Lectures on the Fourier Transform and Its Applications

The Sally Series
Pure and Applied UNDERGRADUATE TEXTS · 33

Lectures on the Fourier Transform and Its Applications

Brad G. Osgood

AMS AMERICAN MATHEMATICAL SOCIETY
Providence, Rhode Island USA

2010 *Mathematics Subject Classification*. Primary 42A38, 42B10, 65T50, 94A20, 46F10.

For additional information and updates on this book, visit
www.ams.org/bookpages/amstext-33

Library of Congress Cataloging-in-Publication Data

Names: Osgood, Brad, author.
Title: Lectures on the Fourier transform and its applications / Brad G. Osgood.
Description: Providence, Rhode Island : American Mathematical Society, [2019] | Series: Pure and applied undergraduate texts ; volume 33 | Includes an index.
Identifiers: LCCN 2017061807 | ISBN 9781470441913 (alk. paper)
Subjects: LCSH: Fourier transformations. | Transformations (Mathematics) | AMS: Harmonic analysis on Euclidean spaces – Harmonic analysis in one variable – Fourier and Fourier-Stieltjes transforms and other transforms of Fourier type. msc | Harmonic analysis on Euclidean spaces – Harmonic analysis in several variables – Fourier and Fourier-Stieltjes transforms and other transforms of Fourier type. msc | Numerical analysis – Numerical methods in Fourier analysis – Discrete and fast Fourier transforms. msc | Information and communication, circuits – Communication, information – Sampling theory. msc | Functional analysis – Distributions, generalized functions, distribution spaces – Operations with distributions. msc
Classification: LCC QC20.7.F67 O84 2019 | DDC 515/.723–dc23
LC record available at https://lccn.loc.gov/2017061807

∞ The paper used in this book is acid-free and falls within the guidelines
established to ensure permanence and durability.
Visit the AMS home page at `https://www.ams.org/`

10 9 8 7 6 5 4 3 2 1 24 23 22 21 20 19

To Becky, Miles and Maddie, and Lynn
And to the memory of Ruth, Herb, and Cassie

Contents

Preface

Greetings to all! I always begin my classes with that exclamation, and since this book grew out of lecture notes for a real class with real students, let me begin the same way. The class is EE261: The Fourier Transform and Its Applications, at Stanford. It's considered a course for beginning graduate students and advanced undergraduates in engineering. I get a crowd from all over engineering and science, and even a few math people. As of my last count I've had the pleasure of sharing the subject with around 2,200 students. Not a huge number, but not a small one either.

I've worked on the notes over a number of years but I've been slow to publish them, not the least because there are many books for such a course, including some by friends and colleagues. However, I can honestly say that generations of students have found the notes to be helpful. I hear from them, and I have greatly benefitted from their suggestions. Many books are buffed and polished to an extent that the author sees only his own reflection. Not the case here. I hope that these pages still carry the voice of a teacher, enthusiastic and certainly unperfected. Me talking to you.

So here you go. If you want (virtual) live action, the lectures from one version of the course are available on YouTube through the Stanford Engineering Everywhere program. I've never watched them. The rest of this introduction is pretty much what I hand out to the students as a quick summary of what to expect.

The Fourier transform has it all. As a tool for applications it is used in virtually every area of engineering and science. In electrical engineering, methods based on the Fourier transform are used in all varieties of signal processing, from communications and circuit design to the analysis of imaging systems. In materials science, physics, chemistry, and molecular biology it is used to understand diffraction and the crystal structures of compounds. There's even a field called Fourier optics. In mathematics, Fourier series and the Fourier transform are cornerstones of the broad area known as harmonic analysis, and applications range from number theory to the modern formulation of the theory of partial differential equations. Lurking not so far beneath the surface are deep connections to groups and symmetry. Particularly widely used in practical applications is the discrete Fourier transform computed via the FFT (Fast Fourier Transform) algorithms. It is not an exaggeration to say that much of the increasingly digital world depends on the FFT.

Historically, Fourier analysis developed from employing sines and cosines to model periodic physical phenomena. This is the subject matter of Fourier series (our first topic), and here we learn that a complicated signal can be written as a sum of sinusoidal components, the simplest periodic signals.[1] Borrowing from musical terminology, where pure tones are single sinusoids and the frequency is the pitch, the component sinusoids are often called the harmonics. Complicated tones are sums of harmonics. In this way, with a periodic signal we associate a discrete set of frequencies — its spectrum — and the amount that each harmonic contributes to the total signal. If we know the spectrum and the amount that each harmonic contributes, then we know the signal, and vice versa. We analyze the signal into its component parts, and we synthesize the signal from its component parts.

The Fourier transform arose as a way of analyzing and synthesizing nonperiodic signals. The spectrum becomes a continuum of frequencies rather than a discrete set. Through the Fourier transform, and its inverse, we now understand that *every signal has a spectrum, and that the spectrum determines the signal.* This maxim surely ranks as one of the major secrets of the universe. A signal (often a time varying function, thus a representation in the "time domain") and its Fourier transform (a function depending on the spectrum, thus a representation in the "frequency domain") are equivalent in that one determines the other and we can pass back and forth between the two. The signal appears in different guises in the time domain and in the frequency domain, and this enhances the usefulness of both representations.

"Two representations for the same quantity" will be a steady refrain in our work. In signal processing, "filtering in the frequency domain" is an example of this, where operations are carried out on the spectrum to, in turn, produce a signal in the time domain having desired properties. Another important example of the use of the dual representations is the sampling theorem, which is fundamental in passing from an analog to a digital signal. In optics, examples are diffraction and interference phenomena; in physics, an example is the Heisenberg uncertainty principle. In mathematics, celebrated identities in number theory come from Rayleigh's identity, which, in physics, says that the energy of a signal can be computed in either the time domain or the frequency domain representation.

Underlying much of this development and its wide applicability is the notion of *linearity.* The operations of analyzing a signal into its component parts (taking the Fourier transform) and synthesizing a signal from its component parts (taking the inverse Fourier transform) are *linear operations*, namely integration. The principle of superposition applies to linear systems — the sum of inputs produces a sum of outputs — and one can thus work with a complicated signal by working with

[1] Electrical engineers speak in terms of signals and mathematicians in terms of functions. As far as I'm concerned, the two terms are interchangeable. I'll use both.

linear combinations, sums or integrals, of simpler signals. This is fundamental to all signal processing. The desire to extend the applicability of Fourier series and the Fourier transform led not only to an increasing array of real-world applications but also to purely mathematical questions. Investigations on many fronts led to a better understanding of approximation and of limiting processes, the nature of integration, linear operators, eigenfunctions, and orthogonality. In extending the range of the mathematical methods it became necessary to move beyond classical ideas about functions and to develop the theory and practice of distributions, also known as generalized functions. The need for such objects was recognized earlier by engineers for very practical applications and by physicists for computations in quantum mechanics.

In short, in this class there's a little magic every day. I want you to see that.

A few comments on audience and backgrounds. The majority of the students in my class are EEs but not everyone is, and collectively they all bring a great variety of preparation and interests. Students new to the subject are introduced to a set of tools and ideas that are used by engineers and scientists in an astonishing assortment of areas. Students who have seen bits and pieces of Fourier techniques in other courses benefit from a course that puts it all together. I frequently hear this from people who have taken the class. I urge the students to keep in mind the different backgrounds and interests and to keep an open mind to the variety of the material, while recognizing that with such a rich panorama we have to be quite selective.

The goals for the course are to gain a facility with *using* the Fourier transform, both specific techniques and general principles, and learning to recognize when, why, and how it is used. Together with a great variety, the subject also has a great coherence, and my hope is that students will appreciate both.

A few comments on math. This is a very mathematical subject, no doubt about it, and I am a mathematician, but this is not a mathematics textbook. What are the distinctions? One is the role of applications. It is true that much of the mathematical development of Fourier analysis has been independent of the growing list of applications in engineering and science, but the mathematicians are missing out on at least some of the fun. Moreover, the applications have proliferated largely because of the computational power now available, and the tilt toward computation is also viewed suspiciously by maybe a few (not all!) mathematicians. While we will be able to see only a fraction of the problems that call on Fourier analysis, for (almost) every mathematical idea we develop there will be significant applications. This is a feature not always found in courses and textbooks outside of engineering. For example, when speaking of applications in the preface to his well-known book *Fourier Integrals* (from 1937), the British mathematician E. C. Titchmarsh wrote:

> As exercises in the theory I have written out a few of these as it seemed to me that an analyst should. I have retained, as having a certain

> picturesqueness, some references to "heat", "radiation", and so forth; but the interest is purely analytical, and the reader need not know whether such things exist.

Bad attitude.

It is important, however, to at least touch on some of the mathematical issues that arise. Engineering, particularly electrical engineering, draws more and more on mathematical ideas and does so with increasing sophistication. Those ideas may have come from questions once considered to be solely in the realm of pure mathematics or may have had their start in real-world problems. This back-and-forth is not new, but the exchange between mathematical ideas and engineering or scientific applications has become more circuitous and sometimes strained. Consider this comment of the Russian mathematician V. I. Arnold in his graduate-level book on differential equations, *Geometric Methods in the Theory of Ordinary Differential Equations*:

> The axiomatization and algebraization of mathematics, after more than 50 years, has lead to the illegibility of such a large number of mathematical texts that the threat of complete loss of contact with physics and the natural sciences has been realized.

It can be hard for engineers, both students and working professionals, to open an advanced math book and get anywhere. That's understandable, as per Arnold, but it is limiting. If you look at the current engineering and scientific literature where the Fourier transform is used, it looks pretty mathematical. To take one fairly current example, the subject of *wavelets* (which we won't do) has become quite important in various parts of signal processing. So, too, *compressive sensing*. You'll understand these and other new ideas much better if you know the mathematical infrastructure that we'll develop. For a list of mathematical topics that we'll call on see Appendix A.

You do need mathematics. You need to understand better the mathematics you've already learned, and you need the confidence to learn new things and then to use those new things. If you're looking for a learning goal, as everyone seems to be doing, that's a big one. Beyond the specific formulas, facts, etc., I hope that this book offers a dose of mathematical know-how, honest and helpful, and with a light touch.

Thanks

I'd like to thank my colleagues in the electrical engineering department for their good cheer and good advice liberally dispensed, particularly Stephen Boyd, Tom Cover, Bob Dutton, Abbas El Gamal, Bob Gray, Tom Kailath, David Miller, Dwight Nishimura, and John Pauly. Special thanks to John Gill who has taught from the notes on multiple occasions, thus offering an existence proof that someone else can deliver the goods. Thanks also to Ronald Bracewell for many engaging conversations, sometimes even about the Fourier transform. I was in the dean's office for N years while a lot of this scribbling was going on, and it helped that the dean, Jim Plummer, along with the whole group in the office, had a sense of humor and recognized mine. So, too, did other friends and colleagues at Stanford, particularly John Bravman, Judy Goldstein, and Stephanie Kalfayan.

Among my teachers and friends in mathematics, all fine expositors as well as fine mathematicians, I want to mention Lars Ahlfors, Peter Duren, Fred Gehring, Yitzhak Katznelson, Zeev Nehari, Walter Noll, Peter Sarnak, and Juan Schäffer. I had the benefit of working on a series of math books with very good writers, and I hope some of their skill rubbed off, so thanks to Andrew Gleason, Deborah Hughes Hallett, William McCallum, and the rest of the working group. Gerald Folland read an early version of this book and offered many helpful suggestions and corrections. Evelin Sullivan, a star instructor in Stanford's Technical Communications Program, read every line and coached me through the perils of commas, colons and semicolons, and choppy sentences. I'd also like to thank my friends and former students Martin Chuaqui, Ethan Hamilton, Aditya Siripuram, and William Wu, and for something completely different, a shout out to the sixteen talented jazz musicians whom I play with every Tuesday night.

The organization of my version of the course was strongly influenced by Jospeh Goodman's course notes, which in turn drew on Ron Bracewell's book. Joe and Bob Gray together also wrote a textbook that's been featured in the course. One fine summer, Almir Mutapcic started to turn my many TeX files into something that began to look like this book, and then John Gill felt compelled to introduce macros, correct grammar, tinker with notation, change spacing, all of that. I could never have made it this far without their efforts. Ina Mette of the American Mathematical Society took over from there and has been very encouraging and very patient. I hope she believes it's been worth the wait.

I dedicate the book to my loving family, and I'm giving the EE261 TAs an acknowledgement of their own. That leaves one more special group: the students who took my course. You worked hard and you have my thanks. I hope you are flourishing.

The TAs. What a great group over the years. I'd like to recognize and thank them for their many contributions to the class and to this book: Eduardo Abeliuk, Rajiv Agarwal, Sercan Arik, Leighton Barnes, Neda Beheshti, Raj Bhatnagar, Ryan Cassidy, Damien Cerbelaud, Panu Chaichanavong, Albert Chen, David Choi, Aakanksha Chowdhery, Eva Enns, Irena Fischer-Hwang, John Haller, Sangoh Jeong, Jieyang Jia, Thomas John, Max Kamenetsky, Eric Keller, Hye Ji Kim, Youngsik Kim, Zahra Koochak, Matt Kraning, Hongquan Li, Liz Li, Lykomidis Mastroleon, Tom McGiffen, Almir Mutapcic, Deirdre O'Brien, Seonghyun Paik, Stephanie Pancoast, Cristina Puig, Shalini Ranmuthu, Prashant Ramanathan, Paul Reynolds, Vatsal Sharan, Aditya Siripuram, Joelle Skaff, Nikola Stikov, Logi Vidarsson, Adam Wang, Moosa Zaida, and Kai Zang.

Chapter 1

Fourier Series

1.1. Choices: Welcome Aboard

The very first choice is naturally where to start, and my choice is to start with a brief treatment of Fourier series.[1] The kind of analysis introduced by Fourier was originally concerned with representing periodic phenomena via what came to be known as Fourier series and then later with extending those insights to nonperiodic phenomena via the Fourier transform (an integral). In fact, one way of getting from Fourier series to the Fourier transform is to consider nonperiodic phenomena as a limiting case of periodic phenomena as the period tends to infinity. That's what we'll do, in the next chapter.

Associated with the Fourier series of a periodic signal is a discrete set of frequencies. These becomes a continuum of frequencies for the Fourier transform of a nonperiodic signal. In either case, the set of frequencies constitutes the *spectrum*, and with the spectrum comes the most important principle of the subject. To repeat from the preface:

- Every signal has a spectrum and the spectrum determines the signal.

Catchy, but I should say that it's the frequencies together with how much each frequency contributes to the signal that one needs to know.

Ideas drawing on Fourier series and Fourier transforms were thoroughly grounded in physical applications. Most often the phenomena to be studied were modeled by the fundamental differential equations of physics (heat equation, wave equation, Laplace's equation), and the solutions were usually constrained by boundary conditions.

[1] Many books for engineers launch right into Fourier transforms. The popular book *The Fourier Transform and Its Applications* by R. Bracewell, for example, does that, picking up a little on Fourier series later. It's a defensible choice, but it's not my choice.

At first the idea was to use Fourier series to find explicit solutions. This work raised hard and far-reaching questions that led in different directions. For example, setting up Fourier series (in sines and cosines) was later recast in the more general framework of orthogonality, linear operators, and eigenfunctions. That led to the general idea of working with *eigenfunction expansions* of solutions of differential equations, a ubiquitous line of attack in many areas and applications. In the modern formulation of partial differential equations, the Fourier transform has become the basis for *defining* the objects of study, while still remaining a tool for solving specific equations. Much of this development depends on the remarkable relation between Fourier transforms and convolution, something that was also seen earlier in the use of Fourier series. In an effort to apply the methods with increasing generality, mathematicians were pushed (to some extent by engineers and physicists) to reconsider how general the notion of "function" can be and what kinds of functions can be — and should be — admitted into the operating theater of calculus. Differentiation and integration were both generalized in the service of Fourier analysis.

Other directions combine tools from Fourier analysis with symmetries of the objects being analyzed. This might make you think of crystals and crystallography, and you'd be right, while mathematicians think of number theory and Fourier analysis on groups. Finally, I have to mention that in the purely mathematical realm the question of convergence of Fourier series, believe it or not, led G. Cantor near the turn of the 20th century to investigate and invent the theory of infinite sets and to distinguish different cardinalities of infinite sets.

1.2. Periodic Phenomena

To begin the course with Fourier series is to begin with periodic functions, those functions that exhibit a regularly repeating pattern. It shouldn't be necessary to pitch periodicity as an important physical (and mathematical) phenomenon — you've most likely seen examples and applications of periodic behavior in almost every class you've taken. I would only remind you that periodicity often shows up in two varieties, sometimes related, sometimes not. Generally speaking, we think about periodic phenomena according to whether they are *periodic in time* or *periodic in space*.

1.2.1. Time and space. In the case of periodicity in time the phenomenon comes to you. For example, you stand at a fixed point in the ocean (or at a fixed point on an AC electrical circuit) and the waves (or the electrical current) wash over you with a regular, recurring pattern of crests and troughs. The height of the wave is a periodic function of time. Sound is another example. Sound reaches your ear as a longitudinal pressure wave, a periodic compression and rarefaction of the air.

In the case of periodicity in space, you come to the phenomenon. You take a picture, say, and you observe repeating patterns.

Temporal and spatial periodicity come together most naturally in wave motion. Take the case of one spatial dimension and consider, for example, a single sinusoidal wave traveling along a string. For such a wave the periodicity in time is measured by the frequency[2] ν with dimension 1/time and units Hz (Hertz = cycles/second)[3], and the periodicity in space is measured by the wavelength λ with dimension length and units whatever is convenient for the particular setting. If we fix a point in space and let the time vary (take a video of the wave motion at that point), then successive crests of the wave come past that point at a rate of ν times per second, and so do successive troughs. If we fix the time and examine how the wave is spread out in space (take a snapshot instead of a video), we see that the distance between successive crests is a constant λ, as is the distance between successive troughs.

The frequency and wavelength are related through the equation $v = \lambda\nu$, where v is the speed of propagation. This fundamental equation is nothing but the wave version of another fundamental equation: speed = distance/time. If the speed is fixed, like the speed of electromagnetic waves in a vacuum, then the frequency determines the wavelength and vice versa; if you can measure one, you can find the other. For sound, we identify the physical property of frequency with the perceptual property of pitch. For visible light, frequency is perceived as color. The higher the frequency, the shorter the wavelength, and the lower the frequency, the longer the wavelength. With this simple observation we've already encountered what will be a persistent and unifying theme for us, a reciprocal relationship between defining quantities.

More on spatial periodicity. Another way spatial periodicity occurs is when there is a repeating pattern or some kind of symmetry in a spatial region and physically observable quantities associated with that region have a repeating pattern that reflects this. For example, a crystal has a regular, repeating pattern of atoms in space; the arrangement of atoms is called a *lattice*. The function that describes the electron density distribution of the crystal is then a periodic function of the spatial variables, in three dimensions, that describes the crystal. I mention this example, which we'll return to much later, because in contrast to the usual one-dimensional examples you might think of, here the function has three independent periods corresponding to the three directions that describe the crystal lattice.

Here's another example, this time in two dimensions, that is very much a natural subject for Fourier analysis. Consider these stripes of dark and light:

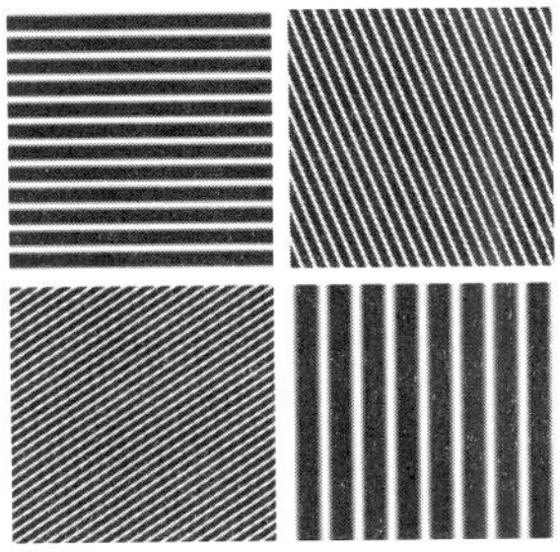

No doubt there's some kind of spatially periodic behavior going on in the respective images. Even without stating a precise definition, it's reasonable to say

[2]A lot of engineering books use f for frequency, naturally enough. Since that conflicts with using f for a generic function, we'll use ν, which is also a common choice.

[3]"Cycles" is neither a dimension nor a unit. It's one complete pattern of the pattern that repeats.

that some of the patterns are lower frequency and that others are higher frequency, meaning that there are fewer stripes per unit length in some than in others. For periodicity in two dimensions, and higher, there's an extra subtlety: "spatial frequency," however we ultimately define it, must be a vector quantity, not a number. We have to say that the stripes occur with a certain spacing in a certain direction.

Such periodic stripes and the functions that generate them are the building blocks of general two-dimensional images. When there's no color, an image is a two-dimensional array of varying shades of gray, and this can be realized as a synthesis — a Fourier synthesis — of just such alternating stripes. There are interesting perceptual questions in constructing images this way, and color is more complicated still.

1.2.2. Definitions, examples, and things to come. To be certain we all know what we're talking about, a function $f(t)$, $-\infty < t < \infty$, is *periodic* of period T if there is a number $T > 0$ such that

$$f(t+T) = f(t)$$

for all t. If there is such a T, then the smallest one for which the equation holds is called the *fundamental period* of the function $f(t)$.[4] Every *integer* multiple of the fundamental period is also a period:[5]

$$f(t+nT) = f(t)\,, \quad n = 0, \pm 1, \pm 2, \ldots .$$

I'm calling the variable t here because I have to call it something, but the definition is general and is not meant to imply periodicity in time.

The graph of $f(t)$ over *any* interval of length T is one *cycle*, also called one period. Geometrically, the periodicity condition means that the shape of one cycle (any cycle) determines the graph everywhere: the shape is repeated over and over. A problem asks you to turn this idea into a formula (periodizing a function). If you know the function on one period, you know everything.

This is all old news to everyone, but, by way of example, there are a few more points I'd like to make. Consider the function

$$f(t) = \cos 2\pi t + \tfrac{1}{2}\cos 4\pi t\,,$$

whose graph is shown below.

[4]Sometimes when people say simply "period," they mean the smallest or fundamental period. (I usually do, for example.) Sometimes they don't. Ask them what they mean.

[5]It's clear from the geometric picture of a repeating graph that this is true. To show it algebraically, if $n \geq 1$, then we see inductively that $f(t+nT) = f(t+(n-1)T+T) = f(t+(n-1)T) = f(t)$. Then to see algebraically why negative multiples of T are also periods, once we've established it for positive multiples, we have, for $n \geq 1$, $f(t-nT) = f(t-nT+nT) = f(t)$.

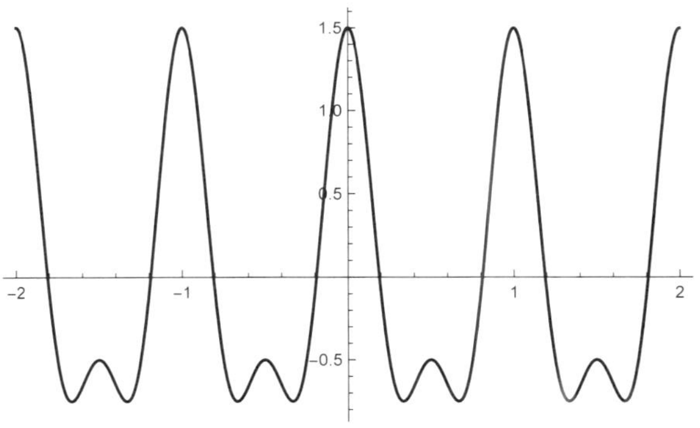

The individual terms are periodic with periods 1 and 1/2, respectively, but the sum is periodic with period 1:

$$\begin{aligned} f(t+1) &= \cos 2\pi(t+1) + \tfrac{1}{2}\cos 4\pi(t+1) \\ &= \cos(2\pi t + 2\pi) + \tfrac{1}{2}\cos(4\pi t + 4\pi) = \cos 2\pi t + \tfrac{1}{2}\cos 4\pi t = f(t)\,. \end{aligned}$$

There is no smaller value of T for which $f(t+T) = f(t)$ for all t. The overall pattern repeats every 1 second, but if this function represented some kind of wave, would you say it had frequency 1 Hz? Somehow I don't think so. It has one *period* but you'd probably say that it has, or contains, *two* frequencies, one cosine of frequency 1 Hz and one of frequency 2 Hz.

The subject of adding up periodic functions is worth a general question:

- Is the sum of two periodic functions periodic?

The answer is no, and that's because of irrational numbers. For example, $\cos t$ and $\cos(\sqrt{2}t)$ are each periodic, with periods 2π and $2\pi/\sqrt{2}$, respectively, but the sum $\cos t + \cos(\sqrt{2}t)$ is not periodic.[6]

Here are plots of $f_1(t) = \cos t + \cos 1.4t$ (left), which is periodic, and of $f_2(t) = \cos t + \cos(\sqrt{2}t)$ (right), which isn't. They are not identical. Look carefully.

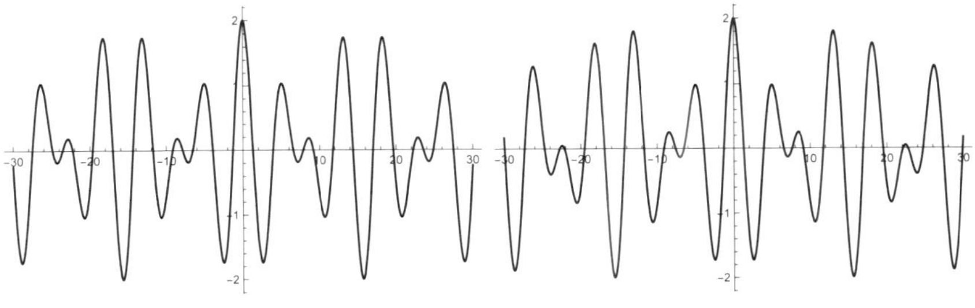

What's the period of $f_1(t)$? One of the problems will allow you to find it.

I'm aware of the irony in making a show of computer plots depending on an irrational number when the computer has to take a rational approximation to draw the picture. I plotted these functions using Mathematica®, and I don't know what

[6]This is left as a problem.

it uses as an approximation to $\sqrt{2}$. How artificial an example is this? Not artificial at all. We'll see why, below.

1.2.3. The building blocks: A few more examples. The classic example of temporal periodicity is the harmonic oscillator, whether it's a mass on a spring (no friction) or current in an LC circuit (no resistance). The harmonic oscillator is treated in exhaustive detail in just about every physics class. This is so because it is the *only* problem that can be treated in exhaustive detail. Well, not quite true — as has been pointed out to me, the two-body problem has also been thoroughly worked over.

The state of the system is described by a *single* sinusoid, say of the form

$$A\sin(2\pi\nu t + \phi)\,.$$

The parameters in this expression are the *amplitude* A, the *frequency* ν, and the *phase* ϕ. The period of this function is $1/\nu$ since

$$\begin{aligned}A\sin\left(2\pi\nu\left(t+\frac{1}{\nu}\right)+\phi\right) &= A\sin\left(2\pi\nu t + 2\pi\nu\frac{1}{\nu}+\phi\right)\\ &= A\sin(2\pi\nu t + 2\pi + \phi) = A\sin(2\pi\nu t+\phi)\,.\end{aligned}$$

The classic example of spatial periodicity, the example that started the whole subject, is the distribution of heat in a circular ring. A ring is heated up, somehow, and the heat then distributes itself, somehow, through the material. In the long run we expect all points on the ring to be of the same temperature, but they won't be in the short run. At each fixed time, how does the temperature vary around the ring?

In this problem the periodicity comes from the coordinate description of the ring. Think of the ring as a circle. Then a point on the ring is determined by an angle θ and quantities which depend on position are functions of θ. Since θ and $\theta + 2\pi$ are the same point on the circle, any continuous function describing a physical quantity on the circle, e.g., temperature, is a periodic function of θ with period 2π.

The distribution of temperature is not given by a simple sinusoid. It was Fourier's hot idea to consider a *sum* of sinusoids as a model for the temperature distribution:

$$\sum_{n=1}^{N} A_n\sin(n\theta+\phi_n)\,.$$

The dependence on time is in the coefficients A_n. We'll study this problem more completely later.

Regardless of the physical context, the individual terms in a trigonometric sum such as the one above are called *harmonics*, terminology that comes from the mathematical representation of musical pitch. The terms contribute to the sum in varying amplitudes and phases, and these can have any values. The frequencies of the terms, on the other hand, are integer multiples of the fundamental frequency. For the sum above that's $1/2\pi$. Because the frequencies are integer multiples of the fundamental frequency, the sum is also periodic of period 2π. The term

$A_n \sin(n\theta + \phi_n)$ has period $2\pi/n$, but the whole sum can't have a shorter cycle than the longest cycle that occurs, and that's 2π.

Musical pitch and tuning. Musical pitch and the production of musical notes is a periodic phenomenon of the same general type as we've been considering. Notes can be produced by vibrating strings or other objects that can vibrate regularly (like lips, reeds, or the bars of a xylophone). The engineering problem is how to tune musical instruments. The subject of tuning has a fascinating history, from the natural tuning of the Greeks, based on ratios of integers, to the theory of the equal tempered scale, which is the system of tuning used today. That system is based on $2^{1/12}$, an irrational number.

There are 12 notes in the equal tempered scale, going from any given note to the same note an octave up, and two adjacent notes have frequencies with ratio $2^{1/12}$. The notes in the scale are $\cos(2\pi \cdot 440 \cdot 2^{n/12}t)$ from $n = 0$ to $n = 12$. If an A of frequency 440 Hz (concert A) is described by

$$\mathrm{A} = \cos(2\pi \cdot 440\, t)\,,$$

then 6 notes up from A in a well-tempered scale is a D♯ given by

$$\mathrm{D}\sharp = \cos(2\pi \cdot 440\sqrt{2}\, t)\,.$$

Playing the A and the D♯ together gives essentially the signal we had earlier, $\cos t + \cos 2^{1/2}t$. I'll withhold judgment whether it sounds any good — it's the tritone, the augmented fourth, the "devil in music."

Of course, when you tune a piano, you don't tighten the strings irrationally. The art is to make the right approximations. There are many discussions of this; see, for example

`http://www.precisionstrobe.com/`.

To read more about tuning in general try

`http://www.wikipedia.org/wiki/Musical_tuning`.

Here's a quote from the first reference describing the need for well-tempered tuning:

> Two developments occurred in music technology which necessitated changes from the just toned temperament. With the development of the fretted instruments, a problem occurs when setting the frets for just tuning, that octaves played across two strings around the neck would produce impure octaves. Likewise, an organ set to a just tuning scale would reveal chords with unpleasant properties. A compromise to this situation was the development of the mean toned scale. In this system several of the intervals were adjusted to increase the number of usable keys. With the evolution of composition technique in the 18th century increasing the use of harmonic modulation a change was advocated to the equal tempered scale. Among these advocates was J. S. Bach who published two entire works entitled *The Well-tempered Clavier*. Each of these works contain 24 fugues written in each of twelve major and twelve minor keys and demonstrated that using an equal tempered scale, music could be written in, and shifted to any key.

1.3. It All Adds Up

From simple, single sinusoids we can build up much more complicated periodic functions by taking sums of sinusoids, and for now just think of finite sums. To highlight the essential ideas it's convenient to standardize a little and consider functions with period 1. This simplifies some of the writing and it will be easy to modify the formulas if the period is not 1. The basic function of period 1 is $\sin 2\pi t$, and so the Fourier-type sum we considered briefly above now looks like

$$\sum_{n=1}^{N} A_n \sin(2\pi nt + \phi_n)\,.$$

Whatever psychological advantage there is in displaying the amplitude and phase of each harmonic, it turns out to be a somewhat awkward expression to calculate with. It's more common to write a general trigonometric sum as

$$\sum_{n=1}^{N} (a_n \cos 2\pi nt + b_n \sin 2\pi nt)\,.$$

Using the addition formulas for the sine function you can pass between the two expressions — they are equivalent. (Treat that as an exercise.)

If we include a constant term ($n = 0$), we write

$$\frac{a_0}{2} + \sum_{n=1}^{N} (a_n \cos 2\pi nt + b_n \sin 2\pi nt)\,.$$

The reason for writing the constant term with the fraction 1/2 is because, as you will see below, it matches better with still another expression for such a sum. In electrical engineering the constant term is often referred to as the *DC component* as in "direct current." The other terms, being periodic, alternate, as in "alternating current," or AC. Aside from the DC component, the harmonics have periods $1, 1/2, 1/3, \ldots, 1/N$, respectively, or frequencies $1, 2, 3, \ldots, N$. Because the frequencies of the individual harmonics are integer multiples of the lowest frequency, the period of the sum is 1.

Using complex exponentials. Algebraic work on such trigonometric sums is made incomparably easier if we use complex exponentials to represent the sine and cosine. If you're rusty, see Appendix B on complex numbers where there is a discussion of complex exponentials and how they can be used without fear to represent real signals, and an answer to the question of what is meant by negative frequencies. The latter are about to make an appearance.

I remind you that

$$\cos t = \frac{e^{it} + e^{-it}}{2}, \quad \sin t = \frac{e^{it} - e^{-it}}{2i}\,.$$

Hence

$$\cos 2\pi nt = \frac{e^{2\pi int} + e^{-2\pi int}}{2}, \quad \sin 2\pi nt = \frac{e^{2\pi int} - e^{-2\pi int}}{2i}\,.$$

(By the way, here and everywhere else in this book, $i = \sqrt{-1}$. Not j. See the declaration of principles in Appendix B.) Let's use these to rewrite the sum

$$\frac{a_0}{2} + \sum_{n=1}^{N} (a_n \cos 2\pi nt + b_n \sin 2\pi nt)\,.$$

We have

$$\begin{aligned}
&\frac{a_0}{2} + \sum_{n=1}^{N} (a_n \cos 2\pi nt + b_n \sin 2\pi nt) \\
&\quad = \frac{a_0}{2} + \sum_{n=1}^{N} \left(a_n \frac{e^{2\pi int} + e^{-2\pi int}}{2} - ib_n \frac{e^{2\pi int} - e^{-2\pi int}}{2} \right) \\
&\quad\quad \text{(using } 1/i = -i \text{ in the sine term)} \\
&\quad = \frac{a_0}{2} + \sum_{n=1}^{N} \left(\frac{1}{2}(a_n - ib_n)e^{2\pi int} + \frac{1}{2}(a_n + ib_n)e^{-2\pi int} \right) \\
&\quad = \frac{a_0}{2} + \sum_{n=1}^{N} \frac{1}{2}(a_n - ib_n)e^{2\pi int} + \sum_{n=1}^{N} \frac{1}{2}(a_n + ib_n)e^{-2\pi int}\,.
\end{aligned}$$

We want to write this as a single sum that runs over positive and negative n. For this, put

$$c_n = \frac{1}{2}(a_n - ib_n) \quad \text{and} \quad c_{-n} = \frac{1}{2}(a_n + ib_n)\,, \qquad n = 1, \dots, N\,,$$

and also

$$c_0 = \frac{a_0}{2}\,.$$

Leave the first sum alone, and write the second sum with n going from -1 to $-N$, so that the expression becomes

$$\frac{a_0}{2} + \sum_{n=1}^{N} c_n e^{2\pi int} + \sum_{n=-1}^{-N} c_n e^{2\pi int}\,.$$

Thus everything combines to produce

$$\frac{a_0}{2} + \sum_{n=1}^{N} (a_n \cos 2\pi nt + b_n \sin 2\pi nt) = \sum_{n=-N}^{N} c_n e^{2\pi int}\,.$$

The negative frequencies are the negative values of n. And now you can see why we wrote the DC term as $a_0/2$.

In this final form of the sum the coefficients c_n are *complex* numbers. If we assume that the signal is real — which we didn't have to in the preceding calculations — then their definition in terms of the a's and b's entails the relationship

$$c_{-n} = \overline{c_n}\,.$$

In particular, for any value of n the magnitudes of c_n and c_{-n} are equal:

$$|c_n| = |c_{-n}|\,.$$

Notice that when $n = 0$ we have

$$c_0 = \overline{c_0}\,,$$

which implies that c_0 is a real number; this jibes with c_0 equalling the real number $a_0/2$.

The conjugate symmetry property $c_{-n} = \overline{c_n}$ for real signals is important. If we start out with the representation

$$f(t) = \sum_{n=-N}^{N} c_n e^{2\pi int}$$

where the c_n are complex numbers, then $f(t)$ is a real signal *if and only if* the coefficients satisfy $c_{-n} = \overline{c_n}$. For if $f(t)$ is real, then

$$\sum_{n=-N}^{N} c_n e^{2\pi int} = \overline{\sum_{n=-N}^{N} c_n e^{2\pi int}} = \sum_{n=-N}^{N} \overline{c_n}\,\overline{e^{2\pi int}} = \sum_{n=-N}^{N} \overline{c_n}\,e^{-2\pi int}\,,$$

and equating like terms gives the relation $c_{-n} = \overline{c_n}$. Conversely, suppose the relation is satisfied. For $n = 0$ we have $c_0 = \overline{c_0}$, so c_0 is real. For each $n \neq 0$ we can group $c_n e^{2\pi int}$ with $c_{-n}e^{-2\pi int}$, and then

$$c_n e^{2\pi int} + c_{-n}e^{-2\pi int} = c_n e^{2\pi int} + \overline{c_n e^{2\pi int}} = 2\,\mathrm{Re}\left(c_n e^{2\pi int}\right).$$

Then

$$\sum_{n=-N}^{N} c_n e^{2\pi int} = c_0 + \sum_{n=1}^{N} 2\,\mathrm{Re}\left(c_n e^{2\pi int}\right) = c_0 + 2\,\mathrm{Re}\left\{\sum_{n=1}^{N} c_n e^{2\pi int}\right\},$$

and so the sum of complex exponentials produces a real signal.

1.3.1. Lost at *c*. Suppose you're confronting a complicated looking periodic signal. You can think of the signal as varying in time but, again and always, the reasoning to follow applies to any sort of one-dimensional periodic phenomenon. In any event, we can scale to assume that the period is 1. Might we express the signal as a sum of simpler periodic signals of period 1?

Call the signal $f(t)$. *Suppose* we can write $f(t)$ as a sum

$$f(t) = \sum_{n=-N}^{N} c_n e^{2\pi int}\,.$$

The unknowns in this expression are the coefficients c_n. Can we solve for them?[7]

Solve in terms of what? We're assuming that $f(t)$ is known, so we want an expression for the coefficients in terms of $f(t)$. Let's take a direct approach with algebra, trying to isolate the coefficient c_k for a fixed k. First, pull the kth term out of the sum to get

$$c_k e^{2\pi ikt} = f(t) - \sum_{n=-N,n\neq k}^{N} c_n e^{2\pi int}\,.$$

[7]See the story *Lost at c* by Jean Sheppard in the collection *A Fistful of Fig Newtons*.

Now multiply both sides by $e^{-2\pi ikt}$ to get c_k alone:

$$\begin{aligned}
c_k &= e^{-2\pi ikt}f(t) - e^{-2\pi ikt} \sum_{n=-N,n\neq k}^{N} c_n e^{2\pi int} \\
&= e^{-2\pi ikt}f(t) - \sum_{n=-N,n\neq k}^{N} c_n e^{-2\pi ikt}e^{2\pi int} \\
&= e^{-2\pi ikt}f(t) - \sum_{n=-N,n\neq k}^{N} c_n e^{2\pi i(n-k)t} .
\end{aligned}$$

Well, brilliant, we have managed to solve for c_k in terms of all of the other unknown coefficients.

This is as far as algebra will take us, and when algebra is exhausted, the desperate mathematician will turn to calculus, i.e., to differentiation or integration. Here's a hint: differentiation won't get you anywhere.

Another idea is needed, and that idea is integrating both sides from 0 to 1; if you know what happens on an interval of length 1, you know everything — that's periodicity working for you. We take the interval from 0 to 1 as base period for the function, but any interval of length 1 would work, as we'll explain below (that's periodicity working for you, again). We have

$$\begin{aligned}
c_k = \int_0^1 c_k\,dt &= \int_0^1 e^{-2\pi ikt}f(t)\,dt - \int_0^1 \left(\sum_{n=-N,n\neq k}^{N} c_n e^{2\pi i(n-k)t} \right) dt \\
&= \int_0^1 e^{-2\pi ikt}f(t)\,dt - \sum_{n=-N,n\neq k}^{N} c_n \int_0^1 e^{2\pi i(n-k)t}\,dt .
\end{aligned}$$

Just as in calculus, we can evaluate the integral of a complex exponential by

$$\begin{aligned}
\int_0^1 e^{2\pi i(n-k)t}\,dt &= \frac{1}{2\pi i(n-k)} e^{2\pi i(n-k)t} \Big]_{t=0}^{t=1} \\
&= \frac{1}{2\pi i(n-k)} (e^{2\pi i(n-k)} - e^0) \\
&= \frac{1}{2\pi i(n-k)} (1-1) \quad (\text{remember } e^{2\pi i\cdot\text{integer}} = 1) \\
&= 0 .
\end{aligned}$$

Note that $n \neq k$ is needed here.

All of the terms in the sum integrate to zero! We have found a formula for the kth coefficient:

$$c_k = \int_0^1 e^{-2\pi ikt}f(t)\,dt .$$

Let's summarize and be careful to note what we've done here, and what we haven't done. We've shown that *if* we can write a periodic function $f(t)$ of period

1 as a sum

$$f(t) = \sum_{n=-N}^{N} c_n e^{2\pi int},$$

then the coefficients c_n must be given by

$$c_n = \int_0^1 e^{-2\pi int} f(t)\, dt\,.$$

We have *not* shown that every periodic function *can* be expressed this way.

By the way, in none of the preceding calculations did we have to assume that $f(t)$ is a real signal. If, however, we do assume that $f(t)$ is real, then let's see how the formula for the coefficients jibes with $\overline{c_n} = c_{-n}$. We have

$$\begin{aligned}
\overline{c_n} &= \overline{\left(\int_0^1 e^{-2\pi int} f(t)\, dt\right)} = \int_0^1 \overline{e^{-2\pi int}}\; \overline{f(t)}\, dt \\
&= \int_0^1 e^{2\pi int} f(t)\, dt \quad \text{(because } f(t) \text{ is real, as are } t \text{ and } dt\text{)} \\
&= c_{-n} \quad \text{(by definition of } c_n\text{)}.
\end{aligned}$$

1.3.2. Fourier coefficients. The c_n are called the *Fourier coefficients* of $f(t)$ because it was Fourier who introduced these ideas into mathematics and science (but working with the sine and cosine form of the expression). The sum

$$\sum_{n=-N}^{N} c_n e^{2\pi int}$$

is called a (finite) *Fourier series.*

If you want to be mathematically hip and impress your friends at cocktail parties, use the notation

$$\hat{f}(n) = \int_0^1 e^{-2\pi int} f(t)\, dt$$

for the Fourier coefficients. Read this as "f hat of n." Always conscious of social status, I will use this notation.

Note in particular that the zeroth Fourier coefficient is the *average value* of the function:

$$\hat{f}(0) = \int_0^1 f(t)\, dt\,.$$

As promised, let's now check that any interval of length 1 will do to calculate $\hat{f}(n)$. To integrate over an interval of length 1 is to integrate from a to $a+1$, where a is any number. We can examine how the integral varies as a function of a

by differentiating with respect to a:

$$\begin{aligned}\frac{d}{da}\Big(\int_a^{a+1} e^{-2\pi int} f(t)\,dt\Big) &= e^{-2\pi in(a+1)} f(a+1) - e^{-2\pi ina} f(a)\\ &= e^{-2\pi ina} e^{-2\pi in} f(a+1) - e^{-2\pi ina} f(a)\\ &= e^{-2\pi ina} f(a) - e^{-2\pi ina} f(a)\\ &\quad \text{(using } e^{-2\pi in} = 1 \text{ and } f(a+1) = f(a))\\ &= 0\,.\end{aligned}$$

In other words, the integral

$$\int_a^{a+1} e^{-2\pi int} f(t)\,dt$$

is independent of a. So, in particular,

$$\int_a^{a+1} e^{-2\pi int} f(t)\,dt = \int_0^1 e^{-2\pi int} f(t)\,dt = \hat{f}(n)\,.$$

A common instance of this is to use

$$\hat{f}(n) = \int_{-1/2}^{1/2} e^{-2\pi int} f(t)\,dt\,.$$

There are times when such a choice is helpful — integrating over a symmetric interval, that is.

Symmetry relations. When $f(t)$ is a real signal, we've seen the symmetry property

$$\hat{f}(-n) = \overline{\hat{f}(n)}$$

of the Fourier coefficients. From this,

$$|\hat{f}(-n)| = |\hat{f}(n)|\,.$$

In words, the *magnitude* of the Fourier coefficients is an even function of n.

If $f(t)$ itself has symmetry, then so do the Fourier coefficients. For example, suppose that $f(t)$ is even, i.e., $f(-t) = f(t)$, but don't suppose necessarily that $f(t)$ is real. For the nth Fourier coefficient we have

$$\begin{aligned}\hat{f}(-n) &= \int_0^1 e^{-2\pi i(-n)t} f(t)\,dt = \int_0^1 e^{2\pi int} f(t)\,dt\\ &= -\int_0^{-1} e^{-2\pi ins} f(-s)\,ds \text{ (using } t=-s \text{ and changing limits accordingly)}\\ &= \int_{-1}^{0} e^{-2\pi ins} f(s)\,ds\\ &\quad \text{(flipping the limits and using the assumption that } f(t) \text{ is even)}\\ &= \hat{f}(n) \quad \text{(you can integrate over any period, in this case from } -1 \text{ to } 0).\end{aligned}$$

This shows that $\hat{f}(n)$ itself is an even function of n. Combining this with $\hat{f}(-n) = \overline{\hat{f}(n)}$ *for a real function*, we can further conclude that then $\hat{f}(n) = \overline{\hat{f}(n)}$, and putting the two together:

- If $f(t)$ is even, then the Fourier coefficients $\hat{f}(n)$ are also even.
- If $f(t)$ is real and even, then the Fourier coefficients $\hat{f}(n)$ are also real and even.

What if $f(t)$ is odd, i.e., $f(-t) = -f(t)$? Repeat the calculations (and comments) above, but this time using the oddness:

$$\begin{aligned}\hat{f}(-n) &= \int_0^1 e^{-2\pi i(-n)t} f(t)\,dt = \int_0^1 e^{2\pi i n t} f(t)\,dt \\ &= -\int_0^{-1} e^{-2\pi i n s} f(-s)\,ds \\ &\quad \text{(substituting } t = -s \text{ and changing limits accordingly)} \\ &= -\int_{-1}^{0} e^{-2\pi i n s} f(s)\,ds \\ &\quad \text{(flipping the limits and using that } f(t) \text{ is odd)} \\ &= -\hat{f}(n) \\ &\quad \text{(because you can integrate over any period, in this case from } -1 \text{ to } 0).\end{aligned}$$

This shows that $\hat{f}(n)$ is an odd function of n. Now if in addition we assume that $f(t)$ is real, then this combines with $\hat{f}(-n) = \overline{\hat{f}(n)}$ to give $\overline{\hat{f}(n)} = -\hat{f}(n)$. Meaning:

- If $f(t)$ is odd, then the Fourier coefficients $\hat{f}(n)$ are also odd.
- If $f(t)$ is real and odd, then the Fourier coefficients $\hat{f}(n)$ are odd and purely imaginary.

These can be helpful consistency checks for your calculations. Suppose you're calculating Fourier coefficients for a particular function — it could happen — then symmetry properties of the function should be inherited by the Fourier coefficients. Check.

In a story that we'll spin out over the course of this book, we think of

$$\hat{f}(n) = \int_0^1 e^{-2\pi i n t} f(t)\,dt$$

as a *transform* of f to a new function $\hat{f}$. While $f(t)$ is defined for a continuous variable t, the transformed function $\hat{f}$ is defined on the integers. There are reasons for this that are much deeper than simply solving for the unknown coefficients in a Fourier series.

1.3.3. Period, frequencies, and spectrum. Now we give a few more general observations and some terminology. And in the spirit of generality I'm now allowing infinite Fourier series into the arena, never mind questions of convergence. In the following we're assuming that $f(t)$ is a real, periodic signal of period 1 with a representation as the series

$$f(t) = \sum_{n=-\infty}^{\infty} \hat{f}(n) e^{2\pi i n t} .$$

For such a function there may be only a finite number of nonzero coefficients; or maybe all but a finite number of coefficients are nonzero; or maybe none of the coefficients are zero; or maybe there are an infinite number of nonzero coefficients but also an infinite number of coefficients that are zero — I think that's everything. Any such possibility may and can occur. Also interesting, and useful to know for some applications, is that one can say something about the *size* of the coefficients. Along with convergence, we'll come back to that, too.

Earlier, I have often used the more geometric term *period* instead of the more physical term *frequency*. It's natural to talk about the period for a Fourier series as above. The period is 1. The values of the function repeat according to $f(t+1) = f(t)$ and so do the values of all the individual terms, though the terms for $n \neq 1$, $\hat{f}(n)e^{2\pi i n t}$, have the strictly shorter period $1/n$. (By convention, we sort of ignore the constant term c_0 when talking about periods or frequencies. It's obviously periodic of period 1, or any other period for that matter.) As mentioned earlier, it doesn't seem natural to talk about "the frequency" (should it be 1 Hz?). That misses the point. Rather, being able to write $f(t)$ as a Fourier series means that it is synthesized from many harmonics, many frequencies, positive and negative, perhaps an infinite number. The set of frequencies n that are present in a given periodic signal is the *spectrum* of the signal. These are the numbers n for which $\hat{f}(n) \neq 0$. While strictly speaking it's the frequencies, like $n = \pm 2, \pm 7, \pm 325$, that make up the spectrum, people often use the word "spectrum" to refer also to the (nonzero) values $\hat{f}(n)$, like $\hat{f}(\pm 2)$, $\hat{f}(\pm 7)$, $\hat{f}(\pm 325)$. This ambiguity should not cause any problems.

Let's recall the symmetry relation for a real signal, $\hat{f}(-n) = \overline{\hat{f}(n)}$. Because of this relation, the coefficients $\hat{f}(n)$ and $\hat{f}(-n)$ are either both zero or both nonzero. If the coefficients are *all* zero from some point on, say $\hat{f}(n) = 0$ for $|n| > N$, then it's common to say that the signal has *no* spectrum from that point on. One also says in this case that the signal is *bandlimited* and that the *bandwidth* is $2N$.

The square magnitude of the nth coefficient, $|\hat{f}(n)|^2$, is said to be the *energy* of the (positive and negative) harmonic $e^{\pm 2\pi i n t}$. (More on this later.) The sequence of squared magnitudes $|\hat{f}(n)|^2$ is called the *energy spectrum* or the *power spectrum* of the signal (different names in different fields). It's a fundamental fact, known as Rayleigh's identity,[8] that

$$\int_0^1 |f(t)|^2 \, dt = \sum_{n=-\infty}^{\infty} |\hat{f}(n)|^2 .$$

We'll come back to this in Section 1.7.

[8] It's actually Lord Rayleigh, the third Baron Rayleigh.

Viewing a signal: Why do musical instruments sound different? The frequencies and the Fourier coefficients provide the recipe for the signal, and it has proved so useful to view a signal through the lens of its spectrum that the *spectrum analyzer* has been invented to display it. These days you can get an app for your smart phone that does this, often coupled with an oscilloscope to display the picture of the signal in time. You really should pick one up and do your own experiments. Certainly every EE meets these devices in the course of his or her studies. Let's see what the pictures look like for some actual signals

As to the question in the heading, more precisely, why do two instruments sound different even when they are playing the same note? It's because the note that an instrument produces is not a single sinusoid of a single frequency, not a pure A at 440 Hz, for example, but a sum of many sinusoids each of its own frequency and each contributing its own amount of energy. The complicated wave that reaches your ear is the combination of many ingredients, and two instruments sound different because of the different harmonics they produce and because of the different strengths of the harmonics.

Here I am playing an A 440 on my trombone, that most romantic of all instruments.

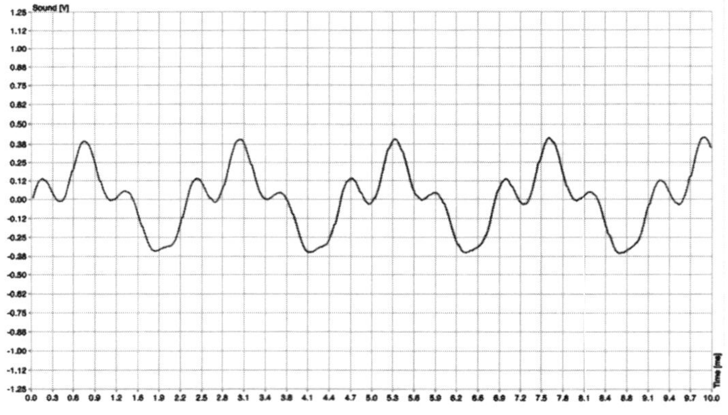

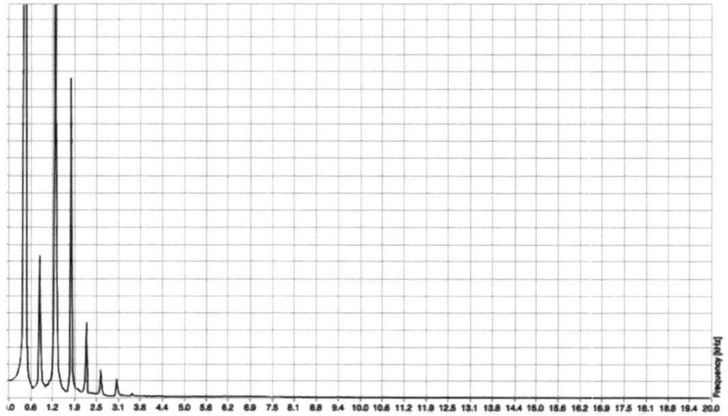

The first picture is what an oscilloscope shows you — the time-varying waveform. The horizontal axis is time marked off in milliseconds, and the vertical is a varying voltage tracking a varying air pressure.

The second picture is a plot of the spectrum. The app is computing Fourier coefficients corresponding to different frequencies and plotting, as a bar chart, their *magnitudes* as a function of frequency.[9] Since $|\hat{f}(-n)| = |\hat{f}(n)|$ only the positive frequencies are shown. The horizontal axis is frequency in kHz, up to 20 kHZ, which is about the upper limit of human hearing. I didn't show units on the vertical axis because different apps have different conventions for displaying the spectrum. It's really the *relative* heights of the bars that are interesting. Some harmonics are contributing more energy relative to others,

I asked my friend Randy Smith, physicist and trumpet player, to play an A 440 on his trumpet and show me the wave form and the spectrum. He obliged:

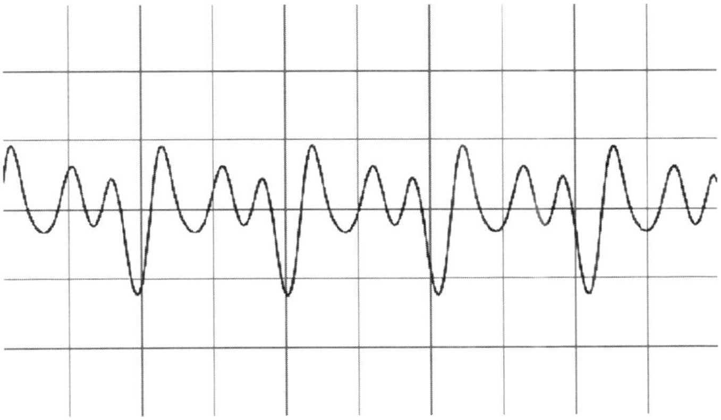

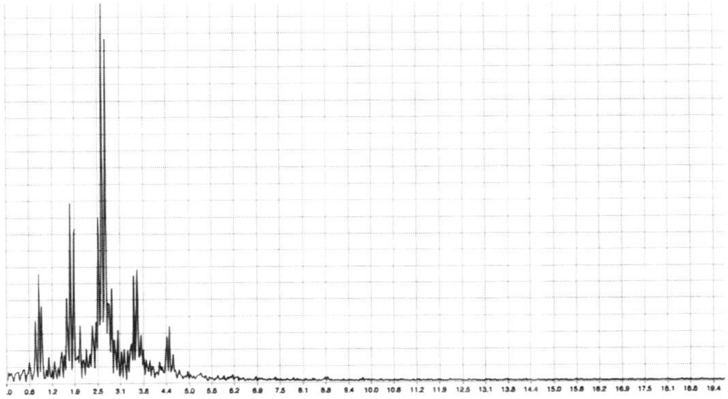

[9]Remember, the Fourier coefficients themselves are complex. The plot does not capture information on the phases. The algorithm used to find the coefficients is the famous Fast Fourier Transform, FFT, which we'll study later. It is applied to a sampled, hence discrete, version of the signal.

For Randy's trumpet there's more energy in the higher harmonics than in the fundamental. (Also, his mic was picking up a certain amount of ambient noise.) Here are Randy's comments:

> The trumpet I play is a Benge, which has the reputation of being a lead horn. It gives a kind of bright tone or sizzle to the lead voice in a big band. I believe the main parameter determining this in trumpet design is the rate of flair of the bell. The Benge flares out more towards the very end, whereas other trumpets flair out more gradually along the last horizontal run to the bell. This is a subtle effect and is difficult to see even in side-by-side comparison.
>
> I have to say I was surprised (or had forgotten) that there seems to be more energy in the upper partials than in the fundamental! Why does the ear even know to perceive this as an A 440, when to my eye I think I would call these notes an A 2640!

Certainly the wave forms look different, but the spectral representation — the frequency domain — gives a much clearer explanation than does the time domain portrait of why the instruments sound different. In the frequency domain you can see how the ingredients differ and by how much. I'd say (I would) that the trombone has a more pure sound in that more harmonics are concentrated around the fundamental. Once in class we did this comparison between my trombone and a singer who had perfect pitch. When she sang an A, it was almost one pure spike at 440 Hz.

The spectral representation also offers opportunities for varieties of signal processing that would not be so easy to do, much less to imagine, in the time domain. Thinking of the spectrogram as a bar chart, it's easy to imagine pushing some bars down, pulling others up, or eliminating whole blocks of bars, operations whose actions in the time domain are far from clear.

As an aside, I've never found a satisfactory answer to the question of why orchestras tune to an oboe playing an A. I thought it might be because the oboe produces a very pure note, mostly a perfect 440 with very few other harmonics, and this would be desirable. In fact, it seems just the opposite is the case. The spectrum of the oboe is very rich, with plenty of harmonics. I wonder if this means that whatever instrument you happen to play there's a little bit of you in the oboe and a little bit of the oboe in you. Maybe subconsciously that helps you tune?

Finally, do a little reading on how hearing works. To oversimplify, your inner ear (the cochlea) finds the harmonics and their amounts (the key is resonance) and passes that data on to your brain to do the synthesis. Nature knew Fourier analysis all along!

1.3.4. Changing the period, and another reciprocal relationship. We've standardized to assume that our signals have period 1, and the formulas reflect that choice. Changing to a base period other than 1 does not present too stiff a challenge, and it brings up a very important phenomenon.

Suppose $f(t)$ has period T. If we scale the time and let

$$g(t) = f(Tt),$$

then $g(t)$ has period 1,

$$g(t+1) = f(T(t+1)) = f(Tt+T) = f(Tt) = g(t).$$

Now suppose we can represent $g(t)$ as the Fourier series

$$g(t) = \sum_{n=-\infty}^{\infty} c_n e^{2\pi int}.$$

Write $s = Tt$, so that $g(t) = f(s)$. Then

$$f(s) = g(t) = \sum_{n=-\infty}^{\infty} c_n e^{2\pi int} = \sum_{n=-\infty}^{\infty} c_n e^{2\pi ins/T}.$$

The harmonics for $f(s)$ are now $e^{2\pi ins/T}$.

What about the coefficients? For the expansion of $g(t)$, which has period 1,

$$c_n = \hat{g}(n) = \int_0^1 e^{-2\pi int} g(t)\, dt.$$

Then making the same change of variable $s = Tt$, the integral becomes, in terms of $f(s)$,

$$\frac{1}{T}\int_0^T e^{-2\pi ins/T} f(s)\, ds.$$

These are the Fourier coefficients for the expansion of $f(s)$ in the harmonics $e^{2\pi ins/T}$.

To summarize, and calling the variable t again, the Fourier series for a function $f(t)$ of period T is

$$f(t) = \sum_{n=-\infty}^{\infty} c_n e^{2\pi int/T}$$

where the coefficients are given by

$$c_n = \frac{1}{T}\int_0^T e^{-2\pi int/T} f(t)\, dt.$$

We could again use the notation $\hat{f}(n)$ for the coefficients, but if you do so just make sure to remind yourself that the period is now T. The coefficient c_0 is still the average value of the function because we divide the integral by the length of the period interval.

As in the case of period 1, we can integrate over any interval of length T to find c_n. For example,

$$c_n = \frac{1}{T}\int_{-T/2}^{T/2} e^{-2\pi int/T} f(t)\, dt.$$

Remark. As we'll see later, there are reasons to take the harmonics to be
$$\frac{1}{\sqrt{T}}e^{2\pi int/T}$$
and the Fourier coefficients then to be
$$c_n = \frac{1}{\sqrt{T}}\int_0^T e^{-2\pi int/T} f(t)\,dt\,.$$
This makes no difference in the final formula for the series because we have two factors of $1/\sqrt{T}$ coming in, one from the differently normalized Fourier coefficient and one from the differently normalized complex exponential.

Time domain – frequency domain reciprocity. We can read a little more into the expansion
$$f(t) = \sum_{n=-\infty}^{\infty} c_n e^{2\pi int/T}\,.$$
In the time domain the signal repeats after T seconds, while the points in the spectrum are 0, $\pm 1/T$, $\pm 2/T$, $\dots$, which are spaced $1/T$ apart. (Of course for period $T = 1$ the spacing in the spectrum is also 1.) This is worth elevating to an aphorism:

- The larger the period in time, the smaller the spacing in the spectrum. The smaller the period in time, the larger the spacing in the spectrum.

We'll see other instances of this aphorism.[10]

Thinking, loosely, of long periods as slow oscillations and short periods as fast oscillations, convince yourself that the statment makes intuitive sense. If you allow yourself to imagine letting $T \to \infty$, you can also allow yourself to imagine the discrete set of frequencies becoming a continuum of frequencies. We'll allow ourselves to imagine this when we introduce the Fourier transform in the next chapter.

1.4. Two Examples and a Warning

All of this is fine, but does it really work? Given a periodic function, can we expect to write it as a sum of exponentials in the way we have described? Let's look at an example.

Consider a *square wave* of period 1, such as illustrated below.

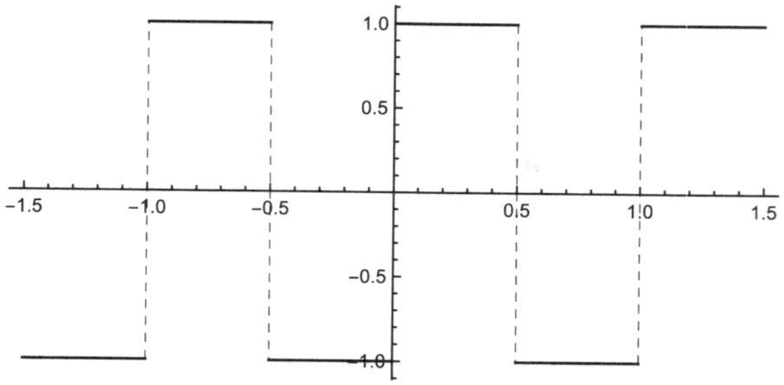

[10]We'll also have other such aphorisms — they're meant to help you organize your understanding and intuition for the subject and for the applications.

Let's calculate the Fourier coefficients. The function is

$$f(t) = \begin{cases} +1, & 0 \le t < \frac{1}{2}, \\ -1, & \frac{1}{2} \le t < 1, \end{cases}$$

and then extended to be periodic of period 1 by repeating the pattern. The zeroth coefficient is the average value of the function on $0 \le t \le 1$, and this is zero. For the other coefficients we have

$$\begin{aligned} \hat{f}(n) &= \int_0^1 e^{-2\pi int} f(t)\, dt \\ &= \int_0^{1/2} e^{-2\pi int}\, dt - \int_{1/2}^1 e^{-2\pi int}\, dt \\ &= \left[-\frac{1}{2\pi in} e^{-2\pi int} \right]_0^{1/2} - \left[-\frac{1}{2\pi in} e^{-2\pi int} \right]_{1/2}^1 = \frac{1}{\pi in}\big(1 - e^{-\pi in}\big). \end{aligned}$$

We should thus consider the *infinite* Fourier series

$$\sum_{n \ne 0} \frac{1}{\pi in}\big(1 - e^{-\pi in}\big) e^{2\pi int}\,.$$

We can write this in a simpler form by first noting that

$$1 - e^{-\pi in} = \begin{cases} 0, & n \text{ even}, \\ 2, & n \text{ odd}, \end{cases}$$

so the series becomes[11]

$$\sum_{n \text{ odd}} \frac{2}{\pi in} e^{2\pi int}\,.$$

Now combine the positive and negative terms and use

$$e^{2\pi int} - e^{-2\pi int} = 2i \sin 2\pi nt\,.$$

Substituting this into the series and writing $n = 2k+1$, our final answer is[12]

$$\frac{4}{\pi} \sum_{k=0}^{\infty} \frac{1}{2k+1} \sin 2\pi(2k+1)t\,.$$

What kind of series is this? In what sense does it converge, if at all, and to what does it converge? Can we represent $f(t)$ as a Fourier series through the sum?

[11] Consistency check: The signal is real and odd and the Fourier coefficient, $\hat{f}(n) = 2/\pi in$, is an odd function of n and purely imaginary.

[12] Another consistency check: The signal is odd and this jibes with the Fourier series having only sine terms.

The graphs below are sums of terms up to frequencies 11 (k from 0 to 5) and 31 (k from 0 to 15), respectively.

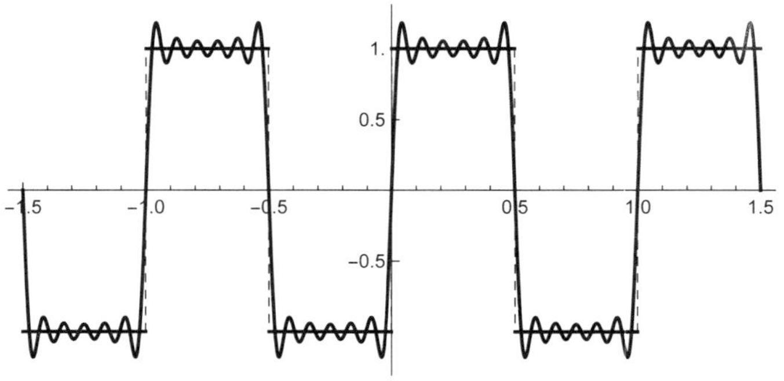

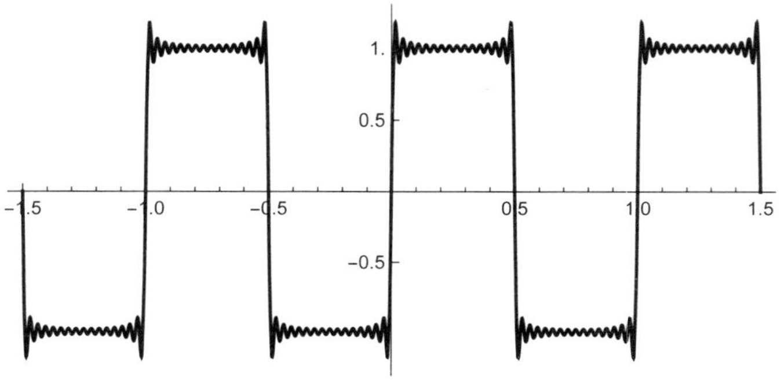

We see a strange phenomenon. We certainly see the general shape of the square wave, but there's trouble at the corners. We'll return to this in Section 1.5.

In retrospect, we shouldn't expect to represent a function like the square wave by a *finite* sum of complex exponentials. Why? Because a finite sum of continuous functions is continuous and the square wave has jump discontinuities. For maybe the first time in your life, one of those theorems from calculus that seemed so pointless at the time makes an appearance: The sum of two (or a finite number) of continuous functions is continuous. Whatever else we may be able to conclude about a Fourier series representation for a square wave, it must contain arbitrarily high frequencies.

I picked the example of a square wave because it's easy to carry out the integrations needed to find the Fourier coefficients. However, it's not only a discontinuity that forces high frequencies. Take a triangle wave, say defined for $|t| \leq 1/2$ by

$$g(t) = \frac{1}{2} - |t| = \begin{cases} \frac{1}{2} + t, & -\frac{1}{2} \leq t \leq 0, \\ \frac{1}{2} - t, & 0 \leq t \leq +\frac{1}{2}, \end{cases}$$

and then extended to be periodic of period 1 by repeating the pattern. Here's a plot of five cycles.

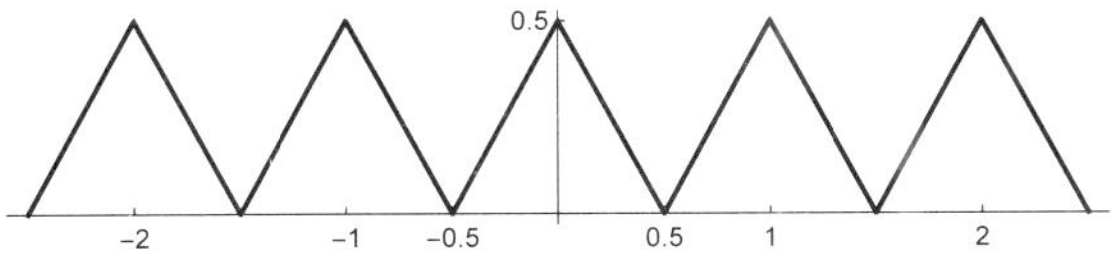

The signal is continuous. There are no jumps, though there are corners.

It takes a little more work than for the square wave to find the Fourier coefficients. The zeroth coefficient is the area of the triangle, so $\hat{g}(0) = 1/4$. For the others, when $n \neq 0$,

$$\begin{aligned}\hat{g}(n) &= \int_{-1/2}^{1/2} e^{-2\pi int}\Big(\frac{1}{2} - |t|\Big)\,dt \\ &= \frac{1}{2}\int_{-1/2}^{1/2} e^{-2\pi int}\,dt - \int_{-1/2}^{1/2} e^{-2\pi int}|t|\,dt\,.\end{aligned}$$

The first integral is zero, leaving us with

$$\begin{aligned}\hat{g}(n) &= -\int_{-1/2}^{1/2} e^{-2\pi int}|t|\,dt \\ &= \int_{-1/2}^{0} e^{-2\pi int}t\,dt - \int_{0}^{1/2} e^{-2\pi int}t\,dt\,.\end{aligned}$$

For each integral, integration by parts is called for with $u = t$ and $dv = e^{-2\pi int}\,dt$.[13] The integrals don't directly combine, but they are related in a way that can save us a little work (something you should always look to do) through the following observation. Give the first integral a name, say,

$$A(n) = \int_{-1/2}^{0} e^{-2\pi int}t\,dt\,.$$

Then

$$\begin{aligned}A(-n) &= \int_{-1/2}^{0} e^{2\pi int}t\,dt \\ &= \int_{1/2}^{0} e^{-2\pi ins}(-s)(-ds) \quad \text{(substituting } s = -t\text{)} \\ &= -\int_{0}^{1/2} e^{-2\pi ins}s\,ds\,;\end{aligned}$$

that's the second integral. Thus

$$\hat{g}(n) = A(n) + A(-n)$$

[13]On the off chance that you've blotted this out of memory, here's what the formula looks like, as it's usually written:

$$\int_a^b u\,dv = [uv]_a^b - \int_a^b v\,du\,.$$

To apply integration by parts in a given problem is to decide which part of the integrand is u and which part is dv.

and it remains to find $A(n)$ alone. Notice also that this expression for $\hat{g}(n)$ implies that $\hat{g}(n) = \hat{g}(-n)$, which in turn implies that $\hat{g}$ is both even and real as a function of n *just as it should be* (consistency check!).

I'm inclined not to integrate by parts in public, or in print, but I'm pleased to report the result:

$$\int_{-1/2}^{0} e^{-2\pi i n t} t \, dt = \frac{1}{4\pi^2 n^2}(1 + e^{\pi i n}(\pi i n - 1)).$$

Finding $\hat{g}(n)$ is then pretty easy:

$$\hat{g}(n) = \frac{1}{2\pi^2 n^2}(1 - n\pi \sin n\pi - \cos n\pi) = \frac{1}{2\pi^2 n^2}(1 - \cos n\pi).$$

This is even and real (one more check) and, actually,

$$\hat{g}(n) = \begin{cases} 0, & n \text{ even}, \\ \frac{1}{\pi^2 n^2}, & n \text{ odd}. \end{cases}$$

As with the square wave, we combine the $+n$ and $-n$ terms, for n odd:

$$\hat{g}(n)e^{2\pi i n t} + \hat{g}(-n)e^{-2\pi i n t} = \frac{2}{\pi^2 n^2} \cos 2\pi n t$$

and we can then write the Fourier series as

$$\tfrac{1}{4} + \sum_{k=0}^{\infty} \frac{2}{\pi^2 (2k+1)^2} \cos(2\pi(2k+1)t).$$

This time the series involves only cosines, a reflection of the fact that the triangle wave is an even function.

Here, as for the square wave, there are infinitely many terms. This is yet another occurrence of one of those calculus theorems: the sines and cosines are differentiable to any order, so any finite sum of them is also differentiable to any order. We cannot expect a finite Fourier series to represent the triangle wave exactly because the triangle wave has corners. Let's state an aphorism:

- A discontinuity in any order derivative of a periodic function will force an infinite Fourier series.

This is not to say that if the function is differentiable to any order, then the Fourier series is necessarily a finite sum. It isn't.

Consider that a warning: finite Fourier series are an exception. Convergence has to be an issue.

How good a job do the finite sums do in approximating the triangle wave? Here are plots of two cycles showing approximations up to the first 7 and 15 frequencies, respectively.

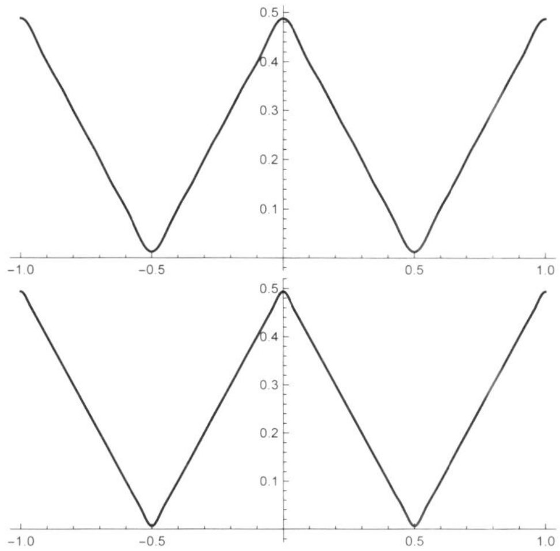

Pretty good. No strange wiggling as with the discontinuous square wave. The corners are rounded, as they must be for a finite sum.

Note also that for the triangle wave the coefficients decrease like $1/n^2$ while for the square wave they decrease like $1/n$. This has exactly to do with the fact that the square wave is discontinuous while the triangle wave is continuous but its *derivative* is discontinuous. If you can wait that long, we'll explain this, and more, in Section 1.8.2.

One thing common to these two examples might be stated as another aphorism:

- It takes high frequencies to make sharp corners.

This particular aphorism is important, for example, in questions of filtering, a topic we'll consider in detail later. For now:

- Filtering (usually) means cutting off, either in time or in frequency.
- Cutting off means sharp corners.
- Sharp corners in time means high frequencies.

This comes up in signal processing for music, for example. If you're not careful to avoid discontinuities in filtering the signal (the music), you'll hear clicks — symptoms of high frequencies — when the signal is played back. A sharp cutoff will inevitably yield an unsatisfactory result, so you have to design your filters to minimize this problem.

1.5. The Math, Part 1: A Convergence Result

At this stage we have only demonstrated that if it is possible to write a periodic function as a sum of simple harmonics,

$$f(t) = \sum_{n=-N}^{N} c_n e^{2\pi int} ,$$

then the coefficients are

$$c_n = \hat{f}(n) = \int_0^1 e^{-2\pi int} f(t)\, dt\,.$$

We also have some examples that indicate the possible difficulties in this sort of representation; i.e., an infinite series may be required and then convergence questions inevitably loom.

In this short section we'll provide a quotable result on convergence, enough to (mostly) soothe any unease you may be feeling. We'll provide a few further details later, as well as a much different take on what we've seen so far on the mathematical issues raised by Fourier series.

1.5.1. The theorem you want. Start with a periodic function $f(t)$ of period 1. Find the Fourier coefficients, assuming that those integrals are fine, and form the Fourier series

$$\sum_{n=-\infty}^{\infty} \hat{f}(n) e^{2\pi int}\,.$$

You want to know two things:

(1) If you plug a point t_0 into the series, as in $\sum_{n=-\infty}^{\infty} \hat{f}(n) e^{2\pi int_0}$, does it converge?

(2) If so, does it converge to $f(t_0)$? That is, do you have

$$f(t_0) = \sum_{n=-\infty}^{\infty} \hat{f}(n) e^{2\pi int_0}\ ?$$

You might have thought that the answer to the first question was enough, but the questions are separate, as we'll see, and you want to know answers to both.

You want answers to these questions because you're a practical, busy person who computes using finite approximations, as I did to plot approximations to the square wave and the triangle wave.[14] You might not use the mathematical terminology, but what you use for your computations are the (finite) *partial sums*, denoted by

$$S_N(t) = \sum_{n=-N}^{N} \hat{f}(n) e^{2\pi int}\,.$$

You want to know if these serve as a good approximation to $f(t)$, better and better as N increases. So you slide into wanting to know:

$1'$ For a point t_0 does $\lim_{N\to\infty} S_N(t_0)$ exist?

$2'$ If so, do we have $f(t_0) = \lim_{N\to\infty} S_N(t_0)$?

You may have seen limits of partial sums in a calculus course, but it may also have been awhile.

All questions about convergence of Fourier series are questions about limits of the corresponding partial sums. Have no fear, there's a fine result that answers the questions. Let me give an easily quotable version, followed by a comment on terminology.

[14]Or, if you're a mathematician, you just want to know.

Theorem. Suppose $f(t)$ is a periodic, piecewise continuous function with a piecewise continuous derivative. If $f(t)$ is continuous at t_0, then

$$f(t_0) = \lim_{N\to\infty} S_N(t_0) = \sum_{n=-\infty}^{\infty} \hat{f}(n)e^{2\pi i n t_0}.$$

If $f(t)$ has a jump discontinuity at t_0 with right and left limits $f(t_0^+)$ and $f(t_0^-)$, respectively, then

$$\frac{1}{2}(f(t_0^+) + f(t_0^-)) = \lim_{N\to\infty} S_N(t_0) = \sum_{n=-\infty}^{\infty} \hat{f}(n)e^{2\pi i n t_0}.$$

The terminology: To say that $f(t)$ is piecewise continuous is to say that it has at most a finite number of finite jump discontinuities on any period interval.[15] The same is also assumed to hold for $f'(t)$. The type of convergence identified in the theorem is often referred to as *pointwise convergence*, meaning that if you plug in a point, you get convergence. Yes.

This theorem is generally attributed, in various forms, to the German mathematician Lejeune Dirichlet, and the hypotheses are often referred to as the Dirichlet conditions. A little more on this in Section 1.8.

It's an eminently reasonable result, and in particular it applies to the square and triangle waves. For the triangle wave, the Fourier series converges at each point to the function. For the square wave, the Fourier series converges either to 1 or to -1 away from the jumps, and to $1/2$ at the jumps.

The weird jiggling at the jumps, as seen in the plots, is called *Gibbs's Phenomenon*, named after J. W. Gibbs.[16]. It persists even as N gets larger. It was observed experimentally by Michelson[17] and Stratton who designed a mechanical device to draw finite Fourier series. Michelson and Stratton assumed that the extra wiggles they were seeing at jumps came from a mechanical problem with the machine. But Gibbs (who used a sawtooth wave as an example) showed that the phenomenon is real and *does not go away* even in the limit. The oscillations may become more compressed, but they don't go away. A standard way to formulate Gibbs's phenomenon precisely is for a square wave that jumps from -1 to $+1$ at $t = 0$ when t goes from negative to positive. Away from the single jump discontinuity, $S_N(t)$ tends uniformly to the values, $+1$ or -1 as the case may be, as $N \to \infty$. Hence the precise statement of Gibbs's phenomenon will be that the maximum of $S_N(t)$ remains greater than 1 as $N \to \infty$. And that's what is proved:

$$\lim_{N\to\infty} \max S_N(t) = 1.1789797\ldots.$$

[15] So the jumps in the values of $f(t)$, if any, are finite and in particular $f(t)$ is bounded — it doesn't become infinite at any point in a period interval. At a point where $f(t)$ is continuous we have $f(t_0) = \frac{1}{2}(f(t_0^+) + f(t_0^-))$, because the left and right limits agree. So the two parts of the theorem could actually be combined and stated simply as $\frac{1}{2}(f(t_0^+) + f(t_0^-)) = \sum_{n=-\infty}^{\infty} \hat{f}(n)e^{2\pi i n t_0}$ at any t_0. But nobody thinks of it that way.

[16] Gibbs was a very eminent American mathematician, physicist, and engineer. Yale awarded him the first American doctorate in engineering, and he continued as a professor there for his entire career.

[17] That's the Albert Michelson of the famous Michelson and Morley experiment proving that there is no luminiferous aether.

So the overshoot is almost 18% — quite noticeable! If you're curious, and it's mostly a curiosity, you can find a discussion in most books on Fourier series.[18] There's something here that may bother you. We have the theorem on pointwise convergence that says at a jump discontinuity the partial sums converge to the average of the values at the jump. We also have Gibbs's phenomenon and the picture of an overshooting oscillation that doesn't go away. How can these two pictures coexist? If you're confused, it's because you're thinking that convergence of $S_N(t)$ is the same as convergence of the graphs of the $S_N(t)$ to the graph of the sawtooth function. But they are *not* the same thing. It's the distinction between pointwise and uniform convergence. We'll return to this in Section 1.8 and specifically in Section 1.8.4.

1.6. Fourier Series in Action

We'll return to the math, which is both interesting and important, but I want first to present a few model cases of how Fourier series can be applied. The range of applications is vast, so the principle of selection has been to choose examples that are both interesting in themselves and have connections with different areas.

The first applications are to heat flow. These are classical, celebrated problems and should be in your storehouse of general knowledge. Another reason for including them is the *form* that one of the solutions takes as a convolution integral, anticipating things to come. We'll also look briefly at how the differential equation governing heat flow comes up in other areas. The key word is *diffusion*.

The next application is not classical at all. It has to do, on the one hand, with synthesizing sound, and on the other, as we'll see later, with sampling theory. Later, when we do higher-dimensional Fourier analysis, we'll have a final application of higher-dimensional Fourier series to random walks on a lattice. With a little probability thrown in, the analysis of the problem is not beyond what we know to this point, but enough is enough for now.

1.6.1. Hot enough for ya? The study of how temperature varies in time and in space over a heated region was the first use by Fourier in the 1820s of the method of expanding a function into a series of trigonometric functions. The physical phenomenon is described, at least approximately, by a partial differential equation, and Fourier series can be used to write down solutions.

I'll give a brief, standard derivation of the differential equation in one spatial dimension, so the configuration to think of is a one-dimensional rod — straight or bent, it doesn't matter. The argument involves a number of common but difficult, practically undefined, terms, first among them the term "heat," followed closely by the term "temperature."

As it is usually stated, heat is a transfer of energy (another undefined term, thank you) due to temperature difference. The transfer process is called "heat." What gets transferred is energy. Because of this, heat is usually identified as a form of energy and has units of energy. We speak of heat as a transfer of energy, and hence we speak of heat flow, because, like so many other physical quantities, heat

[18] I am indebted to Gerald Folland for pointing out that originally I had quoted a wrong figure of about 9% and that I had joined a long line of distinguished people who had done the same.

is only interesting if it's associated with a change. Temperature, more properly called thermodynamic temperature (formerly absolute temperature), is a derived quantity. The temperature of a substance is proportional to the kinetic energy of the atoms in the substance.[19] A substance at temperature 0 (absolute zero) cannot transfer energy — it's not "hot." The principle at work, according to Newton, is:

> A temperature difference between two substances in contact with each other causes a *transfer* of energy from the substance of higher temperature to the substance of lower temperature, and that's heat, or heat flow. No temperature difference, no heat.

Back to the rod. However the rod got it, its temperature is a function of both the spatial variable x, giving the position along the rod, and of the time t. We let $u(x,t)$ denote the temperature, and the problem is to find it. The description of heat, just above, with a little amplification, is enough to propose a partial differential equation that $u(x,t)$ should satisfy.[20]

To derive the equation, we introduce $q(x,t)$, the amount of heat that "flows" per second at x and t, meaning that $q(x,t)$ is the rate at which energy is transferred at x and t. Newton's law of cooling says that this is proportional to the gradient of the temperature:

$$q(x,t) = -ku_x(x,t)\,, \quad k > 0\,.$$

The reason for the minus sign is that if $u_x(x,t) > 0$, i.e., if the temperature is increasing at x, then the rate at which heat flows at x is negative — from hotter to colder, hence back from x. The constant k can be identified with the reciprocal of *thermal resistance* of the substance. For a given temperature gradient, the higher the resistance, the smaller the heat flow per second, and similarly the smaller the resistance, the greater the heat flow per second.

As the heat flows from hotter to colder, the temperature rises in the colder part of the substance. The rate at which the temperature rises at x, given by the derivative $u_t(x,t)$, is proportional to the rate at which heat accumulates (for lack of a better term) per unit length. Now $q(x,t)$ is already a rate — the heat flow per second — so the rate at which heat accumulates per unit length is the "rate in" minus the "rate out" per length, which is (if the heat is flowing from left to right)

$$\frac{q(x,t) - q(x+\Delta x,t)}{\Delta x}\,.$$

Thus in the limit

$$u_t(x,t) = -k'q_x(x,t)\,, \quad k' > 0\,.$$

The constant k' can be identified with the reciprocal of the *thermal capacity* per unit length. Thermal resistance and thermal capacity are not the standard terms, but they can be related to standard terms, e.g., specific heat. They are used here because of the similarity of heat flow to electrical phenomena — see the discussion, below, of the mathematical analysis of telegraph cables.

---

[19]With this (partial) definition, the unit of temperature is the Kelvin.

[20]This follows Bracewell's presentation.

Next, differentiate the first equation with respect to x to get

$$q_x(x,t) = -ku_{xx}(x,t)\,,$$

and substitute this into the second equation to obtain an equation involving $u(x,t)$ alone:

$$u_t(x,t) = kk'u_{xx}(x,t)\,.$$

This is the *heat equation*.

To summarize, in whatever particular context it's applied, the setup for a problem based on the heat equation involves:

- A region in space.
- An initial distribution of temperature on that region.

It's natural to think of fixing one of the variables and letting the other change. Then the solution $u(x,t)$ tells you:

- For each fixed time t how the temperature is distributed on the region.
- At each fixed point x how the temperature is changing over time.

We want to look at two examples of using Fourier series to solve such a problem: heat flow on a circle and, more dramatically, the temperature of the earth. These are nice examples because they show different aspects of how the methods can be applied and, as mentioned above, they exhibit forms of solutions, especially for the circle problem, of a type we'll see frequently.

Why a circle, why the earth — and why Fourier methods? Because in each case the function $u(x,t)$ will be *periodic* in one of the variables. In one case we work with periodicity in space and in the other with periodicity in time.

Heating a circle. Suppose a circle (e.g., a wire) is heated up. This provides the circle with an initial distribution of temperature, heat then flows around the circle, and the temperature changes over time. At any fixed time the temperature must be a periodic function of the position on the circle, for if we specify points on the circle by an angle θ, then the temperature, as a function of θ, is the same at θ and at $\theta + 2\pi$.

We can imagine a circle as an interval with the endpoints identified, say the interval $0 \le x \le 1$. We let $f(x)$ be the initial temperature, and we let $u(x,t)$ be the temperature as a function of position and time. Then

$$u(x,0) = f(x)\,.$$

We choose units so that the heat equation takes the form

$$u_t = \tfrac{1}{2}u_{xx}\,,$$

that is, so the constant depending on physical attributes of the wire is 1/2.[21]

The functions $f(x)$ and $u(x,t)$ are periodic in the spatial variable x with period 1, and we can try expanding $u(x,t)$ as a Fourier series with coefficients that

[21] Nothing particularly special to this normalization, but I'm used to it. It's not hard to incorporate the constant if you'd like to.

depend on time:

$$u(x,t) = \sum_{n=-\infty}^{\infty} c_n(t)e^{2\pi inx} \quad \text{where} \quad c_n(t) = \int_0^1 e^{-2\pi inx}u(x,t)\,dx\,.$$

The unknowns are the coefficients $c_n(t)$. Note that

$$c_n(0) = \int_0^1 e^{-2\pi inx}u(x,0)\,dx = \int_0^1 e^{-2\pi inx}f(x)\,dx = \hat{f}(n)\,.$$

We want to plug the Fourier series for $u(x,t)$ into the heat equation, and for this we allow ourselves the possibility of differentiating the series term by term.[22] So

$$u_t(x,t) = \sum_{n=-\infty}^{\infty} c_n'(t)e^{2\pi inx}\,,$$

$$u_{xx}(x,t) = \sum_{n=-\infty}^{\infty} c_n(t)(2\pi in)^2 e^{2\pi inx} = \sum_{n=-\infty}^{\infty} (-4\pi^2n^2)c_n(t)e^{2\pi inx}\,,$$

and

$$\sum_{n=-\infty}^{\infty} c_n'(t)e^{2\pi inx} = u_t(x,t) = \frac{1}{2}u_{xx}(x,t) = \sum_{n=-\infty}^{\infty} (-2\pi^2n^2)c_n(t)e^{2\pi inx}\,.$$

Equate coefficients[23] to conclude that

$$c_n'(t) = -2\pi^2n^2c_n(t)\,.$$

We have found that $c_n(t)$ satisfies a simple *ordinary* differential equation

$$c_n'(t) = -2\pi^2n^2\,c_n(t)\,,$$

whose solution is

$$c_n(t) = c_n(0)e^{-2\pi^2n^2t} = \hat{f}(n)e^{-2\pi^2n^2t}\,.$$

The solution to the heat equation is therefore

$$u(x,t) = \sum_{n=-\infty}^{\infty} \hat{f}(n)e^{-2\pi^2n^2t}e^{2\pi inx}\,.$$

Pretty impressive. Thank you, Fourier.

Having come this far mathematically, let's look at the form of the solution and see what properties it captures, physically. If nothing else, this can serve as a reality check on where our reasoning has led. Let's look at the two extreme cases, $t = 0$ and $t = \infty$. We know what should happen at $t = 0$, but taking off from the final

[22] Yes, this can be justified.

[23] See previous footnote.

formula we have

$$u(x,0) = \sum_{n=-\infty}^{\infty} \hat{f}(n)e^{-2\pi^2 n^2 \cdot 0} e^{2\pi i n x} = \sum_{n=-\infty}^{\infty} \hat{f}(n)e^{2\pi i n x},$$

and this is the Fourier series for $f(x)$,

$$u(x,0) = \sum_{n=-\infty}^{\infty} \hat{f}(n)e^{2\pi i n x} = f(x).$$

Perfect.

What happens as $t \to \infty$? The exponential factors $e^{-2\pi^2 n^2 t}$ tend to zero *except* when $n = 0$. Thus all terms with $n \neq 0$ die off in the limit and

$$\lim_{t\to\infty} u(x,t) = \hat{f}(0).$$

This, too, has a natural physical interpretation, for

$$\hat{f}(0) = \int_0^1 f(x)\,dx$$

is the average value of $f(x)$, or, physically in this application, the *average of the initial temperature*. We see, as a consequence of the mathematical form of the solution, that as $t \to \infty$ the temperature $u(x,t)$ tends to a constant value, the average of the initial temperature distribution. You might very well have reasoned your way to that conclusion thinking physically, but it's good to see it play out mathematically as well.[24]

The formula for $u(x,t)$ is a neat way of writing the solution, and we could leave it at that, but in anticipation of things to come it's useful to bring back the integral definition of $\hat{f}(n)$ and write the expression differently. Write the formula for $\hat{f}(n)$ as

$$\hat{f}(n) = \int_0^1 f(y)e^{-2\pi i n y}\,dy.$$

(Don't use x as the variable of integration since it's already in use in the formula for $u(x,t)$.) Then

$$\begin{aligned} u(x,t) &= \sum_{n=-\infty}^{\infty} e^{-2\pi^2 n^2 t} e^{2\pi i n x} \int_0^1 f(y)e^{-2\pi i n y}\,dy \\ &= \int_0^1 f(y) \sum_{n=-\infty}^{\infty} e^{-2\pi^2 n^2 t} e^{2\pi i n (x-y)}\,dy, \end{aligned}$$

or, with

$$g(x-y,t) = \sum_{n=-\infty}^{\infty} e^{-2\pi^2 n^2 t} e^{2\pi i n (x-y)},$$

[24]What happens to the solution if the initial temperature itself is constant?

we have

$$u(x,t) = \int_0^1 f(y)g(x-y,t)\,dy\,.$$

The function

$$g(x,t) = \sum_{n=-\infty}^{\infty} e^{-2\pi^2 n^2 t} e^{2\pi i n x}$$

is called *Green's function*,[25] or the *fundamental solution* for the heat equation for a circle. Note that $g(x,t)$ is a periodic function of period 1 in the spatial variable. The expression for the solution $u(x,t)$ is a *convolution integral*, a term you have probably heard in earlier classes, but new here. In words, $u(x,t)$ is given by the convolution of the initial temperature $f(x)$ with Green's function $g(x,t)$. This is a very important fact.

Convolution for periodic functions. Convolution gets a chapter of its own once we have introduced the Fourier transform. A few kind words here for this operation in the setting of periodic functions.

In general (not assuming any extra time dependence, as above), if $f(x)$ and $g(x)$ are periodic, the integral

$$\int_0^1 f(y)g(x-y)\,dy$$

is called the *convolution* of f and g. There's a special notation,

$$(f * g)(x) = \int_0^1 f(y)g(x-y)\,dy\,,$$

read: "f convolved with g."

The integral makes sense because of periodicity. That is, for a given x between 0 and 1 and for y varying from 0 to 1 (as the variable of integration), $x-y$ will assume values outside the interval $[0,1]$. If $g(x)$ were not periodic, it wouldn't make sense to consider $g(x-y)$, but the periodicity is just what allows us to do that. Furthermore, though the definition makes it looks like $f(x)$ and $g(x)$ do not enter symmetrically in forming $f * g$, as long as $f(x)$ is also periodic we have

$$(f * g)(x) = \int_0^1 f(y)g(x-y)\,dy = \int_0^1 f(x-y)g(y)\,dy = (g * f)(x)\,.$$

[25]Named in honor of the British mathematician George Green. You may have encountered *Green's theorem* on line integrals in a vector calculus class. Same Green.

To see this, make a change of variable $u = x - y$ in the first integral. Then

$$\begin{aligned}
(f * g)(x) &= \int_0^1 f(y)g(x-y)\,dy \\
&= \int_x^{x-1} f(x-u)g(u)(-du) \\
&\quad \text{(makes sense because } f(x) \text{ is periodic)} \\
&= \int_{x-1}^{x} f(x-u)g(u)\,du \\
&= \int_0^1 f(x-u)g(u)\,du \\
&\quad \text{(integrand is periodic: any interval of length 1 will do)} \\
&= (g * f)(x)\,.
\end{aligned}$$

There are some problems on convolution and Fourier series at the end of the chapter. Foremost among them is what happens to Fourier coefficients under convolution. The identity is

$$\widehat{f * g}(n) = \hat{f}(n)\hat{g}(n)\,.$$

The Fourier coefficients *multiply.* Another way of putting this is that if

$$f(t) = \sum_{n=-\infty}^{\infty} a_n e^{2\pi int}, \; g(t) = \sum_{n=-\infty}^{\infty} b_n e^{2\pi int}, \quad \text{then} \quad (f*g)(t) = \sum_{n=-\infty}^{\infty} a_n b_n e^{2\pi int}\,.$$

We'll use this property of convolution in Section 1.8, and we'll see this phenomenon again when we do convolution and Fourier transforms. Among other applications, convolution is the basis for using one signal to filter another.

There's an interesting and important property of the solution of the heat equation as a convolution that has to do with shifts. Suppose we shift the initial temperature to $f(x - x_0)$, for some x_0 on the circle — it's like the initial temperature distribution around the circle has just been rotated by an amount x_0. Our sense is that the solution to the heat equation should also be shifted by the same amount, so $u(x, t)$ should become $u(x - x_0, t)$. Let's see how that follows from the convolution form of the solution.

Our arguments have shown that the temperature at x and t arising from an initial temperature $f(x - x_0)$ is

$$\int_0^1 g(x-y,t)f(x-x_0)\,dy\,.$$

Now substitute $u = y - x_0$ to get

$$\begin{aligned}
\int_0^1 g(x-y,t)f(x-x_0)\,dy &= \int_{-x_0}^{1-x_0} g(x-(u+x_0),t)f(u)\,du \\
&= \int_{-x_0}^{1-x_0} g((x-x_0)-u,t)f(u)\,du\,.
\end{aligned}$$

Since the integrand is periodic of period 1,

$$\int_{-x_0}^{1-x_0} g((x-x_0)-u,t)f(u)\,du = \int_0^1 g((x-x_0)-y,t)f(y)\,dy$$

(calling the variable of integration y again).

The latter integral is exactly $u(x-x_0,t)$ where $u(x,t)$ is the solution corresponding to an unshifted initial temperature $f(x)$.

To think more in EE terms, if you know the terminology coming from linear systems, Green's function $g(x,t)$ is the *impulse response* associated with the linear system "heat flow on a circle," meaning:

- Inputs go in: the initial heat distribution $f(x)$.
- Outputs come out: the temperature $u(x,t)$.
- The system from inputs to outputs is linear; i.e., a superposition of inputs produces a superposition of outputs. (You can think about why that's true.)
- The output is the convolution of the input with g:

$$u(x,t) = \int_0^1 f(y)g(x-y,t)\,dy\,.$$

In fact, convolution *has* to come in because of the shifting property. This will be a central topic in the chapter on linear systems.

Convolutions occur absolutely everywhere in Fourier analysis, and we'll be spending a lot of time with them. In our example, as a formula for the solution, the convolution may be interpreted as saying that for each time t the temperature $u(x,t)$ at a point x is a kind of smoothed average of the initial temperature distribution $f(x)$. In other settings a convolution may have different interpretations.

Heating the earth, storing your wine. The wind blows, the rain falls, and the temperature at any particular place on earth changes over the course of a year. Let's agree that the way the temperature varies is pretty much the same year after year, so that the temperature at any particular place on earth is approximately a periodic function of time, where the period is 1 year. What about the temperature x-meters *under* that particular place? How does the temperature depend on x and t?[26]

Fix a place on earth and let $u(x,t)$ denote the temperature x-meters underground at time t. We assume again that u satisfies the heat equation, $u_t = \frac{1}{2}u_{xx}$. This time we try a solution of the form

$$u(x,t) = \sum_{n=-\infty}^{\infty} c_n(x)e^{2\pi int}\,,$$

reflecting the periodicity in time. From this,

$$u_{xx}(x,t) = \sum_{n=-\infty}^{\infty} c_n''(x)e^{2\pi int} \quad \text{and} \quad u_t(x,t) = \sum_{n=-\infty}^{\infty} c_n(x)(2\pi in)e^{2\pi int}\,.$$

[26]This example is taken from *Fourier Series and Integrals* by H. Dym and H. McKean, who credit A. Sommerfeld.

Plugging u_{xx} and u_t into the heat equation then gives

$$c_n''(x) = 4\pi i n c_n(x)\,,$$

a second-order ordinary differential equation for $c_n(x)$.

We can solve this second-order differential equation in x easily by noting that

$$(4\pi i n)^{1/2} = \pm(2\pi|n|)^{1/2}(1 \pm i)\,,$$

where in the second $\pm$ we take $1+i$ when $n > 0$ and $1-i$ when $n < 0$. I'll leave it to you to decide that the root to take in the first $\pm$ is $-(2\pi|n|)^{1/2}(1 \pm i)$. Thus

$$c_n(x) = A_n e^{-(2\pi|n|)^{1/2}(1\pm i)x}\,.$$

What is the initial value $A_n = c_n(0)$? We assume that at $x = 0$ there is a periodic function of t that models the temperature (at the fixed spot on earth) over the course of the year. Call this $f(t)$. Then $u(0,t) = f(t)$, and

$$c_n(0) = \int_0^1 u(0,t)e^{-2\pi i n t}\,dt = \hat{f}(n)\,.$$

Our solution is

$$u(x,t) = \sum_{n=-\infty}^{\infty} \hat{f}(n)e^{-(2\pi|n|)^{1/2}(1\pm i)x}e^{2\pi i n t}\,.$$

Care to express this as a convolution?

The expression for $u(x,t)$ is not beautiful, but it becomes more interesting if we rearrange the exponentials to isolate the periodic parts (the ones that have an i in them) from the nonperiodic part that remains. The latter is $e^{-(2\pi|n|)^{1/2}x}$. The terms then look like

$$\hat{f}(n)\,e^{-(2\pi|n|)^{1/2}x}\,e^{2\pi i n t \mp (2\pi|n|)^{1/2} i x}\,.$$

What's interesting here? The dependence on the depth, x. Each term is *damped* by the exponential

$$e^{-(2\pi|n|)^{1/2}x}$$

and *phase shifted* by the amount $(2\pi|n|)^{1/2}x$.

Take a simple case. Suppose that the temperature at the surface $x = 0$ is given just by $\sin 2\pi t$ and that the mean annual temperature is 0 (in some non-Kelvin scale of temperature); i.e.,

$$\int_0^1 f(t)\,dt = \hat{f}(0) = 0\,.$$

All Fourier coefficients other than the first (and minus first) are zero, and the solution reduces to

$$u(x,t) = e^{-(2\pi)^{1/2}x}\sin(2\pi t - (2\pi)^{1/2}x)\,.$$

Take the depth x so that $(2\pi)^{1/2}x = \pi$. Then the temperature is damped by $e^{-\pi} = 0.04$, quite a bit, and it is half a period (six months) out of phase with the temperature at the surface. The temperature x-meters below stays pretty constant because of the damping, and because of the phase shift it's cool in the summer and warm in the winter. There's a name for a place like that. It's called a cellar.

The first shot in the second industrial revolution. Many types of diffusion processes are in principle similar enough to the flow of heat that they are modeled by the heat equation, or a variant of the heat equation, and Fourier analysis is often used to find solutions. One celebrated example of this was the paper by William Thomson (later Lord Kelvin): "On the theory of the electric telegraph" published in 1855 in the Proceedings of the Royal Society.

The high-tech industry of the mid to late 19th century was submarine telegraphy, and there were challenges. Sharp pulses representing the dots and dashes of Morse code were sent at one end of the cable and in transit, if the cable was very long and if the pulses were sent in too rapid a succession, the pulses were observed to smear out and overlap to the degree that at the receiving end it was impossible to resolve them. The commercial success of telegraph transmissions between continents depended on undersea cables reliably handling a large volume of traffic. What was causing the problem? How should cables be designed? The stakes were high and a quantitative analysis was needed.

A qualitative explanation of signal distortion was offered by Michael Faraday, who was shown the phenomenon by Latimer Clark. Clark, an official of the Electric and International Telegraph Company, had observed the blurring of signals on the Dutch-Anglo line. Faraday surmised that a cable immersed in water became in effect an enormous capacitor, consisting as it does of two conductors — the wire and the water — separated by insulating material (gutta-percha in those days). When a signal was sent, part of the energy went into charging the capacitor, which took time, and when the signal was finished, the capacitor discharged and that also took time. The delay associated with both charging and discharging distorted the signal and caused signals sent too rapidly to overlap.

Thomson took up the problem in two letters to G. Stokes (of Stokes's theorem fame), which became the published paper. We won't follow Thomson's analysis at this point, because, with the passage of time, it is more easily understood via Fourier transforms rather than Fourier series. However, here are some highlights. Think of the whole cable as a (flexible) cylinder with a wire of radius a along the axis and surrounded by a layer of insulation of radius b (thus of thickness $b-a$). To model the electrical properties of the cable, Thomson introduced the "electrostatic capacity per unit length" depending on a, b, and ϵ, the permittivity of the insulator. His formula was

$$C = \frac{2\pi\epsilon}{\ln(b/a)}\,.$$

(You may have done this calculation in an EE or physics class.) He also introduced the "resistance per unit length," denoting it by K. Imagining the cable as a series of infinitesimal pieces and using Kirchhoff's circuit law and Ohm's law on each piece, he argued that the voltage $v(x,t)$ at a distance x from the end of the cable and at a time t must satisfy the partial differential equation

$$v_t = \frac{1}{KC} v_{xx}\,.$$

Thomson states: "This equation agrees with the well-known equation of the linear motion of heat in a solid conductor, and various forms of solution which Fourier has given are perfectly adapted for answering practical questions regarding the use of the telegraph wire."

After the fact, the basis of the analogy is that charge diffusing through a cable may be described in the same way as heat through a rod, with a gradient in electric potential replacing the gradient of temperature, etc. (Keep in mind, however, that the electron was not discovered till 1897.) Here we see K and C playing the role of thermal resistance and thermal capacity in the derivation of the heat equation.

The result of Thomson's analysis that had the greatest practical consequence was his demonstration that "...the time at which the maximum electrodynamic effect of connecting the battery for an instant ..." [sending a sharp pulse, that is] occurs for

$$t_{\max} = \tfrac{1}{6}KCx^2 .$$

The number $t_{\max}$ is what's needed to understand the delay in receiving the signal. It's the fact that the distance from the end of the cable, x, comes in *squared* that's so important. This means, for example, that the delay in a signal sent along a 1,000 mile cable will be 100 times as large as the delay along a 100 mile cable, and not 10 times as large, as was thought. This was Thomson's "Law of squares."

Thomson's work has been called "The first shot in the second industrial revolution."[27] This was when electrical engineering became decidedly mathematical. His conclusions did not go unchallenged, however. Consider this quote of Edward Whitehouse, chief electrician for the Atlantic Telegraph Company, speaking in 1856:

> I believe nature knows no such application of this law [the law of squares] and I can only regard it as a fiction of the schools; a forced and violent application of a principle in Physics, good and true under other circumstances, but misapplied here."

Thomson's analysis did not prevail and the first transatlantic cable was built without regard to his specifications. Thomson said the cable had to be designed to make KC small. The telegraph company thought they could just crank up the power. The continents were joined on August 5, 1858, after four previous failed attempts. (Laying an ocean's worth of cable presented its own difficulties.) The first successful sent message was on August 16. The cable failed three weeks later. Too high a voltage. They fried it.

Rather later, in 1876, Oliver Heaviside greatly extended Thomson's work by including the effects of induction. He derived a more general differential equation for the voltage $v(x,t)$ in the form

$$v_{xx} = KCv_t + SCv_{tt} ,$$

where S denotes the inductance per unit length and, as before, K and C denote the resistance and capacitance per unit length. The significance of this equation, though not realized till later still, is that it allows for solutions that represent propagating *waves*. Indeed, from a partial differential equations point of view the

[27] See *Getting the Message: A History of Communications* by L. Solymar.

equation looks like a mix of the heat equation and the wave equation. (We'll see the wave equation later.) It is Heaviside's equation that is now usually referred to as the "telegraph equation."

The last shot in the Second World War. Speaking of high stakes diffusion processes, in the early stages of the theoretical analysis of atomic explosives, it was necessary to study the diffusion of neutrons produced by fission as they worked their way through a mass of uranium. The question: How much mass is needed so that enough uranium nuclei will fission in a short enough time to produce an explosion? An analysis of this problem was carried out by Robert Serber and some students at Berkeley in the summer of 1942, preceding the opening of the facilities at Los Alamos (where the bulk of the work was done and the bomb was built). They found that the so-called critical mass needed for an explosive chain reaction was about 60 kg of U^{235}, arranged in a sphere of radius about 9 cm (together with a tamper surrounding the uranium). A less careful model of how the diffusion works gives a critical mass of 200 kg. As the story goes, in the development of the German atomic bomb project (which predated the American efforts), Werner Heisenberg worked with a less careful model and obtained too high a number for the critical mass. This set Germany's program back.

For a fascinating and accessible account of this and more, see Robert Serber's *The Los Alamos Primer*. These are the notes of the first lectures given by Serber at Los Alamos on the state of knowledge of atomic bombs, annotated by him for this edition. For a dramatized account of Heisenberg's role in the German atomic bomb project — including the misunderstanding of diffusion — try Michael Frayn's play *Copenhagen*.

Incidentally, as Serber explains:

> ...the origin of the energy released in fission is exactly the same as the origin of the energy released when two atoms or molecules react chemically. It's the electrostatic energy between two similarly charged particles. Two similarly charged particles repel each other. There's an electrical force pushing them apart. Work has to be done to overcome the repulsion and push them together from a large distance ... [He goes on to apply this to the protons in a uranium nucleus.]
>
> Somehow the popular notion took hold long ago that Einstein's theory of relativity, in particular his famous equation $E = mc^2$, plays some essential role in the theory of fission. ...his theory of relativity is not required in discussing fission. The theory of fission is what physicists call a nonrelativistic theory, meaning that relativistic effects are too small to affect the dynamics of the fission process significantly.

1.6.2. A nonclassical example: What's the buzz? A musical tone is a periodic wave and we model it by sinusoids. A pure tone is a single sinusoid, while more complicated tones are sums of sinusoids. The frequencies of the higher harmonics are integer multiples of the fundamental harmonic, and the harmonics will typically have different energies. As a model of the most complete and uniform tone, we might take a sum of *all* harmonics, each present with the same energy, say 1. If we further assume that the period is 1 (i.e., that the fundamental harmonic has frequency 1), then we're looking at the signal

$$f(t) = \sum_{n=-\infty}^{\infty} e^{2\pi int}.$$

What does this sound like? Not very pleasant, depending on your tastes. Let's call it a buzz, with some further comments later.

The sum $\sum_{n=-\infty}^{\infty} e^{2\pi int}$ is not a classical Fourier series in any way, shape, or form. It does not represent a signal with finite energy and the series does not converge in any easily defined sense. Setting aside these inconvenient truths, the buzz is an important signal for several reasons.

A formula can help us to get more of a sense of its properties. In Appendix C, on geometric sums, you're asked to consider

$$D_N(t) = \sum_{n=-N}^{N} e^{2\pi int},$$

a partial sum for the buzz, and to derive the expression

$$D_N(t) = \frac{\sin(2\pi t(N+1/2))}{\sin(\pi t)}.$$

The letter "D" here is for Dirichlet (again) who introduced this expression to study convergence questions. He wasn't thinking about buzzers. In his honor, $D_N(t)$ is called the *Dirichlet kernel*. We see that $D_N(t)$ is an even function; always look for symmetry, you never know when it will come up.

There's another form of the sum $\sum_{n=-N}^{N} e^{2\pi int}$ that's also useful. Isolating the $n = 0$ term and combining positive and negative terms we get

$$\sum_{n=-N}^{N} e^{2\pi int} = 1 + \sum_{n=1}^{N}(e^{2\pi int} + e^{-2\pi int}) = 1 + 2\sum_{n=1}^{N} \cos 2\pi nt.$$

The latter expression makes it easy to see that the value at the origin is $1+2N$ and that $1+2N$ is the maximum of $D_N(t)$, since the individual terms have a maximum of 1 at $t = 0$. The maximum gets bigger with N. By periodicity, $1+2N$ is also the value at each of the integers. Note one other interesting fact, namely that

$$\int_0^1 D_N(t)\,dt = 1.$$

It follows quickly either from the definition of $D_N(t)$ as the sum of complex exponentials or from the second expression in terms of cosines. The complex exponentials $e^{2\pi int}$ integrate to 0 except when $n = 0$; likewise the cosines integrate away. (It's not so easy to see the integral property from the formula for $D_N(t)$ as a ratio of sines.) In any event, this turns out to be an important property of $D_N(t)$, as we'll see in Section 1.8.

Here are plots of $D_N(t)$ for $N = 5$, 10, and 20.

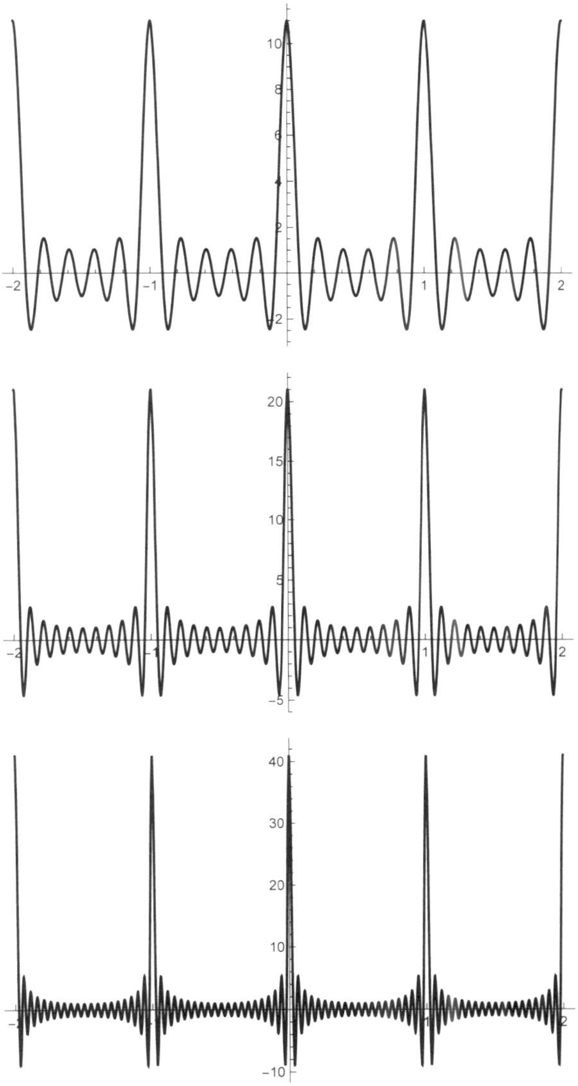

The signal becomes more and more concentrated at the integers, with higher and higher peaks. In fact, as we'll see later, the sequence of signals $D_N(t)$ tends to a sum of δ's at the integers as $N \to \infty$:

$$D_N(t) \to \sum_{n=-\infty}^{\infty} \delta(t-n)\,.$$

In what sense the convergence takes place will also have to wait till later.

The sum of infinitely many, regularly spaced δ's is sometimes called an *impulse train*, and we'll have other descriptive names for it. It will be a real workhorse for us, starting in Chapter 5.

In digital signal processing, particularly computer music, it's the finite, discrete form of an impulse train that's used. See Chapter 7. Rather than creating a sound by adding (sampled) sinusoids, one works in the frequency domain and synthesizes the sound from its spectrum. Start with the discrete impulse train that has all frequencies with equal energy — the discrete time buzz. This is easy to generate. Then shape the spectrum by increasing or decreasing the energies of the various harmonics, perhaps decreasing some to zero. The sound is synthesized, in the time domain, from this shaped spectrum. Additional operations are also possible. For an interesting discussion I recommend *A Digital Signal Processing Primer* by Ken Steiglitz. This is also where I got the terminology "buzz."[28]

One final look back at heat. Green's function for the heat equation had the form

$$g(x,t) = \sum_{n=-\infty}^{\infty} e^{-2\pi^2 n^2 t} e^{2\pi i n x}\,.$$

As $t \to 0$ the function $g(x,t)$ tends to

$$\sum_{n=-\infty}^{\infty} e^{2\pi i n x}\,.$$

Just thought you'd find that provocative.

1.7. The Math, Part 2: Orthogonality and Square Integrable Functions

Those who cared gradually realized that there's much more to the structure of the Fourier coefficients and to the idea of writing a periodic function as a sum of complex exponentials than might appear from our simple derivation. There are:

- Algebraic and geometric aspects
  - These are extensions of the algebra and geometry of vectors in Euclidean space to apply to functions. The key ideas are the inner product (dot product), orthogonality, and norm. The extensions are straightforward in the sense that the formal operations look (almost) the same for functions as they do for geometric vectors. We can pretty much cover the whole thing.

    This will give you the opportunity to reflect on things you already know from your more advanced perspective. Always to be welcomed. Your job is to transfer your geometric intuition from vectors, say in two dimensions, to a more general setting where the "vectors" are signals and "geometry" is by analogy, meaning that the words transfer in some kind of analogous way even if the pictures do not.

[28] "Buzz" is probably not the best word to describe the signal in the continuous-time case, but it will do for a shorthand reference.

- Analytic aspects
  - The analytic aspects require new ideas on using limits and on the nature of integration. We really can't cover the whole thing, and it's not appropriate to attempt to. But I'll say a little. Similar issues will come up when we work with the Fourier transform.

I'd like to add a third item, without the bullet: eigenfunctions of linear operators (especially differential operators). I'll say very little about this, though it will figure in our discussion of linear systems. I mention it because it's this perspective together with orthogonality that are still in force when one wants to work with expansions of a signal using functions other than sine and cosine, i.e., beyond considering solely periodic functions. It also provides another reason why only integer multiples of the fundamental frequency enter in Fourier series. That has to do with the harmonics being eigenfunctions of a differential operator and the boundary conditions that go with the problem. Do a search on Sturm-Liouville theory, for example, to see what this is about. This is covered in many engineering courses on partial differential equations.

This section is long, but you'll see the terms and the ideas in many areas of applications. Well worth your effort. The plan is to keep an eye on the big picture and save some details for Section 1.8, which, though it hurts me to say so, you could regard as optional.

It will help to ground the discussion by beginning with geometric vectors, where the ideas should be familiar to you. On with the show.

1.7.1. Vectors and orthogonality. In its origin, geometry is about measuring things — real, physical things. The aspect of Euclidean geometry that sets it apart from geometries that share most of its other features is how *perpendicularity* (the right angle) comes in. Consider the famous fifth postulate on parallel lines: "That, if a straight line falling on two straight lines makes the interior angles on the same side less than two right angles, the two straight lines, if produced indefinitely, meet on that side on which are the angles less than the two right angles."

Perpendicularity, such a familiar term, becomes austere when mathematicians start referring to it as *orthogonality*. They mean the same thing. I'm used to "orthogonal" and it's another term you can throw around to impress your friends.

Perpendicularity (orthogonality) meets measurement in one place, the Pythagorean theorem: Three numbers a, b, and c are sides of a right triangle if and only if

$$c^2 = a^2 + b^2 .$$

You surely remember the "only if" part, but may have forgotten the "if" part, and that's an important part of our story.

Speaking of perpendicularity and measurement, how do you lay out a big rectangular field of specified dimensions? You use the Pythagorean theorem. I had an encounter with this a few summers ago when I volunteered to help lay out soccer fields. I was only asked to assist, because evidently I could not be trusted with the details. Put two stakes in the ground to determine one side of the field. That's one leg of what is to become a right triangle — half the field. I hooked a tape measure on one stake and walked off in a direction generally perpendicular to the first leg, stopping when I had gone the regulation distance for that side of the field. The chief of the crew hooked another tape measure on the other stake and walked approximately along the diagonal of the field — the hypotenuse. We adjusted our positions, but not the length we had walked off, to meet up, so that the Pythagorean theorem was satisfied; he had a chart showing what this distance should be. Hence at our meeting point the leg I determined must be perpendicular to the first leg we laid out. This was my first practical use of the Pythagorean theorem, and so began my transition from a pure mathematician to an engineer.

To set up a notion of perpendicularity that incorporates the Pythagorean theorem in settings other than the familiar Euclidean plane, or three-dimensional space, is to try to copy the Euclidean properties that go with it. That's what we'll do for signals.

Vectors allow us to do geometry by way of algebra, and as well may allow us to give geometric interpretations of algebraic expressions. Both aspects are on display in the problem of identifying algebraically when two vectors are orthogonal. That's the problem we want to solve.

To fix some notation, which you probably know, we let $\mathbb{R}$ denote the set of real numbers and we write vectors in $\mathbb{R}^n$ as n-tuples of real numbers:

$$\underline{v} = (v_1, v_2, \ldots, v_n)\,.$$

The v_i are called the components of $\underline{v}$. The underline notation for vectors is easy to write by hand, easier than boldface which is often used, and less awkward looking than drawing a little arrow over the symbol. We'll use this notation again in working with discrete signals and the discrete Fourier transform, and with Fourier transforms in two and higher dimensions. If you want to use your own notation, that's fine, but pick something — you really will need some notation to distinguish a vector from a scalar.

The length, or *norm*, of $\underline{v}$ is

$$\|\underline{v}\| = (v_1^2 + v_2^2 + \cdots + v_n^2)^{1/2}\,.$$

Say $\underline{v}$ is a *unit vector* if $\|\underline{v}\| = 1$. The distance between two vectors $\underline{v}$ and $\underline{w}$ is $\|\underline{v} - \underline{w}\|$. The triangle inequality says that

$$\|\underline{v} + \underline{w}\| \leq \|\underline{v}\| + \|\underline{w}\|\,,$$

with equality if and only if $\underline{u}$ and $\underline{v}$ are in the same direction. We'll see a proof of this in Section 1.9. The inequality is a way of saying, analytically, that the shortest distance between two points is a straight line; convince yourself that's what's being said.

What does the Pythagorean theorem say in terms of vectors? Let's just work in $\mathbb{R}^2$. Let $\underline{u} = (u_1, u_2)$, $\underline{v} = (v_1, v_2)$, so $\underline{u} + \underline{v} = (u_1 + v_1, u_2 + v_2)$ is the hypotenuse of the triangle with sides $\underline{u}$ and $\underline{v}$. The vectors $\underline{u}$, $\underline{v}$, and $\underline{u}+\underline{v}$ form a right triangle if and only if

$$\|\underline{u}+\underline{v}\|^2 = \|\underline{u}\|^2 + \|\underline{v}\|^2 , \quad \text{which we expand as}$$
$$(u_1+v_1)^2 + (u_2+v_2)^2 = (u_1^2+u_2^2) + (v_1^2+v_2^2)$$
$$(u_1^2+2u_1v_1+v_1^2) + (u_2^2+2u_2v_2+v_2^2) = u_1^2+u_2^2+v_1^2+v_2^2 .$$

The squared terms cancel and we conclude that the algebraic condition

$$u_1v_1 + u_2v_2 = 0$$

is necessary and sufficient for $\underline{u}$ and $\underline{v}$ to be perpendicular. Not news to you. Without any further argument — the calculation is similar — let me also state that the same thing happens in higher dimensions: $\underline{u} = (u_1, u_2, \dots, u_n)$ and $\underline{v} = (v_1, v_2, \dots, v_n)$ are orthogonal if and only if

$$u_1v_1 + u_2v_2 + \cdots + u_nv_n = 0 .$$

One of the great strengths of vector methods in geometry is that while your initial intuition may be confined to situations where you can draw a picture (two and three dimensions, say), most often the calculations work even when you can't draw a picture (n dimensions). Follow the words. Follow your pencil.[29] Don't underestimate the power of this.

A remark on trade secrets. We've solved our problem. We've found a way to express algebraically the geometric property of orthogonality. Allow me to take the occasion to let you in on a trade secret of mathematics and mathematicians.

Mathematics progresses more from making definitions than from proving theorems. A definition is motivated by the solution of an interesting problem. A really good definition finds applications beyond the original problem. Think "derivative" as capturing the idea of instantaneous rate of change (or slope of a tangent line to a curve, if you prefer) or "integral" as a precise expression of area under a curve or "group" as a way of capturing symmetry. Less grandly, think of what we just did with vectors and orthogonality.

The mathematician turns around the solution of a problem and makes a definition out of it. Then a mathematician often covers up any trace of there ever having been an original motivating problem, except to announce to the world their great discovery that *by means of the elegant new definition* they can solve the problem they actually wanted to solve. This is one of the things that makes it hard for the uninitiated to read math books. Math books (often) start with definitions and leave the examples — the motivating problems — till later.

We'll see this pattern over and over. We'll have our share of definitions, but I'll try to keep them problem driven.

[29]Supposedly the great mathematician Leonhard Euler, whom we met in connection with complex exponentials, said when explaining his astonishing insight into formulas: "I was just following my pencil." Or words to that effect. I have bought many pencils looking for just the right one.

Inner product defined. And so we introduce the definition of the *inner product* or *dot product* of two vectors. We give this in $\mathbb{R}^n$:

- If $\underline{u} = (u_1, u_2, \ldots, u_n)$ and $\underline{v} = (v_1, v_2, \ldots, v_n)$, then the inner product is

$$\underline{u} \cdot \underline{v} = u_1 v_1 + u_2 v_2 + \cdots + u_n v_n \,.$$

Other notations for the inner product are $(\underline{u}, \underline{v})$ (just parentheses; we'll be using this notation) and $\langle \underline{u}, \underline{v} \rangle$ (angle brackets for those who think parentheses are not fancy enough; the use of angle brackets is especially common in physics).

Notice that

$$(\underline{v}, \underline{v}) = v_1^2 + v_2^2 + \cdots + v_n^2 = \|\underline{v}\|^2 \,.$$

Thus

$$\|\underline{v}\| = (\underline{v}, \underline{v})^{1/2} \,.$$

We now announce to the world the most excellent theorem that

- $(\underline{u}, \underline{v}) = 0$ if and only if $\underline{u}$ and $\underline{v}$ are orthogonal.

This was the point, after all. The game is rigged. We invented $(\underline{u}, \underline{v})$ to make the result true. Nevertheless, it is a truly helpful result, especially because it's so easy to verify whether $(\underline{u}, \underline{v}) = 0$.

Let me also remind you of the familiar formula

$$(\underline{u}, \underline{v}) = \|\underline{u}\| \, \|\underline{v}\| \cos\theta \,,$$

where θ is the angle between $\underline{u}$ and $\underline{v}$. This is sometimes taken as an alternate definition of the inner product. It's the law of cosines,

$$c^2 = a^2 + b^2 - 2ab\cos\theta \,,$$

written in terms of vectors; work it out. For further comments on this see Section 1.9.

Projections. The inner product does more than identify orthogonal vectors, and that makes it a really good definition. When it's nonzero, it tells you how much of one vector is in the direction of another. The vector

$$\frac{(\underline{v}, \underline{w})}{\|\underline{w}\|} \frac{\underline{w}}{\|\underline{w}\|} \quad \text{also written as} \quad \frac{(\underline{v}, \underline{w})}{(\underline{w}, \underline{w})} \underline{w}$$

is the projection of $\underline{v}$ onto the unit vector $\underline{w}/\|\underline{w}\|$, or, if you prefer, $(\underline{v}, \underline{w})/\|\underline{w}\|$ is the (scalar) component of $\underline{v}$ in the direction of $\underline{w}$. Simplest of all, if $\underline{w}$ is a unit vector, then $(\underline{v}, \underline{w})\underline{w}$ is the amount of $\underline{v}$ in the $\underline{w}$ direction. If we scale $\underline{v}$ by α, then the projection scales by α.

I think of the inner product as measuring how much one vector knows another; two orthogonal vectors don't know each other.

Algebraic properties of the inner product. Finally, I want to list the main algebraic properties of the inner product and what they're called in mathland. I won't give the proofs: they are straightforward verifications. We'll see these properties again — modified slightly to allow for complex numbers — a little later.

(1) $(\underline{v}, \underline{v}) \geq 0$ and $(\underline{v}, \underline{v}) = 0$ if and only if $\underline{v} = 0$ (positive definiteness).

(2) $(\underline{v}, \underline{w}) = (\underline{w}, \underline{v})$ (symmetry).

(3) $(\alpha\underline{v}, \underline{w}) = \alpha(\underline{v}, \underline{w})$ for any scalar α (homogeneity).
By the symmetry property we also have $(\underline{v}, \alpha\underline{w}) = \alpha(\underline{v}, \underline{w})$.

(4) $(\underline{v} + \underline{w}, \underline{u}) = (\underline{v}, \underline{u}) + (\underline{w}, \underline{u})$ (additivity).

In fact, these are exactly the properties that ordinary multiplication has.

Orthonormal basis. The natural basis for $\mathbb{R}^n$ is the vectors of length 1 in the n-coordinate directions:

$$\underline{e}_1 = (1, 0, \ldots, 0), \ \underline{e}_2 = (0, 1, \ldots, 0), \ \ldots, \ \underline{e}_n = (0, 0, \ldots, 1).$$

These vectors are called the "natural" basis because a vector $\underline{v} = (v_1, v_2, \ldots, v_n)$ is expressed "naturally" in terms of its components as

$$\underline{v} = v_1\underline{e}_1 + v_2\underline{e}_2 + \cdots + v_n\underline{e}_n .$$

The natural basis $\underline{e}_1, \underline{e}_2, \ldots, \underline{e}_n$ is an *orthonormal basis* for $\mathbb{R}^n$, meaning

$$(\underline{e}_i, \underline{e}_j) = \delta_{ij} ,$$

where δ_{ij} is the *Kronecker delta*,[30] defined by

$$\delta_{ij} = \begin{cases} 1, & i = j , \\ 0, & i \neq j . \end{cases}$$

Notice that

$$(\underline{v}, \underline{e}_k) = v_k$$

and hence that

$$\underline{v} = \sum_{k=1}^{n} (\underline{v}, \underline{e}_k)\underline{e}_k .$$

In words:

- When $\underline{v}$ is decomposed as a sum of vectors in the directions of the orthonormal basis vectors, the components are given by the inner product of $\underline{v}$ with the basis vectors.
- Since the $\underline{e}_k$ have length 1, the inner products $(\underline{v}, \underline{e}_k)$ are the projections of $\underline{v}$ onto the basis vectors.
  Put that the other way I like so much: the inner product $(\underline{v}, \underline{e}_k)$ is how much $\underline{v}$ and $\underline{e}_k$ know each other.
- Furthermore, we have

$$\|\underline{v}\|^2 = \sum_{k=1}^{n} (\underline{v}, \underline{e}_k)^2 .$$

[30] Named after the German mathematician Leopold Kronecker. He worked mainly in number theory. Every math major learns the famous quote: "God made the integers; all else is the work of man."

Regarding the last formula, it doesn't say anything you didn't already know, but it's worth the trip to see how it follows from the orthonormality of the basis vectors and the algebraic properties of the inner product. To wit:

$$\begin{aligned}
\|\underline{v}\|^2 &= (\underline{v}, \underline{v}) \\
&= \left(\sum_{k=1}^{n} (\underline{v}, \underline{e}_k)\underline{e}_k, \sum_{\ell=1}^{n} (\underline{v}, \underline{e}_\ell)\underline{e}_\ell \right) \\
&\qquad \text{(We have to use different indices of summation.} \\
&\qquad\quad \text{See Appendix C, on geometric sums, if you don't know why.)} \\
&= \sum_{k,\ell=1}^{n} (\underline{v}, \underline{e}_k)(\underline{v}, \underline{e}_\ell)(\underline{e}_k, \underline{e}_\ell) \quad \text{(using homogeneity)} \\
&= \sum_{k,\ell=1}^{n} (\underline{v}, \underline{e}_k)(\underline{v}, \underline{e}_\ell)\delta_{k\ell} \\
&= \sum_{k=1}^{n} (\underline{v}, \underline{e}_k)(\underline{v}, \underline{e}_k)1 \\
&\qquad \text{(only when } k = \ell \text{ does a term survive)} \\
&= \sum_{k=1}^{n} (\underline{v}, \underline{e}_k)^2 .
\end{aligned}$$

Done. Make sure you understand each step in this derivation. And also make sure you understand why this result holds for *any* orthonormal basis, not just the natural basis.

There's a catch: Complex vectors. There's always a catch. The definition of the inner product for real vectors has to be modified when the vectors are complex. Recall that we denote the set of complex numbers by $\mathbb{C}$ and the n-dimensional space, of n-tuples from $\mathbb{C}$, by $\mathbb{C}^n$. The complex inner product is making an appearance now primarily as motivation for what's coming in the next section, but we'll be using it again very strongly when we cover the discrete Fourier transform in Chapter 7.

Here's the issue. It takes a minute to explain. Again just working in two dimensions, suppose that $\underline{u} = (u_1, u_2)$ and $\underline{v} = (v_1, v_2)$ where u_1, u_2, v_1, v_2 are complex numbers. First, the norm of a vector is

$$\|u\| = (|u_1|^2 + |u_2|^2)^{1/2} .$$

Those are the magnitudes of the complex numbers u_1, u_2, not absolute values. Fine. Next, as we did for real vectors, expand the vector form of the Pythagorean theorem:

$$\begin{aligned}
&\|\underline{u} + \underline{v}\|^2 = \|\underline{u}\|^2 + \|\underline{v}\|^2 , \\
&|u_1 + v_1|^2 + |u_2 + v_2|^2 = |u_1|^2 + |u_2|^2 + |v_1|^2 + |v_2|^2 , \\
&(|u_1|^2 + 2\,\mathrm{Re}\{u_1\overline{v_1}\} + |v_1|^2) + (|u_2|^2 + 2\,\mathrm{Re}\{u_2\overline{v_2}\} + |v_2|^2) \\
&\qquad = |u_1|^2 + |u_2|^2 + |v_1|^2 + |v_2|^2 , \\
&\mathrm{Re}\{u_1\overline{v_1} + u_2\overline{v_2}\} = 0 .
\end{aligned}$$

(See Appendix B, on complex numbers, if you don't recognize what happened on the left-hand side with the real parts coming in.) So if we want to make a definition of orthogonality for complex vectors, why not define the inner product to be

$$(\underline{u}, \underline{v}) = \mathrm{Re}\{u_1\overline{v_1} + u_2\overline{v_2}\}$$

and say that this should be 0? Note that if $\underline{v} = \underline{u}$, then we do recover the norm from this definition,

$$\mathrm{Re}\{u_1\overline{u_1} + u_2\overline{u_2}\} = \mathrm{Re}\{|u_1|^2 + |u_2|^2\} = |u_1|^2 + |u_2|^2 .$$

We also have symmetry,

$$(\underline{u}, \underline{v}) = \mathrm{Re}\{u_1\overline{v_1} + u_2\overline{v_2}\} = \mathrm{Re}\{v_1\overline{u_1} + v_2\overline{u_2}\} = (\underline{v}, \underline{u}) ,$$

and additivity checks out as well. The problem is with homogeneity. If α is a complex number, then with this (tentative) definition,

$$(\alpha\underline{u}, \underline{v}) = \mathrm{Re}\{\alpha u_1\overline{v_1} + \alpha u_2\overline{v_2}\} \neq \alpha\,\mathrm{Re}\{u_1\overline{v_1} + u_2\overline{v_2}\} = \alpha(\underline{u}, \underline{v}) .$$

Similar problem with $(\underline{u}, \alpha\underline{v})$.

Homogeneity is too much to give up. For one thing, you can't work effectively with the projection of one vector onto another without homogeneity, because without it you can't scale the projection. To preserve homogeneity — actually, to almost preserve homogeneity — the elders have decided that the inner product of two complex vectors (in n dimensions) should be

$$(\underline{u}, \underline{v}) = u_1\overline{v_1} + u_2\overline{v_2} + \cdots + u_n\overline{v_n} .$$

With this definition the inner product of two complex vectors is a complex number. Nevertheless, we can still declare $\underline{u}$ and $\underline{v}$ to be orthogonal if $(\underline{u}, \underline{v}) = 0$, and we still have the expression for the norm

$$(\underline{u}, \underline{u}) = |u_1|^2 + |u_2|^2 + \cdots + |u_n|^2 = \|\underline{u}\|^2.$$

But there are consequences of adopting this definition. Here's a list of the algebraic properties showing the similarities and the differences with the real case. As before, I'll leave these to you to verify.

(1) $(\underline{v}, \underline{v}) \geq 0$ and $(\underline{v}, \underline{v}) = 0$ if and only if $\underline{v} = 0$ (positive definiteness).

(2) $(\underline{v}, \underline{w}) = \overline{(\underline{w}, \underline{v})}$ (conjugate symmetry, also called Hermitian[31] symmetry).

(3) $(\alpha\underline{v}, \underline{w}) = \alpha(\underline{v}, \underline{w})$ for any scalar α (homogeneity).
This time, by the conjugate symmetry property we have $(\underline{v}, \alpha\underline{w}) = \bar{\alpha}(\underline{v}, \underline{w})$.

(4) $(\underline{v} + \underline{w}, \underline{u}) = (\underline{v}, \underline{u}) + (\underline{w}, \underline{u})$ (additivity).

Order matters in the symmetry property. It's the slot, first or second, that matters in the homogeneity property.

The projection of $\underline{v}$ onto a unit vector $\underline{w}$ is $(\underline{v}, \underline{w})\underline{w}$ (even if you can't draw a picture of the complex vectors). If α is a complex number, then the projection of

[31]Named after the French mathematician Charles Hermite. Among other mathematical topics, he worked on quadratic forms, of which the inner product is an example. He is also the father of a family of eponymous polynomials that play a role in properties of the Fourier transform. Saved for a later problem.

$\alpha\underline{v}$ onto $\underline{w}$ is $(\alpha\underline{v}, \underline{w})\underline{w} = \alpha(\underline{v}, \underline{w})\underline{w}$ because the inner product is homogeneous in the first slot. Projection scales, albeit allowing for scaling by a complex number (which you can't draw).

From now on, the default assumption is that we use the inner product for complex vectors. The real vectors $\underline{e}_1, \underline{e}_2, \ldots, \underline{e}_n$ are an orthonormal basis for $\mathbb{C}^n$ (because we allow for complex scalars) and any vector $\underline{v}$ in $\mathbb{C}^n$ can be written in terms of this basis as

$$\underline{v} = \sum_{k=1}^{n} (\underline{v}, \underline{e}_k)\underline{e}_k \,.$$

1.7.2. Back to Fourier series. Let me introduce the notation and basic terminology and state the important results so you can see the point. Then I'll explain where these ideas come from and how they fit together.

Once again, to be definite, we'll work with periodic functions of period 1. We can consider such a function already to be defined for all real numbers and satisfying the identity $f(t+1) = f(t)$ for all t, or we can consider $f(t)$ to be defined initially only on the interval from 0 to 1, say, and then extended to be periodic and defined on all of $\mathbb{R}$ by repeating the graph. In either case, once we know what we need to know about the function on $[0, 1]$, or on any interval of length 1, we know everything. All of the action in the following discussion takes place on $[0, 1]$.

Square integrable functions and their norm. When $f(t)$ is a signal defined on $[0, 1]$, its *energy* is defined to be the integral

$$\int_0^1 |f(t)|^2 \, dt \,.$$

This definition of energy comes up in other physical contexts; we don't have to be talking about functions of time. (In some areas the integral of the square is identified with power.) Thus

$$\int_0^1 |f(t)|^2 \, dt < \infty$$

means that the signal has *finite energy*, a reasonable condition to expect or to impose. One also says that the signal is *square integrable*. Any signal that is continuous, or even piecewise continuous with finite jumps, is square integrable. That covers a lot.

I'm writing the definition in terms of the integral of the magnitude squared, $|f(t)|^2$, rather than just $f(t)^2$ because we'll want to consider the definition to apply to complex-valued functions. For real-valued functions it doesn't matter whether we integrate $|f(t)|^2$ or $f(t)^2$; they're the same.

One further point before we go on. Although our purpose is to use the finite energy condition to work with periodic functions and though you think of periodic functions as defined for all time, you can see why we have to restrict attention to one period (any period). An integral of the square of a (nonzero) periodic function from $-\infty$ to ∞, for example

$$\int_{-\infty}^{\infty} \sin^2 2\pi t \, dt \,,$$

is infinite.

For mathematical reasons, primarily, it's best to take the square root of the integral and to define the *norm* (length) of a function $f(t)$ by

$$\|f\| = \left(\int_0^1 |f(t)|^2 \, dt \right)^{1/2} .$$

With this definition one has that

$$\|\alpha f\| = |\alpha| \, \|f\| \,,$$

whereas if we didn't take the square root, the constant would come out to the second power. If we've really defined a length, then scaling $f(t)$ to $\alpha f(t)$ should scale the length of $f(t)$ by $|\alpha|$. If we didn't take the square root in defining $\|f\|$, the length wouldn't scale to the first power. (Same thing as for geometric vectors and their norm.)

One can also show, though the proof is not so obvious (see Section 1.9), that the triangle inequality holds:

$$\|f + g\| \leq \|f\| + \|g\| \,.$$

Write that out in terms of integrals if you think it's obvious:

$$\left(\int_0^1 |f(t) + g(t)|^2 \, dt \right)^{1/2} \leq \left(\int_0^1 |f(t)|^2 \, dt \right)^{1/2} + \left(\int_0^1 |g(t)|^2 \, dt \right)^{1/2} .$$

We can measure the distance between two functions via

$$\|f - g\| = \left(\int_0^1 |f(t) - g(t)|^2 \, dt \right)^{1/2} .$$

Then $\|f - g\| = 0$ if and only if $f = g$.

This is a test. We jumped right into some definitions — sorry! — but you're meant to make the connection to geometric vectors. The length of a vector is the square root of the sum of the squares of its components. The norm defined by an integral (being a limit of sums) is the continuous analog of that. The definition of distance is likewise analogous to the distance between vectors. We'll make the analogy to geometric vectors even closer when we introduce the corresponding inner product.

We let $L^2([0,1])$ be the set of complex-valued functions $f(t)$ on $[0,1]$ for which

$$\int_0^1 |f(t)|^2 \, dt < \infty \,.$$

The "L" stands for Henri Lebesgue, the French mathematician who introduced a new definition of the integral that underlies the analytic aspects of the results we're about to talk about. His work was around the turn of the 20th century. The length we've just introduced, $\|f\|$, is then called the *square norm* or the L^2-*norm* of the function. When we want to distinguish this from other norms that might (and will) come up, we write $\|f\|_2$.

Any geometric vector has finite length, so in the geometric setting we don't need to make finite norm a separate assumption. But not every function on $[0,1]$ is square integrable (find some examples!), and we do need to assume square integrability of a signal to get anything done. In other words, if it's not square integrable, we're not interested. Fortunately, as mentioned above, there are plenty of functions to be interested in.

The inner product on $L^2([0,1])$. Having passed the test of recognizing the analogy between lengths of functions in $L^2([0,1])$ and lengths of geometric vectors, you will be very comfortable defining the (complex!) inner product of two functions f and g in $L^2([0,1])$ as

$$(f,g) = \int_0^1 f(t)\overline{g(t)}\,dt\,.$$

Order matters, for

$$(g,f) = \int_0^1 g(t)\overline{f(t)}\,dt = \int_0^1 \overline{\overline{g(t)}f(t)}\,dt = \overline{\int_0^1 \overline{g(t)}f(t)\,dt} = \overline{(f,g)}\,.$$

This is the same Hermitian symmetry that we observed for the inner product of geometric vectors. Note that

$$(f,f) = \|f\|^2.$$

The rest of the algebraic properties of the inner product also hold. We list them for your convenience.

(1) $(f,f) \geq 0$ and $(f,f) = 0$ if and only if $f = 0$.

(2) $(\alpha f, g) = \alpha(f,g)$, $(f, \alpha g) = \bar{\alpha}(f,g)$ (In the second slot α comes out as a complex conjugate.)

(3) $(f, g+h) = (f,g) + (f,h)$.

For those keeping mathematical score, it's necessary to show that if $f(t)$ and $g(t)$ are square integrable, then $f(t)\overline{g(t)}$ is integrable, so that the inner product makes sense. I'll leave this nugget till Section 1.9.

Armed with an inner product, orthogonality is but a definition away. Namely, f and g in $L^2([0,1])$ are *orthogonal* if

$$(f,g) = \int_0^1 f(t)\overline{g(t)}\,dt = 0\,.$$

For this definition it doesn't matter which order we take, (f,g) or (g,f).

If $f(t)$ and $g(t)$ are orthogonal, then the Pythagorean theorem holds,

$$\|f+g\|^2 = \|f\|^2 + \|g\|^2\,.$$

This follows from a frequently employed identity for the complex inner product:

$$\|f+g\|^2 = \|f\|^2 + 2\,\mathrm{Re}(f,g) + \|g\|^2\,.$$

In turn, the verification of this is an exercise in using the algebraic properties of the inner product, and it is worth going through:

$$\begin{aligned}\|f+g\|^2 &= (f+g, f+g) = (f, f+g) + (g, f+g)\\ &= (f,f) + (f,g) + (g,f) + (g,g)\\ &= (f,f) + (f,g) + \overline{(f,g)} + (g,g) = \|f\|^2 + 2\,\mathrm{Re}(f,g) + \|g\|^2 .\end{aligned}$$

A similar identity is

$$\|f-g\|^2 = \|f\|^2 - 2\,\mathrm{Re}(f,g) + \|g\|^2 .$$

1.7.3. The complex exponentials are orthonormal, and more! Allowing yourself the geometric perspective, we can now recast the fundamental calculation we did when solving for the Fourier coefficients:

- The complex exponentials $e^{2\pi int}$, $n = 0, \pm1, \pm2, \ldots$, are orthonormal.

Let's see that calculation again. Write

$$e_n(t) = e^{2\pi int} .$$

The inner product of two of them, $e_n(t)$ and $e_m(t)$, when $n \neq m$, is

$$\begin{aligned}(e_n, e_m) &= \int_0^1 e^{2\pi int}\overline{e^{2\pi imt}}\,dt = \int_0^1 e^{2\pi int}e^{-2\pi imt}\,dt = \int_0^1 e^{2\pi i(n-m)t}\,dt\\ &= \frac{1}{2\pi i(n-m)}e^{2\pi i(n-m)t}\Big]_0^1\\ &= \frac{1}{2\pi i(n-m)}\left(e^{2\pi i(n-m)} - e^0\right) = \frac{1}{2\pi i(n-m)}(1-1) = 0 .\end{aligned}$$

They are *orthogonal*. And when $n = m$,

$$\begin{aligned}\|e_n\|^2 = (e_n, e_n) &= \int_0^1 e^{2\pi int}\overline{e^{2\pi int}}\,dt = \int_0^1 e^{2\pi int}e^{-2\pi int}\,dt\\ &= \int_0^1 e^{2\pi i(n-n)t}\,dt = \int_0^1 1\,dt = 1 .\end{aligned}$$

Therefore the functions $e_n(t)$ are *orthonormal*:

$$(e_n, e_m) = \delta_{nm} = \begin{cases}1, & n = m,\\ 0, & n \neq m.\end{cases}$$

Go easy on yourself. Let me relieve you of a burden that you may believe you must carry. Just because we use the geometric terminology, for good reason, does not mean that you should be able to visualize the orthogonality of the complex exponentials. It's an analogy; it's not an identification. Ditto for any other family of orthogonal functions (like sines and cosines or Legendre polynomials, etc; see the problems). So don't lie in bed staring at the ceiling, imagining graphs, trying to imagine orthogonality. Just don't.

The Fourier coefficients are projections. What is the component of a function $f(t)$ "in the direction" $e_n(t)$? By analogy to the Euclidean case, it is given by the inner product

$$(f, e_n) = \int_0^1 f(t)\overline{e_n(t)}\, dt = \int_0^1 f(t)e^{-2\pi int}\, dt\,,$$

precisely the nth Fourier coefficient $\hat{f}(n)$. (Note that e_n really does have to be in the second slot here.)

Thus writing the Fourier series

$$f(t) = \sum_{n=-\infty}^{\infty} \hat{f}(n)e^{2\pi int}$$

is exactly like the decomposition in terms of an orthonormal basis and associated inner product:

$$f = \sum_{n=-\infty}^{\infty} (f, e_n)e_n\,.$$

The complex exponentials are an orthonormal basis. What we haven't done is to show that the complex exponentials are a *basis* as well as being orthonormal. *They are!* This result is the culmination of the algebraic/geometric/analytic approach we've just been developing.

- If $f(t)$ is in $L^2([0,1])$, then

$$f = \sum_{n=-\infty}^{\infty} (f, e_n)e_n = \sum_{n=-\infty}^{\infty} \hat{f}(n)e^{2\pi int}\,.$$

For this we are required to show that

$$\lim_{N\to\infty} \left\| f - \sum_{n=-N}^{N} (f, e_n)e_n \right\| = 0\,.$$

Writing

$$f = \sum_{n=-\infty}^{\infty} (f, e_n)e_n\,,$$

as above, means the equals sign is interpreted in terms of the limit. It's the convergence issue again, but recast. More on this in Section 1.8.3.

Bessel's inequality. Here's a nice (math) application of the orthonormality of the complex exponentials and the properties of the inner product. Using what we've done, you can show that

$$\left\| f - \sum_{n=-N}^{N} \hat{f}(n)e^{2\pi int} \right\|^2 = \|f\|^2 - \sum_{n=-N}^{N} |\hat{f}(n)|^2\,.$$

Since, as a norm, $\|f - \sum_{n=-N}^{N} \hat{f}(n)\|^2 \geq 0$, this in turn implies that

$$\sum_{n=-N}^{N} |\hat{f}(n)|^2 \leq \|f\|^2\,.$$

This is known as *Bessel's inequality.*[32] You get an inequality because on the left-hand side you don't have all the components of f in the sum, so you don't get the whole L^2-norm of f. However, to push this a little further, Bessel's inequality is true for all N, no matter how large, so we can safely let $N \to \infty$ and conclude that

$$\sum_{n=-\infty}^{\infty} |\hat{f}(n)|^2 \leq \|f\|^2 < \infty\,,$$

i.e., that $\sum_{n=-\infty}^{\infty} |\hat{f}(n)|^2$ converges. This has as a consequence that

$$|\hat{f}(n)|^2 \to 0 \text{ as } n \to \pm\infty, \text{ hence also that } \hat{f}(n) \to 0 \text{ as } n \to \infty.$$

Here we have invoked a general result on convergent series from good old calculus days; if a series converges, then the general term must tend to zero.[33]

However, from just the Fourier coefficients tending to 0, one cannot conclude conversely that the Fourier series converges. Remember the example from those same good old calculus days: $1/n \to 0$ but $\sum_{n=1}^{\infty} 1/n$ diverges. Moreover, the Fourier series, involving complex exponentials, is a sum of oscillatory terms (cosines and sines if you split it into real and imaginary parts), so convergence is even trickier.

The fact that the complex exponentials are an orthonormal *basis* for $L^2([0,1])$ means that, in fact, equality holds in Bessel's inequality:

$$\sum_{n=-\infty}^{\infty} |\hat{f}(n)|^2 = \|f\|^2 < \infty\,.$$

This is Rayleigh's identity and it gets a section of its own.

Rayleigh's identity. As a further application of these ideas, let's derive Rayleigh's identity,

$$\int_0^1 |f(t)|^2\,dt = \sum_{n=-\infty}^{\infty} |\hat{f}(n)|^2\,.$$

Physically, Rayleigh's identity says that we can compute the energy of the signal by adding up the energies of the individual harmonics, for the magnitude $|\hat{f}(n)|^2$ is the energy contributed by the nth harmonic. That's quite a satisfactory state of affairs and an extremely useful result. You'll see examples of its use in the problems.

We really have equal contributions from the positive and negative harmonics, $e^{2\pi int}$ and $e^{-2\pi int}$, since $|\hat{f}(-n)| = |\hat{f}(n)|$ (note the absolute values here). In passing between the complex exponential form

$$\sum_{n=-\infty}^{\infty} c_n e^{2\pi int}\,, \quad c_n = \hat{f}(n)\,,$$

[32]That's Friedrich Bessel, whose name is also attached to a family of special functions. You'll meet the functions in some problems.

[33]In particular, $\sum_{n=-\infty}^{\infty} e^{2\pi int}$, the buzz example, cannot converge for any value of t since $|e^{2\pi int}| = 1$.

and the sine-cosine form,

$$\tfrac{1}{2}a_0 + \sum_{n=1}^{\infty} a_n \cos 2\pi nt + \sum_{n=1}^{\infty} b_n \sin 2\pi nt\,,$$

of the Fourier series, we have $|c_n| = \frac{1}{2}\sqrt{a_n^2 + b_n^2}$, so $\hat{f}(n)$ and $\hat{f}(-n)$ together contribute a total energy of $\sqrt{a_n^2 + b_n^2}$.

The derivation of Rayleigh's identity is a cinch! Happily accepting convergence, it's an algebraic calculation. In fact, we gave the argument in the context of geometric vectors for the real inner product, and for emphasis let's do it in the present setting.

Expand $f(t)$ as a Fourier series:

$$f(t) = \sum_{n=-\infty}^{\infty} \hat{f}(n)e^{2\pi int} = \sum_{n=-\infty}^{\infty} (f, e_n)e_n(t)\,.$$

Then

$$\begin{aligned}\int_0^1 |f(t)|^2\,dt &= \|f\|^2 = (f, f)\\ &= \Big(\sum_{n=-\infty}^{\infty} (f, e_n)e_n, \sum_{m=-\infty}^{\infty} (f, e_m)e_m\Big)\\ &= \sum_{n,m} (f, e_n)\overline{(f, e_m)}(e_n, e_m) = \sum_{n,m=-\infty}^{\infty} (f, e_n)\overline{(f, e_m)}\delta_{nm}\\ &= \sum_{n=-\infty}^{\infty} (f, e_n)\overline{(f, e_n)} = \sum_{n=-\infty}^{\infty} |(f, e_n)|^2 = \sum_{n=-\infty}^{\infty} |\hat{f}(n)|^2\,.\end{aligned}$$

The derivation used

(1) the algebraic properties of the complex inner product,

(2) the fact that the $e_n(t) = e^{2\pi int}$ are orthonormal with respect to this inner product,

(3) know-how in working with sums.

Do not go to sleep until you can follow every line.[34]

Writing Rayleigh's identity as

$$\|f\|^2 = \sum_{n=-\infty}^{\infty} |(f, e_n)|^2$$

again highlights the parallels between the geometry of L^2 and the geometry of vectors: how do you find the squared length of a vector? By adding the squared magnitudes of its components with respect to an orthonormal basis. That's exactly what Rayleigh's identity is saying. It's the completion of Bessel's inequality, which we saw earlier.

[34]If you derived Bessel's inequality earlier, you could probably doze off.

Best L^2-approximation by finite Fourier series. There's another take on the Fourier coefficients being projections onto the complex exponentials $e_n(t) = e^{2\pi i n t}$. It says that a finite Fourier series of degree N gives the best (trigonometric) approximation of that order in $L^2([0,1])$ to a function. More precisely:

- If $f(t)$ is in $L^2([0,1])$ and $\alpha_1, \alpha_2, \ldots, \alpha_N$ are any complex numbers, then
$$\left\| f - \sum_{n=-N}^{N} (f, e_n) e_n \right\| \leq \left\| f - \sum_{n=-N}^{N} \alpha_n e_n \right\| .$$
  Furthermore, equality holds only when $\alpha_n = (f, e_n)$ for every n.

It's this last statement, on the case of equality, that leads to the Fourier coefficients in a different way than solving for them directly as we did originally. Another way of stating the result is that the *orthogonal projection* of f onto the subspace of $L^2([0,1])$ spanned by the e_n, $n = -N, \ldots, N$, is
$$\sum_{n=-N}^{N} \hat{f}(n) e^{2\pi i n t} .$$
This identifies the Fourier coefficients as solutions of an optimization problem, and engineers love optimization problems. Mathematicians love optimization problems, too, though sometimes for different reasons. They love this result because it's an essential step in proving that the complex exponentials are an orthonormal basis of $L^2([0,1])$. We'll discuss this in Section 1.8.3.

Here comes the proof. Hold on. Write
$$\left\| f - \sum_{n=-N}^{N} \alpha_n e_n \right\|^2 = \left\| f - \sum_{n=-N}^{N} (f, e_n) e_n + \sum_{n=-N}^{N} (f, e_n) e_n - \sum_{n=-N}^{N} \alpha_n e_n \right\|^2$$
$$= \left\| \left(f - \sum_{n=-N}^{N} (f, e_n) e_n \right) + \sum_{n=-N}^{N} ((f, e_n) - \alpha_n) e_n \right\|^2 .$$
We squared all the norms because we want to use the properties of inner products to expand the last line. Using the identity we derived a few pages back,[35] the last line equals
$$\left\| \left(f - \sum_{n=-N}^{N} (f, e_n) e_n \right) + \sum_{n=-N}^{N} ((f, e_n) - \alpha_n) e_n \right\|^2$$
$$= \left\| f - \sum_{n=-N}^{N} (f, e_n) e_n \right\|^2$$
$$+ 2 \operatorname{Re} \left(f - \sum_{n=-N}^{N} (f, e_n) e_n, \sum_{m=-N}^{N} ((f, e_m) - \alpha_m) e_m \right) + \left\| \sum_{n=-N}^{N} ((f, e_n) - \alpha_n) e_n \right\|^2 .$$

[35]The identity on the Pythagorean theorem and orthogonality for the complex inner product.

This looks complicated, but the middle term is just a sum of multiples of terms of the form

$$\left(f - \sum_{n=-N}^{N} (f, e_n)e_n, e_m\right) = (f, e_m) - \sum_{n=-N}^{N} (f, e_n)(e_n, e_m) = (f, e_m) - (f, e_m) = 0\,,$$

so the whole thing drops out! The final term is

$$\left\| \sum_{n=-N}^{N} \big((f, e_n) - \alpha_n\big)e_n \right\|^2 = \sum_{n=-N}^{N} \left|(f, e_n) - \alpha_n\right|^2.$$

We are left with

$$\left\| f - \sum_{n=-N}^{N} \alpha_n e_n \right\|^2 = \left\| f - \sum_{n=-N}^{N} (f, e_n)e_n \right\|^2 + \sum_{n=-N}^{N} |(f, e_n) - \alpha_n|^2.$$

This completely proves the result, for the right-hand side is the sum of two positive terms and hence

$$\left\| f - \sum_{n=-N}^{N} \alpha_n e_n \right\|^2 \geq \left\| f - \sum_{n=-N}^{N} (f, e_n)e_n \right\|^2$$

with equality holding if and only if

$$\sum_{n=-N}^{N} |(f, e_n) - \alpha_n|^2 = 0\,.$$

The latter holds if and only if $\alpha_n = (f, e_n)$ for all n.

This argument may have seemed labor intensive, but it was *all algebra* based on the properties of the inner product. Imagine trying to write everything out in terms of integrals.

What if the period isn't 1? Remember how we modified the Fourier series when the period is T rather than 1. We were led to the expansion

$$f(t) = \sum_{n=-\infty}^{\infty} c_n e^{2\pi i n t/T}\,,$$

where

$$c_n = \frac{1}{T}\int_0^T e^{-2\pi i n t/T} f(t)\,dt\,.$$

The whole setup we've just been through can easily be modified to cover this case. We work in the space $L^2([0, T])$ of square integrable functions on the interval $[0, T]$. The (complex) inner product is

$$(f, g) = \int_0^T f(t)\overline{g(t)}\,dt\,.$$

What happens with the T-periodic complex exponentials $e^{2\pi int/T}$? If $n \neq m$, then, much as before,

$$\begin{aligned}
(e^{2\pi int/T}, e^{2\pi imt/T}) &= \int_0^T e^{2\pi int/T}\,\overline{e^{2\pi imt/T}}\,dt = \int_0^T e^{2\pi int/T}\,e^{-2\pi imt/T}\,dt \\
&= \int_0^T e^{2\pi i(n-m)t/T}\,dt = \frac{1}{2\pi i(n-m)/T}\,e^{2\pi i(n-m)t/T}\Big]_0^T \\
&= \frac{1}{2\pi i(n-m)/T}\,(e^{2\pi i(n-m)} - e^0) = \frac{1}{2\pi i(n-m)/T}\,(1-1) = 0\,.
\end{aligned}$$

And when $n = m$,

$$\begin{aligned}
(e^{2\pi int/T}, e^{2\pi int/T}) &= \int_0^T e^{2\pi int/T}\,\overline{e^{2\pi int/T}}\,dt \\
&= \int_0^T e^{2\pi int/T}\,e^{-2\pi int/T}\,dt = \int_0^T 1\,dt = T\,.
\end{aligned}$$

Aha — it's not 1; it's T. The complex exponentials with period T are orthogonal but not orthonormal. To get the latter property to hold, we scale the complex exponentials to

$$e_n(t) = \frac{1}{\sqrt{T}}e^{2\pi int/T}\,,$$

for then

$$(e_n, e_m) = \begin{cases} 1, & n = m, \\ 0, & n \neq m. \end{cases}$$

This is where the factor $1/\sqrt{T}$ comes from, the factor mentioned earlier in this chapter. The inner product of f with e_n is

$$(f, e_n) = \frac{1}{\sqrt{T}}\int_0^T f(t)e^{-2\pi int/T}\,dt\,.$$

Then

$$\begin{aligned}
\sum_{n=-\infty}^{\infty} (f, e_n)e_n &= \sum_{n=-\infty}^{\infty}\left(\frac{1}{\sqrt{T}}\int_0^T f(s)e^{-2\pi ins/T}\,ds\right)\frac{1}{\sqrt{T}}e^{2\pi int/T} \\
&= \sum_{n=-\infty}^{\infty} c_n e^{2\pi int/T}\,,
\end{aligned}$$

where

$$c_n = \frac{1}{T}\int_0^T e^{-2\pi int/T}f(t)\,dt\,,$$

as above. We're back to our earlier formula.

Here ends Section 1.7. A lot has happened, let us say. What follows to round out the chapter are sections providing some details on some of the mathematical points and references where you can find out more. Read as your interests may guide you.

1.8. Appendix: Notes on the Convergence of Fourier Series

My first comment is, "Don't go there." Or at least proceed cautiously. Ever since Fourier scandalized French society with his claims of the universality of his methods, questions on the convergence of Fourier series have been among the thorniest in mathematical analysis.[36] Here we're going to be very circumspect, whereas a little googling will quickly turn up as much or more than you want to know. We'll give a few explanations, a few references, but nothing like complete proofs or even complete introductions to proofs. Think of this section rather as an introduction to terminology that you may encounter elsewhere and with some results on display that you can use in self-defense if a mathematician sneaks up on you.

The convergence of infinite series is all about the partial sums. In the case of Fourier series that means about

$$S_N(t) = \sum_{n=-N}^{N} \hat{f}(n)e^{2\pi int},$$

for a function $f(t)$. The action takes place for $0 \le t \le 1$. The question is what happens as $N \to \infty$. Roughly speaking, there are three types of good things:

(1) *Pointwise convergence.* Plugging in a point t, we have $\lim_{N\to\infty} S_N(t) = f(t)$, but we can't say much more, e.g., about the rate of convergence from one point to the next.

(2) *Uniform convergence.* This is stronger than pointwise convergence. Beyond saying that $\lim_{N\to\infty} S_N(t) = f(t)$ for a $t \in [0,1]$, uniform convergence says that $\max_{0\le t\le 1} |S_N(t) - f(t)| \to 0$ as $N \to \infty$. In words, when N is large, the partial sum $S_N(t)$ stays close to $f(t)$ over the whole interval $0 \le t \le 1$, from which comes the descriptor "uniform," as in "uniformly close." See Section 1.8.4 for more on this.

(3) *Convergence in $L^2([0,1])$.* This is what we just finished talking about in the preceding section. The condition is

$$\lim_{N\to\infty} \int_0^1 |S_N(t) - f(t)|^2 \, dt \to 0 .$$

More complete definitions require the seasoned use of ϵ's and δ's, and you will be pummeled by these if you take a math course on analysis.[37]

Of the three modes, uniform convergence is the strongest in that it implies the other two. Pointwise convergence is the kind of convergence you get from the theorem in Section 1.5.1, and at a jump discontinuity the partial sums actually converge to the average of the jump. Uniform convergence needs more smoothness than just piecewise continuity (the hypothesis in that theorem). There will be a theorem to this effect in Section 1.8.2.

---

[36] Dramatized accounts would have you believe that this way leads to madness. Poor Georg Cantor suffered terribly, but not because of the convergence of Fourier series. Really, not because of Fourier series.

[37] But it can be a worthwhile pummeling. My colleague Stephen Boyd, a former math major, wants all of his PhD students to take analysis, not because they'll ever use the specific results, but because of the clarity of thought that the course engenders. I have some hopes for this course, too.

It would seem that L^2-convergence, as a sort of convergence in the mean, has nothing to do with what happens to the partial sums if you plug in a particular value of t. However, there are theorems that do address just this connection, and they are among the most difficult and subtle in the business. Let's move on.

1.8.1. Studying partial sums via the Dirichlet kernel: The buzz is back.

Look back to Section 1.6, to the separate discussions of the buzz signal and convolution. They're going to appear together in a most interesting way, especially to electrical engineers.

If you have some background in signal processing, you might recognize that the partial sum

$$S_N(t) = \sum_{n=-N}^{N} \hat{f}(n)e^{2\pi int}$$

is the result of applying a low-pass filter to the full Fourier series,

$$f(t) = \sum_{n=-\infty}^{\infty} \hat{f}(n)e^{2\pi int}.$$

The low frequencies, those from $-N$ to N, are "passed" unchanged, and the frequencies above and below $\pm N$ are zeroed out.

How is this done? By convolution with the Dirichlet kernel

$$D_N(t) = \sum_{n=-N}^{N} e^{2\pi int}.$$

Recall that when two periodic signals are convolved, their Fourier coefficients multiply. In this case, for $D_N(t)$ and $f(t)$,

$$\widehat{(D_N * f)}(n) = \widehat{D_N}(n)\hat{f}(n).$$

But

$$\widehat{D_N}(n) = \begin{cases} 1, & n = 0, \pm 1, \pm 2, \dots, \pm N, \\ 0, & \text{otherwise}, \end{cases}$$

and so the Fourier series for $(D_N * f)(t)$ is

$$(D_N * f)(t) = \sum_{n=-\infty}^{\infty} \widehat{D_N}(n)\hat{f}(n)e^{2\pi int} = \sum_{n=-N}^{N} \hat{f}(n)e^{2\pi int} = S_N(t).$$

That's low-pass filtering. I never saw this description in my math major days. We'll see low-pass filters again, involving the Fourier transform.

As an integral,[38]

$$S_N(t) = (D_N * f)(t) = \int_0^1 D_N(t-\tau)f(\tau)\,d\tau.$$

[38]If you've wondered why $D_N(t)$ is called a "kernel," it's because of this equation. Right at this juncture $D_N(t-\tau)$ appears in the integrand, integrated against $f(\tau)$ to produce a function of t. In some areas of mathematics, where functions of t are defined by integral expressions of the form $\int_a^b k(t,\tau)f(\tau)\,d\tau$ (of which convolution is an example), the function $k(t,\tau)$ is referred to as a *kernel*. We'll see this again in Chapter 7. Now you know. On the other hand, I don't know where the particular word "kernel" comes from.

There's another way to write this that turns out to be more helpful in analyzing the convergence of $S_N(t)$.[39] Since $D_N(t)$ is even, $D_N(t-\tau) = D_N(\tau-t)$, and using the substitution $u = \tau - t$ gives

$$\begin{aligned}\int_0^1 D_N(t-\tau)f(\tau)\,d\tau &= \int_0^1 D_N(\tau-t)f(\tau)\,d\tau \\ &= \int_{-t}^{1-t} D_N(u)f(t+u)\,du \\ &= \int_0^1 D_N(u)f(t+u)\,du\,.\end{aligned}$$

The final equality holds because, as $D_N(u)f(u+t)$ is periodic of period 1, integrating over any interval of length 1 gives the same answer.

Now, remember, it's not just that we want to know that $S_N(t)$ converges as $N \to \infty$, we want to know that it converges to $f(t)$ (or to an average in the case of a jump discontinuity). To help with that there's another trick. Recalling that

$$\int_0^1 D_N(u)\,du = 1,$$

we have

$$f(t) = f(t)\int_0^1 D_N(u)\,du = \int_0^1 f(t)D_N(u)\,du\,,$$

so that

$$\begin{aligned}S_N(t) - f(t) &= \int_0^1 D_N(u)f(t+u)\,du - \int_0^1 f(t)D_N(u)\,du \\ &= \int_0^1 D_N(u)(f(t+u) - f(t))\,du\,.\end{aligned}$$

Finally, bring back the closed form expression for D_N as a ratio of sines, change the variable u back to τ because it looks better, and change the limits of integration to be from $-1/2$ to $1/2$ (possible by periodicity) because it makes the subsequent analysis easier:

$$S_N(t) - f(t) = \int_{-1/2}^{1/2} \frac{\sin(2\pi\tau(N+1/2))}{\sin(\pi\tau)}(f(t+\tau) - f(\tau))\,d\tau\,.$$

It is this expression that's used to prove results on $\lim_{N\to\infty} S_N(t)$.

Very briefly, here's why. The hypotheses in the convergence theorems, the theorem in Section 1.5.1, and a theorem to come in the next section involve *local* properties of $f(t)$, e.g., continuity at a point, a jump discontinuity, and differentiability at a point. For a fixed point t, that means looking at $f(t+\tau) - f(t)$ when τ

[39] I didn't say that any of this is the obvious thing to do, just that "it turns out to be more helpful," as we'll see. Go along for the ride; it will end soon.

is small. The idea is to take a small $\delta > 0$ and split the integral into pieces:

$$\begin{aligned}\int_{-1/2}^{1/2} &\frac{\sin(2\pi\tau(N+1/2))}{\sin(\pi\tau)}(f(t+\tau)-f(\tau))\,d\tau \\ &= \int_{|\tau|\leq\delta} \frac{\sin(2\pi\tau(N+1/2))}{\sin(\pi\tau)}(f(t+\tau)-f(\tau))\,d\tau \\ &\quad + \int_{\delta\leq|\tau|\leq 1} \frac{\sin(2\pi\tau(N+1/2))}{\sin(\pi\tau)}(f(t+\tau)-f(\tau))\,d\tau\,.\end{aligned}$$

Reasonable hypotheses on $f(t)$, like those we've seen on piecewise continuity and differentiability, together with the explicit form of the Dirichlet kernel make it possible to estimate the integrands and show convergence as $N \to \infty$. You use different estimates for the different integrals. Depending on the hypotheses, one may deduce just pointwise convergence or the stronger uniform convergence.

If this were a regulation math course, I'd give the full arguments and make the estimates, which I've always liked. But this is not a regulation math course so at this point I am sending you off to the literature if you want to find out more. Aside from the many books specializing in Fourier series, many books on introductory mathematical analysis include a chapter on Fourier series and will pick up where we left off.[40] A particularly lucid discussion is in Peter Duren's *An Invitation to Classical Analysis*. There are a few more mathematical points that I want to cover instead.

1.8.2. Rates of convergence and smoothness: How big are the Fourier coefficients? Suppose we have a Fourier series of a square integrable function $f(t)$,

$$\sum_{n=-\infty}^{\infty} \hat{f}(n)e^{2\pi int}\,.$$

As noted earlier, from Bessel's inequality,

$$|\hat{f}(n)|^2 \to 0 \quad \text{as } n \to \pm\infty\,.$$

Knowing that the coefficients tend to zero, can we say how fast?

Here's a simple-minded approach that gives some sense of the answer and shows how the answer depends on discontinuities in the function or in its derivatives. All of this discussion is based only on integration by parts with definite integrals.

Suppose, as always, that $f(t)$ is periodic of period 1. Let's assume for this discussion that the function doesn't jump at the endpoints 0 and 1 and that any problem points are inside the interval.[41] That is, we're imagining that there may be trouble at a point t_0 with $0 < t_0 < 1$, as in maybe $f(t)$ jumps there, or maybe $f(t)$ is continuous at t_0 but there's a corner, so $f'(t)$ jumps at t_0, and so on.

---

[40] If you do look things up, be aware that different normalizations, i.e., a different choice of period, will affect the Dirichlet kernel. If the author is working with functions that are periodic of period 2π, for example, then the corresponding Dirichlet kernel is $D_N(t) = \frac{\sin(n+\frac{1}{2})t}{\sin\frac{1}{2}t}$.

[41] This really isn't a restriction. I just want to deal with a single discontinuity for the argument to follow.

The nth Fourier coefficient is

$$\hat{f}(n) = \int_0^1 e^{-2\pi int} f(t)\, dt\,.$$

To analyze the situation near t_0, write this as the sum of two integrals:

$$\hat{f}(n) = \int_0^{t_0} e^{-2\pi int} f(t)\, dt + \int_{t_0}^1 e^{-2\pi int} f(t)\, dt\,.$$

Apply integration by parts to each of these integrals. In doing so, we're going to suppose that at least away from t_0 the function has as many derivatives as we want. Then, on a first pass,

$$\int_0^{t_0} e^{-2\pi int} f(t)\, dt = \left[\frac{e^{-2\pi int} f(t)}{-2\pi in}\right]_0^{t_0} - \int_0^{t_0} \frac{e^{-2\pi int} f'(t)}{-2\pi in}\, dt\,,$$

$$\int_{t_0}^1 e^{-2\pi int} f(t)\, dt = \left[\frac{e^{-2\pi int} f(t)}{-2\pi in}\right]_{t_0}^{1} - \int_{t_0}^{1} \frac{e^{-2\pi int} f'(t)}{-2\pi in}\, dt\,.$$

Add these together. Using $f(0) = f(1)$, this results in

$$\hat{f}(n) = \left[\frac{e^{-2\pi int} f(t)}{-2\pi in}\right]_{t_0^+}^{t_0^-} - \int_0^1 \frac{e^{-2\pi int} f'(t)}{-2\pi in}\, dt\,,$$

where the notations t_0^- and t_0^+ indicate that we're looking at the values of $f(t)$ as we take left-hand and right-hand limits at t_0. If $f(t)$ is continuous at t_0, then the terms in brackets cancel and we're left with just the integral as an expression for $\hat{f}(n)$. But if $f(t)$ is not continuous at t_0 — if it jumps, for example — then we don't get cancellation, and we expect that the Fourier coefficient will be of order $1/n$ in magnitude.[42]

Now suppose that $f(t)$ *is* continuous at t_0, and integrate by parts a second time. In the same manner as above, this gives

$$\hat{f}(n) = \left[\frac{e^{-2\pi int} f'(t)}{(-2\pi in)^2}\right]_{t_0^-}^{t_0^+} - \int_0^1 \frac{e^{-2\pi int} f''(t)}{(-2\pi in)^2}\, dt\,.$$

If $f'(t)$ (the *derivative*) is continuous at t_0, then the bracketed part disappears. If $f'(t)$ is not continuous at t_0, for example if there is a corner at t_0, then the terms do not cancel and we expect the Fourier coefficient to be of size $1/n^2$.

We can continue in this way. The rough rule of thumb may be stated as follows:

- If $f(t)$ is not continuous, then the Fourier coefficients should have some terms like $1/n$.

[42]If we had more jump discontinuities, we'd split the integral up going over several subintervals and we'd have several terms of order $1/n$. The combined result would still be of order $1/n$. This would also be true if the function jumped at the endpoints 0 and 1.

- If $f(t)$ is differentiable except for corners ($f(t)$ is continuous but $f'(t)$ is not), then the Fourier coefficients should have some terms like $1/n^2$.
- If $f''(t)$ exists but is not continuous, then the Fourier coefficients should have some terms like $1/n^3$.

A discontinuity in $f''(t)$ is harder to visualize; in terms of graphs it's typically a discontinuity in the curvature. For example, imagine a curve consisting of an arc of a circle and a line segment that meets the arc tangent to the circle at their endpoints. Something like

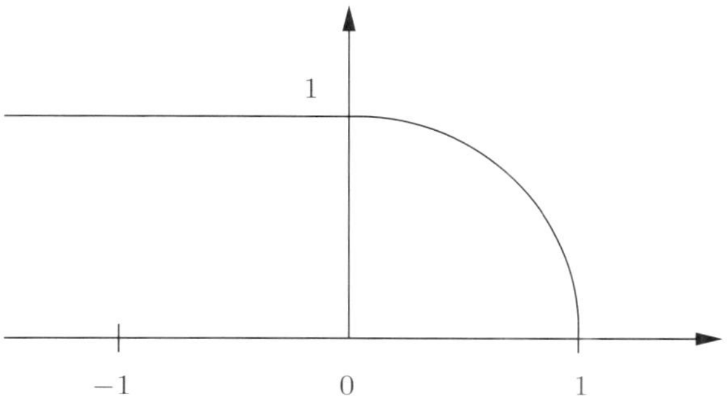

The curve and its first derivative are continuous at the point of tangency, but the second derivative has a jump. If you rode along this path at constant speed, you'd feel a jerk — a discontinuity in the acceleration — when you passed through the point of tangency.

Obviously this pattern extends to continuities/discontinuities in higher-order derivatives. It also jibes with some examples we had earlier. The square wave

$$f(t) = \begin{cases} +1, & 0 \le t < \frac{1}{2}, \\ -1, & \frac{1}{2} \le t < 1, \end{cases}$$

has jump discontinuities, and its Fourier series is

$$\sum_{n \text{ odd}} \frac{2}{\pi i n} e^{2\pi i n t} = \frac{4}{\pi} \sum_{k=0}^{\infty} \frac{1}{2k+1} \sin 2\pi(2k+1)t\,.$$

The triangle wave

$$g(t) = \begin{cases} \frac{1}{2} + t, & -\frac{1}{2} \le t \le 0, \\ \frac{1}{2} - t, & 0 \le t \le +\frac{1}{2}, \end{cases}$$

is continuous but the derivative is discontinuous. (In fact, the derivative is the square wave.) Its Fourier series is

$$\tfrac{1}{4} + \sum_{k=0}^{\infty} \frac{2}{\pi^2(2k+1)^2} \cos(2\pi(2k+1)t)\,.$$

The smoothness (degree of differentiability) of a periodic function is closely related to the rate of convergence and size of the Fourier coefficients. In their book, *Fourier Series and Integrals*, mentioned earlier, Dym and McKean state the result this way:

> **Theorem.** Let $f(t)$ be periodic of period 1. Suppose that $f(t)$ is p-times continuously differentiable, where p is at least 1. (This includes the derivatives matching up continuously at the endpoints 0 and 1.) Then the partial sums
>
> $$S_N(t) = \sum_{n=-N}^{N} \hat{f}(n) e^{2\pi i n t}$$
>
> converge to $f(t)$ pointwise and uniformly on $[0,1]$ as $N \to \infty$. Furthermore
>
> $$\max |S_N(t) - f(t)| \leq \text{constant} \frac{1}{N^{p-\frac{1}{2}}}$$
>
> for $0 \leq t \leq 1$.

The fact that there is a bound on the maximum of $|S_N(t) - f(t)|$ that tends to 0 is what makes the convergence uniform.

We won't prove the theorem; the proof uses the Dirichlet kernel representation of $S_N(t) - f(t)$ from Section 1.8.1. Once again, an interesting aspect of this result has to do with how local properties of the function are reflected in global properties of its Fourier series. In the present setting, "local properties" of a function refers to how smooth it is, i.e., how many times it's continuously differentiable. About the only kind of "global question" one can ask about series is how fast they converge, and that's what is estimated here. The essential point is that the error in the approximation, and indirectly the rate at which the coefficients decrease, is governed by the smoothness (the degree of differentiability) of the signal. The smoother the function, a "local" statement, the better the approximation, and this is not just in the mean L^2 sense but uniformly over the interval, a "global" statement. Moreover, the Fourier coefficients themselves, being defined by an integral, are global aspects of the function.

One final comment about the differentiability and the size of the Fourier coefficients. The extreme case of smoothness is when $f(t)$ is continuously differentiable to *any* order. The notation to capture this is that $f(t)$ is of class C^∞ on $[0,1]$. (This means that all derivatives also match up at the endpoints, $f^{(k)}(0) = f^{(k)}(1)$ for all $k \geq 0$.) In that case, one can show that the Fourier coefficients are *rapidly decreasing*. This means that

$$n^k \hat{f}(n) \to 0 \text{ as } n \to \pm\infty \text{ for any } k \geq 0\,.$$

In words, $\hat{f}(n)$ tends to zero faster than any nonnegative power of n. In fact, the implication goes the other way, too: if the Fourier coefficients are rapidly decreasing, then $f(t)$ is C^∞.

You may be getting tired of all this foreshadowing, but we will also see the local versus global interplay at work in properties of the Fourier transform, particularly in the relation between smoothness and rate of decay. This is one reason why I wanted us to see this for Fourier series. And wouldn't you know it: we'll also encounter rapidly decreasing functions again.

1.8.3. The complex exponentials really are an orthonormal basis for $L^2([0,1])$.

The one remaining point in our hit parade of the L^2-theory of Fourier series is that the complex exponentials are a basis, meaning that

$$\lim_{N\to\infty}\left\| \sum_{n=-N}^{N} \hat{f}(n)e^{2\pi int} - f(t)\right\| = 0\,.$$

I said earlier that we wouldn't attempt a complete proof of this, and we won't. But with the discussion just preceding, we can say more precisely how the proof goes and what the issues are that we cannot get into. The argument is in three steps.

Let $f(t)$ be a square integrable function and let $\epsilon > 0$.

Step 1. Any function in $L^2([0,1])$ can be approximated in the L^2-norm by a continuously differentiable function. Starting with a given f in $L^2([0,1])$ and any $\epsilon > 0$ we can find a function $g(t)$ that is continuously differentiable on $[0,1]$ for which

$$\|f - g\| < \epsilon\,.$$

This is exactly the step we cannot do! It's here, in proving this statement, that one needs the more general theory of Lebesgue integration and the limiting processes that go with it.[43] Let it rest.

Step 2. From the discussion in the preceding section we now know that the partial sums for a continuously differentiable function ($p = 1$ in the statement of the theorem) converge uniformly to the function. Thus with $g(t)$ as in Step 1 we can choose N so large that

$$\max_{0\le t\le 1}\left| g(t) - \sum_{n=-N}^{N} \hat{g}(n)e^{2\pi int}\right| < \epsilon\,.$$

Then for the L^2-norm,

$$\begin{aligned}\int_0^1 \left| g(t) - \sum_{n=-N}^{N} \hat{g}(n)e^{2\pi int}\right|^2 dt &\le \int_0^1 \left(\max\left| g(t) - \sum_{n=-N}^{N} \hat{g}(n)e^{2\pi int}\right|\right)^2 dt\\ &< \int_0^1 \epsilon^2\, dt = \epsilon^2\,.\end{aligned}$$

Hence

$$\left\| g(t) - \sum_{n=-N}^{N} \hat{g}(n)e^{2\pi int}\right\| < \epsilon\,.$$

[43] Actually, it's true that any function in $L^2([0,1])$ can be approximated by an *infinitely* differentiable function. We will come back to this in Chapter 4 in the context of convolution being a smoothing operation.

Step 3. Remember that the Fourier coefficients provide the best finite approximation in L^2 to the function. As we'll need it:

$$\left\| f(t) - \sum_{n=-N}^{N} \hat{f}(n)e^{2\pi int} \right\| \le \left\| f(t) - \sum_{n=-N}^{N} \hat{g}(n)e^{2\pi int} \right\| .$$

Then at last

$$\begin{aligned}\left\| f(t) - \sum_{n=-N}^{N} \hat{f}(n)e^{2\pi int} \right\| &\le \left\| f(t) - \sum_{n=-N}^{N} \hat{g}(n)e^{2\pi int} \right\| \\ &= \left\| f(t) - g(t) + g(t) - \sum_{n=-N}^{N} \hat{g}(n)e^{2\pi int} \right\| \\ &\le \| f(t) - g(t) \| + \left\| g(t) - \sum_{n=-N}^{N} \hat{g}(n)e^{2\pi int} \right\| < 2\epsilon .\end{aligned}$$

This shows that

$$\left\| f(t) - \sum_{n=-N}^{N} \hat{f}(n)e^{2\pi int} \right\|$$

can be made arbitrarily small by taking N large enough, which is what we were required to do.

To complete the picture, let me add a final point that's a sort of converse to what we've done. We won't use this, but it ties things up nicely.

If $\{c_n : n = 0, \pm 1, \pm 2, \dots\}$ is any sequence of complex numbers for which

$$\sum_{n=-\infty}^{\infty} |c_n|^2 < \infty ,$$

then the function

$$f(t) = \sum_{n=-\infty}^{\infty} c_n e^{2\pi int}$$

is in $L^2([0,1])$, meaning the limit of the partial sums converges to a function in $L^2([0,1])$, and $c_n = \hat{f}(n)$.

This is often referred to as the Riesz-Fischer theorem.

Farewell to all that. Had enough? There's a lot of material that just passed your way and, I repeat, you are likely to encounter the ideas and the terminology in other venues. But, honestly, the results are more mathematical issues than they are of day-to-day, practical concern. To paraphrase John Tukey, an eminent applied mathematician who helped to invent the Fast Fourier Transform algorithm (a coming attraction!): "I wouldn't want to fly in a plane whose design depended on whether a function was Riemann or Lebesgue integrable."

1.8.4. Appendix: Pointwise convergence vs. uniform convergence. Here's an example, a classic of its type, to show that pointwise convergence is not the same as uniform convergence, or what amounts to the same thing, that we can have a sequence of functions $f_n(t)$ with the property that $f_n(t) \to f(t)$ for every value of t as $n \to \infty$ but the graphs of the $f_n(t)$ *do not* ultimately look like the graph of $f(t)$. Let me describe such a sequence of functions in words, draw a few pictures, and leave it to you to write down a formula.

The functions $f_n(t)$ will all be defined on $0 \le t \le 1$. For each n the graph of $f_n(t)$ is zero from $1/n$ to 1 and is an isosceles triangle with height n^2 for $0 \le t \le 1/n$. Here are pictures of $f_1(t)$, $f_5(t)$, and $f_{10}(t)$.

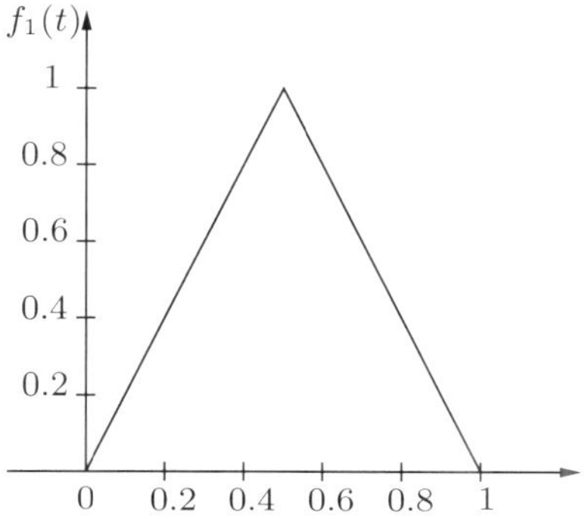

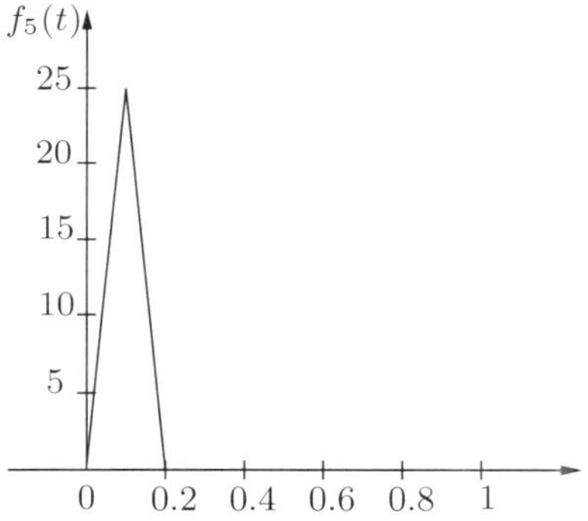

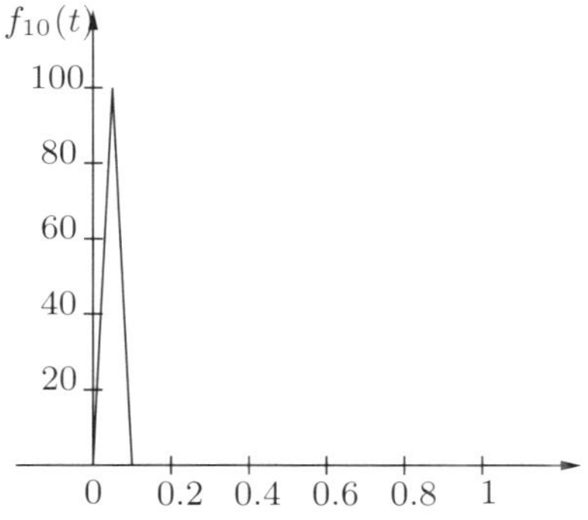

The peak slides to the left and gets higher and higher as n increases. It's clear that for each t the sequence $f_n(t)$ tends to 0. This is so because $f_n(0) = 0$ for all n, and for any $t \neq 0$ eventually (that is, for large enough n) the peak is going to slide to the left of t and $f_n(t)$ will be zero from that n onward. Thus $f_n(t)$ converges pointwise to the constant 0. But the graphs of the $f_n(t)$ certainly are not uniformly close to 0!

Because uniform convergence is much to be prized, you might be pleased to know that there's a fairly flexible result that guarantees it. It's called the Weierstrass M-test[44] and it goes like this.

> **Weierstrass M-test.** Let $f_1(t), f_2(t), \ldots$ be a sequence of functions on $0 \le t \le 1$ and suppose there are numbers $M_1, M_2, \ldots$ with the following properties:
> (1) For each n we have $|f_n(t)| \le M_n$ for any $0 \le t \le 1$.
> (2) The series of numbers $\sum_{n=1}^{\infty} M_n$ converges.
> Then the series of functions $\sum_{n=1}^{\infty} f_n(t)$ converges uniformly for $0 \le t \le 1$.

The theorem doesn't say anything about what the series converges to.

Harking back to Section 1.4, the Weierstrass M-test is all we need in order to show that the Fourier series for the triangle wave converges uniformly for $0 \le t \le 1$. Why? But the test does not allow us to draw any conclusion for the square wave. Why?

1.9. Appendix: The Cauchy-Schwarz Inequality

Such a famous and useful inequality. It had to go somewhere, so here it is. The *Cauchy-Schwarz inequality* is a relationship between the inner product of two vectors and their norms. It states that

$$|(\underline{v}, \underline{w})| \le \|\underline{v}\| \, \|\underline{w}\| .$$

It's a real workhorse and you should know it. You'll even see it in action in some problems scattered through the book.

For geometric vectors this is trivial to see from the geometric formula for the inner product,

$$|(\underline{v}, \underline{w})| = \|\underline{v}\| \, \|\underline{w}\| \, |\cos\theta| \le \|\underline{v}\| \, \|\underline{w}\| ,$$

because $|\cos\theta| \le 1$. Actually, the rationale for the geometric formula of the inner product will *follow* from the Cauchy-Schwarz inequality.

It's certainly not obvious how to derive the inequality from the algebraic definition of the inner product of geometric vectors. Written out in components, the inequality says (for real vectors) that

$$\left| \sum_{k=1}^{n} v_k w_k \right| \le \left(\sum_{k=1}^{n} v_k^2 \right)^{1/2} \left(\sum_{k=1}^{n} w_k^2 \right)^{1/2} .$$

Sit down and try that one out sometime.

The derivation of the Cauchy-Schwarz inequality in general uses only the four algebraic properties of the inner product listed earlier. Consequently the same

[44] Named after K. Weierstrass. He had an interesting career, being largely self-taught and teaching (German) high school for a number of years. He was known for an emphasis on rigor, thus causing endless worrying by many mathematics students and mathematics professors.

argument applies to any product satisfying these properties, the inner product on $L^2([0,1])$, for example. It's such an elegant argument (due to John von Neumann, I believe) that I'd like to show it to you. We'll give this for the real inner product here, with comments on the complex case to follow.

Any inequality can ultimately be written in a way that says that some quantity is positive, or, at least, nonnegative. Examples of things that we know are positive are the square of a real number, the area of something, and the length of something. More subtle inequalities sometimes rely on convexity, e.g., the center of gravity of a system of masses is contained within the convex hull of the masses. This little riff on the nature of inequalities qualifies as a minor secret of the universe.

For the proof of the Cauchy-Schwarz inequality we use that the norm of a vector is nonnegative, but we throw in a parameter.[45] Let r be any real number. Then $\|\underline{v} - r\underline{w}\|^2 \geq 0$. Write this in terms of the inner product and expand using the algebraic properties; because of homogeneity, symmetry, and additivity, this is just like multiplication — that's important to realize:

$$\begin{aligned} 0 &\leq \|\underline{v} - r\underline{w}\|^2 \\ &= (\underline{v} - r\underline{w}, \underline{v} - r\underline{w}) \\ &= (\underline{v}, \underline{v}) - 2r(\underline{v}, \underline{w}) + r^2(\underline{w}, \underline{w}) \\ &= \|\underline{v}\|^2 - 2r(\underline{v}, \underline{w}) + r^2\|\underline{w}\|^2 . \end{aligned}$$

This is a quadratic equation in r, of the form $ar^2 + br + c$, where $a = \|\underline{w}\|^2$, $b = -2(\underline{v}, \underline{w})$, and $c = \|\underline{v}\|^2$. The first inequality, and the chain of equalities that follow, says that this quadratic is *always nonnegative*. Now a quadratic that's always nonnegative has to have a *nonpositive* discriminant: the discriminant, $b^2 - 4ac$ determines the nature of the roots of the quadratic. If the discriminant is positive, then there are two real roots, and if there are two real roots, then the quadratic must be negative somewhere.

Therefore $b^2 - 4ac \leq 0$, which translates to

$$4(\underline{v}, \underline{w})^2 - 4\|\underline{w}\|^2\,\|\underline{v}\|^2 \leq 0 \quad \text{or} \quad (\underline{v}, \underline{w})^2 \leq \|\underline{w}\|^2\,\|\underline{v}\|^2 .$$

Take the square root of both sides to obtain

$$|(\underline{v}, \underline{w})| \leq \|\underline{v}\|\,\|\underline{w}\| ,$$

as desired. Amazing, isn't it — a nontrivial application of the *quadratic formula*![46]

To get back to geometry, we now know that

$$-1 \leq \frac{(\underline{v}, \underline{w})}{\|\underline{v}\|\,\|\underline{w}\|} \leq 1 .$$

[45] "Throwing in a parameter" goes under the heading of dirty tricks of the universe.

[46] As a slight alternative to this argument, if the quadratic $ar^2 + br + c$ is everywhere nonnegative, then, in particular, its minimum value is nonnegative. This minimum occurs at $t = -b/2a$ and leads to the same inequality, $4ac - b^2 \geq 0$.

Therefore there is a unique angle θ with $0 \le \theta \le \pi$ such that

$$\cos\theta = \frac{(\underline{v}, \underline{w})}{\|\underline{v}\| \, \|\underline{w}\|},$$

i.e.,

$$(\underline{v}, \underline{w}) = \|\underline{v}\| \, \|\underline{w}\| \cos\theta .$$

Identifying θ as the angle between $\underline{v}$ and $\underline{w}$, we have now reproduced the geometric formula for the inner product. What a relief.

This also demonstrates when equality holds in the Cauchy-Schwarz inequality, namely, when the vectors are in the same (or opposite) direction.

The triangle inequality,

$$\|\underline{v} + \underline{w}\| \le \|\underline{v}\| + \|\underline{w}\| ,$$

follows directly from the Cauchy-Schwarz inequality. Here's the argument.

$$\begin{aligned}
\|\underline{v} + \underline{w}\|^2 &= (\underline{v} + \underline{w}, \underline{v} + \underline{w}) \\
&= (\underline{v}, \underline{v}) + 2(\underline{v}, \underline{w}) + (\underline{w}, \underline{w}) \\
&\le (\underline{v}, \underline{v}) + 2|(\underline{v}, \underline{w})| + (\underline{w}, \underline{w}) \\
&\le (\underline{v}, \underline{v}) + 2\|\underline{v}\| \, \|\underline{w}\| + (\underline{w}, \underline{w}) \quad \text{(by Cauchy-Schwarz)} \\
&= \|\underline{v}\|^2 + 2\|\underline{v}\| \, \|\underline{w}\| + \|\underline{w}\|^2 = (\|\underline{v}\| + \|\underline{w}\|)^2.
\end{aligned}$$

Now take the square root of both sides to get $\|\underline{v} + \underline{w}\| \le \|\underline{v}\| + \|\underline{w}\|$. In coordinates this says that

$$\left(\sum_{k=1}^{n} (v_k + w_k)^2\right)^{1/2} \le \left(\sum_{k=1}^{n} v_k^2\right)^{1/2} + \left(\sum_{k=1}^{n} w_k^2\right)^{1/2}.$$

Here's how to get the Cauchy-Schwarz inequality for complex inner products from what we've already done. The inequality states

$$|(\underline{v}, \underline{w})| \le \|\underline{v}\| \, \|\underline{w}\|$$

for complex vectors $\underline{v}$ and $\underline{w}$. On the left-hand side we have the magnitude of the complex number $(\underline{v}, \underline{w})$. As a slight twist on what we did in the real case, let $\alpha = re^{i\theta}$ be a complex number (r real) and consider

$$\begin{aligned}
0 \le \|\underline{v} - \alpha\underline{w}\|^2 &= \|\underline{v}\|^2 - 2\operatorname{Re}(\underline{v}, \alpha g) + \|\alpha\underline{w}\|^2 \\
&= \|\underline{v}\|^2 - 2\operatorname{Re}\left(\overline{\alpha}(\underline{v}, \underline{w})\right) + \|\alpha\underline{w}\|^2 \\
&= \|\underline{v}\|^2 - 2r\operatorname{Re}\left(e^{-i\theta}(\underline{v}, \underline{w})\right) + r^2\|\underline{w}\|^2 .
\end{aligned}$$

Now, we can choose θ to be anything we want, and we do so to make

$$\operatorname{Re}\left(e^{-i\theta}(\underline{v}, \underline{w})\right) = |(\underline{v}, \underline{w})| .$$

Multiplying $(\underline{v}, \underline{w})$ by $e^{-i\theta}$ rotates the complex number $(\underline{v}, \underline{w})$ clockwise by θ, so choose θ to rotate $(\underline{v}, \underline{w})$ to be real and positive. From here the argument is the same as it was in the real case.

Cauchy-Schwarz for $L^2([0,1])$. Let me emphasize again that the proof of the Cauchy-Schwarz inequality depended only on the algebraic properties of the inner product and therefore holds for the (complex) inner product on $L^2([0,1])$. It takes the impressive form

$$\left|\int_0^1 f(t)\overline{g(t)}\,dt\right| \le \left(\int_0^1 |f(t)|^2\,dt\right)^{1/2}\left(\int_0^1 |g(t)|^2\,dt\right)^{1/2}.$$

We now also know that the triangle inequality holds:

$$\|f+g\| \le \|f\| + \|g\|,$$

i.e.,

$$\left(\int_0^1 |f(t)+g(t)|^2\,dt\right)^{1/2} \le \left(\int_0^1 |f(t)|^2\,dt\right)^{1/2} + \left(\int_0^1 |g(t)|^2\,dt\right)^{1/2}.$$

By the way, the inner product does make sense. Since this is a mathematical section, I should point out that I have skipped over a mathematical something. If $f(t)$ and $g(t)$ are square integrable, then in order to get the Cauchy-Schwarz inequality working, one has to know that the inner product (f,g) makes sense, i.e., that

$$\left|\int_0^1 f(t)\overline{g(t)}\,dt\right| < \infty.$$

Have no fear. To deduce this you can use

$$2|f(t)||g(t)| \le |f(t)|^2 + |g(t)|^2.$$

This is the inequality between the arithmetic and geometric mean — look it up![47] Then

$$\begin{aligned}\left|\int_0^1 f(t)\overline{g(t)}\,dt\right| \le \int_0^1 |f(t)||\overline{g(t)}|\,dt &= \int_0^1 |f(t)||g(t)|\,dt\\ &\le \frac{1}{2}\left(\int_0^1 |f(t)|^2\,dt + \int_0^1 |g(t)|^2\,dt\right) < \infty,\end{aligned}$$

since we started by assuming that $|f(t)|$ and $|g(t)|$ are square integrable. The Cauchy-Schwarz inequality is well founded, and we are pleased.

The Fourier coefficients are fine. Now that we know that the Cauchy-Schwarz inequality is really OK, one consequence is the fortunate fact that the Fourier coefficients of a function in $L^2([0,1])$ exist. That is, we might be wondering about the existence of

$$(f, e_n) = \int_0^1 f(t)e^{-2\pi int}\,dt.$$

[47] It's positivity again: $0 \le (a-b)^2 = a^2 - 2ab + b^2$.

We wonder no more:[48]

$$|(f, e_n)| \leq \|f\| \|e^{2\pi int}\| = \|f\| \,.$$

And convolution is also OK. We didn't make an issue of being sure that the convolution $(f * g)(t)$ of two periodic functions $f(t)$, $g(t)$ is really defined, i.e., that the integral

$$\int_0^1 f(t-\tau)g(\tau)\, d\tau$$

exists. Assuming that $f(t)$ and $g(t)$ are L^2-functions, we have

$$\left|\int_0^1 f(t-\tau)g(\tau)\, d\tau\right| \leq \left(\int_0^1 |f(t-\tau)|^2\, d\tau\right)^{1/2} \left(\int_0^1 |g(\tau)|^2\, d\tau\right)^{1/2} .$$

Now you can finish the argument that the right-hand side is finite by invoking periodicity:

$$\int_0^1 |f(t-\tau)|^2\, d\tau = \int_0^1 |f(\tau)|^2\, d\tau \,.$$

Square integrable implies integrable. Another consequence of Cauchy-Schwarz that's sometimes under-appreciated is that a square integrable function is also absolutely integrable, meaning that if

$$\int_0^1 |f(t)|^2\, dt < \infty \,,$$

then

$$\int_0^1 |f(t)|\, dt < \infty \,.$$

To see this, apply the Cauchy-Schwarz inequality to $|f(t)| \cdot 1$, producing

$$\int_0^1 |f(t)|\, 1\, dt \leq \|f\| \|1\| < \infty \,,$$

since both $f(t)$ and the constant function 1 are square integrable on $[0, 1]$.

> **Warning:** This simple, almost offhand argument *would not work* if the interval $[0, 1]$ were replaced by the entire real line. The constant function 1 has an infinite integral on $\mathbb{R}$. You may think we can get around this little inconvenience, but it is *exactly* the sort of trouble that comes up sometimes in trying to apply Fourier *series* ideas (where functions are defined on finite intervals) to Fourier *transform* ideas (where functions are defined on all of $\mathbb{R}$).

[48] You can also deduce that (f, e_n) makes sense by using the arithmetic-geometric mean inequality.

Problems and Further Results

Since this is the first section of problems, let me make a few general comments that apply to the whole book. All the problems are from problem sets or exams so I think they're pretty well debugged. Read through the whole section; you'll find interesting material.

The problems originate from many sources and have arrived here through many people. Many problems we made up (myself, colleagues, TAs), and many were, in the finest academic tradition, ~~stolen~~ borrowed from the many resources that are available: books, websites, etc.[49] I've tried to give credit when I knew it, but I also know I've fallen far short. I ask for forgiveness and immunity from prosecution. I do want to single out the book *A First Course in Fourier Analysis* by D. Kammler for its wonderful collection of problems (among other attributes), a number of which I've reproduced. The classic signals and systems text by Oppenheim and Willsky was also a good source of problems, some of which arrived here via the TAs who took their course!

Downloading files for problems. Some problems rely on software or ask you to write some code. We used MATLAB® for the class, I tend to use Mathematica, and any standard package should be sufficient. For problems calling for files (some scripts in MATLAB, for example), these can be found via Box in a folder called "Fourier Transform Problem Files". Problems will say something like: "Download `filename` …". You can download the files from

`https://stanford.box.com/s/gv9kltoukmpr36xefwa6hcp8x1h12vm5.`

(You should not need a Box account, but you might want to open one anyway.)

1.1. *Playing with Fourier series using* MATLAB

This problem is based on the MATLAB application in the 'sinesum' folder on the above website. Go there to get the files. It's a tool to plot sums of sinusoids of the form

$$\sum_{n=1}^{N} A_n \sin(2\pi n t + \phi_n)\,.$$

Using the sinesum application, generate (approximately) the signal plotted below. You should be able to do this with, say, five harmonics. Along with the overall shape, the things to get right are the approximate shift and orientation of the signal. To get started, the signal looks like a sawtooth signal. The coefficient A_n is of order $1/n$ and that gives you the relative sizes of the A_n. However, you'll see that additional shifting and flipping needs to be done. This can be accomplished by changing the phases ϕ_n.

[49] I refer to T. Körner's invocation of Kipling in the introduction to his book *Exercises in Fourier Analysis*. I was tempted to ~~steal~~ borrow it.

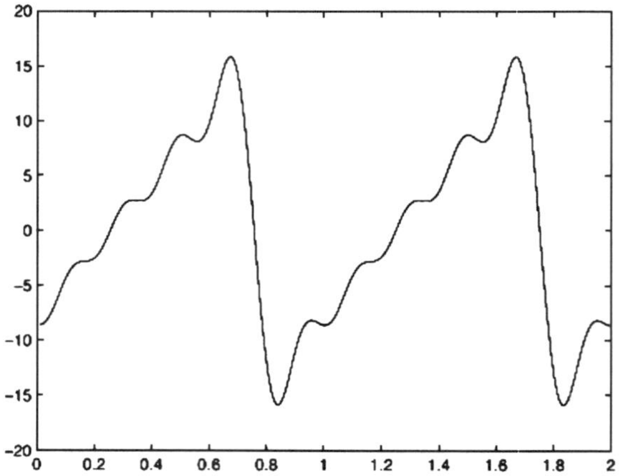

The point of the problem is to get a little hands-on experience with making complicated periodic signals, with MATLAB doing the dirty work. Some of us even think this is fun.

1.2. *Adding periodic functions*

You can sometimes be misled by thinking too casually about modifying or combining periodic functions: scaling a periodic function results in a periodic function; shifting a periodic function results in a periodic function. What about adding?

(a) Let $f(x) = \sin(2\pi mx) + \sin(2\pi nx)$ where n and m are positive integers. Is $f(x)$ periodic? If so, what is its period?

(b) Let $g(x) = \sin(2\pi px) + \sin(2\pi qx)$ where p and q are positive rational numbers (say $p = m/r$ and $q = n/s$, as fractions in lowest terms). Is $g(x)$ periodic? If so, what is its period?

(c) It's not true that the sum of two periodic functions is periodic. For example, show that $f(t) = \cos t + \cos\sqrt{2}t$ is not periodic. (*Hint*: Suppose by way of contradiction that there is some T such that $f(t+T) = f(t)$ for all t. In particular, the maximum value of $f(t)$ repeats. This will lead to a contradiction.)

1.3. *Periods of sums and products*[50]

Let $f(t) = \sin 3t + \cos 5t$ and $g(t) = \sin 3t \cdot \cos 5t$.

(a) What is the period of $f(t)$? Find the Fourier series for $f(t)$.

(b) Find the Fourier series for $g(t)$. What is the period of $g(t)$?
(The period of the product is more interesting. The product repeats every 2π, that is, $g(t+2\pi) = g(t)$, so the period of $g(t)$ is a divisor of 2π. To determine the fundamental frequency of $g(t)$, we find its Fourier series.)

[50]From John Gill.

1.4. *Different definitions of periodicity*

(a) Show that $f(t)$ is periodic of period p if and only if $f(t-p) = f(t)$ for all t. The upshot is that it doesn't matter if we define periodicity as $f(t+p) = f(t)$ or as $f(t-p) = f(t)$.

(b) Show that $f(t)$ is periodic of period p if and only if $f(t+p/2) = f(t-p/2)$ for all t.

1.5. *Overheard at a problem session ...*

Suppose two sinusoids have the same frequency but possibly different amplitudes and phases. What about their sum? Each of the following answers was proposed at a problem session:

(a) Has twice the frequency of the original sinusoids.
(b) Is either zero or has exactly the same frequency as the original sinusoids.
(c) May exhibit beats.
(d) Is not necessarily periodic at all.

Which is correct, and why? (*Hint*: Think in terms of complex exponentials.)

1.6. *Low voltage*[51]

A periodic voltage is given by $v(t) = 3\cos(2\pi\nu_1 t - 1.3) + 5\cos(2\pi\nu_2 t + 0.5)$. Regardless of the frequencies ν_1, ν_2, the maximum voltage is always less than 8, but it can be much smaller. Use MATLAB (or another program) to find the maximum voltage if $\nu_1 = 2$ Hz and $\nu_2 = 1$ Hz.

1.7. *Periodizing a triangle*[52]

The triangle function with a parameter $p > 0$ is

$$\Lambda_p(t) = \begin{cases} 1 - \frac{1}{p}|t|\,, & |t| \le p, \\ 0\,, & |t| \ge p. \end{cases}$$

We'll be seeing it a lot. When $p = 1$ (the model case in many instances), we simplify the notation to just $\Lambda(t)$. Note that $\Lambda_p(t) = \Lambda(t/p)$. The graph of $\Lambda_p(t)$ is

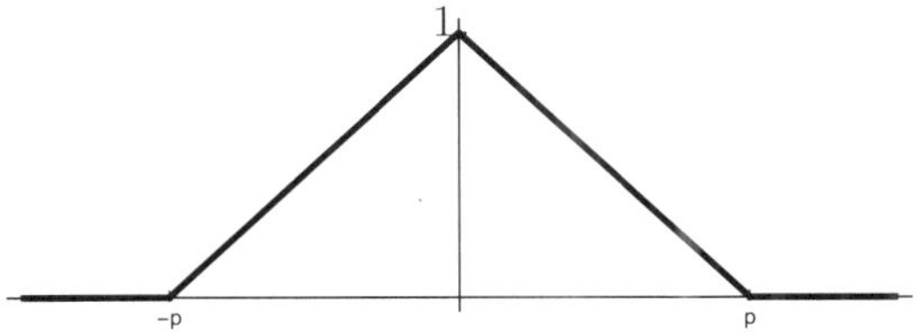

The parameter p specifies the length of the base, namely $2p$. Alternately, p determines the slopes of the sides: the left side has slope $1/p$ and the right side has slope $-1/p$.

[51] From Paul Nahim, *The science of radio*.

[52] Do this problem and also the next problem! We'll be periodizing a fair amount.

Now for $T > 0$ define

$$g(t) = \sum_{n=-\infty}^{\infty} \Lambda_p(t - nT)\,.$$

(a) For $p = 1/2$ and $T = 1/2$, $T = 3/4$, $T = 1$, $T = 2$, sketch the graphs of $g(t)$. Periodic? In each case what is the period?
(b) In which of these cases is it possible to recover the original signal $\Lambda_{1/2}(t)$ from the sum $g(t)$? What operation would you apply to $g(t)$ to do this?
(c) In general, what condition on p and T will guarantee that $\Lambda_p(t)$ can be recovered from the sum $g(t)$?

1.8. *Periodization in general*[53]
Let $f(t)$ be a function, defined for all t, and let $T > 0$. Define

$$g(t) = \sum_{n=-\infty}^{\infty} f(t - nT)\,,$$

(as we did for Λ in the previous problem).
(a) Provided the sum converges, show that $g(t)$ is periodic with period T. One says that $g(t)$ is the *periodization* of $f(t)$ of period T. (Later we'll learn to express such a sum as a convolution.)
(b) Give a (simple) condition on $f(t)$ and T that guarantees that the sum $g(t)$ converges *and* that it is possible to recover $f(t)$ from $g(t)$.
(c) If a function $f(t)$ is already periodic, is it equal to its own periodization? Explain.

1.9. *Integrals over a period*
Let $f(t)$ be a function of period 1.
(a) Any period will do: show that

$$\int_a^{a+1} f(t)\,dt = \int_0^1 f(t)\,dt$$

for any a.
(b) Shifts don't matter: show that

$$\int_0^1 f(x-y)\,dx = \int_0^1 f(x)\,dx$$

for any y.
(c) Suppose in addition to being periodic that $f(t)$ is odd. Show that

$$\int_0^1 f(t)\,dt = 0.$$

[53] See the previous footnote.

1.10. Let $g(x)$ be periodic of period T with Fourier coefficients c_k. Show that

$$c_k = \frac{1}{mT}\int_0^{mT} g(x)e^{-2\pi ikx/T}\,dx, \quad m \text{ a positive integer.}$$

1.11. *Signal generation via Fourier series*

Consider the function $f(t)$ sketched below.

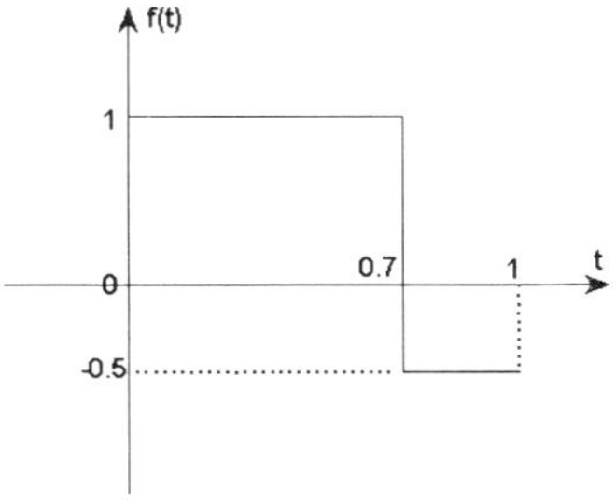

We periodize $f(t)$ to have period 1 and expand it in a Fourier series,

$$f(t) = \sum_{n=-\infty}^{\infty} c_n e^{2\pi int}.$$

You do not need to calculate the c_n's for this problem.

By modifying the Fourier coefficients c_n of $f(t)$ we can generate new signals $g(t)$ that are related to $f(t)$. Say the Fourier series for $g(t)$ is

$$g(t) = \sum_{n=-\infty}^{\infty} d_n e^{2\pi int}.$$

For each case below express $g(t)$ in terms of $f(t)$ and draw $g(t)$ for $0 < t < 1$. If necessary, draw both real and imaginary parts of $g(t)$ in separate graphs.

(a) $d_n = \mathrm{Re}(c_n) = \frac{1}{2}(c_n + \overline{c_n})$.

(b) $d_n = (-1)^n c_n = e^{\pi in} c_n$.

(c) $d_n = \begin{cases} c_n & n \text{ even,} \\ 0 & n \text{ odd.} \end{cases}$

Hint: Use part (b).

1.12. If $f(t)$ is a differentiable function of period T, show that $f'(t)$ also has period T. What about an antiderivative of $f(t)$?

1.13. *Half-wave symmetry*

Let $f(t)$ be a periodic signal of period 1.

We say that $f(t)$ has *half-wave symmetry* if

$$f\left(t - \frac{1}{2}\right) = -f(t).$$

(a) Sketch an example of a signal that has half-wave symmetry.

(b) If $f(t)$ has half-wave symmetry and its Fourier series is

$$f(t) = \sum_{n=-\infty}^{\infty} c_n e^{2\pi i n t},$$

show that $c_n = 0$ if n is even.
Hint: $-c_n = -\int_0^1 e^{-2\pi i n t} f(t)\,dt = \int_0^1 e^{-2\pi i n t} f(t - \frac{1}{2})\,dt$.

1.14. *Downsampling the spectrum*
Define a signal $f(t)$ by

$$f(t) = \begin{cases} t & \text{for } 0 \le t < 0.25, \\ -t + 0.5 & \text{for } 0.25 \le t \le 0.5, \\ 0 & \text{for } 0.5 \le t \le 1, \end{cases}$$

and periodize $f(t)$ to have period 1. See the plot below.

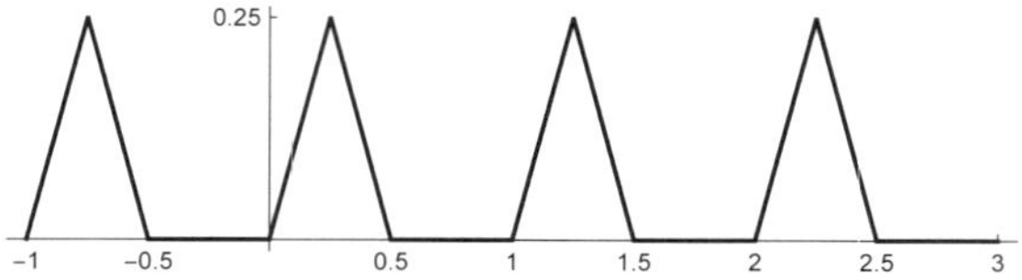

Let $\hat{f}(n)$ be the nth Fourier coefficient of $f(t)$. We define a new periodic signal $h(t)$ of period 1 whose Fourier coefficients are given by

$$\hat{h}(n) = \begin{cases} \hat{f}(n) & \text{for } n \text{ even}, \\ 0 & \text{for } n \text{ odd}; \end{cases}$$

i.e., we have thrown away all the odd-indexed Fourier coefficients from f to obtain h. The Fourier series for $h(t)$ is

$$h(t) = \sum_{n=-\infty}^{\infty} \hat{h}(n) e^{2\pi i n t}.$$

Find h in terms of f and sketch a graph of $h(t)$ for $0 \le t \le 1$. What do you observe?

This is an example of *downsampling*. We will come back to downsampling when we learn about the discrete Fourier transform.

Hint: Write

$$\hat{h}(n) = (1/2)\hat{f}(n)\,(1 + (-1)^n),$$

and express $(-1)^n$ as a complex exponential.

1.15. *Upsampling the spectrum*
In Problem 1.14 we considered the effect of downsampling the spectrum of a periodic signal $f(t)$ of period 1. There we defined a new function $h(t)$ also of period 1 by setting the odd Fourier coefficients of $f(t)$ equal to zero and keeping only the even Fourier coefficients. A related operation is called *upsampling*. For upsampling we keep all the original Fourier coefficients of $f(t)$ and increase the total spectrum by *adding* extra zeros between the original Fourier coefficients to define a new function.

To be precise, let $f(t)$ be a periodic function with period 1 and with Fourier coefficients $\hat{f}(n)$. Define a periodic function $h(t)$ *of period* 2 by setting

$$\hat{h}(n) = \begin{cases} 0, & n \text{ odd}, \\ \hat{f}(n/2), & n \text{ even}. \end{cases}$$

(a) Find $h(t)$ in terms of $f(t)$ on the interval $0 \leq t \leq 2$.
(b) How does the energy of the signal $h(t)$ compare to that of $f(t)$?
(c) If (one period of) $f(t)$ is the function in Problem 1.14 , what is the graph of $h(t)$ for $0 \leq t \leq 2$?
(d) How does your answer to (b) jibe with Rayleigh's identity and the graph you plotted in part (c)?

We will see upsampling again when we study the discrete Fourier transform.

1.16. *Upsampling, again*

In the preceding problem you showed how upsampling Fourier coefficients — inserting zeros between the Fourier coefficients of a given function — leads to squeezing that function in the time domain. Here you will show that the reverse also holds, that squeezing a function in the time domain leads to upsampling its Fourier coefficients.

Let $f(x)$ be a T-periodic function with Fourier series

$$f(x) = \sum_{k=-\infty}^{\infty} c_k e^{2\pi ikx/T}.$$

(a) For a constant b consider the shifted function

$$f(x-b) = \sum_{k=-\infty}^{\infty} d_k e^{2\pi ikx/T}.$$

Find the Fourier coefficients d_k in terms of the c_k.

(b) Suppose that $f(x)$ also has a period smaller than T; say that $f(x)$ is T/m-periodic for some integer m. Show that $c_k = 0$ when $k \neq 0, \pm m, \pm 2m, \ldots$.
Hint: Use that $f(x - \frac{T}{m}) - f(x) = 0$ and your result in part (a).

(c) Finally, suppose that $f(x)$ is a squeezed version of a T-periodic function $g(x)$, that is,

$$f(x) = g(mx),$$

so that, as above, $f(x)$ has periods T and T/m. Suppose also that $g(x)$ has the Fourier series expansion

$$g(x) = \sum_{k=-\infty}^{\infty} c'_k e^{2\pi ikx/T}.$$

Find the Fourier coefficients c_k of $f(x)$ in terms of the c'_k.
Hint: Since $g(x)$ is T-periodic, its Fourier coefficients satisfy

$$c'_k = \frac{1}{mT}\int_0^{mT} g(x)e^{-2\pi ikx/T}\,dx.$$

Explain this formula, and then use it.

1.17. *Walsh functions*[54]

The set of Walsh functions is a simple and useful set of orthonormal functions (using the inner product on $L^2([0,1])$). They come up in various areas of digital signal processing. They take only the values 1 and -1 and these values on subintervals of $[0,1]$ determined by *dyadic fractions*, meaning fractions of the form $1/2^n$. The first five Walsh functions, $y_0(t)$, $y_1(t)$, $y_2(t)$, $y_3(t)$, $y_4(t)$, are illustrated in the figures below.

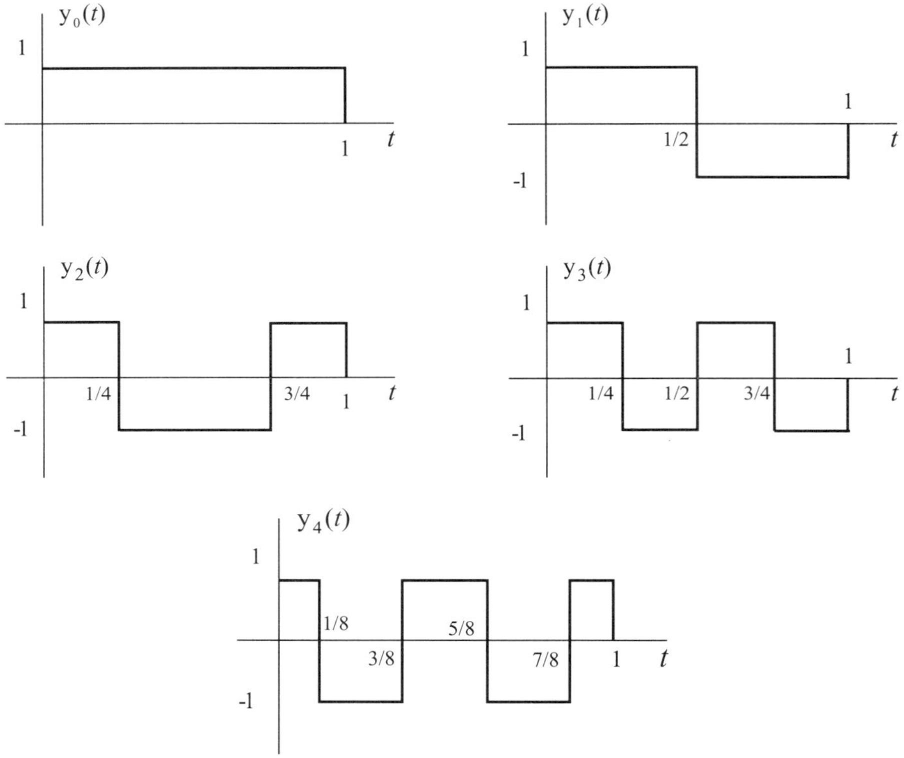

Let $x(t) = \sin \pi t$. Find an approximation to $x(t)$ of the form

$$X(t) = \sum_{i=0}^{4} a_i y_i(t)$$

such that $\int_0^1 |x(t) - \hat{x}(t)|^2 \, dt$ is minimized. Follow the reasoning in the section on best L^2-approximation by finite Fourier series to determine the a_i.

[54]Named for the American mathematician Joseph Walsh

1.18. *Haar functions*[55]

Define a sequence of functions $E_n^k(x)$ on $0 \le x \le 1$ by

$$E_n^k(x) \quad = \quad \begin{cases} 2^{n/2}\,, & \frac{k-1}{2^n} \le x < \frac{k-\frac{1}{2}}{2^n}\,, \\ -2^{n/2}\,, & \frac{k-\frac{1}{2}}{2^n} \le x < \frac{k}{2^n}\,, \\ 0\,, & \text{otherwise}, \end{cases}$$

where k and n are integers, $1 \le k \le 2^n$, and $n \ge 0$.

Sketch the graphs of E_n^k for $n \le 2$, and show that in general the E_n^k are orthonormal with respect to the usual inner product on $L^2([0,1])$.

In fact, the E_n^k form an orthonormal basis of $L^2([0,1])$, though you are not asked to show this. They are called the *Haar basis* and they are used in the construction of *wavelets*. (Since they form a basis, any function in L^2 — including smooth functions — can be approximated in the L^2-norm by the discontinuous Haar functions. Sort of the opposite of using the smooth functions sine and cosine to approximate discontinuous functions like the square wave.)

1.19. *Legendre polynomials*[56]

The *Legendre polynomials* $P_n(x)$ are, by definition, solutions of the second-order differential equation (the *Legendre equation*)

$$((1-x^2)y')' + n(n+1)y = (1-x^2)y'' - 2xy' + n(n+1)y = 0\,, \qquad n = 0, 1, \ldots\,.$$

The usual approach to solving this equation is via an infinite series. Special arguments are needed to show that the series terminates at a finite number of terms and hence that $P_n(x)$ is a polynomial. (Here the fact that n is a nonnegative integer is essential.) In fact, $P_n(x)$ is of degree n, and the further normalization $P_n(1) = 1$ determines it uniquely. With this normalization the first four Legendre polynomials are $P_0(x) = 1$, $P_1(x) = x$, $P_2(x) = (1/2)(3x^2 - 1)$, and $P_3(x) = (1/2)(5x^3 - 3x)$. Many mathematical software packages, Mathematica for example, have Legendre polynomials built in, but check the normalization in use.

The Legendre equation and its solutions come up in a variety of applications (potential theory, Laplace's equation in a spherical region in 3D, solving the Schrödinger equation for the hydrogen atom). In Section 1.7, I mentioned Sturm-Liouville theory and the Legendre equation is an example. Why bring this up in a chapter on Fourier series? Orthogonality and the possible usefulness of expanding a function $f(x)$ as a series

$$f(x) = \sum_{n=0}^{\infty} a_n P_n(x)\,.$$

[55] Named for the Hungarian mathematician Alfréd Haar.

[56] Named for the French mathematician Adrien-Marie Legendre.

The action takes place on $L^2([-1,1])$, the space of square integrable functions on $-1 \le x \le 1$ with the inner product

$$\int_{-1}^{1} f(x)g(x)\,dx\,.$$

(a) Let's show that the Legendre polynomials are orthogonal. We have

$$((1-x^2)P_n'(x))' + n(n+1)P_n(x) = 0\,,$$
$$((1-x^2)P_m'(x))' + m(m+1)P_m(x) = 0\,.$$

Multiply the first equation by $P_m(x)$, multiply the second equation by $P_n(x)$, and subtract the two. Now integrate this from -1 to 1 — integration by parts can help — and conclude that if $n \neq m$, then $P_n(x)$ and $P_m(x)$ are orthogonal.[57]

The Legendre polynomials are not orthonormal. Rather, they satisfy

$$\int_{-1}^{1} P_n(x)^2\,dx = \frac{2}{2n+1}\,.$$

Leave that as something to look up; it can be deduced, with some effort, from Rodrigues's formula

$$P_n(x) = \frac{1}{2^n n!}\frac{d^n}{dx^n}(x^2-1)^n\,,$$

which is also something to look up.

The headline result is that the functions $\sqrt{\frac{2n+1}{2}}P_n(x)$, $n = 0, 1, 2, \ldots$, are an orthonormal basis for $L^2([-1,1])$.

(b) In the expansion $f(x) = \sum_{n=0}^{\infty} a_n P_n(x)$, find a formula for a_n.

1.20. *Rayleigh's identity and a famous sum*

It doesn't matter if you're an engineer, a scientist, or a mathematician, you cannot go through life knowing about Fourier series and not know the beautiful application of Rayleigh's identity to evaluating a *very* famous sum. It is a leading example of the maxim: if you have two ways of computing the same thing, you have something significant.

Let $S(t)$ be the sawtooth function, that is, $S(t) = t$ for $0 \le t < 1$, and then periodized to have period 1. Show that the Fourier series for $S(t)$ is

$$\frac{1}{2} - \sum_{n\neq 0} \frac{1}{2\pi i n} e^{2\pi i n t}$$

and use it to show that

$$\sum_{n=1}^{\infty} \frac{1}{n^2} = \frac{\pi^2}{6}.$$

[57] There's another reason that the Legendre polynomials are orthogonal. For those who are familiar with the Gram-Schmidt process, the Legendre polynomials are obtained by applying it to the simple powers $1, x, x^2, \ldots$.

1.21. If you liked Problem 1.20, you'll love this one.

Let $f(t) = t^2$ for $0 \le t < 2$ and periodize $f(t)$ to have period 2. Here's the picture.

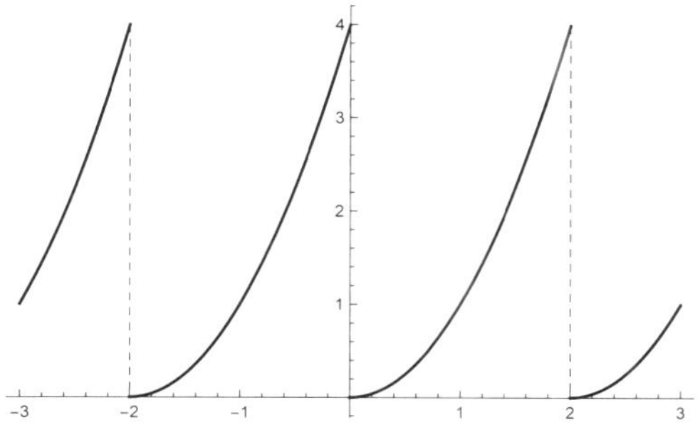

(a) Find the Fourier series coefficients, c_n, of $f(t)$.

(b) Using your result from part (a), obtain the following:

$$\sum_{n=1}^{\infty} \frac{1}{n^2} = \frac{\pi^2}{6},$$
$$\sum_{n=1}^{\infty} \frac{(-1)^{n+1}}{n^2} = \frac{\pi^2}{12},$$
$$\sum_{n=0}^{\infty} \frac{1}{(2n+1)^2} = \frac{\pi^2}{8}.$$

1.22. *Rayleigh's identity for period T*

What will Rayleigh's identity be for periodic functions of period T (when $T \neq 1$)? Try starting with the usual identity, for the $T = 1$ case, and scale the variable to convert a period T function to a period 1 function.

1.23. *A generalization of Rayleigh's identity*

Suppose $f(t)$ and $g(t)$ are square integrable periodic functions of period 1 with Fourier series

$$f(t) = \sum_{n=-\infty}^{\infty} a_n e^{2\pi i n t}, \quad g(t) = \sum_{n=-\infty}^{\infty} b_n e^{2\pi i n t}.$$

Show that

$$(f, g) = \sum_{n=-\infty}^{\infty} a_n \overline{b_n},$$

where

$$(f, g) = \int_0^1 f(t)\overline{g(t)}\, dt.$$

1.24. Can two periodic, square integrable functions $f(t)$ and $g(t)$ satisfy $\hat{f}(n) = \hat{g}(n) + 1$?

1.25. *Orthogonality of sines and cosines*

Though we use complex exponentials for Fourier series, we could have developed the whole thing with sines and cosines, and you often see this approach. The key fact is the orthogonality of the functions $\sin 2\pi nt$ and $\cos 2\pi nt$, $n = 0, 1, 2, \dots$.

(a) Verify the following:

$$(i)\quad \int_0^1 \sin 2\pi nt \cos 2\pi mt \, dt = 0.$$

$$(ii)\quad \int_0^1 \sin 2\pi nt \sin 2\pi mt \, dt = \begin{cases} 0, & m \neq n, \\ 1/2, & m = n. \end{cases}$$

$$(iii)\quad \int_0^1 \cos 2\pi nt \cos 2\pi mt \, dt = \begin{cases} 0, & m \neq n, \\ 1/2, & m = n. \end{cases}$$

Notice that the sines and cosines are not orthonormal, but they are orthogonal.

You can do this with some trig identities, but my advice, for (ii) and (iii), is to use a modified version of orthogonality of the complex exponentials including a $\pm$, as in

$$\int_0^1 e^{2\pi int} e^{\pm 2\pi imt} = \begin{cases} 0, & n \neq \mp m, \\ 1, & n = \mp m. \end{cases}$$

Split this into real and imaginary parts using Euler's formula and see how that implies the orthogonality statements about sines and cosines.

(b) If we write a Fourier series as

$$f(t) = \frac{a_0}{2} + \sum_{n=1}^{\infty} a_n \cos 2\pi nt + b_n \sin 2\pi nt,$$

what is a formula for the coefficients in terms of $f(t)$?

1.26. *Wirtinger's inequality*[58]

Let $f(t)$ be periodic of period 1 and suppose that the average value of $f(t)$ is zero; i.e.,

$$\int_0^1 f(t)\, dt = 0.$$

(a) Deduce Wirtinger's inequality:

$$\int_0^1 |f(t)|^2 \, dt \leq \frac{1}{4\pi^2} \int_0^1 |f'(t)|^2 \, dt.$$

Hint: Expand $f(t)$ in a Fourier series, and allow yourself to differentiate term by term to also get the Fourier series for $f'(t)$.

(b) When does equality hold?

[58]Named for the German mathematician W. Wirtinger. Among his contributions was the idea of the spectrum of a linear operator as a generalization of eigenvalue.

(c) If instead $f(t)$ is periodic of period 2π and satisfies $\int_0^{2\pi} f(t)\,dt = 0$, show that Wirtinger's inequality becomes

$$\int_0^{2\pi} |f(t)|^2\,dt \le \int_0^{2\pi} |f'(t)|^2\,dt.$$

By the way, notice what happens when $f(t)$ is constant. In this case $f'(t) = 0$, so the right-hand side of the inequality is zero. But if $f(t)$ is constant *and* its average value is zero, then $f(t)$ must be identically zero, so the left-hand side of the inequality is zero as well.

1.27. *Isoperimetric inequality*

Wirtinger's inequality, from the previous problem, can be used along with Fourier series, to prove the *isoperimetric inequality*, which states, informally:

Among all closed curves of a given length, the circle encloses the maximum area.

Stated more formally:

Let C be a simple closed curve (no self-intersections) of length L and enclosing an area A. Then

$$L^2 \ge 4\pi A,$$

with equality holding exactly for a circle of length L.

See, for example, the article by R. Osserman, The isoperimetric inequality, Bulletin of the American Mathematical Society **84** (1978), 1182–1238. Let's see how it's done.

To establish this we'll need some notions and formulas from calculus, which you can look up. First, suppose the curve C is given by parametric equations $(x(t), y(t))$, traced out counterclockwise. If we use the arclength parameter s, then $0 \le s \le L$ and one has

$$\left(\frac{dx}{ds}\right)^2 + \left(\frac{dy}{ds}\right)^2 = 1.$$

(Physically, C is traced out at unit speed.) However, we're going to appeal to Wirtinger's inequality, the version for 2π-periodic functions, and to do that we scale s, setting $t = (2\pi/L)s$. Thus in describing C by $(x(t), y(t))$ the parameter t varies from 0 to 2π, and $x(t)$ and $y(t)$ are periodic of period 2π. Furthermore, for this parametrization:

(a) Show that

$$\int_0^{2\pi} \left(\frac{dx}{dt}\right)^2 + \left(\frac{dy}{dt}\right)^2 dt = \frac{L^2}{2\pi}.$$

Next, from Green's theorem the area enclosed by C can be expressed as

$$A = -\int_0^{2\pi} y\frac{dx}{dt}\,dt.$$

(b) Show that

$$L^2 - 4\pi A = 2\pi \int_0^{2\pi} \left(\frac{dx}{dt} + y\right)^2 dt + 2\pi \int_0^{2\pi} \left[\left(\frac{dy}{dt}\right)^2 - y^2\right] dt.$$

Now, we can choose coordinates so that

$$\int_0^{2\pi} y(t)\,dt = 0.$$

This is the case if we shift the curve so that the x-axis passes through the center of gravity of the curve. Then the y-coordinate of the center of gravity is 0, which is the integral equation. Finally, then:

(c) Show that $L^2 - 4\pi A \geq 0$ with equality only for a circle of length L.

1.28. *Convolution, autocorrelation, and Fourier series*

Recall that the *convolution* of two functions $f(t)$ and $g(t)$ of period 1 is defined by

$$(f * g)(t) = \int_0^1 f(\tau)g(t-\tau)\,d\tau\,.$$

(a) Show that $f * g$ is periodic of period 1.

(b) Suppose $f(t)$ and $g(t)$ have Fourier series

$$f(t) = \sum_{n=-\infty}^{\infty} a_n e^{2\pi int}, \qquad g(t) = \sum_{n=-\infty}^{\infty} b_n e^{2\pi int},$$

respectively. Find the Fourier series of $(f * g)(t)$ and explain why this implies that $f * g = g * f$.

Let $f(x)$ be a real, periodic function of period 1. The autocorrelation of f with itself is the function

$$(f \star f)(x) = \int_0^1 f(y)f(y+x)\,dy\,.$$

Note the difference in notation: "$\star$" for autocorrelation and "$*$" for convolution.

(c) Show that $f \star f$ is also periodic of period 1.

(d) If

$$f(x) = \sum_{n=-\infty}^{\infty} \hat{f}(n)e^{2\pi inx},$$

show that the Fourier series of $(f \star f)(x)$ is

$$(f \star f)(x) = \sum_{n=-\infty}^{\infty} |\hat{f}(n)|^2 e^{2\pi inx}\,.$$

1.29. *Convolution of Fourier coefficients*

There's a dual result to the problem on convolution. (We'll talk more formally about "duality" when we learn about Fourier transforms. For now just consider it an overused word.) Suppose

$$f(t) = \sum_{n=-\infty}^{\infty} a_n e^{2\pi int}, \quad g(t) = \sum_{n=-\infty}^{\infty} b_n e^{2\pi int}$$

are periodic functions of period 1, expressed as Fourier series. Their product $f(t)g(t)$ is also periodic of period 1 with, say,

$$f(t)g(t) = \sum_{n=-\infty}^{\infty} c_n e^{2\pi int}.$$

Show that

$$c_n = \sum_{k=-\infty}^{\infty} a_k b_{n-k}.$$

1.30. *Convolving with a constant*

Let $f(t)$ be a periodic signal of period 1. Show that $(c * f)(t) = c\hat{f}(0)$ where c is a constant. (We regard a constant as a periodic function, in this case of period 1; it's not much of a repeating pattern, but you have to admit that if you know it on an interval, you know it everywhere.)

1.31. *More on autocorrelation: Pitch detection*

Recall that the autocorrelation of a real-valued, periodic function (period 1) $f(t)$ is

$$(f \star f)(x) = \int_0^1 f(y)f(y+x)\,dy$$

and that if

$$f(x) = \sum_{n=-\infty}^{\infty} \hat{f}(n)e^{2\pi inx},$$

then the Fourier series of $(f \star f)(x)$ is

$$(f \star f)(x) = \sum_{n=-\infty}^{\infty} |\hat{f}(n)|^2 e^{2\pi inx}.$$

Show that for any $x \in [0, 1]$,

$$|(f \star f)(x)| \leq (f \star f)(0) = \sum_{n=-\infty}^{\infty} |\hat{f}(n)|^2.$$

Hint: Use the Cauchy-Schwarz inequality.

Observe that, by periodicity,

$$|(f \star f)(x)| \leq (f \star f)(m)$$

for any integer m.

Autocorrelation is used to detect the fundamental (lowest) frequency in a musical sound. If the sound is $f(t)$, the method is to plot the autocorrelation $(f \star f)(x)$ and look for the biggest values. That this works follows from a version of the inequality above; modify the inequality to apply to a signal of lowest period T instead of 1. (What would that modification be?)

See, for example, `http://cnx.org/content/m11714/latest/`.

1.32. *The Dirichlet problem, convolution, and the Poisson kernel*

When modeling physical phenomena by partial differential equations it is frequently necessary to solve a *boundary value problem.* One of the most famous and important of these is associated with Laplace's equation:

$$\Delta u = \frac{\partial^2 u}{\partial x^2} + \frac{\partial^2 u}{\partial y^2} = 0\,,$$

where $u(x, y)$ is defined on a region R in the plane. The operator

$$\Delta = \frac{\partial^2}{\partial x^2} + \frac{\partial^2}{\partial y^2}$$

is called the *Laplacian* and a real-valued function $u(x, y)$ satisfying $\Delta u = 0$ is called *harmonic.* The *Dirichlet problem* for Laplace's equation is this:

Given a function $f(x, y)$ defined on the boundary of a region R, find a function $u(x, y)$ defined on R that is harmonic in R and equal to $f(x, y)$ on the boundary.

Fourier series and convolution combine to solve this problem when R is a disk.

As with many problems where circular symmetry is involved, in this case where the functions are defined on a circular disk, it is helpful to introduce polar coordinates (r, θ), with $x = r\cos\theta$, $y = r\sin\theta$. Writing $U(r, \theta) = u(r\cos\theta, r\sin\theta)$, so regarding u as a function of r and θ, Laplace's equation becomes

$$\frac{\partial^2 U}{\partial r^2} + \frac{1}{r}\frac{\partial U}{\partial r} + \frac{1}{r^2}\frac{\partial^2 U}{\partial \theta^2} = 0\,.$$

(You need not derive this.)

(a) Let $\{c_n\}$, $n = 0, 1, \dots$, be a bounded sequence of complex numbers, let $r < 1$, and define $u(r, \theta)$ by the series

$$u(r, \theta) = \operatorname{Re}\left\{c_0 + 2\sum_{n=1}^{\infty} c_n r^n e^{in\theta}\right\}.$$

From the assumption that the coefficients are bounded and comparison with a geometric series, it can be shown that the series converges, but you need not do this.

Show that $u(r, \theta)$ is a harmonic function.

[*Hint*: Superposition in action — the Laplacian of the sum is the sum of the Laplacians, and the c_n comes out, too. Also, the derivative of the real part of something is the real part of the derivative of the thing.]

(b) Suppose that $f(\theta)$ is a real-valued, continuous, periodic function of period 2π and let

$$f(\theta) = \sum_{n=-\infty}^{\infty} c_n e^{in\theta}$$

be its Fourier series. Now form the harmonic function $u(r, \theta)$ as above, with these coefficients c_n. This solves the Dirichlet problem of finding a harmonic function on the unit disk $x^2 + y^2 < 1$ with boundary values $f(\theta)$ on the unit circle $x^2 + y^2 = 1$; precisely,

$$\lim_{r \to 1} u(r, \theta) = f(\theta)\,.$$

You are not asked to show this — it requires a fair amount of work — but assuming that all is well with convergence, explain why one has

$$u(1, \theta) = f(\theta).$$

[This uses the symmetry property of Fourier coefficients. Remember also that $\mathrm{Re}\{z\} = (1/2)(z + \bar{z})$.]

(c) The solution can also be written as a convolution! Show that

$$u(r, \theta) = \frac{1}{2\pi} \int_0^{2\pi} f(\phi) P(r, \theta - \phi)\, d\phi,$$

where

$$P(r, \theta) = \frac{1 - r^2}{1 - 2r\cos\theta + r^2}.$$

(Introduce the Fourier coefficients of $f(\theta)$ — meaning their definition as an integral. You'll also need a formula for the sum of a geometric series.)

(d) The function $P(r, \theta)$ is called the *Poisson*[59] *kernel*. Show that it is a harmonic function. (This is a special case of the result you established in part (a) for the particular case $c_n = 1$.)

1.33. *Convolution in terms of cosines and sines*

As further evidence that it is easier to work with complex exponentials when computing with Fourier series, consider convolution. Suppose

$$f(t) = \frac{a_0}{2} + \sum_{n=1}^{\infty} a_n \cos 2\pi nt + b_n \sin 2\pi nt,$$

$$g(t) = \frac{c_0}{2} + \sum_{n=1}^{\infty} c_n \cos 2\pi nt + d_n \sin 2\pi nt.$$

Show that the Fourier series, in terms of cosines and sines, for the convolution

$$(f * g)(t) = \int_0^1 f(\tau) g(t - \tau)\, d\tau$$

is

$$(f * g)(t) = \frac{a_0 c_0}{4} + \sum_{n=1}^{\infty} \frac{1}{2}(a_n c_n - b_n d_n) \cos 2\pi nt + \frac{1}{2}(a_n d_n + b_n c_n) \sin 2\pi nt.$$

Use the results of the problem on the orthogonality of the trig functions. You'll need the addition formulas for the sine and cosine. Sorry.

1.34. *Periodic in time and space?*[60]

A friend asks you about periodicity of the function of two variables $f(x, t) = \cos(2\pi xt)$. She says that the cosine itself is a periodic function, so this function should be periodic in both time (t) and space (x). How would you describe the situation? What about the function $g(x, t) = \cos(2\pi(x + t))$? You might also make 3D plots of the functions.

[59] S. Poisson, a French polymath, will make several appearances. We'll see his name in connection with an important result known as the Poisson summation formuls.

[60] Suggested by a student.

1.35. *Bessel functions and Fourier series*[61]

(a) The Bessel equation of order n is

$$x^2y'' + xy' + (x^2 - n^2)y = 0\,.$$

It arises in many applications, often — but not always! — in finding solutions of partial differential equations by the technique of separation of variables into radial and angular parts. A solution of the equation is the *Bessel function of the first kind of order* n, given by the integral

$$J_n(x) = \frac{1}{2\pi}\int_0^{2\pi} \cos(x\sin\theta - n\theta)\,d\theta\,.$$

Show that

$$e^{ix\sin\theta} = \sum_{n=-\infty}^{\infty} J_n(x)e^{in\theta}\,, \qquad e^{ix\cos\theta} = \sum_{n=-\infty}^{\infty} i^n J_n(x)e^{in\theta}\,.$$

These formulas come up in quantum scattering theory, diffraction, and other areas *not* related to separation of variables.

(b) The function $I_0(a)$, a a real number, defined by the integral

$$I_0(a) = \frac{1}{2\pi}\int_0^{2\pi} e^{a\cos t}\,dt,$$

is called the *modified Bessel function of the first kind of order* 0. It arises in the study of FM radio and wireless communication, *e.g.*, in the model of Rician fading of signals.

Show that

$$I_0(a) = \sum_{n=0}^{\infty} \frac{a^{2n}}{2^{2n}(n!)^2}\,.$$

Hint: Consider the function $f(t) = e^{(1/2)ae^{it}}$. This is periodic (in t) of period 2π. Find its Fourier series by using a *Taylor series* expansion of e^x, for $x = (1/2)ae^{it}$, and then apply Rayleigh's identity for periodic functions of period 2π.

1.36. *Fitting a Fourier series — approximately — if the frequencies aren't integer multiples of a fundamental frequency*

Let $\nu_1, \nu_2, \ldots, \nu_N$ be distinct positive numbers and let

$$f(t) = a_0 + \sum_{n=1}^{N}(a_n\cos 2\pi\nu_n t + b_n\sin 2\pi\nu_n t).$$

A sum like this might come up if $f(t)$ models a phenomenon that is the additive result of effects from independent sources that are each periodic but possibly with periods that aren't (reciprocal) integer multiples. To take a real example, $f(t)$ might be the height of the tide at a point on the earth, a periodic function that is the additive effect of the rotation of the earth relative to the moon and the rotation of the earth relative to the sun. (In general, $f(t)$ might not even

[61] From a Caltech problem set.

be periodic, considering what we learned in an earlier problem. And tuning musical chords presents similar problems!)

(a) Show that the sum can be written in the form

$$f(t) = \sum_{n=-N}^{N} c_n e^{2\pi i \nu_n t};$$

i.e., what is c_n in terms of a_n and b_n, and what is ν_n for negative n?

(b) Show that

$$\lim_{T\to\infty} \left\{ \frac{1}{T} \int_{t_0}^{t_0+T} e^{-2\pi i \nu_m t} f(t)\, dt \right\} = c_m.$$

for any m, $-N \leq m \leq N$.

In evaluating the integral $\int_{t_0}^{t_0+T} e^{-2\pi i \nu_m t} f(t)\, dt$, term by term, you will get Tc_m plus a sum that you'd like to do something with. The terms in the sum won't generally be zero, as they are when all the frequencies are integers, but you'd like to think maybe you get some kind of geometric series (because there always seem to be geometric series in this subject). Not this time. Instead, get the result indicated by showing that the terms in the sum can be (easily) bounded above.

(c) From part (b), finally show that, for any t_0,

$$\lim_{T\to\infty} \left\{ \frac{2}{T} \int_{t_0}^{t_0+T} f(t) \cos 2\pi \nu_n t\, dt \right\} = a_n,$$

and

$$\lim_{T\to\infty} \left\{ \frac{2}{T} \int_{t_0}^{t_0+T} f(t) \sin 2\pi \nu_n t\, dt \right\} = b_n,$$

$1 \leq n \leq N$, and that

$$\lim_{T\to\infty} \left\{ \frac{1}{T} \int_{t_0}^{t_0+T} f(t)\, dt \right\} = a_0.$$

Practically, the result means that if we know measurements for $f(t)$ over long time intervals (starting at any time t_0 and extending to $t_0 + T$ where T is large), then we can calculate approximately the coefficients a_n and b_n.

1.37. *Frequencies of a piano chord*

Dividing a musical scale into steps, and the allied problem of tuning musical chords, has a long and fascinating history; a quick web search will turn up many references. For a sound wave — the periodic compression and rarefaction of air — the physical characteristic of frequency is perceived as pitch, and all systems of tuning start from the fact that raising a note one octave in pitch corresponds to doubling its frequency. The *well-tempered* scale divides the steps in pitch of a scale covering one octave into twelve equal steps in frequency. Thus the ratio of the frequency of one note in the scale to the note just below it is $2^{1/12}$, an irrational number. Twelve steps double the frequency.

A piano chord is therefore an example of the sum of several frequencies that are not integral multiples of a fundamental frequency. This problem, coupled with the preceding problem (on fitting a Fourier series approximately), Problem 1.36, asks you to explore this.

Download the MATLAB data file `PianoChords.mat`. Make sure the file is in your current directory, load the data, and listen to the chords:

```
load PianoChords.mat
soundsc(chord1, fs);
soundsc(chord2, fs);
```

Note that the"piano notes" are pure frequencies and do not contain any harmonics, so the chords don't necessarily sound like a piano. The loaded variables are the waveforms of two piano chords (`chord1, chord2`), the time vector (`t`), the sound duration (`dur`), the sampling frequency (`fs`), and the frequencies of the piano keys from Middle C to Tenor C (`freq`). These keys are called Middle C, C♯/D♭, D, D♯/E♭, E, F, F♯/G♭, G, G♯/A♭, A, A♯/B♭, B, and Tenor C, respectively. (MATLAB inherently stores signals as discrete-time signals. The sampling frequency represents how many times the sound wave is sampled per second.)

(a) Chord 1 is a C major chord, produced by playing three keys: Middle C - E - G. Borrowing from the notation of Problem 1.36, find coefficients a_n, b_n for the three frequencies. Assume that the duration (5 sec.) is long enough to approximate the limit to infinity.
Hint: The magnitude of each frequency is 1. In other words, $|a_n|^2 + |b_n|^2 = 1$.

(b) Chord 2 is a minor chord, again produced by playing three keys. Write a MATLAB script that analyzes a chord and identifies any frequencies present (among those in `freq`) and their coefficients a_n, b_n. You can assume that the magnitude of any frequency present is 1. What are the keys played to produce Chord 2?

(c) (Optional) Create your own major or minor chords. For a major chord, key2 is 4 keys higher than key1, and key3 is 3 keys higher than key2, as is evident by Middle C - E - G. For a minor chord, key2 is 3 keys higher than key1, and key3 is 4 keys higher than key2. By changing the relative weights between a_n and b_n (while keeping the magnitude 1), we are essentially changing the phase of each note. Form two chords that are identical except for different phases in the notes. Can you hear a difference? Surprisingly (or perhaps not), our ears are quite insensitive to phase.

1.38. *Averaging operator*

Let $f(t)$ be a periodic signal of period 1 and define the *averaging operator* depending on a parameter $h > 0$ by

$$\mathcal{A}_h f(x) = \frac{1}{2h} \int_{x-h}^{x+h} f(t)\, dt\,.$$

Thus $\mathcal{A}_h f(x)$ is a new signal.

(a) Show that $\mathcal{A}_h f(x)$ is periodic of period 1 as a function of x; i.e.,

$$\mathcal{A}_h f(x+1) = \mathcal{A}_h f(x)\,.$$

Hint:

$$\mathcal{A}_h f(x+1) = \frac{1}{2h} \int_{x+1-h}^{x+1+h} f(t)\,dt\,.$$

Now make a change of variable $t = u + 1$.

(b) Find the Fourier series for $\mathcal{A}_h f(x)$ in terms of the Fourier series for $f(t)$.

1.39. *Identities for inner products and norms*

Using the algebraic properties of inner products, show that:

(a) $\|f+g\|^2 + \|f-g\|^2 = 2\|f\|^2 + 2\|g\|^2$.
What does this mean geometrically for the usual inner product and vectors in $\mathbb{R}^2$?

(b) $\|f - ig\|^2 = \|f\|^2 - 2\,\mathrm{Im}(f,g) + \|g\|^2$.

(c) $4(f,g) = \left(\|f+g\|^2 - \|f-g\|^2\right) + i\left(\|f+ig\|^2 - \|f-ig\|^2\right)$.
(This is sometimes called the *polarization identity*. It says that the norm determines the inner product.)

1.40. *Reversals, delays, scaling, and inner products*

Let $f(t)$ and $g(t)$ be two signals defined for all t. Their inner product is

$$(f,g) = \int_{-\infty}^{\infty} f(t)\overline{g(t)}\,dt\,.$$

Define the *reversed signal* to be

$$f^-(t) = f(-t)\,.$$

Define the *delay operator*, or *shift operator*, by

$$\tau_b f(t) = f(t-b)\,, \quad b \in \mathbb{R}\,.$$

Define the *scaling operator* to be

$$\sigma_a f(t) = f(at)\,, \quad a \in \mathbb{R}\,, a \neq 0\,.$$

(a) If both $f(t)$ and $g(t)$ are reversed, what happens to their inner product?

(b) If both $f(t)$ and $g(t)$ are delayed by the same amount, what happens to their inner product?

(c) If both $f(t)$ and $g(t)$ are scaled by the same amount, what happens to their inner product?

1.41. *Reproducing kernel*

Recall that the inner product of two functions on $[0,1]$ is

$$(f,g) = \int_0^1 f(x)\overline{g(x)}\,dx.$$

Let $\{\varphi_n(t)\}$ be an orthonormal basis for $L^2([0,1])$, meaning that

$$(\varphi_n, \varphi_m) = \begin{cases} 1, & n = m, \\ 0, & n \neq m, \end{cases}$$

and that we can write any function $f(t)$ in $L^2([0,1])$ as

$$f(t) = \sum_{n=1}^{\infty} (f, \varphi_n)\varphi_n(t).$$

Define a function of two variables

$$K(t,\tau) = \sum_{n=1}^{\infty} \varphi_n(t)\overline{\varphi_n(\tau)}.$$

Show that

$$(f(t), K(t,\tau)) = \int_0^1 f(t)\overline{K(t,\tau)}\, dt = f(\tau)\,.$$

(Ignore all questions of convergence of series, *etc.*)

Because of this property, $K(t,\tau)$ is called a *reproducing kernel*; taking the inner product of $f(t)$ with $K(t,\tau)$ "reproduces" the value of f at τ. In electrical engineering literature, sampling theorems are often expressed in terms of reproducing kernels.

1.42. *Tiling a rectangle*

Consider the following geometry problem:[62]

Suppose a rectangle is tiled by subrectangles, as in the illustration. (The shading has no special meaning. It's just to make the tiling easier to see.) Suppose also that each subrectangle has at least one side of integer length. Show that the big rectangle has at least one side of integer length.

Sounds reasonable, but maybe not so obvious.

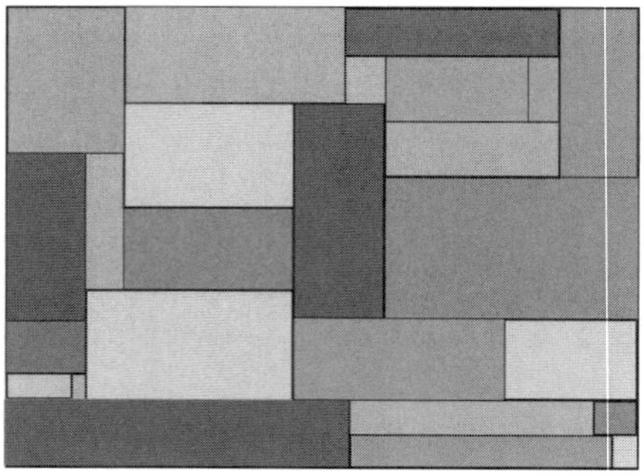

In fact, there's a sneaky approach to this problem using properties of complex exponentials, the point being that you never know when considerations typically used in Fourier analysis can be brought to bear on a problem seemingly from a much different area. The hint is:

Let R be the rectangle

$$R = \{(x,y) : a \le x \le b,\ c \le y \le d\}.$$

Show that

$$\iint_R e^{2\pi i(x+y)}\, dxdy = 0 \quad \text{if and only if } b-a \text{ is an integer or } d-c \text{ is an integer.}$$

[62]From *Proofs from THE BOOK*, by Martin Aigner, Günter M. Ziegler, Karl H. Hofmann.

1.43. *Edge detection and Fourier coefficients*

Many applications involve the direct numerical approximation of the Fourier coefficients of a signal $f(t)$. Moreover, in many cases the signal may have jump discontinuities (e.g., square wave or sawtooth) or its derivative may have jump discontinuities (e.g., corners as in a triangle wave), and we may not know of this ahead of time.

How much information can be learned about edges in a signal from its Fourier coefficients? The following edge detector has been proposed.[63] Suppose $f(t)$ is periodic of period 1 and has Fourier coefficients c_n. For all points t where $f(t)$ is continuous, we can write

$$f(t) = \sum_{n=-\infty}^{\infty} c_n e^{2\pi int}.$$

Likewise, at points where the derivative of the signal, $f'(t)$, is continuous, we can write

$$f'(t) = \sum_{n=-\infty}^{\infty} 2\pi inc_n e^{2\pi int}.$$

If we let

$$s_n = -inc_n,$$

we define the edge detector of order N to be

$$E_N(t) = \frac{\pi}{N} \sum_{n=-N}^{N} s_n e^{2\pi int}.$$

For values of t where $f(t)$ or $f'(t)$ is continuous, we expect that $E_N(t)$ tends to 0 as N increases due to the convergence of the Fourier series at these points (the coefficients have to be tending to zero). However, $E_N(t)$ will be *nonzero* as N increases for values of t where either $f(t)$ or $f'(t)$ has a jump discontinuity. In fact, the value of $E_N(t)$ at these points approximates (depending on N) the height of the jump in $f(t)$ or $f'(t)$.

(a) Calculate the Fourier coefficients c_n for the following functions, periodic with period 1:

$$f(t) = \begin{cases} 1, & 0 < t < \frac{1}{2}, \\ 0, & \frac{1}{2} < t < 1, \end{cases}$$

$$g(t) = t, \quad 0 < t < 1.$$

(b) In MATLAB, write a script that calculates $E_N(t)$ for these two functions for different values of N. Plot the resulting $E_N(t)$ for each function with $N = 10$, 100, and 1,000.

(c) Comment on your results: What features of the initial function are detected by $E_N(t)$? How does the amplitude of $E_N(t)$ relate to the initial function? How does $E_N(t)$ change as N increases?

[63] S. Engelberg, Edge detection using Fourier coefficients, Amer. Math. Monthly **116** (6) (2008), 499–513.

(d) Download the file `edge.m`. Typing `edge` in MATLAB will define a vector t and a vector $f(t)$, representing a signal containing jump discontinuities. Write a MATLAB script that computes the Fourier coefficients c_n of $f(t)$ and calculates the resulting $E_N(t)$. (*Hint*: Use the MATLAB functions fft and fftshift). Plot $f(t)$ and $E_N(t)$ for $N = 10$, 100, and 1,000. How do your results compare with your previous observations?

Chapter 2

Fourier Transform

2.1. A First Look at the Fourier Transform

We're about to make the transition from Fourier series to the Fourier transform. "Transition" is the appropriate word, for we've chosen a path by which the Fourier transform emerges in passing from periodic to nonperiodic functions. To make the trip, we'll view a nonperiodic function (which can be just about anything) as a limiting case of a periodic function as the period becomes longer and longer. Actually, this process doesn't immediately produce the desired result. It takes a little extra tinkering to coax the Fourier transform out of the Fourier coefficient, but we'll land comfortably and it's an interesting excursion.

2.1.1. An example: The rectangle function and its Fourier transform. Let's take a specific, simple, and important example. Consider the rectangle function, or "rect" for short, defined by

$$\Pi(t) = \begin{cases} 1, & |t| < 1/2, \\ 0, & |t| \geq 1/2. \end{cases}$$

Here's the graph, which is not very complicated.

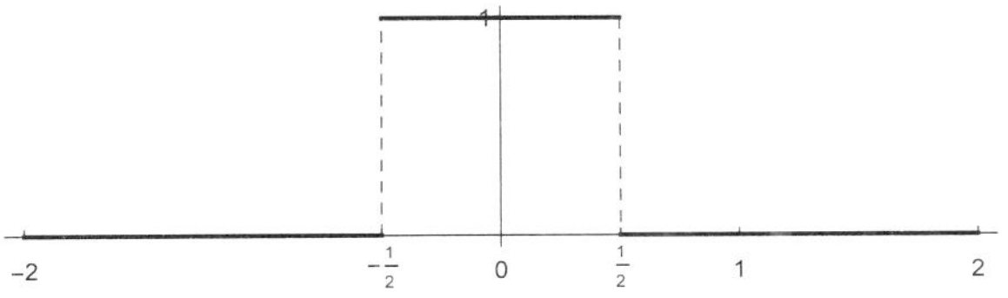

$\Pi(t)$ is even — centered at the origin — and has width 1. Later we'll consider shifted and scaled versions. You can think of $\Pi(t)$ as modeling a switch that is on for one second and off for the rest of the time. Π is also called, variously, the *top hat* function (because of its graph), the *indicator* function, or the *characteristic* function for the interval $(-1/2, 1/2)$.

While we have defined $\Pi(\pm 1/2) = 0$, other common conventions are either to have $\Pi(\pm 1/2) = 1$ or $\Pi(\pm 1/2) = 1/2$. And some people don't define Π at $\pm 1/2$ at all, leaving two holes in the domain. I don't want to get dragged into this dispute. It almost never matters, though for some purposes the choice $\Pi(\pm 1/2) = 1/2$ makes the most sense. We'll deal with this, exceptionally, when it comes up.[1]

$\Pi(t)$ is not periodic. It doesn't have a Fourier series. In problems, you experimented a little with periodizations and I again want to do that with Π. As a periodic version of $\Pi(t)$ we repeat the nonzero part of the function at regular intervals, separated by (long) intervals where the function is zero. We can think of such a function arising when we flip a switch on for a second at a time, doing so repeatedly, and we keep it off for a long time in between the times it's on. (One often hears the term *duty cycle* associated with this sort of thing.) Here's a plot of $\Pi(t)$ periodized to have period 16.

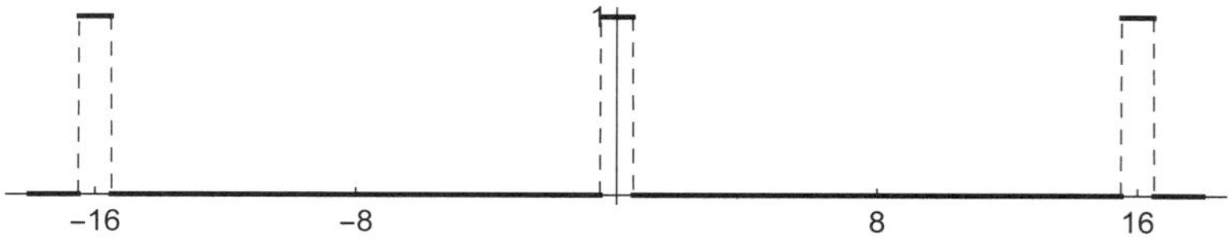

Here are plots of some Fourier coefficients of periodized rectangle functions with periods 4, 16, and 32; the zeroth coefficient and, respectively for each of these, 10 coefficients, 60 coefficients, and 200 coefficients.[2] The spectrum is symmetric, and both positive and negative frequencies are shown. The periodized functions are real and even, so in each case the Fourier coefficients are real. These are plots of the actual coefficients not their magnitudes.

Actually, not quite. I'll explain in a moment the scaling on the horizontal axis, i.e., why the Fourier coefficients c_n aren't just plotted at the integers n. There's also an important issue of vertical scaling. But keep the overall shape in mind — that's the point.

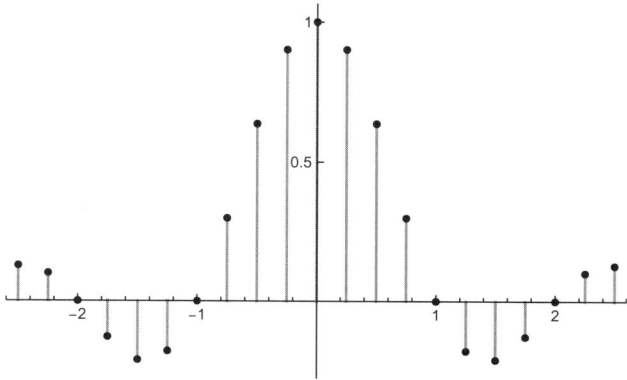

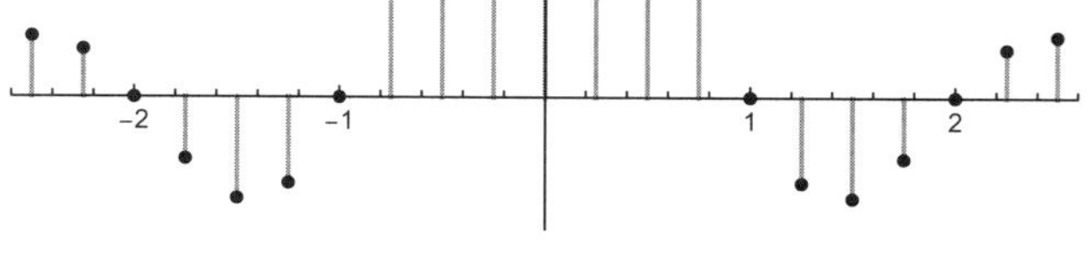

[1]It's actually more of an issue for discrete signals and for a discrete version of Π. This will come up in Chapter 7.

[2]Since Π is discontinuous, the Fourier coefficients go on forever.

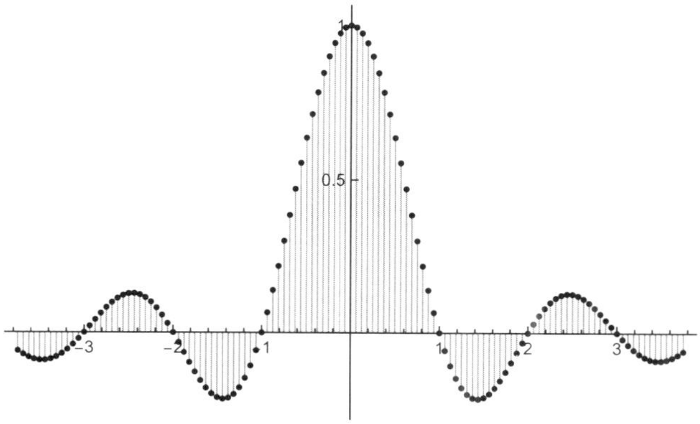

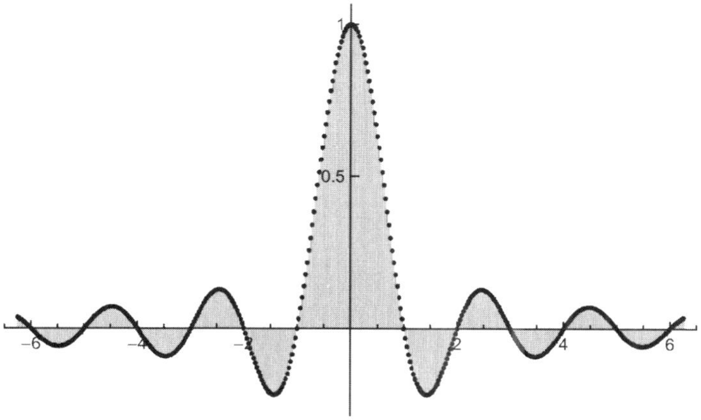

As plotted, as the period increases, the frequencies are getting closer and closer together and it sure looks as though the coefficients are tracking some definite curve. We can analyze what's going on in this particular example and combine that with some general statements to lead us forward.

Recall that for a function $f(t)$ of period T the Fourier series has the form

$$f(t) = \sum_{n=-\infty}^{\infty} c_n e^{2\pi int/T}$$

and that the frequencies are $0, \pm 1/T, \pm 2/T, \ldots$. Points in the spectrum are spaced $1/T$ apart and, indeed, in the pictures above the spectrum is getting more tightly packed as the period T increases. The nth Fourier coefficient is given by

$$\begin{aligned} c_n &= \frac{1}{T}\int_0^T e^{-2\pi int/T} f(t)\,dt \\ &= \frac{1}{T}\int_{-T/2}^{T/2} e^{-2\pi int/T} f(t)\,dt\,. \end{aligned}$$

We can calculate this Fourier coefficient for $\Pi(t)$:

$$\begin{aligned} c_n &= \frac{1}{T}\int_{-T/2}^{T/2} e^{-2\pi int/T}\Pi(t)\,dt = \frac{1}{T}\int_{-1/2}^{1/2} e^{-2\pi int/T}\cdot 1\,dt \\ &= \frac{1}{T}\left[\frac{1}{-2\pi in/T}e^{-2\pi int/T}\right]_{t=-1/2}^{t=1/2} = \frac{1}{2\pi in}\left(e^{\pi in/T} - e^{-\pi in/T}\right) = \frac{1}{\pi n}\sin\left(\frac{\pi n}{T}\right). \end{aligned}$$

Now, although the spectrum is *indexed* by n, the points in the spectrum are n/T ($n = 0, \pm 1, \pm 2, \ldots$), and it's more helpful to think of the spectral information (the value of c_n) as a transform of Π *evaluated* at the points n/T. Write this, provisionally, as

$$(\text{transform of periodized } \Pi)\left(\frac{n}{T}\right) = \frac{1}{\pi n}\sin\left(\frac{\pi n}{T}\right).$$

We're almost there, but not quite. If you're dying to just take a limit as $T \to \infty$, consider that, for each n, if T is very large, then n/T is very small and

$$\frac{1}{\pi n}\sin\left(\frac{\pi n}{T}\right) \quad \text{is about size } \frac{1}{T} \quad (\text{remember } \sin\theta \approx \theta \text{ if } \theta \text{ is small}).$$

In other words, for each n this so-called transform,

$$\frac{1}{\pi n}\sin\left(\frac{\pi n}{T}\right),$$

tends to 0 like $1/T$. The Fourier coefficients are all tending to 0 as $T \to \infty$. To compensate for this we scale up by T, considering instead

$$(\text{scaled transform of periodized } \Pi)\left(\frac{n}{T}\right) = T\,\frac{1}{\pi n}\sin\left(\frac{\pi n}{T}\right) = \frac{\sin(\pi n/T)}{\pi n/T}.$$

In fact, the plots of the *scaled* transforms are what I showed you, above.

Next, if T is large, then we can think of replacing the closely packed discrete points n/T by a continuous variable, say s, so that with $s = n/T$ we would then write, approximately,

$$(\text{scaled transform of periodized } \Pi)(s) = \frac{\sin \pi s}{\pi s}.$$

What does this procedure look like in terms of the integral formula? Simply

$$\begin{aligned} (\text{scaled transform of periodized } \Pi)\left(\frac{n}{T}\right) &= T\cdot c_n \\ &= T\cdot\frac{1}{T}\int_{-T/2}^{T/2} e^{-2\pi int/T} f(t)\,dt \\ &= \int_{-T/2}^{T/2} e^{-2\pi int/T} f(t)\,dt. \end{aligned}$$

We now think of $T \to \infty$ as having the effect of replacing the discrete variable n/T by the continuous variable s, as well as pushing the limits of integration to $\pm\infty$.

Then we may write for the (limiting) transform of Π the integral expression

$$\widehat{\Pi}(s) = \int_{-\infty}^{\infty} e^{-2\pi i s t}\, \Pi(t)\, dt\,.$$

Behold, the Fourier transform is born! Or it soon will be. We use the notation $\widehat{\Pi}$ in tribute to the notation for Fourier coefficients, but more on notation later.

Let's calculate the integral (we know what the answer is because we saw the discrete form of it earlier):

$$\widehat{\Pi}(s) = \int_{-\infty}^{\infty} e^{-2\pi i s t}\Pi(t)\, dt = \int_{-1/2}^{1/2} e^{-2\pi i s t} \cdot 1\, dt = \frac{\sin \pi s}{\pi s}\,.$$

Here's a graph. You can now certainly see the continuous curve that is tracked by the plots of the discrete, scaled Fourier coefficients.

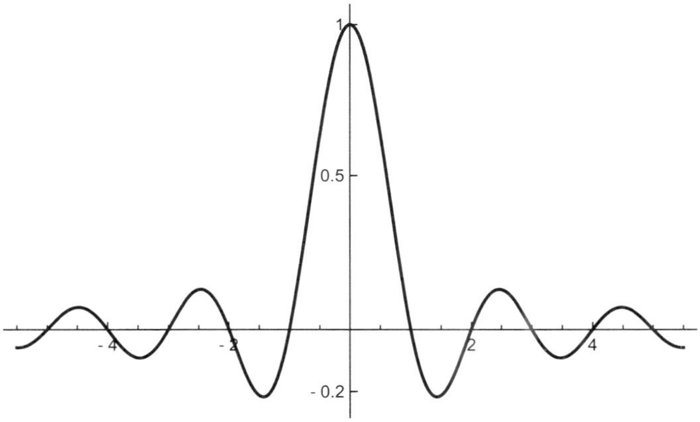

The function $\sin \pi x / \pi x$ (written now with a generic variable x) comes up so often in this subject that it's given a name, sinc:

$$\operatorname{sinc} x = \frac{\sin \pi x}{\pi x}\,,$$

pronounced "sink." Note that

$$\operatorname{sinc} 0 = 1$$

by virtue of the famous limit

$$\lim_{x \to 0} \frac{\sin x}{x} = 1\,.$$

No doubt you saw this limit when you took calculus, and you probably thought you'd never see it again. But the sinc function is very important in EE applications, and elsewhere, *because* it's the Fourier transform of the rectangle function. In fact, it's fair to say that many EE's see the sinc function in their dreams.

Fry's Electronics is a famous electronics store in Silicon Valley and beyond. They certainly know their customers. I have personally done my bit to help their business. I am using these pictures with the kind permission of Fry's Electronics Inc. and the American Institute of Mathematics.

2.1.2. In general. We would be led to the same idea — scale the Fourier coefficients by T — if we had started off periodizing just about any function with the intention of letting the period T tend to infinity. Suppose $f(t)$ is zero outside of $|t| \leq 1/2$. (Any interval will do; we just want to suppose that a function is zero outside some interval so we can periodize.) We periodize $f(t)$ to have period T and compute the Fourier coefficients:

$$c_n = \frac{1}{T}\int_{-T/2}^{T/2} e^{-2\pi int/T} f(t)\,dt = \frac{1}{T}\int_{-1/2}^{1/2} e^{-2\pi int/T} f(t)\,dt\,.$$

How big is this? We can estimate

$$\begin{aligned}|c_n| &= \frac{1}{T}\left|\int_{-1/2}^{1/2} e^{-2\pi int/T} f(t)\,dt\right| \\ &\le \frac{1}{T}\int_{-1/2}^{1/2} |e^{-2\pi int/T}|\,|f(t)|\,dt = \frac{1}{T}\int_{-1/2}^{1/2} |f(t)|\,dt = \frac{A}{T}\,,\end{aligned}$$

where

$$A = \int_{-1/2}^{1/2} |f(t)|\,dt\,.$$

This A is some *fixed number independent* of n and T. Again we see that c_n tends to 0 like $1/T$, and so again we scale back up by T and consider

$$(\text{scaled transform of periodized } f)\left(\frac{n}{T}\right) = Tc_n = \int_{-T/2}^{T/2} e^{-2\pi int/T} f(t)\,dt\,.$$

In the limit as $T \to \infty$ we replace n/T by s and consider

$$\hat{f}(s) = \int_{-\infty}^{\infty} e^{-2\pi ist} f(t)\,dt\,.$$

We're back to the same integral formula.

2.1.3. Fourier transform defined.

There you have it. We now *define* the *Fourier transform* of a function $f(t)$ to be

$$\hat{f}(s) = \int_{-\infty}^{\infty} e^{-2\pi ist} f(t)\,dt\,.$$

For now, just take this as a formal definition; later we'll discuss conditions for such an integral to exist. We assume that $f(t)$ is defined for all real numbers t. We're not assuming that $f(t)$ is zero outside some interval, and we're not periodizing. This definition is general. Take note: we didn't *derive* the Fourier transform — we *motivated the definition.*

For any $s \in \mathbb{R}$, integrating $f(t)$ against the complex-valued function $e^{-2\pi ist}$ with respect to t produces, generally, a complex-valued function of s. Remember that the Fourier transform $\hat{f}(s)$ is a complex-valued function of $s \in \mathbb{R}$. There are cases when $\hat{f}(s)$ is real (like $\widehat{\Pi}(s) = \operatorname{sinc} s$) for reasons of symmetry, and we'll discuss this.

The spectrum of a periodic function is a discrete set of frequencies, possibly an infinite set (when there's a discontinuity of some order, for example) but always a discrete set. By contrast, the Fourier transform of a nonperiodic signal produces a continuous spectrum, or a continuum of frequencies. It may be that the transform $\hat{f}(s)$ is identically zero for $|s|$ sufficiently large — an important class of signals called *bandlimited* — or it may be that the nonzero values of $\hat{f}(s)$ extend to $\pm\infty$ or it may be that $\hat{f}(s)$ is zero for just a few values of s.

While the Fourier transform takes flight from the desire to find spectral information on nonperiodic functions with periodic functions as the model, the extra complications and extra richness of what results will soon make it seem like we're in a much different world. The definition just given is a good one *because* of the richness and *despite* the complications. Periodic functions are great, but there's a lot more in the world to analyze.

Still by analogy to Fourier series, the Fourier transform analyzes a signal $f(t)$ into its frequency components $\hat{f}(s)$. We haven't yet considered how the corresponding synthesis goes. How can we recover $f(t)$ in the time domain from $\hat{f}(s)$ in the frequency domain?

Recovering $f(t)$ from $\hat{f}(s)$. We can push the ideas on nonperiodic functions as limits of periodic functions a little further and discover how we might obtain $f(t)$ from its transform $\hat{f}(s)$. Again suppose $f(t)$ is zero outside some interval and periodize it to have (large) period T. We expand $f(t)$ in a Fourier series,

$$f(t) = \sum_{n=-\infty}^{\infty} c_n e^{2\pi int/T}\,.$$

The Fourier *coefficients* can be written via the Fourier *transform* of f evaluated at the points $s_n = n/T$:

$$\begin{aligned} c_n &= \frac{1}{T}\int_{-T/2}^{T/2} e^{-2\pi int/T} f(t)\,dt = \frac{1}{T}\int_{-\infty}^{\infty} e^{-2\pi int/T} f(t)\,dt \\ &\qquad \text{(we can extend the limits to } \pm\infty \text{ since } f(t) \text{ is zero outside of } [-T/2, T/2]) \\ &= \frac{1}{T}\hat{f}\left(\frac{n}{T}\right) = \frac{1}{T}\hat{f}(s_n)\,. \end{aligned}$$

Plug this into the expression for $f(t)$:

$$f(t) = \sum_{n=-\infty}^{\infty} \frac{1}{T}\hat{f}(s_n)e^{2\pi is_n t}\,.$$

The points $s_n = n/T$ are spaced $1/T$ apart, so we can think of $1/T$ as, say, Δs, and the sum above as a Riemann sum approximating an integral

$$\sum_{n=-\infty}^{\infty} \frac{1}{T}\hat{f}(s_n)e^{2\pi is_n t} = \sum_{n=-\infty}^{\infty} \hat{f}(s_n)e^{2\pi is_n t}\Delta s \approx \int_{-\infty}^{\infty} \hat{f}(s)e^{2\pi ist}\,ds\,.$$

The limits on the integral go from $-\infty$ to ∞ because the sum, and the points s_n, go from $-\infty$ to ∞. Thus as the period $T \to \infty$ we would expect to have

$$f(t) = \int_{-\infty}^{\infty} \hat{f}(s)e^{2\pi ist}\,ds$$

and we have recovered $f(t)$ from $\hat{f}(s)$. We have found the *inverse Fourier transform* and *Fourier inversion*.

The inverse Fourier transform defined, and Fourier inversion, too. The integral we've just come up with can stand on its own as a transform, and so we define the *inverse Fourier transform* of a function $g(s)$ to be

$$\check{g}(t) = \int_{-\infty}^{\infty} e^{2\pi ist} g(s)\, ds \quad \text{(upside down hat — cute; read: "check").}$$

The inverse Fourier transform looks just like the Fourier transform except the latter has a minus sign in the complex exponential. Later we'll say more about the symmetry between the Fourier transform and its inverse.

Again, we're treating this formally for the moment, withholding a discussion of conditions under which the integral makes sense. In the same spirit we've also produced the *Fourier inversion theorem*, namely,

$$f(t) = \int_{-\infty}^{\infty} e^{2\pi ist} \hat{f}(s)\, ds\,.$$

This applies (when it applies) to a general function and its transforms.

Written very compactly,

$$(\hat{f})\check{} = f\,.$$

By the way, we could have gone through the whole argument, above, starting with $\hat{f}$ as the basic function instead of f. If we did that, we'd be led to the complementary result on Fourier inversion,

$$(\check{g})\hat{} = g\,.$$

2.1.4. A quick summary and a look ahead.

Let's summarize what we've done, partly to fix ideas and partly as a guide to what we'd like to do next. There's so much involved, all of importance, and it will take some time before everything is in place.

- The Fourier transform of the signal $f(t)$ is

  $$\hat{f}(s) = \int_{-\infty}^{\infty} f(t) e^{-2\pi ist}\, dt\,.$$

  This is a complex-valued function of s.

- The domain of $\hat{f}(s)$ is the set of real numbers s where the integral exists. One says that $\hat{f}(s)$ is defined on the *frequency domain* and that the original signal $f(t)$ is defined on the *time domain* (or the *spatial domain*, depending on the context). We've already slipped into using this terminology.

  For a (nonperiodic) signal, defined on the whole real line, we do not have a discrete set of frequencies, as in the periodic case, but rather a *continuum* of frequencies.[3] We still do call the individual s's "frequencies" however, and the set of all frequencies s where $\hat{f}(s)$ exists is the *spectrum* of $f(t)$.

  As with Fourier series, people often refer to the *values* $\hat{f}(s)$ as the spectrum. This ambiguity shouldn't cause confusion, but it points up a difference

[3] A periodic function *does* have a Fourier transform, but it's a sum of δ-functions. We'll have to do that, too, and it will take some effort.

between Fourier series and Fourier transforms that I'll discuss briefly, below. And again as with Fourier series, we refer to the complex exponentials $e^{2\pi ist}$ as the *harmonics*.

One particular value of $\hat{f}(s)$ is worth calling attention to; namely, for $s = 0$ we have

$$\hat{f}(0) = \int_{-\infty}^{\infty} f(t)\, dt\,.$$

If $f(t)$ is real, as it most often is in applications, then $\hat{f}(0)$ is real even though other values of the Fourier transform may be complex. In calculus terms $\hat{f}(0)$ is the area under the graph of $f(t)$.

- The inverse Fourier transform is defined by

$$\check{g}(t) = \int_{-\infty}^{\infty} e^{2\pi ist} g(s)\, ds\,.$$

Fourier inversion says that

$$(\hat{f})\check{} = f\,, \quad (\check{g})\hat{} = g\,.$$

Taken together, the Fourier transform and its inverse provide a way of passing between two (equivalent) representations of a signal.

The functions $f(t)$ and $\hat{f}(s)$ may be "equivalent" via Fourier inversion, but they may also have quite different properties; e.g., one may be real valued and the other complex valued. Is it really true that when $\hat{f}(s)$ exists, we can just pop it into the formula for the inverse Fourier transform — which is also an improper integral that looks the same as the forward transform except for the minus sign — and really get back $f(t)$? Really? Not obvious! Worth wondering about.

We note one consequence of Fourier inversion, namely that

$$f(0) = \int_{-\infty}^{\infty} \hat{f}(s)\, ds\,.$$

There is no quick calculus interpretation of this result. The right-hand side is an integral of a complex-valued function (generally), and the result is real (if $f(0)$ is real).

- If t has dimension time, then to make st dimensionless in the exponential $e^{\pm 2\pi ist}$ the variable s must have dimension 1/time, i.e., the dimension of frequency, of course. In general, whatever the variables mean in $e^{\pm 2\pi ist}$ their dimensions must be reciprocal.

  This is the first example, of many to come, of a *reciprocal relationship* between the two domains. We'll take note of such relationships when they occur. Expect them to occur. They will help you to organize your understanding of the subject.

- The square magnitude $|\hat{f}(s)|^2$ is called the *power spectrum* (especially in connection with its use in communications) or the *spectral power density* (especially in connection with its use in optics) or the *energy spectrum* (especially in every other connection).

An important relation between the energy of the signal in the time domain and the energy spectrum in the frequency domain is Parseval's identity for Fourier transforms:[4]

$$\int_{-\infty}^{\infty} |f(t)|^2 \, dt = \int_{-\infty}^{\infty} |\hat{f}(s)|^2 \, ds \, .$$

This is the Fourier transform version of Rayleigh's identity and is also a future attraction.

A warning on notations: None is perfect; all are in use. Depending on the operation to be performed, or on the context, it's useful to have alternate notations for the Fourier transform. But here's a warning, which is the start of a complaint, which is the prelude to a full blown rant. Diddling with notation seems to be an unavoidable hassle in this subject. Flipping back and forth between a transform and its inverse, naming the variables in the different domains (even writing or not writing the variables), changing plus signs to minus signs, taking complex conjugates, these are all routine day-to-day operations, and they can cause endless muddles if you are not careful and sometimes even if you are careful. You will believe me when we have some examples, and you will hear me complain about it frequently.

Here's one example of a common notational convention:

> If the function is called f, then one often uses the corresponding capital letter, F, to denote the Fourier transform. So one sees a and A, z and Z, and everything in between. Note, however, that one typically uses different names for the variable for the two functions, as in $f(x)$ (or $f(t)$) and $F(s)$. This "capital letter notation" is very common in engineering but often confuses people when duality is invoked, to be explained below.

And then there's this:

> Since taking the Fourier transform is an operation that is applied to a function to produce a new function, it's also sometimes convenient to indicate this by a kind of operational notation. For example, it's common to write $\mathcal{F}f(s)$ for $\hat{f}(s)$, and so, to repeat the full definition,
>
> $$\mathcal{F}f(s) = \int_{-\infty}^{\infty} e^{-2\pi i s t} f(t) \, dt \, .$$
>
> This is often the most unambiguous notation.
>
> For those who believe in the power of parentheses, it would be even more proper to write
>
> $$(\mathcal{F}f)(s) = \int_{-\infty}^{\infty} e^{-2\pi i s t} f(t) \, dt \, ,$$
>
> indicating that we take the transform of f and we evaluate this transform *at* s by means of the given formula. But the extra parentheses are (probably) too much of a good thing and we won't use them. Just be careful.

---

[4]For the French mathematician Marc-Antoine Parseval. The identity is also referred to as Plancherel's theorem, though usually in a different connection.

The operation of taking the inverse Fourier transform is then denoted by $\mathcal{F}^{-1}$, and so

$$\mathcal{F}^{-1}g(t) = \int_{-\infty}^{\infty} e^{2\pi i s t} g(s)\, ds\,.$$

Fourier inversion looks like

$$\mathcal{F}^{-1}\mathcal{F}f = f, \quad \mathcal{F}\mathcal{F}^{-1}f = f.$$

We will use the notations $\mathcal{F}f$ and $\mathcal{F}^{-1}f$ more often than not. This notation, too, is far from ideal, but overall it offers fewer problems.

Finally, a function and its Fourier transform are said to constitute a *Fourier pair*. There have been various notations devised to indicate this sibling relationship. One is

$$f(t) \rightleftharpoons F(s)\,.$$

Bracewell advocated the use of

$$F(s) \supset f(t),$$

and others also use it. I . . . don't like it.

A warning on definitions. Our definition of the Fourier transform is a common one, but it's not the only one. One question is where to put the 2π: in the exponential, as we have done; or perhaps as a factor out front; or perhaps left out completely. There's also a question of which is the Fourier transform and which is the inverse, i.e., which transform gets the minus sign in the exponential. All of the various conventions are in day-to-day use in the professions. At the end of the chapter I'll give a summary of what happens to which formulas under the various conventions. I only mention this now because when you're talking with a friend about the Fourier transform, be sure you both know which conventions are being followed. Friendships have broken up over this.

A comment on the spectrum. For Fourier series, we considered a point n to be in the spectrum only when the corresponding Fourier coefficient $\hat{f}(n) \neq 0$. That was the proper thing to do since, in the Fourier series

$$\sum_{n=-\infty}^{\infty} \hat{f}(n)^{2\pi i n t}\,,$$

whether a coefficient is zero or nonzero changes the signal.

Not so for the Fourier transform. Changing the value of $\mathcal{F}f(s)$ at a point, or at several points, or more generally on a set of *measure zero*, does not affect the recovery of $f(t)$ from $\mathcal{F}f(s)$ via Fourier inversion. In the formula

$$f(t) = \int_{-\infty}^{\infty} e^{2\pi i s t} \mathcal{F}f(s)\, ds$$

the integral is not changed by altering the values of $\mathcal{F}f(s)$ on a set of measure zero. That's the key fact about sets of measure zero: they don't affect the value of an integral. By the same token, altering the values of f on a set of measure zero leaves its Fourier transform $\mathcal{F}f$ unchanged.

And what is a set of measure zero? Fair question, but a question that goes beyond what we need.[5] Suffice it to say that finite sets are of measure zero, and so too are some infinite sets.

So, with an exception, the criterion for a point belonging to the spectrum is just the existence of the Fourier transform. The exception is the case when $\mathcal{F}f(s) = 0$ not just at isolated points, but on, say, a (finite) collection of intervals. If $\mathcal{F}f(s) = 0$ on an interval[6] I, then

$$\int_I e^{2\pi i s t} \mathcal{F}f(s)\, ds = 0\,,$$

and so we can leave out those intervals in the integral:

$$f(t) = \int_{-\infty}^{\infty} e^{2\pi i s t} \mathcal{F}f(s)\, ds = \int_{\text{complement of all such } I\text{'s}} e^{2\pi i s t} \mathcal{F}f(s)\, ds\,.$$

In other words, intervals where $\mathcal{F}f(s) = 0$ don't contribute to recovering $f(t)$ from $\mathcal{F}f(s)$ through Fourier inversion, and I'd say those values of s are *not* in the spectrum.

Important examples are the *bandlimited signals* arising in sampling theory (Chapter 6). These are the signals $f(t)$ for which

$$\mathcal{F}f(s) = 0 \quad \text{for } |s| \geq s_0\,,$$

for some number s_0. We wouldn't consider points with $|s| > s_0$ to be in the spectrum (but there are arguments about the endpoints).

We're done with warnings and comments. On to more interesting things.

2.2. Getting to Know Your Fourier Transform

In one way, at least, our study of the Fourier transform will run the same course as your study of calculus. When you learned calculus, it was necessary to learn the derivative and integral formulas for specific functions and for types of functions (powers, exponentials, trig functions — the functions society needs) and also to learn the general principles and rules of differentiation and integration that allow you to work with combinations of functions (product rule, chain rule, inverse functions). It will be the same thing for us now. We'll need to have a storehouse of specific functions and their transforms that we can call on, and we'll need to develop general principles and results on how the Fourier transform operates.

[5] Sets of measure zero are important in generalizing from the Riemann integral to the Lebesgue integral. There's a little more discussion of this in Section 4.3. It's unlikely you'll need to worry about such things in typical applications.

[6] Or more generally on a set of *positive* measure.

2.2.1. Some specific transforms. We've already seen the example

$$\widehat{\Pi}(s) = \operatorname{sinc} s, \quad \text{or} \quad \mathcal{F}\Pi(s) = \operatorname{sinc} s \quad \text{using the } \mathcal{F} \text{ notation.}$$

Let's do a few more examples.

The triangle function. Consider next the triangle function, defined by

$$\Lambda(x) = \begin{cases} 1 - |x|, & |x| \leq 1, \\ 0, & \text{otherwise.} \end{cases}$$

Here's the graph.

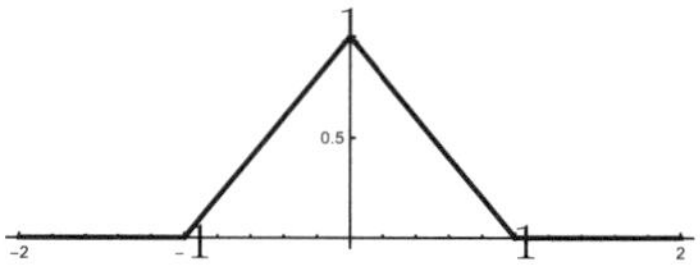

You worked with the triangle function and a scaled version in Problem 1.7 on periodization.

For the Fourier transform,

$$\mathcal{F}\Lambda(s) = \int_{-\infty}^{\infty} \Lambda(x)e^{-2\pi isx}\, dx = \int_{-1}^{0} (1+x)e^{-2\pi isx}\, dx + \int_{0}^{1} (1-x)e^{-2\pi isx}\, dx\,.$$

Integration by parts is called for, and if you look back to our calculation of the Fourier coefficients for the triangle wave in Chapter 1, a similar observation can save us a little work. Let $A(s)$ be the first integral,

$$A(s) = \int_{-1}^{0} (1+x)e^{-2\pi isx}\, dx\,.$$

Then

$$A(-s) = \int_{-1}^{0} (1+x)e^{2\pi isx}\, dx$$

and a change of variable $u = -x$ turns this into

$$A(-s) = \int_{1}^{0} (1-u)e^{-2\pi isu}(-du) = \int_{0}^{1} (1-u)e^{-2\pi isu} du\,.$$

Thus

$$\mathcal{F}\Lambda(s) = A(s) + A(-s)\,,$$

and we only have to subject ourselves to integration by parts to find $A(s)$. For this, with $u = 1 + x$ and $dv = e^{-2\pi i s x}\,dx$ we get

$$A(s) = -\frac{1}{2\pi i s} + \frac{1}{4\pi^2 s^2}(1 - e^{2\pi i s})\,.$$

Then

$$\begin{aligned}
\mathcal{F}\Lambda(s) &= A(s) + A(-s) \\
&= \frac{2}{4\pi^2 s^2} - \frac{e^{2\pi i s}}{4\pi^2 s^2} - \frac{e^{2\pi i s}}{4\pi^2 s^2} \\
&= \left(\frac{e^{\pi i s}}{2\pi i s}\right)^2 - 2\frac{1}{2\pi i s}\frac{1}{2\pi i s} + \left(\frac{e^{-\pi i s}}{2\pi i s}\right)^2 \\
&= \left(\frac{e^{\pi i s}}{2\pi i s} - \frac{e^{-\pi i s}}{2\pi i s}\right)^2 \qquad (\text{using } (a - a^{-1})^2 = a^2 - 2 + a^{-2}) \\
&= \left(\frac{1}{\pi s}\left(\frac{e^{\pi i s} - e^{-\pi i s}}{2i}\right)\right)^2 \\
&= \left(\frac{\sin \pi s}{\pi s}\right)^2 = \operatorname{sinc}^2 s\,.
\end{aligned}$$

Tricky. It's no accident that the Fourier transform of the triangle function turns out to be the square of the Fourier transform of the rect function. It has to do with convolution, an operation we have seen for Fourier series and will see anew for Fourier transforms in the next chapter.

The graph of $\operatorname{sinc}^2 s$ looks like

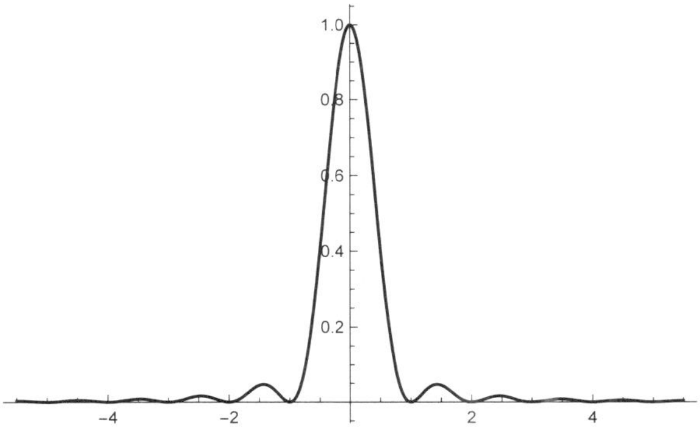

The exponential decay. Another commonly occurring function is the one-sided exponential decay, defined by

$$f(t) = \begin{cases} 0, & t \le 0, \\ e^{-at}, & t > 0, \end{cases}$$

where a is a positive constant. This function models a signal that is zero, switched on, and then decays exponentially. Here are graphs for $a = 2,\ 1.5,\ 1.0,\ 0.5,\ 0.25$.

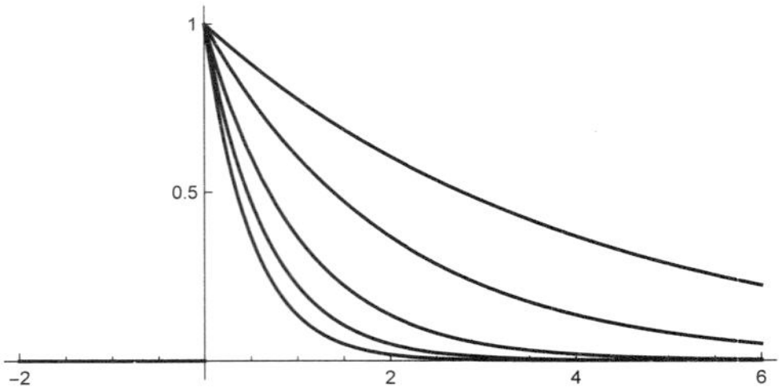

Which is which? If you can't say, see the discussion on scaling the independent variable at the end of this section.

Back to the exponential decay, we can calculate its Fourier transform directly:

$$\begin{aligned}\mathcal{F}f(s) &= \int_0^\infty e^{-2\pi i s t} e^{-at}\,dt = \int_0^\infty e^{-2\pi i s t - at}\,dt \\ &= \int_0^\infty e^{(-2\pi i s - a)t}\,dt = \left[\frac{e^{(-2\pi i s - a)t}}{-2\pi i s - a}\right]_{t=0}^{t=\infty} \\ &= \frac{e^{(-2\pi i s)t}}{-2\pi i s - a} e^{-at}\Bigg|_{t=\infty} - \frac{e^{(-2\pi i s - a)t}}{-2\pi i s - a}\Bigg|_{t=0} = \frac{1}{2\pi i s + a}.\end{aligned}$$

In this case, unlike the results for the rect function and the triangle function, the Fourier transform is complex. The fact that $\mathcal{F}\Pi(s)$ and $\mathcal{F}\Lambda(s)$ are real is because $\Pi(x)$ and $\Lambda(x)$ are even functions; we'll go over this shortly. There is no such symmetry for the exponential decay.

The power spectrum of the exponential decay is

$$|\mathcal{F}f(s)|^2 = \frac{1}{|2\pi i s + a|^2} = \frac{1}{a^2 + 4\pi^2 s^2}\,.$$

Here are graphs of this function for the same values of a as in the graphs of the exponential decay function.

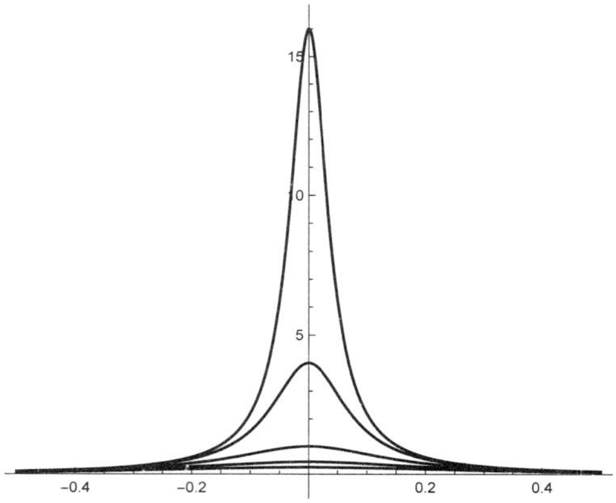

Which is which? You'll soon learn to spot that immediately, relative to the pictures in the time domain, and it's important. Also note that $|\mathcal{F}f(s)|^2$ *is* an even function of s even though $\mathcal{F}f(s)$ is not. We'll see why later. The shape of $|\mathcal{F}f(s)|^2$ is that of a bell curve, though this is *not* a Gaussian, a function we'll discuss just below. The curve is known as a *Lorentz profile*[7] and comes up in analyzing the transition probabilities and lifetime of the excited state in atoms.

How does the graph of $f(ax)$ compare with the graph of $f(x)$? Let me remind you of some elementary lore on scaling the independent variable in a function. The question is how the graph of $f(ax)$ compares with the graph of $f(x)$ when $0 < a < 1$ and when $a > 1$; I'm talking about any generic function $f(x)$ here. This is very simple, especially compared to what we've done and what we're going to do, but you'll want it at your fingertips and *everyone* has to think about it for a few seconds. Here's how to spend those few seconds.

Consider, for example, the graph of $f(2x)$. The graph of $f(2x)$, compared with the graph of $f(x)$, is squeezed. Why? Think about what happens when you plot the graph of $f(2x)$ over, say, $-1 \leq x \leq 1$. When x goes from -1 to 1, $2x$ goes from -2 to 2, so while you're plotting $f(2x)$ over the interval from -1 to 1, you have to compute the values of $f(x)$ from -2 to 2. That's more of the function in less space, as it were, so the graph of $f(2x)$ is a squeezed version of the graph of $f(x)$. Clear?

Similar reasoning shows that the graph of $f(x/2)$ is stretched. If x goes from -1 to 1, then $x/2$ goes from $-1/2$ to $1/2$, so while you're plotting $f(x/2)$ over the interval -1 to 1, you have to compute the values of $f(x)$ from $-1/2$ to $1/2$. That's less of the function in more space, so the graph of $f(x/2)$ is a stretched version of the graph of $f(x)$.

2.2.2. For Whom the Bell Curve Tolls. Let's next consider the Gaussian function and its Fourier transform. We'll need this for many examples and problems. This function, the famous bell shaped curve, was used by Gauss for statistical problems. It has some striking properties with respect to the Fourier transform that, on the one hand, give it a special role within Fourier analysis and, on the other hand, allow Fourier methods to be applied to other areas where the function comes up. We'll see an application to probability in Chapter 3.

The basic Gaussian is $f(x) = e^{-x^2}$. The shape of the graph is familiar to you:

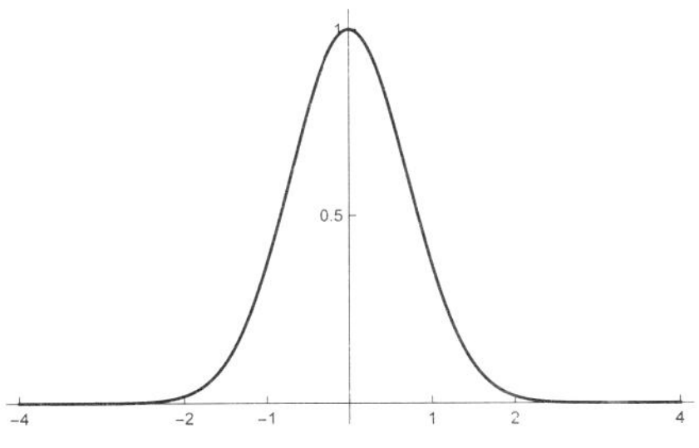

For various applications one throws in extra factors to modify particular properties of the function. We'll do this too, and there's not a complete agreement on what's best. There is an agreement that before anything else happens, one has to

[7]Named after the Dutch physicist H. Lorentz. He was known for work in electrodynamics and (before it was so named) special relativity. You may have run across the Lorentz contraction.

know the amazing equation[8]

$$\int_{-\infty}^{\infty} e^{-x^2}\,dx = \sqrt{\pi}.$$

Now, the function $f(x) = e^{-x^2}$ does not have an elementary antiderivative, so this integral cannot be found directly by an appeal to the Fundamental Theorem of Calculus. The fact that it *can* be evaluated exactly is one of the most famous tricks in mathematics. It's due to Euler, and you shouldn't go through life not having seen it. And even if you have seen it, it's worth seeing again; see the discussion following this section. First, though, the Fourier transform.

The Fourier transform of a Gaussian. In whatever subject it's applied, it seems always to be useful to normalize the Gaussian so that the total area is 1. This can be done in several ways, but for Fourier analysis the best choice, as we shall see, is

$$f(x) = e^{-\pi x^2}.$$

You can check using the result for the integral of e^{-x^2} that

$$\int_{-\infty}^{\infty} e^{-\pi x^2}\,dx = 1.$$

Let's compute the Fourier transform

$$\mathcal{F}f(s) = \int_{-\infty}^{\infty} e^{-\pi x^2} e^{-2\pi i s x}\,dx.$$

Differentiate with respect to s:

$$\frac{d}{ds}\mathcal{F}f(s) = \int_{-\infty}^{\infty} e^{-\pi x^2}(-2\pi i x)e^{-2\pi i s x}\,dx.$$

This is set up perfectly for an integration by parts, where $dv = -2\pi i x e^{-\pi x^2}\,dx$ and $u = e^{-2\pi i s x}$. Then $v = ie^{-\pi x^2}$, and evaluating the product uv at the limits $\pm\infty$ gives 0. Thus

$$\begin{aligned}\frac{d}{ds}\mathcal{F}f(s) &= -\int_{-\infty}^{\infty} ie^{-\pi x^2}(-2\pi i s)e^{-2\pi i s x}\,dx\\ &= -2\pi s\int_{-\infty}^{\infty} e^{-\pi x^2}e^{-2\pi i s x}\,dx\\ &= -2\pi s\mathcal{F}f(s).\end{aligned}$$

So $\mathcal{F}f(s)$ satisfies the simple differential equation

$$\frac{d}{ds}\mathcal{F}f(s) = -2\pi s\mathcal{F}f(s)$$

whose unique solution, incorporating the initial condition, is

$$\mathcal{F}f(s) = \mathcal{F}f(0)e^{-\pi s^2}.$$

---

[8]Speaking of this equation, William Thomson, after he became Lord Kelvin, said: "A mathematician is one to whom *that* is as obvious as that twice two makes four is to you." I'd like to think he was kidding.

But

$$\mathcal{F}f(0) = \int_{-\infty}^{\infty} e^{-\pi x^2}\, dx = 1\,.$$

Hence

$$\mathcal{F}f(s) = e^{-\pi s^2}\,.$$

We have found the remarkable fact that the Gaussian $f(x) = e^{-\pi x^2}$ is its own Fourier transform!

Evaluation of the Gaussian integral. We want to evaluate

$$I = \int_{-\infty}^{\infty} e^{-x^2}\, dx\,.$$

It doesn't matter what we call the variable of integration, so we can also write the integral as

$$I = \int_{-\infty}^{\infty} e^{-y^2}\, dy\,.$$

Therefore

$$I^2 = \left(\int_{-\infty}^{\infty} e^{-x^2}\, dx\right)\left(\int_{-\infty}^{\infty} e^{-y^2}\, dy\right).$$

Because the variables aren't coupled, we can combine this into a double integral[9]

$$\int_{-\infty}^{\infty}\left(\int_{-\infty}^{\infty} e^{-x^2}\, dx\right) e^{-y^2}\, dy = \int_{-\infty}^{\infty}\int_{-\infty}^{\infty} e^{-(x^2+y^2)}\, dx\, dy\,.$$

Now we make a change of variables, introducing polar coordinates, (r, θ). First, what about the limits of integration? To let both x and y range from $-\infty$ to ∞ is to describe the entire plane, and to describe the entire plane in polar coordinates is to let r go from 0 to ∞ and θ go from 0 to 2π. Next, $e^{-(x^2+y^2)}$ becomes e^{-r^2} and the area element $dx\, dy$ becomes $r\, dr\, d\theta$. It's the extra factor of r in the area element that makes all the difference. With the change to polar coordinates we have

$$I^2 = \int_{-\infty}^{\infty}\int_{-\infty}^{\infty} e^{-(x^2+y^2)}\, dx\, dy = \int_0^{2\pi}\int_0^{\infty} e^{-r^2}\, r\, dr\, d\theta.$$

Because of the factor r, the inner integral can be done directly:

$$\int_0^{\infty} e^{-r^2}\, r\, dr = -\tfrac{1}{2}e^{-r^2}\Big]_0^{\infty} = \tfrac{1}{2}\,.$$

The double integral then reduces to

$$I^2 = \int_0^{2\pi} \tfrac{1}{2}\, d\theta = \pi\,,$$

from which

$$\int_{-\infty}^{\infty} e^{-x^2}\, dx = I = \sqrt{\pi}\,.$$

Wonderful.

[9]We will see the same sort of thing in other settings, for example when we work with the product of two Fourier transforms on our way to defining convolution in the next chapter.

A final word on the examples. You might have noticed that our list so far does not include many of the basic functions that society needs. For example the sine and cosine haven't appeared, and you can't get much more basic than that. This will have to wait. There's no way to make sense, classically, of the integrals

$$\int_{-\infty}^{\infty} e^{-2\pi ist} \cos 2\pi t \, dt \,, \quad \int_{-\infty}^{\infty} e^{-2\pi ist} \sin 2\pi t \, dt \,.$$

A whole different approach will be necessary.

Nonetheless, there are more examples that can be done. You can find compilations on the web, and a very attractive "Pictorial Dictionary of Fourier Transforms" is in Bracewell's book.

2.3. Getting to Know Your Fourier Transform, Better

We've started to build a storehouse of specific transforms. Let's now proceed for awhile along the other path and develop some general properties. For this discussion, and indeed for much of our work over the next many pages, we are going to abandon all worries about transforms existing, integrals converging, and whatever other worries you might be carrying. Relax and enjoy the ride. The rigor police are off duty.

2.3.1. Fourier transform pairs and duality. One striking feature of the Fourier transform and the inverse Fourier transform is the symmetry between the two formulas, something you *don't* see for Fourier series. For Fourier series the coefficients are given by an integral (a transform of $f(t)$ into $\hat{f}(n)$), but the inverse transform is the series itself. For Fourier transforms, $\mathcal{F}$ and $\mathcal{F}^{-1}$ look the same except for the minus sign in the complex exponential.[10] In words, we can say that if you replace s by $-s$ in the formula for the Fourier transform, then you're taking the inverse Fourier transform. Likewise, if you replace t by $-t$ in the formula for the inverse Fourier transform, then you're taking the Fourier transform. That is,

$$\mathcal{F}f(-s) = \int_{-\infty}^{\infty} e^{-2\pi i(-s)t} f(t) \, dt = \int_{-\infty}^{\infty} e^{2\pi ist} f(t) \, dt = \mathcal{F}^{-1} f(s) \,,$$
$$\mathcal{F}^{-1} f(-t) = \int_{-\infty}^{\infty} e^{2\pi is(-t)} f(s) \, ds = \int_{-\infty}^{\infty} e^{-2\pi ist} f(s) \, ds = \mathcal{F}f(t) \,.$$

This might cause some consternation because you generally want to think of the two variables, s and t, as somehow associated with separate and different domains, one domain for the forward transform and one for the inverse transform, one for time and one for frequency, while in each of these formulas a single variable is used in both domains. You have to get over this kind of consternation, because it's going

[10]Here's the reason that the formulas for the Fourier transform and its inverse appear so symmetric; it's quite a deep mathematical fact. As the general theory goes, if the original function is defined on a group, then the transform (also defined in generality) is defined on the *dual group*, which I won't define for you here. In the case of Fourier series the function is periodic and so its natural domain is the circle; think of the circle as $[0, 1]$ with the endpoints identified, and the elements of the group are rotations of the circle. It turns out that the dual of the circle group is the integers, and that's why $\hat{f}$ is evaluated at integers n. It also turns out that when the group is $\mathbb{R}$, a group under addition, the dual group is again $\mathbb{R}$. Thus the Fourier transform of a function defined on $\mathbb{R}$ is itself defined on $\mathbb{R}$. Working through the general definitions of the Fourier transform and its inverse in this case produces the symmetric result that we have before us. Kick that one around over dinner some night.

to come up again. Think purely in terms of the math: the transform is an operation on a function that produces a new function. To write down the formula, I have to evaluate the transform at a variable, but the variable is only a placeholder and it doesn't matter what I call it as long as I keep its roles in the formulas straight.

Also be observant of what the notation in the formula says and, just as important, what it doesn't say. The first formula, for example, says what happens when you *first* take the Fourier transform of f and *then* evaluate it at $-s$; it's *not* a formula for $\mathcal{F}(f(-s))$ as in "first change s to $-s$ in the formula for f and then take the transform." I could have written the first displayed equation as $(\mathcal{F}f)(-s) = \mathcal{F}^{-1}f(s)$, with an extra set of parentheses around the $\mathcal{F}f$ to emphasize this, but I thought that looked too clumsy. It cannot be said too often: be careful, please.

The equations

$$\mathcal{F}f(-s) = \mathcal{F}^{-1}f(s),$$
$$\mathcal{F}^{-1}f(-t) = \mathcal{F}f(t)$$

are sometimes referred to as the *duality* properties of the transforms. They look like different statements but you can get from one to the other. We'll set this up a little differently in the next section.

Here's an example of how duality is used. We know that

$$\mathcal{F}\Pi = \operatorname{sinc}$$

and hence that

$$\mathcal{F}^{-1}\operatorname{sinc} = \Pi\,.$$

"By duality" we can find $\mathcal{F}\operatorname{sinc}$:

$$\mathcal{F}\operatorname{sinc}(t) = \mathcal{F}^{-1}\operatorname{sinc}(-t) = \Pi(-t)\,.$$

Troubled by the variables? Remember, the left-hand side is $(\mathcal{F}\operatorname{sinc})(t)$. Now with the *additional* knowledge that Π is an even function ($\Pi(-t) = \Pi(t)$) we can conclude that

$$\mathcal{F}\operatorname{sinc} = \Pi\,.$$

Let's apply the same argument to find $\mathcal{F}\operatorname{sinc}^2$. Remember the triangle function Λ and the result

$$\mathcal{F}\Lambda = \operatorname{sinc}^2\,,$$

so

$$\mathcal{F}^{-1}\operatorname{sinc}^2 = \Lambda\,.$$

But then

$$\mathcal{F}\operatorname{sinc}^2(t) = (\mathcal{F}^{-1}\operatorname{sinc}^2)(-t) = \Lambda(-t)$$

and since Λ is even,

$$\mathcal{F}\operatorname{sinc}^2 = \Lambda\,.$$

Duality and reversed signals. There's a slightly different take on duality that I prefer because it suppresses the variables. I find it easier to remember. Starting with a signal $f(t)$, define the *reversed signal* f^- by

$$f^-(t) = f(-t)\,.$$

Note that a double reversal gives back the original signal,

$$(f^-)^- = f\,.$$

Note also that the conditions defining when a function is even or odd are easy to write in terms of the reversed signals:

$$f \text{ is even if } f^- = f,$$
$$f \text{ is odd if } f^- = -f.$$

In words, a signal is even if reversing the signal doesn't change it, and a signal is odd if reversing the signal changes its sign. We'll pick up on this in the next section.

Simple enough — to reverse the signal is just to reverse the time. This is a general operation, whatever the nature of the signal and whether or not the variable is time. Using this notation, we can rewrite the first duality equation, $\mathcal{F}f(-s) = \mathcal{F}^{-1}f(s)$, as

$$(\mathcal{F}f)^- = \mathcal{F}^{-1}f$$

and we can rewrite the second duality equation, $\mathcal{F}^{-1}f(-t) = \mathcal{F}f(t)$, as

$$(\mathcal{F}^{-1}f)^- = \mathcal{F}f\,.$$

This makes it very clear that the two equations are saying the same thing. One is just the reverse of the other.

Furthermore, using this notation, the result $\mathcal{F}\,\mathrm{sinc} = \Pi$, for example, goes a little more quickly:

$$\mathcal{F}\,\mathrm{sinc} = (\mathcal{F}^{-1}\,\mathrm{sinc})^- = \Pi^- = \Pi\,.$$

Likewise

$$\mathcal{F}\mathrm{sinc}^2 = (\mathcal{F}^{-1}\mathrm{sinc}^2)^- = \Lambda^- = \Lambda\,.$$

A natural variation on the preceding duality results is to ask what happens with $\mathcal{F}f^-$, the Fourier transform of the reversed signal. Let's work this out. By definition,

$$\mathcal{F}f^-(s) = \int_{-\infty}^{\infty} e^{-2\pi ist} f^-(t)\,dt = \int_{-\infty}^{\infty} e^{-2\pi ist} f(-t)\,dt\,.$$

There's only one thing to do at this point, and we'll be doing it a lot: make a change of variable in the integral. Let $u = -t$ so that $du = -dt$, or $dt = -du$. Then as t goes from $-\infty$ to ∞, the variable $u = -t$ goes from ∞ to $-\infty$ and we have

$$\int_{-\infty}^{\infty} e^{-2\pi ist} f(-t)\, dt = \int_{\infty}^{-\infty} e^{-2\pi is(-u)} f(u)\,(-du)$$
$$= \int_{-\infty}^{\infty} e^{2\pi isu} f(u)\, du$$
(the minus sign, on the du, flips the limits back)
$$= \mathcal{F}^{-1}f(s).$$

Thus, quite neatly,

$$\mathcal{F}f^- = \mathcal{F}^{-1}f.$$

Even more neatly, if we now substitute $\mathcal{F}^{-1}f = (\mathcal{F}f)^-$ from earlier, we have

$$\mathcal{F}f^- = (\mathcal{F}f)^-.$$

Note carefully where the parentheses are here. In words:

- The Fourier transform of the reversed signal is the reversal of the Fourier transform of the signal.

That one I can remember.

To finish off these questions, we have to know what happens to $\mathcal{F}^{-1}f^-$. But we don't have to do a separate calculation here. Using our earlier duality result,

$$\mathcal{F}^{-1}f^- = (\mathcal{F}f^-)^- = (\mathcal{F}^{-1}f)^-.$$

In words, the inverse Fourier transform of the reversed signal is the reversal of the inverse Fourier transform of the signal. We can also take this one step farther and get back to $\mathcal{F}^{-1}f^- = \mathcal{F}f$.

And so, the list of duality relations to this point boils down to

$$\mathcal{F}f = (\mathcal{F}^{-1}f)^-,$$
$$\mathcal{F}f^- = \mathcal{F}^{-1}f.$$

Learn these. There's one more:

$$\mathcal{F}(\mathcal{F}f)(s) = f(-s) \quad \text{or} \quad \mathcal{F}(\mathcal{F}f) = f^- \quad \text{without the variable.}$$

This follows from Fourier inversion:

$$\mathcal{F}(\mathcal{F}f)(s) = \int_{-\infty}^{\infty} e^{-2\pi ist}\mathcal{F}f(t)\, dt = \mathcal{F}^{-1}(\mathcal{F}f)(-s) = f(-s) = f^-(s).$$

It's common to drop the one set of parentheses and write the result as

$$\mathcal{F}\mathcal{F}f = f^-.$$

Of course we also have

$$\mathcal{F}\mathcal{F}f^- = f.$$

Based on this and the earlier duality results, you can then check that

$$\mathcal{F}\mathcal{F}\mathcal{F}\mathcal{F}f = f\,,$$

written as $\mathcal{F}^4 f = f$, not $\mathcal{F}$ to the fourth power, but $\mathcal{F}$ applied four times. Thus $\mathcal{F}^4$ is the identity transformation. Some people attach mystical significance to this fact.

An example of $\mathcal{F}\mathcal{F}f = f^-$ in action is yet one more derivation of $\mathcal{F}\,\text{sinc} = \Pi$, for

$$\mathcal{F}\,\text{sinc} = \mathcal{F}\mathcal{F}\Pi = \Pi^- = \Pi\,.$$

Even and odd symmetries and the Fourier transform. We've already had a number of occasions to use even and odd symmetries of functions. In the case of real-valued functions the conditions have obvious interpretations in terms of the symmetries of the graphs; the graph of an even function is symmetric about the y-axis and the graph of an odd function is symmetric through the origin. However, the algebraic definitions of even and odd apply to complex-valued as well as to real-valued functions, though the geometric picture is lacking when the function is complex valued because we can't draw the graph. A function can be even, odd, or neither, but it can't be both even and odd unless it's identically zero.

How are symmetries of a function reflected in properties of its Fourier transform? I won't give a complete accounting, but here are a few important cases.

- If $f(x)$ is even or odd, respectively, then so is its Fourier transform.

Working with reversed signals, we have to show that $(\mathcal{F}f)^- = \mathcal{F}f$ if f is even and $(\mathcal{F}f)^- = -\mathcal{F}f$ if f is odd. It's lighting fast using the equations that we derived above:

$$(\mathcal{F}f)^- = \mathcal{F}f^- = \begin{cases} \mathcal{F}f & \text{if } f \text{ is even, } f^- = f\,, \\ \mathcal{F}(-f) = -\mathcal{F}f & \text{if } f \text{ is odd, } f^- = -f\,. \end{cases}$$

Because the Fourier transform of a function is complex valued there are other symmetries we can consider for $\mathcal{F}f(s)$, namely what happens under complex conjugation. Take this one:

- If $f(t)$ is real valued, then $(\mathcal{F}f)^- = \overline{\mathcal{F}f}$ and $\mathcal{F}(f^-) = \overline{\mathcal{F}f}$.

Here it's worthwhile to bring the variable back in,

$$\mathcal{F}f(-s) = \overline{\mathcal{F}f(s)}.$$

In particular,

$$|\mathcal{F}f(-s)| = |\mathcal{F}f(s)|.$$

For a real-valued signal the magnitude of the Fourier transform is the same at corresponding positive and negative frequencies. This comes up all the time and is analogous to the conjugate symmetry property possessed by the Fourier coefficients for a real-valued periodic function.

The derivation is essentially the same as it was for Fourier coefficients, but it may be helpful to repeat it for practice and to see the similarities:

$$\begin{aligned}\mathcal{F}f(-s) &= \int_{-\infty}^{\infty} e^{2\pi ist} f(t)\,dt \\ &= \overline{\left\{\int_{-\infty}^{\infty} e^{-2\pi ist} f(t)\,dt\right\}} \\ &\qquad (\overline{e^{-2\pi ist}} = e^{2\pi ist}, \text{ and } \overline{f(t)} = f(t) \text{ since } f(t) \text{ is real}) \\ &= \overline{\mathcal{F}f(s)}.\end{aligned}$$

We can refine this if the function $f(t)$ itself has symmetry. For example, combine the previous results, and remember that a complex number is real if it's equal to its conjugate and is purely imaginary if it's equal to minus its conjugate. We then have:

- If f is a real-valued and even function, then its Fourier transform is real valued and even.
- If f is a real-valued and odd function, then its Fourier transform is purely imaginary and odd.

We saw this first point in action for Fourier transforms of the rect function $\Pi(t)$ and the triangle function $\Lambda(t)$. Both functions are even and their Fourier transforms, sinc and sinc^2, respectively, are even and real. Good thing it worked out that way.

2.3.2. Linearity. One of the simplest and most frequently invoked properties of the Fourier transform is that it is linear (operating on functions). This means

$$\begin{aligned}\mathcal{F}(f+g)(s) &= \mathcal{F}f(s) + \mathcal{F}g(s), \\ \mathcal{F}(\alpha f)(s) &= \alpha\mathcal{F}f(s) \quad \text{for any number } \alpha \text{ (real or complex).}\end{aligned}$$

The linearity properties are easy to check from the corresponding properties for integrals. For example,

$$\begin{aligned}\mathcal{F}(f+g)(s) &= \int_{-\infty}^{\infty} (f(x)+g(x))e^{-2\pi isx}\,dx \\ &= \int_{-\infty}^{\infty} f(x)e^{-2\pi isx}\,dx + \int_{-\infty}^{\infty} g(x)e^{-2\pi isx}\,dx = \mathcal{F}f(s) + \mathcal{F}g(s)\,.\end{aligned}$$

Linearity also holds for $\mathcal{F}^{-1}$.

We used (without comment) the property on multiples when we wrote $\mathcal{F}(-f) = -\mathcal{F}f$ in talking about odd functions and their transforms. I bet it didn't bother you that we hadn't yet stated the property formally.

2.3.3. Shift theorem. A shift of the variable t (a delay in time) has a simple effect on the Fourier transform. We would expect the magnitude of the Fourier transform $|\mathcal{F}f(s)|$ to stay the same, since shifting the original signal in time should not change the energy at any point in the spectrum. Hence the only change should be a phase shift in $\mathcal{F}f(s)$, and that's exactly what happens.

To compute the Fourier transform of $f(t-b)$, for a constant b, we have

$$\begin{aligned}
\int_{-\infty}^{\infty} f(t-b)e^{-2\pi ist}\,dt &= \int_{-\infty}^{\infty} f(u)e^{-2\pi is(u+b)}\,du \\
&\qquad \text{(substituting } u=t-b\text{; the limits still go from } -\infty \text{ to } \infty) \\
&= \int_{-\infty}^{\infty} f(u)e^{-2\pi isu}e^{-2\pi isb}\,du \\
&= e^{-2\pi isb}\int_{-\infty}^{\infty} f(u)e^{-2\pi isu}\,du \\
&\qquad (e^{-2\pi isb} \text{ comes out of the integral because it doesn't depend on } u) \\
&= e^{-2\pi isb}\hat{f}(s).
\end{aligned}$$

The best notation to capture this property is probably the pair notation, $f \rightleftharpoons F$.[11] Thus:

- If $f(t) \rightleftharpoons F(s)$, then $f(t-b) \rightleftharpoons e^{-2\pi isb}F(s)$.

Notice that, as promised, the magnitude of the Fourier transform has not changed under a time shift because the factor out front has magnitude 1:

$$\left|e^{-2\pi isb}F(s)\right| = \left|e^{-2\pi isb}\right|\,|F(s)| = |F(s)|\,.$$

The shift theorem for the inverse Fourier transform looks like this:

- If $F(s) \rightleftharpoons f(t)$, then $F(s-b) \rightleftharpoons e^{2\pi itb}f(t)$.

You can derive this from the integral defining $\mathcal{F}^{-1}$ or by using duality.

2.3.4. Modulation theorem.

First cousin to the shift theorem is the *modulation theorem*, which states:

- If $f(t) \rightleftharpoons F(s)$, then $e^{2\pi is_0t}f(t) \rightleftharpoons F(s-s_0)$.

In words, a phase change in time corresponds to a shift in frequency. That's a modulation of the spectrum.

A real-valued version of changing the phase in time is:

- If $f(t) \rightleftharpoons F(s)$, then $f(t)\cos(2\pi s_0t) \rightleftharpoons \frac{1}{2}(F(s-s_0)+F(s+s_0))$.

I'm going to leave these as exercises.

For the inverse Fourier transform:

- If $F(s) \rightleftharpoons f(t)$, then $e^{2\pi it_0s}F(s) \rightleftharpoons f(t+t_0)$.
- If $F(s) \rightleftharpoons f(t)$, then $F(s)\cos(2\pi t_0s) \rightleftharpoons \frac{1}{2}(f(t-t_0)+f(t+t_0))$.

[11] This is, however, an excellent opportunity to complain about notational matters. Writing $\mathcal{F}f(t-b)$ invites the same anxieties that some of us had when changing signs. What's being transformed? What's being plugged in? There's no room to write an s. The hat notation is even worse — there's no place for the s, again, and do you really want to write $\widehat{f(t-b)}$ with such a wide hat?

2.3.5. Stretch (similarity) theorem. How does the Fourier transform change if we stretch or shrink the variable in the time domain? If we scale t to at, what happens to the Fourier transform of $f(at)$?

We assume that $a \neq 0$. First suppose $a > 0$. Then

$$\begin{aligned}\int_{-\infty}^{\infty} f(at)e^{-2\pi ist}\,dt &= \int_{-\infty}^{\infty} f(u)e^{-2\pi is(u/a)}\frac{1}{a}\,du \\ &\qquad \text{(substituting } u = at\text{; the limits go the same way because } a > 0\text{)} \\ &= \frac{1}{a}\int_{-\infty}^{\infty} f(u)e^{-2\pi i(s/a)u}\,du = \frac{1}{a}\mathcal{F}f\left(\frac{s}{a}\right).\end{aligned}$$

If $a < 0$, the limits of integration are reversed when we make the substitution $u = at$:

$$\begin{aligned}\int_{-\infty}^{\infty} f(at)e^{-2\pi ist}\,dt &= \frac{1}{a}\int_{+\infty}^{-\infty} f(u)e^{-2\pi is(u/a)}\,du \\ &= -\frac{1}{a}\int_{-\infty}^{+\infty} f(u)e^{-2\pi i(s/a)u}\,du \\ &\qquad \text{(flipping the limits back introduces a minus sign)} \\ &= -\frac{1}{a}\mathcal{F}f\left(\frac{s}{a}\right).\end{aligned}$$

Since $-a$ is positive when a is negative ($-a = |a|$), we can combine the two cases and present the stretch theorem in its full glory:

- If $f(t) \rightleftharpoons F(s)$, then $f(at) \rightleftharpoons \dfrac{1}{|a|}F\left(\dfrac{s}{a}\right)$.

This is also sometimes called the *similarity theorem* because changing the variable from x to ax, as a change of scale, is also known as a similarity.

The result for the inverse Fourier transform looks the same:

- If $F(s) \rightleftharpoons f(t)$, then $F(as) \rightleftharpoons \dfrac{1}{|a|}f\left(\dfrac{t}{a}\right)$.

Some important observations go with the stretch theorem. First, it's clearly a reciprocal relationship between the two domains — there it is again.

Let's say a little more, and let's take a to be positive, which is the most common situation. If a is large (bigger than 1, at least), then the graph of $f(at)$ is squeezed horizontally compared to $f(t)$. Something different is happening in the frequency domain, in fact in two ways. The Fourier transform is $(1/a)F(s/a)$. If a is large, then $F(s/a)$ is *stretched out* compared to $F(s)$, rather than squeezed in. Furthermore, multiplying by $1/a$, since the transform is $(1/a)F(a/s)$, also squashes down the values of the transform. The opposite happens if a is small (less than 1). In that case the graph of $f(at)$ is stretched out horizontally compared to $f(t)$, while the Fourier transform is compressed *horizontally* and stretched *vertically*.

To sum up, a function stretched out in the time domain is squeezed in the frequency domain, and vice versa.[12] The phrase that's often used to describe this phenomenon is that a signal cannot be *localized* (meaning becoming concentrated at a point) in *both* the time domain and the frequency domain. We will see more precise formulations of this principle.[13]

This is somewhat analogous to what happens to the spectrum of a periodic function for long or short periods. Say the period is T, and recall that the points in the spectrum are spaced $1/T$ apart, a fact we've used several times. If T is large, then it's fair to think of the function as spread out in the time domain — it goes a long time before repeating. But then since $1/T$ is small, the spectrum is squeezed. On the other hand, if T is small, then the function is squeezed in the time domain — it goes only a short time before repeating — while the spectrum is spread out, since $1/T$ is large.

> **Careful here.** In the discussion just above I tried not to talk in terms of properties of the *graph of the transform* — though you may have reflexively thought in those terms and I slipped into it a little — because the transform is generally *complex valued.* You do see this squeezing and spreading phenomenon geometrically by looking at the graphs of $f(t)$ in the time domain (if $f(t)$ is real) and the *magnitude* of the Fourier transform $|\mathcal{F}f(s)|$ in the frequency domain.[14]

Example: The stretched rect. Hardly a felicitous phrase, "stretched rect," but the function comes up often in applications. Let $p > 0$ and define

$$\Pi_p(t) = \begin{cases} 1, & |t| < p/2\,, \\ 0, & |t| \geq p/2\,. \end{cases}$$

Thus Π_p is a rect function of width p. We can find its Fourier transform by direct integration, but we can also find it by means of the stretch theorem if we observe that

$$\Pi_p(t) = \Pi(t/p)\,.$$

Just to be sure, write down the definition of Π and follow through:

$$\Pi(t/p) = \begin{cases} 1, & |t/p| < 1/2, \\ 0, & |t/p| \geq 1/2 \end{cases} = \begin{cases} 1, & |t| < p/2, \\ 0, & |t| \geq p/2 \end{cases} = \Pi_p(t)\,.$$

By the stretch theorem applied to $\Pi(t/p)$ we have

$$\mathcal{F}\Pi_p(s) = p\mathcal{F}\Pi(ps) = p\,\mathrm{sinc}\,ps.$$

This comes up often enough to memorize.

Here are plots of the Fourier transform pairs for $p = 1/5$ and $p = 5$, respectively. Note the scales on the axes.

---

[12]It's the same sort of phenomenon if $a < 0$, but there's an additional reversal. Talk yourself through that.

[13]In fact, the famous Heisenberg uncertainty principle in quantum mechanics is an example.

[14]We observed this for the one-sided exponential decay and its Fourier transform, and you should now go back to that example and match up the graphs of $|\mathcal{F}f|$ with the various values of the parameter.

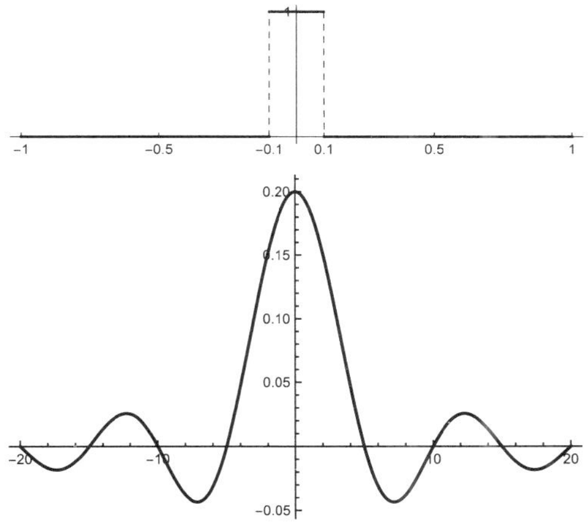

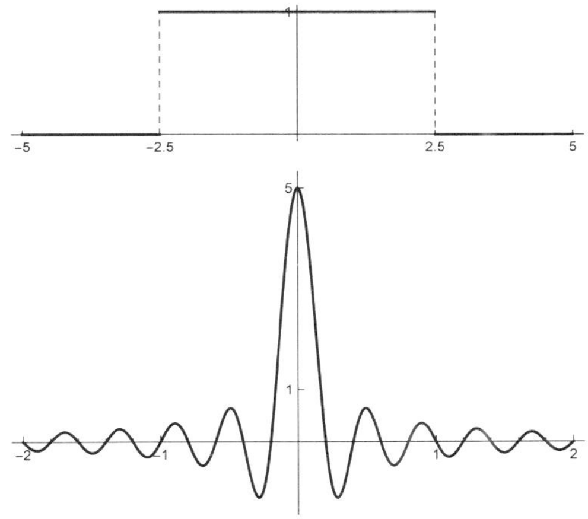

Example: Stretched triangle. Let's throw in the stretched triangle function and its Fourier transform; these will likewise come up often. For $p > 0$ let

$$\Lambda_p(t) = \Lambda(t/p) = \begin{cases} 1 - |t/p|, & |t| \le p, \\ 0, & |t| \ge p. \end{cases}$$

Note that the width of the triangle is $2p$.

Here are plots of $\Lambda_p(t)$ for $p = 0.5, 1, 2$:

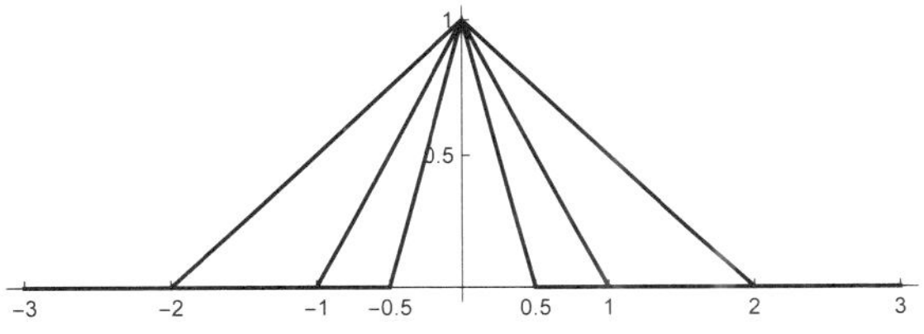

From the stretch theorem applied to $\Lambda(t/p)$ we get

$$\mathcal{F}\Lambda_p(s) = p\mathcal{F}\Lambda(ps) = p\,\text{sinc}^2(ps).$$

Once again, worth memorizing.

Here are plots of the corresponding Fourier transforms. Match them!

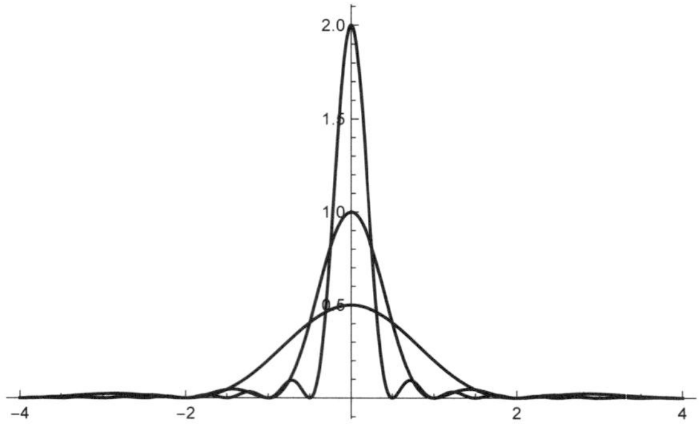

Example: Two-sided exponential decay. Here's an example of how you might combine the properties we've developed to find the transform of another commonly occurring signal. Let's find the Fourier transform of the *two*-sided exponential decay

$$g(t) = e^{-a|t|}, \quad a \text{ a positive constant.}$$

Here are plots of $g(t)$ for $a = 0.5, 1, 2$. Match them!

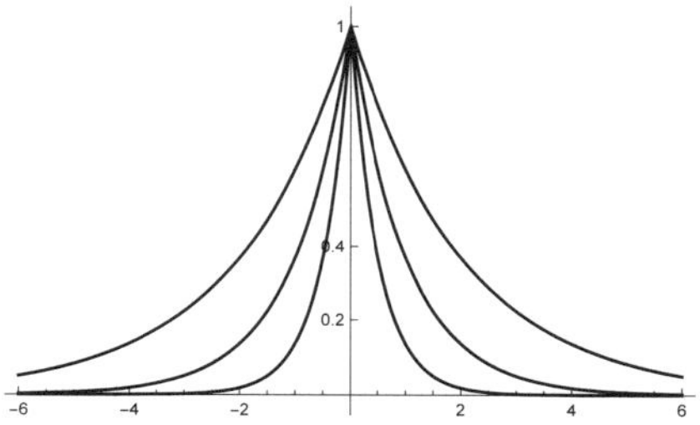

We could compute the transform directly; plugging into the formula for the Fourier transform would give us integrals we could do. However, we've already done half the work when we found the Fourier transform of the one-sided exponential decay. Recall that

$$f(t) = \begin{cases} 0, & t < 0, \\ e^{-at}, & t \geq 0 \end{cases} \quad \Rightarrow \quad F(s) = \mathcal{F}f(s) = \frac{1}{2\pi i s + a}$$

and now realize

$$g(t) \quad \text{is almost equal to} \quad f(t) + f(-t)\,.$$

They agree except at the origin, where $g(0) = 1$ and $f(t)$ and $f(-t)$ are both one. But two functions that agree except for one point will clearly give the same result when integrated[15] against $e^{-2\pi i s t}$. Therefore

$$\begin{aligned} G(s) = \mathcal{F}g(s) &= F(s) + F(-s) \\ &= \frac{1}{2\pi i s + a} + \frac{1}{-2\pi i s + a} = \frac{2a}{a^2 + 4\pi^2 s^2}\,. \end{aligned}$$

[15]See, e.g., the earlier comments on how the spectrum is defined.

Note that $g(t)$ is even and $G(s)$ is real. These sorts of quick checks on correctness and consistency (evenness, oddness, real or purely imaginary, etc.) are useful when you're doing calculations. Here are plots of $G(s)$ for the $a = 0.5, 1, 2$. Match them!

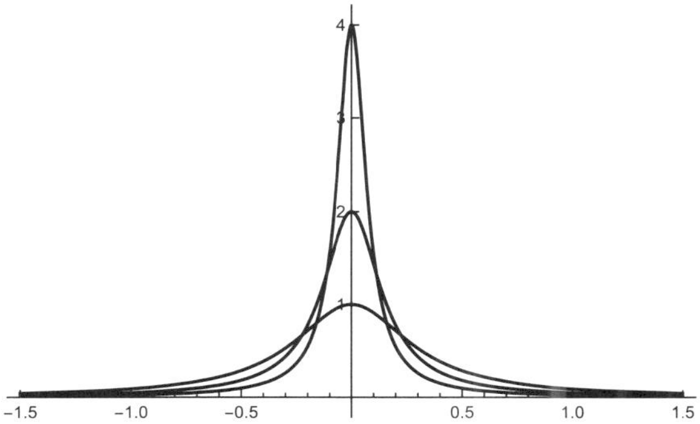

Down the road we'll see an application of the two-sided exponential decay to solving a second-order ordinary differential equation.

Example: Other Gaussians. As mentioned, there are other ways of normalizing a Gaussian. For example, instead of $e^{-\pi x^2}$ we can take

$$g(x) = \frac{1}{\sigma\sqrt{2\pi}} e^{-x^2/2\sigma^2} .$$

You might recognize this from applications to probability and statistics as the Gaussian with *mean* zero and *standard deviation* σ (or *variance* σ^2). The Gaussian with mean μ and standard deviation σ is the shifted version of this:

$$g(x, \mu, \sigma) = \frac{1}{\sigma\sqrt{2\pi}} e^{-(x-\mu)^2/2\sigma^2} .$$

Geometrically, σ is a measure of how peaked or spread out the curve is about the mean. In terms of the graph, the inflection points occur at $\mu \pm \sigma$; thus if σ is large, the curve is spread out, and if σ is small, the curve is sharply peaked. The area under the graph is still 1. We'll meet Gaussians again in the next chapter when we talk about the Central Limit Theorem.

The question here for us is what happens to the Fourier transform when the Gaussian is modified in this way. This can be answered by the results on shifts and stretches. Take the case of $\mu = 0$, for simplicity. To find the Fourier transform we can apply the similarity theorem: $f(ax) \rightleftharpoons (1/|a|)F(s/a)$. With $a = 1/\sigma\sqrt{2\pi}$ this gives

$$g(t) = \frac{1}{\sigma\sqrt{2\pi}} e^{-x^2/2\sigma^2} \quad \Rightarrow \quad \mathcal{F}g(s) = e^{-2\pi^2\sigma^2 s^2} ,$$

still a Gaussian, but not an exact replica of what we started with. Note that with $\mu = 0$ here the Gaussian is even and the Fourier transform is real and even.

2.3.6. The shift operator and the scaling operator. There are times when it's helpful to think about shifting and scaling as operations on a signal. We define the delay operator, or shift, or translation operator τ_b operating on a signal $f(x)$ by

$$\tau_b f(x) = f(x - b)$$

("τ" for "translation"). We define the scaling operator, or stretch operator, σ_a by

$$\sigma_a f(x) = f(ax)$$

("σ" for "scaling"[16]). You can also think in system terminology, where the input is a signal $f(x)$ and the output is the delayed signal $\tau_b f(x) = f(x-b)$, or the scaled signal $\sigma_a f(x) = f(ax)$.[17]

If nothing else, introducing these operators allows maybe a cleaner formulation of the shift theorem and the scaling theorem, i.e., less of a conflict over the variables:

$$\mathcal{F}(\tau_b f)(s) = e^{-2\pi i s b}\mathcal{F}f(s)\,,$$
$$\mathcal{F}(\sigma_a f)(s) = \frac{1}{|a|}\mathcal{F}f\Big(\frac{s}{a}\Big)\,.$$

We can even take the stretch theorem one step further, writing

$$\frac{1}{|a|}\mathcal{F}f\Big(\frac{s}{a}\Big) = \frac{1}{|a|}\sigma_{1/a}(\mathcal{F}f)(s)\,,$$

and doing so allows us to suppress the s variable completely, if you like that sort of thing:

$$\mathcal{F}(\sigma_a f) = \frac{1}{|a|}\sigma_{1/a}(\mathcal{F}f).$$

For the inverse Fourier transform,

$$\mathcal{F}^{-1}(\tau_b f)(t) = e^{2\pi i t b}\mathcal{F}^{-1}f(t)\,.$$

The stretch theorem for $\mathcal{F}^{-1}$ in terms of σ_a just swaps $\mathcal{F}^{-1}$ for $\mathcal{F}$.

Confusion often reigns — I have seen it — when shifting and scaling are combined, as in $\tau_b(\sigma_a f)$ versus $\sigma_a(\tau_b f)$. These are *not* the same. Let's ease into that, next.

Combining shifts and stretches. We can combine the shift theorem and the stretch theorem to find the formula for the Fourier transform of $f(ax-b)$, but let's have an example at the ready.

It's easy to find the Fourier transform of $f(x) = \Pi((x-3)/2) = \Pi(\frac{1}{2}x - \frac{3}{2})$ by direct integration. The shifted, scaled rect function is nonzero only for $2 < t < 4$, and thus

$$\begin{aligned} F(s) &= \int_2^4 e^{-2\pi i s x}\,dx \\ &= -\frac{1}{2\pi i s}e^{-2\pi i s x}\Big]_{x=2}^{x=4} = -\frac{1}{2\pi i s}(e^{-8\pi i s} - e^{-4\pi i s})\,. \end{aligned}$$

We can still bring in the sinc function, but the factoring is a little trickier:

$$e^{-8\pi i s} - e^{-4\pi i s} = e^{-6\pi i s}(e^{-2\pi i s} - e^{2\pi i s}) = e^{-6\pi i s}(-2i)\sin 2\pi s\,.$$

[16]This has nothing to do with σ as a standard deviation, above. Sorry, there are only so many letters.

[17]Not to be pedantic about it, but you could write $(\tau_b f)(x) = f(x-b)$, $(\sigma_a f)(x) = f(ax)$, the placement of parentheses indicating that you are acting on a signal f and evaluating the result at x according to the formulas. Like writing $(\mathcal{F}f)(s)$ instead of just $\mathcal{F}f(s)$.

Going through that is part of the point of the example. Plugging in gives

$$F(s) = e^{-6\pi is}\frac{\sin 2\pi s}{\pi s} = 2e^{-6\pi is}\operatorname{sinc} 2s\,.$$

The Fourier transform has become complex because shifting the rect function has destroyed its symmetry.

Here's a plot of $\Pi((x-3)/2)$ and of $4\operatorname{sinc}^2 2s$, the square of the *magnitude* of its Fourier transform. Once again, looking at the latter gives you no information on the phases in the spectrum, only on the energies.

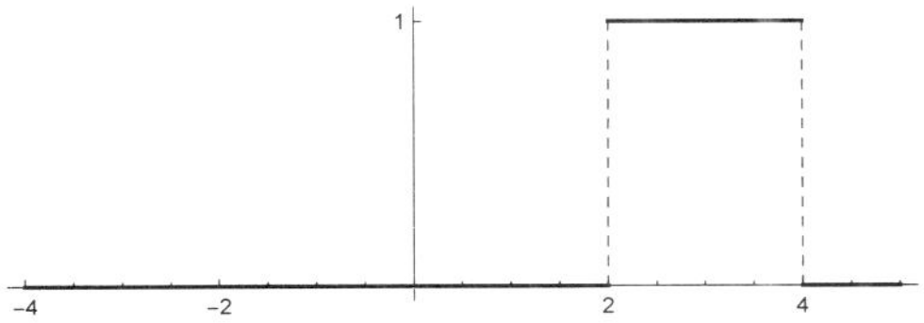

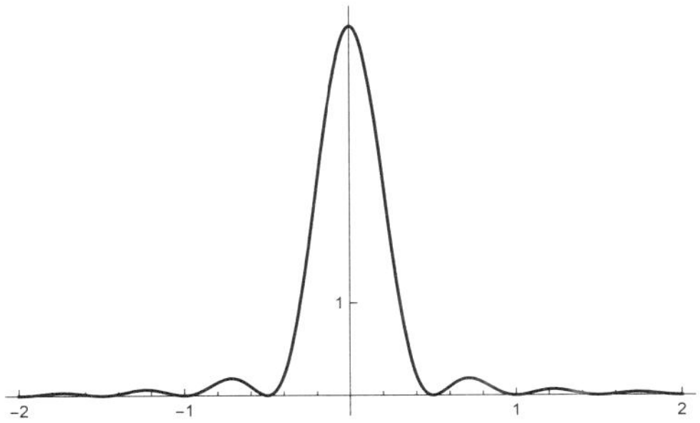

Now let's do the general formula. For the Fourier transform of $f(ax-b)$ it's a matter of making a substitution in the Fourier transform integral. We're now used to this so we'll be less detailed:

$$\begin{aligned}\int_{-\infty}^{\infty} e^{-2\pi isx} f(ax-b) &= \frac{1}{|a|}\int_{-\infty}^{\infty} e^{-2\pi i(\frac{1}{a}u+\frac{b}{a})s} f(u)\,du \quad (\text{letting } u = ax-b)\\ &= \frac{1}{|a|}e^{-2\pi isb/a}\int_{-\infty}^{\infty} e^{-2\pi isu/a} f(u)\,du\\ &= \frac{1}{|a|}e^{-2\pi isb/a}\mathcal{F}f\left(\frac{s}{a}\right).\end{aligned}$$

In its full glory:

- If $f(x) \rightleftharpoons F(s)$, then $f(ax-b) \rightleftharpoons \dfrac{1}{|a|}e^{-2\pi isb/a}F\left(\dfrac{s}{a}\right)$.

Try this on $\Pi((x-3)/2) = \Pi\left(\frac{1}{2}x - \frac{3}{2}\right)$. With $a = 1/2$ and $b = 3/2$ we get

$$\mathcal{F}\left(\Pi\left(\tfrac{1}{2}x - \tfrac{3}{2}\right)\right) = 2e^{-6\pi is}\mathcal{F}\Pi(2s) = 2e^{-6\pi is}\operatorname{sinc} 2s$$

just like before.

And for $F = \mathcal{F}^{-1} f$:

- If $F(s) \rightleftharpoons f(x)$, then $F(as-b) \rightleftharpoons \dfrac{1}{|a|} e^{2\pi i t b/a} f\Big(\dfrac{t}{a}\Big)$,

the difference being no minus sign in the complex exponential phase factor.

Finally, let's bring in the shift and scaling operators. There are two ways to use σ and τ, depending on the order; yes, they lead to the same answer. On the one hand,

$$f(ax-b) = \sigma_a(\tau_b f)(x),$$

because (note the parentheses and the order of operations!)

$$\sigma_a(\tau_b f)(x) = (\tau_b f)(ax) = f(ax-b).$$

Then

$$\mathcal{F}(\sigma_a(\tau_b f))(s) = \frac{1}{|a|}\mathcal{F}(\tau_b f)\Big(\frac{s}{a}\Big) = \frac{1}{|a|} e^{-2\pi i b(s/a)} \mathcal{F} f\Big(\frac{s}{a}\Big).$$

So far so good. On the other hand,

$$f(ax-b) = \tau_{b/a}(\sigma_a f)(x),$$

because

$$\tau_{b/a}(\sigma_a f)(x) = (\sigma_a f)\Big(x - \frac{b}{a}\Big) = f\Big(a\Big(x - \frac{b}{a}\Big)\Big) = f(ax-b).$$

Then

$$\mathcal{F}(\tau_{b/a}(\sigma_a f))(s) = e^{-2\pi i (b/a)s}\mathcal{F}(\sigma_a f)(s) = e^{-2\pi i (b/a)s}\frac{1}{|a|}\mathcal{F} f\Big(\frac{s}{a}\Big).$$

It works. Was there any doubt?

2.3.7. Parseval's identity. We mentioned Parseval's identity earlier, in passing. It says

$$\int_{-\infty}^{\infty} |f(t)|^2\,dt = \int_{-\infty}^{\infty} |\mathcal{F} f(s)|^2\,ds$$

and is the analog for Fourier transforms of Rayleigh's identity for Fourier series. Here we'll derive a more general statement (also referred to as Parseval's identity):

$$\int_{-\infty}^{\infty} f(t)\overline{g(t)}\,dt = \int_{-\infty}^{\infty} \mathcal{F} f(s)\overline{\mathcal{F} g(s)}\,ds.$$

Setting $f = g$ gives the first version.

Hold on; here we go:

$$\int_{-\infty}^{\infty} f(t)\overline{g(t)}\,dt = \int_{-\infty}^{\infty} f(t)\overline{\left(\int_{-\infty}^{\infty} e^{2\pi ist}\mathcal{F}g(s)\,ds\right)}\,dt$$

(applying Fourier inversion to g and $\mathcal{F}g$; that's the key)

$$= \int_{-\infty}^{\infty} f(t)\left(\int_{-\infty}^{\infty} e^{-2\pi ist}\overline{\mathcal{F}g(s)}\,ds\right)dt$$

(complex conjugation doing what it does)

$$= \int_{-\infty}^{\infty}\int_{-\infty}^{\infty} f(t)\overline{\mathcal{F}g(s)}e^{-2\pi ist}\,ds\,dt$$

$$= \int_{-\infty}^{\infty}\left(\int_{-\infty}^{\infty} f(t)e^{-2\pi ist}\,dt\right)\overline{\mathcal{F}g(s)}\,ds$$

(change the order of integration and swap double and iterated integrals)

$$= \int_{-\infty}^{\infty} \mathcal{F}f(s)\overline{\mathcal{F}g(s)}\,ds.$$

Beautiful. Needless to say, there's a whole assortment of mathematical hypotheses needed for the various steps. But not needed by us. Not for now, at least.

Parseval's identity for the inverse Fourier transform is the same as for the Fourier transform:

$$\int_{-\infty}^{\infty} f(t)\overline{g(t)}\,dt = \int_{-\infty}^{\infty} \mathcal{F}^{-1}f(s)\overline{\mathcal{F}^{-1}g(s)}\,ds.$$

If Parseval's identity seems reminiscent to you of inner products and norms, your memory serves you well. The inner product of two functions, defined on $\mathbb{R}$, is

$$(f,g) = \int_{-\infty}^{\infty} f(x)\overline{g(x)}\,dx\,,$$

when the integral makes sense, and the norm is

$$\|f\| = \left(\int_{-\infty}^{\infty} |f(x)|^2\,dx\right)^{1/2}.$$

Thus Parseval's identity is often written as

$$(f,g) = (\mathcal{F}f, \mathcal{F}g)\,,$$

and then also

$$\|f\| = \|\mathcal{F}f\|.$$

We'll discuss this further, later.

2.3.8. Derivative theorems. Some might say that I've saved the best property for last. There are two derivative theorems, each of which speaks to the important and surprising property that the Fourier transform connects differentiation to multiplication:

$$\frac{d}{ds}\mathcal{F}f(s) = \mathcal{F}(-2\pi itf(t))\,,$$

$$\mathcal{F}(f')(s) = 2\pi is\mathcal{F}f(s)\,.$$

For the first,

$$\begin{aligned}\frac{d}{ds}\mathcal{F}f(s) &= \frac{d}{ds}\int_{-\infty}^{\infty} e^{-2\pi ist} f(t)\,dt \\ &= \int_{-\infty}^{\infty} \left(\frac{\partial}{\partial s} e^{-2\pi ist}\right) f(t)\,dt \\ &= \int_{-\infty}^{\infty} (-2\pi it\, e^{-2\pi ist}) f(t)\,dt \\ &= \int_{-\infty}^{\infty} e^{-2\pi ist}(-2\pi itf(t))\,dt \\ &= \mathcal{F}(-2\pi itf(t)).\end{aligned}$$

The mathematical issue at stake is interchanging differentiation with integration. The notational issue at stake is that there's no good way of including the variable s on the right-hand side. For both issues you just have to know what you're doing. Such are the challenges.

For the second, take the Fourier transform of the difference quotient of $f(t)$:

$$\begin{aligned}\int_{-\infty}^{\infty} e^{-2\pi ist}\frac{f(t+h)-f(t)}{h}\,dt &= \frac{1}{h}\left(\int_{-\infty}^{\infty} e^{-2\pi ist} f(t+h)\,dt - \int_{-\infty}^{\infty} e^{-2\pi ist} f(t)\,dt\right) \\ &= \frac{1}{h}(e^{2\pi ish}\mathcal{F}f(s) - \mathcal{F}f(s)) \\ &\quad \text{(using the shift theorem!)} \\ &= \left(\frac{e^{2\pi ish}-1}{h}\right)\mathcal{F}f(s).\end{aligned}$$

To bring the derivative in we want to take the limit of the difference quotient. Now, what is

$$\lim_{h\to 0}\left(\frac{e^{2\pi ish}-1}{h}\right)?$$

It is

$$\left.\frac{d}{dt}e^{2\pi ist}\right|_{t=0} = 2\pi is\, e^{2\pi is0} = 2\pi is.$$

So,

$$\begin{aligned}\mathcal{F}(f')(s) &= \int_{-\infty}^{\infty} e^{-2\pi ist} \lim_{h\to 0}\left(\frac{f(t+h)-f(t)}{h}\right) dt \\ &= \lim_{h\to 0}\int_{-\infty}^{\infty} e^{-2\pi ist}\frac{f(t+h)-f(t)}{h}\,dt \\ &= \lim_{h\to 0}\left(\frac{e^{2\pi ish}-1}{h}\right)\mathcal{F}f(s) \\ &= 2\pi is\mathcal{F}f(s).\end{aligned}$$

The mathematical issue at stake this time is interchanging $\lim_{h\to 0}$ with integration. Such are the challenges.

We can apply each of these formulas repeatedly to obtain versions for higher derivatives:

$$\frac{d^k}{ds^k}\mathcal{F}f(s) = \mathcal{F}((-2\pi it)^k f(t)),$$
$$\mathcal{F}(f^{(k)})(s) = (2\pi is)^k \mathcal{F}f(s).$$

What are the formulas for the inverse Fourier transform? It's just a matter of changing the sign:

$$\frac{d}{ds}\mathcal{F}^{-1}f(s) = \mathcal{F}^{-1}(2\pi itf(t)),$$
$$\mathcal{F}^{-1}(f')(s) = -2\pi is\mathcal{F}^{-1}f(s).$$

2.4. Different Definitions of the Fourier Transform, and What Happens to the Formulas

I mentioned earlier that different fields may adopt different definitions of the Fourier transform and its inverse, perhaps better adapted to the applications, or maybe just to be difficult. Following the very helpful summary provided by T. W. Körner in his book *Fourier Analysis*,[18] I will indicate some of the many irritating variations. You can also find these sorts of lists on the web.

To be general, write

$$\mathcal{F}f(s) = \frac{1}{A}\int_{-\infty}^{\infty} e^{iBst} f(t)\, dt.$$

The choices that are most often found in practice are

$$\begin{aligned} A &= \sqrt{2\pi}, & B &= \pm 1, \\ A &= 1, & B &= \pm 2\pi, \\ A &= 1, & B &= \pm 1. \end{aligned}$$

The definition we've chosen has $A = 1$ and $B = -2\pi$.

For the general $\mathcal{F}$, with A and B unspecified, we have

$$\begin{aligned} \mathcal{F}\mathcal{F}f &= \frac{2\pi}{A^2|B|} f^-, \\ \mathcal{F}(f(t-b)) &= e^{iBbs}\mathcal{F}f(s), \\ \mathcal{F}(f(t/a)) &= \frac{1}{|a|}\mathcal{F}f\Big(\frac{s}{a}\Big) \quad \text{(unchanged)}, \\ \mathcal{F}(f') &= -iB\mathcal{F}f, \\ (\mathcal{F}f)' &= iB\mathcal{F}(tf(t)). \end{aligned}$$

[18]There called "Chase the Constant." Such an interesting book in many ways.

For Parseval's identity, let me just write out some cases:

$$A = 1,\ B = \pm 1 \ \Rightarrow \int_{-\infty}^{\infty} f(x)\overline{g(x)}\, dx = \frac{1}{2\pi}\int_{-\infty}^{\infty} \mathcal{F}f(s)\overline{\mathcal{F}g(s)}\, ds\,,$$

$$A = \sqrt{2\pi},\ B = \pm 1 \ \Rightarrow \int_{-\infty}^{\infty} f(x)\overline{g(x)}\, dx = \int_{-\infty}^{\infty} \mathcal{F}f(s)\overline{\mathcal{F}g(s)}\, ds\,,$$

$$A = \sqrt{2\pi},\ B = \pm 2\pi \ \Rightarrow \int_{-\infty}^{\infty} f(x)\overline{g(x)}\, dx = 2\pi\int_{-\infty}^{\infty} \mathcal{F}f(s)\overline{\mathcal{F}g(s)}\, ds\,.$$

There's also one for convolution, which we'll meet in the next chapter. The result is listed here for convenience:

$$\mathcal{F}(f * g) = A(\mathcal{F}f)(\mathcal{F}g)\,.$$

Realize, too, that changing the constants in the definition of $\mathcal{F}f$ will then change the formulas for particular Fourier transforms. For example,

$$\mathcal{F}\Pi(s) = \frac{1}{A}\frac{\sin(\frac{Bs}{2})}{\frac{Bs}{2}}\,.$$

Happy hunting and good luck out there.

Problems and Further Results

2.1. *Some practice combining scaled triangle functions*

Express each of the following as a sum of two shifted, scaled triangle functions $b_1\Lambda_{a_1}(t - c_1) + b_2\Lambda_{a_2}(t - c_2)$. Think of the sum as a "left-triangle" plus a "right-triangle" ("right" meaning to the right, not having an angle of 90°). For part (d), the values x_1, x_2, and x_3 cannot be arbitrary. Rather, to be able to express the plot as the sum of two Λ's they must satisfy a relationship that you should determine.

Why do this? It's superposition waiting to happen — the possibility of operating on a complicated signal by operating on simpler constituents and adding up the results. Here you are called upon to find the constituents.

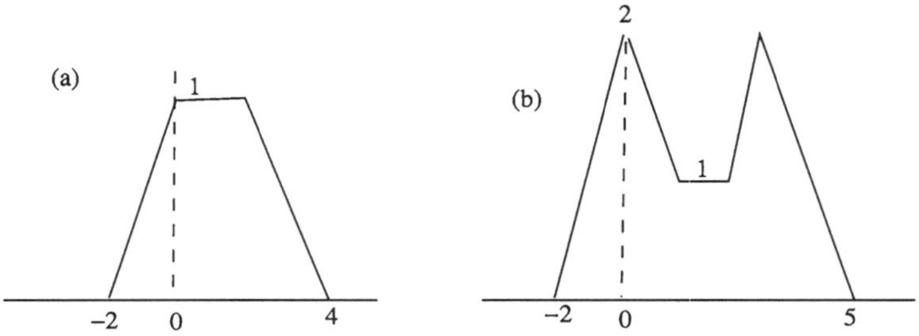

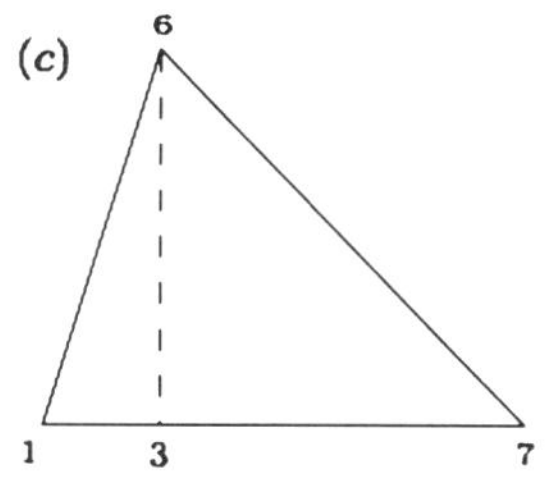

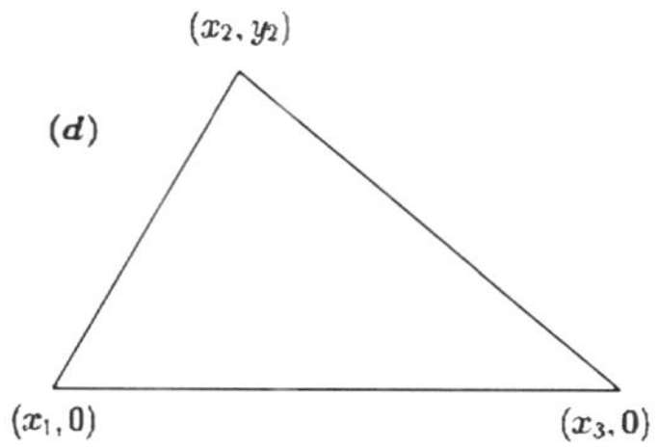

2.2. *From the graphs*

Find the Fourier transforms of $f(x)$ and $g(x)$ for each graph.

(a) A sinc function

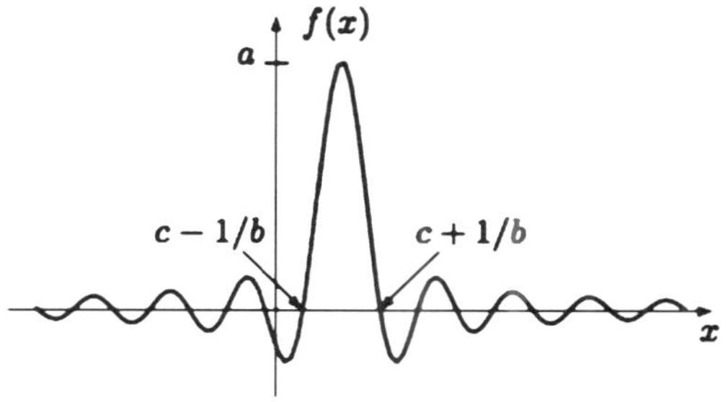

(b) $g(x)$

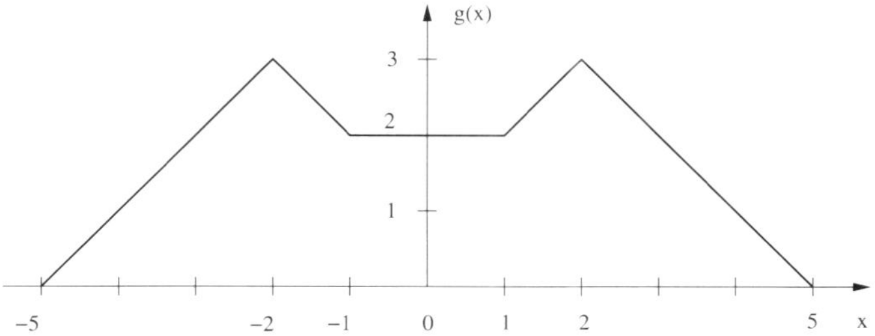

2.3. *Letter M*

Find the Fourier transform of the following signal.

(*Hint*: You can solve this problem without computing any integrals.)

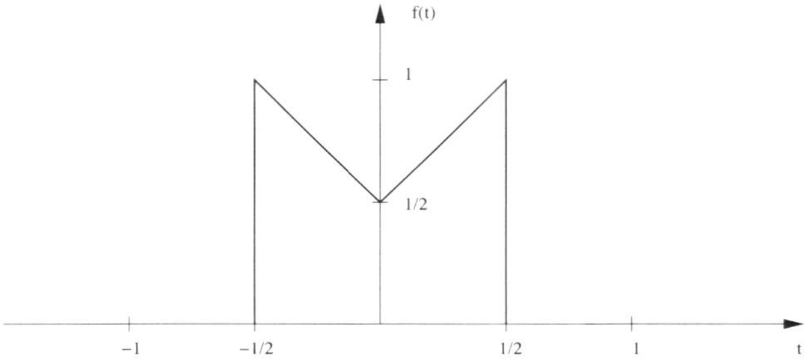

2.4. Find the Fourier transform of the house function, shown below.

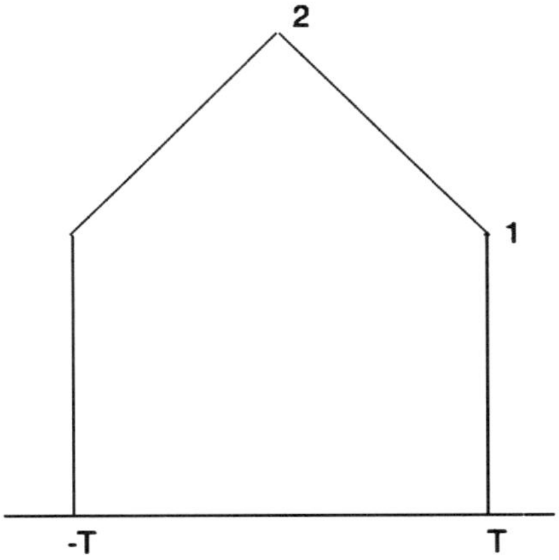

2.5. *Linearity and shifting properties of the Fourier transform*[19]
Suppose we are given the signal

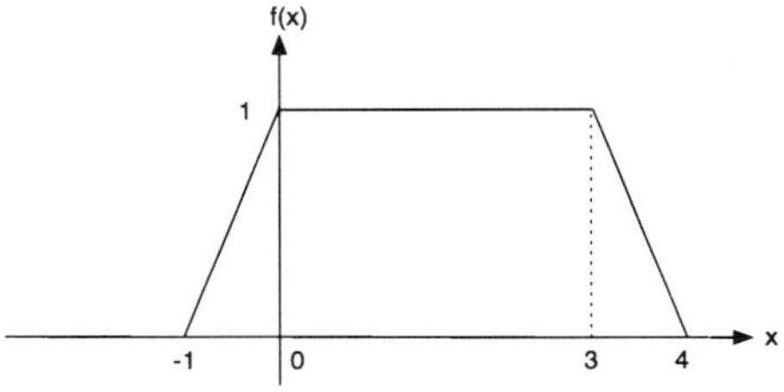

and we are told that its Fourier transform is

$$\mathcal{F}f(s) = -4\,\mathrm{sinc}^2(s)\sin(\pi s)\sin(2\pi s)e^{-i3\pi s}.$$

Using *ONLY* this information, find the Fourier transform of the following signals:
(a) $g(x)$

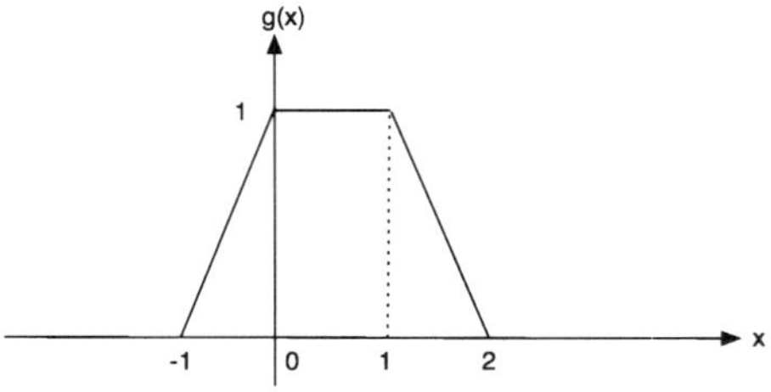

[19]From T. John.

(b) $h(x)$

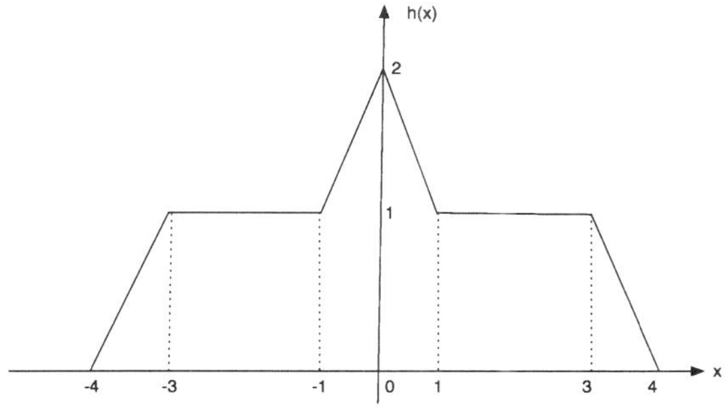

2.6. *Piecewise linear approximations and Fourier transforms*

(a) Find the Fourier transform of the following signal.

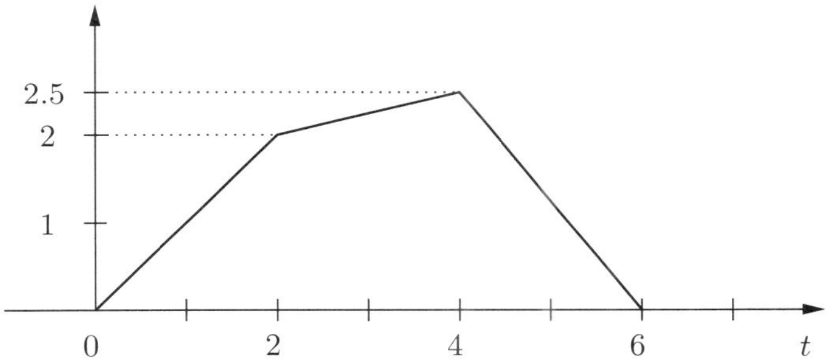

Hint: Think Λ's.

(b) Consider a signal $f(t)$ defined on an interval from 0 to D with $f(0) = 0$ and $f(D) = 0$. We get a uniform, piecewise linear approximation to $f(t)$ by dividing the interval into n equal subintervals of length $T = D/n$ and then joining the values $0 = f(0), f(T), f(2T), \ldots, f(nT) = f(D) = 0$ by consecutive line segments. Let $g(t)$ be the linear approximation of a signal $f(t)$, obtained in this manner, as illustrated in the following figure where $T = 1$ and $D = 6$.

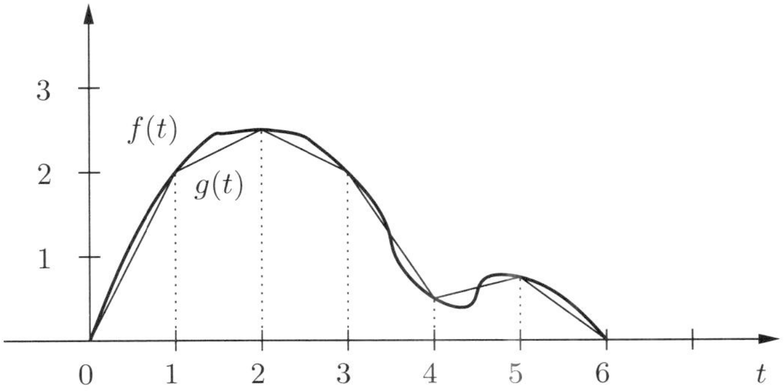

Find $\mathcal{F}g(s)$ for the general problem (*not* for the example given in the figure above) using any necessary information about the signal $f(t)$ or its Fourier transform $\mathcal{F}f(s)$. Think Λ's, again.

2.7. Two signals are plotted below. Without computing the Fourier transforms, determine if one can be the Fourier transform of the other. Explain your reasoning.

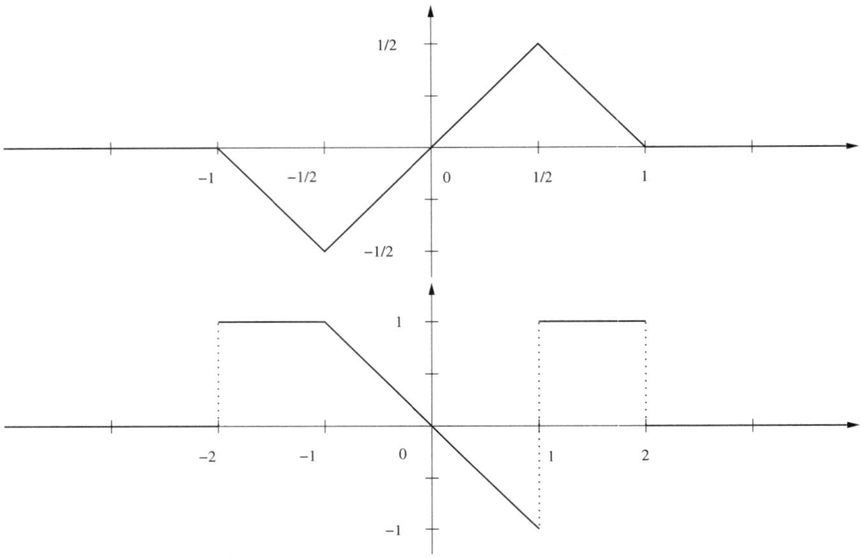

2.8. Let $f(x)$ be a signal whose Fourier transform $F(s)$ is plotted below.[20]

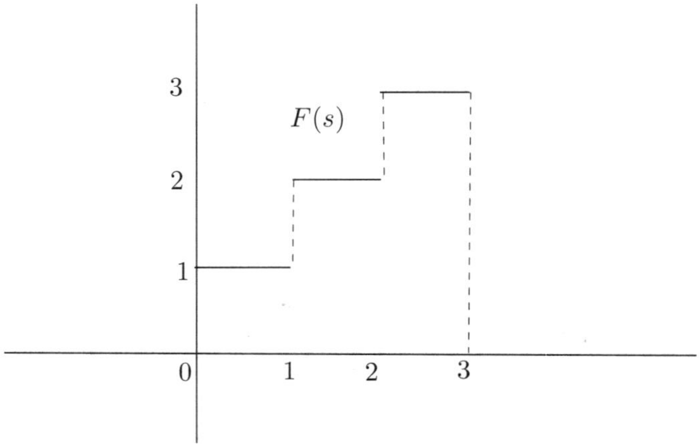

Sketch the graph of the Fourier transform of the following signals:
(a) $f(-x)$,
(b) $f(2x)$,
(c) $e^{4\pi i x} f(x)$,
(d) $(f * f)(x)$,
(e) $\frac{1}{2\pi i} f'(x)$.

[20] From D. Kammler.

2.9. *Derived signals*[21]

The figures below show a signal $f(t)$ (the first one, on top) and six other signals derived from $f(t)$. Note the scales on the axes. Suppose $f(t)$ has Fourier transform $F(s)$.

Express the other six signals, (a)–(f), in terms of $f(t)$ and express the Fourier transforms of these signals, (a)–(f), in terms of $F(s)$.

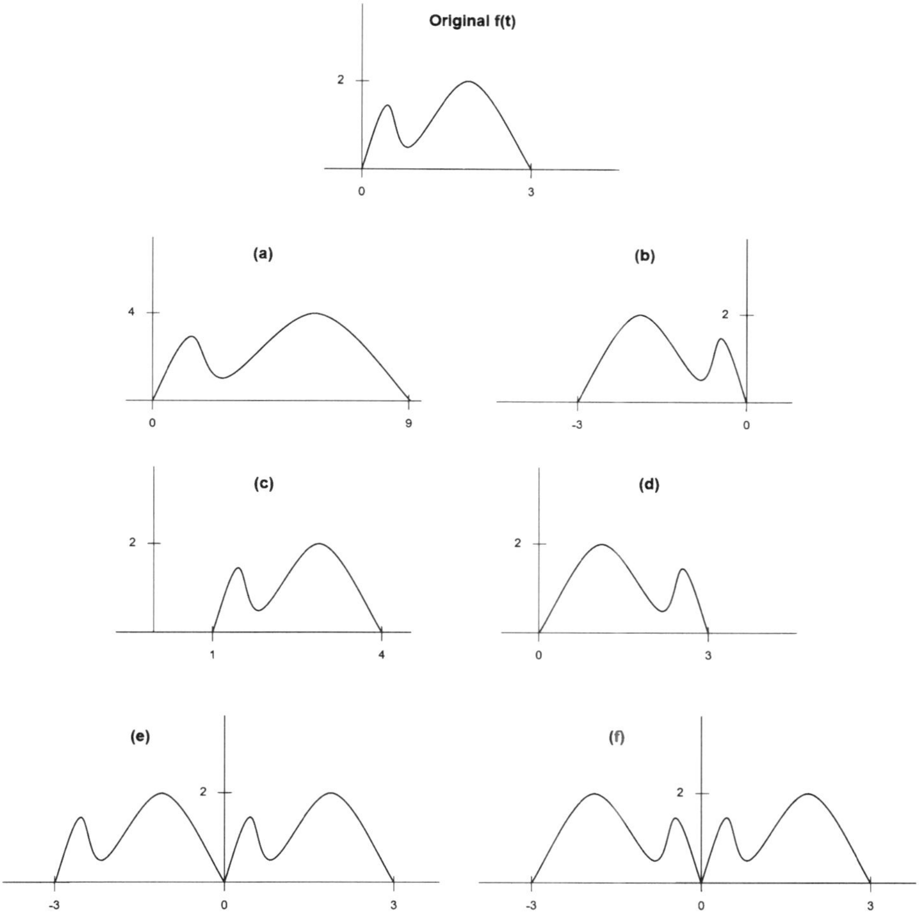

2.10. Suppose we have the real and odd function $f(t)$ and the magnitude of its Fourier transform $|\mathcal{F}f(s)|$. Can $f(t)$ have a $|\mathcal{F}f(s)|$ as shown below? Explain your reasoning.

[21]From D. Kammler, redrawn by Shalini Ranmuthu.

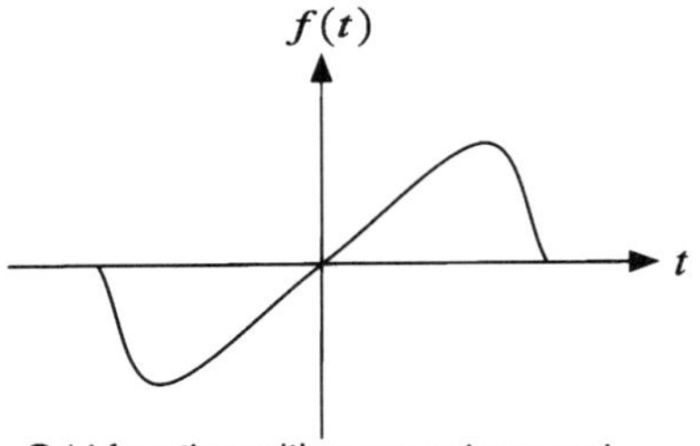

Odd function with compact support

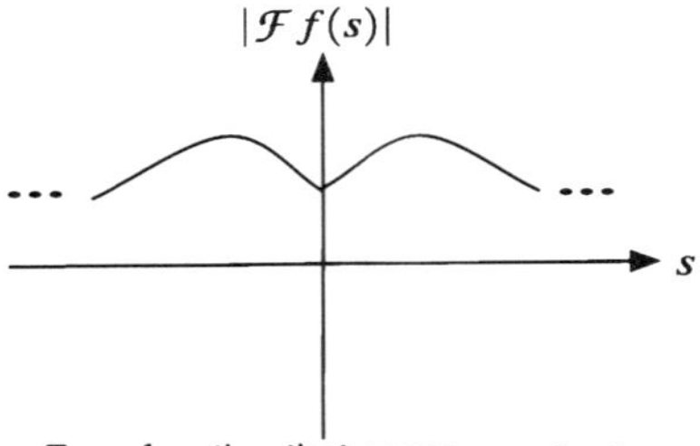

Even function that spans $-\infty < s < \infty$

("Compact support" means that the function is identically zero outside a finite interval. We'll see this condition later in other contexts.)

2.11. *Phase profile possible?*[22]

Suppose we have a complex function $g(t)$ and we construct a new function $f(t) = |g(t)|^2$. Can the phase profile $\angle \mathcal{F}f(s)$ be of the form shown below? Explain your reasoning.

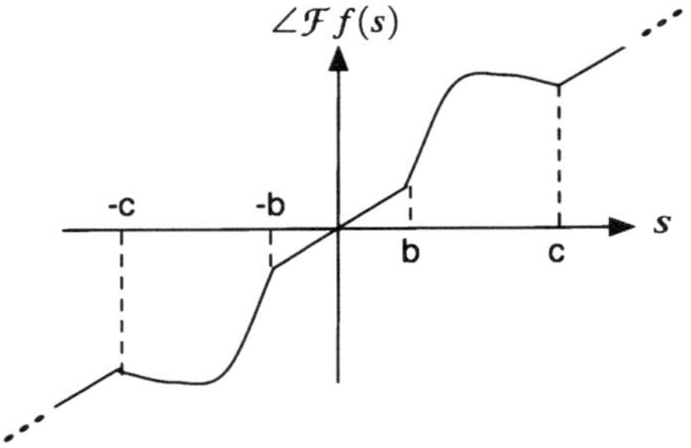

Odd function that spans $-\infty < s < \infty$

2.12. Suppose $f(t)$ is a real and even function. Can the phase profile $\angle \mathcal{F}f(s)$ be of the form shown below? If so, provide an example of such a function $f(t)$.

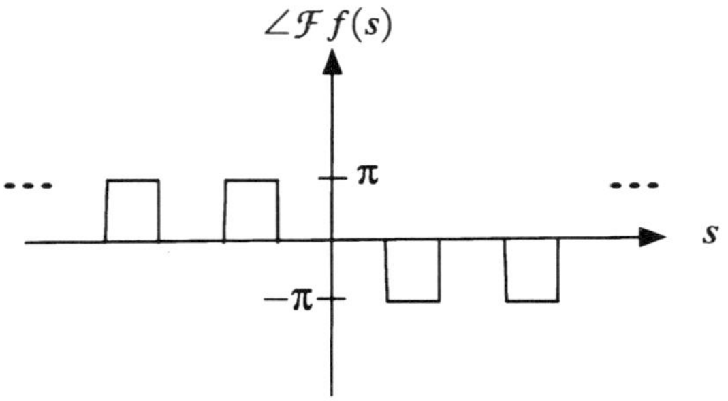

[22] From T. John.

2.13. Let $y(t)$ be a signal with Fourier transform $Y(s)$. Suppose we are given the following facts:
(a) $y(t)$ is real.
(b) $y(t) = 0$ for $t \leq 0$.
(c) $\mathcal{F}^{-1}\{\mathrm{Re}[Y(s)]\} = |t|e^{-|t|}$, where $\mathrm{Re}[Y(s)]$ denotes the real part of $Y(s)$.
Find $y(t)$.

2.14. *The modulation property of the Fourier transform*
(a) Let $f(t)$ be a signal, let s_0 be a number, and define

$$g(t) = f(t)\cos(2\pi s_0 t).$$

Show that

$$\mathcal{F}g(s) = \frac{1}{2}\mathcal{F}f(s - s_0) + \frac{1}{2}\mathcal{F}f(s + s_0).$$

(No δ-functions, please, for those who know about them.)
(b) Find the signal (in the time domain) whose Fourier transform is pictured below.

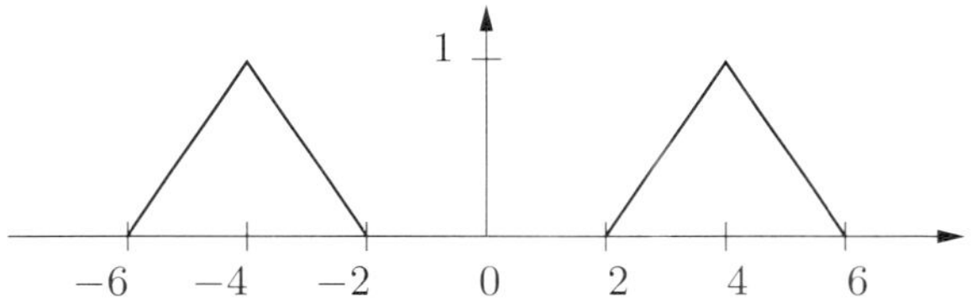

2.15. Let $x(t)$ be a signal whose spectrum is identically zero outside the range $-S_0 \leq s \leq S_0$, for example:

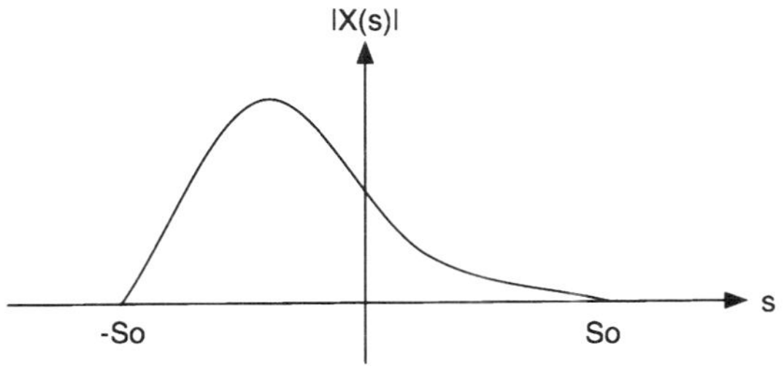

For the following signals, determine the range over which their spectrum is nonzero:
(a) $x(t) + x(t-1)$,
(b) $\frac{dx(t)}{dt}$,
(c) $x(t)\cos(2\pi S_0 t)$,
(d) $x(t)e^{i2\pi b_0 t}$, where b_0 is a positive real constant.

2.16. *Fourier transforms and Fourier coefficients*

Suppose the function $f(t)$ is zero outside the interval $-1/2 \leq t \leq 1/2$. We periodize $f(t)$ to have period 1 by forming

$$g(t) = \sum_{k=-\infty}^{\infty} f(t-k) .$$

Now take the Fourier series representation of $g(t)$:

$$g(t) = \sum_{n=-\infty}^{\infty} \hat{g}(n) e^{2\pi i n t} .$$

Find the relationship between the Fourier transform $\mathcal{F}f(s)$ and the Fourier series coefficients $\hat{g}(n)$.

2.17. Consider the functions $g(x)$ and $h(x)$ shown below. These functions have Fourier transforms $\mathcal{F}g(s)$ and $\mathcal{F}h(s)$, respectively.

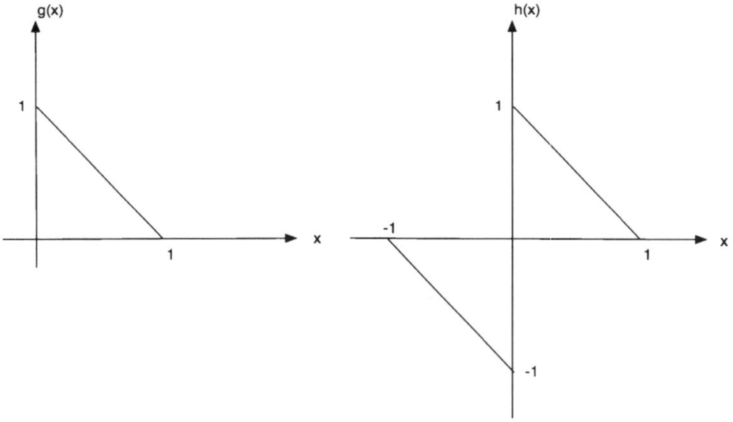

You are told that the imaginary part of $\mathcal{F}g(s)$ is $(\sin(2\pi s) - 2\pi s) / (4\pi^2 s^2)$.

(a) What are the two possible values of $\angle \mathcal{F}h(s)$, i.e., the phase of $\mathcal{F}h(s)$? Express your answer in radians.

(b) Evaluate $\int_{-\infty}^{\infty} \mathcal{F}g(s) \cos(\pi s)\, ds$.

(c) Evaluate $\int_{-\infty}^{\infty} \mathcal{F}h(s) e^{i4\pi s}\, ds$.

(d) Without performing any integration, what is the real part of $\mathcal{F}g(s)$? Explain your reasoning.

(e) Without performing any integration, what is $\mathcal{F}h(s)$? Explain your reasoning.

(f) Suppose $h(x)$ is periodized to have period $T = 2$. Without performing any integration, what are the Fourier coefficents, c_k, of this periodic signal?

2.18. *Periodizations and boundary conditions*[23]

Let $f(x)$ be a real continuously differentiable function defined on the interval $(0,1)$ and zero everywhere else. Recall from the Problems and Further Results section in Chapter 1 that the *periodization* of $f(x)$ is

$$f_{\text{per}}(x) = \sum_{n=-\infty}^{\infty} f(x-n),$$

so that $f_{\text{per}}(x)$ is periodic with period 1.

We now define the *even periodization* of $f(x)$ and the *odd periodization* of $f(x)$ by

$$f_{\text{even per}}(x) = \sum_{n=-\infty}^{\infty} f(x-2n) + f(-x-2n),$$

$$f_{\text{odd per}}(x) = \sum_{n=-\infty}^{\infty} f(x-2n) - f(-x-2n).$$

Notice that both $f_{\text{even per}}(x)$ and $f_{\text{odd per}}(x)$ are periodic functions with period 2. Also note that in the interval $0 < x < 1$, all three periodizations are identical.

Assume that $f(x)$ has Fourier transform $F(s)$ and that the following Fourier series representations exist:

$$f_{\text{even per}}(x) = \sum_{n=0}^{\infty} a_n \cos(\pi n x),$$

$$f_{\text{odd per}}(x) = \sum_{n=0}^{\infty} b_n \sin(\pi n x),$$

$$f_{\text{per}}(x) = \sum_{n=-\infty}^{\infty} c_n e^{2\pi i n x}.$$

(a) Express b_n in terms of $F(s)$.

(b) A guitar string of unit length is stretched along the x-axis. Let $u(x,t)$ represent the height of the string as a function of length x and time t. Assume that the guitar string is fastened at its ends so that $u(0,t) = 0$ and $u(1,t) = 0$. The guitar string is plucked at $t = 0$ and then released. For $t > 0$, the height of the guitar string satisfies the wave equation

$$u_{tt} = u_{xx}.$$

Explain why, to solve this equation, we should substitute the odd periodization

$$u(x,t) = \sum_{n=0}^{\infty} d_n(t) \sin(\pi n x)$$

into the wave equation and not one of the other series. What differential equation should $d_n(t)$ satisfy? (You do not need to actually solve the differential equation.)

[23] From Raj Bhatnagar.

(c) Now assume $f(x)$ is as before and additionally that $f(0) \neq f(1)$. Consider the following approximations to $f(x)$:

$$\text{(i)} \quad \sum_{n=0}^{N} a_n \cos(\pi n x),$$

$$\text{(ii)} \quad \sum_{n=0}^{N} b_n \sin(\pi n x),$$

$$\text{(iii)} \quad \sum_{n=-N}^{N} c_n e^{2\pi i n x}.$$

where a_n, b_n, and c_n are as before and N is a fixed big integer. Which approximation will be best in terms of minimizing error?

MP3 compression commonly begins by dividing an audio signal into chunks of 12 ms, taking the (discrete) cosine transform of each chunk, and passing each result through a low-pass filter. Can you explain why audio engineers prefer to use the cosine transform instead of the Fourier transform?

2.19. *Energy of a bandlimited signal*

The energy of a signal $g(t)$ is the integral

$$\int_{-\infty}^{\infty} |g(t)|^2 \, dt \, .$$

Suppose that $g(t)$ satisfies

$$\mathcal{F}g(s) = 0 \, , \quad |s| \geq \frac{1}{2} \, .$$

One says that $g(t)$ is *bandlimitied*, a class of signals we shall study extensively when we talk about sampling and interpolation. Express the energy of $g(t)$, in terms of the values of $g(t)$ at the integers, $g(n)$, $n = 0, \pm 1, \pm 2, \ldots$. *Hint*: Use the Fourier series of the periodization of $\mathcal{F}g(s)$ of period 1.

2.20. *Voice scrambling/descrambling*[24]

Eva and Albert are designing an encryption circuit for voice communications. One day, Albert comes to Eva with an encryption scheme.

Albert: Eva, I've got a new interesting encryption scheme.

Eva: What is it?

Albert: It's a scrambler! It divides the entire frequency band into several equal sub-bands and interchanges the sub-bands according to a predetermined key. For example, suppose we have 8 sub-bands. Let sub-band A correspond to frequencies between 0 and 1 kHz. Then, sub-band B corresponds to frequencies between 1 and 2 kHz, sub-band C corresponds to frequencies between 2 and 3 kHz, and so on. At the end, sub-band H corresponds to frequencies between 7 and 8 kHz. The original order of the sub-bands is ABCDEFGH. This scrambling technique will be to reorder the sub-bands

[24] From M. Kamenetsky.

to FBHECDGA or HGFEDCBA or any other predetermined order. It is difficult to recover the original signal without knowing the predetermined key.

Eva: Interesting. But can you just try all possible sequences?

Albert: It is not that simple. Suppose we have scrambled a speech signal with 8 sub-bands. The number of possible orders is 8! = 40,320. Do you really want to create 40,320 sound files and listen to them?

Eva: No way. That's too many trials. By the way, isn't it a well-known fact that the overwhelming majority of the power spectrum of human voice is concentrated in the low frequency bands? Maybe I can descramble your signal before trying all 8! cases.

(a) The goal of this problem is for you to design a MATLAB program that will descramble their message. You may obtain the scrambled signal (`scramble.mat`) from the website. It has been scrambled in the fashion described above by rearranging the equally divided 8 sub-bands. Explain how to reconstruct the original signal without listening to 8! sound files. Do you really need to know the exact predetermined key in order to get the original signal? *Hint*: Eva's last argument and a plot of the signal in the frequency domain will be helpful.

(b) Descramble this signal and transcribe the first and last 5 words. *Hint*: You will need to use MATLAB's `fft` and `ifft` (Fast Fourier Transform and inverse Fast Fourier Transform, which we will learn about later). The way MATLAB computes the `fft`, when you take the `fft` of the scrambled signal, you will get a DC component, $\mathcal{F}f(0)$, assuming the signal is $f(t)$, and 16 bands: $\mathcal{F}f(0)$ AB...H $\bar{\text{H}}\cdots\bar{\text{B}}\bar{\text{A}}$. You want to reorder this, for example, $\mathcal{F}f(0)$ HG...A $\bar{\text{A}}\cdots\bar{\text{G}}\bar{\text{H}}$. Please be aware that the DC component is not a part of a sub-band, and each band has to have a complex conjugated pair on the other side. You might have a better idea of what's going on if you plot the magnitude (`abs`), real part, or imaginary part of the scrambled/descrambled spectrum.

(c) Is this a good encryption scheme (i.e., could an eavesdropper who has intercepted the scrambled message but who does not know the scrambling sequence still decode the message)? If so, give a brief explanation of why. If not, explain its deficiencies and suggest a way to improve its resistance to attacks.

2.21. *Rectangular wave with an arbitrary duty cycle*[25]

Consider a periodic rectangular wave with period $T = 1$. Its *duty cycle* is the ratio between its high time T_{high} (time when the rectangular wave is equal to its maximum value during the period) and its overall period T. The duty cycle is represented using parameter a given by

$$a = T_{\text{high}}/T.$$

Note that $0 < a < 1$.

See the figure below for a rectangular wave with $a = 0.5$.

[25] From A. Mutapcic.

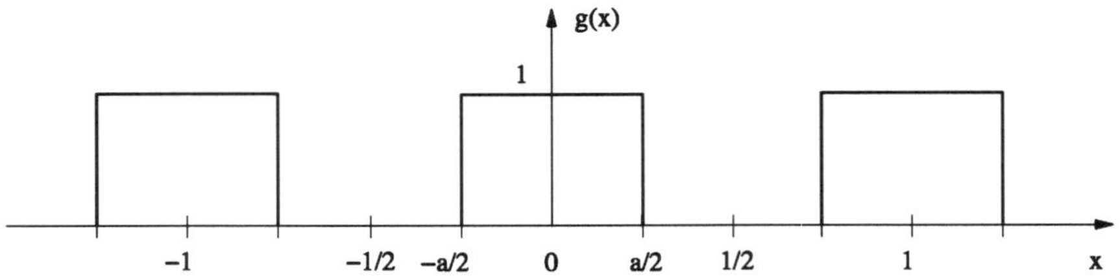

(a) Find the Fourier series coefficients, $\hat{f}(n)$, for a rectangular wave with an arbitrary duty cycle as determined by parameter a.
You can solve this problem using a direct Fourier series approach or by sampling the Fourier transform of a single cycle of the given periodic function (as derived in Problem 2.16).

(b) Find the energy in all the harmonics excluding the constant (DC) term; i.e., compute the quantity

$$E_h = \sum_{\substack{n=-\infty \\ n\neq 0}}^{\infty} |\hat{f}(n)|^2.$$

2.22. *Periodizing and half-wave symmetry*
Let $f(t)$ be a function that is identically 0 outside of the interval $(0, T)$. As a variant of the usual way to periodize $f(t)$ consider the alternating sum

(*) $$g(t) = \sum_{n=-\infty}^{\infty} (-1)^n f(t - nT).$$

(a) For the function $f(t)$ shown below, sketch $g(t)$ when $0 \leq t \leq 3T$.

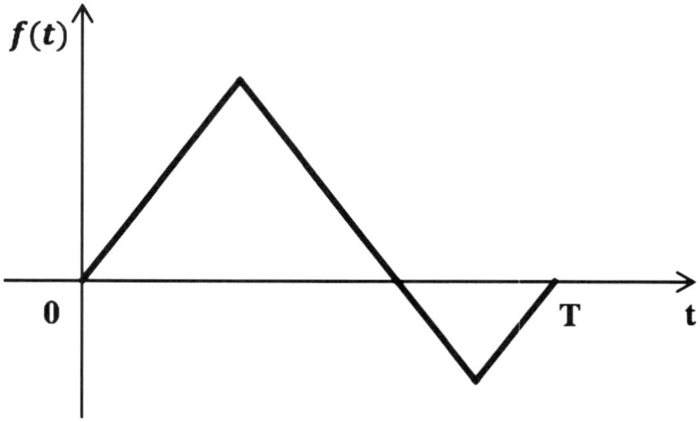

The following questions pertain to a general function $g(t)$ as defined in (*), not the particular example in part (a).

(b) For $g(t)$ as in equation (*), show that $g(t)$ has half-wave symmetry,

$$g(t + T) = -g(t).$$

(c) Show that $g(t)$ has period $2T$,

$$g(t + 2T) = g(t).$$

You can either use part (b) or show this directly from the definition in equation (*).

(d) What are the Fourier coefficients of $g(t)$ in terms of $\mathcal{F}f(s)$? As a check on your work, you should get that half of the Fourier coefficients are 0.

2.23. *Fourier transform properties and graphs, again*

A signal $f(t)$ has the *Fourier transform* $F(s)$ shown below. Answer the following questions for the signals $f_1(t)$, $f_2(t)$, and $f_3(t)$, also shown below. *Note*: $f_3(t)$ is a *periodic* signal.

$F(s)$

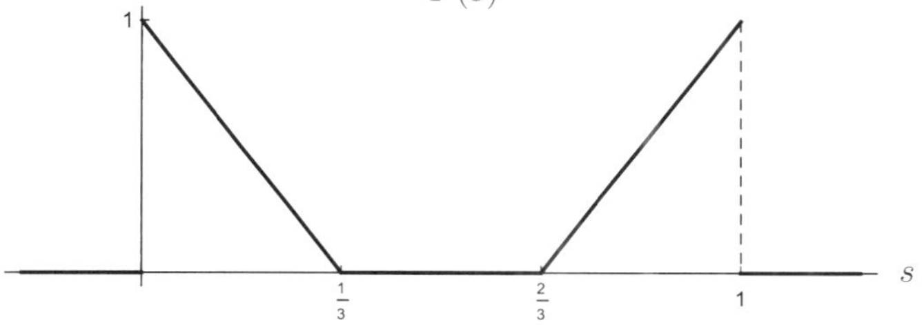

(a) Find the Fourier transform of $f_1(t)$ in terms of f.
(b) Find the Fourier transform of $f_2(t)$ in terms of f.
(c) Find the Fourier coefficients of the periodic signal $f_3(t)$ in terms of f.

$f_1(t)$

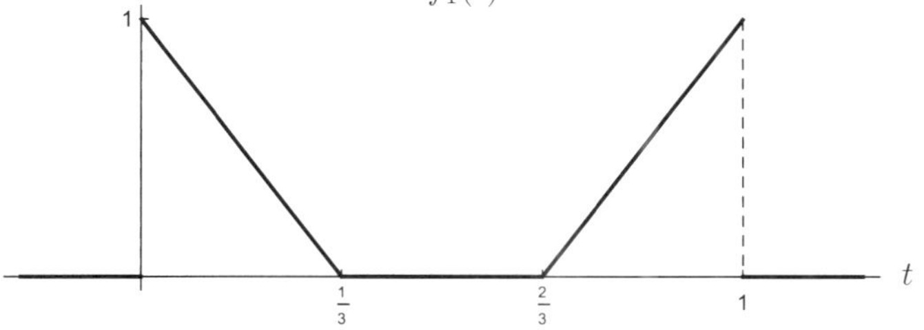

$f_2(t)$

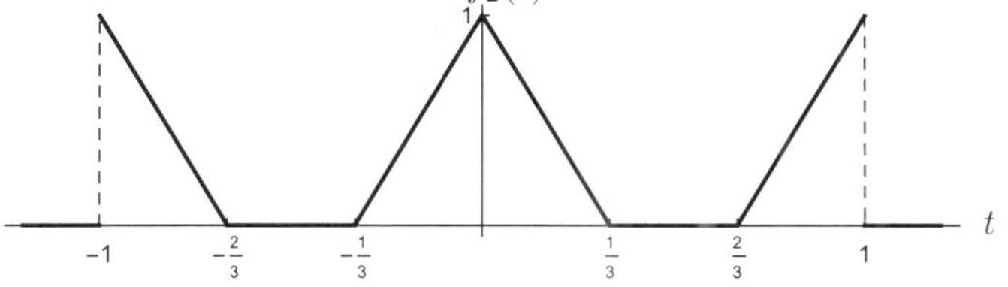

$f_3(t)$

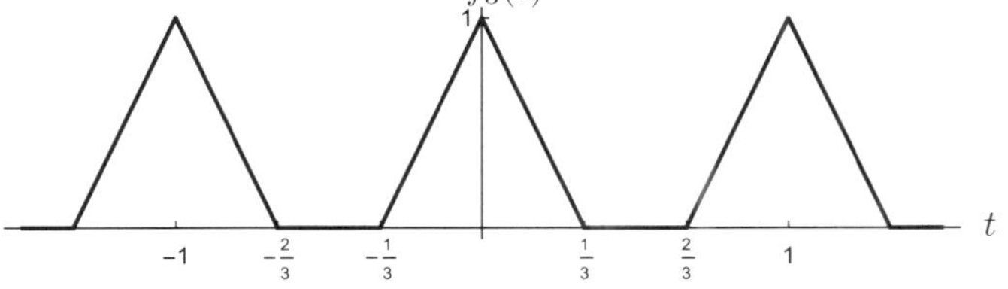

2.24. *Fourier transform values from a graph*

The function $F(s)$ sketched below is the Fourier transform of an unknown function $f(x)$.

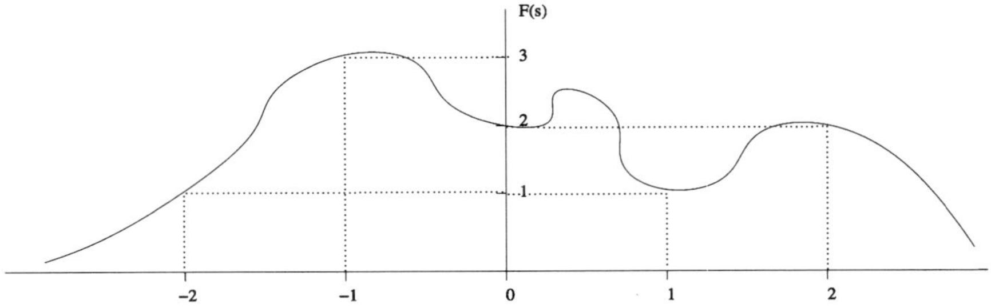

Evaluate the following integrals. Your answers should be numbers, not functions!

(a)
$$\int_{-\infty}^{\infty} f(x)\, dx.$$

(b)
$$\int_{-\infty}^{\infty} f(x)e^{-2\pi i x}\, dx.$$

(c)
$$\int_{-\infty}^{\infty} \operatorname{Re}\{f(x)\}e^{4\pi i x} dx.$$

(d)
$$\int_{-\infty}^{\infty} (f * \Pi)(x) dx.$$

(e)
$$\int_{-\infty}^{\infty} f(x)\cos(2\pi x)\, dx.$$

(f)
$$\int_{-\infty}^{\infty} f(x)\cos(2\pi x)e^{2\pi i x}\, dx.$$

2.25. Sketch the derivative of the function $f(x)$ shown below. Then using the derivative theorem for Fourier transforms, find the Fourier transform of f.

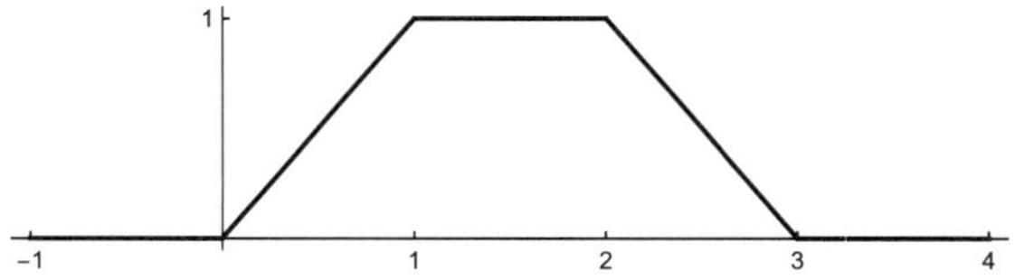

2.26. *Scale invariance*[26]

A function f is called *scale invariant* if there exists a number $\lambda > 0$, called the *scaling factor*, and a number d, called the *scaling dimension*, such that

$$f(\lambda x) = \lambda^d f(x).$$

(Google "scale invariance.")

(a) Is the Fourier transform of a scale invariant function f scale invariant? If so, how are the scaling dimension and scaling factors of f related to those of $\mathcal{F}f$?

(b) If f and g are scale invariant with the same scaling factor λ, is $f * g$ scale invariant? If so, what are its scaling factors and scaling dimension?

2.27. Evaluate the following integrals with the help of Fourier transform theorems:

(a)

$$\int_{-\infty}^{\infty} \operatorname{sinc}^4(t)\, dt,$$

(b)

$$\int_{-\infty}^{\infty} \frac{2}{1+(2\pi t)^2} \operatorname{sinc}(2t)\, dt.$$

2.28. *Alternate Fourier transform conventions*

Our definition of the Fourier transform is

$$F(s) = \mathcal{F}f(s) = \int_{-\infty}^{\infty} f(t)e^{-2\pi i s t} dt,$$

and the corresponding inverse Fourier transform is

$$f(t) = \mathcal{F}^{-1}F(t) = \int_{-\infty}^{\infty} F(s)e^{2\pi i s t} ds.$$

At an international conference you meet a physicist who defines a different Fourier transform:

$$G(\omega) = \mathcal{G}f(\omega) = \int_{-\infty}^{\infty} f(t)e^{i\omega t} dt.$$

(Note the sign of the exponent!) Derive a formula for the inverse transform of $\mathcal{G}$.

2.29. *Cosine transform*

Let $g(x)$ be a real-valued function defined for $0 \le x < \infty$. The function

$$\mathcal{C}g(s) = 2\int_0^{\infty} g(x)\cos 2\pi s x\, dx$$

is called the *cosine transform* of $g(x)$. It is also a real-valued function. A discrete version of the cosine transform (called the DCT) is used in lossy compression of audio (e.g., MP3) and of images (e.g., JPEG).

[26] From A. Siripram.

Let

$$f(x) = \begin{cases} g(x), & x \geq 0, \\ g(-x), & x \leq 0. \end{cases}$$

Note that $f(x)$ is even, as is then $\mathcal{F}f(s)$.

(a) Show that

$$\mathcal{F}f(s) = \mathcal{C}g(s).$$

(b) *Inverse cosine transform.* Using part (a) and considerations of evenness, show that

$$f(x) = 2\int_0^\infty \mathcal{C}g(s)\cos 2\pi sx\, ds.$$

From this we conclude that when $x \geq 0$,

$$g(x) = 2\int_0^\infty \mathcal{C}g(s)\cos 2\pi sx\, ds.$$

2.30. *Sine transform*

Your friend, who also worked on Problem 2.29 on the cosine transform, jumps in and says, "We should also define a *sine transform* by

$$\mathcal{S}f(s) = 2i\int_0^\infty f(t)\sin(2\pi st)\, dt.$$

I'm sure that if $f(t)$ is odd, then $\mathcal{S}f(s) = \mathcal{F}f(s)$ is odd, too. Nature is like that, or it ought to be." You're not so sure about nature. What do you get?

2.31. *Fourier transform equals the signal*

Find signals $f(t)$'s such that:

(a) The Fourier transform of $f(t)$ is equal to itself; i.e., $\mathcal{F}f(s) = f(s)$.

(b) $f(0) = 2$.

Can you think of a general rule that generates infinitely many $f(t)$'s?

2.32. *Short-time Fourier transform*

To handle time variations in spectral analysis one often works within a local time section of the signal using a window function $w(t)$. For this, the *short-time Fourier transform* (STFT) is defined to be

$$F(\tau, s) = \int_{-\infty}^\infty f(t)w(t-\tau)e^{-2\pi ist}dt.$$

The window function $w(t)$ is real valued, has a Fourier transform $W(s)$, and satisfies the scaling condition

$$W(0) = \int_{-\infty}^\infty w(t)dt = 1.$$

(a) Express $\mathcal{F}f(s)$, the Fourier transform of $f(t)$, in terms of $F(\tau, s)$.

(b) Derive an expression for the inverse STFT; i.e., express $f(t)$ in terms of $F(\tau, s)$.

(c) Find the STFT of $f(t) = 1$.

(d) Find the STFT of $f(t) = e^{2\pi i s_0 t}$.

2.33. *Equivalent width: Still another reciprocal relationship*

The *equivalent width* of a signal $f(t)$, with $f(0) \neq 0$, is the width of a rectangle having height $f(0)$ and area the same as under the graph of $f(t)$. Thus

$$W_f = \frac{1}{f(0)} \int_{-\infty}^{\infty} f(t)\, dt\,.$$

This is a measure for how spread out a signal is.

Show that $W_f W_{\mathcal{F}f} = 1$. Thus, the equivalent widths of a signal and its Fourier transform are reciprocal.

From the Internet Encyclopedia of Science: *Equivalent width*: A measure of the strength of a spectral line. On a plot of intensity against wavelength, a spectral line appears as a curve with a shape defined by the line profile. The equivalent width is the width of a rectangle centered on a spectral line that, on a plot of intensity against wavelength, has the same area as the line.

2.34. *Some symmetry properties*

Let $f(t)$ be a real signal, with Fourier transform

$$\mathcal{F}f(s) = A(s) + iB(s),$$

where $A(s)$ and $B(s)$ are the real and imaginary parts, respectively, of $\mathcal{F}f(s)$.

(a) Show that $A(s)$ is even and $B(s)$ is odd.

(b) One says that $f(t)$ is *causal* if $f(t) = 0$ for $t \leq 0$. If $f(t)$ is causal, show that $A(s)$ and $B(s)$ each contain half the total energy:

$$\int_{-\infty}^{\infty} A(s)^2 ds = \frac{1}{2} \int_{-\infty}^{\infty} f(t)^2 dt = \int_{-\infty}^{\infty} B(s)^2 ds.$$

Hint: Use $A = \operatorname{Re} \mathcal{F}f = (\mathcal{F}f + \overline{\mathcal{F}f})/2$ and apply Parseval's identity to A and its inverse Fourier transform.

2.35. *Analytic signal*

In communications theory the *analytic signal* $f_a(t)$ of a signal $f(t)$ is defined, via the Fourier transform, by

$$\mathcal{F}f_a(s) = \begin{cases} \mathcal{F}f(s), & s \geq 0, \\ 0, & \text{otherwise}\,. \end{cases}$$

For a real-valued signal $f(t)$ that is not identically zero, could the corresponding analytic signal $f_a(t)$ also be real? Why or why not?

2.36. *Instantaneous frequency*

Let $f(t)$ be a real signal and let

$$F(s) = \int_{-\infty}^{\infty} e^{-2\pi i s t} f(t)\, dt$$

be its Fourier transform. Write

$$F(s) = A(s) e^{i\phi(s)};$$

i.e., $A(s)$ is the magnitude of $F(s)$ and $\phi(s)$ is the phase. Assume that both $A(s)$ and $\phi(s)$ are differentiable.

(a) Show that $A'(0) = 0$ and that $\phi(0) = 0$. (Use symmetry.)

(b) The derivative $\phi'(s)$ is sometimes called the instantaneous frequency. Show that

$$\phi'(0) = -2\pi \frac{\int_{-\infty}^{\infty} t f(t)\, dt}{\int_{-\infty}^{\infty} f(t)\, dt}.$$

You may assume the results in part (a).

2.37. *Dispersion*

Propagation in dispersive optical media, such as an optical fiber, is governed by

$$\frac{\partial E(z,t)}{\partial z} + \frac{i\beta}{2}\frac{\partial^2 E(z,t)}{\partial t^2} = 0$$

where $E(z,t)$ is the electric field as a function of the longitudinal coordinate z and time t and β is the dispersion constant. The observed phenomenon is that for an input that is a narrow impulse, the output is spread out in frequency after propagating a distance d because of dispersion.

Let $\widetilde{E}(z,s)$ be the Fourier transform of $E(z,t)$ with respect to t. This is the frequency domain representation of the electric field at any given z. What is the relationship between $\widetilde{E}(z,s)$ and $\widetilde{E}(0,s)$. What about $|\widetilde{E}(z,s)|$ and $|\widetilde{E}(0,s)|$?

2.38. *Fourier transforms of chirp signals and the principle of stationary phase*[27]

Consider a complex-valued function $f(t)$, and denote the phase of $f(t)$ by $\phi(t)$. Also assume that the amplitude $|f(t)|$ is a relatively slowly varying function of t. Write the Fourier transform of f as

$$\begin{aligned} F(s) &= \int_{-\infty}^{\infty} f(t)e^{-i2\pi st}dt \\ &= \int_{-\infty}^{\infty} |f(t)|e^{i(\phi(t)-2\pi st)}dt, \end{aligned}$$

and define

$$\mu(t) \equiv \frac{1}{s}\phi(t) - 2\pi t.$$

The *principle of stationary phase* says that the value of this Fourier transform integral is dominated by the region in which the phase of the integrand, $\mu(t)$, is nearly stationary, i.e., the region surrounding

$$\frac{d\mu(t)}{dt} = 0.$$

In the other regions, the areas under the positive and negative lobes of the integrand tend to (approximately) cancel. Using a Taylor series expansion of

[27]From R. Bhatnagar.

$\mu(t)$ around the point t_0 where $\frac{d\mu(t)}{dt} = 0$, it can be shown that

$$F(s) \approx |f(t_0)|\sqrt{\frac{2\pi}{|s\mu''(t_0)|}}e^{i\pi/4}e^{is\mu(t_0)}$$

where $\mu''(t)$ is the second derivative of $\mu(t)$.

A *chirp signal* is of the form

$$f(t) = \Pi(t) \cdot e^{i\pi a t^2}.$$

Its *instantaneous frequency* is

$$s_{\text{inst}} = \frac{1}{2\pi}\frac{d}{dt}[\pi a t^2] = at.$$

Assume $a > 0$.

(a) Choose $a = 200$. For $s = 0$ and $s = 50$, plot the real and imaginary parts of the integrand $|f(t)|e^{is\mu(t)}$. Which region contributes most to the value of the Fourier transform integral?

(b) Using the formula provided, find the approximate Fourier transform of $f(t)$.

(c) Using MATLAB, calculate the Fourier transform of $f(t)$ using the fft() function. Be sure to choose a sampling rate of at least several hundred Hz. Plot the real and imaginary parts of this on top of the approximate Fourier transform given by the expression you found in part (b). How well do they compare?

2.39. *Conservation of energy*

Suppose $u(x,t)$ satisfies the wave equation on the real line

$$\frac{\partial^2 u}{\partial t^2} = \frac{\partial^2 u}{\partial x^2}$$

with the initial conditions $u(x,0) = f(x)$ and $u_t(x,0) = 0$. Define the energy of the wave at time t by

$$E(t) = \int_{-\infty}^{\infty} \left|\frac{\partial u}{\partial x}(x,t)\right|^2 dx + \int_{-\infty}^{\infty} \left|\frac{\partial u}{\partial t}(x,t)\right|^2 dx.$$

The first term represents potential energy and the second term kinetic energy.

Prove that the wave conserves energy:

$$E(t) = E(0).$$

Hint: The Fourier transform $U(s,t)$ of $u(x,t)$ with respect to x has the form

$$U(s,t) = \frac{1}{2}F(s)(e^{2\pi i s t} + e^{-2\pi i s t}) = F(s)\cos 2\pi s t \quad \text{where } F(s) = \mathcal{F}f(s).$$

2.40. *The Hartley transform: An exercise in duality*

Let $f(t)$ be a *real* signal. The Hartley[28] transform of $f(t)$ is defined by

$$\mathcal{H}f(s) = \int_{-\infty}^{\infty} (\cos 2\pi s t + \sin 2\pi s t) f(t)\, dt \quad \textit{Note}\text{: There is no } i \text{ in the integrand.}$$

[28] Named after Ralph Hartley. He also did important work on the early ideas of information theory.

Other definitions (e.g., where the 2π goes) are in use. This transform shares a number of properties with the Fourier transform, but it is real for a real signal and this provides some computational advantages for the discrete form of the transform.

(a) We can get an expression for $\mathcal{H}f$ in terms of the Fourier transform of f. Recall that the real and imaginary parts of a complex number z are given by

$$\operatorname{Re} z = \frac{z+\bar{z}}{2}, \quad \operatorname{Im} z = \frac{z-\bar{z}}{2i}.$$

From the definition of $\mathcal{H}f$ and

$$e^{-2\pi i s t} = \cos 2\pi st - i \sin 2\pi st$$

we see that

$$\mathcal{H}f = \operatorname{Re}\mathcal{F}f - \operatorname{Im}\mathcal{F}f.$$

Using this, show that

$$\mathcal{H}f = \frac{1+i}{2}\mathcal{F}f + \frac{1-i}{2}\mathcal{F}f^- \quad \text{and} \quad \mathcal{H}f^- = \frac{1+i}{2}\mathcal{F}f^- + \frac{1-i}{2}\mathcal{F}f.$$

Remember that $1/i = -i$.
It follows that $\mathcal{H}f^- = (\mathcal{H}f)^-$, just like the Fourier transform.

(b) Next, show that

$$\mathcal{F}\mathcal{H}f = \frac{1+i}{2}f^- + \frac{1-i}{2}f \quad \text{and} \quad \mathcal{F}\mathcal{H}f^- = \frac{1+i}{2}f + \frac{1-i}{2}f^-.$$

(c) Using part (b) show that $\mathcal{H}$ is its own inverse, i.e., that

$$\mathcal{H}\mathcal{H}f = f.$$

(d) If $f(t)$ is real and even, show that $\mathcal{F}$ and $\mathcal{H}$ are inverses of each other, namely,

$$\mathcal{H}\mathcal{F}f = \mathcal{F}\mathcal{H}f = f.$$

Why do we need the assumption that f is even?

2.41. *Hartley conserves energy*

Let $f(t)$ be a real signal, with Fourier transform given by

$$\mathcal{F}f(s) = A(s) + iB(s);$$

so $A(s)$ and $B(s)$ are the real and imaginary parts, respectively, of $\mathcal{F}f(s)$. As in Problem 2.40, the Hartley transform of the real function f, written $\mathcal{H}f$, is defined in terms of $A(s)$ and $B(s)$ as

$$\mathcal{H}f(s) = A(s) - B(s).$$

Show that the Hartley transform conserves energy, i.e.,

$$\int_{-\infty}^{\infty} \mathcal{H}f(s)^2 \, ds = \int_{-\infty}^{\infty} (A(s) - B(s))^2 \, ds = \int_{-\infty}^{\infty} f(t)^2 \, dt.$$

You'll find Problem 2.34 helpful.

2.42. *Duality implies Fourier inversion*

Suppose you know the duality formula $\mathcal{F}\mathcal{F}f = f^-$. From this deduce the formula for $\mathcal{F}^{-1}$.

2.43. *Hubbard-Stratonovich formula*

Show that

$$\int_{-\infty}^{\infty} e^{-\pi x^2} e^{-2\pi s x}\,dx = e^{\pi s^2}\,.$$

Use the same approach we took in finding the Fourier transform of a Gaussian.

In physics, this equation and higher-dimensional generalizations are called Hubbard-Stratonovich formulas; do a google search. (You might recognize it as the two-sided Laplace transform of $e^{-\pi x^2}$. Note that there's no i in the $e^{-2\pi s x}$.)

2.44. *Hermite polynomials as eigenfunctions of the Fourier transform*

The Hermite polynomials $H_n(x)$ are defined by the formula

$$H_n(x) = (-1)^n e^{x^2} \frac{d^n}{dx^n}(e^{-x^2})\,.$$

They satisfy various recurrence relationships, for example

$$H_n'(x) = 2xH_n(x) - H_{n+1}(x)\,.$$

The first several Hermite polynomials are

$$H_0(x) = 1\,, \quad H_1(x) = 2x\,, \quad H_2(x) = 4x^2 - 2\,, \quad H_3(x) = 8x^3 - 12x\,.$$

The Hermite polynomials arise in many areas of applications, one being the study of the quantum oscillator; they are solutions of the differential equation

$$y'' + 2xy' + 2ny = 0\,.$$

(The definition given here is the one commonly used in areas of physics. There is an alternate definition, $H_n(x) = (-1)^n e^{x^2/2}(d^n/dx^n)(e^{-x^2/2})$, favored for applications to questions in probability.) Define

$$\psi_n(x) = H_n(\sqrt{2\pi}x)e^{-\pi x^2}\,.$$

In particular, $\psi_0(x) = e^{-\pi x^2}$ is our model Gaussian. Show that

$$\mathcal{F}\psi_n(s) = (-i)^n \psi_n(s)\,.$$

That is, the functions $\psi_0, \psi_1, \psi_2, \ldots$ are eigenfunctions of the Fourier transform with eigenvalues $1, -i, -1, i, \ldots$. *Hint*: You know that the Fourier transform of the Gaussian is itself. Use the recursion relationship and argue inductively.

Chapter 3

Convolution

3.1. A $*$ Is Born

How can we use one signal to modify another? That's what signal processing is all about, and some of the properties of the Fourier transform that we have already covered can be thought of as addressing this question. The easiest is the result on additivity, according to which

$$\mathcal{F}(f+g) = \mathcal{F}f + \mathcal{F}g.$$

Adding the signal $g(t)$ to the signal $f(t)$ adds the amounts $\mathcal{F}g(s)$ to the frequency components $\mathcal{F}f(s)$. (Symmetrically, $f(t)$ modifies $g(t)$ in the same way.) The spectrum of $f+g$ may be more or less complicated than the spectrum of f and g alone, and it's an elementary operation in *both* the time domain and the frequency domain that produces or eliminates the complications. It's also an operation that's easily undone. See some frequencies you don't like in the spectrum (a bad buzz)? Then try adding something in or subtracting something out and see what the signal looks like.

We can view the question of using one signal to modify another in either the time domain or the frequency domain, sometimes with equal ease and sometimes with one point of view preferred. We just looked at sums. What about products? The trivial case is multiplying by a constant, as in $\mathcal{F}(af)(s) = a\mathcal{F}f(s)$. The energies of the harmonics are all affected by the same amount, so, thinking of music for example, the signal sounds the same, only louder or softer.

It's much less obvious how to scale the harmonics *separately*. We ask:

> Is there some combination of the signals $f(t)$ and $g(t)$ in the time domain so that in the frequency domain the Fourier transform of the combination is
>
> $$\mathcal{F}g(s)\mathcal{F}f(s)\,?$$

In other words, in the time domain can we combine the signal $g(t)$ with the signal $f(t)$ so that the frequency components $\mathcal{F}f(s)$ of $f(t)$ are scaled by the frequency components $\mathcal{F}g(s)$ of $g(t)$? (Once again this is symmetric — we could say that the frequency components $\mathcal{F}g(s)$ are scaled by the frequency components $\mathcal{F}f(s)$.)

Let's check this out, and remember that the rigor police are still off duty. No arrests will be made for unstated assumptions, integrals that are not known to converge, etc.

The product of the Fourier transforms of $f(t)$ and $g(t)$ is

$$\mathcal{F}g(s)\,\mathcal{F}f(s) = \int_{-\infty}^{\infty} e^{-2\pi ist} g(t)\,dt \int_{-\infty}^{\infty} e^{-2\pi isx} f(x)\,dx\,.$$

We used different variables of integration in the two integrals because we're going to combine the product into an iterated integral[1]:

$$\begin{aligned}
\int_{-\infty}^{\infty} e^{-2\pi ist} g(t)\,dt \int_{-\infty}^{\infty} e^{-2\pi isx} f(x)\,dx &= \int_{-\infty}^{\infty}\int_{-\infty}^{\infty} e^{-2\pi ist} e^{-2\pi isx} g(t) f(x)\,dt\,dx \\
&= \int_{-\infty}^{\infty}\int_{-\infty}^{\infty} e^{-2\pi is(t+x)} g(t) f(x)\,dt\,dx \\
&= \int_{-\infty}^{\infty}\left(\int_{-\infty}^{\infty} e^{-2\pi is(t+x)} g(t)\,dt\right) f(x)\,dx.
\end{aligned}$$

Now make the change of variable $u = t + x$ in the inner integral. Then $t = u - x$, $du = dt$, and the limits are the same. The result is

$$\int_{-\infty}^{\infty}\left(\int_{-\infty}^{\infty} e^{-2\pi is(t+x)} g(t)\,dt\right) f(x)\,dx = \int_{-\infty}^{\infty}\left(\int_{-\infty}^{\infty} e^{-2\pi isu} g(u-x)\,du\right) f(x)\,dx.$$

Next, switch the order of integration:

$$\begin{aligned}
\int_{-\infty}^{\infty}\left(\int_{-\infty}^{\infty} e^{-2\pi isu} g(u-x)\,du\right) f(x)\,dx &= \int_{-\infty}^{\infty}\int_{-\infty}^{\infty} e^{-2\pi isu} g(u-x) f(x)\,du\,dx \\
&= \int_{-\infty}^{\infty}\int_{-\infty}^{\infty} e^{-2\pi isu} g(u-x) f(x)\,dx\,du \\
&= \int_{-\infty}^{\infty} e^{-2\pi isu}\left(\int_{-\infty}^{\infty} g(u-x) f(x)\,dx\right) du.
\end{aligned}$$

Look at what's happened here. The inner integral is a function of u. Let's set it up on its own:

$$h(u) = \int_{-\infty}^{\infty} g(u-x) f(x)\,dx\,.$$

Then the outer integral produces the Fourier transform of h:

$$\int_{-\infty}^{\infty} e^{-2\pi isu}\left(\int_{-\infty}^{\infty} g(u-x) f(x)\,dx\right) du = \int_{-\infty}^{\infty} e^{-2\pi isu} h(u)\,du = \mathcal{F}h(s).$$

[1] If you're uneasy with this (never mind issues of convergence), you might convince yourself that it's correct by working your way backwards from the double integral to the product of the two single integrals.

Switching the variable name for h from $h(u)$ to $h(t)$ (solely for psychological comfort), we have discovered that the signals $f(t)$ and $g(t)$ are combined into a signal

$$h(t) = \int_{-\infty}^{\infty} g(t-x)f(x)\,dx\,,$$

and comparing where we started to where we finished, we have

$$\mathcal{F}h(s) = \mathcal{F}g(s)\mathcal{F}f(s)\,.$$

Remarkable.

We have solved our problem. The only thing to do is to realize what we've done and declare it to the world. We make the following definition:

- **Convolution defined.** The *convolution* of two functions $g(t)$ and $f(t)$ is the function

$$h(t) = \int_{-\infty}^{\infty} g(t-x)f(x)\,dx\,.$$

  We use the notation

$$(g * f)(t) = \int_{-\infty}^{\infty} g(t-x)f(x)\,dx\,.$$

We can now proudly announce:

- **Convolution Theorem.** $\mathcal{F}(g * f)(s) = \mathcal{F}g(s)\mathcal{F}f(s)$.
  - In other notation: If $f(t) \rightleftharpoons F(s)$ and $g(t) \rightleftharpoons G(s)$, then $(g * f)(t) \rightleftharpoons G(s)F(s)$.
- In words: Convolution in the time domain corresponds to multiplication in the frequency domain.

The convolution theorem is indeed remarkable, but the game was rigged! We defined convolution to make it work. What we've just gone through is the same sort of thing we did when we found the formula for the Fourier coefficients for a periodic function. Or when we defined the inner product of two vectors by first asking for an algebraic condition that guaranteed orthogonality. That game was rigged, too. Remember the principle, going back to Archimedes: first suppose the problem is solved and see what the answer must be. The second step, assuming the first one works, is to turn that solution into a definition and then announce to the world that you have solved your original problem based on your brilliant definition.[2]

Recall that when we studied Fourier series, convolution came up in the form

$$(g * f)(t) = \int_0^1 g(t-x)f(x)\,dx\,.$$

In that setting, for the integral to make sense, i.e., to be able to evaluate $g(t-x)$ at points outside the interval from 0 to 1, we had to assume that g was periodic. That's not an issue in the present setting, where we assume that $f(t)$ and $g(t)$ are defined for all t so the factors in the integrand, $g(t-x)f(x)$, are defined everywhere.

---

[2]Mathematicians, in particular, are very good at presenting their results and writing their books in this way — do step one in secret and tell the world only step two. It's extremely irritating.

Questions may be raised as to whether the integral converges, but at least the setup makes sense.

Let's see a quick application of our brilliant new discovery. As an exercise you can show (by hand, using the definition as an integral) that

$$(\Pi * \Pi)(x) = \Lambda(x).$$

Recall that Λ is the triangle function. Applying the convolution theorem, we find that

$$\mathcal{F}\Lambda(s) = \mathcal{F}(\Pi * \Pi)(s) = \operatorname{sinc} s \cdot \operatorname{sinc} s = \operatorname{sinc}^2 s,$$

just like before. To ask again, was there any doubt?

Remark on notation. It's common to see people write the convolution as $g(t) * f(t)$, putting the variable t in each of the functions g and f. There are times when that's OK, even sometimes preferable to introducing a lot of extra notation, but in general I think it's a bad idea because it can lead to all sorts of abuses and possible mistakes. For example, what's $g(2t) * f(t)$? If you plugged in too casually, you might write this as the integral

$$\int_{-\infty}^{\infty} g(2t - x) f(x)\, dx.$$

That's wrong. The right answer in convolving $g(2t)$ and $f(t)$ is

$$\int_{-\infty}^{\infty} g(2(t-x)) f(x)\, dx = \int_{-\infty}^{\infty} g(2t - 2x) f(x)\, dx.$$

Make sure you understand why the first is wrong and second is right.

One way to be unambiguous about this is to say something like: let's define $h(t) = g(2t)$; then

$$(h * f)(t) = \int_{-\infty}^{\infty} h(t-x) f(x)\, dx = \int_{-\infty}^{\infty} g(2(t-x)) f(x)\, dx.$$

Or, what amounts to the same thing, one can also use the scaling operator σ_a from the last chapter. For example, with $(\sigma_2 g)(x) = g(2x)$ we're looking at

$$(\sigma_2 g * f)(t) = \int_{-\infty}^{\infty} (\sigma_2 g)(t-x) f(x)\, dx = \int_{-\infty}^{\infty} g(2(t-x)) f(x)\, dx.$$

I concede that this may be too much of a hassle in most cases. Just be careful.

One more comment. In the definition

$$(g * f)(x) = \int_{-\infty}^{\infty} g(y) f(x-y)\, dy$$

(you should also practice using different variables), how do you remember if it's $x - y$ or $y - x$? The answer is that the arguments in the integrand add up to x, as in $y + (x - y) = x$, the variable where $g * f$ is evaluated. I've seen people stumble over this.

Convolving in the frequency domain. If you look at the argument for the convolution theorem $\mathcal{F}(g * f) = \mathcal{F}g \cdot \mathcal{F}f$, you'll see that we could have carried the whole thing out for the inverse Fourier transform and given the symmetry between the Fourier transform and its inverse that's not surprising. So we also have

$$\mathcal{F}^{-1}(g * f) = \mathcal{F}^{-1}g \cdot \mathcal{F}^{-1}f\,.$$

What's more interesting and doesn't follow without a little additional argument is this:

$$\mathcal{F}(gf)(s) = (\mathcal{F}g * \mathcal{F}f)(s)\,.$$

In words:

- Multiplication in the time domain corresponds to convolution in the frequency domain.

Here's how the derivation goes. We'll need one of the duality formulas, the one that says

$$\mathcal{F}(\mathcal{F}f)(s) = f(-s) \quad \text{or} \quad \mathcal{F}\mathcal{F}f = f^- \quad \text{without the variable.}$$

To derive the identity $\mathcal{F}(gf) = \mathcal{F}g * \mathcal{F}f$, we write, for convenience, $h = \mathcal{F}f$ and $k = \mathcal{F}g$. Then we're to show

$$\mathcal{F}(gf) = k * h\,.$$

The one thing we know is how to take the Fourier transform of a convolution, so, in the present notation, $\mathcal{F}(k * h) = (\mathcal{F}k)(\mathcal{F}h)$. But now $\mathcal{F}k = \mathcal{F}\mathcal{F}g = g^-$, from the identity above, and likewise $\mathcal{F}h = \mathcal{F}\mathcal{F}f = f^-$. So $\mathcal{F}(k * h) = g^- f^- = (gf)^-$, or

$$gf = \mathcal{F}(k * h)^-\,.$$

Now, finally, take the Fourier transform of both sides of this last equation and appeal to the $\mathcal{F}\mathcal{F}$ identity again:

$$\mathcal{F}(gf) = \mathcal{F}(\mathcal{F}(k * h)^-) = k * h = \mathcal{F}g * \mathcal{F}f\,.$$

We're done.

Remark. You may wonder why we didn't start this chapter by trying to prove $\mathcal{F}(gf)(s) = (\mathcal{F}g * \mathcal{F}f)(s)$ rather than $\mathcal{F}(g * f) = (\mathcal{F}f)(\mathcal{F}g)$, as we did. Perhaps it seems more natural to multiply signals in the time domain and see what effect this has in the frequency domain, so why not work with $\mathcal{F}(fg)$ directly? You can write the integral for $\mathcal{F}(gf)$ but you'll probably find that you can't do much.

Fourier coefficients. In Problem 1.28 you were asked to show that if

$$f(t) = \sum_{n=-\infty}^{\infty} \hat{f}(n)e^{2\pi int} \quad \text{and} \quad g(t) = \sum_{n=-\infty}^{\infty} \hat{g}(n)e^{2\pi int},$$

then the nth Fourier coefficient of the convolution,

$$(f * g)(t) = \int_0^1 f(\tau)g(t-\tau)\,d\tau,$$

is

$$\widehat{(f * g)}(n) = \hat{f}(n)\hat{g}(n).$$

Thus

$$(f * g)(t) = \sum_{n=-\infty}^{\infty} \hat{f}(n)\hat{g}(n)e^{2\pi int}.$$

Think that was an accident?

Convolution with different normalizations. Back in Section 2.4 we reproduced a handy guide to Fourier transform formulas as they depend on the particular convention being used, meaning that if someone's Fourier transform has

$$\mathcal{F}f(s) = \frac{1}{A}\int_{-\infty}^{\infty} e^{iBst} f(t)\, dt\,,$$

then various formulas also depend on A and B. Let me remind you that the convolution theorem likewise has this dependence. The rule is

$$\mathcal{F}(f * g) = A(\mathcal{F}f)(\mathcal{F}g).$$

3.2. What Is Convolution, Really?

There's not a single answer to that question. Those of you who have had a course in "Signals and Systems" probably saw convolution in connection with linear time invariant systems (LTI systems) and in particular with the *impulse response* for such a system. That's a very natural setting for convolution, and we'll consider it later, after we have the machinery of δ-functions et al.[3]

The fact is that convolution is used in many ways and for many reasons, and it can be a mistake to try to attach to it one particular meaning or interpretation. This multitude of interpretations and applications is somewhat like the situation with the definite integral. When you learned about the integral, chances are that it was introduced via an important motivating problem, typically recovering the distance traveled by integrating velocity, or the problem of finding the area under a curve. That's fine, but the integral is really a much more general and flexible concept than those two sample problems alone might suggest. You do yourself no service if every time you think to use an integral you think only of one of those problems. Likewise, you do yourself no service if you insist on one particular interpretation of convolution.

Still, to pursue the analogy with the integral a little bit further, in pretty much all applications of the integral there *is* a general method at work: cut the problem into small pieces where it can be solved approximately, sum up the solution for the pieces, and pass to a limit.[4] Likewise, there is also often a general method to working with, or seeking to work with, convolutions. Usually there's something in the picture that has to do with *smoothing* and *averaging*, understood broadly. You see this in both the continuous case (which we're doing in the present chapter) and the discrete case (which we'll do in a later chapter).

For example, in using Fourier series to solve the heat equation on a circle, we saw that the solution was expressed as a convolution of the initial heat distribution with

[3]This already came up in connection with our solution of the heat equation via Fourier series.

[4]This goes back to Archimedes, who called his paper on the subject "The Method."

the Green's function (or fundamental solution). That's a smoothing and averaging interpretation (both!) of the convolution — the initial temperature is smoothed out, toward its average, as time increases.[5]

In brief, we'll get to know convolution by seeing it in action:

- Convolution *is* what convolution *does*.

As tautological as that is, it's probably the best answer to the question in the heading to this section.

But can I visualize convolution? Or "Flip this, buddy". I'm tempted to say don't bother. Again for those of you who have seen convolution in earlier courses, you've probably heard the expression "flip and drag." For

$$(g * f)(t) = \int_{-\infty}^{\infty} g(t-x)f(x)\,dx,$$

here's what this means.

- Fix a value t. The graph of the function $g(x-t)$, as a function of x, has the same shape as $g(x)$ but is shifted to the right by t. Then forming $g(t-x)$ flips the graph (left-right) about the line $x = t$. If the most interesting or important features of $g(x)$ are near $x = 0$, e.g., if it's sharply peaked there, then those features are shifted to $x = t$ for the function $g(t-x)$, but there's the extra "flip" to keep in mind.
- Multiply the two functions $f(x)$ and $g(t-x)$ and integrate with respect to x. Right. Remember that the value of the convolution $(g * f)(t)$ is not just the product of the values of f and the flipped and shifted g, it's the *integral* of the product — much harder to visualize. Integrating the product sums up these values; that's the "dragging" part.

Back to smoothing and averaging. I like to think of the convolution operation as using one function to smooth and average the other. To fix ideas, say g is used to smooth f in forming $g * f$. In many common applications $g(x)$ is a positive function, concentrated near 0, with total area 1,

$$\int_{-\infty}^{\infty} g(x)\,dx = 1,$$

like a sharply peaked Gaussian, for example (as we'll see). Then $g(t-x)$ is concentrated near t and still has area 1. For a fixed t, the integral

$$\int_{-\infty}^{\infty} g(t-x)f(x)\,dx$$

is a weighted average of the values of $f(x)$ near $x = t$, weighted by the values of (the flipped and shifted) g. It's an average because $\int_{-\infty}^{\infty} g(x)\,dx = 1$. Computing the convolution $g*f$ at t replaces the value $f(t)$ by a weighted average of the values of f near t.

[5]It's also a linear systems interpretation of convolution, where the system is described by the heat equation. The input is the initial temperature; the output is the solution of the heat equation, which is a convolution.

Where does the smoothing come in? Here's where:

- Changing t ("dragging" $g(t-x)$ through different values of t) repeats this operation.

Again take the case of an averaging-type function $g(t)$, as above. At a given point t, $(g*f)(t)$ is a weighted average of values of f near t. Move t a little to a nearby point t'. Then $(g*f)(t')$ is a weighted average of values of f near t', which will include values of f that entered into the average near t. Thus the values of the convolutions $(g*f)(t)$ and $(g*f)(t')$ will likely be closer to each other than are the values $f(t)$ and $f(t')$. In this sense, $(g*f)(t)$ is smoothing f as t varies — there's less of a change between values of the convolution $g*f$ than between values of f.

We'll study this in more detail later, but you've already seen at least one example of smoothing. The rectangle function $\Pi(x)$ is discontinuous — it has jumps at $\pm 1/2$. The convolution $\Pi * \Pi$ is the triangle function Λ, which is *continuous* — the jumps at the endpoints have been smoothed out. There's still a corner, but there's *no* discontinuity.

In fact, as an aphorism we can state the following:

- The convolution $g*f$ is at least as good a function as g and f are separately, maybe better.

We'll come back to this in Section 4.8

A smear job, too. Now, be a little careful in how you think about this averaging and smoothing process. Computing any value of $(g*f)(t)$ involves *all* of the values of g and *all* of the values of f and involves adding the products of corresponding values of g and f with one of the functions flipped and dragged. If *both* $f(t)$ and $g(t)$ become identically zero after a while, then the convolution $g*f$ will also be identically zero outside of some interval. But if $f(t)$ or $g(t)$ does not become identically zero, then neither will the convolution. In addition to averaging and smoothing, the convolution also "smears" out the factors — not a becoming description, but that's what everyone says.

Definitely keep the general description we've just gone through in mind, but as far as visualizing the convolution of any two old functions, I think it's of dubious value to beat yourself up trying to do that. It's hard geometrically, and it's hard computationally, in the sense that you have to calculate some tedious integrals. You do have to do a few of these in your life — hence some problems to work out — but only a few.

There are plenty of web resources out there to help. For developing further intuition, I particularly like the Johns Hopkins webpage on signals, systems, and control:

`http://www.jhu.edu/~signals/.`

There you'll find a Java applet called "Joy of Convolution" (and many other things). It will allow you to select sample curves $f(t)$ and $g(t)$, or to draw your own curves with a mouse, and then produce the convolution $(g*f)(t)$.

Of course you can try to get some intuition for how the convolution looks by thinking of what's happening in the frequency domain. It may not be so far-fetched to imagine the Fourier transforms $\mathcal{F}f$, $\mathcal{F}g$, and their product and then imagine the inverse transform to get you $g * f$.

3.3. Properties of Convolution: It's a Lot Like Multiplication

Convolution behaves in many ways like multiplication, though not in all ways. For example, it is commutative:

$$f * g = g * f\,.$$

So although it looks like the respective roles of f and g are different — one is "flipped and dragged," and the other isn't — in fact they share equally in the end result.

Do we have to prove this? Not among friends. After all, we *defined* the convolution so that the convolution theorem holds, that is, so that $\mathcal{F}(g * f) = \mathcal{F}g\mathcal{F}f$. But g and f enter symmetrically on the right-hand side, multiplication being commutative, from which also $g * f = f * g$. In the language of signal processing one might say $g(t)$ can be used to modify $f(t)$ or $f(t)$ can be used to modify $g(t)$.

However, the commutativity property is easy to check from the definition:

$$\begin{aligned}(f * g)(t) &= \int_{-\infty}^{\infty} f(t-u)g(u)\,du \\ &= \int_{-\infty}^{\infty} g(t-v)f(v)\,dv \quad \text{(making the substitution } v = t-u\text{)} \\ &= (g * f)(t)\,.\end{aligned}$$

The same idea, a change of variable, but with much more bookkeeping, establishes that convolution is associative:[6]

$$(f * g) * h = f * (g * h)\,.$$

Using the convolution theorem reduces this to the associativity of multiplication.

Nothing gets some people more incensed than a casual attitude toward the associative property of convolution. The issue is the convergence of the various iterated integrals one encounters in forming $(f * g) * h$ and $f * (g * h)$. So, to be honest, one can find examples where $(f * g) * h \neq f * (g * h)$. It won't be an issue. If someone comes after you on this, wait them out and then go about your work.

Much more easily one gets that

$$f * (g + h) = f * g + f * h\,.$$

The corresponding statements are easily verified in the frequency domain.

[6]An exercise for you in integrals. Set aside more than a few minutes.

How about a "1"? Is there a function that is to convolution as 1 is to multiplication? Is there a function g such that

$$(g * f)(t) = f(t), \quad \text{for } \textit{all} \text{ functions } f?$$

What property would such a g have? Take Fourier transforms of both sides:

$$\mathcal{F}f(s)\mathcal{F}g(s) = \mathcal{F}f(s)\,.$$

Then $g(x)$ must be such that

$$\mathcal{F}g(s) = 1\,.$$

Is there such a g? Applying the inverse Fourier transform would lead to

$$g(x) = \int_{-\infty}^{\infty} e^{2\pi isx}\,ds\,,$$

and, emphatically, that integral does not exist — even I wouldn't try to slip that by the rigor police. Something is up here. Maybe Fourier inversion doesn't work in this case, or else there's no classical function whose Fourier transform is 1, or something. In fact, though the integral does not exist in any sense, the problem of a "1 for convolution" leads exactly to the δ-function. Next chapter.

How about division? Suppose we know h and g in

$$h = f * g$$

and we want to solve for f. Again, taking Fourier transforms, we would say

$$\mathcal{F}h = \mathcal{F}f \cdot \mathcal{F}g \Rightarrow \mathcal{F}f = \frac{\mathcal{F}h}{\mathcal{F}g}\,.$$

We'd like the convolution quotient to be the inverse Fourier transform of $\mathcal{F}h/\mathcal{F}g$. But there are problems caused by places where $\mathcal{F}g = 0$, along with the usual problems with the existence of the integral for the inverse Fourier transform.

Solving for $f(t)$ is the *deconvolution* problem, which is extremely important in applications. Many times a noisy signal comes to you in the form $h = f * g$; the signal is f, the noise is g, you receive h. You make some assumptions about the nature of the noise, usually statistical assumptions, and you want to separate the signal from the noise. You want to deconvolve. Poke around for some guidance on this; it's an important though somewhat specialized subject.

Other algebraic identities. It's not hard to combine the various rules and develop an algebra of convolutions. Such identities can be of great use; it sure beats calculating integrals. Here's an assortment. (Lowercase and uppercase letters are Fourier pairs.)

$$\begin{aligned}
\big((f \cdot g) * (h \cdot k)\big)(t) &\rightleftharpoons \big((F * G) \cdot (H * K)\big)(s), \\
\big(f(t) + g(t)\big) \cdot \big(h(t) + k(t)\big) &\rightleftharpoons ((F + G) * (H + K))(s), \\
\big(f(t) \cdot (g * h)\big)(t) &\rightleftharpoons \big(F * (G \cdot H)\big)(s).
\end{aligned}$$

You can write down others. Be confident. Be careful, but be confident.

Derivatives. Moving on from algebra to calculus, a result on derivatives combines the two maxims of smoothing and $f * g$ being as good as or better than f and g separately. Without stating all the assumptions: If f is differentiable, then $f * g$ is differentiable, even if g is not, and

$$(f * g)'(x) = (f' * g)(x) .$$

So here's an example of convolution *not* being like multiplication. For pursuing the analogy you might think there's some kind of product rule, as in

$$(f * g)'(x) = (f' * g)(x) + (f * g')(x) .$$

No, no.

Here's how to enjoy the derivation:

$$\begin{aligned}(f * g)'(x) &= \frac{d}{dx}\int_{-\infty}^{\infty} f(x-y)g(y)\,dy \\ &= \int_{-\infty}^{\infty} \frac{d}{dx} f(x-y)g(y)\,dy \\ &\qquad \text{(we joyfully differentiate under the integral sign)} \\ &= \int_{-\infty}^{\infty} f'(x-y)g(y)\,dy \\ &\qquad \text{(the } d/dx \text{ hits only } f(x-y)\text{, and use the chain rule)} \\ &= (f' * g)(x) .\end{aligned}$$

We'll return to this, with a few more details, in the next chapter. The results also extend to higher derivatives, for example

$$(f * g)''(x) = (f'' * g)(x) .$$

If both f and g are differentiable, then you can put the derivative on either f or g:

$$(f * g)'(x) = (f' * g)(x) = (f * g')(x) .$$

3.4. Convolution in Action I: A Little Bit on Filtering

"Filtering" is a generic term for just about any operation one might want to apply to a signal, within reason of course. There's usually some feature of the signal that one wants to enhance or eliminate, and one expects *something* of the original signal to be recognizable or recoverable after it's been filtered. Most filters are described as modifying the spectral content of a signal, and they are thus set up as operations on the Fourier transform of a signal. We'll take up this topic in more detail in a later chapter when we discuss linear time invariant systems, but it's worthwhile saying a little bit now because the most common filters operate simply through multiplication in the frequency domain, hence through convolution in the time domain.

The features are:

- An input signal $v(t)$.
- An output signal $w(t)$.

- The operation from input to output, in the time domain, is convolution with a function $h(t)$:

$$w(t) = (h * v)(t) = \int_{-\infty}^{\infty} h(t-x)v(x)\,dx.$$

With this description the Fourier transforms of the input and output are related by multiplication in the frequency domain,

$$W(s) = H(s)V(s)\,,$$

where, following tradition, we denote the Fourier transforms by the corresponding capital letters.

Here $h(t)$, hence $H(s)$, is fixed. Specifying $h(t)$ or $H(s)$ defines the system. It's wired into the circuit or coded into the software, and it does what it does to *any* input $v(t)$ you may give it. In this context $h(t)$ is usually called the *impulse response* and $H(s)$ is called the *transfer function*. The term "impulse response" comes from how the system "responds" to a unit impulse (a δ-function), as we'll see later. The term "transfer function" is probably more due to lack of imagination, "transferring" information from $V(s)$ to $W(s)$. It seems to be a matter of course to denote a generic impulse response by $h(t)$ and to denote the corresponding transfer function by $H(s)$. Who am I to do otherwise?

Since filters based on convolution are usually designed to have a specific effect on the spectrum of an input, to design a filter is to design a transfer function. The operations, for which you're invited to draw a block diagram, are thus

input $\to$ Fourier transform $\to$ multiply by H $\to$ inverse Fourier transform = output.

We want to see some examples of this, filters that are in day-to-day use and the principles that go into their design.

One preliminary comment about how the spectra of the input and output are related. Write

$$V(s) = |V(s)|e^{i\phi_V(s)}, \quad \phi_V(s) = \tan^{-1}\left(\frac{\operatorname{Im} V(s)}{\operatorname{Re} V(s)}\right),$$

so the phase of $V(s)$ is $\phi_V(s)$, with similar notations for the phases of $W(s)$ and $H(s)$. Then

$$\begin{aligned}|W(s)|e^{i\phi_W(s)} &= |H(s)|\,e^{i\phi_H(s)}\,|V(s)\,|e^{i\phi_V(s)} \\ &= |H(s)|\,|V(s)|\,e^{i(\phi_H(s)+\phi_V(s))}.\end{aligned}$$

Thus the *magnitudes multiply* and the *phases add*:

$$\begin{aligned}|W(s)| &= |H(s)|\,|V(s)|, \\ \phi_W(s) &= \phi_V(s) + \phi_H(s).\end{aligned}$$

Multiplying $V(s)$ by $H(s)$ can't make the spectrum of $V(s)$ any bigger, in s that is. The spectrum of the output takes up no more of $\mathbb{R}$ than the spectrum of the input. One says that no new frequencies are added to the spectrum. But multiplying $V(s)$

by $H(s)$ *can* make the spectrum smaller by zeroing out parts of $V(s)$. Furthermore, there is no phase change when $\phi_H(s) = 0$, and this happens when $H(s)$ is real. In this case only the amplitude is changed when the signal goes through the filter.

Common examples of filters that do both of these things — modify some part of the magnitude of the spectrum with no phase change — are low-pass, band-pass, high-pass, and notch filters, to which we now turn.

3.4.1. Designing filters.

Low-pass filters. An *ideal low-pass filter* cuts off all frequencies *above* a certain amount ν_c ("c" for "cutoff") and lets all frequencies *below* ν_c pass through unchanged. Your signal has some unwanted high frequency components, e.g., hiss in audio? A low-pass filter may be just the thing.

If we write

$$w(t) = (\text{low} * v)(t) \rightleftharpoons W(s) = \text{Low}(s)V(s)\,,$$

then the transfer function we want is

$$\text{Low}(s) = \begin{cases} 1, & |s| < \nu_c, \\ 0, & |s| \geq \nu_c. \end{cases}$$

Multiplying $V(s)$ by $\text{Low}(s)$ leaves unchanged the spectrum of v for $|s| < \nu_c$ and eliminates the other frequencies. The transfer function is just a scaled rect function, and we can write it (to remind you) as

$$\text{Low}(s) = \Pi_{2\nu_c}(s) = \Pi(s/2\nu_c) = \begin{cases} 1, & |s/2\nu_c| < \frac{1}{2}, \\ 0, & |s/2\nu_c| \geq \frac{1}{2} \end{cases} = \begin{cases} 1, & |s| < \nu_c\,, \\ 0, & |s| \geq \nu_c\,. \end{cases}$$

In the time domain the impulse response is the inverse Fourier transform of $\Pi_{2\nu_c}$, and this is

$$\text{low}(t) = 2\nu_c \operatorname{sinc}(2\nu_c t)\,.$$

By the way, why is this called just a "low-pass filter"? Aren't the frequencies below $-\nu_c$ also eliminated and so not "passed" by the filter? The frequencies below $-\nu_c$ aren't really "low." Remember that for real signals $v(t)$ (which is where this is applied) one has the symmetry relation $V(-s) = \overline{V(s)}$. The positive and negative frequencies combine in reconstructing the real signal in the inverse Fourier transform, much like what happens with Fourier series. Thus one wants to pass the frequencies with $-\nu_c < s < \nu_c$, called the *pass-band*, and eliminate the frequencies with $s \geq \nu_c$ *and* $s \leq -\nu_c$, called the *stop-band*.

And why is this called an *ideal* low-pass filter? Because the cutoff is a sharp one, right at a particular frequency ν_c. In practice this cannot be achieved, and much of the original art of filter design is concerned with useful approximations to a sharp cutoff. Perhaps the best known is the *Butterworth filter*. Too specialized to go into here, but you can easily find the basic facts and appreciate them.

Band-pass filters. Another very common filter passes a particular *band* of frequencies through unchanged and eliminates all others. More precisely, the filter passes symmetric bands of positive and negative frequencies and eliminates all others. This is the *ideal band-pass filter*. Its transfer function, $\text{Band}(s)$, can be constructed by shifting and adding transfer functions for the low-pass filter, i.e., adding two shifted and scaled rectangle functions.

We center our band-pass filter at $\pm\nu_0$ and cutoff frequencies more than ν_c above and below ν_0 and $-\nu_0$. Simple enough, the transfer function of a band-pass filter is

$$\begin{aligned}\text{Band}(s) &= \begin{cases} 1, & \nu_0 - \nu_c < |s| < \nu_0 + \nu_c, \\ 0, & \text{otherwise} \end{cases} \\ &= \Pi_{2\nu_c}(s - \nu_0) + \Pi_{2\nu_c}(s + \nu_0) \quad \text{(shifted low-pass filters).}\end{aligned}$$

Here's what it looks like.

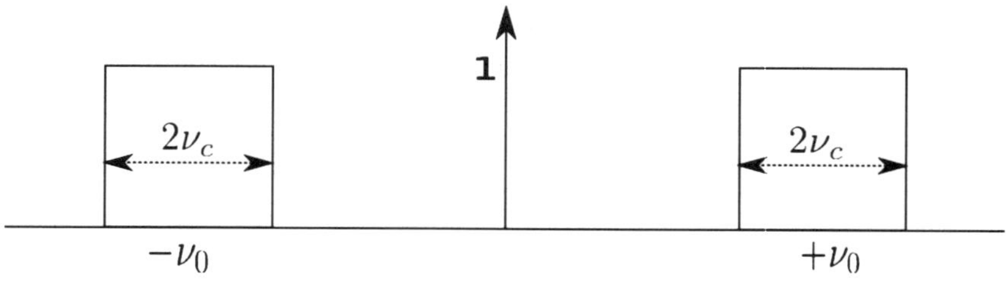

From the representation of Band(s) in terms of shifted and scaled rectangle functions, it's easy to find the impulse response, band(t). That's given by

$$\begin{aligned}\text{band}(t) &= h(t)e^{2\pi i\nu_0 t} + h(t)e^{-2\pi i\nu_0 t} \\ &\qquad \text{(using the shift theorem or the modulation theorem)} \\ &= 4\nu_c \cos(2\pi\nu_0 t)\,\text{sinc}(2\nu_c t).\end{aligned}$$

Here's a plot of band(t) for $\nu_0 = 10$ and $\nu_c = 2$.

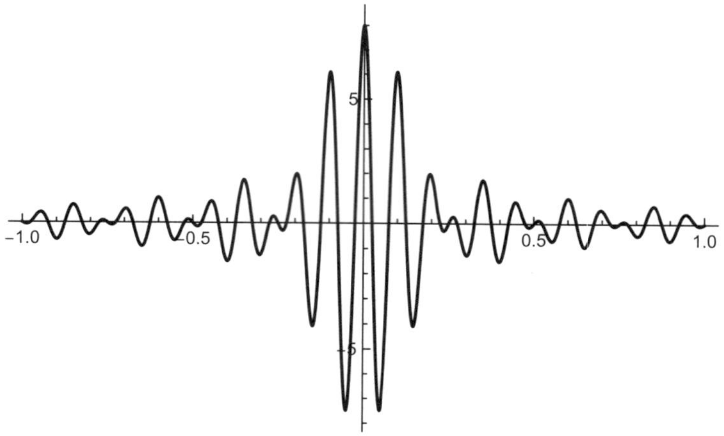

Now, tell the truth, do you really think you could just flip and drag and figure out what the convolution looks like of that thing with some other thing?

High-pass filters. The twin to an ideal low-pass filter is an ideal high-pass filter, where all frequencies beyond a cutoff frequency $\pm\nu_c$ are passed through unchanged and all frequencies in between $\pm\nu_c$ are eliminated. You might use this, for example, if there's a slow "drift" in your data that suggests a low frequency disturbance or noise that you may want to eliminate. High-pass filters are used on images to sharpen edges and details (associated with high spatial frequencies) and eliminate blurring (associated with low spatial frequencies).

The graph of the transfer function for an ideal high-pass filter looks like

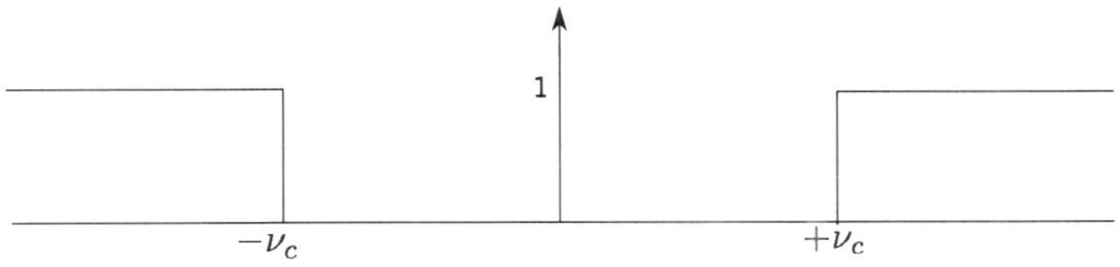

It's easy to write a formula for this; it's just

$$\mathrm{High}(s) = 1 - \Pi_{2\nu_c}(s)$$

where ν_c is the cutoff frequency.[7] At this point we're stuck in finding the impulse response because we haven't yet gained the knowledge that the inverse Fourier transform of 1 is the δ-function, not to mention how to convolve with a δ-function.[8] But, clearly, the filtered version of a signal $v(t)$ is

$$w(t) = v(t) - 2\nu_c \operatorname{sinc}(2\nu_c t) * v(t),$$

because taking the Fourier transform,

$$W(s) = V(s) - \Pi_{2\nu_c}(s)V(s) = (1 - \Pi_{2\nu_c}(s))V(s),$$

exactly eliminates the middle of the spectrum of $v(t)$ from $-\nu_c$ to ν_c.

Notch filters. The evil twin of a band-pass filter is a *notch filter*. The effect of a notch filter is to *eliminate* frequencies within a given band (the "notch," also known again as the stop-band) and to pass frequencies outside that band. To get the transfer function we just subtract a band-pass transfer function from 1. Using the one we already have,

$$\mathrm{Notch}(s) = 1 - \mathrm{Band}(s) = 1 - (\Pi_{2\nu_c}(s - \nu_0) + \Pi_{2\nu_c}(s + \nu_0))\,.$$

This will eliminate the positive frequencies between $\nu_0 - \nu_c$ and $\nu_0 + \nu_c$ and the symmetric corresponding negative frequencies between $-\nu_0 - \nu_c$ and $-\nu_0 + \nu_c$ and will pass all frequencies outside of these two bands. The picture is

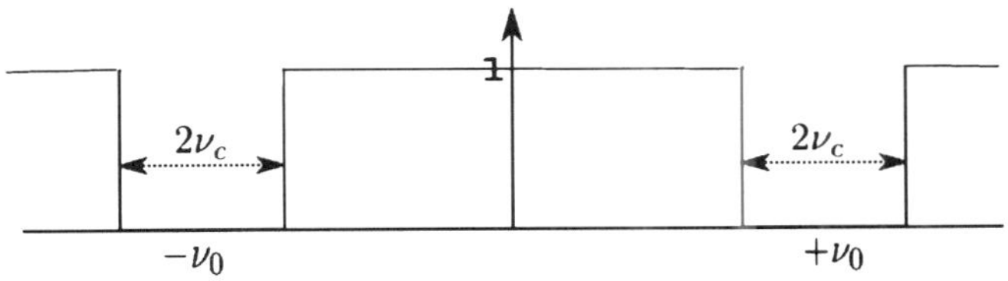

[7]OK, this High(s) is 1 at the endpoints $\pm\nu_c$ instead of 0, but that makes no practical difference. On the other hand, this is a further argument for defining Π to have value 1/2 at the endpoints, for then the transfer functions for the low- and high-pass filters agree in how they cut off.

[8]So think of the high-pass filter as the evil twin of the low-pass filter.

For the impulse response we're in the same position here as we were for the high-pass filter for we cannot take the inverse Fourier transform to find the impulse response without recourse to δ's. But again, as with the band-pass filter, we can write down the filtered function directly:

$$w(t) = v(t) - 4\nu_c \cos(2\pi\nu_0 t)\,\mathrm{sinc}(2\nu_c t) * v(t).$$

3.5. Convolution in Action II: Differential Equations

Solving differential equations was Fourier's original motivation for introducing his series, and the use of the Fourier transform to this end has continued to exercise a strong influence on the theory and the applications. We'll consider several illustrations, from a simple ordinary differential equation to problems associated with the heat equation. We'll also revisit the problem of a signal propagating along a cable.

The derivative formula, again. To put the Fourier transform to work we need a formula for the Fourier transform of the derivative, and as we found in the previous chapter,

$$(\mathcal{F}f')(s) = 2\pi i s\, \mathcal{F}f(s)\,.$$

For higher derivatives,

$$(\mathcal{F}f^{(n)})(s) = (2\pi i s)^n\, \mathcal{F}f(s)\,.$$

In general, a *differential operator* can be thought of as a polynomial in d/dx, say of the form

$$P\Big(\frac{d}{dx}\Big) = a_n\Big(\frac{d}{dx}\Big)^n + a_{n-1}\Big(\frac{d}{dx}\Big)^{n-1} + \cdots + a_1\frac{d}{dx} + a_0\,,$$

and when applied to a function $f(x)$ the result is

$$a_n f^{(n)} + a_{n-1}f^{(n-1)} + \cdots + a_1 f' + a_0 f\,.$$

If we now take the Fourier transform of this expression, we wind up with the Fourier transform of f multiplied by the corresponding nth degree polynomial evaluated at $2\pi i s$:

$$\begin{aligned}\Big(\mathcal{F}\Big(P\Big(\frac{d}{dx}\Big)f\Big)\Big)(s) &= P(2\pi i s)\,\mathcal{F}f(s)\\ &= \big(a_n(2\pi i s)^n + a_{n-1}(2\pi i s)^{n-1} + \cdots + a_1(2\pi i s) + a_0\big)\,\mathcal{F}f(s)\,.\end{aligned}$$

Don't underestimate how important this is.

3.5.1. A simple ordinary differential equation and how to solve it.

You might like starting off with a classic second-order, ordinary differential equation[9]

$$u'' - u = -f.$$

Maybe you've looked at a different form of this equation, but I'm writing it this way to make the subsequent calculations a little easier. $f(t)$ is a given function and you want to find $u(t)$.

[9]Adapted from Dym and McKean.

Take the Fourier transform of both sides:

$$(2\pi is)^2\mathcal{F}u - \mathcal{F}u = -\mathcal{F}f,$$
$$-4\pi^2 s^2\mathcal{F}u - \mathcal{F}u = -\mathcal{F}f,$$
$$(1 + 4\pi^2 s^2)\mathcal{F}u = \mathcal{F}f.$$

Since $1 + 4\pi^2 s^2 \neq 0$, we can solve for $\mathcal{F}u$ as

$$\mathcal{F}u = \frac{1}{1 + 4\pi^2 s^2}\mathcal{F}f\,,$$

and, with a little struggle, we recognize $1/(1 + 4\pi^2 s^2)$ as the Fourier transform of $\frac{1}{2}e^{-|t|}$. Thus

$$\mathcal{F}u = \mathcal{F}\left(\tfrac{1}{2}e^{-|t|}\right) \cdot \mathcal{F}f\,.$$

The right-hand side is the product of two Fourier transforms, and then according to the convolution theorem,

$$u(t) = \tfrac{1}{2}e^{-|t|} * f(t)\,.$$

Written out in full this is

$$u(t) = \tfrac{1}{2}\int_{-\infty}^{\infty} e^{-|t-\tau|} f(\tau)\, d\tau\,.$$

And there you have the two-sided exponential decay in action, as well as convolution.

Those of you paying close attention might wonder about solutions to the homogeneous equation, when $f = 0$ and the ODE is just $u'' = u$. The argument above produces $u = 0$ as the solution, which is fine, but it misses two other solutions, the hyperbolic sine and cosine. If you've seen them:

$$\sinh t = \frac{e^t - e^{-t}}{2}, \qquad \cosh t = \frac{e^t + e^{-t}}{2}\,.$$

You can check that

$$\frac{d}{dt}\sinh t = \cosh t \quad \text{and} \quad \frac{d}{dt}\cosh t = \sinh t\,,$$

from which it follows that both $\sinh t$ and $\cosh t$ satisfy $u'' = u$. What gives? The argument using the Fourier transform should assume that any solution u is twice continuously differentiable *and integrable* (and that f is continuous and integrable), and that's the kind of solution the method produces. Neither $\sinh t$ nor $\cosh t$ is integrable on the whole real line.

3.5.2. The heat equation. Remember the heat equation? In one spatial dimension, the equation that describes the rates of change of the temperature $u(x,t)$ of the body at a point x and time t (with some normalization of the constants associated with the material) is the partial differential equation

$$u_t = \tfrac{1}{2}u_{xx}\,.$$

In our earlier work on Fourier series we considered heat flow on a circle, and we looked for solutions that are periodic functions of x with period 1, so $u(x,t)$ was to satisfy $u(x+1,t) = u(x,t)$. This time we want to consider the problem of heat flow on the infinite rod. A rod of great length (effectively of infinite length) is provided with an initial temperature distribution $f(x)$, and we want to find a solution $u(x,t)$ of the heat equation with

$$u(x,0) = f(x)\,.$$

Both $f(x)$ and $u(x,t)$ are defined for $-\infty < x < \infty$, and there is no assumption of periodicity. Knowing the Fourier transform of the Gaussian is essential for the treatment we're about to give.

The idea is to take the Fourier transform of both sides of the heat equation with respect to x. The Fourier transform of the right-hand side of the equation, $\frac{1}{2}u_{xx}(x,t)$, is

$$\tfrac{1}{2}\mathcal{F}u_{xx}(s,t) = \tfrac{1}{2}(2\pi i s)^2\mathcal{F}u(s,t) = -2\pi^2 s^2\mathcal{F}u(s,t)\,,$$

from the derivative formula. Observe that the frequency variable (still using that terminology) s is now in the first slot of the transformed function and that the time variable t is just going along for the ride. For the left-hand side, $u_t(x,t)$, we do something different. We have

$$\begin{aligned}
\mathcal{F}u_t(s,t) &= \int_{-\infty}^{\infty} u_t(x,t)e^{-2\pi i s x}\,dx \quad \text{(Fourier transform in } x\text{)}\\
&= \int_{-\infty}^{\infty} \frac{\partial}{\partial t}u(x,t)e^{-2\pi i s x}\,dx\\
&= \frac{\partial}{\partial t}\int_{-\infty}^{\infty} u(x,t)e^{-2\pi i s x}\,dx = \frac{\partial}{\partial t}\mathcal{F}u(s,t).
\end{aligned}$$

Thus taking the Fourier transform with respect to x of both sides of the equation

$$u_t = \tfrac{1}{2}u_{xx}$$

leads to

$$\frac{\partial \mathcal{F}u(s,t)}{\partial t} = -2\pi^2 s^2\mathcal{F}u(s,t)\,.$$

This is a differential equation in t — an ordinary differential equation, despite the partial derivative symbol — and we can solve it:

$$\mathcal{F}u(s,t) = \mathcal{F}u(s,0)e^{-2\pi^2 s^2 t}\,.$$

What is the initial condition $\mathcal{F}u(s,0)$?

$$\begin{aligned}
\mathcal{F}u(s,0) &= \int_{-\infty}^{\infty} u(x,0)e^{-2\pi i s x}\,dx\\
&= \int_{-\infty}^{\infty} f(x)e^{-2\pi i s x}\,dx = \mathcal{F}f(s)\,.
\end{aligned}$$

Putting it all together,

$$\mathcal{F}u(s,t) = \mathcal{F}f(s)e^{-2\pi^2 s^2 t}\,.$$

We recognize (we are good) that the exponential factor on the right-hand side is the Fourier transform of the Gaussian,

$$g(x,t) = \frac{1}{\sqrt{2\pi t}}e^{-x^2/2t}\,.$$

We then have a product of two Fourier transforms,

$$\mathcal{F}u(s,t) = \mathcal{F}g(s,t)\,\mathcal{F}f(s)$$

and we invert this to obtain a convolution in the spatial domain:

$$u(x,t) = g(x,t) * f(x) = \left(\frac{1}{\sqrt{2\pi t}} e^{-x^2/2t}\right) * f(x) \quad \text{(convolution in } x\text{)}$$

or, written out,

$$u(x,t) = \int_{-\infty}^{\infty} \frac{1}{\sqrt{2\pi t}} e^{-(x-y)^2/2t} f(y)\, dy\,.$$

It's reasonable to believe that the temperature $u(x,t)$ of the rod at a point x at a time $t > 0$ is some kind of averaged, smoothed version of the initial temperature $f(x) = u(x,0)$. That's convolution at work.

The function

$$g(x,t) = \frac{1}{\sqrt{2\pi t}} e^{-x^2/2t}$$

is called the *heat kernel* (or Green's function or the fundamental solution) for the heat equation for the infinite rod. Here are plots of $g(x,t)$, as a function of x, for $t = 1, 0.5, 0.1, 0.05, 0.01$; the more peaked curves corresponding to the smaller values of t.

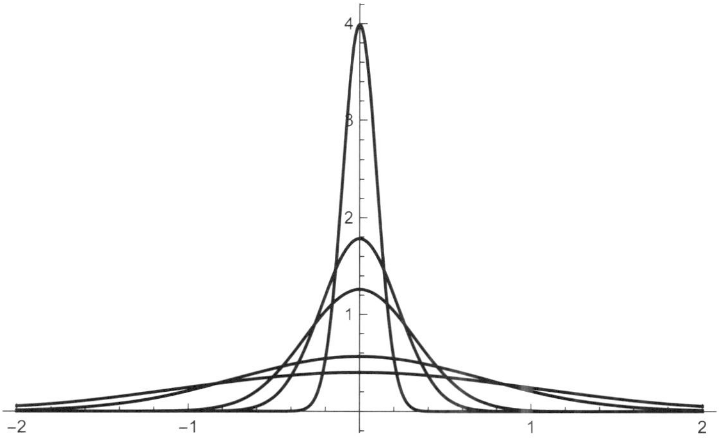

You can see that the curves are becoming more concentrated near $x = 0$. However, they are doing so in a way that keeps the area under each curve 1, for

$$\begin{aligned}\int_{-\infty}^{\infty} \frac{1}{\sqrt{2\pi t}} e^{-x^2/2t}\, dx &= \frac{1}{\sqrt{2\pi t}} \int_{-\infty}^{\infty} e^{-\pi u^2} \sqrt{2\pi t}\, du \\ &\qquad \text{(making the substitution } u = x/\sqrt{2\pi t}\text{)} \\ &= \int_{-\infty}^{\infty} e^{-\pi u^2}\, du = 1\,.\end{aligned}$$

We'll see later that the $g(x,t)$ serve as an approximation to the δ-function as $t \to 0$.

You might ask at this point: Didn't we already solve the heat equation? Is what we did in Chapter 1 related to what we did just now? Indeed we did and indeed they are: see Section 3.5.4.

3.5.3. More on diffusion; back to the cable. When last we left the problem of signals propagating down a cable, William Thomson had appealed to the heat equation to study the delay in a signal sent along a long, undersea telegraph cable. The physical intuition, as of the mid 19th century, was that charge diffused along the cable. To reconstruct part of Thomson's solution (essentially) we must begin with a slightly different setup. The equation is the same':

$$u_t = \tfrac{1}{2} u_{xx}\,,$$

so we're choosing constants as above and not explicitly incorporating physical parameters such as resistance per length, capacitance per length, etc. More importantly, the initial and boundary conditions are different.

We consider a *semi-infinite* rod, having one end (at $x = 0$) but effectively extending infinitely in the positive x-direction. Instead of an initial distribution of temperature along the entire rod, we suppose there is a source of heat (or voltage) $f(t)$ at the end $x = 0$. Thus we have the initial condition

$$u(0, t) = f(t)\,.$$

We also suppose that

$$u(x, 0) = 0\,,$$

meaning that at $t = 0$ there's no temperature (or charge) in the rod. We also assume that $u(x, t)$ and its derivatives tend to zero as $x \to \infty$. Finally, we set

$$u(x, t) = 0 \quad \text{for } x < 0$$

so that we can regard $u(x, t)$ as defined for all x. We want a solution that expresses $u(x, t)$, the temperature (or voltage) at a position $x > 0$ and time $t > 0$ in terms of the initial temperature (or voltage) $f(t)$ at the endpoint $x = 0$.

The analysis of this is *really* involved. It's quite a striking formula that works out in the end, but, be warned, the end is a way off. Proceed only if interested.

First take the Fourier transform of $u(x, t)$ with respect to x (the notation $\hat{u}$ seems more natural here):

$$\hat{u}(s, t) = \int_{-\infty}^{\infty} e^{-2\pi i s x} u(x, t)\, dx\,.$$

Then, using the heat equation,

$$\frac{\partial}{\partial t}\hat{u}(s, t) = \int_{-\infty}^{\infty} e^{-2\pi i s x} \frac{\partial}{\partial t} u(x, t)\, dx = \int_{-\infty}^{\infty} e^{-2\pi i s x} \frac{1}{2}\frac{\partial^2}{\partial x^2} u(x, t)\, dx\,.$$

We need to integrate only from 0 to ∞ since $u(x,t)$ is identically 0 for $x < 0$. We integrate by parts once:

$$\begin{aligned}
&\int_0^\infty e^{-2\pi i s x}\frac{1}{2}\frac{\partial^2}{\partial x^2}u(x,t)\,dx \\
&\quad = \frac{1}{2}\left(\left[e^{-2\pi i s x}\frac{\partial}{\partial x}u(x,t)\right]_{x=0}^{x=\infty} + 2\pi i s\int_0^\infty \frac{\partial}{\partial x}u(x,t)\,e^{-2\pi i s x}\,dx\right) \\
&\quad = -\tfrac{1}{2}u_x(0,t) + \pi i s\int_0^\infty \frac{\partial}{\partial x}u(x,t)\,e^{-2\pi i s x}\,dx\,,
\end{aligned}$$

taking the boundary conditions on $u(x,t)$ into account. Now integrate by parts a second time:

$$\begin{aligned}
\int_0^\infty \frac{\partial}{\partial x}u(x,t)\,e^{-2\pi i s x}\,dx &= \left[e^{-2\pi i s x}\,u(x,t)\right]_{x=0}^{x=\infty} + 2\pi i s\int_0^\infty e^{-2\pi i s t}\,u(x,t)\,dx \\
&= -u(0,t) + 2\pi i s\int_0^\infty e^{-2\pi i s t}\,u(x,t)\,dx \\
&= -f(t) + 2\pi i s\int_{-\infty}^\infty e^{-2\pi i s t}\,u(x,t)\,dx
\end{aligned}$$

(we drop the bottom limit back to $-\infty$ to bring back the Fourier transform)

$$= -f(t) + 2\pi i s\,\hat{u}(s,t).$$

Putting these calculations together yields

$$\frac{\partial}{\partial t}\hat{u}(s,t) = -\tfrac{1}{2}u_x(0,t) - \pi i s f(t) - 2\pi^2 s^2\hat{u}(s,t)\,.$$

This is a linear, first-order, ordinary differential equation (in t) for $\hat{u}$. It's of the general type

$$y'(t) + P(t)y(t) = Q(t)\,.$$

If you cast your mind back to courses from the dim and distant past, you will recall that to solve such an equation you multiply both sides by the integrating factor

$$e^{\int_0^t P(\tau)\,d\tau}\,,$$

which produces

$$\left(y(t)e^{\int_0^t P(\tau)\,d\tau}\right)' = e^{\int_0^t P(\tau)\,d\tau}Q(t)\,.$$

From here you get $y(t)$ by direct integration. For our particular application we have

$P(t) = 2\pi^2 s^2$ (that's a constant as far as we're concerned because there's no t),

$Q(t) = -\frac{1}{2}u_x(0,t) - \pi i s f(t)$.

The integrating factor is $e^{2\pi^2 s^2 t}$ and we're to solve

$$(e^{2\pi^2 s^2 t}\hat{u}(t))' = e^{2\pi^2 s^2 t}\left(-\tfrac{1}{2}u_x(0,t) - \pi i s f(t)\right)\,.$$

I want to carry this out so you don't miss anything.

Write τ for t and integrate both sides from 0 to t with respect to τ:

$$e^{2\pi^2 s^2 t}\hat{u}(s,t) - \hat{u}(s,0) = \int_0^t e^{2\pi^2 s^2 \tau}\left(-\tfrac{1}{2}u_x(0,\tau) - \pi i s f(\tau)\right)\,d\tau\,.$$

But $\hat{u}(s,0) = 0$ since $u(x,0)$ is identically 0, so

$$\begin{aligned}\hat{u}(s,t) &= e^{-2\pi^2 s^2 t}\int_0^t e^{2\pi^2 s^2 \tau}\left(-\tfrac{1}{2}u_x(0,\tau) - \pi i s f(\tau)\right)d\tau\\ &= \int_0^t e^{-2\pi^2 s^2 (t-\tau)}\left(-\tfrac{1}{2}u_x(0,\tau) - \pi i s f(\tau)\right)d\tau.\end{aligned}$$

We need to take the inverse transform of this to get $u(x,t)$. Be not afraid:

$$\begin{aligned}u(x,t) &= \int_{-\infty}^{\infty} e^{2\pi i s x}\hat{u}(s,t)\,ds\\ &= \int_{-\infty}^{\infty} e^{2\pi i s x}\Big(\int_0^t e^{-2\pi^2 s^2 (t-\tau)}\left(-\tfrac{1}{2}u_x(0,\tau) - \pi i s f(\tau)\right)d\tau\Big)\,ds\\ &= \int_0^t\int_{-\infty}^{\infty} e^{2\pi i s x}e^{-2\pi^2 s^2 (t-\tau)}\left(-\tfrac{1}{2}u_x(0,\tau) - \pi i s f(\tau)\right)ds\,d\tau\,.\end{aligned}$$

Appearances to the contrary, this is not hopeless. Let's pull out the inner integral for further examination:

$$\begin{aligned}&\int_{-\infty}^{\infty} e^{2\pi i s x}(e^{-2\pi^2 s^2 (t-\tau)}\left(-\tfrac{1}{2}u_x(0,\tau) - \pi i s f(\tau)\right))\,ds\\ &\quad= -\tfrac{1}{2}u_x(0,\tau)\int_{-\infty}^{\infty} e^{2\pi i s x}e^{-2\pi^2 s^2 (t-\tau)}\,ds - \pi i f(\tau)\int_{-\infty}^{\infty} e^{2\pi i s x}\,s\,e^{-2\pi^2 s^2 (t-\tau)}\,ds\,.\end{aligned}$$

The first integral is the inverse Fourier transform of a Gaussian; we want to find $\mathcal{F}^{-1}\big(e^{-2\pi s^2 (t-\tau)}\big)$. Recall the formulas

$$\mathcal{F}\Big(\frac{1}{\sigma\sqrt{2\pi}}e^{-x^2/2\sigma^2}\Big) = e^{-2\pi^2\sigma^2 s^2}\,, \quad \mathcal{F}(e^{-x^2/2\sigma^2}) = \sigma\sqrt{2\pi}\,e^{-2\pi^2\sigma^2 s^2}.$$

Apply this with

$$\sigma = \frac{1}{2\pi\sqrt{(t-\tau)}}\,.$$

Then, using duality and evenness of the Gaussian, we have

$$\int_{-\infty}^{\infty} e^{2\pi i s x}e^{-2\pi s^2 (t-\tau)}\,ds = \mathcal{F}^{-1}\big(e^{-2\pi s^2 (t-\tau)}\big) = \frac{e^{-x^2/2(t-\tau)}}{\sqrt{2\pi(t-\tau)}}\,.$$

In the second integral we want to find $\mathcal{F}^{-1}(s\,e^{-2\pi^2 s^2 (t-\tau)})$. For this, note that

$$s e^{-2\pi^2 s^2 (t-\tau)} = -\frac{1}{4\pi^2(t-\tau)}\frac{d}{ds}e^{-2\pi^2 s^2 (t--\tau)}$$

and hence

$$\begin{aligned}\int_{-\infty}^{\infty} e^{2\pi i s x}\,s\,e^{-2\pi^2 s^2 (t-\tau)}\,ds &= \mathcal{F}^{-1}\Big(-\frac{1}{4\pi^2(t-\tau)}\frac{d}{ds}e^{-2\pi^2 s^2 (t-\tau)}\Big)\\ &= -\frac{1}{4\pi^2(t-\tau)}\mathcal{F}^{-1}\Big(\frac{d}{ds}e^{-2\pi^2 s^2 (t-\tau)}\Big)\,.\end{aligned}$$

We know how to take the inverse Fourier transform of a derivative, or rather we know how to take the (forward) Fourier transform, and that's all we need to do by another application of duality. Let me remind you: we use, for a general function f,

$$\mathcal{F}^{-1}f' = (\mathcal{F}f')^- = (2\pi i x \mathcal{F}f)^- = -2\pi i x(\mathcal{F}f)^- = -2\pi i x \mathcal{F}^{-1}f\,.$$

Apply this to

$$\begin{aligned}\mathcal{F}^{-1}\Big(\frac{d}{ds}e^{-2\pi^2 s^2(t-\tau)}\Big) &= -2\pi i x\mathcal{F}^{-1}\big(e^{-2\pi^2 s^2(t-\tau)}\big)\\ &= -2\pi i x\frac{1}{\sqrt{2\pi(t-\tau)}}\,e^{-x^2/2(t-\tau)}\end{aligned}$$

(from our earlier calculation, fortunately).

Then

$$-\frac{1}{4\pi^2(t-\tau)}\mathcal{F}^{-1}\Big(\frac{d}{ds}e^{-2\pi^2 s^2(t-\tau)}\Big) = \frac{2\pi i x}{4\pi^2(t-\tau)}\,\frac{e^{-x^2/2(t-\tau)}}{\sqrt{2\pi(t-\tau)}} = \frac{i}{2\pi}\,\frac{x\,e^{-x^2/2(t-\tau)}}{\sqrt{2\pi(t-\tau)^3}}\,.$$

That is,

$$\mathcal{F}^{-1}\Big(s\,e^{-2\pi^2 s^2(t-\tau)}\Big) = \frac{i}{2\pi}\,\frac{x\,e^{-x^2/2(t-\tau)}}{\sqrt{2\pi(t-\tau)^3}}\,.$$

Finally getting back to the expression for $u(x,t)$, we can combine what we've calculated for the inverse Fourier transforms and write

$$\begin{aligned}u(x,t) &= -\tfrac{1}{2}\int_0^t u_x(0,\tau)\mathcal{F}^{-1}(e^{-2\pi s^2(t-\tau)})\,d\tau \;-\; \pi i\int_0^t f(\tau)\mathcal{F}^{-1}(s\,e^{-2\pi^2 s^2(t-\tau)})\,d\tau\\ &= -\tfrac{1}{2}\int_0^t u_x(0,\tau)\frac{e^{-x^2/2(t-\tau)}}{\sqrt{2\pi(t-\tau)}}\,d\tau \;+\; \tfrac{1}{2}\int_0^t f(\tau)\,\frac{x\,e^{-x^2/2(t-\tau)}}{\sqrt{2\pi(t-\tau)^3}}\,d\tau.\end{aligned}$$

We're almost there. We'd like to eliminate $u_x(0,\tau)$ from this formula and express $u(x,t)$ in terms of $f(t)$ only. This can be accomplished by a very clever, and I'd say highly nonobvious, observation. We know that $u(x,t)$ is zero for $x<0$; we have defined it to be so. Hence the integral expression for $u(x,t)$ is zero for $x<0$. Because of the evenness and oddness *in* x of the two integrands this has a consequence for the values of the integrals when x is positive. (The first integrand is even in x and the second is odd in x.) In fact, the integrals are equal!

Let me explain what happens in a general situation, stripped down, so you can see the idea. Suppose we have

$$\Phi(x,t) = \int_0^t \phi(x,\tau)\,d\tau + \int_0^t \psi(x,\tau)\,d\tau$$

where we know that: $\Phi(x,t)$ is zero for $x<0$; $\phi(x,\tau)$ is even in x; $\psi(x,\tau)$ is odd in x. Take $a>0$. Then $\Phi(-a,\tau)=0$; hence using the evenness of $\phi(x,\tau)$ and the oddness of $\psi(x,\tau)$,

$$0 = \int_0^t \phi(-a,\tau)\,d\tau + \int_0^t \psi(-a,\tau)\,d\tau = \int_0^t \phi(a,\tau)\,d\tau - \int_0^t \psi(a,\tau)\,d\tau\,.$$

We conclude that for all $a > 0$,

$$\int_0^t \phi(a,\tau) = \int_0^t \psi(a,\tau)\,d\tau\,,$$

and hence for $x > 0$ (writing x for a)

$$\begin{aligned}\Phi(x,t) &= \int_0^t \phi(x,\tau)\,d\tau + \int_0^t \psi(x,\tau)\,d\tau \\ &= 2\int_0^t \psi(x,\tau)\,d\tau = 2\int_0^t \phi(x,\tau)\,d\tau \quad \text{(either } \phi \text{ or } \psi \text{ could be used).}\end{aligned}$$

We apply this in our situation with

$$\phi(x,\tau) = -\tfrac{1}{2}u_x(0,\tau)\frac{e^{-x^2/2(t-\tau)}}{\sqrt{2\pi(t-\tau)}}\,,\qquad \psi(x,\tau) = \tfrac{1}{2}f(\tau)\,\frac{x\,e^{-x^2/2(t-\tau)}}{\sqrt{2\pi(t-\tau)^3}}\,.$$

The result is that we can eliminate the integral with the $u_x(0,\tau)$ and write the solution — the final solution — as

$$u(x,t) = \int_0^t f(\tau)\,\frac{x\,e^{-x^2/2(t-\tau)}}{\sqrt{2\pi(t-\tau)^3}}\,d\tau\,.$$

This form of the solution was the one given by Stokes. He wrote to Thomson:

> In working out myself various forms of the solution of the equation $dv/dt = d^2v/dx^2$ [*Note*: He puts a 1 on the right-hand side instead of a 1/2, and he uses an ordinary "*d*".] under the condition $v = 0$ when $t = 0$ from $x = 0$ to $x = \infty$; $v = f(t)$ when $x = 0$ from $t = 0$ to $t = \infty$ I found the solution ... was ...
>
> $$v(x,t) = \frac{x}{2\sqrt{\pi}}\int_0^t (t-t')^{-3/2}e^{-x^2/4(t-t')}f(t')\,dt'\,.$$

3.5.4. Didn't we already solve the heat equation? Our first application of Fourier series (*the* first application of Fourier series) was to solve the heat equation. Let's recall the setup and the form of the solution. We heat a circle, which we consider to be the interval $0 \le x \le 1$ with the endpoints identified. If the initial distribution of temperature is the function $f(x)$, then the temperature $u(x,t)$ at a point x at time $t > 0$ is given by

$$u(x,t) = \int_0^1 g(x-y,t)f(y)\,dy\,,$$

where

$$g(x,t) = \sum_{n=-\infty}^{\infty} e^{-2\pi^2n^2t}e^{2\pi inx}\,.$$

So, explicitly,

$$u(x,t) = g(x,t) * f(x) = \int_0^1 \sum_{n=-\infty}^{\infty} e^{-2\pi^2n^2t}e^{2\pi in(x-y)}f(y)\,dy\,.$$

This was our first encounter with convolution, a convolution in the spatial variable but with limits of integration just from 0 to 1. Here $f(x)$, $g(x,t)$, and $u(x,t)$ are periodic of period 1 in x.

How does this compare to what we did for the rod? If we imagine initially heating up a circle as heating up an infinite rod by a *periodic* function $f(x)$, then shouldn't we be able to express the temperature $u(x,t)$ for the circle as we did for the rod? We will show that the solution for a circle does have the *same form* as the solution for the infinite rod by means of the remarkable identity:

$$\sum_{n=-\infty}^{\infty} e^{-(x-n)^2/2t} = \sqrt{2\pi t} \sum_{n=-\infty}^{\infty} e^{-2\pi^2 n^2 t} e^{2\pi i n x}$$

Needless to say, this is *not* obvious.

As an aside, for general interest, a special case of this identity is particularly famous. The *Jacobi theta function*[10] is defined by

$$\vartheta(t) = \sum_{n=-\infty}^{\infty} e^{-\pi n^2 t},$$

for $t > 0$. It comes up in surprisingly diverse pure and applied fields, including number theory and statistical mechanics! (In the latter it is used to study partition functions.) Jacobi's identity is

$$\vartheta(t) = \frac{1}{\sqrt{t}} \vartheta\left(\frac{1}{t}\right).$$

It follows from the identity above, with $x = 0$ and replacing t by $1/2\pi t$.

We'll show later why the general identity holds. But first, assuming that it does, let's work with the solution of the heat equation for a circle and see what we get. Applying the identity to Green's function $g(x,t)$ for heat flow on the circle we have

$$g(x,t) = \sum_{n=-\infty}^{\infty} e^{-2\pi^2 n^2 t} e^{2\pi i n x} = \frac{1}{\sqrt{2\pi t}} \sum_{n=-\infty}^{\infty} e^{-(x-n)^2/2t}.$$

[10] Named after C. Jacobi. He contributed a staggering amount of work to mathematics, many ideas finding applications in very diverse areas. He is perhaps second only to Euler for pure analytic skill.

Regard the initial distribution of heat $f(x)$ as being defined on all of $\mathbb{R}$ and having period 1. Then

$$\begin{aligned}
u(x,t) &= \int_0^1 \sum_{n=-\infty}^{\infty} e^{-2\pi^2 n^2 t} e^{2\pi i n(x-y)} f(y)\, dy \\
&= \frac{1}{\sqrt{2\pi t}} \int_0^1 \sum_{n=-\infty}^{\infty} e^{-(x-y-n)^2/2t} f(y)\, dy \\
&\qquad \text{(using the identity for Green's function)} \\
&= \frac{1}{\sqrt{2\pi t}} \sum_{n=-\infty}^{\infty} \int_0^1 e^{-(x-y-n)^2/2t} f(y)\, dy \\
&= \frac{1}{\sqrt{2\pi t}} \sum_{n=-\infty}^{\infty} \int_n^{n+1} e^{-(x-u)^2/2t} f(u-n)\, du \quad \text{(substituting } u = y+n\text{)} \\
&= \frac{1}{\sqrt{2\pi t}} \sum_{n=-\infty}^{\infty} \int_n^{n+1} e^{-(x-u)^2/2t} f(u)\, du \quad \text{(using that } f \text{ has period 1)} \\
&= \frac{1}{\sqrt{2\pi t}} \int_{-\infty}^{\infty} e^{-(x-u)^2/2t} f(u)\, du\,.
\end{aligned}$$

Voilà! We are back to the solution of the heat equation on the line.

Incidentally, since the problem was originally formulated for heating a circle, the function $u(x,t)$ is periodic in x. Can we see that from this form of the solution? Yes, for

$$\begin{aligned}
u(x+1,t) &= \frac{1}{\sqrt{2\pi t}} \int_{-\infty}^{\infty} e^{-(x+1-u)^2/2t} f(u)\, du \\
&= \frac{1}{\sqrt{2\pi t}} \int_{-\infty}^{\infty} e^{-(x-w)^2/2t} f(w+1)\, dw \quad \text{(substituting } w = u-1\text{)} \\
&= \frac{1}{\sqrt{2\pi t}} \int_{-\infty}^{\infty} e^{-(x-w)^2/2t} f(w)\, dw \quad \text{(using the periodicity of } f(x)\text{)} \\
&= u(x,t)\,.
\end{aligned}$$

Now let's derive the identity

$$\sum_{n=-\infty}^{\infty} e^{-(x-n)^2/2t} = \sqrt{2\pi t} \sum_{n=-\infty}^{\infty} e^{-2\pi^2 n^2 t} e^{2\pi i n x}\,.$$

This is a great combination of many of the things we've developed to this point, and it will come up again.[11] Consider the left-hand side as a function of x, say

$$h(x) = \sum_{n=-\infty}^{\infty} e^{-(x-n)^2/2t}\,.$$

[11]It's worth your effort to go through this. The calculations in this special case will come up more generally when we do the Poisson summation formula. That formula is the basis of the sampling theorem.

This is a periodic function of period 1; it's the *periodization* of the Gaussian $e^{-x^2/2t}$. (It's even not hard to show that the series converges, etc., but we won't go through that.) What are its Fourier coefficients? We can calculate them:

$$\begin{aligned}
\hat{h}(k) &= \int_0^1 h(x)e^{-2\pi ikx}\,dx \\
&= \int_0^1 \left(\sum_{n=-\infty}^{\infty} e^{-(x-n)^2/2t}\right)e^{-2\pi ikx}\,dx \\
&= \sum_{n=-\infty}^{\infty} \int_0^1 e^{-(x-n)^2/2t}e^{-2\pi ikx}\,dx \\
&= \sum_{n=-\infty}^{\infty} \int_{-n}^{-n+1} e^{-u^2/2t}e^{-2\pi iku}\,du \\
&\qquad \text{(substituting } u = x - n \text{ and using periodicity of } e^{-2\pi ikx}\text{)} \\
&= \int_{-\infty}^{\infty} e^{-u^2/2t}e^{-2\pi iku}\,du\,.
\end{aligned}$$

But this last integral is exactly the Fourier transform of the Gaussian $e^{-x^2/2t}$ at $s = k$. We know how to do that. The answer is $\sqrt{2\pi t}\,e^{-2\pi^2k^2t}$.

We have shown that the Fourier coefficients of $h(x)$ are

$$\hat{h}(k) = \sqrt{2\pi t}\,e^{-2\pi^2k^2t}\,.$$

Since the function is equal to its Fourier series (really equal here because all the series converge), we conclude that

$$\begin{aligned}
h(x) &= \sum_{n=-\infty}^{\infty} e^{-(x-n)^2/2t} \\
&= \sum_{n=-\infty}^{\infty} \hat{h}(n)e^{2\pi inx} = \sqrt{2\pi t}\sum_{n=-\infty}^{\infty} e^{-2\pi^2n^2t}e^{2\pi inx}\,,
\end{aligned}$$

and there's the identity we wanted to prove.

3.6. Convolution in Action III: The Central Limit Theorem

Several times we've met the idea that convolution is a smoothing operation. Let me begin with some graphical examples of this, convolving a discontinuous or rough function repeatedly with itself.

In a problem you computed, by hand, the convolution of the rectangle function Π with itself a few times. Here are plots of this, up to $\Pi * \Pi * \Pi * \Pi$, displayed clockwise from Π.

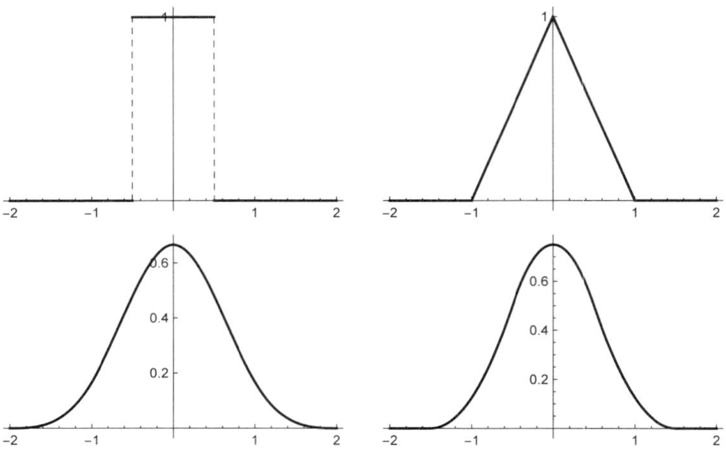

Not only are the convolutions becoming smoother, the unmistakable shape of a Gaussian is emerging. Is this a coincidence, based on the particularly simple nature of the function Π, or is something more going on?

Here is a plot of, literally, a random function $f(x)$ — the values $f(x)$ are just randomly chosen numbers between 0 and 1 — and its self-convolutions up to the four-fold convolution $f * f * f * f$, again displayed clockwise.

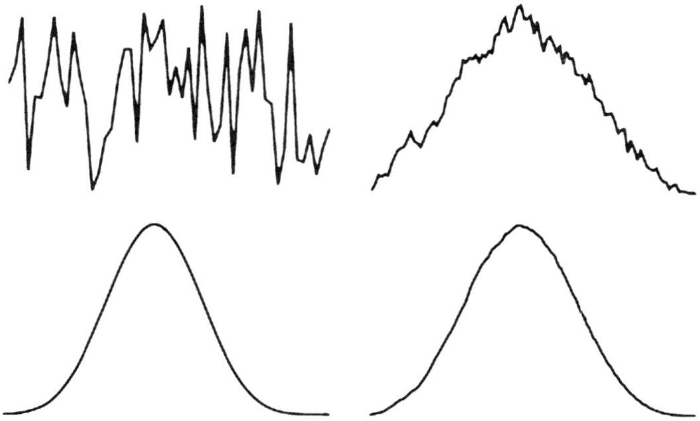

From seeming chaos, again we see a Gaussian emerging. This is the spookiest thing I know.

The object of this section is to explain this phenomenon, to give substance to the famous quotation:

> Everyone believes in the normal approximation, the experimenters because they think it is a mathematical theorem, the mathematicians because they think it is an experimental fact.
>
> G. Lippman, French Physicist, 1845–1921

The "normal approximation" (or normal distribution) is the Gaussian. The "mathematical theorem" here is the *Central Limit Theorem*. To understand the theorem and to appreciate the "experimental fact," we have to develop some ideas from probability.

3.6.1. Random variables.

In whatever field of science or engineering you pursue, you will certainly use probabilistic ideas, and you will certainly use the Gaussian. I'm going under the assumption that you probably already know some probability and probably some statistics, too, even if only in a casual way. For our present work we don't need complete generality based on hard won, abstract definitions. Just the basics will do. Every mature field settles on its fundamental notions (or tries to), sometimes clear at the outset and sometimes slow to emerge. One of the fundamental notions for probability, maybe *the* fundamental notion, is the *random variable*. It was slow to emerge, or at least a suitable definition was slow to emerge. We're going to be informal, even cavalier about this: a random variable is a number you don't know yet.[12] By that I mean that a random variable, or rather *its value*, is the numerical result of some process, like a measurement or the result of an experiment. The assumption is that you can make the measurement, you can perform the experiment, but until you do you don't know the value of the random variable. It's called "random" because a particular object to be measured is thought of as being drawn "at random" from a collection of all such objects. For example:

Random variable	Value of random variable
Height of people in US population	Height of a particular person
Length of pins produced	Length of a particular pin
Momentum of atoms in a gas	Momentum of a particular atom
Resistance of resistors off a production line	Resistance of a particular resistor
Toss of coin	0 or 1 (head or tail)
Roll of dice	The numbers that come up

A common notation is to write X for the name of the random variable and x for its value. If you thus think that a random variable X is just a function, you're right, but deciding what the *domain* of such a function should be and what mathematical structure to require of both the domain and the function demands the kind of care that we don't want to get into. As I said, this was a long time in coming. Consider, for example, Mark Kac's comment: "Independent random variables were to me (and others, including my teacher Steinhaus) shadowy and not really well-defined objects." Kac was one of the most eminent probabilists of the 20th century.

3.6.2. Probability distributions and probability density functions.

"Random variable" is the fundamental notion, but not the fundamental object of study. For a given random variable, what we're most interested in is how its values are distributed. For this it's helpful to distinguish between two types of random variables.

---

[12] I think this phrase to describe a random variable is due to Sam Savage in the Management Science and Engineering department at Stanford.

- A random variable is *discrete* if its values are among only a discrete set of possibilities, for us, a finite set.
  - For example, "roll the die" is a discrete random variable with values 1, 2, 3, 4, 5, or 6. "Toss the coin" is a discrete random variable with values "head" and "tail," or 0 and 1. (A random variable with values 0 and 1 is the basic random variable in coding and information theory.)
- A random variable is *continuous* if its values do not form a discrete set, for us typically filling up one or more intervals of real numbers.
  - For example, "length of a pin" is a continuous random variable since, in theory, the length of a pin can vary continuously.

For a discrete random variable we are used to the idea of displaying the distribution of values as a *histogram*. We set up bins, each bin corresponding to each of the possible values (or sometimes a range of values), we run the random process however many times we please, and for each bin we draw a bar with height indicating the *percentage* that the particular value occurs among all actual outcomes of the runs.[13] Since we plot percentages, or fractions, the total area of the histogram is 100%, or just 1.

A series of runs of the same experiment or of the same measurement will produce histograms of varying shapes.[14] We often expect some kind of limiting shape as we increase the number of runs, or we may *suppose* that the ideal distribution has some shape and then compare the actual data from a series of runs to the ideal, theoretical answer.

- The theoretical histogram is called the *probability distribution*.
- The function that describes the histogram (the shape of the distribution) is called the *probability density function*, or *pdf*, of the random variable.

Is there a difference between the probability distribution and the probability density function? No, not really — it's like distinguishing between the graph of a function and the function. Both terms are in common use, more or less interchangeably.

- The *probability* that any particular value comes up is the area of its bin in the probability distribution, which is therefore a number between 0 and 1.

If the random variable is called X and the value we're interested in is x, we write this as

$$\text{Prob}(X = x) = \text{area of the bin over } x\,.$$

Also

$$\text{Prob}(a \le X \le b) = \text{areas of the bins from } a \text{ to } b\,.$$

Thus probability is the percentage of the occurrence of a particular outcome, or range of outcomes, among all possible outcomes. We *must* base the definition of probability on what we presume or assume is the distribution function for a given

[13] I have gotten into heated arguments with physicist friends who insist on plotting frequencies of values rather than percentages. Fools.

[14] A run is like: "Do the experiment 10 times and make a histogram of your results for those 10 trials." A series of runs is like: "Do your run of 10 times, again. And again."

random variable. A statement about probabilities for a run of experiments is then a statement about long term trends, thought of as an approximation to the ideal distribution.[15]

One can also introduce probability distributions and probability density functions for continuous random variables. You can think of this — in fact you probably should think of this — as a continuous version of a probability histogram. It's a tricky business, however, to take a limit of the distributions for a discrete random variable, which have bins of a definite size, to produce a distribution for a continuous random variable, imagining the latter as having infinitely many infinitesimal bins.

It's easiest, and best, to define the distribution for a continuous random variable directly. In fact, we're used to this. It's another example of turning the solution of a problem into a definition. Years of bitter experience taught probabilists what properties they needed of probability distributions to make their theory work, to prove their theorems. Not so many. They started over with the following definition and haven't looked back:

- A *probability density function* is a nonnegative function $p(x)$ with area 1; i.e.,

$$\int_{-\infty}^{\infty} p(x)\, dx = 1\,.$$

Remember, x is the measured value of some experiment. By convention, we take x to go from $-\infty$ to ∞ so we don't constantly have to say how far the values extend.

In this context, over the years I have been taught to say the phrase: "Let X be a random variable *drawn* from a distribution $p(x)$ ..." and to write $X \sim p(x)$.

Here's one quick and important property of pdfs:

- If $p(x)$ is a pdf and $a > 0$, then $ap(ax)$ is also a pdf.

To show this, we have to check that the integral of $ap(ax)$ is 1. But

$$\int_{-\infty}^{\infty} ap(ax)\, dx = \int_{-\infty}^{\infty} ap(u)\frac{1}{a}\, du = \int_{-\infty}^{\infty} p(u)\, du = 1\,,$$

making the change of variable $u = ax$. We'll soon see this property in action.

- We think of a pdf as being associated with a random variable X whose values are x, and we write p_X if we want to emphasize this. The (probability) distribution of X is the graph of p_X, but, again, the terms *probability density function* and *probability distribution* are used interchangeably.
- *Probability* is defined by

$$\begin{aligned}\text{Prob}(X \le a) &= \text{area under the curve for } x \le a\\ &= \int_{-\infty}^{a} p_X(x)\, dx.\end{aligned}$$

---

[15]We are in dangerous territory here. Look up "Bayesian."

Also,

$$\text{Prob}(a \le X \le b) = \int_a^b p_X(x)\,dx\,.$$

For continuous random variables it really only makes sense to talk about the probability of a range of values occurring, not the probability of the occurrence of a single value. Think of the pdf as describing a limit of a (discrete) histogram: if the bins are becoming infinitely thin, what kind of event could land in an infinitely thin bin?[16]

Finally, for variable t, say, we can view

$$P(t) = \int_{-\infty}^t p(x)\,dx$$

as the probability function. It's also called the *cumulative probability* or the *cumulative density function*.[17] We then have

$$\text{Prob}(X \le t) = P(t)$$

and

$$\text{Prob}(a \le X \le b) = P(b) - P(a)\,.$$

According to the Fundamental Theorem of Calculus we can recover the probability density function from $P(t)$ by differentiation:

$$\frac{d}{dt}P(t) = p(t)\,.$$

In short, to know $p(t)$ is to know $P(t)$ and vice versa. You might not think this news is of any particular practical importance, but you're about to see that it is.

3.6.3. Mean, variance, and standard deviation. Suppose X is a random variable with pdf $p(x)$. The x's are the values assumed by X, so the *mean* μ of X is the weighted average of these values, weighted according to p. That is,

$$\mu(X) = \int_{-\infty}^{\infty} xp(x)\,dx\,.$$

The mean is also called the *expected value*, usually written $\mathbb{E}(X)$. "Expected value" is maybe more in keeping with the value of a random variable being determined by a measurement, as in, "What number were you expecting?"

Be careful here — the mean of X is *not* the average value of the function $p(x)$. Also, it might be that $\mu(X) = \infty$, i.e., that the integral of $xp(x)$ does not converge. This has to be checked for any particular example.

If $\mu(X) < \infty$, then we can always subtract off the mean to assume that X has mean zero. Here's what this means, no pun intended. In fact, let's do something slightly more general. What do we mean by $X - a$, when X is a random variable

---

[16] There's also the familiar integral identity

$$\int_a^a p_X(x)\,dx = 0$$

to contend with. In this context we would interpret this as saying that $\text{Prob}(X = a) = 0$.

[17] Cumulative density function is the preferred term because it allows for a three letter acronym: *cdf*.

and a is a constant? Nothing deep — you do the experiment to get a value of X (X is a number you don't know yet) and then you subtract a from it. What is the pdf of $X - a$? To figure that out, we have

$$\begin{aligned}\text{Prob}(X - a \le t) &= \text{Prob}(X \le t + a)\\ &= \int_{-\infty}^{t+a} p(x)\,dx\\ &= \int_{-\infty}^{t} p(u+a)\,du \quad (\text{substituting } u = x - a).\end{aligned}$$

This identifies the pdf of $X - a$ as $p(x + a)$, the shifted pdf of X.[18]

Next, what is the mean of $X - a$? It must be $\mu(X) - a$. (Common sense, please.) Let's check this now knowing what pdf to integrate:

$$\begin{aligned}\mu(X - a) &= \int_{-\infty}^{\infty} xp(x+a)\,dx\\ &= \int_{-\infty}^{\infty} (u-a)p(u)\,du \quad (\text{substituting } u = x + a)\\ &= \int_{-\infty}^{\infty} up(u)\,du - a\int_{-\infty}^{\infty} p(u)\,du = \mu(X) - a\,.\end{aligned}$$

Note that translating the pdf $p(x)$ to $p(x+a)$ does nothing to the shape, or areas, of the distribution, hence does nothing to calculating any probabilities based on $p(x)$. As promised, the mean is $\mu(X) - a$. We are also happy to be certain now that "subtracting off the mean," as in $X - \mu(X)$, really does result in a random variable with mean 0. This normalization is often a convenient one to make in deriving formulas.

Continue to suppose that the mean $\mu(X)$ is finite. The *variance* $\sigma^2(X)$ is a measure of the amount that the values of the random variable deviate from the mean, *on average*, i.e., as weighted by the pdf $p(x)$. Since some values are above the mean and some are below, we weight the *square* of the differences, $(x - \mu(X))^2$, by $p(x)$ and define

$$\sigma^2(X) = \int_{-\infty}^{\infty} (x - \mu(X))^2 p(x)\,dx\,.$$

If we have normalized so that the mean is zero, this becomes simply

$$\sigma^2(X) = \int_{-\infty}^{\infty} x^2 p(x)\,dx\,.$$

The *standard deviation* is $\sigma(X)$, the square root of the variance. Even if the mean is finite, it might be that $\sigma^2(X)$ is infinite. This, too, has to be checked for any particular example.

We've just seen that we can normalize the mean of a random variable to be 0. Assuming that the variance is finite, can we normalize it in some helpful way?

[18]This is an illustration of the practical importance of going *from* the probability function *to* the pdf. We identified the pdf by knowing the probability function. This won't be the last time we do this.

Suppose X has pdf p and let a be a positive constant. Then

$$\begin{aligned}\operatorname{Prob}\left(\frac{1}{a}X \le t\right) &= \operatorname{Prob}(X \le at)\\ &= \int_{-\infty}^{at} p(x)\,dx\\ &= \int_{-\infty}^{t} ap(au)\,du \quad \left(\text{making the substitution } u = \frac{1}{a}x\right).\end{aligned}$$

This says that the random variable $\frac{1}{a}X$ has pdf $ap(ax)$. (Here in action is the scaled pdf $ap(ax)$, which we had as an example of operations on pdf's.) Suppose that we've normalized the mean of X to be 0. Then the variance of $\frac{1}{a}X$ is

$$\begin{aligned}\sigma^2\left(\frac{1}{a}X\right) &= \int_{-\infty}^{\infty} x^2 ap(ax)\,dx\\ &= a\int_{-\infty}^{\infty} \frac{1}{a^2}u^2 p(u)\frac{1}{a}\,du \quad (\text{making the substitution } u = ax)\\ &= \frac{1}{a^2}\int_{-\infty}^{\infty} u^2 p(u)\,du = \frac{1}{a^2}\sigma^2(X)\,.\end{aligned}$$

In particular, if we choose $a = \sigma(X)$, then the variance of $\frac{1}{a}X$ is one. This is also a convenient normalization for many formulas.

In summary:

- Given a random variable X with $\mu(X) < \infty$ and $\sigma(X) < \infty$, it is possible to normalize and assume that $\mu(X) = 0$ and $\sigma^2(X) = 1$.

You see these assumptions a lot.

3.6.4. Two examples.

Let's have two leading examples of pdfs to refer to. Additional examples are in the problems.

The uniform distribution. "Uniform" refers to a random process where all possible outcomes are equally likely. In the discrete case, tossing a coin or throwing a die are examples. All bins in the ideal histogram have the same height, two bins of height 1/2 for the toss of a coin, six bins of height 1/6 for the throw of a single die, and N bins of height $1/N$ for a discrete random variable with N values.

For a continuous random variable, the uniform distribution is identically 1 on an interval of length 1 and zero elsewhere. We've seen such a graph before. If we shift to the interval to go from $-1/2$ to $1/2$, it's the graph of the ever versatile rectangle function: $\Pi(x)$ is now starring in yet another role, that of the uniform distribution.

The mean is 0, obviously,[19] but to verify this formally:

$$\mu = \int_{-\infty}^{\infty} x\Pi(x)\,dx = \int_{-1/2}^{1/2} x\,dx = \tfrac{1}{2}x^2\Big]_{-1/2}^{+1/2} = 0\,.$$

[19]. . . the mean of the random variable with pdf $p(x)$ is *not* the average value of $p(x)$ . . . .

(Also, we're integrating an odd function x, over a symmetric interval.) The variance is then

$$\sigma^2 = \int_{-\infty}^{\infty} x^2 \Pi(x)\, dx = \int_{-1/2}^{1/2} x^2\, dx = \tfrac{1}{3}x^3\Big]_{-1/2}^{+1/2} = \tfrac{1}{12},$$

perhaps not quite so obvious.

The normal distribution. This whole production is about getting to Gaussians, so it seems appropriate that at some point I mention:

- The Gaussian is a pdf.

Indeed, to borrow information from earlier work in this chapter, the Gaussian

$$g(x, \mu, \sigma) = \frac{1}{\sigma\sqrt{2\pi}} e^{-(x-\mu)^2/2\sigma^2}$$

is a pdf with mean μ and variance σ^2. The distribution associated with such a Gaussian is called a *normal distribution*. There, it's official. But why is it "normal"? You're soon to find out.

3.6.5. Independence. An important extra property that two or more random variables may have is *independence*. The plain English description of independence is that one event or measurement doesn't influence another event or measurement. Each flip of a coin, roll of a die, or measurement of a resistor is a new event, not influenced by previous events.

Operationally, independence implies that the probabilities multiply: if two random variables X_1 and X_2 are independent, then

$$\text{Prob}(X_1 \le a \text{ and } X_2 \le b) = \text{Prob}(X_1 \le a) \cdot \text{Prob}(X_2 \le b)\,.$$

In words, if $X_1 \le a$ occurs r percent and $X_2 \le b$ occurs s percent, then, if the events are independent, the percent that $X_1 \le a$ occurs *and* $X_2 \le b$ occurs is r percent of s percent (or the other way around).

3.6.6. Convolution appears. Using the terminology we've encountered, we can begin to be more precise about the content of the Central Limit Theorem. That result — the ubiquity of the bell-shaped curve — has to do with sums of independent random variables and with the distributions of those sums. The two notions are linked through convolution.

While we'll ultimately work with continuous random variables, let's look at the discrete random variable X = "roll the die" as an example. The ideal histogram for the toss of a single die is uniform; each number 1 through 6 comes up with equal probability. We might represent it pictorially like this:

I don't mean to think just of a picture of dice here, I mean to think of the distribution as six bins of equal height 1/6, each bin corresponding to one of the six possible tosses.

What about the *sum* of the tosses of two dice? What is the distribution, theoretically, of the sums? The possible values of the sum are 2 through 12, but the values do not occur with equal probability. There's only one way of making 2 and one way of making 12, but there are more ways of making the other possible sums. In fact, 7 is the most probable sum, with six ways it can be achieved. We might represent the distribution for the sum of two dice pictorially like this:

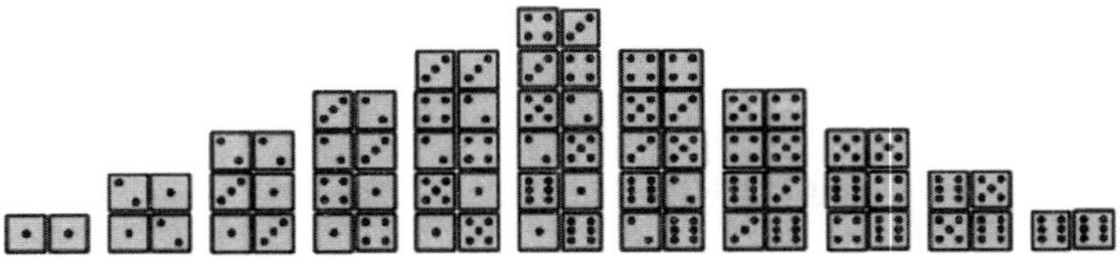

It's triangular. Now let's see ... for the single random variable $X =$ "roll one die" we have a distribution like a rect function. For the sum, say random variables $X_1 + X_2 =$ "roll of die 1 plus roll of die 2", the distribution looks like the triangle function....

The key discovery is this:

- **Convolution and probability density functions.** The probability density function of the sum of two independent random variables is the convolution of the probability density functions of each.

What a beautiful, elegant, and useful statement! Let's see why it works.

We can get a good intuitive sense of why this result might hold by looking again at the discrete case and at the example of tossing two dice. To ask about the distribution of the sum of two dice is to ask about the probabilities of particular numbers coming up, and these we can compute directly, using the rules of probability. Take, for example, the probability that the sum is 7. Count the ways, distinguishing which throw is first:

$$\begin{aligned}
&\text{Prob}(\text{Sum} = 7)\\
&\quad = \text{Prob}(\{1 \text{ and } 6\} \text{ or } \{2 \text{ and } 5\} \text{ or } \{3 \text{ and } 4\} \text{ or } \{4 \text{ and } 3\} \text{ or } \{5 \text{ and } 2\} \text{ or } \{6 \text{ and } 1\})\\
&\quad = \text{Prob}(1 \text{ and } 6) + \text{Prob}(2 \text{ and } 5) + \text{Prob}(3 \text{ and } 4)\\
&\qquad + \text{Prob}(4 \text{ and } 3) + \text{Prob}(5 \text{ and } 2) + \text{Prob}(6 \text{ and } 1)\\
&\qquad\quad \text{(probabilities add when events are mutually exclusive)}\\
&\quad = \text{Prob}(1)\,\text{Prob}(6) + \text{Prob}(2)\,\text{Prob}(5) + \text{Prob}(3)\,\text{Prob}(4)\\
&\qquad + \text{Prob}(4)\,\text{Prob}(3) + \text{Prob}(5)\,\text{Prob}(2) + \text{Prob}(6)\,\text{Prob}(1)\\
&\qquad\quad \text{(probabilities multiply when events are independent)}\\
&\quad = 6\left(\frac{1}{6}\right)^2 = \frac{1}{6}\,.
\end{aligned}$$

The particular answer, Prob(sum $= 7) = 1/6$, is not important here;[20] it's the form of the expression for the solution that should catch your eye. We can write it as

$$\text{Prob}(\text{sum} = 7) = \sum_{k=1}^{6} \text{Prob}(k)\,\text{Prob}(7-k)\,,$$

which is visibly a discrete convolution of Prob with itself; it has the same form as an integral convolution with the sum replacing the integral.

We can extend this observation by introducing

$$p(n) = \begin{cases} \frac{1}{6}, & n = 1, 2, \ldots, 6\,, \\ 0, & \text{otherwise.} \end{cases}$$

This is the discrete uniform density for the random variable "throw one die." Then, by the same reasoning as above,

$$\text{Prob}(\text{sum of two dice} = n) = \sum_{k=-\infty}^{\infty} p(k)p(n-k)\,.$$

You can check that this gives the right answers, including the answer 0 for n bigger than 12 or n less than 2:

n	Prob(sum $= n$)
2	1/36
3	2/36
4	3/36
5	4/36
6	5/36
7	6/36
8	5/36
9	4/36
10	3/36
11	2/36
12	1/36

Now let's turn to the case of continuous random variables, and in the following argument look for similarities to the example we just treated. Let X_1 and X_2 be independent random variables with probability density functions $p_1(x_1)$ and $p_2(x_2)$. Because X_1 and X_2 are independent,

$$\text{Prob}(a_1 \le X_1 \le b_1 \text{ and } a_2 \le X_2 \le b_2) = \left(\int_{a_1}^{b_1} p_1(x_1)\,dx_1\right)\left(\int_{a_2}^{b_2} p_2(x_2)\,dx_2\right).$$

Using what has now become a familiar trick, we write this as a double integral:

$$\left(\int_{a_1}^{b_1} p_1(x_1)\,dx_1\right)\left(\int_{a_2}^{b_2} p_2(x_2)\,dx_2\right) = \int_{a_2}^{b_2}\int_{a_1}^{b_1} p_1(x_1)p_2(x_2)\,dx_1\,dx_2\,;$$

[20]But do note that it agrees with what we can observe from the graphic of the sum of two dice. We see that the total number of possibilities for two throws is 36 and that 7 comes up $6/36 = 1/6$ of the time.

that is,

$$\operatorname{Prob}(a_1 \le X_1 \le b_1 \text{ and } a_2 \le X_2 \le b_2) = \int_{a_2}^{b_2} \int_{a_1}^{b_1} p_1(x_1)p_2(x_2)\,dx_1\,dx_2\,.$$

If we let a_1 and a_2 drop to $-\infty$, then

$$\operatorname{Prob}(X_1 \le b_1 \text{ and } X_2 \le b_2) = \int_{-\infty}^{b_2} \int_{-\infty}^{b_1} p_1(x_1)p_2(x_2)\,dx_1\,dx_2\,.$$

Since this holds for any b_1 and b_2, we can conclude that

$$\operatorname{Prob}(X_1 + X_2 \le t) = \iint_{x_1+x_2\le t} p_1(x_1)p_2(x_2)\,dx_1\,dx_2$$

for every t. In words, the probability that $X_1 + X_2 \le t$ is computed by integrating the *joint probability density* $p_1(x_1)p_2(x_2)$ over the region in the (x_1, x_2)-plane where $x_1 + x_2 \le t$.

We're going to make a change of variable in this double integral. We let

$$\begin{aligned} x_1 &= u\,, \\ x_2 &= v - u\,. \end{aligned}$$

Notice that $x_1 + x_2 = v$. Thus under this transformation the (oblique) line $x_1 + x_2 = t$ becomes the horizontal line $v = t$, and the region $x_1 + x_2 \le t$ in the (x_1, x_2)-plane becomes the half-plane $v \le t$ in the (u, v)-plane.

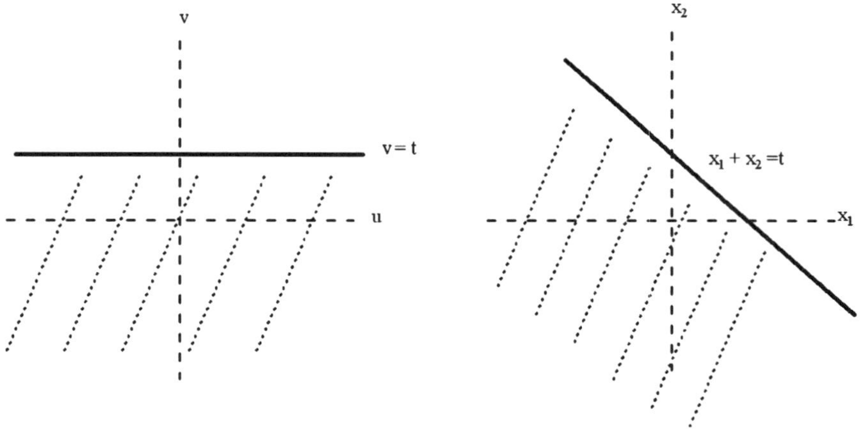

The integral then becomes

$$\begin{aligned} \iint_{x_1+x_2\le t} p_1(x_1)p_2(x_2)\,dx_1\,dx_2 &= \int_{-\infty}^{t} \int_{-\infty}^{\infty} p_1(u)p_2(v-u)\,du\,dv \\ &\qquad \text{(the convolution of } p_1 \text{ and } p_2 \text{ is inside!)} \\ &= \int_{-\infty}^{t} (p_2 * p_1)(v)\,dv\,. \end{aligned}$$

To summarize, we now see that the probability $\text{Prob}(X_1 + X_2 \le t)$ for any t is given by

$$\text{Prob}(X_1 + X_2 \le t) = \int_{-\infty}^{t} (p_2 * p_1)(v)\, dv\,.$$

Therefore the probability density function of $X_1 + X_2$ is $(p_2 * p_1)(t)$.

This extends to the sum of any finite number of random variables: if $X_1, X_2, \ldots, X_n$ are independent random variables with probability density functions $p_1, p_2, \ldots, p_n$, respectively, then the probability density function of $X_1 + X_2 + \cdots + X_n$ is $p_1 * p_2 * \cdots * p_n$. Cool. Cool. ... Cool.

For a single probability density $p(x)$ we'll write

$$p^{*n}(x) = (p*p*\cdots*p)(x) \quad (n \text{ factors of } p, \text{ i.e., } n-1 \text{ convolutions of } p \text{ with itself}).$$

3.6.7. The Central Limit Theorem: The bell curve tolls for thee. People say things like, "The Central Limit Theorem[21] says the sum of n independent random variables is well approximated by a Gaussian if n is large." They mean to say that the sum is *distributed* like a Gaussian. To make a true statement we have to begin with a few assumptions — but not many — on how the random variables themselves are distributed. Call the random variables X_1, $X_2, \ldots$, $X_n, \ldots$. We assume first of all that the X's are independent. We also assume that all of the X's have the same probability density function. So for practical purposes you can think of the random variables as being the same, like making the same measurements in different trials, or throwing a die hundreds of times, recording the results, and then doing it again. Naturally, there's some terminology and an acronym that goes along with this. One says that the X's are *independent and identically distributed*, or *iid*. In particular, the X's all have the same mean, say μ, and they all have the same standard deviation, say σ. We assume that both μ and σ are finite.

Consider the sum

$$S_n = X_1 + X_2 + \cdots + X_n\,.$$

We want to say that S_n is distributed like a Gaussian as n increases. But which Gaussian? The mean and standard deviation for the X's are all the same, but for S_n they are changing with n. It's not hard to show, though, that for S_n the mean scales by n and the standard deviation scales by $\sqrt{n}$:

$$\mu(S_n) = n\mu,$$
$$\sigma(S_n) = \sqrt{n}\,\sigma\,.$$

We'll derive these later, in Section 3.6.8, so as not to interrupt the action.

To make sense of S_n approaching a particular Gaussian, we should therefore recenter and rescale the sum, say fix the mean to be zero and fix the standard deviation to be 1. We should work with

$$\frac{S_n - n\mu}{\sqrt{n}\,\sigma}$$

[21] Abbreviated, of course, as CLT.

and ask what happens as $n \to \infty$. One form of the Central Limit Theorem says that

$$\lim_{n\to\infty} \text{Prob}\left(a < \frac{S_n - n\mu}{\sqrt{n}\sigma} < b\right) = \frac{1}{\sqrt{2\pi}} \int_a^b e^{-x^2/2}\,dx\,.$$

On the right-hand side is the Gaussian $(1/\sqrt{2\pi})e^{-x^2/2}$ with mean 0 and standard deviation 1. The theorem says that probabilities for the *normalized* sum of the random variables approach those based on this Gaussian.

We'll focus on the convergence of the pdf's for S_n, sort of an unintegrated form of the way the CLT is stated above. Let $p(x)$ be the common probability density function for the $X_1, X_2, \dots, X_n$ (the pdf for the iid X's, for those who like to compress their terminology). We'll start by assuming already that $\mu = 0$ and $\sigma = 1$ for the X's. This means that

$$\int_{-\infty}^{\infty} xp(x)\,dx = 0 \quad \text{and} \quad \int_{-\infty}^{\infty} x^2 p(x)\,dx = 1\,,$$

in addition to

$$\int_{-\infty}^{\infty} p(x)\,dx = 1\,,$$

which is true for every pdf.

Now, the mean of S_n is zero, but the standard deviation is $\sqrt{n}$, so we want to work $S_n/\sqrt{n}$. What is the pdf of this? We've shown that the pdf for $S_n = X_1 + \cdots + X_n$ is

$$p^{*n}(x) = (p * p * \cdots * p)(x)\,;$$

hence the probability density function for $S_n/\sqrt{n}$ is

$$p_n(x) = \sqrt{n}\, p^{*n}(\sqrt{n}\, x)\,.$$

(Careful here: It's $(p * p * \cdots * p)(\sqrt{n}\, x)$, not $p(\sqrt{n}\, x) * p(\sqrt{n}\, x) * \cdots p(\sqrt{n}\, x)$.)

We're all set to show:

- **Central Limit Theorem.** Let $X_1, X_2, \dots, X_n, \dots$ be independent, identically distributed random variables with mean 0 and standard deviation 1. Let $p_n(x)$ be the probability density function for

  $$S_n/\sqrt{n} = (X_1 + X_2 + \cdots + X_n)/\sqrt{n}.$$

  Then

  $$p_n(x) \to \frac{1}{\sqrt{2\pi}} e^{-x^2/2} \quad \text{as } n \to \infty\,.$$

The idea is to take the Fourier transform of p_n, which, by the convolution theorem, will essentially be the *product* of the Fourier transforms of p. Products are easier than convolutions, and the hope is to use the assumptions on p to get some information on the form of this product as $n \to \infty$. For dramatic effect, and it's very dramatic, I'm going to run through this a little quickly, supplying some further details afterward.

Begin with the Fourier transform of

$$p_n(x) = \sqrt{n}\, p^{*n}(\sqrt{n}\, x)\,.$$

We'll use the capital letter notation and write $P(s) = \mathcal{F}p(s)$. Then the Fourier transform of $p_n(x)$ is

$$P^n\left(\frac{s}{\sqrt{n}}\right) \quad \text{(ordinary } n\text{th power here).}$$

The normalization of mean zero and standard deviation allows us to do something with $P(s/\sqrt{n})$. Using a Taylor series approximation for the exponential function,[22] we have

$$\begin{aligned}
P\left(\frac{s}{\sqrt{n}}\right) &= \int_{-\infty}^{\infty} e^{-2\pi i s x/\sqrt{n}}\, p(x)\, dx \\
&= \int_{-\infty}^{\infty} \left(1 - \frac{2\pi i s x}{\sqrt{n}} + \frac{1}{2}\left(\frac{2\pi i s x}{\sqrt{n}}\right)^2 + \text{small}\right) p(x)\, dx \\
&= \int_{-\infty}^{\infty} \left(1 - \frac{2\pi i s x}{\sqrt{n}} - \frac{2\pi^2 s^2 x^2}{n} + \text{small}\right) p(x)\, dx \\
&= \underbrace{\int_{-\infty}^{\infty} p(x)\, dx}_{=1} - \frac{2\pi i s}{\sqrt{n}} \underbrace{\int_{-\infty}^{\infty} x p(x)\, dx}_{=0} \\
&\qquad - \frac{2\pi^2 s^2}{n} \underbrace{\int_{-\infty}^{\infty} x^2 p(x)\, dx}_{=1} + \int_{-\infty}^{\infty} (\text{small})\, p(x)\, dx \\
&= 1 - \frac{2\pi^2 s^2}{n} + \text{small}\,.
\end{aligned}$$

See how the normalizations came in:

$$\int_{-\infty}^{\infty} p(x)\, dx = 1\,, \quad \int_{-\infty}^{\infty} x p(x)\, dx = 0\,, \quad \int_{-\infty}^{\infty} x^2 p(x)\, dx = 1\,.$$

That "small" term comes from the remainder in the Taylor series for the exponential and it tends to 0 faster than $1/n$ as $n \to \infty$; see below. Using $\big(1 + x/n\big)^n \to e^x$, from calculus days, we have for large n

$$P^n\left(\frac{s}{\sqrt{n}}\right) \approx \left(1 - \frac{2\pi^2 s^2}{n}\right)^n \approx e^{-2\pi^2 s^2}\,.$$

Taking the inverse Fourier transform of $e^{-2\pi^2 s^2}$ and knowing what happens to the Gaussian, taking the limit as $n \to \infty$, taking the rest of the day off for a job well done, we conclude that

$$p_n(x) \to \frac{1}{\sqrt{2\pi}} e^{-x^2/2}\,.$$

Catch your breath and relax.

---

[22] You remember that $e^x = 1 + x + \frac{x^2}{2!} + \frac{x^3}{3!} + \cdots$. This works when x is complex, too.

What about that "small" part? Only for those who are interested — you have to be somewhat comfortable with the remainder in Taylor series. To see more carefully what's going on with the "small" part, write

$$e^{-2\pi i(s/\sqrt{n})x} = 1 - \frac{2\pi isx}{\sqrt{n}} - \frac{2\pi s^2x^2}{n} + R_n(x).$$

The remainder $R_n(x)$ also depends on s and n and for fixed x (and s) it tends to 0 as $n \to \infty$ like $1/n^{3/2}$. Split the Fourier transform into two pieces:

$$\int_{-\infty}^{\infty} e^{-2\pi i(s/\sqrt{n})x}p(x)\,dx = \int_{|x|\leq 1} e^{-2\pi i(s/\sqrt{n})x}p(x)\,dx + \int_{|x|\geq 1} e^{-2\pi i(s/\sqrt{n})x}p(x)\,dx\,.$$

Then,

$$\begin{aligned}
&\int_{|x|\leq 1} e^{-2\pi i(s/\sqrt{n})x}p(x)\,dx + \int_{|x|\geq 1} e^{-2\pi i(s/\sqrt{n})x}p(x)\,dx \\
&\quad = \int_{|x|\leq 1}\left(1 - \frac{2\pi isx}{\sqrt{n}} - \frac{2\pi^2s^2x^2}{n} + R_n(x)\right)p(x)\,dx \\
&\qquad + \int_{|x|\geq 1}\left(1 - \frac{2\pi isx}{\sqrt{n}} - \frac{2\pi^2s^2x^2}{n} + \frac{1}{x^2}x^2R_n(x)\right)p(x)\,dx \\
&\quad = \int_{-\infty}^{\infty}\left(1 - \frac{2\pi isx}{\sqrt{n}} - \frac{2\pi^2s^2x^2}{n}\right)p(x)\,dx \\
&\qquad + \int_{|x|\leq 1} R_n(x)p(x)\,dx + \int_{|x|\geq 1}\left(\frac{1}{x^2}R_n(x)\right)x^2p(x)\,dx.
\end{aligned}$$

Since $\int_{-\infty}^{\infty} p(x)\,dx = 1$ and $R_n(x)$ is bounded for $|x| \leq 1$, we have

$$\int_{|x|\leq 1} R_n(x)p(x)\,dx = o(1), \quad n \to \infty.$$

Here $o(1)$ stands for a quantity that tends to 0 as $n \to \infty$. Likewise, since

$$\frac{1}{x^2}R_n(x) = \frac{1}{x^2}\left(e^{-2\pi i(s/\sqrt{n})x} - \left(1 - \frac{2\pi isx}{\sqrt{n}} - \frac{2\pi^2s^2x^2}{n}\right)\right)$$

is bounded for $|x| \geq 1$ and $\int_{-\infty}^{\infty} x^2p(x)\,dx = 1$, we also have

$$\int_{|x|\geq 1}\left(\frac{1}{x^2}R_n(x)\right)x^2p(x)\,dx = o(1), \quad n \to \infty.$$

Putting these together and appealing to the normalizations as earlier, we can conclude that

$$\int_{-\infty}^{\infty} e^{-2\pi i(s/\sqrt{n})x}p(x)\,dx = 1 - \frac{2\pi^2s^2}{n} + o(1).$$

That's where we ended up before, when we swept all the details into "small." All is well.

3.6.8. The mean and standard deviation of the sum of random variables. The setup for the Central Limit Theorem involves the sum

$$S_n = X_1 + X_2 + \cdots + X_n$$

of n independent random variables, all having the same pdf $p(x)$. Thus all of the X's have the same mean and the same variance:

$$\mu = \int_{-\infty}^{\infty} xp(x)\,dx\,, \quad \sigma^2 = \int_{-\infty}^{\infty} x^2 p(x)\,dx\,.$$

We needed to know that the mean and the standard deviation of S_n are

$$\mu(S_n) = n\mu\,, \quad \sigma(S_n) = \sqrt{n}\,\sigma\,.$$

Take the first of these. The pdf for $S_2 = X_1 + X_2$ is $p * p$, and hence

$$\begin{aligned}
\mu(S_2) &= \int_{-\infty}^{\infty} x(p*p)(x)\,dx \\
&= \int_{-\infty}^{\infty} x\left(\int_{-\infty}^{\infty} p(x-y)p(y)\,dy\right)dx \\
&= \int_{-\infty}^{\infty}\left(\int_{-\infty}^{\infty} xp(x-y)\,dx\right)p(y)\,dy \\
&= \int_{-\infty}^{\infty}\left(\int_{-\infty}^{\infty} (u+y)p(u)\,du\right)p(y)\,dy \quad (\text{using } u = x-y) \\
&= \int_{-\infty}^{\infty}\left(\int_{-\infty}^{\infty} up(u)\,du + y\int_{-\infty}^{\infty} p(u)\,du\right)p(y)\,dy \\
&= \int_{-\infty}^{\infty} (\mu+y)p(y)\,dy \quad \left(\text{using } \textstyle\int_{-\infty}^{\infty} up(u)\,du = \mu \text{ and } \int_{-\infty}^{\infty} p(u)du = 1\right) \\
&= \mu\int_{-\infty}^{\infty} p(u)\,du + \int_{-\infty}^{\infty} yp(y)\,dy \\
&= \mu + \mu\,.
\end{aligned}$$

Inductively, we get $\mu(S_n) = n\mu$.

How about the variance, or standard deviation? Again let's do this for $S_2 = X_1 + X_2$. We first assume that the mean of the X's is 0, and hence the mean of S_2 is 0 as well:

$$\int_{-\infty}^{\infty} xp(x)\,dx = 0 \quad \text{and} \quad \int_{-\infty}^{\infty} x(p*p)(x)\,dx = 0\,.$$

Then the variance of S_2 is

$$\begin{aligned}
\sigma^2(S_2) &= \int_{-\infty}^{\infty} x^2 (p*p)(x)\,dx \\
&= \int_{-\infty}^{\infty} x^2 \left(\int_{-\infty}^{\infty} p(x-y)p(y)\,dy\right) dx \\
&= \int_{-\infty}^{\infty} \left(\int_{-\infty}^{\infty} x^2 p(x-y)\,dx\right) p(y)\,dy \\
&= \int_{-\infty}^{\infty} \left(\int_{-\infty}^{\infty} (u+y)^2 p(u)\,du\right) p(y)\,dy \quad (\text{using } u = x-y) \\
&= \int_{-\infty}^{\infty} \left(\int_{-\infty}^{\infty} (u^2+2uy+y^2) p(u)\,du\right) p(y)\,dy \\
&= \int_{-\infty}^{\infty} \left(\int_{-\infty}^{\infty} u^2 p(u)\,du + 2y\int_{-\infty}^{\infty} up(u)\,du + y^2\int_{-\infty}^{\infty} p(u)\,du\right) p(y)\,dy \\
&= \int_{-\infty}^{\infty} (\sigma^2+y^2)p(y)\,dy \ (\text{using } \textstyle\int_{-\infty}^{\infty} u^2p(u)du = \sigma^2 \text{ and } \int_{-\infty}^{\infty} up(u)du = 0) \\
&= \sigma^2 \int_{-\infty}^{\infty} p(y)\,dy + \int_{-\infty}^{\infty} y^2 p(y)\,dy \\
&= \sigma^2 + \sigma^2 \;=\; 2\sigma^2\,.
\end{aligned}$$

So the variance of S_2 is $2\sigma^2$ and the standard deviation is $\sigma(S_2) = \sqrt{2}\sigma$. Once again, inductively,

$$\sigma(S_n) = \sqrt{n}\,\sigma\,.$$

Pretty nice, really. I'll let you decide what to do if the mean is not zero at the start.

3.7. Heisenberg's Inequality

Since we've gone to the trouble of introducing some of the terminology from probability and statistics (mean, variance, etc.), I thought you might appreciate seeing another application.

Consider the stretch theorem, which reads:

- If $f(t) \rightleftharpoons F(s)$, then $f(at) \rightleftharpoons \dfrac{1}{|a|}F\left(\dfrac{s}{a}\right)$.

If a is large, then $f(at)$ is squeezed and $(1/|a|)F(s/a)$ is stretched. Conversely, if a is small, then $f(at)$ is stretched and $(1/|a|)F(s/a)$ is squeezed.

A more quantitative statement of the trade-off between the spread of a signal and the spread of its Fourier transform is related to (equivalent to) that most famous inequality in quantum mechanics, the Heisenberg uncertainty principle.

Suppose $f(x)$ is a signal with finite energy,

$$\int_{-\infty}^{\infty} |f(x)|^2\,dx < \infty\,.$$

Normalize the signal by dividing f by the square root of its energy, and thus assume that

$$\int_{-\infty}^{\infty} |f(x)|^2 \, dx = 1 \, .$$

We can then regard $|f(x)|^2$ as defining a probability density function, and it has a mean and a variance. Now, by Parseval's identity,

$$\int_{-\infty}^{\infty} |\hat{f}(s)|^2 \, ds = \int_{-\infty}^{\infty} |f(x)|^2 \, dx = 1 \, .$$

Thus $|\hat{f}(s)|^2$ also defines a probability distribution, and it too has a mean and variance. How do they compare to those of $|f(x)|^2$?

As earlier, we shift $f(x)$, or rather $|f(x)|^2$, to assume that the mean is 0. The effect on $\hat{f}(s)$ of shifting $f(x)$ is to multiply by a complex exponential, which has absolute value 1 and hence does not affect $|\hat{f}(s)|^2$. In the same manner, we can shift $\hat{f}(s)$ so that it has zero mean, and again there will be no effect on $|f(x)|^2$.

To summarize, we assume that the probability distributions $|f(x)|^2$ and $|\hat{f}(s)|^2$ each have mean 0, and we are interested in comparing their variances:

$$\sigma^2(f) = \int_{-\infty}^{\infty} x^2 |f(x)|^2 \, dx \quad \text{and} \quad \sigma^2(\hat{f}) = \int_{-\infty}^{\infty} s^2 |\hat{f}(s)|^2 \, ds \, .$$

The Heisenberg uncertainty principle states that

$$\sigma(f)\sigma(\hat{f}) \geq \frac{1}{4\pi} \, .$$

In words, this says that not both of $\sigma(f)$ and $\sigma(\hat{f})$ can be small. If one is tiny, the other has to be big enough so that their product is at least $1/4\pi$.

After all the setup, the argument to deduce the lower bound is pretty easy, except for a little trick right in the middle. It's also helpful to assume that we're working with complex-valued functions — I think the trick that comes up is actually a little easier to verify in that case. Finally, we're going to assume that $|f(x)|$ decreases rapidly enough at $\pm\infty$. You'll see what's needed. The result can be proved for more general functions via approximation arguments. Here we go:

$$\begin{aligned}
4\pi^2\sigma(f)^2\sigma(\hat{f})^2 &= 4\pi^2 \int_{-\infty}^{\infty} x^2 |f(x)|^2 \, dx \int_{-\infty}^{\infty} s^2 |\hat{f}(s)|^2 \, ds \\
&= \int_{-\infty}^{\infty} x^2 |f(x)|^2 \, dx \int_{-\infty}^{\infty} |2\pi i s|^2 |\hat{f}(s)|^2 \, ds \\
&= \int_{-\infty}^{\infty} |x f(x)|^2 \, dx \int_{-\infty}^{\infty} |\widehat{f'}(s)|^2 \, ds \\
&= \int_{-\infty}^{\infty} |x f(x)|^2 \, dx \int_{-\infty}^{\infty} |f'(x)|^2 \, dx \\
&\qquad \text{(by Parseval's identity applied to } f'(x)\text{)} \\
&\geq \left(\int_{-\infty}^{\infty} |x\overline{f(x)} f'(x)| \, dx \right)^2 \quad \text{(by the Cauchy-Schwarz inequality).}
\end{aligned}$$

Here comes the trick. In the integrand, we have $|\overline{f(x)}f'(x)|$. The magnitude of any complex number is always greater than its real part.[23] Hence

$$\begin{aligned}|x\overline{f(x)}f'(x)| &\geq x\,\mathrm{Re}\{\overline{f(x)}f'(x)\}\\ &= x\,\frac{1}{2}\big(\overline{f(x)}f'(x) + f(x)\overline{f'(x)}\big) = \frac{1}{2}x\frac{d}{dx}\big(\overline{f(x)}f(x)\big) = \frac{1}{2}x\frac{d}{dx}|f(x)|^2 .\end{aligned}$$

Use this in the last line, above:

$$\Big(\int_{-\infty}^{\infty} |x\overline{f(x)}f'(x)|\,dx\Big)^2 \geq \Big(\int_{-\infty}^{\infty} x\frac{d}{dx}\left(\frac{1}{2}|f(x)|^2\right)dx\Big)^2$$

Now integrate by parts with $u = x$, $dv = \frac{d}{dx}\frac{1}{2}|f(x)|^2\,dx$. The term $[uv]_{-\infty}^{\infty}$ drops out because *we assume it does*; i.e., we assume that $x|f(x)|$ goes to zero as $x \to \pm\infty$. Therefore we're left with the integral of $v\,du$ (and the whole thing is squared). That is,

$$\Big(\int_{-\infty}^{\infty} x\frac{d}{dx}\left(\frac{1}{2}|f(x)|^2\right)dx\Big)^2 dx = \frac{1}{4}\Big(\int_{-\infty}^{\infty} |f(x)|^2\,dx\Big)^2 = \frac{1}{4}.$$

To summarize, we have shown that

$$4\pi^2\sigma(f)^2\sigma(\hat{f})^2 \geq \frac{1}{4} \quad \text{or} \quad \sigma(f)\sigma(\hat{f}) \geq \frac{1}{4\pi}.$$

Remark. Using the case of equality in the Cauchy-Schwarz inequality, one can show that equality holds in Heisenberg's inequality exactly for constant multiples of $f(x) = e^{-kx^2}$. Yet another spooky appearance of the Gaussian.

Is this quantum mechanics? I'm told it is. The quantum mechanics of a particle moving in one dimension that goes along with this inequality runs as follows, in skeletal form, with no attempt at motivation:

The *state* of a particle moving in one dimension is given by a complex-valued function ψ in $L^2(\mathbb{R})$, the square integrable functions on the real line. L^2 plays a big role in quantum mechanics — you need a space to work in, and L^2 is the space. Really.[24] Probabilities are done with complex quantities in this business, and the first notion is that the probability of finding the particle in the interval $a \leq x \leq b$ is given by

$$\int_a^b \psi(x)^*\psi(x)\,dx\,,$$

where in this field it's customary to write the complex conjugate of a quantity using an asterisk instead of an overline.

An *observable* is a symmetric linear operator A operating on some subset of functions (states) in $L^2(\mathbb{R})$.[25] The *average* of A in the state ψ is defined to be

$$\int_{-\infty}^{\infty} \psi(x)^*(A\psi)(x)\,dx.$$

[23] Draw a picture. The complex number is a vector, which is always longer than its x-component.

[24] It's orthogonality and all that goes with it. The inner product is $(f, g) = \int_{-\infty}^{\infty} f(x)\overline{g(x)}\,dx$. We'll come back to $L^2(\mathbb{R})$ in the next chapter.

[25] "Symmetric" means that A is equal to its transpose, which, in terms of inner products, translates to $(Af, g) = (f, Ag)$.

One important observable is the "position of the particle," and this, as it turns out, is associated to the operator "multiplication by x." Thus the average position is

$$\int_{-\infty}^{\infty} \psi(x)^*(A\psi)(x)\,dx = \int_{-\infty}^{\infty} \psi(x)^* x\psi(x)\,dx = \int_{-\infty}^{\infty} x|\psi(x)|^2\,dx\,.$$

Another important observable is *momentum*, and this is associated with the operator

$$B = \frac{1}{2\pi i}\frac{d}{dx}\,.$$

The average momentum is then

$$\begin{aligned}\int_{-\infty}^{\infty} \psi(x)^*(B\psi)(x)\,dx &= \int_{-\infty}^{\infty} \psi(x)^*\frac{1}{2\pi i}\psi'(x)\,dx \\ &= \int_{-\infty}^{\infty} \hat{\psi}(s)^* s\hat{\psi}(s)\,ds \\ &\qquad \text{(using Parseval's identity for products of functions)} \\ &= \int_{-\infty}^{\infty} s|\psi(s)|^2\,ds\,.\end{aligned}$$

The position and momentum operators do *not* commute:

$$(AB - BA)(\psi) = \frac{1}{2\pi i}\left(x\frac{d}{dx} - \frac{d}{dx}x\right)(\psi) = -\frac{1}{2\pi i}\psi\,.$$

In quantum mechanics this means that the position and momentum cannot *simultaneously* be measured with arbitrary accuracy. The Heisenberg inequality, as a lower bound for the product of the two variances, is a quantitative way of stating this.

"Uncertainty principles" have assumed an important role in Fourier analysis. Start with the fundamental paper of Donaho and Stark, Uncertainty principles and signal recovery, SIAM J. Appl. Math. **49** (1989), 906–931. Work your way forward from there.

Problems and Further Results

3.1. (a) Show that $\Pi * \Pi = \Lambda$ using the definition of convolution, i.e.,

$$(\Pi * \Pi)(x) = \int_{-\infty}^{\infty} \Pi(y)\Pi(x-y)\,dy.$$

(b) What about $\Pi_a * \Pi_a$? (Use any method you wish.)

(c) *Sercan and Aditya discuss convolution:*

Sercan: You know, I think this problem suggests something general about how convolution spreads out a signal.

Aditya: How so?

Sercan: Well, we showed that $\Pi * \Pi = \Lambda$. I know that Λ has a different shape than Π, but notice that while $\Pi(x) = 0$ for $|x| \geq 1/2$, we have $\Lambda(x) = 0$ outside the bigger interval $|x| \geq 1$. So $\Pi * \Pi$ is more spread out.

Aditya: I agree, but do you have any more examples?

Sercan: The second example $\Pi_a * \Pi_a$ exhibits the same phenomenon.

Aditya: Is there a general pattern there?

Sercan: I'm not sure, but based on how we showed $\Pi * \Pi = \Lambda$, I think I'll conjecture that if $f(x) = 0$ for $|x| \geq a/2$, then $(f * f)(x) = 0$ for $|x| \geq a$, so $(f * f)(x)$ is twice as spread out as $f(x)$.

Is Sercan on to something or too optimistic?

3.2. Let $x(t)$ be a signal whose spectrum is identically zero outside the range $-S_0 \leq s \leq S_0$. An example of such a spectrum is shown below.

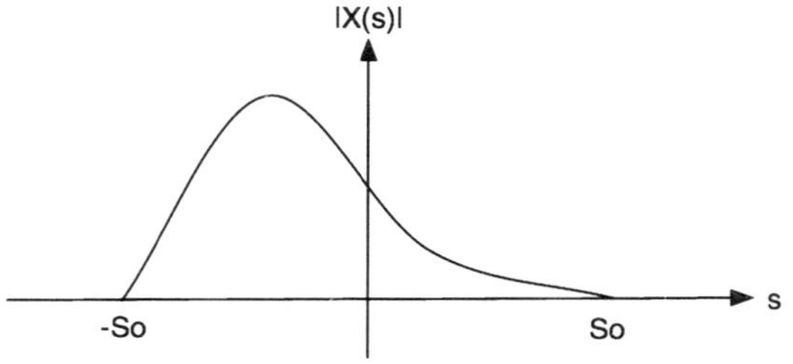

For the signal $x^3(t) * x^8(t)$, determine the range over which the spectrum is nonzero. See Problem 3.1.

3.3. *Some sample convolutions*[26]

(a) It is often useful to represent operations on signals as convolutions. For each of the following, find a function $h(t)$ such that $y(t) = (x * h)(t)$.

(i) $y(t) = \displaystyle\int_{-\infty}^{t-1} x(\tau)d\tau.$

(ii) $y(t) = \displaystyle\int_{t-T/2}^{t+T/2} x(\tau)d\tau.$

(b) Let $h(t) = \Pi(t - 1/2)$ and $x(t) = \sin(2\pi t)u(t)$, where $u(t) = 1$ for $t \geq 0$ and 0 for $t < 0$. Sketch $h(t)$ and $x(t)$ and then find $y(t) = (x * h)(t)$.

3.4. *Reversals, shifts, stretches, and convolution*

(a) If both $f(t)$ and $g(t)$ are reversed, what happens to their convolution? If one of $f(t)$ and $g(t)$ is reversed, what happens to their convolution?

(b) Show that

$$(\tau_b f) * g = \tau_b(f * g) = f * (\tau_b g)\,.$$

[26] From Raj Bhatnagar.

Use this result to deduce that if either f or g is periodic of period T, then $f * g$ is periodic of period T.

(c) Show that

$$(\sigma_a f) * g = \frac{1}{|a|}\sigma_a(f * (\sigma_{1/a} g)), \quad (\sigma_a f) * (\sigma_a g) = \frac{1}{|a|}\sigma_a(f * g).$$

3.5. Eva and Rajiv converse about convolution:

Rajiv: You know, convolution really is a remarkable operation, the way it imparts properties of one function onto the convolution with another. Take periodicity: if $f(t)$ is periodic, then $(f * g)(t)$ is periodic with the same period as f. That was an earlier problem.

Eva: There's a problem with that statement. You want to say that if $f(t)$ is a periodic function of period T, then $(f * g)(t)$ is also periodic of period T.

Rajiv: Right.

Eva: What if $g(t)$ is also periodic, say of period R? Then doesn't $(f * g)(t)$ have two periods, T and R?

Rajiv: I suppose so.

Eva: But wouldn't this lead right to a contradiction? I mean, for example, you can't have a function with two periods, can you?

Rajiv: I think we've found a fundamental contradiction in mathematics.

Eva: Why don't we look at a simple, special case first. What happens if you convolve $\sin 2\pi t$ with itself?

Rajiv: OK, both functions have period 1 so for the convolution you get a function that's periodic of period 1, no problem.

Eva: No, you don't. Something goes wrong.

What's going on? With whom do you agree and why? What do you think about that statement: "If $f(t)$ is periodic, then $f * g$ is periodic."

3.6. *Some practice with convolution*

(a) Let $f(x) = e^{-|x|}$, $-\infty < x < \infty$. Find $(f * f)(x)$.

(b) Let $g(x) = e^{-\pi x^2}$, $-\infty < x < \infty$. Show that $(g * g)(x) = \frac{1}{\sqrt{2}}e^{-\pi x^2/2}$. From this, deduce the result of the n-fold convolution of g, i.e., $g * g * \ldots * g$ (with n factors of g).

3.7. *Convolving sincs*

(a) Show that $\operatorname{sinc} t * \operatorname{sinc} t = \operatorname{sinc} t$.

(b) A little more generally, show that $\operatorname{sinc}(pt) * \operatorname{sinc}(p't) = \frac{p''}{pp'}\operatorname{sinc}(p''s)$ where $p'' = \min\{p, p'\}$. Assume that $p, p' > 0$.

(c) Scaling equally in the two sincs, show that

$$\operatorname{sinc}(pt - a) * \operatorname{sinc}(pt - b) = \frac{1}{p}\operatorname{sinc}(pt - (a + b)).$$

There's still a more general formula allowing for independent scaling of the sincs. You can work that out.

3.8. *Areas multiply under convolution*
Let $h = f * g$. Show that

$$\int_{-\infty}^{\infty} h(t)\,dt = \left(\int_{-\infty}^{\infty} f(t)\,dt\right)\left(\int_{-\infty}^{\infty} g(t)\,dt\right).$$

Joelle says: "This can be used to show that the convolution of two probability distributions is a probability distribution." Is Joelle right?

3.9. *Convolution and derivative*
Use the convolution theorem and the derivative theorem $\mathcal{F}f'(s) = 2\pi is\mathcal{F}f(s)$ to show $(f * g)' = f' * g = f * g'$.

3.10. Convolution with the unit step function

$$u(x) = \begin{cases} 1, & x \geq 0\,, \\ 0, & x < 0\,, \end{cases}$$

can be used as an integrator. Show for a signal $f(x)$ that

$$\int_0^x f(y)\,dy = (u * (uf))(x)\,.$$

(uf is the product of u and f; i.e., $(uf)(x) = u(x)f(x)$.)

3.11. *Periodizing and convolution*
Let f and g be functions that are identically 0 outside the interval $(-1/2, 1/2)$. Let f_T and g_T denote the periodizations of f and g with period T (with $T \geq 1$); i.e.,

$$f_T(t) = \sum_{n=-\infty}^{\infty} f(t - nT), \quad g_T(t) = \sum_{n=-\infty}^{\infty} g(t - nT).$$

Let p be the convolution of f and g, and let q be the convolution of f_T and g_T, the latter taken as the convolution of periodic functions:

$$p(t) = \int_{-\infty}^{\infty} f(x)g(t - x)dx, \quad q(t) = \int_{-T/2}^{T/2} f_T(y)g_T(t - y)dy.$$

Note that q is periodic of period T. Show that q is a periodization of p with period T; i.e.,

$$q(t) = \sum_{n=-\infty}^{\infty} p(t - nT).$$

Hint: Work your way from the right-hand side to the left-hand side.

3.12. *Divided time signal*

Consider a time signal $f(t)$ with Fourier transform $F(s)$. We divide this signal into equal nonoverlapping time segments, denoting segments as

$$f_n(t) = \begin{cases} f(t), & nT < t < (n+1)T, \\ 0, & \text{otherwise}, \end{cases}$$

where n is an integer.

For an illustration, please see the figure below for division of an arbitrary time signal. The signal $f_0(t)$ is the portion of $f(t)$ in the interval $(0, T)$, $f_1(t)$ is the portion in the interval $(T, 2T)$, and so on.

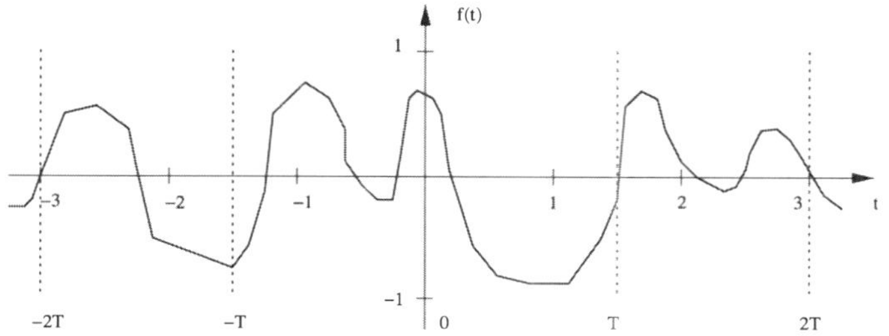

(a) Find the formula for the Fourier transform of a segment function $f_n(t)$, where n is an integer, in terms of the original function's Fourier transform $F(s)$.

(b) Now let $f(t)$ be *periodic* with period T and divided equally into segment functions, $f_n(t)$, as described above. Let $F_0(s)$ be the Fourier transform of the segment $f_0(t)$. Find $F_n(s)$, the Fourier transform of other segments, in terms of $F_0(s)$.

3.13. Let $f(x)$ be a signal and for $h > 0$ let $A_h f(x)$ be the averaging operator,

$$A_h f(x) = \frac{1}{2h} \int_{x-h}^{x+h} f(y)\, dy.$$

(a) Show that we have the alternate expressions for $A_h f(x)$:

$$A_h f(x) = \frac{1}{2h} \int_{-h}^{h} f(x+y)\, dy = \frac{1}{2h} \int_{-h}^{h} f(x-y)\, dy\,.$$

(b) Express $A_h f(x)$ as a convolution and find the Fourier transform $\mathcal{F}(A_h f)(s)$.

3.14. Let $f(x)$ be a signal and for $h > 0$ let $A_h f(x)$ be the averaging operator from Problem 3.13. The signal

$$f(t) = 3\sin 4\pi t + 2\cos 8\pi t - 4\sin 24\pi t$$

is plotted below.

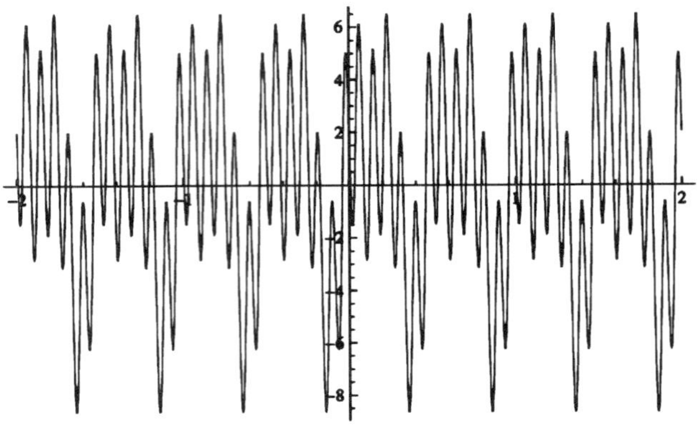

It's complicated. The signal is then averaged with $h = 1/8$ to obtain

$$g(t) = A_{1/8} f(t).$$

The plot of $g(t)$, which looks like

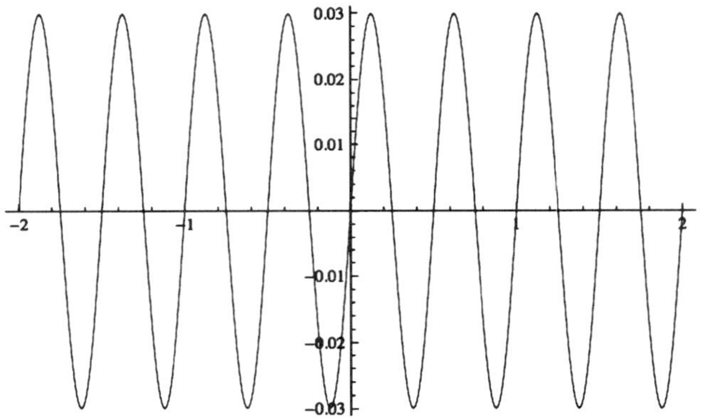

shows only the lowest frequency of the original signal $f(t)$. Why is this?

3.15. *Frequency-hopping spread spectrum*

Per Wikipedia: "Frequency-hopping spread spectrum (FHSS) is a method of transmitting radio signals by rapidly switching a carrier among many frequency channels, using a pseudorandom sequence known to both transmitter and receiver." To an outside party without knowledge of this frequency hopping sequence, the sequence appears random; on the other hand, the intended receiver can determine the sequence with his knowledge of the seed used to generate the pseudorandom sequence. As we will see, this makes FHSS robust against eavesdropping and jamming.

Imagine that your favorite FM radio station employs FHSS. Then for instance it would transmit at 101.3 MHz for 1 minute, then switch to 88.5 MHz for 1 minute, then 94.5 MHz, 97.3 MHz, 90.1 Mhz, and so forth. If you did not have knowledge of this hopping sequence, then you would have trouble listening in! However, if you knew the seed used to generate the sequence of hops, you could anticipate each hop to the next station and dial in appropriately every minute. Because at any given time each signal only occupies a small part of the spectrum, multiple users could transmit messages simultaneously, each hopping about the spectrum.

In this MATLAB exercise, you will decode two messages that were simultaneously transmitted, each with its own pseudorandom hopping sequence. The messages $m_1(t), m_2(t)$ are bandlimited to 4 kHz (i.e., $\mathcal{F}m_1(s), \mathcal{F}m_2(s) = 0$ for $|s| > 2$ kHz) and modulated to their carrier frequency by multiplication with a cosine (see Problem 2.7). The carrier frequencies, $f_1(t), f_2(t)$, change every 1 second, but are constant for each second of transmission. Because you (as the receiver) know the pseudorandom seeds, the hopping sequences of $f_1(t)$ and $f_2(t)$ across 20 channels are known to you. These 20 channels have carrier frequencies of 12, 16, 20, ..., 88 kHz. The transmitted message is

$$Y(t) = m_1(t)\cos(2\pi f_1(t)t) + m_2(t)\cos(2\pi f_2(t)t).$$

Download

`genSpreadSpectrum.mat`,
`genSpreadSpectrum.m`,
`decodeSpreadSpectrum.m`.

This exercise will parallel the `decodeSpreadSpectrum.m` script. The script will run as is, but your job is to complete the code inside `decodeSpreadSpectrum.m` that decodes the two messages for each scenario.

(a) For each second of transmission, show how multiplying $Y(t)$ by $\cos(2\pi f_1(t)t)$ can be used to recover $m_1(t)$ when $f_1(t) \neq f_2(t)$. What happens if $f_1(t) = f_2(t)$?

(b) Let's take a look at the frequency spectrum over time, using `spectrogram`. Does the spectrogram make sense given the channel sequence? Suppose the hopping sequence of the two messages are independent of each other. Do you see a potential problem with this? What is the probability of the two messages NOT colliding during these 10 seconds if the hopping sequences are chosen independently, without memory of its previous hops and with equal likelihood at each of the 20 channels?

(c) Decode both messages, using your knowledge of the encoding seeds. Do you hear something unexpected midway through message 1? What is the phrase, and why do we hear it?

(d) *Eavesdropping (fixed channel).* Assume that an eavesdropper is listening in on a fixed channel. For instance, listen in on channel 19. Qualitatively, what do you hear?

(e) *Eavesdropping (wrong seed).* Assume that as an eavesdropper, you have the wrong seed. So while you are demodulating some frequency hopping pattern, it does not correspond to either message. Qualitatively, what do you hear?

(f) *Jamming (fixed channel).* `Jam1` is a signal designed to jam channel 11. Supposed your enemy is trying to impede your communications by transmitted noise on one of your channels. Look at the spectrogram. What do the two decoded messages sound like? What is the effect of fixed channel jamming?

(g) *Jamming (spread spectrum).* `Jam2` is a signal whose energy is spread across all channels. However, the energy required is proportional to the bandwidth (number of channels), so if we have a fixed amount of energy, the jamming energy per channel drops. Check that `Jam2` and `Jam1` have the same energy (what is the energy?). What do the two decoded messages sound like? What is the effect of spread spectrum jamming? (Is it really as bad as it looks in the spectrogram?)

Note: While we strive to make this a realistic exercise, there are several issues you should consider. First, the sampling frequency `fs` is absurdly high (200 kHz) for a voice transmission. We needed this so that we could emulate a continuous-time operation (demodulation by a cos) in MATLAB. In practice, the sampling would only happen after demodulation and low-pass filtering; this would only require an 8 kHz sampling rate. Second, the message bandwidth is usually much smaller relative to the spread across the spectrum, so it is much harder to observe the entire spectrum to track the hopping pattern (as we could determine from the spectrogram). Third, in this exercise, you could conceivably try all 500 pseudorandom seeds, but in practice you might use a 128-bit seed (as opposed to a 9-bit seed) and a predefined algorithm that operates on the seed (as opposed to a lookup table). Fourth, the message could be encrypted (such as in the voice scrambling problem earlier) prior to transmission for truly secure communcations. Finally, another interpretation of the spread spectrum jamming result is that if there is no jamming, we could transmit at much lower power ("stealth mode") to the point where the message is almost hidden by the ambient, background noise, making it much more difficult to recognize that any transmission is occurring unless you happen to be decoding correctly.

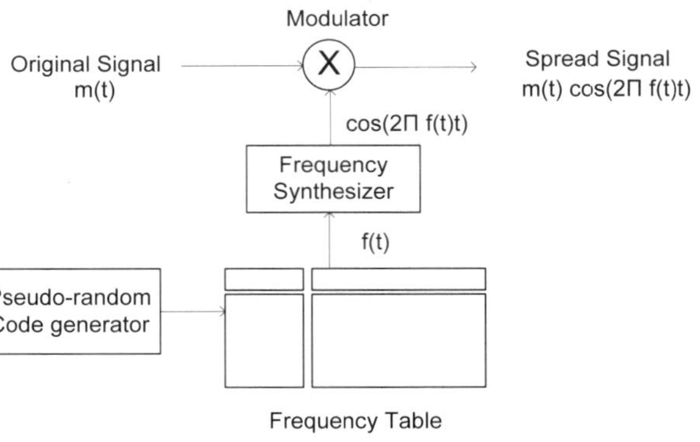

3.16. *Reproducing function*

A real function $f(t)$ is said to reproduce itself under convolution if

$$f(t/a) * f(t/b) = kf(t/c),$$

where k and c in general depend on a and b but the form of the relation is to be valid for any a and b. In words, if such a function is convolved with a similar function having the same shape (i.e., identical except for time scale and possibly amplitude), then the result also has the same shape.

(a) Find an equivalent condition in terms of $\hat{f}(s)$, the Fourier transform of $f(t)$.

(b) Assuming that $\hat{f}(0) = 1$, $\hat{f}'(0) = 0$ and that $\hat{f}''(0)$ exists but is not zero, show that $c^2 = a^2 + b^2$ and $k = |ab/c|$.

(c) Show that the function $f(t) = e^{-\pi t^2}$ reproduces itself under convolution. Are the relationships derived in (b) satisfied for $f(t)$?

3.17. *Applications of the Fourier transform to differential equations*

Let $H(t)$ be the unit step function,

$$H(t) = \begin{cases} 0, & t \leq 0, \\ 1, & t > 0. \end{cases}$$

(a) Using the Fourier transform properties, find a solution for the differential equation $x''(t) + 2x'(t) + 2x(t) = e^{-t}H(t)$. (Thus the right-hand side is a one-sided exponential decay.)

(b) The function $g(t) = e^{-a|t|}$ has the following Fourier transform:

$$G(s) = \frac{2a}{a^2 + 4\pi^2 s^2}.$$

Knowing this transform, demonstrate that one possible solution for the equation $x''(t) - x(t) = -2e^{-t}H(t)$ is

$$x(t) = \int_0^\infty e^{-|t-\tau|}e^{-\tau}\,d\tau.$$

3.18. *Exponential decay and a common differential equation*[27]

(a) Let $a > 0$ and let $g(t) = e^{-a|t|}$ be the two-sided exponential decay. Find $\mathcal{F}g(s)$. *Hint*: Use the one-sided exponential decay and a reversal.

(b) Let $u(t)$ be the unit step function

$$u(t) = \begin{cases} 1, & t > 0, \\ 0, & t \leq 0. \end{cases}$$

Show that a solution of the differential equation

$$x''(t) - x(t) = -2e^{-t}u(t)$$

is

$$x(t) = \int_0^\infty e^{-|t-\tau|}e^{-\tau}\,d\tau.$$

[27] From S. Arik.

3.19. *Solving the wave equation*

An infinite string is stretched along the x-axis and is given an initial displacement described by a function $f(x)$. It is then free to vibrate. The displacement $u(x, t)$ at a time $t > 0$ and at a point x on the string is described by the *wave equation*

$$\frac{\partial^2 u}{\partial t^2} = \frac{\partial^2 u}{\partial x^2}.$$

One often includes physical constants in the equation, e.g., the speed of the wave, but these are suppressed to keep things simple.
Assume that $u(x, 0) = f(x)$ and $u_t(x, 0) = 0$ (zero initial velocity) and use the Fourier transform to show that

$$u(x, t) = \frac{1}{2}(f(x + t) + f(x - t)).$$

This is d'Alembert's (famous) solution to the wave equation.

3.20. *An application to quantum mechanics*

In quantum mechanics, the state of a free particle is entirely described by its wavefunction $\psi(x, t)$, which evolves according to the time-dependent Schrödinger equation:

$$\frac{\partial \psi(x,t)}{\partial t} = \frac{i}{2m} \frac{\partial^2 \psi(x,t)}{\partial x^2}.$$

(a) If the wavefunction of a particle begins in state $\phi(x)$ (i.e., $\psi(x, 0) = \phi(x)$), show that

$$\psi(x, t) = \phi(x) * \sqrt{\frac{m}{2\pi i t}} e^{\frac{imx^2}{2t}}$$

for $t \geq 0$. (You've done problems like this before — treat the Schrödinger equation as a heat equation with a complex constant of proportionality.)

(b) At any given time t, we would like to interpret the square magnitude of the wavefunction, $|\psi(x, t)|^2$, to be the probability density function (pdf) associated with finding the particle at position x. Recall that a function $p(x)$ is a probability density function if
   (i) $p(x) \geq 0$,
   (ii) $\int_{-\infty}^{\infty} p(x)\, dx = 1$.

Using the result in part (a), show that if $|\phi(x)|^2$ is a pdf, then $|\psi(x, t)|^2$ is a pdf for $t \geq 0$.

3.21. *Cross correlation*

The *cross-correlation* (sometimes just called correlation) of two real-valued signals $f(t)$ and $g(t)$ is defined by

$$(f \star g)(x) = \int_{-\infty}^{\infty} f(y) g(x + y)\, dy\,.$$

Note the notation: $\star$ for cross-correlation. $(f \star g)(x)$ is often described as a measure of how well the values of g, when shifted by x, correlate with the values of f. It depends on x; some shifts of g may correlate better with f than other shifts.

To get a sense of this, think about when $(f \star g)(x)$ is positive (and large) or negative (and large) or zero (or near zero). If, for a given x, the values $f(y)$ and $g(x+y)$ are tracking each other — both positive or both negative — then the integral will be positive and so the value $(f \star g)(x)$ will be positive. The closer the match between $f(x)$ and $g(x+y)$ (as y varies) the larger the integral and the larger the cross-correlation.

In the other direction, if, for example, $f(y)$ and $g(x+y)$ maintain opposite signs as y varies (so are negatively correlated), then the integral will be negative and $(f \star g)(x) < 0$. The more negatively they are correlated, the more negative $(f \star g)(x)$.

Finally, it might be that the values of $f(y)$ and $g(x+y)$ jump around as y varies, sometimes positive and sometimes negative, and it may then be that in taking the integral the values cancel out, making $(f * g)(x)$ near zero. One might say — one does say — that f and g are uncorrelated if $(f \star g)(x) = 0$ for all x.

(a) Cross-correlation is similar to convolution, with some important differences. Show that $f \star g = f^- * g = (f * g^-)^-$. Is it true that $f \star g = g \star f$?

(b) *Cross-correlation and delays.* Show that

$$f \star (\tau_b g) = \tau_b (f \star g)\,,$$

where $(\tau_b f)(t) = f(t-b)$. Why does this make sense, intuitively? What about $(\tau_b f) \star g$?

3.22. *Autocorrelation*

The *autocorrelation* of a real-valued signal $f(t)$ with itself is defined to be

$$(f \star f)(x) = \int_{-\infty}^{\infty} f(y) f(x+y)\, dy\,.$$

This is a measure of how much the values of f are correlated, so how much they match up under shifts. For example, *ideal white noise* has uncorrelated values; thus one definition of white noise is a signal with $f \star f = 0$.

(a) Find $\Pi \star \Pi$, without much work. (Use the relationship between correlation and convolution.)

(b) Show that

$$(f \star f)(x) \le (f \star f)(0).$$

You saw this for Fourier series. To show this, use the *Cauchy-Schwarz* inequality. See Chapter 1 for a discussion of this in the context of general inner products. In the present context the Cauchy-Schwarz inequality states that

$$\int_{-\infty}^{\infty} f(t) g(t)\, dt \le \left\{ \int_{-\infty}^{\infty} f(t)^2\, dt \right\}^{1/2} \left\{ \int_{-\infty}^{\infty} g(t)^2\, dt \right\}^{1/2},$$

with equality holding if and only if $f(t)$ and $g(t)$ are proportional.

(c) Show that

$$\mathcal{F}(f \star f) = |\mathcal{F}f|^2 .$$

You also saw a version of this for Fourier series. For Fourier transforms the result is sometimes known as the *Wiener-Khintchine Theorem*, though they actually proved a more general, limiting result that applies when the transform may not exist.

The quantity $|\mathcal{F}f|^2$ is often called the power spectrum, and the result, $\mathcal{F}(f \star f) = |\mathcal{F}f|^2$, is stated for short as: "The Fourier transform of the autocorrelation is the power spectrum." Its many uses include the study of noise in electronic systems.

(d) *Correlation and radar detection.* Here's an application of the maximum value property, $(f \star f)(t) \leq (f \star f)(0)$, above. A radar signal $f(t)$ is sent out, takes a time T to reach an object, and is reflected back and received by the station. In total, the signal is delayed by a time $2T$, attenuated by an amount, say α, and subject to noise, say given by a function $n(t)$. Thus the received signal, $f_r(t)$, is modeled by

$$f_r(t) = \alpha f(t - 2T) + n(t) = \alpha(\tau_{2T} f)(t) + n(t) .$$

You know $f(t)$ and $f_r(t)$, but because of the noise you are uncertain about T, which is what you want to know to determine the distance of the object from the radar station. You also don't know much about the noise, but one assumption that is made is that the cross-correlation of f with n is constant,

$$(f \star n)(t) = C, \quad \text{a constant.}$$

You can compute $f \star f_r$ (so do that) and you can tell (in practice, given the data) where it takes its maximum, say at t_0. Find T in terms of t_0.

3.23. Compute the autocorrelation function $\Gamma_f(\tau) = (f \star f)(\tau)$ for the one-sided exponential decay function, $f(t) = e^{-at}\, u(t)$. (Here, $u(t)$ is a unit step function.)

3.24. *Signal phase.*

Let $h(t)$ be a real signal with Fourier transform, $H(s)$. Write $H(s) = |H(s)|e^{i\theta(s)}$, where $\theta(s) = \angle H(s)$ is the *phase*. Assume that $\angle H(s)$ is not identically zero.

Let $x(t)$ be a real signal whose Fourier transform, $X(s)$, has a phase $\angle X(s)$. Let $x(t)$ undergo the following set of transformations, resulting in the signal $y(t)$:

$$\begin{aligned} g(t) &= h(t) * x(t) , \\ f(t) &= g(t) \star h(t) , \\ y(t) &= f(-t) . \end{aligned}$$

Let $Y(s)$ be the Fourier transform of $y(t)$. What can you say about the phase of $Y(s)$, $\angle Y(s)$, as it relates to $\angle X(s)$? Explain your work.

3.25. *Signal transmission*

You wish to transmit a secret signal $x(t)$ whose Fourier transform is shown in the figure below, centered at the origin.

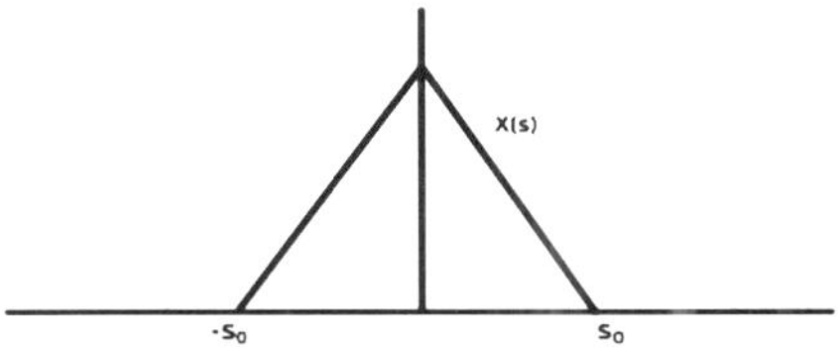

Unfortunately, the entire frequency spectrum is jammed by Dr. Evil, except the regions from $2s_0$ to $2.5s_0$, $4s_0$ to $4.5s_0$, and the mirror images, $-2.5s_0$ to $-2s_0$ and $-4.5s_0$ to $-4s_0$, as shown in the figure below (the shaded regions indicate parts of the spectrum jammed by Dr. Evil):

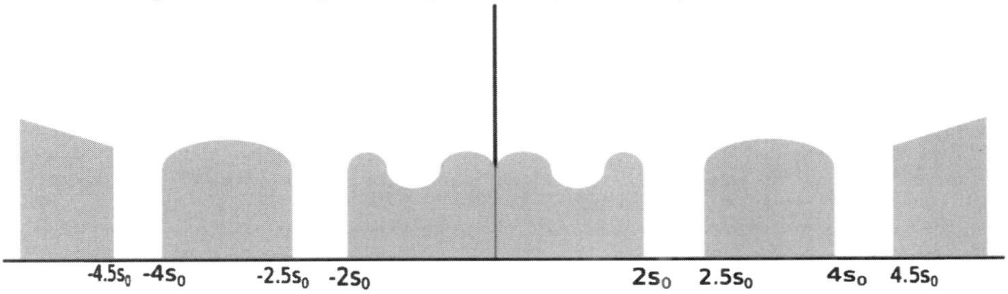

(a) Assuming that any mathematical operation can be performed on the signal on the receiver side and assuming that you can also perform any mathematical operation on your side (the transmitter), what must you do (in mathematical terms) to the signal for it to not be affected by the jamming during the transmission? What must the receiver do to reconstruct the original signal? Your answer may be in either the time or the frequency domain.

(b) You learn that Dr. Evil has stopped jamming the frequencies he was previously jamming but is now eavesdropping on those same frequencies, i.e., the shaded frequencies in the above figure. Is your transmission network of part (a) adequate to hide the signal from Dr. Evil? If not, what modifications would you make to it in order to hide the transmission from Dr. Evil's eavesdropping? Your answer may be in either the time or the frequency domain.

3.26. *Convolution theorem for the Hartley transform*

Recall the Hartley transform that we studied previously. It is defined for a real function $f(t)$ as

$$\mathcal{H}f(s) = \int_{-\infty}^{\infty} (\cos 2\pi st + \sin 2\pi st) f(t)\, dt.$$

(a) Find the convolution theorem for the Hartley transform; i.e., find an expression for $\mathcal{H}(f * g)(s)$ in terms of $\mathcal{H}f(s)$ and $\mathcal{H}g(s)$.

Hint: You may find the following identities you derived useful:

$$\mathcal{H}f = \frac{1+i}{2}\mathcal{F}f + \frac{1-i}{2}\mathcal{F}f^- \quad \text{and} \quad \mathcal{H}f^- = \frac{1+i}{2}\mathcal{F}f^- + \frac{1-i}{2}\mathcal{F}f.$$

(It's not as simple as for the Fourier transform!)

(b) If the functions f and g have even symmetry, what can you say about $\mathcal{H}(f * g)(s)$?

3.27. *Characteristic functions*

Recall that a probability density function (pdf) satisfies the following.

(1) $p(x)$ is real valued.
(2) $p(x) \geq 0$ for all x.
(3) $\int_{-\infty}^{\infty} p(x)\,dx = 1$.

In probability theory one uses the term *characteristic function* to refer to the *Fourier transform* of a probability density function.

If $\psi(s)$ is a characteristic function corresponding to pdf $p(x)$, show that the following are also characteristic functions. Brief explanations are sufficient. The key is to show that $p_i(x) = \mathcal{F}^{-1}\psi_i(x)$ is also a pdf. If $p_i(x)$ is a pdf, then $\psi_i(s)$ is a characteristic function.

(a) $\psi_1(s) = e^{-2\pi i b s}\psi(s)$, where b is a real constant.
(b) $\psi_2(s) = \psi(as)$, where $a > 0$.
(c) $\psi_3(s) = \overline{\psi}(s)$.
(d) $\psi_4(s) = \text{Re}[\psi(s)]$.
(e) $\psi_5(s) = |\psi(s)|^2$.

3.28. *Fourier transforms of probability densities*

We saw that the Gaussian is an example of a probability density whose Fourier transform is identical to the function; i.e., it satisfies $\mathcal{F}f(s) = f(s)$. In this problem we investigate if it is possible that $\mathcal{F}f(s) = af(s)$ for some complex number a.

Recall that a probability density p is a nonnegative function which integrates to 1.

(a) Does there exist a probability density p whose Fourier transform is a scaling of itself, i.e., $\mathcal{F}p(s) = ap(s)$, for $a \neq 1$?
(b) Does there exist a probability density p whose Fourier transform is the negative of itself, i.e., $\mathcal{F}p(s) = -p(s)$?

Thus, the only possible value of a is 1.

3.29. *Gaussians and convolution*

A continuous random variable X is said to be *normally distributed* with mean μ and variance σ^2 if its probability density function (pdf) is given by

$$f(x) = \frac{1}{\sigma\sqrt{2\pi}} e^{\frac{-(x-\mu)^2}{2\sigma^2}}.$$

A common shorthand notation for this is $X \sim N(\mu, \sigma)$, read as "X is drawn from a normal distribution of mean μ and variance σ^2."

Given two independent random variables $X_1 \sim N(\mu_1, \sigma_1)$ and $X_2 \sim N(\mu_2, \sigma_2)$, find the pdf of their sum $X_1 + X_2$ by finding the convolution of their pdfs.

How are the mean and variance of $X_1 + X_2$ related to the mean and variance of X_1 and X_2?

3.30. *Infinitely divisible probability distributions*

Let $f(x)$ be a probability distribution function for a random variable X. We say that $f(x)$ is *infinitely divisible* if for every positive integer n there are n independent, identically distributed random variables $X_1, X_2, \ldots, X_n$ such that $X_1 + X_2 + \cdots + X_n$ has the same distribution as X, i.e., has pdf $f(x)$. (Note that this does *not* mean that the individual X_i's have distribution $f(x)$.)

This may seem to be a very strong condition, but many pdf's that come up in practice are infinitely divisible.

Recall that if we add two independent random variables, the distribution of the sum is the convolution of their distributions.

(a) Show that if $f(x)$ is infinitely divisible, then for every n, there exists a distribution $f_{(n)}(x)$ such that

$$\mathcal{F}f(s) = \left(\mathcal{F}f_{(n)}(s)\right)^n .$$

(b) The Gaussian distribution, with mean μ and variance σ^2, is

$$f(x) = \frac{1}{\sigma\sqrt{2\pi}} e^{-\frac{(x-\mu)^2}{2\sigma^2}} .$$

Use part (a) to show that the Gaussian distribution is infinitely divisible. What are the $f_{(n)}(x)$? Is the mean and variance of $f_{(n)}(x)$ what you might expect?

Hint: You may find it easier to first consider $\mu = 0$ and $\sigma^2 = 1$ and then use the scaling and shift theorems for arbitrary μ, σ^2.

(c) The Cauchy distribution is

$$f(x) = \frac{1}{\pi\gamma\left(1 + \left(\frac{x - x_0}{\gamma}\right)^2\right)};$$

it involves the parameters x_0 and $\gamma > 0$. This distribution is symmetric about x_0, but neither its mean nor its variance is finite. Still, we can show that the Cauchy distribution is infinitely divisible. What are the $f_{(n)}(x)$, and how are these distributions related to the original $f(x)$?

Hint: Again, you may find it easier to first consider $x_0 = 0, \gamma = 1$. Have we seen this expression before? Look at the two-sided exponential decay.

3.31. *Plotting distributions using* MATLAB.

Suppose you roll a fair, six-sided die many times.

(a) What do you think the average of the numbers should be as the number of trials increases?

(b) Let X_i be the random variable denoting the ith roll of the die. Then the empirical average after N rolls of the die is equal to

$$\frac{1}{N}\sum_{i=1}^{N} X_i.$$

Use MATLAB to plot the distribution of the empirical average, for $N = 2$, 10, and 100.

3.32. *Approximating π: Brad learns a lesson*

Brad was doing some exercises to learn the Julia programming language. One exercise was to use Monte Carlo techniques to approximate π via the ratio of two areas. Consider the square with $-1 \le x \le 1$, $-1 \le y \le 1$, and the inscribed unit circle $x^2 + y^2 \le 1$. Then

$$\pi = 4(\text{area of the circle})/\text{area of the square}.$$

This ratio is approximated by generating a large number of points in the square, randomly chosen and uniformly distributed, and computing

$$4(\text{number of points landing in the circle})/\text{total points in the square}.$$

Here's the code that Brad wrote, followed by the code that the Julia people suggested.

```
N=100000 # Specifies number of random points in the unit square
M=0
# Initialize the variable that counts the points in the square
 that are also in the unit circle
for k=1:N
point=rand(2)-rand(2)
# A random point (2-vector) whose entries can be between -1
 and 1.
t=dot(point,point)  # Its length, squared
if  t <=1
M=M+1
end
end
println(4M/N)
```

```
N=100000
M=0
for k=1:N
 point=[2*rand()-1,2*rand()-1]
 # A different way of generating a random point in the square.
t=dot(point,point)
if  t <=1
M=M+1
end
end
println(4M/N)
```

The function

```
rand(2)
```

returns pair of random numbers between 0 and 1. The function

```
rand()
```

returns a single random number between 0 and 1. The function

```
dot
```

returns the dot product.

A few trials of Julia's code gave the numbers 3.14084, 3.13944, 3.13608, 3.13676 whereas a few trials of Brad's code gave 3.90108, 3.89736, 3.89872, 3.89532. Even bumping up to a million points didn't help Brad; Julia's code gave 3.141872 and Brad's code gave 3.898568. What did Brad do wrong?

3.33. *An example of the Cauchy distribution*

Stand a fixed distance from a wall, hold a laser horizontally, and point it at the wall at a random angle. The locations of the burn marks on the wall, along any line, are a random variable X whose values are distributed according to

$$p(x) = \frac{1}{\pi(1+x^2)}\,.$$

This is a *Cauchy distribution* and arises in many applications.

(a) Find the distribution for the average of two (independent) series of such burn marks, i.e., $(1/2)(X_1 + X_2)$, where X_1 and X_2 are distributed as above. For this you'll likely need the Fourier transform of a two-sided decaying exponential and the fact that if $p(x)$ is the distribution for a random variable X, then $(1/a)X$ has distribution $ap(ax)$.

(b) Without any further work, what happens if we average N series of such burn marks? Note that the Central Limit Theorem does *not* apply because the mean and variance are infinite.

People's intuition is often that if they perform the same experiment many times and average the results, then they're getting close to the "actual" or "ideal" answer. What do your results above say about this intuition?

3.34. *More on convolution and probability*[28]

Let $p(x)$ be the probability density of a uniform random variable on $[0,1]$; i.e.,

$$p(x) = \begin{cases} 1, & 0 \le x \le 1\,, \\ 0, & \text{otherwise.} \end{cases}$$

Let $p^{(n)}$ be the n-fold convolution of p given by

$$p^{(n)} = \underbrace{p * p * p * \ldots * p}_{n \text{ times}}\,.$$

[28]From A. Siripuram.

(a) Sketch a graph of $p^{(2)}(x)$.
Note that repeated convolutions keep widening the signal: $p(x)$ is nonzero on $[0, 1]$, $p^{(2)}(x)$ is nonzero on $[0, 2]$, and in general $p^{(n)}(x)$ is nonzero on $[0, n]$. For this problem, we are only interested in computing $p^{(n)}(x)$ on $[0, 1]$, even though it is nonzero on $[0, n]$, a much larger interval.

(b) Show that

$$p^{(n)}(x) = \frac{x^{n-1}}{(n-1)!} \quad \text{for } 0 \le x \le 1 .$$

We will use this later in the problem. Now suppose we have a container of capacity 1 unit. Water is poured into the container each second (the amount of water poured is random), and we want to determine when the container overflows. Let V_i be the amount of water added to the container in the ith second. Assume that V_i is a random variable distributed uniformly in $[0, 1]$; i.e., it has a probability density(pdf) of $p(x)$. The container overflows at the Nth second if $V_1 + V_2 + \cdots + V_N > 1$. We want to find $\mathbb{E}(N)$, the expected time after which the container overflows. Now, we have

$$\begin{aligned} \mathbb{E}(N) &= \sum_{n=0}^{\infty} \text{Prob}(N > n) \\ &= 1 + \sum_{n=1}^{\infty} \text{Prob}(V_1 + V_2 + \cdots + V_n \le 1) \\ (*) \qquad &= 1 + \sum_{n=1}^{\infty} \int_0^1 \phi_n(x)dx, \text{ where } \phi_n(x) \text{ is the pdf of } V_1 + V_2 + \cdots + V_n. \end{aligned}$$

(You are not asked to prove any of the equalities above.) You can assume that the amounts of water added in any second are independent of each other; i.e., the V_i are independent.

(c) Complete the argument from (*) to show that $\mathbb{E}(N) = e$; i.e., the container overflows on average after e seconds.

3.35. *Estimating parameters with minimum spectral samples*[29]
You are given a signal f of the form

$$f(x) = \frac{1}{\sqrt{2\pi\sigma^2}} e^{\frac{(x-\mu)^2}{2\sigma^2}}$$

and want to estimate the parameters μ and σ. The constraint is that we are only allowed to take measurements of the spectrum $\mathcal{F}f$.

Assume you are given a black box which, on input s_0, outputs $\mathcal{F}f(s_0)$. The idea is to take measurements at as few s_0 as possible to find μ and σ. For the purposes of this problem, you can assume $\mathcal{F}f(s_0)$ is given to you with infinite precision. You can also assume that μ is between -1 and 1, even though you do not know its exact value.

[29] From A. Siripuram.

(a) How will you estimate μ and σ with the procedure described above? What is the least number of measurements required?
(b) How many measurements would be needed if you were *not* given that $\mu \in (-1, 1)$?

3.36. *Benford's Law: A logarithmic central limit theorem?* (This problem is not as long as it looks!)

Benford's Law is an empirical observation about the distribution of leading digits in a list of numbers. You might think that the leading digits of naturally occurring measurements are uniformly distributed, but in a very large number of very diverse situations this is not the case. The number 1 occurs as the leading digit about 30% of the time while the number 9 occurs as the leading digit only about 5% of the time.

Benford's Law says that the leading digits follow a logarithmic distribution, that for $n = 1, 2, \ldots, 9$ the fraction of numbers with leading digit n is approximately $\log(n + 1) - \log n$ (log base 10 here). Thus if $n = 1$, this is $\log 2 - \log 1 = .30$ and for $n = 9$ it is $\log 9 - \log 8 = .05$, hence the numbers 30% and 5% mentioned above.

There are many fascinating discussions of this observation and even some explanations. In recent years it has been used as a tool in discovering fraud, for peoples' tendency in fabricating lists of numbers is to think that the leading digits should be uniformly distributed, and this can be checked against what Benford's Law indicates the distribution should be. A broadcast of the NPR program Radiolab (highly recommended) had a segment on this use of Benford's Law; see their webpage for a podcast:

`http://www.radiolab.org/2009/nov/30/from-benford-to-erdos/`

See also

`http://www.washingtonpost.com/wp-dyn/content/article/2009/06/20/AR2009062000004.html`

for a recent article. The mathematician Terrence Tao has also compared several such empirical laws:

`http://terrytao.wordpress.com/2009/07/03/benfords-law-zipfs-law-and-the-pareto-distribution/`.

The Central Limit Theorem tells us that the sum of independent, identically distributed random variables will be distributed like a Gaussian. Why a log in Benford's Law? An approach to Benford's Law via Fourier series has been given in a paper by J. Boyle[30] upon which this discussion is based. In his paper, Boyle gives as motivation the following statement:

> Frequently, naturally occurring data can be thought of as products or quotients on random variables. For example ... if a city has an initial population P_0 and grows by $r_i\%$ in year i, then the population in n years is $P_n = P_0(1+r_1)(1+r_2)\cdots(1+r_n)$, a product of a number of random variables.

[30] J. Boyle, An application of Fourier series to the most significant digit problem, Amer. Math. Monthly **101** (1994), 879–886.

The setup for the study goes as follows. We write numbers in scientific notation:

$$x = M(x) \times 10^{N(x)}$$

where the *mantissa* $M(x)$ satisfies $1 \le M(x) < 10$. The leading digit of x is then $\lfloor M(x) \rfloor$ so we want to understand the distribution of $M(x)$. For a given collection of data to say that the $M(x)$ has a log distribution is to say

$$\text{Prob}(M(x) \le M) = \log M.$$

Then

$$\text{Prob}(M \le M(x) \le M+1) = \log(M+1) - \log M.$$

In turn, $M(x)$ has a log distribution if and only if $\log M(x)$ has a uniform distribution. The goal is thus to understand when $\log M(x)$ has a uniform distribution.

If we multiply two numbers, their mantissas multiply and

$$M(xy) = \begin{cases} M(x)M(y), & 1 \le M(x)M(y) < 10, \\ \frac{M(x)M(y)}{10}, & 10 \le M(x)M(y) < 100. \end{cases}$$

If we take the log, this can be expressed more conveniently as

$$\log M(xy) = \log M(x) + \log M(y) \mod 1.$$

The following theorem of Boyle gives a justification for Benford's Law for the distribution of the leading digit when continuous random variables are multiplied:

Theorem 3.1. *Let $X_1, X_2, \ldots, X_N$ be independent, identically distributed continuous random variables on the interval $[0, 1]$. Let*

$$Z = \sum_{k=1}^{N} X_k \mod 1.$$

Then as $N \to \infty$ the probability distribution of Z approaches a uniform distribution.

Boyle goes on to extend Theorem 3.1 to quotients and powers of random variables, but we will not pursue this.

We're going to step through the proof, providing explanations for some steps and asking you for explanations in others.

The probability density functions we work with are defined on $[0, 1]$, and we assume that they are continuous and also square integrable. Moreover, since we use arithmetic mod 1 (to define Z), it is natural to assume they are periodic of period 1, and the machinery of Fourier series is then available.

Prove the following general lemma, which will be used in the proof of the theorem.

Lemma 3.1. *If $p(x)$ is a continuous, square integrable probability density on $[0, 1]$, periodic of period 1, then*

(a) $\hat{p}(0) = 1$, (b) $|\hat{p}(n)| \le 1$, $n \ne 0$.

Actually, a stronger statement than (b) holds — and that's what we'll really need, and you may use it — but we will not ask you for the proof. Namely,

$$\sup_{n \neq 0} |\hat{p}(n)| \leq A < 1$$

for some constant A. Here sup is the supremum, or least upper bound, of the Fourier coefficients for $n \neq 0$. The statement means that other than for the zeroth coefficient the $\hat{p}(n)$ stay strictly less than 1 (uniformly).

We turn to the proof of the theorem. Suppose that $p(x)$ is the common probability density function for the X_i and let $f(x)$ be the probability distribution for $Z = \sum_{k=1}^{N} X_k \mod 1$. What we (you) will show is that

$$\|f - 1\|^2 = \int_0^1 (f(x) - 1)^2 \, dx \to 0 \quad \text{as } N \to \infty.$$

This says that $f(x)$ tends to the uniform distribution, 1, with respect to the norm on the space $L^2([0,1])$ of square integrable functions. It turns out that this is sufficient, as far as probabilities are concerned, to conclude that Z has the uniform distribution in the limit.

The steps in establishing the statement above, with prompts for you to fill in the reasons, are

$$\begin{aligned}
\int_0^1 (f(x) - 1)^2 \, dx &= \sum_{n \neq 0} |\hat{f}(n)|^2 \quad \text{(why?)} \\
&= \sum_{n \neq 0} |\hat{p}(n)|^{2N} \quad \text{(why?)} \\
&= \sum_{n \neq 0} |\hat{p}(n)|^2 \, |\hat{p}(n)|^{2N-2} \\
&\leq A^{2N-2} \sum_{n \neq 0} |\hat{p}(n)|^2 \quad \text{(why?)}.
\end{aligned}$$

Now explain why

$$\int_0^1 (f(x) - 1)^2 \, dx \to 0 \quad \text{as } N \to \infty.$$

3.37. *Poisson transform*[31]

The *Poisson distribution* (in probability) is used to model the probability of a given number of events occurring over a period of time. For example, the number of phone calls arriving at a call center per minute is modeled using the Poisson distribution:

$$\text{Prob(number of calls} = k) = \frac{\lambda^k}{k!} e^{-\lambda}, \quad k = 0, 1, 2, \dots,$$

where λ is the number of phone calls per minute. Now, the rate of arrivals of phone calls typically varies throughout the day, so λ itself should be considered

[31] From A. Siripuram.

to come from a random variable, Λ, say with probability density function $p(\lambda)$. In that case, the probability Prob(number of calls $= k$) is more realistically modeled by the function

$$P(k) = \int_0^\infty \frac{\lambda^k}{k!} e^{-\lambda} p(\lambda)\, d\lambda, \quad k = 0, 1, 2, \dots.$$

We will call $P(k)$ the Poisson transform of $p(\lambda)$.

It only makes sense to consider $\lambda \geq 0$, but it is convenient for the discussion below to allow $\lambda < 0$ by declaring $p(\lambda) = 0$ when $\lambda < 0$. Then we can write

$$P(k) = \int_{-\infty}^\infty \frac{\lambda^k}{k!} e^{-\lambda} p(\lambda)\, d\lambda, \quad k = 0, 1, 2, \dots.$$

Note that unlike the Fourier transform, the Poisson transform takes a function of a real variable and produces a function defined on the natural numbers $0, 1, 2, \dots$.

(a) We would like to know that the Poisson transform is invertible; i.e., given the numbers $P(0), P(1), P(2), \dots$, it is possible to find $p(\lambda)$. This is a relevant question, as the values $P(k)$ can be measured experimentally from a large set of observations.
For this, given the values $P(k)$ define the function

$$Q(s) = \sum_{k=0}^\infty (-2\pi i s)^k P(k).$$

Using the definition of $P(k)$ show that $Q(s)$ is the Fourier transform of $q(\lambda) = e^{-\lambda} p(\lambda)$, and so find $p(\lambda)$ in terms of $Q(s)$.
Hint: Recall the Taylor series for the exponential function

$$e^a = \sum_{k=0}^\infty \frac{a^k}{k!}.$$

(b) Suppose there are two independent factors contributing to the arrival rate of phone calls, so that $\Lambda = \Lambda_1 + \Lambda_2$. Recall that for the corresponding probability density functions we have

$$p = p_1 * p_2.$$

We want to find the relationship between the Poisson transforms, $P(k)$ for p, and $P_1(k), P_2(k)$ for p_1, p_2.

(i) As above, let $q(\lambda) = e^{-\lambda} p(\lambda)$ and, correspondingly, $q_1(\lambda) = e^{-\lambda} p_1(\lambda)$, $q_2(\lambda) = e^{-\lambda} p_2(\lambda)$. Show that

$$q_1 * q_2 = q \quad \text{(use the definition of convolution)}.$$

(ii) Using $Q(s) = Q_1(s) Q_2(s)$ for the Fourier transforms, deduce that

$$P(k) = \sum_{m=0}^k P_1(m) P_2(k-m),$$

a discrete convolution of P_1 and P_2!

3.38. *Deriving the solution of the Black-Scholes equation for financial derivatives*[32] (This problem is also not as long as it looks.)

As per Wikipedia: A call option is a financial contract between two parties, the buyer and the seller of this type of option. The buyer of the call option has the right, but not the obligation, to buy an agreed quantity of a particular commodity (the underlying) from the seller of the option at the expiration date for a certain price (the strike price). The seller is obligated to sell the commodity should the buyer so decide. The buyer pays a fee (called a premium) for this right. The buyer of a call option wants the price of the underlying instrument to rise in the future while the seller expects that it will not. A European call option allows the buyer to buy the option only on the option expiration date.

The partial differential equation to obtain a Black-Scholes European call option pricing formula is

$$(*)\qquad rC = C_t + \frac{1}{2}\sigma^2 S^2 C_{SS} + rS\, C_S \quad \text{for } t \in (0,T), S \in (0,\infty).$$

Here $C(S,t)$ is the value of the call option at time t with expiration time T, strike price K, and risk-free interest rate r. Let S be the price of the underlying call option and let the volatility of the price of the stock be σ^2. Note that $C_t = \frac{\partial C(S,t)}{\partial t}$, $C_S = \frac{\partial C(S,t)}{\partial S}$, and $C_{SS} = \frac{\partial^2 C(S,t)}{\partial S^2}$. The side conditions of this partial differential equation are

$$\begin{aligned} &C(S,T) = \max(0, S-K) \quad \text{for } S \in (0,\infty),\\ &C(0,t) = 0,\\ &C(S,t) \to S - Ke^{-r(T-t)} \quad \text{as } S \to \infty \text{ for } t \in (0,T). \end{aligned}$$

A judicious change of variables allows us to find the solution to the Black-Scholes equation using a more well-known partial differential equation, namely the heat equation on an infinite rod. First, the Black-Scholes equation is reduced to a (parabolic) differential equation

$$v_\tau = v_{xx} + (k-1)v_x - kv \quad \text{for } x \in (-\infty,\infty), \tau \in (0, T\sigma^2/2)$$

using the change of variables

$$C(S,t) = Kv(x,\tau), \quad S = Ke^x, \quad t = T - \frac{2\tau}{\sigma^2}, \quad k = 2r/\sigma^2.$$

Next, this partial differential equation is reduced to the heat equation on an infinite rod using the change of variables $v(x,\tau) = e^{\alpha x + \beta\tau}u(x,\tau)$:

$$(**)\qquad \begin{aligned} &u_\tau = u_{xx} \quad \text{for } x \in (-\infty,\infty) \text{ and } \tau \in (0, T\sigma^2/2),\\ &u(x,0) = \max(0, (e^{(k+1)x/2} - e^{(k-1)x/2})) \quad \text{for } x \in (-\infty,\infty),\\ &u(x,\tau) \to 0 \quad \text{as } x \to \pm\infty \text{ for } \tau \in (0, T\sigma^2/2), \end{aligned}$$

where $\alpha = (1-k)/2$ and $\beta = -(k+1)^2/4$. You do not need to verify that equation (*) reduces to (**) using the given change of variables.

[32] From Raj Bhatnagar.

(a) Verify that the solution of the partial differential equation (**) is given by a function $u(x,\tau)$ of the form $f(x) * \frac{1}{2\sqrt{\pi\tau}}e^{-\frac{x^2}{4\tau}}$. Find the function $f(x)$.

(b) Show that

$$u(x,\tau) = e^{(k+1)x/2+(k+1)^2\tau/4}\Phi\left(\frac{x+(k+1)\tau}{\sqrt{2\tau}}\right) - e^{(k-1)x/2+(k-1)^2\tau/4}\Phi\left(\frac{x+(k-1)\tau}{\sqrt{2\tau}}\right),$$

where $\Phi(x) = \frac{1}{\sqrt{2\pi}}\int_{-\infty}^{x} e^{-w^2/2}dw$ is the cumulative distribution function of a Gaussian random variable with mean 0 and variance 1. (*Hint*: You may use $I_\alpha = \frac{1}{\sqrt{4\pi\tau}}\int_0^\infty e^{-\frac{(x-u)^2}{4\tau}+\alpha u}du = e^{\alpha x+\alpha^2\tau}\Phi(\frac{x+2\tau\alpha}{\sqrt{2\tau}})$.)

(c) Reverse the change of variables to get the solution of the original partial differential equation (*) given by

$$C(S,t) = \Phi(w) - Ke^{-r(T-t)}\Phi(w - \sigma\sqrt{T-t}),$$

where $w = \frac{\log(S/K)+(r+\sigma^2/2)(T-t)}{\sigma\sqrt{T-t}}$. This is the Black-Scholes European call option pricing formula.

Note that the reverse change of the variables u, x, and τ to F, S, and t requires $v(x,\tau) = e^{\alpha x+\beta\tau}u(x,\tau)$, $\alpha = (1-k)/2$, $\beta = -(1+k)^2/4$, $C(S,t) = Kv(x,\tau)$, $x = \log\left(\frac{S}{K}\right)$, $\tau = \frac{\sigma^2(T-t)}{2}$, $k = 2r/\sigma^2$.

(d) What is the price $C(S,t)$ of a European call option on a stock when the stock price S is 62 euros per share, the strike price K is 60 euros per share, the continuously compounded interest rate r is 10%, the stock's volatility σ is 20% per year, and the exercise time is five months ($T = 5/12$)? What is the price at $t = 0$? (*Hint*: In MATLAB, you may use function 'qfunc(x)' to compute $Q(x) = 1 - \Phi(x) = \frac{1}{\sqrt{2\pi}}\int_x^\infty e^{-w^2/2}dw$.)

The Black-Scholes model is widely used in practice to estimate the prices of call options in financial markets. This is mainly due to the existence of a concrete, closed-form solutions for the price of the options given the parameters in the Black-Scholes formulas. The only trouble seems to be the estimation of the parameters, especially the estimation of the volatility σ from historical data. Robert Merton and Myron Scholes were awarded the Nobel Prize in Economics in 1997 to honor their contributions to option pricing!

Chapter 4

Distributions and Their Fourier Transforms

4.1. The Day of Reckoning

We've been playing a little fast and loose with the Fourier transform, applying Fourier inversion, appealing to duality, and all that. "Fast and loose" is an understatement, but it's also true that we haven't done anything wrong. All of our formulas and all of our applications have been correct, if not fully justified. Nevertheless, we have to come to terms with some fundamental questions. It will take us some time, but in the end we will have settled on a very wide class of signals with these properties:

- The allowed signals include Π's, Λ's, δ's, unit steps, ramps, sines, cosines, and other standard signals that the world's economy depends on.
- The Fourier transform and its inverse are defined for all of these signals.
- Fourier inversion works.

These are the three most important features of the development to come, but we'll also reestablish some of our specific results and as an added benefit we'll even finish off differential calculus!

4.1.1. A too simple criterion.

It's not hard to write down an assumption on a function that guarantees the existence of its Fourier transform and even implies a little more than existence.

- If $\displaystyle\int_{-\infty}^{\infty} |f(t)|\,dt < \infty$, then $\mathcal{F}f$ exists and is continuous.

The same result holds for $\mathcal{F}^{-1}f$, and the argument below applies to both $\mathcal{F}f$ and $\mathcal{F}^{-1}f$.

"Existence" means that the integral defining the Fourier transform converges. This follows from

$$
\begin{aligned}
|\mathcal{F}f(s)| &= \left|\int_{-\infty}^{\infty} e^{-2\pi i s t} f(t)\,dt\right| \\
&\le \int_{-\infty}^{\infty} |e^{-2\pi i s t}|\,|f(t)|\,dt = \int_{-\infty}^{\infty} |f(t)|\,dt < \infty\,.
\end{aligned}
$$

Here we've used that the magnitude of the integral is less that the integral of the magnitude.[1] There's actually something to say here, and while it's not complicated, I'd just as soon defer this and other comments on general facts on integrals to Section 4.3. Continuity is the little extra information we get beyond existence. Continuity can be deduced as follows: for any s and s' we have

$$
\begin{aligned}
|\mathcal{F}f(s) - \mathcal{F}f(s')| &= \left|\int_{-\infty}^{\infty} e^{-2\pi i s t} f(t)\,dt - \int_{-\infty}^{\infty} e^{-2\pi i s' t} f(t)\,dt\right| \\
&= \left|\int_{-\infty}^{\infty} (e^{-2\pi i s t} - e^{-2\pi i s' t}) f(t)\,dt\right| \\
&\le \int_{-\infty}^{\infty} |e^{-2\pi i s t} - e^{-2\pi i s' t}|\,|f(t)|\,dt\,.
\end{aligned}
$$

As a consequence of $\int_{-\infty}^{\infty} |f(t)|\,dt < \infty$ we can take the limit as $s' \to s$ inside the integral.[2] If we do that, then $|e^{-2\pi i s t} - e^{-2\pi i s' t}| \to 0$, and hence,

$$
|\mathcal{F}f(s) - \mathcal{F}f(s')| \to 0 \quad \text{as } s' \to s\,.
$$

This says that $\mathcal{F}f(s)$ is continuous.

Functions satisfying

$$
\int_{-\infty}^{\infty} |f(t)|\,dt < \infty
$$

are said to be in $L^1(\mathbb{R})$. One also says simply that $|f(t)|$ is *integrable*, or that $f(t)$ is absolutely integrable. The "L" is for Lebesgue, as for the square integrable functions $L^2([0,1])$ that we met in connection with Fourier series. It's Lebesgue's generalization of the Riemann integral that's intended by invoking the notation $L^1(\mathbb{R})$. The "1" refers to the fact that we integrate $|f(t)|$ to the first power. The L^1-norm of f is defined to be

$$
\|f\|_1 = \int_{-\infty}^{\infty} |f(t)|\,dt\,.
$$

No inner products or orthogonality for $L^1(\mathbb{R})$. If you're wondering about $L^2(\mathbb{R})$, wait till the end of the chapter.

---

[1] Magnitude, not absolute value, because the integral is a complex number.

[2] A general fact. We've done this before and will do it again, and I will say a little more about it in Section 4.3. But save yourself for other things and let some of these general facts ride without insisting on complete justifications — such justifications creep in everywhere once you let the rigor police back on the beat.

In this terminology the criterion says that if $f \in L^1(\mathbb{R})$, then $\mathcal{F}f$ exists and is continuous, and using the norm notation we can write

$$|\mathcal{F}f(s)| \leq \|f\|_1 \quad \text{for } -\infty < s < \infty.$$

This is a handy way to write the inequality.

While the result on existence (and continuity) holds for both $\mathcal{F}f$ and $\mathcal{F}^{-1}f$, we haven't said anything about Fourier inversion or duality, and no such statement appears in the criterion. Let's look right away at a test case.

The very first example we computed, and still an important one, is the Fourier transform of Π. We found directly that

$$\mathcal{F}\Pi(s) = \int_{-\infty}^{\infty} e^{-2\pi i s t}\Pi(t)\,dt = \int_{-1/2}^{1/2} e^{-2\pi i s t}\,dt = \operatorname{sinc} s\,.$$

No problem there, no problem whatsoever. Note that the criterion applies to guarantee existence, for Π is surely in $L^1(\mathbb{R})$:

$$\int_{-\infty}^{\infty} |\Pi(t)|\,dt = \int_{-1/2}^{1/2} 1\,dt = 1\,.$$

Furthermore, the transform $\mathcal{F}\Pi(s) = \operatorname{sinc} s$ is continuous, as it should be.[3] That's worth remarking on: although the signal jumps (Π has a discontinuity), the *Fourier transform* does not. Make this part of your intuition on the Fourier transform vis à vis the signal.

Many of the other examples we worked with are L^1-functions — the triangle function, the exponential decay (one- or two-sided), Gaussians — so our computations of the Fourier transforms in those cases were perfectly justifiable (and correct).

Appealing to duality we then reached the happy conclusion that $\mathcal{F}\operatorname{sinc} s = \Pi(s)$. But without invoking duality, i.e., using directly the formula,

$$\mathcal{F}\operatorname{sinc}(s) = \int_{-\infty}^{\infty} e^{-2\pi i s t}\operatorname{sinc} t\,dt\,,$$

we have a problem. The sinc function *does not* satisfy the integrability criterion. It is my sad duty to inform you that

$$\int_{-\infty}^{\infty} |\operatorname{sinc} t|\,dt = \infty\,.$$

I'll give you two ways of seeing the failure of $|\operatorname{sinc} t|$ to be integrable. First, if sinc did satisfy the criterion $\int_{-\infty}^{\infty} |\operatorname{sinc} t|\,dt < \infty$, then its Fourier transform would be continuous. But its Fourier transform, which *has* to come out to be Π, is *not* continuous. Or, if you don't like that, here's a direct argument. We can find infinitely many intervals where $|\sin \pi t| \geq 1/2$; this happens when t is between $1/6$ and $5/6$, and that repeats for infinitely many intervals, for example on

---

[3]Excuse the mathematical fine point: $\operatorname{sinc} s$ is continuous because of the famous limit $\lim_{s\to 0} \sin \pi s/\pi s = 1$.

$I_n = [\frac{1}{6} + 2n, \frac{5}{6} + 2n]$, $n = 0, 1, 2, \ldots$, because $\sin \pi t$ is periodic of period 2. The I_n all have length 2/3. On I_n we have $|t| \leq \frac{5}{6} + 2n$, so

$$\frac{1}{|t|} \geq \frac{1}{5/6 + 2n}$$

and

$$\int_{I_n} \frac{|\sin \pi t|}{\pi |t|}\, dt \geq \frac{1}{2\pi} \frac{1}{5/6 + 2n} \int_{I_n} dt = \frac{1}{2\pi} \frac{2}{3} \frac{1}{5/6 + 2n}.$$

Then

$$\int_{-\infty}^{\infty} \frac{|\sin \pi t|}{\pi |t|}\, dt \geq \sum_n \int_{I_n} \frac{|\sin \pi t|}{\pi |t|}\, dt = \frac{1}{3\pi} \sum_{n=1}^{\infty} \frac{1}{5/6 + 2n} = \infty\,.$$

Yes, $|\operatorname{sinc} t| = |\sin \pi t / \pi t|$ tends to 0 as $t \to \pm\infty$, but not fast enough to make the integral of $|\operatorname{sinc} t|$ converge.

This is the most basic example in the theory! It's not clear that the integral defining the Fourier transform of sinc exists — at least it doesn't follow from the criterion. Doesn't this bother you? Isn't it a little embarrassing that multibillion dollar industries seem to depend on integrals that don't converge?

In fact, there isn't so much of a problem with either Π or sinc. It is true that

$$\int_{-\infty}^{\infty} e^{-2\pi i s t} \operatorname{sinc} s\, ds = \begin{cases} 1, & |t| < \frac{1}{2}, \\ \frac{1}{2}, & |t| = \frac{1}{2}, \\ 0, & |t| > \frac{1}{2}. \end{cases}$$

However showing this — evaluating the improper integral that defines the Fourier transform — requires special arguments and techniques. The sinc function oscillates, as do the real and imaginary parts of the complex exponential, and integrating $e^{-2\pi i s t} \operatorname{sinc} s$ involves enough cancellations for the limit

$$\lim_{\substack{a \to -\infty \\ b \to \infty}} \int_a^b e^{-2\pi i s t} \operatorname{sinc} s\, ds$$

to exist. This is laid out for you in the problems. Thus Fourier inversion, and duality, can be pushed through in this case.[4]

The truth is that cancellations that occur in the sinc integral or in its Fourier transform are a very subtle and dicey thing. Such risky encounters are to be approached with respect, or maybe to be avoided. We'd like a more robust, trustworthy theory, and it seems that L^1-integrability of a signal is just too simple a criterion on which to build.

There are problems with other, simpler functions. Take, for example, the signal $f(t) = \cos 2\pi t$. As it stands now, this signal does not even *have* a Fourier transform — does not have a spectrum! — for the integral

$$\int_{-\infty}^{\infty} e^{-2\pi i s t} \cos 2\pi t\, dt$$

[4] And I also know full well that I have reignited the debate on how Π should be defined at the endpoints.

does not converge, no way, no how. Or an even more basic example is $f(t) = 1$ and the integral

$$\int_{-\infty}^{\infty} e^{-2\pi i s t} \cdot 1 \, dt \,.$$

No way, no how, again. This is no good.

Before we bury $L^1(\mathbb{R})$ as insufficient for our needs, here's one more good thing about it. There's actually an additional consequence for $\mathcal{F}f$ when $f \in L^1(\mathbb{R})$, namely:

- If $\displaystyle\int_{-\infty}^{\infty} |f(t)| \, dt < \infty$, then $\mathcal{F}f(s) \to 0$ as $s \to \pm\infty$.

This is called the *Riemann-Lebesgue lemma* and it's more difficult to prove than the statements on existence and continuity. I'll comment on it in Section 4.8.3. One might view the result as saying that $\mathcal{F}f(s)$ is at least *trying* to be integrable. It's continuous and it tends to zero as $s \to \pm\infty$. Unfortunately, the fact that $\mathcal{F}f(s) \to 0$ does not imply that it's integrable (think of sinc, again). In fact, a function in $L^1(\mathbb{R})$ need not tend to zero at $\pm\infty$; that's also discussed in Section 4.3. If we knew something, or could insist on something about the *rate* at which a signal or its transform tends to zero at $\pm\infty$, then perhaps we could push on further.

4.1.2. The path, the way. Fiddling around with $L^1(\mathbb{R})$ or substitutes, putting extra conditions on jumps — all have been used. The path to success lies elsewhere. It is well marked and firmly established, but it involves a break with the classical point of view. The outline of how all this is settled goes like this:

1. We single out a collection of functions $\mathcal{S}$ for which convergence of the Fourier integrals is assured, for which a function *and* its Fourier transform are both in $\mathcal{S}$, and for which Fourier inversion works. Furthermore, Parseval's identity holds:
$$\int_{-\infty}^{\infty} |f(x)|^2 \, dx = \int_{-\infty}^{\infty} \mathcal{F}f(s)|^2 \, ds \,.$$

Perhaps surprisingly it's not so hard to find a suitable collection $\mathcal{S}$, at least if one knows what one is looking for. Also perhaps surprisingly, if not ironically, the path to finding Fourier transforms of bad functions, so to speak, starts with the *best* functions.

This much is classical; new ideas with new intentions, yes, but not new *objects*. But what comes next is definitely not classical. Parts of the theory had been anticipated first by Oliver Heaviside and were used effectively by him for solving differential equations that arise in electrical applications. He was steeped in the problems of telegraphy. The conventional formulation of Maxwell's equations of electromagnetics is due to Heaviside. No Fourier transforms, but a lot of δ's (not yet so named).

Heaviside's approach was developed, somewhat, mostly by less talented people and it remained highly suspicious. The ideas were further cultivated by the work of Paul Dirac for use in quantum mechanics, with great success and with the official naming of the Dirac δ.

Mathematicians still found much to be suspicious about, but the rest of the world was warming to the obviously important applications and to the ease with which calculations could be made. Still no generalized Fourier transform. Another influential collection of results originated in the work of Sergei Sobolev on partial differential equations, the use of so-called *weak solutions*.

Finally, in a tour de force, the ideas were expanded and crystalized by Laurent Schwartz. That's where we're headed. $\mathcal{S}$ is for Schwartz, by the way.

2. $\mathcal{S}$ forms a class of *test functions* which, in turn, serve to define a larger class of *generalized functions* or *distributions*. Distributions *operate* on test functions — you pass a distribution a test function, it returns a complex number. The operation is assumed to be linear (so we can invoke superposition when needed) and continuous (so we can take limits when needed). The model case is the δ-function, which we'll define precisely. You've probably seen δ, imprecisely, in other classes, often in connection with the term "impulse response."

"Distribution" is Schwartz's term, and it has nothing to do with probability distributions, etc. There are just so many words to go around, and sometimes they get used more than once and for different purposes. For the class $\mathcal{S}$ of test functions, one uses the term *tempered distributions* and the notation $\mathcal{T}$ for the collection of tempered distributions. The tempered distributions include, for example, L^1- and L^2-functions (which can be wildly discontinuous), the sinc function, and complex exponentials (hence periodic functions). But they include much more, like δ-functions and related objects.

Precisely because $\mathcal{S}$ was chosen to be the ideal Fourier friendly space of classical signals, the tempered distributions are well suited for Fourier methods. *We'll define the Fourier transform of a tempered distribution.*

3. The Fourier transform and its inverse will be defined so as to operate on tempered distributions, and they operate to produce distributions of the same type. The inverse Fourier transform can be applied, and the Fourier inversion theorem holds in this setting.
4. In the case when a tempered distribution comes from a function — in a way we'll explain — the Fourier transform reduces to the usual definition as an integral, when the integral makes sense, so we won't have lost anything in the process. However, tempered distributions are more general than functions, so we really will have done something new.

Our goal is to hit the relatively few main ideas in the outline above, suppressing the considerable mass of technical details. In practical terms this will enable us to introduce δ-functions and the like as tools for computation and to feel a greater measure of confidence in the range of applicability of the formulas. We're taking this path because it works, it's very interesting, *and it's easy to compute with.* I especially want you to come to believe the last point.

Vacuum tubes? We'll touch on some other approaches to defining distributions (as limits), but as far as I'm concerned they are the equivalent of vacuum tube technology. You can do distributions in other ways, and some people really love building things with vacuum tubes, but wouldn't you rather learn something more up to date?

Let us now praise famous men. Heaviside was an odd character for sure. An appreciative biography is *Oliver Heaviside: Sage in Solitude*, by Paul Nahim, published by the IEEE press. Paul Dirac was at least as odd, at least according to the title of a recent biography: *The Strangest Man: The Hidden Life of Paul Dirac, Mystic of the Atom*, by Graham Farmelo. Schwartz's autobiography (no particular oddness known to me) has now been translated into English under the title *A Mathematician Grappling with His Century*. He is an imposing figure. I don't know much about Sobolev, odd or otherwise, beyond what can be found through Wikipedia et al.

4.2. The Best Functions for Fourier Transforms: Rapidly Decreasing Functions

From our discussion of orthogonality in Chapter 1 you might remember a short sermon on the role of definitions. Ditto for the definition of convolution. This bears repeating at the commencement of this new journey: stated as a working principle, mathematics progresses more by making intelligent definitions than by proving theorems.

The hardest work is often in formulating the fundamental concepts in the right way, a way that will then make the deductions from those definitions (relatively) easy and natural. This can take awhile to sort out, and a subject might be reworked several times as it matures. When discoveries accumulate and one sees where things end up, there's a tendency to go back and change the starting point so that the trip becomes easier. I made mention of this sort of thing in connection to probability, where "random variable" ultimately emerged as the starting point. Mathematicians may be especially self-conscious about this process, but there are certainly examples in engineering where close attention to the basic definitions has shaped a field. Think of Shannon's work on information theory for a particularly striking example.

Engineers can find this tiresome, wanting to *do something* and not, so it may seem, just talk about it. "Devices don't have hypotheses" is how one of my colleagues put it. One can also have too much of a good thing. Too many trips back to the starting point to rewrite the rules can make it hard to follow the game, especially if one has already played by the earlier rules. I'm sympathetic to these concerns, and for our present work on the Fourier transform I'll try to steer a course that makes the definitions reasonable and lets us make steady forward progress. But for the wary let me also offer some honest feedback from teaching this topic this way to many engineers of many stripes. They liked it. They appreciated seeing the ideas coalesce and they liked seeing how the computations played out. Read the chapter lightly, skip around, and try to keep those things in sight. If you're more of a mind mostly to hunt for formulas, you should be able to find what you need, and that's perfectly OK.

4.2.1. Smoothness and decay. To ask how fast $\mathcal{F}f(s)$ might tend to zero, depending on what additional assumptions we might make about the function $f(x)$ *beyond* integrability, will lead to our defining rapidly decreasing functions, and this is the key. Integrability is too weak a condition on a signal $f(x)$ to get very far, but it does imply that $\mathcal{F}f(s)$ is continuous and tends to 0 at $\pm\infty$. What we're going to do is study the relationship between the *smoothness* of a function — not just continuity, but how many times it can be differentiated — and the rate at which its Fourier transform decays at infinity.

Before continuing, a word about the word "smooth." As just mentioned, when applied to a function, it's in reference to how many derivatives the function has. More often than not, and certainly in this chapter, the custom is that when one says simply that a function is smooth, one means that it has derivatives of any order. Put another way, a smooth function is, by custom, an *infinitely differentiable* function.

We'll always assume that $f(x)$ is absolutely integrable and so has a Fourier transform. Let's suppose, more stringently, the following:

- $|xf(x)|$ is integrable; i.e., $\displaystyle\int_{-\infty}^{\infty} |xf(x)|\,dx < \infty$.

Then $xf(x)$ has a Fourier transform. So does $-2\pi i x f(x)$, and its Fourier transform is

$$\begin{aligned}
\mathcal{F}(-2\pi i x f(x)) &= \int_{-\infty}^{\infty} (-2\pi i x) e^{-2\pi i s x} f(x)\,dx \\
&= \int_{-\infty}^{\infty} \left(\frac{d}{ds} e^{-2\pi i s x}\right) f(x)\,dx = \frac{d}{ds}\int_{-\infty}^{\infty} e^{-2\pi i s x} f(x)\,dx \\
&\quad \text{(switching } d/ds \text{ and the integral is justified by the integrability of } |xf(x)|\text{)} \\
&= \frac{d}{ds}(\mathcal{F}f)(s)\,.
\end{aligned}$$

This says that the Fourier transform $\mathcal{F}f(s)$ is differentiable and that its derivative is $\mathcal{F}(-2\pi i x f(x))$. When $f(x)$ is merely integrable, we know that $\mathcal{F}f(s)$ is merely continuous, but with the extra assumption on the integrability of $xf(x)$ we conclude that $\mathcal{F}f(s)$ is actually differentiable. (And its derivative is continuous. Why?)

For one more go-round in this direction, what if $|x^2 f(x)|$ is integrable? Then, by the same argument,

$$\begin{aligned}
\mathcal{F}((-2\pi i x)^2 f(x)) &= \int_{-\infty}^{\infty} (-2\pi i x)^2 e^{-2\pi i s x} f(x)\,dx \\
&= \int_{-\infty}^{\infty} \left(\frac{d^2}{ds^2} e^{-2\pi i s x}\right) f(x)\,dx \\
&= \frac{d^2}{ds^2}\int_{-\infty}^{\infty} e^{-2\pi i s x} f(x)\,dx = \frac{d^2}{ds^2}(\mathcal{F}f)(s)\,,
\end{aligned}$$

and we see that $\mathcal{F}f$ is twice differentiable. (And its second derivative is continuous.)

Clearly, we can proceed like this, and as a somewhat imprecise headline we might then announce:

- Faster decay of $f(x)$ at $\pm\infty$ leads to a greater differentiability of the Fourier transform.

Now let's take this in another direction, with an assumption on the differentiability of the signal. Suppose $f(x)$ is differentiable, that its derivative is integrable, and that $f(x) \to 0$ as $x \to \pm\infty$. I've thrown in all the assumptions I need to justify the following calculation:

$$\begin{aligned}
\mathcal{F}f(s) &= \int_{-\infty}^{\infty} e^{-2\pi isx} f(x)\,dx \\
&= \left[f(x)\frac{e^{-2\pi isx}}{-2\pi is} \right]_{x=-\infty}^{x=\infty} - \int_{-\infty}^{\infty} \frac{e^{-2\pi isx}}{-2\pi is} f'(x)\,dx \\
&\qquad \text{(integration by parts with } u = f(x),\ dv = e^{-2\pi isx}dx) \\
&= \frac{1}{2\pi is}\int_{-\infty}^{\infty} e^{-2\pi isx} f'(x)\,dx \quad (\text{using } f(x) \to 0 \text{ as } x \to \pm\infty) \\
&= \frac{1}{2\pi is}(\mathcal{F}f')(s)\,.
\end{aligned}$$

We then have

$$|\mathcal{F}f(s)| = \frac{1}{2\pi|s|}|(\mathcal{F}f')(s)| \leq \frac{1}{2\pi|s|}\|f'\|_1\,.$$

The last inequality is the result: "The Fourier transform is bounded by the L^1-norm of the function." This says that $\mathcal{F}f(s)$ tends to 0 at $\pm\infty$ like $1/s$. (Remember that $\|f'\|_1$ is some fixed number here, independent of s.) Earlier we commented (without proof) that if f is integrable, then $\mathcal{F}f$ tends to 0 at $\pm\infty$, but here with the stronger assumptions, we get the stronger conclusion that $\mathcal{F}f$ tends to zero *at a certain rate.*

Let's go one step further in this direction. Suppose $f(x)$ is *twice* differentiable, that its first and second derivatives are integrable, and that $f(x)$ and $f'(x)$ tend to 0 as $x \to \pm\infty$. The same argument gives

$$\begin{aligned}
\mathcal{F}f(s) &= \int_{-\infty}^{\infty} e^{-2\pi isx} f(x)\,dx \\
&= \frac{1}{2\pi is}\int_{-\infty}^{\infty} e^{-2\pi isx} f'(x)\,dx \quad (\text{picking up on where we were before}) \\
&= \frac{1}{2\pi is}\left(\left[f'(x)\frac{e^{-2\pi isx}}{-2\pi is} \right]_{x=-\infty}^{x=\infty} - \int_{-\infty}^{\infty} \frac{e^{-2\pi isx}}{-2\pi is} f''(x)\,dx \right) \\
&\qquad \text{(integration by parts with } u = f'(x),\ dv = e^{-2\pi isx}dx) \\
&= \frac{1}{(2\pi is)^2}\int_{-\infty}^{\infty} e^{-2\pi isx} f''(x)\,dx \quad (\text{using } f'(x) \to 0 \text{ as } x \to \pm\infty) \\
&= \frac{1}{(2\pi is)^2}(\mathcal{F}f'')(s)\,.
\end{aligned}$$

Thus

$$|\mathcal{F}f(s)| = \frac{1}{|2\pi s|^2}|(\mathcal{F}f'')(s)| \leq \frac{1}{|2\pi s|^2}\|f''\|_1$$

and we see that $\mathcal{F}f(s)$ tends to 0 like $1/s^2$.

The headline:

- Greater differentiability of $f(x)$, plus integrability, leads to faster decay of the Fourier transform at $\pm\infty$.

Remark on the derivative formula for the Fourier transform. The engaged reader will have noticed that in the course of our work we have rederived the derivative formula

$$\mathcal{F}f'(s) = 2\pi i s \mathcal{F}f(s)\,,$$

but here we used the assumption that $f(x) \to 0$, which we didn't use before.[5] What's up? With the assumptions we made, the derivation above via integration by parts is a natural approach. Later, when we develop the generalized Fourier transform, we'll have suitably generalized derivative formulas.

We could go on as we did above, comparing the consequences of higher differentiability, integrability, and decay, bouncing back and forth between the function and its Fourier transform. The great insight in making use of these observations is that the simplest and most useful way to coordinate *all* these phenomena is to allow for *arbitrarily great differentiability* and *arbitrarily fast decay*. We would like to have both phenomena in play. That is the crucial step, and here is the crucial definition.

4.2.2. Rapidly decreasing functions.

A function $f(x)$ is said to be *rapidly decreasing* at $\pm\infty$ if:

(1) It is infinitely differentiable.[6]

(2) For *all* integers $m \geq 0$ and $n \geq 0$,

$$\left|x^m \frac{d^n}{dx^n} f(x)\right| \to 0 \quad \text{as } x \to \pm\infty.$$

In words, *any* nonnegative power of x times *any* order derivative of f tends to zero at $\pm\infty$.[7]

Note that m and n are independent in this definition. That is, we insist that, say, the 5th power of x times the 17th derivative of $f(x)$ tends to zero and that the 100th power of x times the first derivative of $f(x)$ tends to zero, and whatever you want.

[5] In the earlier derivation we interchanged limits and integration to pull the limit of the difference quotient outside the integral. This also requires assumptions on continuity and differentiability of $f(x)$, but not on behavior at $\pm\infty$.

[6] Since differentiability implies continuity, note that all derivatives are continuous.

[7] We follow the convention that the zeroth-order derivative of a function is just the function.

The sum of two rapidly decreasing functions is again rapidly decreasing, as is their product. The latter follows easily from the former and from the product rule.

These are the functions that constitute $\mathcal{S}$. In Schwartz's honor, they are also referred to as *Schwartz functions.*

Are there any such functions? Any infinitely differentiable function that is identically zero outside some finite interval is one example, and I'll even write down a formula for one of these later. Another example is $f(x) = e^{-x^2}$. You may already be familiar with the phrase "the exponential grows faster than any power of x," and likewise with the phrase "e^{-x^2} decays faster than any power of x."[8] In fact, any derivative of e^{-x^2} decays faster than any power of x as $x \to \pm\infty$, as you can check with l'Hôpital's rule, for example. We can express this exactly as in the definition:

$$\left| x^m \frac{d^n}{dx^n} e^{-x^2} \right| \to 0 \quad \text{as } x \to \pm\infty .$$

There are plenty of other rapidly decreasing functions. We also observe that if $f(x)$ is rapidly decreasing, then it is in $L^1(\mathbb{R})$, and also in $L^2(\mathbb{R})$, since it's decaying so fast — faster than it needs to, to guarantee integrability. In fact, the rapid decay actually guarantees pth power integrability for any $p \geq 1$, meaning that $|f(x)|^p$ is integrable. These functions are denoted, naturally enough, by $L^p(\mathbb{R})$.

Alternative definitions. For some derivations (read: proofs) it's helpful to have other conditions that are equivalent to a function being rapidly decreasing. One such is to assume that for any nonnegative integers m and n there is a constant C_{mn} such that

$$\left| x^m \frac{d^n}{dx^n} f(x) \right| \leq C_{mn} \quad \text{as } x \to \pm\infty .$$

In words, the mth power of x times the nth derivative of f remains bounded for all m and n as $x \to \pm\infty$. The constant will depend on which m and n we take, as well as on the particular function, but not on x.

This looks like a weaker condition — boundedness instead of tending to 0 — but it's not. If a function is rapidly decreasing as we've already defined it, then it certainly satisfies the boundedness condition. Conversely, if the function satisfies the boundedness condition, then

$$\left| x^{m+1} \frac{d^n}{dx^n} f(x) \right| \leq C_{(m+1)n} \quad \text{as } x \to \pm\infty ,$$

so

$$\left| x^m \frac{d^n}{dx^n} f(x) \right| \leq \frac{C_{(m+1)n}}{x} \to 0 \quad \text{as } x \to \pm\infty .$$

A third equivalent condition is to assume that for any nonnegative integers m and n there is a constant C'_{mn} (so a different constant than just above) such that

$$(1 + |x|)^m \left| \frac{d^n}{dx^n} f(x) \right| \leq C'_{mn}$$

for all $-\infty < x < \infty$.

[8] I used e^{-x^2} as an example instead of e^{-x} (for which the statement is true as $x \to \infty$) because I wanted to include $x \to \pm\infty$, and I used e^{-x^2} instead of $e^{-|x|}$ because I wanted the example to be smooth. $e^{-|x|}$ has a corner at $x = 0$.

This condition implies and is implied by the preceding boundedness condition. Here's why. If $|x|$ is big, here meaning that $|x| > A$ for some (large, fixed) A, then $(1+|x|)^m$ is about the same size as $|x|^m$, so bounds on $\left|x^m \frac{d^n}{dx^n} f(x)\right|$ are pretty much the same as bounds on $(1+|x|)^m \left|\frac{d^n}{dx^n} f(x)\right|$; if you have one, you have the other. If x is not big, here meaning that $|x| \leq A$, then both $|x|^m$ and $(1+|x|)^m$ are bounded by $(1+A)^m$ so we're wanting $\left|\frac{d^n}{dx^n} f(x)\right|$ to stay bounded for $|x| \leq A$. This follows because $\frac{d^n}{dx^n} f(x)$ is continuous, a point I want to elaborate.

- Here we are using the fact that a continuous function on a closed, bounded interval has a finite maximum and minimum on the interval. That's an important math fact. You may have heard it, and appealed to it, in studying optimization problems, for example.
- There'a a more general statement on the existence of maxima and minima, one that involves some additional terminology that will come up again and in other contexts. A set in $\mathbb{R}$ that is *closed and bounded* is called *compact.* Think of a finite union of closed, bounded intervals. The theorem is that a continuous function on a compact set has a maximum and a minimum on the set.

We'll use these alternative formulations of rapidly decreasing in some arguments. It's a matter of taste which condition one takes as the primary definition.

Just to tell you, an advantage of the final characterization is in some proofs where you might find yourself wanting to bound an integral, something involving

$$\int_{-\infty}^{\infty} \left|\frac{d^n}{dx^n} f(x)\right| dx \leq \int_{-\infty}^{\infty} \frac{C'_{mn}}{(1+|x|)^m}\, dx\,.$$

The integrand on the right has no zeros in the denominator, and that's a good thing. If you instead stuck with the other boundedness condition in your proof, you might instead find yourself working with something like

$$\int_{-\infty}^{\infty} \left|\frac{d^n}{dx^n} f(x)\right| dx \leq \int_{-\infty}^{\infty} \frac{C_{mn}}{|x|^m}\, dx\,.$$

For the integral on the right you would have to say something about avoiding $x = 0$. You would rather not. This will come up in Section 4.8 (which you might gloss over anyway). Just a heads-up.

4.2.3. Rapidly decreasing functions and the classical Fourier transform.

Let's start to see why Schwartz's idea was such a good one.

The Fourier transform of a function in $\mathcal{S}$ is still in $\mathcal{S}$. Let $f(x)$ be a function in $\mathcal{S}$. We want to show that $\mathcal{F}f(s)$ is also in $\mathcal{S}$. The condition involves derivatives of $\mathcal{F}f$, so what comes in are the derivative formulas for the Fourier transform and the version of the formulas for higher derivatives.

Starting with

$$2\pi i s \mathcal{F}f(s) = \Big(\mathcal{F}\frac{d}{dx}f\Big)(s)$$

and

$$\frac{d}{ds}\mathcal{F}f(s) = \mathcal{F}(-2\pi i x f(x)),$$

the higher-order versions of these formulas are

$$(2\pi i s)^n \mathcal{F}f(s) = \Big(\mathcal{F}\frac{d^n}{dx^n}f\Big)(s),$$

$$\frac{d^n}{ds^n}\mathcal{F}f(s) = \mathcal{F}\big((-2\pi i x)^n f(x)\big).$$

Combining these formulas we find, inductively, that for all nonnegative integers m and n,

$$\mathcal{F}\Big(\frac{d^n}{dx^n}\big((-2\pi i x)^m f(x)\big)\Big) = (2\pi i s)^n \frac{d^m}{ds^m}\mathcal{F}f(s).$$

Note how the roles of m and n are flipped in the two sides of the equation.

We use this last identity together with the estimate for the Fourier transform in terms of the L^1-norm of the function. Namely,

$$|s|^n \left|\frac{d^m}{ds^m}\mathcal{F}f(s)\right| = (2\pi)^{m-n}\left|\mathcal{F}\Big(\frac{d^n}{dx^n}(x^m f(x))\Big)\right| \le (2\pi)^{m-n}\left\|\frac{d^n}{dx^n}(x^m f(x))\right\|_1 .$$

The L^1-norm on the right-hand side is finite because f is rapidly decreasing. Since the right-hand side depends on m and n, we have shown that there is a constant C_{mn} with

$$\left|s^n \frac{d^m}{ds^m}\mathcal{F}f(s)\right| \le C_{mn}.$$

This implies that $\mathcal{F}f$ is rapidly decreasing. Done.

Fourier inversion works for functions in $\mathcal{S}$. We first establish the inversion theorem for a *timelimited* function in $\mathcal{S}$; we had this part of the argument back in Chapter 2, but it's been awhile and it's worth seeing it again. Timelimited means that rather than $f(x)$ just tending to zero at $\pm\infty$, we suppose that $f(x)$ is *identically zero* for $|x| \ge T/2$, for some T. Shortly we'll switch to the more common mathematical term *compact support* to refer to such functions.

In this case we can periodize $f(x)$ to get a smooth, periodic function of period T. Expand the periodic function as a *converging* Fourier series. Then for $-T/2 \le t \le T/2$,

$$\begin{aligned}
f(x) &= \sum_{n=-\infty}^{\infty} c_n e^{2\pi i n x/T} \\
&= \sum_{n=-\infty}^{\infty} e^{2\pi i n x/T}\Big(\frac{1}{T}\int_{-T/2}^{T/2} e^{-2\pi i n y/T} f(y)\,dy\Big) \\
&= \sum_{n=-\infty}^{\infty} e^{2\pi i n x/T}\Big(\frac{1}{T}\int_{-\infty}^{\infty} e^{-2\pi i n y/T} f(y)\,dy\Big) \qquad (f(x) = 0 \text{ for } |x| \ge T/2) \\
&= \sum_{n=-\infty}^{\infty} e^{2\pi i x(n/T)}\mathcal{F}f\Big(\frac{n}{T}\Big)\frac{1}{T} = \sum_{n=-\infty}^{\infty} e^{2\pi i x s_n}\mathcal{F}f(s_n)\Delta s,
\end{aligned}$$

with $s_n = n/T$, $\Delta s = 1/T$. Our intention is to let $T \to \infty$. What we see is a Riemann sum for the integral

$$\int_{-\infty}^{\infty} e^{2\pi i x s} \mathcal{F}f(s)\, ds\,,$$

which is $\mathcal{F}^{-1}\mathcal{F}f(x)$, and the Riemann sum converges to the integral because of the smoothness of $f(x)$. I have not slipped anything past you, or the rigor police.

Thus

$$f(x) = \mathcal{F}^{-1}\mathcal{F}f(x)\,,$$

and the Fourier inversion theorem is established for timelimited functions in $\mathcal{S}$.

When f is *not* timelimited, we use *windowing*. The idea is to cut $f(t)$ off *smoothly* by multiplying it by a smooth function — the window — that's 1 over a finite interval and then decreases smoothly down to be identically 0 outside a larger interval. The interesting thing in the present context, for theoretical rather than practical use, is to make the window so smooth that the windowed function is still in $\mathcal{S}$. Some details and further applications are in Section 4.8.1, but here's the setup. To be definite, take a function $c(t)$ that is identically 1 for $-1/2 \leq t \leq 1/2$, that goes *smoothly* (infinitely differentiable) down to zero as t goes from $1/2$ to 1 and from $-1/2$ to -1 and is then identically 0 for $t \geq 1$ and $t \leq -1$. This is a smoothed version of the rectangle function $\Pi(t)$; instead of cutting off sharply at $\pm 1/2$ we bring the function smoothly down to zero. You can certainly imagine drawing such a function, provided your imaginary pencil is infinitely sharp — enough to render curves agreeing to infinite order:

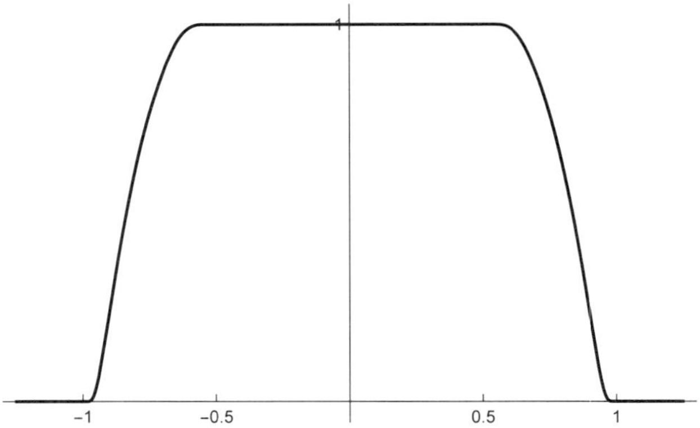

In Section 4.8.1 I'll give an explicit formula for such a window.

Now scale $c(t)$ to $c_n(x) = c(x/n)$. Then $c_n(x)$ is 1 for t between $-n/2$ and $n/2$, goes smoothly down to 0 between $\pm n/2$ and $\pm n$, and is then identically 0 for $|x| \geq n$. Next, the function $f_n(x) = c_n(x)f(x)$ is a timelimited function *in* $\mathcal{S}$. Hence the earlier reasoning shows that the Fourier inversion theorem holds for f_n and $\mathcal{F}f_n$. The window eventually moves past any given x, and consequently $f_n(x) \to f(x)$ as $n \to \infty$. Some estimates based on the properties of the window function, which I won't go through, show that the Fourier inversion theorem also holds in the limit. So far, so good.

Parseval holds for functions in $\mathcal{S}$. Actually, the more general result holds:

If $f(x)$ and $g(x)$ are complex-valued functions in $\mathcal{S}$, then

$$\int_{-\infty}^{\infty} f(x)\overline{g(x)}\,dx = \int_{-\infty}^{\infty} \mathcal{F}f(s)\overline{\mathcal{F}g(s)}\,ds\,.$$

As a special case, if we take $f = g$, then $f(x)\overline{f(x)} = |f(x)|^2$ and the identity becomes

$$\int_{-\infty}^{\infty} |f(x)|^2\,dx = \int_{-\infty}^{\infty} |\mathcal{F}f(s)|^2\,ds\,.$$

The derivation is exactly the one from Chapter 2, with the added confidence that properties of $\mathcal{S}$ justify all steps. I won't repeat it here.

Functions with compact support. Timelimited functions and windowing introduce another category of functions that also come up in the theory of distributions in their own right. These are the *infinitely differentiable functions of compact support* defined on $\mathbb{R}$, the collection of which is often denoted by $C_c^\infty(\mathbb{R})$, or sometimes by $\mathcal{C}$ if you've grown fond of script letters. They are the best functions for partial differential equations in the modern reworking of that subject, and if you take an advanced course on PDE, you will see them all over the place. We'll see them here, too.

To explain the terms, "infinitely differentiable" you already know, and that's the "∞" in $C_c^\infty(\mathbb{R})$. The capital "C" is for "continuous" — all derivatives are continuous. The lowercase, subscript "c" indicates *compact support*, which means that a function is zero in the exterior of a bounded set.

A few technical points, especially for those who have had some topology. The *support* of a function $f(x)$ is the *closure* of the set of points $\{x\colon f(x) \neq 0\}$. That's the smallest closed set containing the set $\{x\colon f(x) \neq 0\}$. This definition is independent of any smoothness property of the function.

As mentioned in Section 4.2.2, a set in $\mathbb{R}$ is said to be *compact* if it is closed and bounded. A bounded set in $\mathbb{R}$ is one that is contained in some finite interval. Thus a function $f(x)$ has compact support if the closure of the set $\{x : f(x) \neq 0\}$ is bounded. (The support is closed by definition — it's the closure of something.) Seems like a much harder way of saying that the function is zero in the exterior of a bounded set, but there are reasons to unpack the terminology.[9] In any event, a consequence of a function $f(x)$ having compact support is then that $f(x)$ is identically 0 for $|x| \geq A$, for some A (generally different A's for different functions). That's the "timelimited" condition we used earlier. The function $c_n(x)$, above, is in $\mathcal{C}$. Its support is $|x| \leq n$. (In particular, there *are* nonzero, smooth functions of compact support. A bit of a relief.)

---

[9]This circle of ideas belongs to the field of *general topology*. Every math major takes some course that includes some aspects of general topology, and you have to be able to toss around the terms. It's good training. One of my teachers in grad school, whose field was partial differential equations, opined that general topology ranked up there with penmanship in importance to science.

Be aware that "functions of compact support" will pop up. In mathematical practice, the property of compact support is often employed in arguments featuring integrals, and integration by parts, to replace an integral over $\mathbb{R}$ by one over a finite interval (the support) or to conclude that terms at $\pm\infty$ are zero. On the other hand, engineers might be more familiar with the term *bandlimited*, referring to a signal whose spectrum is identically zero beyond some point. To reconcile the two terms, a signal is bandlimited if its Fourier transform has compact support. This will be an important class of signals in Chapter 6 on sampling and interpolation.

A function in $\mathcal{C}$ is certainly in $\mathcal{S}$, since by the time x reaches $\pm\infty$ the function has been identically 0 for awhile and not just deceasing to 0. So, symbolically, $\mathcal{C} \subset \mathcal{S}$. The inclusion does not go the other way. A function in $\mathcal{S}$ need not have compact support; for example, the support of the Gaussian e^{-x^2} is all of $\mathbb{R}$ since the Gaussian is never zero.

4.3. A Very Little on Integrals

This section on integrals, more of an early-chapter appendix, is not a short course on integration. It's here to provide a little, but only a little, background explanation for some of the statements on integrals made to this point. It's placed early in the chapter so you might accidentally read it. The star of this section is you.

Integrals are first defined for positive functions. In the general approach to integration (of real-valued functions) you first set out to define the integral for *nonnegative* functions. Why? Because however general a theory you're constructing, an integral is going to be some kind of limit of sums and you'll want to know when that kind of limit exists. If you work with positive, or at least nonnegative, functions, then the issues for limits will be about how big the function gets or about how big the sets are where the function is or isn't big. When the function changes sign, cancellations determine the convergence of the integral, what one of my colleagues called a conspiracy between the positive and negative values. You feel better able to analyze accumulations (for nonnegative functions) than to control conspiratorial cancellations (for functions that change sign). So you first work on defining your integral for functions $f(x)$ with $f(x) \geq 0$.

Integrals are first defined for positive step functions. In fact, having thought about it, you decide to begin with nonnegative *step functions*, functions that look like[10]

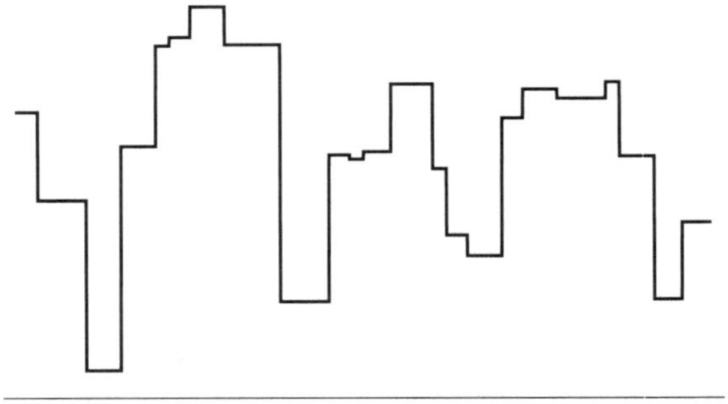

[10]These are the kinds of functions that you get, for example, from a sample-and-hold process that you've probably seen in signal processing.

Step functions are constant on intervals. They're pretty easy to work with. All you do to define an integral is add up the value of the function times the length of the interval for each of the steps.

Then you go on to use step functions to approximate other functions that you want to integrate. Take many, tiny steps. The integral is then again approximated by adding up the values of the function times the lengths of intervals, but for more general functions you find that you really need a more general concept than simply length of an interval. You invent measure theory! You base your theory of the integral on measurable sets and measurable functions. This works fine. You get more than Riemann got with his integral. You call your friend Lebesgue.

Integrals for functions that change sign. Now, backtracking, you know full well that your definition won't be too useful if you can't extend it to functions that can be both positive and negative. Here's how you do this. For any function $f(x)$ you let $f^+(x)$ be its *positive part*:

$$f^+(x) = \max\{f(x), 0\}.$$

Likewise, you let

$$f^-(x) = \max\{-f(x), 0\}$$

be its *negative part*.[11] Tricky: the "negative part" as you've defined it is actually a positive function; taking $-f(x)$ flips over the places where $f(x)$ is negative to be positive. You like that kind of thing. Then

$$f = f^+ - f^-$$

while

$$|f| = f^+ + f^- .$$

You now say that f is integrable if both f^+ and f^- are integrable — a condition which makes sense since f^+ and f^- are both nonnegative functions — and *by definition* you set

$$\int f = \int f^+ - \int f^- .$$

(For complex-valued functions you apply this to the real and imaginary parts.) You follow this approach for integrating functions on a finite interval or on the whole real line. Moreover, according to this definition $|f|$ is integrable if f is because then

$$\int |f| = \int (f^+ + f^-) = \int f^+ + \int f^-$$

and f^+ and f^- are each integrable.[12] It's also true, conversely, that if $|f|$ is integrable, then so is f. You show this by observing that

$$f^+ \le |f| \quad \text{and} \quad f^- \le |f|$$

and this implies that both f^+ and f^- are integrable.

[11] A different use of the notation f^- than we had before, talking about reversal. But we'll never use the negative part of a function again, so be flexible for this section.

[12] Some authors reserve the term "summable" for the case when $\int |f| < \infty$, i.e., for when both $\int f^+$ and $\int f^-$ *are finite*. They still define $\int f = \int f^+ - \int f^-$ but they allow the possibility that one of the integrals on the right may be ∞, in which case $\int f$ is ∞ or $-\infty$ and they don't refer to f as summable.

- You now know where the implication $\int_{-\infty}^{\infty} |f(t)|\, dt < \infty \Rightarrow \mathcal{F}f$ exists comes from.

You get an easy inequality out of this development:

$$\left|\int f\right| \leq \int |f| .$$

In words, the absolute value of the integral is at most the integral of the absolute value. And sure enough that's true, because $\int f$ may involve cancellations of the positive and negative values of f while $\int |f|$ won't have such cancellations. You don't shirk a more formal argument:

$$\begin{aligned}\left|\int f\right| &= \left|\int (f^+ - f^-)\right| = \left|\int f^+ - \int f^-\right| \\ &\leq \left|\int f^+\right| + \left|\int f^-\right| = \int f^+ + \int f^- \quad \text{(since } f^+ \text{ and } f^- \text{ are both nonnegative)} \\ &= \int (f^+ + f^-) = \int |f| .\end{aligned}$$

- You now know where the inequality in

$$\begin{aligned}|\mathcal{F}f(s) - \mathcal{F}f(s')| &= \left|\int_{-\infty}^{\infty} \left(e^{-2\pi i s t} - e^{-2\pi i s' t}\right) f(t)\, dt\right| \\ &\leq \int_{-\infty}^{\infty} \left|e^{-2\pi i s t} - e^{-2\pi i s' t}\right| |f(t)|\, dt\end{aligned}$$

comes from; this came up in showing that $\mathcal{F}f$ is continuous.

sinc stinks. What about the sinc function and trying to make sense of

$$\int_{-\infty}^{\infty} e^{-2\pi i s t} \operatorname{sinc} t\, dt .$$

According to the definitions you just gave, the sinc function is not integrable. In fact, the argument I gave to show that

$$\int_{-\infty}^{\infty} |\operatorname{sinc} t|\, dt = \infty$$

(the second argument) can be easily modified to show that both

$$\int_{-\infty}^{\infty} \operatorname{sinc}^+ t\, dt = \infty \quad \text{and} \quad \int_{-\infty}^{\infty} \operatorname{sinc}^- t\, dt = \infty .$$

So if you wanted to write

$$\int_{-\infty}^{\infty} \operatorname{sinc} t\, dt = \int_{-\infty}^{\infty} \operatorname{sinc}^+ t\, dt - \int_{-\infty}^{\infty} \operatorname{sinc}^- t\, dt,$$

you'd be faced with $\infty - \infty$. Bad. The integral of sinc (and also the integral of $\mathcal{F}$ sinc) *has* to be understood as a limit,

$$\lim_{a\to-\infty,\, b\to\infty} \int_a^b e^{-2\pi i s t} \operatorname{sinc} t\, dt.$$

See the problems. Alternately, evaluating this integral is a classic of contour integration and the residue theorem, which you may have seen in a class on "functions of a complex variable." I won't do it. You may have done it. See *Complex Analysis*, third edition, by Lars Ahlfors, pp. 156–159.

You can relax now. I'll take it from here.

Subtlety vs. cleverness. For the full mathematical theory of Fourier series and Fourier integrals one needs the Lebesgue integral, as I've mentioned before. Lebesgue's approach to defining the integral allows a wider class of functions to be integrated and it allows one to establish very general, very helpful results of the type "the limit of the integral is the integral of the limit," as in

$$f_n \to f \;\Rightarrow\; \lim_{n\to\infty} \int_{-\infty}^{\infty} f_n(t)\,dt = \int_{-\infty}^{\infty} \lim_{n\to\infty} f_n(t)\,dt = \int_{-\infty}^{\infty} f(t)\,dt\,.$$

You probably do things like this routinely, and so do mathematicians but it takes them a year or so of graduate school before they feel good about it. More on this in just a moment.

The definition of the Lebesgue integral is based on a study of the size, or *measure*, of the sets where a function is big or small. Using measure, the sums you write down to define the Lebesgue integral capture different things than the Riemann sums you used in calculus to define the integral. But now take note of the following quote of the mathematician T. Körner from his book *Fourier Analysis*:

> Mathematicians find it easier to understand and enjoy ideas which are clever rather than subtle. Measure theory is subtle rather than clever and so requires hard work to master.

More work than we're willing to do or need to do. Interestingly, the constructions and definitions of measure theory, as Lebesgue and you developed it, were later used in reworking the foundations of probability.

The big theorems. Just a few more things. There's a general result allowing one to push a limit inside the integral sign. It's the *Lebesgue dominated convergence theorem*:

- If f_n is a sequence of integrable functions that converges pointwise to a function f except possibly on a set of measure 0 and if there is an integrable function g with $|f_n| \le g$ for all n (the "dominated" hypothesis), then f is integrable and

$$\lim_{n\to\infty} \int_{-\infty}^{\infty} f_n(t)\,dt = \int_{-\infty}^{\infty} f(t)\,dt\,.$$

There's a variant of this that applies when the integrand depends on a parameter. It goes:

- If $f(x, t_0) = \lim_{t\to t_0} f(x,t)$ for all x and if there is an integrable function g such that $|f(x,t)| \le g(x)$ for all x, then

$$\lim_{t\to t_0} \int_{-\infty}^{\infty} f(x,t)\,dt = \int_{-\infty}^{\infty} f(x,t_0)\,dx\,.$$

The situation described in this last result comes up in many applications, and it's good to know that it holds in great generality. It's used, for example, to justify differentiating under the integral by taking the limit of a difference quotient from outside an integral to inside an integral, as in

$$f(x, \Delta t) = \frac{h(x, t + \Delta t) - h(x, t)}{\Delta t}$$

and

$$\begin{aligned} \frac{\partial}{\partial t} \int_{-\infty}^{\infty} h(x, t)\, dx &= \lim_{\Delta t \to 0} \int_{-\infty}^{\infty} f(x, \Delta t)\, dx \\ &= \int_{-\infty}^{\infty} \lim_{\Delta t \to 0} f(x, \Delta t)\, dx = \int_{-\infty}^{\infty} \frac{\partial}{\partial t} h(x, t)\, dx\,. \end{aligned}$$

If the difference quotient $f(x, \Delta t)$ is bounded by an integrable function, then all is well.

Finally, we've had several occasions to write a double integral as two iterated integrals and also to swap the order of integration. There's a general theorem covering that, too, and it's called *Fubini's theorem.* Without spelling out the hypotheses, which are pretty minimal, it's used to justify statements like

$$\iint_{\mathbb{R}^2} f(x, y)\, dA = \int_{-\infty}^{\infty} \left(\int_{-\infty}^{\infty} f(x, y)\, dx \right) dy = \int_{-\infty}^{\infty} \left(\int_{-\infty}^{\infty} f(x, y)\, dy \right) dx\,.$$

On the left is a double integral using the area measure, followed by iterated integrals with the two orders of integration. The fact that a double integral can be evaluated via iterated single integrals is part of the theorem.

Integrals are not always like sums. Here's one way they're different, and it's important to realize this for our work on Fourier transforms. For sums we have the following result:

$$\text{if } \sum_n a_n \text{ converges, then } a_n \to 0.$$

We used this fact together with Bessel's inequality (or Rayleigh's identity) for Fourier series to conclude that the Fourier coefficients tend to zero. You also know the classic counterexample to the converse of this statement, namely,

$$\frac{1}{n} \to 0 \quad \text{but} \quad \sum_{n=1}^{\infty} \frac{1}{n} \quad \text{diverges}\,.$$

For integrals, however, it is possible that

$$\int_{-\infty}^{\infty} f(x)\, dx$$

exists but $f(x)$ does not tend to zero at $\pm\infty$. Make $f(x)$ nonzero (make it equal to 1 if you want) on thinner and thinner intervals going out toward infinity. Then $f(x)$ doesn't decay to zero, but you can make the intervals thin enough so that the integral converges. Try, for example,

$$\sum_{n=1}^{\infty} n\Pi\left(n^3(x - n)\right)\,.$$

Integrals are sometimes like sums. For a convergent series, $\sum_n a_n$, not only does $a_n \to 0$ but far enough out the *tail* of the series is tiny. Informally, if m is sufficiently large, then $\sum_{n=m}^{\infty} a_n$ (the tail, starting at m) can be made arbitrarily small. It's the same thing for integrals. If, say,

$$\int_{-\infty}^{\infty} |f(x)|\, dx < \infty,$$

then the tails

$$\int_{a}^{\infty} |f(x)|\, dx \quad \text{and} \quad \int_{-\infty}^{-a} |f(x)|\, dx$$

can be made arbitrarily small by taking a sufficiently large.

How shall we test for convergence of integrals? The answer depends on the context, and different choices are possible. Since the convergence of Fourier integrals is at stake, the important thing to measure is the size of a function at infinity — meaning, does it decay fast enough for the integrals to converge?[13] Any kind of measuring requires a standard, and for judging the decay (or growth) of a function the easiest and most common standard is to measure using powers of x. The ruler with which to measure based on powers of x reads

$$\int_{a}^{\infty} \frac{dx}{x^p} \quad \text{is} \quad \begin{cases} \text{infinite} & \text{if } 0 < p \le 1, \\ \text{finite} & \text{if } p > 1. \end{cases}$$

You can check this by direct integration. We take the lower limit a to be positive, but a particular value is irrelevant since the convergence or divergence of the integral depends on the decay near infinity. You can formulate the analogous statements for integrals $-\infty$ to $-a$.

To measure the decay of a function $f(x)$ at $\pm\infty$ we look at

$$\lim_{x \to \pm\infty} |x|^p |f(x)|\,.$$

If, for some $p > 1$, this is *bounded*, then $f(x)$ is integrable. If there is a $0 < p \le 1$ for which the limit is unbounded, i.e., equals ∞, then $f(x)$ is not integrable.

Standards are good only if they're easy to use, and powers of x, together with the conditions on their integrals, are easy to use. You can use these tests to show that every rapidly decreasing function is in both $L^1(\mathbb{R})$ and $L^2(\mathbb{R})$.

Thus ends the happy excursion on integrals, with a few additional points covered here and there when needed.

4.4. Distributions

Our program to extend the applicability of the Fourier transform is a drama in five acts, maybe six. Act I already happened — it was the discovery of the rapidly decreasing functions $\mathcal{S}$. Act II introduces distributions, with a leading role assigned to δ and with particular attention paid to the tempered distributions associated with $\mathcal{S}$.

[13] For now, at least, let's assume that the only cause for concern in convergence of integrals is decay of the function at infinity, not some singularity at a finite point.

4.4.1. What is δ, anyway? The first approach to distributions is to view them as some kind of limit of ordinary functions, which explains the alternate descriptor *generalized functions*. We'll start with δ, viewed in this way, but we'll return to it later from the modern point of view. Also, δ is featured again in the next chapter, highlighting some applications.

I am assuming you may well have seen δ in an undergrad engineering differential equations course or in an undergrad EE signals and systems course. You might not have seen δ in an undergrad math course, because for mathematicians the way it's characterized and used in engineering (or physics) is strictly off limits, even as a limit. In EE applications, starting with Heaviside, you find the *unit impulse* (same thing, different word) used, as an idealization in studying how systems respond to sharp, sudden inputs. We'll come back to this interpretation when we talk about linear systems in Chapter 8. It's a crucial idea.

You probably learned that δ is defined by three properties:

(1) $\delta(x) = 0$ for $x \neq 0$ and $\delta(0) = \infty$.

(2) $\int_{-\infty}^{\infty} \delta(x)\, dx = 1$.

(3) For a function $f(x)$ (whose properties are often left unspecified),

$$\int_{-\infty}^{\infty} \delta(x) f(x)\, dx = f(0)\,.$$

You probably also learned to represent δ graphically as a spike sticking up from 0. The spike is often tagged to have height 1, or strength 1, whatever "height" or "strength" mean for something that's supposed to be ∞ at 0:

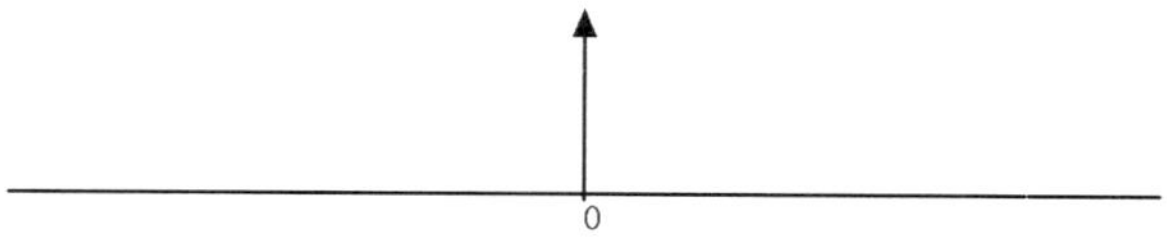

None of (1)–(3) make any sense at all for a classical function, the kind of function you've worked with for years where you plug a number into a formula and get a value. No classical function can satisfy these properties. Nevertheless, it is not my purpose to disabuse you of these ideas, or of the picture. I just want to refine things somewhat.

While the three characteristic properties of δ are bogus, it's quite possible to have a sequence of honest functions approximating the first two properties which, in the limit, yields the third property. And it's the third property that gives δ its operational usefulness — concentration at a point, in this case the point 0.

For example, take the family of scaled Π-functions depending on a parameter $\epsilon > 0$ that will ultimately be tending to 0:

$$\frac{1}{\epsilon}\Pi_\epsilon(x) = \frac{1}{\epsilon}\Pi\left(\frac{x}{\epsilon}\right) = \begin{cases} \frac{1}{\epsilon}, & |x| < \frac{\epsilon}{2}, \\ 0, & |x| \geq \frac{\epsilon}{2}. \end{cases}$$

Here are plots of $\frac{1}{\epsilon}\Pi_\epsilon(x)$ for $\epsilon = 1, 0.5, 0.25$, and 0.1, a family of rectangles getting taller and skinnier.

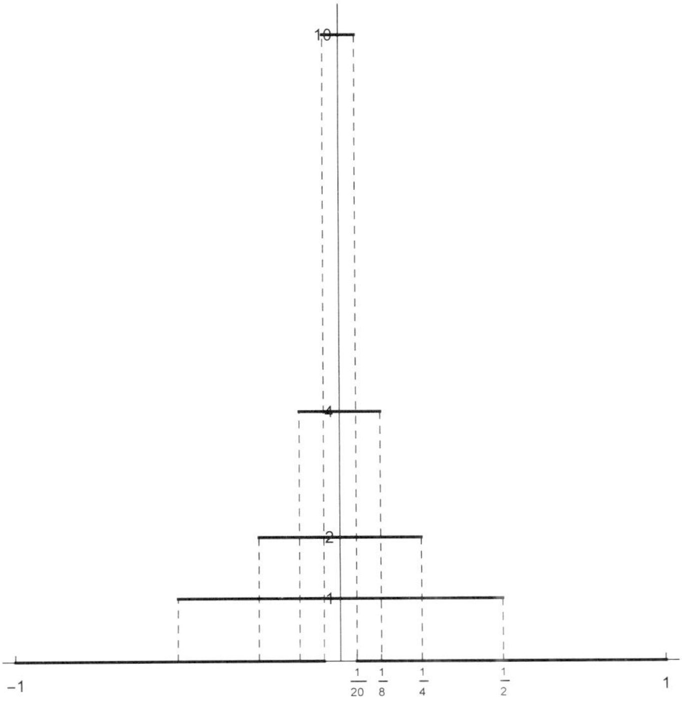

We have $\frac{1}{\epsilon}\Pi_\epsilon(0) = 1/\epsilon$, so the value is getting bigger, while $\frac{1}{\epsilon}\Pi_\epsilon(x) = 0$ for $|x| \geq \epsilon/2$, so the interval where $\frac{1}{\epsilon}\Pi_\epsilon(x)$ is *not* 0 is shrinking. The integral stays at 1:

$$\int_{-\infty}^{\infty} \frac{1}{\epsilon}\Pi_\epsilon(x)\,dx = \frac{1}{\epsilon}\int_{-\epsilon/2}^{\epsilon/2} 1\,dx = \frac{1}{\epsilon}\epsilon = 1\,.$$

Now integrate $\frac{1}{\epsilon}\Pi_\epsilon(x)$ against a function $f(x)$ and use $f(x) = f(0)+f'(0)x+O(x^2)$. Then

$$\begin{aligned}\int_{-\infty}^{\infty} \frac{1}{\epsilon}\Pi_\epsilon(x)f(x)\,dx &= f(0)\frac{1}{\epsilon}\int_{-\epsilon/2}^{\epsilon/2} 1\,dx + f'(0)\frac{1}{\epsilon}\int_{-\epsilon/2}^{\epsilon/2} x\,dx + \frac{1}{\epsilon}\int_{-\epsilon/2}^{\epsilon/2} O(x^2)\,dx \\ &= f(0) + f'(0)\cdot 0 + O(\epsilon^2)\,.\end{aligned}$$

We conclude that

$$\lim_{\epsilon\to 0}\int_{-\infty}^{\infty} \frac{1}{\epsilon}\Pi_\epsilon(x)f(x)\,dx = f(0)\,.$$

This is fine. It captures "concentration at a point" in a limiting sense. What's not fine is to write

$$\lim_{\epsilon\to 0}\int_{-\infty}^{\infty} \frac{1}{\epsilon}\Pi_\epsilon(x)f(x)\,dx = \int_{-\infty}^{\infty}\Big(\lim_{\epsilon\to 0}\frac{1}{\epsilon}\Pi_\epsilon(x)\Big)f(x)\,dx\,.$$

No convergence theorem for integrals will justify this.[14] It's really not OK to blithely write

$$\lim_{\epsilon \to 0} \frac{1}{\epsilon} \Pi_\epsilon(x) = \delta(x)\,,$$

much as you'd like to. Still, convenience and ubiquity make it not worth the sacrifice to abandon

$$\int_{-\infty}^{\infty} \delta(x) f(x)\, dx = f(0)$$

as the way δ *operates*. Embrace it as part of your working day. We'll have more to say, later.

Why do we want to write things in terms of an integral? Aside from the understanding that it represents a reasonable shorthand for the limiting process (keeping that limit outside the integral!), we want to use *operations associated with integration*, chief among them linearity. We want to be able to write, for example,

$$\int_{-\infty}^{\infty} \delta(x)(af(x)+bg(x))\, dx = a \int_{-\infty}^{\infty} \delta(x) f(x)\, dx + b \int_{-\infty}^{\infty} \delta(x) g(x)\, dx = af(0)+bg(0)\,.$$

We also want to be able to make a change of variable, etc., but let's not try for too much just now.

Shifted δ. No surprise that we can shift the approximating rectangles to be centered at a point a and interpret via limits the corresponding properties of $\delta(x-a)$, the shifted δ. As usuallly stated:

(1) $\delta(x-a) = 0$ for $x \neq a$ and $\delta(x-a) = \infty$ for $x = a$.

(2) $\int_{-\infty}^{\infty} \delta(x-a)\, dx = 1$.

(3) For a function $f(x)$,

$$\int_{-\infty}^{\infty} \delta(x-a) f(x)\, dx = f(a)\,.$$

The picture is of a spike sticking up from a.

Convolving with δ. The rectangle function $\Pi(x)$ is even ($\Pi(-x) = \Pi(x)$), and so are the scaled versions $\frac{1}{\epsilon}\Pi_\epsilon(x)$, and from that we get an important convolution property of δ. This is again understood in a limiting sense.

Start with

$$\int_{-\infty}^{\infty} f(x) \frac{1}{\epsilon} \Pi_\epsilon(x-a)\, dx = \int_{-\infty}^{\infty} f(x) \frac{1}{\epsilon} \Pi_\epsilon(a-x)\, dx = \Big(f * \frac{1}{\epsilon}\Pi_\epsilon\Big)(a)\,.$$

Then

$$\lim_{\epsilon \to 0} \Big(f * \frac{1}{\epsilon}\Pi_\epsilon\Big)(a) = \lim_{\epsilon \to 0} \int_{-\infty}^{\infty} f(x) \frac{1}{\epsilon} \Pi_\epsilon(x-a)\, dx = f(a)\,.$$

---

[14] If you read the section on integrals, where the validity of such a result is by appeal to the Lebesgue dominated convergence theorem, what goes wrong here is that $\frac{1}{\epsilon}\Pi_\epsilon(x) f(x)$ is not dominated by an integrable function.

Allowing ourselves (always) the expediency of integrating with δ, this is more often written as

$$f(a) = (f * \delta)(a) = \int_{-\infty}^{\infty} f(x)\delta(a-x)\,dx$$

for any a.

Dropping the a in $f(a) = (f * \delta)(a)$, we have for short

$$f = f * \delta\,.$$

This is the final formula in the parade of properties exhibiting convolution as analogous to multiplication. The number 1 is the *neutral element* for multiplication of numbers, i.e., $x \cdot 1 = x$, and δ is the neutral element for convolution, $f * \delta = f$. We don't get a neutral element for convolution if we stay with classical functions. It's only by allowing for δ and its properties that we can push the (useful!) analogy between multiplication and convolution further. Once we have the generalized Fourier transform we can push it further still.

A smooth approximation to δ: Remember the heat kernel? Depending on the particular application, one can get additional insight by using other sequences of functions to approximate δ. We can do so with a family of smooth functions (even rapidly decreasing functions), and together with convolution this sheds some light on the heat equation.

Consider the family of scaled Gaussians — the heat kernel from the last chapter:

$$g(x,t) = \frac{1}{\sqrt{2\pi t}}e^{-x^2/2t}, \quad t > 0, -\infty < x < \infty\,.$$

Here are plots of functions in the family for $t = 1$, 0.5, 0.1, and 0.01.

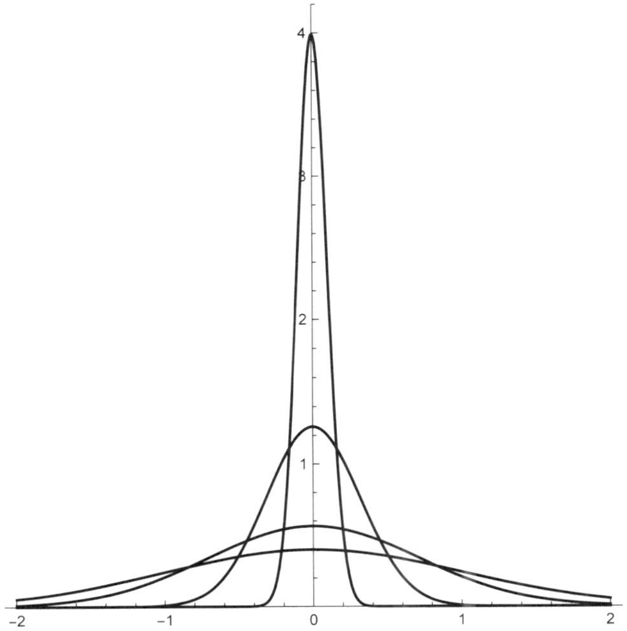

The $g(x,t)$ are rapidly decreasing. The smaller the value of t, the more sharply peaked the function is at 0, and away from 0 the functions are hugging the x-axis more and more closely. Furthermore, for any $t > 0$

$$\int_{-\infty}^{\infty} g(x,t)\,dx = 1\,.$$

We could repeat the same kind of limiting argument we used for the scaled rectangle functions to show that

$$\lim_{t\to 0}\int_{-\infty}^{\infty} g(x,t)f(x)\,dx = f(0)\,,$$

but a little more is involved because the Gaussians are more complicated than simple rectangle functions. Instead, below I'll give a general method for approximating δ, one that applies to both the family of rectangles and the family of Gaussians.

Now recall how we used the functions $g(x,t)$ in solving the heat equation

$$u_t = \tfrac{1}{2}u_{xx}$$

for an infinite rod. When the rod (the real line, in other words) is given an initial temperature $f(x)$, the temperature $u(x,t)$ at a point x on the rod and at a time t is the convolution

$$u(x,t) = g(x,t) * f(x) = \frac{1}{\sqrt{2\pi t}}e^{-x^2/2t} * f(x) = \int_{-\infty}^{\infty} \frac{1}{\sqrt{2\pi t}}e^{-(x-y)^2/2t} f(y)\,dy\,.$$

What I didn't say back in the last chapter, knowing that this day would come, is how one recovers the initial temperature $f(x)$ *from* this formula. The initial temperature is when $t = 0$, so this evidently requires that we take the limit:

$$\lim_{t\to 0^+} u(x,t) = \lim_{t\to 0^+}(g(x,t) * f(x)) = (\delta * f)(x) = f(x)\,.$$

Out pops the initial temperature. Perfect.

A general way to get to δ. There's a general, flexible, and simple approach to getting to δ via a limit. It can be useful to know this if one model approximation might be preferred to another in a particular computation or application. Start with a function $h(x)$ having

$$\int_{-\infty}^{\infty} h(x)\,dx = 1$$

and form

$$h_\epsilon(x) = \frac{1}{\epsilon}h\left(\frac{x}{\epsilon}\right), \quad \epsilon > 0\,.$$

Then

$$\int_{-\infty}^{\infty} h_\epsilon(x)\,dx = \int_{-\infty}^{\infty} h(u)\,du = 1, \quad \text{making the change of variables } u = x/\epsilon.$$

How does $h_\epsilon(x)$ compare with $h(x)$? As ϵ decreases, the scaled functions $h(x/\epsilon)$ concentrate near $x = 0$; that is, the graphs are squeezed in the horizontal direction. Multiplying by $1/\epsilon$ to form $h_\epsilon(x)$ then stretches the values in the vertical direction. We want to show that

$$\lim_{\epsilon\to 0}\int_{-\infty}^{\infty} h_\epsilon(x)f(x)\,dx = f(0)\,.$$

Here we assume that $h(x)f(x)$ is integrable and that $f(x)$ is continuous at $x = 0$.

There is a nice argument for this (in a moment), the lesson being that sometimes it's easier to show something in general than it is to establish particular cases. At least that's what mathematicians say.

But speaking of particular cases. We've seen two applications of this construction, the first using

$$h(x) = \Pi(x),$$

and also

$$h(x) = \frac{1}{\sqrt{2\pi}} e^{x^2/2}, \quad \text{with } \epsilon = \sqrt{t}\,.$$

Another possible choice, believe it or not, is

$$h(x) = \operatorname{sinc} x\,.$$

This works because

$$\int_{-\infty}^{\infty} \operatorname{sinc} x\, dx = 1\,.$$

The integral is the Fourier transform of sinc at 0, and you'll recall that we stated the true fact that

$$\int_{-\infty}^{\infty} e^{-2\pi i s x} \operatorname{sinc} x\, dx = \begin{cases} 1, & |s| < \frac{1}{2}, \\ \frac{1}{2}, & |s| = \frac{1}{2}, \\ 0, & |s| > \frac{1}{2}\,. \end{cases}$$

Back to the argument. Write

$$\begin{aligned} \int_{-\infty}^{\infty} h_\epsilon(x) f(x)\, dx &= \int_{-\infty}^{\infty} h_\epsilon(x)(f(x) - f(0) + f(0))\, dx \\ &= \int_{-\infty}^{\infty} h_\epsilon(x)(f(x) - f(0))\, dx + f(0) \int_{-\infty}^{\infty} h_\epsilon(x)\, dx \\ &= \int_{-\infty}^{\infty} h_\epsilon(x)(f(x) - f(0))\, dx + f(0) \\ &= \int_{-\infty}^{\infty} h(u)(f(\epsilon u) - f(0))\, du + f(0), \end{aligned}$$

where we have used that the integral of h_ϵ is 1 and have made a change of variable $u = x/\epsilon$ in the integral in the last line (note that it's $h(u)$ not $h_\epsilon(u)$ in the integrand).

The object now is to show that the integral of $h(x)(f(\epsilon x) - f(0))$ goes to zero as $\epsilon \to 0$. There are two parts to this. Since the integral of $h(x)(f(\epsilon x) - f(0))$ is finite, the tails of the integral at $\pm\infty$ can be made arbitrarily small, meaning, more formally, that for any $\epsilon' > 0$ there is an $a > 0$ such that

$$\left| \int_{a}^{\infty} h(x)(f(\epsilon x) - f(0))\, dx \right| + \left| \int_{-\infty}^{-a} h(x)(f(\epsilon x) - f(0))\, dx \right| < \epsilon'\,.$$

This didn't involve letting ϵ tend to 0; that comes in now. Fix a as above. It remains to work with the integral

$$\int_{-a}^{a} h(x)(f(\epsilon x) - f(0))\, dx$$

and show that this, too, can be made arbitrarily small. Now

$$\int_{-a}^{a} |h(x)|\, dx$$

is a fixed number, say M, and we can take ϵ so small that $|f(\epsilon x) - f(0)| < \epsilon'/M$ for $|\epsilon x| \leq a$ (continuity of $f(x)$ at 0). With this,

$$\left| \int_{-a}^{a} h(x)(f(\epsilon x) - f(0))\, dx \right| \leq \int_{-a}^{a} |h(x)|\, |f(\epsilon x) - f(0)|\, dx < \epsilon' .$$

Combining the three estimates, we have

$$\left| \int_{-\infty}^{\infty} f(x)(f(x/p) - f(0))\, dx \right| < 2\epsilon' .$$

That's as tiny as we want, and we're done:

$$\lim_{\epsilon \to 0} \int_{-\infty}^{\infty} h_\epsilon(x) f(x)\, dx = f(0) .$$

Notice that the only thing we have to assume about $f(x)$ is that it's continuous at 0.[15]

4.4.2. Distributions as linear functionals: Farewell to vacuum tubes. The path we've just taken leading to δ follows a natural desire to have things look as classical as possible. Other examples of using limiting processes to extend the classical realm, including finding some Fourier transforms, can also be given. Still and all, I maintain that adopting wholesale this approach to defining and working with distributions is using technology from a bygone era. It's time to transistorize.

δ is as δ does. A few more comments about what we didn't do. If you look back, you'll see that we didn't really define δ directly. We defined what we wanted *δ to do* when seen in the company of a function $f(x)$, either writing, loosely,

$$\int_{-\infty}^{\infty} \delta(x) f(x)\, dx = f(0)$$

or replacing this by a limit if guilty feelings are running high.

So what do you do if you're Laurent Schwartz, mathematical heir to much accumulated knowledge from engineering and physics? Surprise! You do what mathematicians always do, some better than others. You turn the solution of a problem into a definition. The problem is to capture concentration at a point, which, *in practice*, means operating on a function to produce the value of the function at a given point. The solution, the definition, is to define δ as the operation — the system, if you prefer that language — that accepts a function $f(x)$ as an

[15] By the way, we didn't do it at the time, but these are the sorts of arguments that are used to prove the theorems on convergence of Fourier series via the Dirichlet kernel, back in Section 1.8.1. Look also at the limiting behavior of the buzz signal, in Section 1.6.2.

input and returns the value $f(0)$ as the output. Write $\delta(f)$ for δ operating on f and define

$$\delta(f) = f(0)\,.$$

(An alternate notation will soon make an appearance.) Furthermore, because of the applications that you know are coming, you, Laurent, are pleased to see that δ is a *linear system*, meaning that for numbers α and β and for functions $f(x)$ and $g(x)$

$$\delta(\alpha f + \beta g) = (\alpha f + \beta g)(0) = \alpha f(0) + \beta g(0) = \alpha\delta(f) + \beta\delta(g)\,.$$

(Linearity is not part of the definition of δ; it's a consequence of the definition.)

While you're at it, you define the shifted δ by

$$\delta_a(f) = f(a)\,.$$

It, too, is a linear system, operating on a function f to return the value $f(a)$.

You're not done. You don't just define one thing, one distribution; you define "distribution" in general.

Test functions and distributions. Distributions operate on functions and return numbers. The kind of functions a distribution operates on, the properties one incorporates into the definition, come from the field of applications that one is aiming to assist. Whatever the particulars, Schwartz chose to use the general term *test functions*, and one speaks of a distribution *on* a class of test functions. For the distributions associated with generalizing the Fourier transform the test functions are $\mathcal{S}$, the rapidly decreasing functions. For the distributions associated with studying partial differential equations the test functions are $\mathcal{C}$, the infinitely differentiable functions of compact support.

Once you've settled on a class of test functions there is a corresponding class of distributions. The definition is in several parts.

(1) Given a test function φ a distribution T operates on φ to produce a complex number $T(\varphi)$. To define a particular distribution is to define how it operates, as δ is defined by $\delta(\varphi) = \varphi(0)$. It's kind of a convention to use Greek letters in denoting test functions.

We often write the action of T on φ as $\langle T, \varphi\rangle$ instead of $T(\varphi)$ and say that T is *paired* with φ. This notation and terminology also go back to Schwartz. Writing $\langle T, \varphi\rangle$, with the distribution in the first slot and the test function in the second slot, is *not* to be interpreted as an inner product.[16]

We would thus write

$$\langle \delta, \varphi\rangle = \varphi(0)$$

as the definition of δ and

$$\langle \delta_a, \varphi\rangle = \varphi(a)$$

as the definition of the shifted δ. Once again, in general, you pass me a test function φ; I have to tell you what $\langle T, \varphi\rangle$ is.

[16]However, the notation is somewhat analogous to the *Bra-ket* notation from quantum mechanics, introduced by Dirac. A Bra-vector $\langle A \mid$ is *paired* with a ket-vector $\mid B\rangle$ to make a Bra-ket (Bracket!) $\langle A \mid B\rangle$. I'll say no more. You're encouraged to read further.

Two distributions T_1 and T_2 are equal if they return the same value when evaluated on a test function:

$$T_1 = T_2 \quad \text{if} \quad \langle T_1, \varphi \rangle = \langle T_2, \varphi \rangle \quad \text{for all test functions } \varphi.$$

Naturally — two functions are equal if they have the same domain and the same values. This isn't part of the definition, but you'll see why I wrote it down separately.

We know, we just know, that superposition is going to come in, so we require it as part of the definition:

(2) A distribution T is linear operating on test functions,

$$T(\alpha_1\varphi_1 + \alpha_2\varphi_2) = \alpha_1 T(\varphi_1) + \alpha_2 T(\varphi_2),$$

or, in the pairing notation,

$$\langle T, \alpha_1\varphi_1 + \alpha_2\varphi_2 \rangle = \alpha_1 \langle T, \varphi_1 \rangle + \alpha_2 \langle T, \varphi_2 \rangle,$$

for test functions φ_1, φ_2 and complex numbers α_1, α_2.

For this to make sense a linear combination of test functions must again be a test function; mathematically speaking, the test functions must form a vector space of functions. That's true for $\mathcal{S}$ and for $\mathcal{C}$.

As part of defining a distribution T, you pass me a linear combination of test functions and, assuming I've already told you how T operates, I have to show you that T satisfies the linearity property above.

For example, δ is linear. You can show that one yourself.

Finally, to have a full and robust theory of distributions one wants to be able to take a limit of a sequence of distributions and to define a distribution via a limit of approximations. We need a condition that speaks to these possibilities and, it turns out, this is best phrased as a kind of continuity condition on how distributions operate on sequences of test functions.

(3) A distribution T is continuous, in the following sense: if φ_n is a sequence of test functions that converges to a test function φ,

$$\lim_{n\to\infty} \varphi_n = \varphi,$$

then the numbers $\langle T, \varphi_n \rangle$ converge to the number $\langle T, \varphi \rangle$,

$$\lim_{n\to\infty} \langle T, \varphi_n \rangle = \langle T, \varphi \rangle \, .$$

So to define a distribution T, after you've passed me a test function φ and I've told you what $\langle T, \varphi \rangle$ is, and after I've shown you that T is linear, finally I have to show you that T satisfies this continuity condition.

This is the hard part of the definition because it's rather involved to define (carefully) what it means for a sequence of test functions to converge to a test function. For either $\mathcal{S}$ or $\mathcal{C}$, for example, to have $\lim_{n\to\infty} \varphi_n = \varphi$ with φ in $\mathcal{S}$ or $\mathcal{C}$ is to control the convergence of φ_n *together with its derivatives of all orders*, and these have to converge to the corresponding derivatives of the limit function φ.

We won't enter into this, and it won't be an issue for us. If you look in standard mathematics books on the theory of distributions, you will find discussions of the appropriate topologies on spaces of functions that must be used to talk about convergence. It's quite technical and without the background and a lot of patience it can be an impediment to going any further — and you'd miss all the fun of what comes next.

For δ, at least, we can be well certain that continuity obtains. Supposing that $\lim_{n\to\infty} \varphi_n = \varphi$, we have to show that

$$\lim_{n\to\infty} \langle \delta, \varphi_n \rangle = \langle \delta, \varphi \rangle .$$

But

$$\langle \delta, \varphi_n \rangle = \varphi_n(0), \quad \langle \delta, \varphi \rangle = \varphi(0) ,$$

so we're asserting that

$$\lim_{n\to\infty} \varphi_n = \varphi \ \Rightarrow \ \lim_{n\to\infty} \varphi_n(0) = \varphi(0) .$$

However complicated the definition of convergence looks in general, for sure it gives us that much in particular. I should say that for many other explicit distributions that we'll define, continuity is not hard to settle (though we still won't often do much about it).[17]

It is now a pleasure to report that δ is a distribution on the classes $\mathcal{S}$ and $\mathcal{C}$ of test functions. Same for the shifted δ. Ah δ, cloaked in mystery, of dubious origin and saved from illegitimacy through taking a limit, now emerges in Schwartz's conception as just about the simplest distribution imaginable: evaluation at a point!

Would we have come upon this simple, direct definition without having gone through the "δ as a limit" approach? Would we have the transistor without first having had vacuum tubes? Perhaps so, perhaps not. The approach to δ via limits provided the basic insights that allowed people, Schwartz in particular, to invent the theory of distributions based on linear operations on test functions. But there's no accounting for genius, and we've barely started to see what the new theory can do for us.

A little additional terminology and notation. There's some terminology that applies to distributions that you ought to be aware of. It comes from the mathematical field of functional analysis.

As we've defined it, a distribution is, for short, a *continuous, linear functional* on a space of test functions. The word "functional," as opposed to the simpler word "function," is used when the values on applying the linear operation are scalars (real or complex numbers). I don't know where this convention comes from, but it's ingrained. The space of all distributions on a given space of test functions is then also known as the *dual space* of the space of test functions.

We didn't make a point of this, but, indeed, distributions themselves form a vector space (the dual space is a vector space): the sum of two distributions is a distribution and a scalar multiple of a distribution is a distribution. In the pairing notation one writes

$$\langle T_1 + T_2, \varphi \rangle = \langle T_1, \varphi \rangle + \langle T_2, \varphi \rangle$$

[17]While the continuity condition may be the hard part of the definition, in truth it's even harder to come up with an example that's *not* continuous. One can show that they exist as a mathematical fact of nature — an existence proof — but nobody can point to an explicit example. So not to worry.

for how the sum of two distributions pairs with a test function φ, and one writes

$$\langle \alpha T, \varphi \rangle = \alpha \langle T, \varphi \rangle$$

for the scalar multiple. I wrote these down because I want you to quietly note that this is not the same as saying that distributions act linearly on test functions. Quietly note that to yourself — I have seen this as a point of confusion because it's easy to confuse a linear combination of distributions with a linear combination of test functions upon which distributions operate.

When the test functions are the rapidly decreasing functions $\mathcal{S}$, the distributions are known as *tempered distributions*, a charming term due to Schwartz. We'll denote them by $\mathcal{T}$. In the language above, $\mathcal{T}$ is the dual space of $\mathcal{S}$.

The space of distributions associated with the test functions $\mathcal{C}$ doesn't have a special adjective, nor a special notation. Some people just write $\mathcal{D}$ for them. In fact, if a mathematician speaks simply of "distributions," without a specific mention of the Fourier transform coming in, he or she usually means the distributions on $\mathcal{C}$.

$\mathcal{D}$ is a bigger collection of distributions than $\mathcal{T}$ because $\mathcal{C}$ is a smaller collection of test functions than $\mathcal{S}$. The latter point should be clear to you: to say that $\varphi(x)$ is smooth and *vanishes identically* outside some interval is a stronger condition than requiring merely that it *decays at* infinity (albeit faster than any power of x). Thus if $\varphi(x)$ is in $\mathcal{C}$, then it's also in $\mathcal{S}$. Then why is $\mathcal{D}$ bigger than $\mathcal{T}$? Since $\mathcal{C}$ is contained in $\mathcal{S}$, a continuous linear functional on $\mathcal{S}$ is automatically a continuous linear functional on $\mathcal{C}$. That is, $\mathcal{T}$ is contained in $\mathcal{D}$. Good old δ is in both $\mathcal{T}$ and $\mathcal{D}$.

Is there a physical analogy for distributions? The word "distribution" was chosen by Schwartz in part because for some examples it might be invoked to refer to a distribution $f(x)$ of a physical quantity, e.g., mass, electrical charge, etc. *And then also* an integral such as

$$\int_{-\infty}^{\infty} f(x)\varphi(x)\,dx$$

might occur as a measurement of $f(x)$ by some testing device modeled by a function $\varphi(x)$. There you see a pairing $\langle f, \varphi \rangle$ of $f(x)$ with $\varphi(x)$ via integration, and it's linear and continuous. (We're about to see that kind of pairing again.)

Personally, I can keep up the charade of physical interpretation for only a short time. There are definitely applications of the mathematical theory of distributions to physics, but that's not the same as saying that distributions have a grounding in physics. As I said earlier, there's no accounting for genius, so let distributions be distributions and enjoy what they can do.

4.4.3. How ordinary functions determine distributions and how distributions include functions society needs. Let us now understand how distributions include the functions we'd like to include. It's important that we don't lose anything along the way, especially for the purposes of working with the Fourier transform. We want distributions to *extend* the applicability of what we've done, not eliminate what we've done.

Functions induce distributions. Suppose $f(x)$ is a function for which the integral

$$\int_{-\infty}^{\infty} f(x)\varphi(x)\,dx$$

exists for all test functions $\varphi(x)$. This is not too high a bar considering that test functions are so good that they're plenty likely to make a product $f(x)\varphi(x)$ integrable even if $f(x)$ isn't so great itself. Think of $\mathcal{S}$, where the functions decrease faster than any power of x, or of $\mathcal{C}$, where the functions are identically zero beyond some point. To make $f(x)\varphi(x)$ integrable it would be enough, say, to suppose that $f(x)$ is piecewise continuous with at most a finite number of finite jump discontinuities and is bounded above by some fixed power of x (even a positive power) at $\pm\infty$.

When $f(x)\varphi(x)$ is integrable, the function $f(x)$ *induces* a distribution T_f *by means of the pairing*

$$\langle T_f, \varphi\rangle = \int_{-\infty}^{\infty} f(x)\varphi(x)\,dx\,.$$

In words, T_f pairs with a test function φ by integration of f against φ.

It's always the same thing:

- To define a distribution T, you pass me a test function φ and I have to tell you what $\langle T, \varphi\rangle$ is. I'm telling you that *in the case of a function f the pairing is by integration.*
  In engineering lingo, T_f is a system that accepts the input $\varphi(x)$ and returns the number $\int_{-\infty}^{\infty} f(x)\varphi(x)\,dx$ as output.

The notation T_f, with the subscript f, just indicates that the distribution is induced by f. One also says that f "defines" or "determines" or "corresponds to" the distribution T_f; pick your preferred descriptive phrase. Using a function $f(x)$ to determine a distribution T_f in this way is a very important way of constructing distributions, and we will employ it frequently.

Let's check that the pairing $\langle T_f, \varphi\rangle$ meets the standard of the definition of a distribution. First of all, $\langle T_f, \varphi\rangle$ is a number, possibly a complex number depending on $f(x)$ or $\varphi(x)$. Next, the pairing is linear because integration is linear:

$$\begin{aligned}\langle T_f, \alpha_1\varphi_1 + \alpha_2\varphi_2\rangle &= \int_{-\infty}^{\infty} f(x)(\alpha_1\varphi_1(x) + \alpha_2\varphi_2(x))\,dx\\ &= \alpha_1\int_{-\infty}^{\infty} f(x)\varphi_1(x)\,dx + \alpha_2\int_{-\infty}^{\infty} f(x)\varphi_2(x)\,dx\\ &= \alpha_1\langle T_f, \varphi_1\rangle + \alpha_2\langle T_f, \varphi_2\rangle\,.\end{aligned}$$

What about continuity? We have to take a sequence of test functions φ_n converging to a test function φ and consider the limit

$$\lim_{n\to\infty}\langle T_f, \varphi_n\rangle = \lim_{n\to\infty}\int_{-\infty}^{\infty} f(x)\varphi_n(x)\,dx\,.$$

Again, we haven't said anything precisely about the meaning of φ_n converging to φ, but the standard results on taking the limit inside the integral[18] will apply for any function f that we're likely to encounter and allow us to conclude that

$$\lim_{n\to\infty}\int_{-\infty}^{\infty} f(x)\varphi_n(x)\,dx = \int_{-\infty}^{\infty} f(x)\lim_{n\to\infty}\varphi_n(x)\,dx = \int_{-\infty}^{\infty} f(x)\varphi(x)\,dx\,,$$

i.e., that

$$\lim_{n\to\infty}\langle T_f,\varphi_n\rangle = \langle T_f,\varphi\rangle\,.$$

This is continuity.

You might well ask whether different functions can give rise to the same distribution. That is, if $T_{f_1} = T_{f_2}$ as distributions, then must we have $f_1(x) = f_2(x)$? Yes, with a caveat. If $T_{f_1} = T_{f_2}$, then for all test functions $\varphi(x)$ we have

$$\int_{-\infty}^{\infty} f_1(x)\varphi(x)\,dx = \int_{-\infty}^{\infty} f_2(x)\varphi(x)\,dx\,;$$

hence

$$\int_{-\infty}^{\infty}(f_1(x)-f_2(x))\varphi(x)\,dx = 0\,.$$

Since this holds for all test functions $\varphi(x)$, we can conclude that $f_1(x) = f_2(x)$. Almost. Actually, changing a function at a finite number of points, or more generally on a set of measure zero, won't change its integral, so instead we should say that $f_1(x) = f_2(x)$ except possibly for a set of measure zero. This is good enough for anything that comes our way.

In other words, distributions contain classical functions. Because a function determines a *unique* distribution, it's natural to identify the function f with the corresponding distribution T_f. Sometimes we then write just f for the corresponding distribution rather than writing T_f, and we write the pairing as

$$\langle f,\varphi\rangle = \int_{-\infty}^{\infty} f(x)\varphi(x)\,dx$$

rather than as $\langle T_f,\varphi\rangle$.

- It is in this sense — identifying a function f with the distribution T_f it determines — that a class of distributions "contains" classical functions.

Let's look at some examples.

---

[18]Lebesgue dominated convergence theorem!

We'll work with distributions on $\mathcal{S}$ or $\mathcal{C}$. The sinc function defines a distribution, because though sinc is not integrable, $\varphi(x)\operatorname{sinc} x$ *is* integrable for any test function $\varphi(x)$. The product dies off fast enough at $\pm\infty$. Likewise, any constant function (like the function 1) defines a distribution, and then so does the unit step function (which is piecewise constant)

$$H(x) = \begin{cases} 0\,, & x < 0\,, \\ 1\,, & x \geq 0\,. \end{cases}$$

Also, any complex exponential $e^{\pm 2\pi i x y}$ defines a distribution (the definition of a distribution allows for complex-valued functions for just this reason), and so do sine and cosine.

For one more example take the unit ramp,

$$u(x) = \begin{cases} 0\,, & x \leq 0\,, \\ x\,, & x \geq 0\,. \end{cases}$$

It tends to ∞ as $x \to \infty$, but it does so only to the first power of x (exactly x). Multiplying by a test function brings $u(x)$ down to make the product integrable, and $u(x)$ determines a distribution:

$$\langle u, \varphi \rangle = \int_{-\infty}^{\infty} u(x)\varphi(x)\, dx\,.$$

The upshot is that sinc, constants, complex exponentials, the unit step, the unit ramp, and many others, can all be considered *to be* distributions. This is a good thing, because we're aiming to define the Fourier transform *of a tempered distribution*.

A note of caution. Not *every* function determines a tempered distribution. For example e^{x^2} doesn't because e^{-x^2} is a rapidly decreasing function and

$$\int_{-\infty}^{\infty} e^{x^2} e^{-x^2}\, dx = \int_{-\infty}^{\infty} 1\, dx = \infty\,.$$

However, e^{x^2} does determine a distribution on $\mathcal{C}$, since multiplying e^{x^2} by a function of compact support makes the product identically zero far enough out, and in particular the product is integrable.

This example shows explicitly that the tempered distributions $\mathcal{T}$, on the test functions $\mathcal{S}$, are strictly contained in the distributions $\mathcal{D}$, on the test functions $\mathcal{C}$. We mentioned this earlier. It's easier to satisfy the integrability condition for $\mathcal{C}$ than for $\mathcal{S}$ because multiplying (a bad) $f(x)$ by a function in $\mathcal{C}$ kills $f(x)$ off completely outside some interval, rather than just bringing it smoothly down to zero at $\pm\infty$ as would happen when multiplying by a function in $\mathcal{S}$. The condition being easier to satisfy makes for more distributions on $\mathcal{C}$ than on $\mathcal{S}$.

But distributions are more general than classical functions. Do all distributions come from functions in the way we've been working with? No. δ does not. As a distribution, δ is defined by the pairing

$$\langle \delta, \varphi \rangle = \varphi(0)\,,$$

but for no classical function $f(x)$ do we have

$$\langle f, \varphi \rangle = \int_{-\infty}^{\infty} f(x)\varphi(x)\,dx = \varphi(0)\,.$$

However, δ, and other, general, distributions, too, *can* be realized as a *limit* of distributions that *do* come from classical functions. In particular, the original approach to δ can be reconciled with the clean, modern definition $\langle \delta, \varphi \rangle = \varphi(0)$. We'll do this in the next section.

The nature of the pairing of a general distribution T with a test function φ is unspecified, so to speak: you give me a test function φ, and *I have to tell you* what the pairing $\langle T, \varphi \rangle$ is. So *don't* think that $\langle T, \varphi \rangle$ is an integral when T is not a classical function; i.e., don't think

$$\langle T, \varphi \rangle = \int_{-\infty}^{\infty} T(x)\varphi(x)\,dx\,.$$

For one thing it doesn't really make sense even to write $T(x)$; you don't pass a distribution a number x to evaluate, you pass it a *function*. The pairing *is* an integral when the distribution comes from a function, but there's more to distributions than that. Nevertheless, there will be an important role played by pairing via integration in figuring out how to define general distributions. We'll make use of this in Section 4.5.

These warnings and promises all duly noted, as I said when we first met δ we will continue to use the notations

$$\int_{-\infty}^{\infty} \delta(x)\varphi(x)\,dx = \varphi(0) \quad \text{and} \quad \int_{-\infty}^{\infty} \delta(x-a)\varphi(x)\,dx = \varphi(a)\,.$$

It's just too much to give this up. Sue me.

4.4.4. Limits and approximation of distributions. There's a very useful general result that allows us to define distributions by means of limits. The statement goes:

- Suppose that T_n is a sequence of distributions and that $\langle T_n, \varphi \rangle$ (a sequence of numbers) converges for every test function φ. Then T_n converges to a distribution T and

$$\langle T, \varphi \rangle = \lim_{n\to\infty} \langle T_n, \varphi \rangle\,.$$

Briefly, distributions can be defined by taking limits of sequences of distributions, and the result says that if the parings converge, then the distributions converge. This is by no means a trivial fact, the key issue being the proper notion of convergence of distributions. That gets mathematically technical and we'll happily be content with the statement and let it go at that.

One instance of this is the ever-present δ. Recall that for a function $h(x)$, continuous at $x = 0$ and satisfying

$$\int_{-\infty}^{\infty} h(x)\,dx = 1\,,$$

we have

$$\lim_{\epsilon \to 0} \int_{-\infty}^{\infty} \frac{1}{\epsilon} h\left(\frac{x}{\epsilon}\right) \varphi(x)\,dx = \varphi(0)\,.$$

In the language we have available now and didn't have before, the integral is the pairing

$$\left\langle \frac{1}{\epsilon} h\left(\frac{x}{\epsilon}\right), \varphi(x) \right\rangle = \int_{-\infty}^{\infty} \frac{1}{\epsilon} h\left(\frac{x}{\epsilon}\right) \varphi(x)\,dx\,,$$

and we can say that the sequence of distributions induced by the functions $(1/\epsilon)h(x/\epsilon)$ converges to the distribution δ. This is the precise way to say

$$\lim_{\epsilon \to 0} \frac{1}{\epsilon} h\left(\frac{x}{\epsilon}\right) = \delta(x).$$

One practical consequence of the theorem is that if, in the limit, different converging sequences of distributions have the same effect on test functions, then they must be converging to the same distribution. More precisely, if $\lim_{n\to\infty}\langle S_n, \varphi\rangle$ and $\lim_{n\to\infty}\langle T_n, \varphi\rangle$ both exist and are equal for every test function φ, then S_n and T_n both converge to the same distribution. That's certainly possible; different sequences can have the same limit, after all. This observation applies, for example, to different choices of the $(1/\epsilon)h(x/\epsilon)$ converging to δ (rectangles, Gaussians, sincs).

Allied to the result on the limit of a sequence of distributions is one on approximation:

- If T is any distribution, then there are test functions ψ_n such that T_{ψ_n} converges to T.

Like the theorem on limits, this is not an easy result.

We're not saying that the T_{ψ_n} converge to T_ψ for some function ψ, i.e., that the limit distribution T is also induced by a function. You *don't* necessarily have $T = T_\psi$ for some function ψ. Again, δ illustrates this point. It's not the test functions ψ_n that are converging to a function; it's the *associated distributions* that are converging to a distribution. The result also doesn't say how you're supposed to find the approximating functions, just that they exist. However, you'll understand how this can be accomplished via convolution if you plow through some of the material at the end of the chapter, on regularizing a distribution. (Optional, but interesting.)

The usefulness of the result is that for the approximating distributions T_{ψ_n} the pairing with a test function φ is given by integration,

$$\langle T_{\psi_n}, \varphi\rangle = \langle \psi_n, \varphi\rangle = \int_{-\infty}^{\infty} \psi_n(x)\varphi(x)\,dx,$$

and so the properties and techniques of integration are available to us. As always, when we define a distribution T, it will be by defining the pairing $\langle T, \varphi\rangle$, like the clean definition $\langle \delta, \varphi \rangle = \varphi(0)$. However, as you will see, we will find our way to the definition of $\langle T, \varphi \rangle$ by first appealing to what happens when the distribution is induced by a function and the pairing is by integration. The approximation theorem justifies the approach, for it says that

$$\lim_{n \to \infty} \langle T_{\psi_n}, \varphi \rangle = \langle T, \varphi \rangle.$$

We'll use this idea over and over.

4.4.5. Principal value: A special function and a special pairing.

For a variety of applications it's important to be able to treat the function

$$f(x) = \frac{1}{x}$$

as a distribution, or more precisely as inducing a distribution. For example, it comes up in defining the *Hilbert transform*, a cornerstone in the analysis of communication systems. See Section 8.9.

The problem is at the origin, of course, not at $\pm\infty$, and the simple pairing we've relied on for functions,

$$\left\langle \frac{1}{x}, \varphi \right\rangle = \int_{-\infty}^{\infty} \frac{1}{x}\, \varphi(x)\, dx\,,$$

can't be used because the integral might not converge. The integral would be OK if $\varphi(x)/x$ stays bounded near $x = 0$, but that might not be the case for an arbitrary test function φ.[19]

To understand the problem, let me pick up some general facts on integration I haven't mentioned. When a function $f(x)$ has a discontinuity (infinite or not), say at 0, then

$$\int_a^b f(x)\, dx, \quad a < 0,\ b > 0,$$

is an *improper integral* and has to be defined via the limits

$$\lim_{\epsilon_1 \to 0} \int_a^{-\epsilon_1} f(x)\, dx + \lim_{\epsilon_2 \to 0} \int_{\epsilon_2}^{b} f(x)\, dx\,, \quad \epsilon_1, \epsilon_2 > 0\,,$$

with ϵ_1 and ϵ_2 tending to zero *separately*. If both limits exist, then so does the integral. That's the definition of $\int_a^b f(x)\, dx$; i.e., you first have to take the separate limits and then add the results. If neither or only one of the limits exists, then the integral does not exist. Harsh, but there you are.

For the simple pairing of $1/x$ with $\varphi(x)$,

$$\int_{-\infty}^{\infty} \frac{1}{x} \varphi(x)\, dx\,,$$

we would thus have to consider

$$\lim_{\epsilon_1 \to 0} \int_{-\infty}^{\epsilon_1} \frac{\varphi(x)}{x}\, dx + \lim_{\epsilon_2 \to 0} \int_{\epsilon_2}^{\infty} \frac{\varphi(x)}{x}\, dx$$

[19] The windowing function that we've used before is an example. It's identically 1 near the origin, so you're integrating $1/x$ near the origin and that blows up.

and these limits *need not exist*. But even if these limits do not exist, what may instead be true is that the *symmetric sum*,

$$\int_{-\infty}^{-\epsilon} \frac{\varphi(x)}{x}\,dx + \int_{\epsilon}^{\infty} \frac{\varphi(x)}{x}\,dx\,,$$

with $\epsilon_1 = \epsilon_2 = \epsilon$, has a limit as $\epsilon \to 0$. This limit is called the *Cauchy principal value* of the improper integral, and one writes

$$\text{pr.v.} \int_{-\infty}^{\infty} \frac{\varphi(x)}{x}\,dx = \lim_{\epsilon\to 0}\left(\int_{-\infty}^{-\epsilon} \frac{\varphi(x)}{x}\,dx + \int_{\epsilon}^{\infty} \frac{\varphi(x)}{x}\,dx\right).$$

(There's no universal agreement on the notation for a principal value integral.)

The principal value does exist for test functions $\varphi(x)$. Here's why. There's no problem away from the origin, so split those integrals off and write, say,

$$\begin{aligned}
\text{pr.v.} \int_{-\infty}^{\infty} \frac{\varphi(x)}{x}\,dx &= \lim_{\epsilon\to 0}\left(\int_{-\infty}^{-\epsilon} \frac{\varphi(x)}{x}\,dx + \int_{\epsilon}^{\infty} \frac{\varphi(x)}{x}\,dx\right) \\
&= \lim_{\epsilon\to 0}\left(\int_{-\infty}^{-1} \frac{\varphi(x)}{x}\,dx + \int_{-1}^{-\epsilon} \frac{\varphi(x)}{x}\,dx + \int_{\epsilon}^{1} \frac{\varphi(x)}{x}\,dx + \int_{1}^{\infty} \frac{\varphi(x)}{x}\,dx\right) \\
&= \int_{-\infty}^{-1} \frac{\varphi(x)}{x}\,dx + \int_{1}^{\infty} \frac{\varphi(x)}{x}\,dx + \lim_{\epsilon\to 0}\left(\int_{-1}^{-\epsilon} \frac{\varphi(x)}{x}\,dx + \int_{\epsilon}^{1} \frac{\varphi(x)}{x}\,dx\right) \\
&= \text{something finite} + \lim_{\epsilon\to 0}\left(\int_{-1}^{-\epsilon} \frac{\varphi(x)}{x}\,dx + \int_{\epsilon}^{1} \frac{\varphi(x)}{x}\,dx\right).
\end{aligned}$$

To investigate the limit use $\varphi(x) = \varphi(0) + \varphi'(0)x + O(x^2)$. Then

$$\begin{aligned}
\int_{-1}^{-\epsilon} \frac{\varphi(x)}{x}\,dx + \int_{\epsilon}^{1} \frac{\varphi(x)}{x}\,dx = \varphi(0)\left(\int_{-1}^{-\epsilon} \frac{1}{x}\,dx + \int_{\epsilon}^{1} \frac{1}{x}\,dx\right) \\
+ \text{ something finite again,}
\end{aligned}$$

and since

$$\int_{-1}^{-\epsilon} \frac{1}{x}\,dx + \int_{\epsilon}^{1} \frac{1}{x}\,dx = 0\,,$$

all is well with letting $\epsilon \to 0$. Computing the principal value gives you something finite.

To conclude, $1/x$ induces a distribution, called a *principal value distribution*, where the pairing is defined by

$$\left\langle \frac{1}{x}, \varphi \right\rangle = \text{pr.v.} \int_{-\infty}^{\infty} \frac{\varphi(x)}{x}\,dx\,.$$

It's clear that the pairing is linear, and to show that it's continuous is left to the imagination (but it is).

4.5. Defining Distributions

Now Act III, where we will define distributions and extend operations on ordinary functions to operations on distributions, including:

(1) The Fourier transform of a tempered distribution!

(2) Fourier transform properties: duality, shifting, and stretching.

(3) Derivatives of distributions, including functions you thought had no business having a derivative.

To define a distribution T is to say what it does to a test function. One more time: you give me a test function φ and I have to tell you $\langle T, \varphi \rangle$. We have done this in four instances, for three particular distributions and for a general class of distributions. In the particular cases, we defined

$$\langle \delta, \varphi \rangle = \varphi(0), \quad \langle \delta_a, \varphi \rangle = \varphi(a), \quad \text{and the principal value } \left\langle \frac{1}{x}, \varphi \right\rangle = \text{pr.v.} \int_{-\infty}^{\infty} \frac{\varphi(x)}{x} \, dx \,.$$

The general class is how a function f determines a distribution T_f by

$$\langle T_f, \varphi \rangle = \int_{-\infty}^{\infty} f(x)\varphi(x) \, dx \,.$$

We identify the distribution T_f with the function f it comes from and we write

$$\langle f, \varphi \rangle = \int_{-\infty}^{\infty} f(x)\varphi(x) \, dx \,.$$

When we want to extend an operation from functions to distributions, e.g., when we want to define the Fourier transform of a distribution or the reverse of a distribution or the shift of a distribution or the derivative of a distribution, we take our cue from the way functions determine distributions and ask how the operation works in the case when the pairing is by integration. What we hope to see is an outcome that suggests a direct definition. It's a procedure we'll follow. It's something to try. It's turning a solution into a definition. Before you dive in, let me offer a few words of guidance and encouragement. There's a lot of material in here — way more than you need to know for your day-to-day working life. We *have* to have this material *in some fashion* but you should probably treat the sections to follow mostly as a reference. Feel free to use the formulas you need when you need them, and remember that our aim is to recover the formulas we know from earlier work in pretty much the same shape as when you first learned them.

At the same time, I do want you to develop confidence in *computing* with distributions and in deriving the formulas. The ideas fit together quite beautifully, and many students, engineers and scientists all, have genuinely appreciated what this approach offers in clarity and certainty. Don't deny yourself the pleasures of the world in solid state. Let's hit it!

4.5.1. The Fourier transform of a tempered distribution. Suppose T is a tempered distribution. Remember this means that the test functions are the Schwartz functions $\mathcal{S}$. Why should such an object have a Fourier transform, and how on earth shall we define it? It can't be an integral, because T isn't a classical function so there's nothing to integrate. If $\mathcal{F}T$ is itself to be a tempered distribution (just as $\mathcal{F}\varphi$ is in $\mathcal{S}$ if φ is in $\mathcal{S}$), then we have to say how $\mathcal{F}T$ pairs with a test function in $\mathcal{S}$, because that's what tempered distributions do. So how?

We have a toehold here. If ψ is a Schwartz function, then $\mathcal{F}\psi$ is again a Schwartz function and we can ask: How does the Schwartz function $\mathcal{F}\psi$ pair with another Schwartz function φ? What is the outcome of $\langle \mathcal{F}\psi, \varphi\rangle$? We know how to pair a distribution that comes from a function ($\mathcal{F}\psi$ in this case) with a Schwartz function; it's

$$\langle \mathcal{F}\psi, \varphi\rangle = \int_{-\infty}^{\infty} \mathcal{F}\psi(x)\varphi(x)\,dx\,.$$

But we can work with the right-hand side:

$$\begin{aligned}
\langle \mathcal{F}\psi, \varphi\rangle &= \int_{-\infty}^{\infty} \mathcal{F}\psi(x)\varphi(x)\,dx \\
&= \int_{-\infty}^{\infty} \left(\int_{-\infty}^{\infty} e^{-2\pi ixy}\psi(y)\,dy\right)\varphi(x)\,dx \\
&= \int_{-\infty}^{\infty}\int_{-\infty}^{\infty} e^{-2\pi ixy}\psi(y)\varphi(x)\,dy\,dx \\
&= \int_{-\infty}^{\infty} \left(\int_{-\infty}^{\infty} e^{-2\pi ixy}\varphi(x)\,dx\right)\psi(y)\,dy \\
&\quad \text{(the interchange of integrals is justified because } \varphi(x)e^{-2\pi isx} \text{ and } \psi(x)e^{-2\pi isx} \text{ are integrable)} \\
&= \int_{-\infty}^{\infty} \mathcal{F}\varphi(y)\psi(y)\,dy \\
&= \langle \psi, \mathcal{F}\varphi\rangle\,.
\end{aligned}$$

In a phrase that will echo through this section and beyond:

- Where did we start; where did we finish?

We started with $\langle \mathcal{F}\psi, \varphi\rangle$ and we finished with $\langle \psi, \mathcal{F}\varphi\rangle$. The outcome of pairing $\mathcal{F}\psi$ with φ is

$$\langle \mathcal{F}\psi, \varphi\rangle = \langle \psi, \mathcal{F}\varphi\rangle\,.$$

This clues us in to how we should make the definition in general:

- Let T be a tempered distribution. The *Fourier transform* of T, denoted by $\mathcal{F}T$, is the tempered distribution defined by

  $$\langle \mathcal{F}T, \varphi\rangle = \langle T, \mathcal{F}\varphi\rangle$$

  for any Schwartz function φ.

Notice that we're using "$\mathcal{F}$" in two ways here. On the left-hand side we're defining $\mathcal{F}T$ for a tempered distribution T. On the right-hand side $\mathcal{F}\varphi$ is the classical Fourier transform of φ defined by an integral. The definition makes sense because when φ is a Schwartz function so is $\mathcal{F}\varphi$; it is only then that the pairing $\langle T, \mathcal{F}\varphi\rangle$ is even defined.

This reasoning, from pairing by integration to pairing in general, will be our modus operandi, so one more word of justification. According to the approximation result in Section 4.4.4, any tempered distribution T is a limit of distributions that come from Schwartz functions, say ψ_n, and we would have

$$\langle T, \varphi\rangle = \lim_{n\to\infty} \langle \psi_n, \varphi\rangle .$$

Then if $\mathcal{F}T$ is to make sense, we might understand it to be given by

$$\langle \mathcal{F}T, \varphi\rangle = \lim_{n\to\infty} \langle \mathcal{F}\psi_n, \varphi\rangle = \lim_{n\to\infty} \langle \psi_n, \mathcal{F}\varphi\rangle = \langle T, \mathcal{F}\varphi\rangle .$$

(The middle equality is how $\mathcal{F}\psi_n$ pairs with φ, from the calculation above.) There's our definition.

We define the inverse Fourier transform by following the same recipe:

- Let T be a tempered distribution. The *inverse Fourier transform* of T, denoted by $\mathcal{F}^{-1}T$, is defined by

$$\langle \mathcal{F}^{-1}T, \varphi\rangle = \langle T, \mathcal{F}^{-1}\varphi\rangle$$

for any Schwartz function φ.[20]

All of a sudden we have:

Fourier inversion. $\mathcal{F}^{-1}\mathcal{F}T = T$ and $\mathcal{F}\mathcal{F}^{-1}T = T$ for any tempered distribution T.

It's a cinch. Watch, it follows directly and immediately from the definitions. For any Schwartz function φ,

$$\begin{aligned}\langle \mathcal{F}^{-1}\mathcal{F}T, \varphi\rangle &= \langle \mathcal{F}T, \mathcal{F}^{-1}\varphi\rangle \quad (\text{definition of } \mathcal{F}^{-1})\\ &= \langle T, \mathcal{F}\mathcal{F}^{-1}\varphi\rangle \quad (\text{definition of } \mathcal{F})\\ &= \langle T, \varphi\rangle \quad (\text{because Fourier inversion works for Schwartz functions}).\end{aligned}$$

We started with $\langle \mathcal{F}^{-1}\mathcal{F}T, \varphi\rangle$ and we finished with $\langle T, \varphi\rangle$, which says that $\mathcal{F}^{-1}(\mathcal{F}T)$ and T have the same value when paired with any Schwartz function. Therefore they are the same distribution: $\mathcal{F}^{-1}\mathcal{F}T = T$. The second identity is derived in the same way.[21]

Done. The most fundamental result in the subject, done, in a few lines. Granted we needed old style Fourier inversion, but remember that the Schwartz functions are the best for Fourier inversion. They are also the best for every other operation with the original Fourier transform.

In Section 4.5.7 we'll show that we've gained and haven't lost. The generalized Fourier transform contains the original, classical Fourier transform in the way tempered distributions contain classical functions.

[20] Mathematicians are fond of writing $\widehat{T}$ and $\check{T}$ for the Fourier transform and inverse Fourier transform, respectively.

[21] Note that in establishing $\mathcal{F}^{-1}\mathcal{F}T = T$ for distributions we used $\mathcal{F}\mathcal{F}^{-1}\varphi = \varphi$ for functions, Fourier inversion in the other order. For showing $\mathcal{F}\mathcal{F}^{-1}T = T$ we'd use $\mathcal{F}^{-1}\mathcal{F}\varphi = \varphi$.

4.5.2. A Fourier transform hit parade, Part 1. With the definition in place it's time to reap the benefits and find some Fourier transforms explicitly. We note one general property:

- $\mathcal{F}$ is linear on tempered distributions.

This means that

$$\mathcal{F}(T_1 + T_2) = \mathcal{F}T_1 + \mathcal{F}T_2 \quad \text{and} \quad \mathcal{F}(\alpha T) = \alpha \mathcal{F}T\,,$$

where α is a number. These follow directly from the definition:

$$\begin{aligned}
\langle \mathcal{F}(T_1 + T_2), \varphi\rangle &= \langle T_1 + T_2, \mathcal{F}\varphi\rangle \\
&= \langle T_1, \mathcal{F}\varphi\rangle + \langle T_2, \mathcal{F}\varphi\rangle \\
&= \langle \mathcal{F}T_1, \varphi\rangle + \langle \mathcal{F}T_2, \varphi\rangle = \langle \mathcal{F}T_1 + \mathcal{F}T_2, \varphi\rangle, \\
\langle \mathcal{F}(\alpha T), \varphi\rangle &= \langle \alpha T, \mathcal{F}\varphi\rangle = \alpha\langle T, \mathcal{F}\varphi\rangle = \alpha\langle \mathcal{F}T, \varphi\rangle = \langle \alpha\mathcal{F}T, \varphi\rangle.
\end{aligned}$$

The Fourier transform of δ. As a first illustration of computing with the generalized Fourier transform, we'll find $\mathcal{F}\delta$. And realize that neither "$\mathcal{F}$" nor "δ" can appear together in any classical sense!

The result is as striking as it is simple; we'll find that

$$\mathcal{F}\delta = 1\,.$$

This should be understood as an equality between distributions, i.e., as saying that $\mathcal{F}\delta$ and 1 produce the same values when paired with any Schwartz function φ. Here "1" is the constant function, and this defines a tempered distribution via the integral

$$\langle 1, \varphi\rangle = \int_{-\infty}^{\infty} 1 \cdot \varphi(x)\, dx\,,$$

which converges.

To find $\mathcal{F}\delta$ we have no recourse but to appeal to the definition of the Fourier transform and the definition of δ. Thus, on the one hand,

$$\langle \mathcal{F}\delta, \varphi\rangle = \langle \delta, \mathcal{F}\varphi\rangle = \mathcal{F}\varphi(0) = \int_{-\infty}^{\infty} \varphi(x)\, dx\,.$$

On the other hand, as we've just noted,

$$\langle 1, \varphi\rangle = \int_{-\infty}^{\infty} 1 \cdot \varphi(x)\, dx = \int_{-\infty}^{\infty} \varphi(x)\, dx\,.$$

We started with $\langle \mathcal{F}\delta, \varphi\rangle$ and we finished with $\langle 1, \varphi\rangle$. We conclude that $\mathcal{F}\delta = 1$ as distributions.

According to the inversion theorem we can also say that $\mathcal{F}^{-1}1 = \delta$.

We can also show that

$$\mathcal{F}1 = \delta\,.$$

First, by definition,

$$\langle \mathcal{F}1, \varphi\rangle = \langle 1, \mathcal{F}\varphi\rangle = \int_{-\infty}^{\infty} \mathcal{F}\varphi(s)\, ds\,.$$

But next we recognize (this takes some practice!) the integral as giving the *inverse* Fourier transform of $\mathcal{F}\varphi$ at 0:

$$\mathcal{F}^{-1}\mathcal{F}\varphi(t) = \int_{-\infty}^{\infty} e^{2\pi i s t}\mathcal{F}\varphi(s)\,ds \quad \text{and, at } t = 0, \quad \mathcal{F}^{-1}\mathcal{F}\varphi(0) = \int_{-\infty}^{\infty} \mathcal{F}\varphi(s)\,ds\,.$$

Now by Fourier inversion on $\mathcal{S}$,

$$\mathcal{F}^{-1}\mathcal{F}\varphi(0) = \varphi(0)\,.$$

Thus

$$\langle \mathcal{F}1, \varphi\rangle = \varphi(0) = \langle \delta, \varphi\rangle\,.$$

We started with $\langle \mathcal{F}1, \varphi\rangle$ and we finished with $\langle \delta, \varphi\rangle$. We conclude that $\mathcal{F}1 = \delta$.

We'll also get this by duality and the evenness of δ once we introduce the reverse of a distribution.

The equations $\mathcal{F}\delta = 1$ and $\mathcal{F}1 = \delta$ are the extreme cases of the trade-off between concentration in time and concentration in frequency, with the Fourier transform turning one into the other. δ is the idealization of the most concentrated function. The function 1, on the other hand, is uniformly spread out over its domain, so it is the least concentrated.

It's rather satisfying that the simplest tempered distribution, δ, has the simplest Fourier transform, 1. (Simplest other than the function that is identically zero.) It's also the simplest argument of its type that gave us the result.

Before there were tempered distributions, however, there was δ, and before there was the Fourier transform of tempered distributions, there was $\mathcal{F}\delta = 1$. In the vacuum tube days this had to be established by limiting arguments, accompanied by an uneasiness (among some) over the nature of the limit and what exactly it produced. Our computation of $\mathcal{F}\delta = 1$ is simple and direct and leaves nothing in question about the meaning of all the quantities involved. Whether it is conceptually simpler than the older approach is something you will have to decide for yourself.

The Fourier transforms of δ_a and $e^{2\pi i x a}$. Recall the shifted delta δ_a, defined by

$$\langle \delta_a, \varphi\rangle = \varphi(a)\,.$$

What is the Fourier transform? One way to obtain $\mathcal{F}\delta_a$ is via a generalization of the shift theorem, which we'll develop later. Even without that we can find $\mathcal{F}\delta_a$ directly from the definition along the same lines as what we did for δ.

We have

$$\langle \mathcal{F}\delta_a, \varphi\rangle = \langle \delta_a, \mathcal{F}\varphi\rangle = \mathcal{F}\varphi(a) = \int_{-\infty}^{\infty} e^{-2\pi i a x}\varphi(x)\,dx\,.$$

This last integral, which is nothing but the definition of the Fourier transform of φ, can also be interpreted as the pairing of the function $e^{-2\pi i a x}$ with the Schwartz function $\varphi(x)$, and again it's *recognizing* the pairing that takes practice. We started with $\langle \mathcal{F}\delta_a, \varphi\rangle$, and we finished with $\langle e^{-2\pi iax}, \varphi\rangle$, hence

$$\mathcal{F}\delta_a = e^{-2\pi isa}\,.$$

To emphasize once again all that is going on here, $e^{-2\pi iax}$ is *not* integrable but it defines a tempered distribution through

$$\int_{-\infty}^{\infty} e^{-2\pi iax}\varphi(x)\,dx\,,$$

which exists because $\varphi(x)$ is rapidly decreasing. And, again, to say $\mathcal{F}\delta_a = e^{-2\pi isa}$ means exactly that

$$\langle \mathcal{F}\delta_a, \varphi\rangle = \langle e^{-2\pi iax}, \varphi\rangle\,.$$

To complete the picture, we can also show that

$$\mathcal{F}e^{2\pi ixa} = \delta_a\,.$$

(There's the usual notational problem here with variables, writing the variable x on the left-hand side. The "variable problem" doesn't go away in this more general setting.) This argument should look familiar: if φ is in $\mathcal{S}$, then

$$\begin{aligned}\langle \mathcal{F}e^{2\pi ixa}, \varphi\rangle &= \langle e^{2\pi ixa}, \mathcal{F}\varphi\rangle\\ &= \int_{-\infty}^{\infty} e^{2\pi ixa}\mathcal{F}\varphi(x)\,dx \quad \text{(the pairing here is with respect to } x\text{).}\end{aligned}$$

But this last integral is the inverse Fourier transform of $\mathcal{F}\varphi$ at a, $\mathcal{F}^{-1}\mathcal{F}\varphi(a)$, and so we get back $\varphi(a)$. Hence

$$\langle \mathcal{F}e^{2\pi ixa}, \varphi\rangle = \varphi(a) = \langle \delta_a, \varphi\rangle\,,$$

whence

$$\mathcal{F}e^{2\pi ixa} = \delta_a\,.$$

Incidentally, recalling the first result in this chapter, $e^{2\pi ixa}$ is not integrable and its Fourier transform is not continuous. Anything but — it's a δ! When you encounter a new result, it's a good idea to touch base with old results. It heightens your appreciation of them and serves as a consistency check. I'm big on consistency checks.

You might be happier using the more traditional notations, $\delta(x)$ for δ and $\delta(x-a)$ for δ_a (and $\delta(x+a)$ for δ_{-a}). I do not object. In this notation the results appear as

$$\mathcal{F}\delta(x \pm a) = e^{\pm 2\pi isa}\,, \quad \mathcal{F}e^{\pm 2\pi ixa} = \delta(s \mp a)\,.$$

Careful how the + and − enter.

The Fourier transform of sine and cosine. We can combine the results above to find the Fourier transform pairs for the cosine and sine. For the cosine,

$$\mathcal{F}\left(\tfrac{1}{2}(\delta_a + \delta_{-a})\right) = \tfrac{1}{2}(e^{-2\pi isa} + e^{2\pi isa}) = \cos 2\pi sa\,.$$

Writing δ with a variable the result looks like

$$\mathcal{F}\left(\tfrac{1}{2}(\delta(x-a) + \delta(x+a))\right) = \cos 2\pi sa\,.$$

Starting instead with a cosine,

$$\mathcal{F}\cos 2\pi ax = \mathcal{F}\left(\tfrac{1}{2}(e^{2\pi ixa} + e^{-2\pi ixa})\right) = \tfrac{1}{2}(\delta_a + \delta_{-a})$$

also written as

$$\mathcal{F}\cos 2\pi ax = \tfrac{1}{2}(\delta(s-a) + \delta(s+a))\,.$$

The Fourier transform of the cosine is often represented graphically as

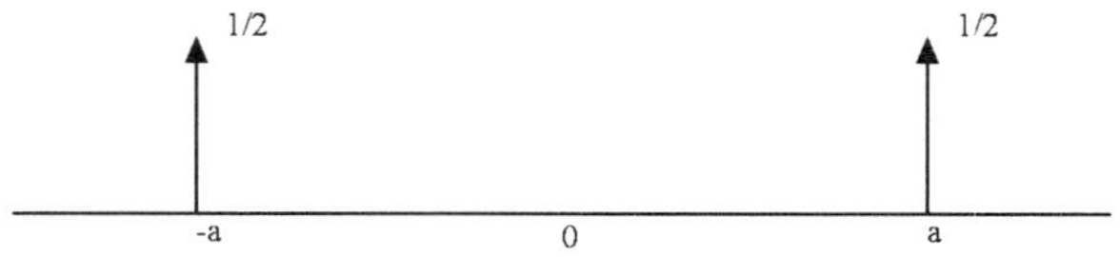

I tagged the spikes with 1/2 to indicate that the shifted δ's have been scaled.

For the sine we have, in a similar way,

$$\mathcal{F}\left(\frac{1}{2i}(\delta_{-a} - \delta_a)\right) = \frac{1}{2i}(e^{2\pi isa} - e^{-2\pi isa}) = \sin 2\pi sa\,,$$

or

$$\mathcal{F}\left(\frac{1}{2i}(\delta(x+a) - \delta(x-a))\right) = \sin 2\pi sa\,.$$

Then

$$\mathcal{F}\sin 2\pi ax = \mathcal{F}\left(\frac{1}{2i}(e^{2\pi ixa} - e^{-2\pi ixa})\right) = \frac{1}{2i}(\delta_a - \delta_{-a})\,,$$

or

$$\mathcal{F}\sin 2\pi ax = \frac{1}{2i}(\delta(s-a) - \delta(s+a))\,.$$

The picture of $\mathcal{F}\sin 2\pi x$ is

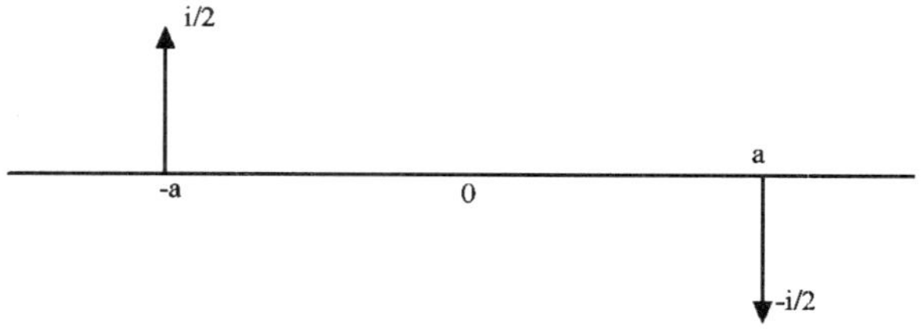

Remember that $1/i = -i$. I've tagged the spike δ_a with $-i/2$ and the spike δ_{-a} with $i/2$.

We'll discuss symmetries of the generalized Fourier transform later, but you can think of $\mathcal{F}\cos 2\pi ax$ as real and even and $\mathcal{F}\sin 2\pi ax$ as purely imaginary and odd.

We should reflect a little on what we've done here and not be too quick to move on. The sine and cosine do not have Fourier transforms in the original, classical sense. It is impossible to do anything with the integrals

$$\int_{-\infty}^{\infty} e^{-2\pi isx}\cos 2\pi x\,dx \quad \text{or} \quad \int_{-\infty}^{\infty} e^{-2\pi isx}\sin 2\pi x\,dx\,.$$

To find the Fourier transform of such basic, important functions we *must* abandon the familiar, classical terrain and plant some spikes in new territory. It is worth the effort.

4.5.3. Duality and symmetries. One of the first things we observed about the Fourier transform and its inverse is that they're pretty much the same thing except for a change in sign in the complex exponential. We recorded the relationships as

$$\begin{aligned}\mathcal{F}f(-s) &= \mathcal{F}^{-1}f(s)\,,\\ \mathcal{F}^{-1}f(-t) &= \mathcal{F}f(t)\,.\end{aligned}$$

We had similar results when we changed the sign of the variable first and then took the Fourier transform:

$$\begin{aligned}\mathcal{F}(f(-t)) &= \mathcal{F}^{-1}f(s)\,,\\ \mathcal{F}^{-1}(f(-s)) &= \mathcal{F}f(s)\,.\end{aligned}$$

(Recall the usual problem with variables. Write these formulas out for yourself to see the roles of s and t.) There was one more:

$$\mathcal{F}\mathcal{F}f(s) = f(-s)\,.$$

We referred to these collectively as the duality between Fourier transform pairs, and we'd like to have similar duality formulas when we take Fourier transforms of distributions.

The problem for distributions is that we don't really have variables to change the sign of. We can't really write $\mathcal{F}T(s)$ or $\mathcal{F}T(-s)$ or $T(-s)$, because distributions don't operate on points s; they operate on test functions. What we can do is define a *reversed distribution*, and once this is set the rest is plain sailing.

Reversed distributions. Recall that for a signal $f(x)$ we introduced the reversed signal defined by

$$f^{-}(x) = f(-x)\,.$$

This helped us to write clean, variable-free versions of the duality results:

$$(\mathcal{F}f)^{-} = \mathcal{F}^{-1}f, \quad (\mathcal{F}^{-1}f)^{-} = \mathcal{F}f\,, \quad \mathcal{F}f^{-} = \mathcal{F}^{-1}f, \quad \mathcal{F}^{-1}f^{-} = \mathcal{F}f\,.$$

Applying $\mathcal{F}$ or $\mathcal{F}^{-1}$ twice reads

$$\mathcal{F}\mathcal{F}f = f^{-}, \quad \mathcal{F}^{-1}\mathcal{F}^{-1}f = f^{-}\,.$$

My personal favorites among formulas of this type are

$$\mathcal{F}f^- = (\mathcal{F}f)^-, \quad \mathcal{F}^{-1}f^- = (\mathcal{F}^{-1}f)^- .$$

What can "sign change," or "reversal," mean for a distribution T? Our standard approach is first to take the case when the distribution comes from a function $f(x)$ and the pairing of f with a test function φ is

$$\langle f, \varphi \rangle = \int_{-\infty}^{\infty} f(x)\varphi(x)\, dx .$$

And if we pair with the reversed signal f^-?

$$\begin{aligned}
\langle f^-, \varphi \rangle &= \int_{-\infty}^{\infty} f(-x)\varphi(x)\, dx \quad \text{(now shift the action to the test function)} \\
&= \int_{\infty}^{-\infty} f(u)\varphi(-u)\,(-du) \quad \text{(by making the change of variable } u = -x\text{)} \\
&= \int_{-\infty}^{\infty} f(u)\varphi(-u)\, du \\
&= \langle f, \varphi^- \rangle .
\end{aligned}$$

We started with $\langle f^-, \varphi \rangle$ and we finished with $\langle f, \varphi^- \rangle$. The general definition is before our eyes:

- If T is a distribution, then the reversed distribution T^- is defined by

$$(T^-, \varphi) = (T, \varphi^-) .$$

Let me add that if T_f denotes the distribution induced by a function f, then with this definition we have, quite agreeably,

$$(T_f)^- = T_{f^-} .$$

If you understand what's just been done you'll understand this last equation. Understand it.

Duality. It's now easy to state the duality relations between the Fourier transform and its inverse. Adopting the notation above, we want to look at $(\mathcal{F}T)^-$ and how it compares to $\mathcal{F}^{-1}T$. For a test function φ,

$$\begin{aligned}
((\mathcal{F}T)^-, \varphi) &= (\mathcal{F}T, \varphi^-) \\
&= (T, \mathcal{F}(\varphi^-)) \quad \text{(that's how the Fourier transform is defined)} \\
&= (T, \mathcal{F}^{-1}\varphi) \quad \text{(because of duality for ordinary Fourier transforms)} \\
&= (\mathcal{F}^{-1}T, \varphi) \quad \text{(that's how the inverse Fourier transform is defined)}
\end{aligned}$$

Pretty slick, really. We can now write simply

$$(\mathcal{F}T)^- = \mathcal{F}^{-1}T .$$

We also then have

$$\mathcal{F}T = (\mathcal{F}^{-1}T)^- .$$

Same formulas as in the classical case.

To take one more example,

$$\langle \mathcal{F}(T^-), \varphi\rangle = \langle T^-, \mathcal{F}\varphi\rangle = \langle T, (\mathcal{F}\varphi)^-\rangle = \langle T, \mathcal{F}^{-1}\varphi\rangle = \langle \mathcal{F}^{-1}T, \varphi\rangle ,$$

and there's the identity

$$\mathcal{F}(T^-) = \mathcal{F}^{-1}T$$

popping out. Finally, we have

$$\mathcal{F}^{-1}(T^-) = \mathcal{F}T .$$

Combining these,

$$\mathcal{F}T^- = (\mathcal{F}T)^-, \quad \mathcal{F}^{-1}T^- = (\mathcal{F}^{-1}T)^- .$$

Applying $\mathcal{F}$ or $\mathcal{F}^{-1}$ twice leads to

$$\mathcal{F}\mathcal{F}T = T^-, \quad \mathcal{F}^{-1}\mathcal{F}^{-1}T = T^- .$$

That's all of them.

*Even and odd distributions: **δ** is even.* Now that we know how to reverse a distribution we can define what it means for a distribution to be even or odd.

- A distribution T is *even* if $T^- = T$. A distribution T is *odd* if $T^- = -T$.

Observe that if $f(x)$ determines a distribution T_f and if $f(x)$ is even or odd, then T_f has the same property. For, as we noted earlier,

$$(T_f)^- = T_{f^-} = T_{\pm f} = \pm T_f .$$

Let's next establish the useful fact:

- δ is even.

This is quick:

$$\langle \delta^-, \varphi\rangle = \langle \delta, \varphi^-\rangle = \varphi^-(0) = \varphi(-0) = \varphi(0) = \langle \delta, \varphi\rangle .$$

So $\delta^- = \delta$. (Where did we start; where did we finish?)

Let's now use this result plus duality to rederive $\mathcal{F}1 = \delta$. This is quick, too:

$$\mathcal{F}1 = (\mathcal{F}^{-1}1)^- = \delta^- = \delta .$$

You can now show that *all* of our old results on evenness and oddness of a signal and its Fourier transform extend in like form to the Fourier transform of distributions. For example, if T is even, then so is $\mathcal{F}T$, for

$$(\mathcal{F}T)^- = \mathcal{F}T^- = \mathcal{F}T ,$$

and if T is odd, then

$$(\mathcal{F}T)^- = \mathcal{F}T^- = \mathcal{F}(-T) = -\mathcal{F}T ;$$

thus $\mathcal{F}T$ is odd. As for functions, any distribution is the sum of an even and an odd distribution.

Finally, what does it mean for a distribution to be real or purely imaginary? It means that the result of applying a distribution to a real-valued test function is a real number or a purely imaginary number, respectively.

Notice how all this works for the cosine (even) and the sine (odd) and their respective Fourier transforms:

$$\mathcal{F}\cos 2\pi ax = \tfrac{1}{2}(\delta_a + \delta_{-a}) \quad \text{(even and real)},$$
$$\mathcal{F}\sin 2\pi ax = \frac{1}{2i}(\delta_a - \delta_{-a}) \quad \text{(odd and purely imaginary)}.$$

Fourier transform of **sinc**. Just for good measure:

$$\begin{aligned}\mathcal{F}\,\mathrm{sinc} &= \mathcal{F}(\mathcal{F}\Pi)\\ &= \Pi^- \quad \text{(one of the duality equations)}\\ &= \Pi \quad (\Pi \text{ is even}).\end{aligned}$$

Looks like our earlier derivation. But to be really careful here, using the generalized Fourier transform, $\mathcal{F}\,\mathrm{sinc}$ is a tempered distribution and the equality $\mathcal{F}\,\mathrm{sinc} = \Pi$ has to be understood as an equation between distributions. You should lose no sleep over this. From now on, write $\mathcal{F}\,\mathrm{sinc} = \Pi$, think in terms of functions, and found your startup.

4.5.4. A function times a distribution makes sense. We interrupt the process of recovering familiar results to say a little bit about multiplication, of all things. We'll need this for the shift theorem, coming up soon, and for other results coming up later. The issue is that there's no way to define the product of two distributions that works consistently with all the rest of the definitions and properties — try as you might, it just won't work. However, it is possible (and easy) to define the product of a function and a distribution, though there are some unexpected consequences.

Say T is a distribution and g is a function. What is gT as a distribution? We take our usual approach and first consider the case when T comes from a function f and the pairing of gf with a test function φ is

$$\langle gf, \varphi\rangle = \int_{-\infty}^{\infty} g(x)f(x)\varphi(x)\,dx = \int_{-\infty}^{\infty} f(x)(g(x)\varphi(x))\,dx$$

(just a little shift of the action in the integrand onto the test function).

As long as $g\varphi$ is still a test function, this last integral is the pairing $\langle f, g\varphi\rangle$. We started with $\langle gf, \varphi\rangle$, finished with $\langle f, g\varphi\rangle$, and we thus make the following general definition:

- Let T be a distribution. If g is a function such that $g\varphi$ is a test function whenever φ is a test function, then gT is the distribution defined by

$$\langle gT, \varphi\rangle = \langle T, g\varphi\rangle\,.$$

This looks as simple as can be, and it is.[22] You may wonder why I even singled out this operation for comment. In fact, some funny things can happen, as we'll now see.

[22] Ah, but there's the question of $g\varphi$ being a test function. Typically one assumes that g is infinitely differentiable, or even a Schwartz function. We'll have occasion to multiply by less well-behaved functions, by a rectangle function, for example, to cut a signal off. In such cases the operations can be justified by a limiting argument, and we won't comment further.

The sampling property of δ. Watch what happens if we multiply δ by $g(x)$:

$$\langle g\delta, \varphi\rangle = \langle \delta, g\varphi\rangle = g(0)\varphi(0)\,.$$

This is the same result as if we had paired $g(0)\delta$ with φ. Thus:

- $g\delta = g(0)\delta$.

Written with the variable,

$$g(x)\delta(x) = g(0)\delta(x).$$

In particular, if $g(0) = 0$, then the result is 0! For example,

$$x\delta = 0$$

or for that matter

$$x^n\delta = 0$$

for any positive power of x.

Along with $g\delta = g(0)\delta$ we have:

- $g\delta_a = g(a)\delta_a$.

To show this:

$$\langle g\delta_a, \varphi\rangle = \langle \delta_a, g\varphi\rangle = g(a)\varphi(a) = g(a)\langle \delta_a, \varphi\rangle = \langle g(a)\delta_a, \varphi\rangle\,.$$

Written with the variable,

$$g(x)\delta(x-a) = g(a)\delta(x-a)\,.$$

The bulleted equations are usually referred to as the *sampling property of δ*. Multiplying a function g by δ_a records the sampled value $g(a)$. But not the sampled value alone — the value $g(a)$ *times* δ_a.

We'll use this property in many applications, for example, as the name implies, when we talk about the famous sampling theorem in Chapter 6.

More on a function times δ. There's a converse to the $x\delta = 0$ property that's interesting in itself and that we'll use later in finding some particular Fourier transforms.

- If T is a distribution and $xT = 0$, then $T = \alpha\delta$ for some constant α.

I'll show you the proof of this, but you can skip it with no hard feelings. The argument is more involved than the simple statement might suggest, though it's a nice example, and a fairly typical example, of the kind of tricks that are used to prove things in this area. Each to his or her own taste.

Knowing where this is going, let me start with an innocent observation.[23] If ψ is a smooth function, then

$$\begin{aligned}\psi(x) &= \psi(0) + \int_0^x \psi'(t)\,dt \\ &= \psi(0) + \int_0^1 x\psi'(xu)\,du \quad \text{(using the substitution } u = t/x) \\ &= \psi(0) + x\int_0^1 \psi'(xu)\,du\,.\end{aligned}$$

Let

$$\Psi(x) = \int_0^1 \psi'(xu)\,du$$

so that

$$\psi(x) = \psi(0) + x\Psi(x)\,.$$

We'll now use this innocent observation in the case when $\psi(0) = 0$, for then

$$\psi(x) = x\Psi(x)\,.$$

It's clear from the definition of Ψ that Ψ is as smooth as ψ is and that if, for example, ψ is rapidly decreasing, then so is Ψ. Put informally, we've shown that if $\psi(0) = 0$, we can factor out an x and still have a function that's as good as ψ.

Now suppose $xT = 0$, meaning that

$$\langle xT, \varphi\rangle = 0$$

for *every* test function φ. Fix a smooth windowing function $\varphi_0(x)$ that is identically 1 on an interval about $x = 0$, goes down to zero smoothly, and is identically zero far enough away from $x = 0$. We mentioned smooth windows earlier, way back in Section 4.2 when we first met Schwartz functions. Here's the plot we had then.

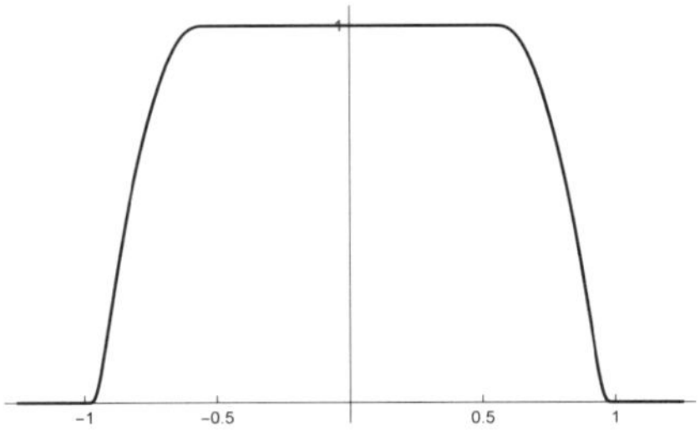

Since φ_0 is fixed in this argument, T operating on φ_0 gives some fixed number, say

$$\langle T, \varphi_0\rangle = \alpha\,.$$

Now write

$$\varphi(x) = \varphi(0)\varphi_0(x) + (\varphi(x) - \varphi(0)\varphi_0(x)) = \varphi(0)\varphi_0(x) + \psi(x)$$

where, by this clever way of writing φ, the function $\psi(x) = \varphi(x) - \varphi(0)\varphi_0(x)$ has the property that

$$\psi(0) = \varphi(0) - \varphi(0)\varphi_0(0) = \varphi(0) - \varphi(0) = 0$$

[23]This innocent observation is actually the beginning of deriving Taylor series "with remainder."

because $\varphi_0(0) = 1$. This means that we can factor out an x and write

$$\psi(x) = x\Psi(x)\,,$$

where Ψ is again a test function, and then

$$\varphi(x) = \varphi(0)\varphi_0(x) + x\Psi(x)\,.$$

But now

$$\begin{aligned}
\langle T, \varphi(x)\rangle &= \langle T, \varphi(0)\varphi_0 + x\Psi\rangle \\
&= \langle T, \varphi(0)\varphi_0\rangle + \langle T, x\Psi\rangle \\
&= \varphi(0)\langle T, \varphi_0\rangle + \langle T, x\Psi\rangle \quad (\varphi(0) \text{ comes out because of linearity}) \\
&= \varphi(0)\langle T, \varphi_0\rangle + \langle xT, \Psi\rangle
\end{aligned}$$

(the x slides over; that's how multiplying T by the smooth function x works)

$$\begin{aligned}
&= \varphi(0)\langle T, \varphi_0\rangle + 0 \quad (\text{because } \langle xT, \Psi\rangle = 0!) \\
&= \alpha\varphi(0) \\
&= \langle \alpha\delta, \varphi\rangle.
\end{aligned}$$

We conclude that

$$T = \alpha\delta\,.$$

4.5.5. Shifts. What should we make of $T(x \pm b)$ for a distribution T when, once again, it doesn't make sense to evaluate T at a point $x \pm b$? We use the familiar strategy, starting by assuming that T comes from a function f and then working with the pairing $\langle f(x-b), \varphi(x)\rangle$:

$$\int_{-\infty}^{\infty} f(x-b)\varphi(x)\,dx = \int_{-\infty}^{\infty} f(u)\varphi(u+b)\,du \quad (\text{making the substitution } u = x - b).$$

Now bring in the shift operator

$$\tau_b f(x) = f(x-b)$$

and write the equation above (using x as a variable of integration in both cases) as

$$\langle \tau_b f, \varphi\rangle = \int_{-\infty}^{\infty} \tau_b f(x)\varphi(x)\,dx = \int_{-\infty}^{\infty} f(x)\tau_{-b}\varphi(x)\,dx = \langle f, \tau_{-b}\varphi\rangle\,.$$

On the left-hand side f is shifted by b, while on the right-hand side φ is shifted by $-b$.

This result guides us in making the general definition:

- If T is a distribution, we define $\tau_b T$ by

$$\langle \tau_b T, \varphi\rangle = \langle T, \tau_{-b}\varphi\rangle\,.$$

(Say that T is delayed or shifted by b.)

You can also check that for a distribution T_f coming from a function f we have

$$\tau_b T_f = T_{\tau_b f}\,.$$

δ_a is a shifted δ, really. To close the loop on some things we said earlier, watch what happens when we delay δ by a:

$$\begin{aligned}\langle \tau_a\delta, \varphi\rangle &= \langle \delta, \tau_{-a}\varphi\rangle \\ &= (\tau_{-a}\varphi)(0) \\ &= \varphi(a) \quad \text{(remember, } \tau_{-a}\varphi(x) = \varphi(x+a)) \\ &= \langle \delta_a, \varphi\rangle\,.\end{aligned}$$

We have shown that

$$\tau_a\delta = \delta_a\,.$$

Writing $\tau_a\delta$ is the variable-free way of writing $\delta(x-a)$.

The shift theorem. We're now ready for the general form of the shift theorem:

- If T is a tempered distribution, then

$$\mathcal{F}(\tau_b T) = e^{-2\pi i b x}\mathcal{F}T\,.$$

The right-hand side makes sense because we have defined what it means to multiply a distribution by a function, in this case the function $e^{-2\pi i b x}$.

To verify the formula, first we have

$$\langle \mathcal{F}(\tau_b T), \varphi\rangle = \langle \tau_b T, \mathcal{F}\varphi\rangle = \langle T, \tau_{-b}\mathcal{F}\varphi\rangle\,.$$

We can evaluate the test function in the last term:

$$\begin{aligned}\tau_{-b}\mathcal{F}\varphi(s) &= \mathcal{F}\varphi(s+b) \\ &= \int_{-\infty}^{\infty} e^{-2\pi i(s+b)x}\varphi(x)\,dx \\ &= \int_{-\infty}^{\infty} e^{-2\pi i s x}e^{-2\pi i b x}\varphi(x)\,dx = \mathcal{F}(e^{-2\pi i b x}\varphi)(s)\,.\end{aligned}$$

Now plug this into what we had before:

$$\begin{aligned}\langle \mathcal{F}(\tau_b T), \varphi\rangle &= \langle T, \tau_{-b}\mathcal{F}\varphi\rangle \\ &= \langle T, \mathcal{F}(e^{-2\pi i b x}\varphi)\rangle \\ &= \langle \mathcal{F}T, e^{-2\pi i b x}\varphi\rangle = \langle e^{-2\pi i b x}\mathcal{F}T, \varphi\rangle\,.\end{aligned}$$

Make sure you understand how in the various steps $\mathcal{F}$ and $e^{-2\pi i b x}$ slide back and forth between slots; that's the crucial aspect of the derivation.

Finally, we started with $\langle \mathcal{F}(\tau_b T), \varphi\rangle$ and finished with $\langle e^{-2\pi i b x}\mathcal{F}T, \varphi\rangle$, which shows that

$$\mathcal{F}(\tau_b T) = e^{-2\pi i b x}\mathcal{F}T\,.$$

As one quick application of this let's see what happens to the shifted δ. By the shift theorem

$$\mathcal{F}\tau_a\delta = e^{-2\pi i a s}\mathcal{F}\delta = e^{-2\pi i s a}\cdot 1 = e^{-2\pi i s a}\,,$$

in accord with what we found earlier for $\mathcal{F}\delta_a$ directly from the definitions of δ_a and $\mathcal{F}$.

4.5.6. Stretches. To obtain the stretch theorem (aka similarity theorem, scaling theorem) we first have to stretch a distribution. Following our now usual procedure, we check what happens with stretches when T comes from a function f, and so we pair $f(ax)$ with a test function $\varphi(x)$. For $a > 0$,

$$\int_{-\infty}^{\infty} f(ax)\varphi(x)\,dx = \int_{-\infty}^{\infty} f(u)\frac{1}{a}\varphi\Big(\frac{u}{a}\Big)\,du\,,$$

making the substitution $u = ax$, and for $a < 0$,

$$\int_{-\infty}^{\infty} f(ax)\varphi(x)\,dx = \int_{\infty}^{-\infty} f(u)\frac{1}{a}\varphi\Big(\frac{u}{a}\Big)\,du = \int_{-\infty}^{\infty} f(u)\left(-\frac{1}{a}\right)\varphi\Big(\frac{u}{a}\Big)\,du\,.$$

Combining the two cases,

$$\int_{-\infty}^{\infty} f(ax)\varphi(x)\,dx = \int_{-\infty}^{\infty} f(u)\frac{1}{|a|}\varphi\Big(\frac{u}{a}\Big)\,du\,.$$

As we expressed shifts via the shift operator τ_b, for stretches we use

$$\sigma_a\varphi(x) = \varphi(ax)$$

to write the pairings and integrals above as

$$\langle\sigma_a f\,\varphi\rangle = \int_{-\infty}^{\infty} \sigma_a f(x)\varphi(x)\,dx = \int_{-\infty}^{\infty} f(x)\frac{1}{|a|}\sigma_{1/a}\varphi(x)\,dx = \left\langle f, \frac{1}{|a|}\sigma_{1/a}\varphi\right\rangle.$$

Thus for a general distribution:

- If T is a distribution, we define $\sigma_a T$ via
$$\langle\sigma_a T,\varphi\rangle = \left\langle T, \frac{1}{|a|}\sigma_{1/a}\varphi\right\rangle.$$

Note also that then

$$\left\langle \frac{1}{|a|}\sigma_{1/a}T, \varphi\right\rangle = \langle T, \sigma_a\varphi\rangle\,.$$

For a distribution T_f coming from a function f the relation is

$$\sigma_a T_f = T_{\sigma_a f}\,.$$

Scaling δ. Since δ is concentrated at a point, however you want to interpret that, you might think that scaling $\delta(x)$ to $\delta(ax)$ (writing the variable) shouldn't have any effect. But it does:

$$\begin{aligned}\langle\sigma_a\delta,\varphi\rangle &= \left\langle\delta, \frac{1}{|a|}\sigma_{1/a}\varphi\right\rangle = \frac{1}{|a|}(\sigma_{1/a}\varphi)(0)\\ &= \frac{1}{|a|}\varphi(0/a) = \frac{1}{|a|}\varphi(0) = \left\langle\frac{1}{|a|}\delta,\varphi\right\rangle.\end{aligned}$$

Hence

$$\sigma_a\delta = \frac{1}{|a|}\delta\,.$$

This is most often written with the variable as

$$\delta(ax) = \frac{1}{|a|}\delta(x)\,.$$

The effect of scaling the variable is to scale the strength of δ by the reciprocal amount. *This is an important result.*

The stretch theorem. With the groundwork laid it's now not difficult to state and derive the general stretch theorem:

- If T is a tempered distribution, then
$$\mathcal{F}(\sigma_a T) = \frac{1}{|a|}\sigma_{1/a}\mathcal{F}T\,.$$
To check this, first observe that
$$\langle \mathcal{F}(\sigma_a T), \varphi\rangle = \langle \sigma_a T, \mathcal{F}\varphi\rangle = \left\langle T, \frac{1}{|a|}\sigma_{1/a}\mathcal{F}\varphi\right\rangle.$$

But now by the stretch theorem for functions
$$\frac{1}{|a|}(\sigma_{1/a}\mathcal{F}\varphi)(s) = \frac{1}{|a|}\mathcal{F}\varphi\left(\frac{s}{a}\right) = \mathcal{F}(\sigma_a\varphi)(s)\,.$$
Plug this back into what we had:
$$\langle \mathcal{F}(\sigma_a T), \varphi\rangle = \left\langle T, \frac{1}{|a|}\sigma_{1/a}\mathcal{F}\varphi\right\rangle = \langle T, \mathcal{F}(\sigma_a\varphi)\rangle = \langle \mathcal{F}T, \sigma_a\varphi\rangle = \left\langle \frac{1}{|a|}\sigma_{1/a}\mathcal{F}T, \varphi\right\rangle.$$
This proves that
$$\mathcal{F}(\sigma_a T) = \frac{1}{|a|}\sigma_{1/a}\mathcal{F}T\,.$$

4.5.7. The generalized Fourier transform includes the classical Fourier transform. Remember that we identify a function f with the induced distribution T_f, and it is in this way that we say that the tempered distributions contain many of the classical functions. Remember also that I said we didn't lose anything by considering the Fourier transform of a tempered distribution. Here's what that means: suppose that a function f defines a distribution and that f has a (classical) Fourier transform $\mathcal{F}f$, which also defines a distribution; i.e., the pairing
$$\langle \mathcal{F}f, \varphi\rangle = \int_{-\infty}^{\infty} \mathcal{F}f(s)\varphi(s)\,ds$$
exists for every Schwartz function φ. This isn't expecting too much. Writing $T_{\mathcal{F}f}$ for the tempered distribution determined by $\mathcal{F}f$,
$$\begin{aligned}
\langle T_{\mathcal{F}f}, \varphi\rangle &= \int_{-\infty}^{\infty} \mathcal{F}f(s)\varphi(s)\,ds\\
&= \int_{-\infty}^{\infty}\left(\int_{-\infty}^{\infty} e^{-2\pi isx}f(x)\,dx\right)\varphi(s)\,ds = \int_{-\infty}^{\infty}\int_{-\infty}^{\infty} e^{-2\pi isx}f(x)\varphi(s)\,ds\,dx\\
&= \int_{-\infty}^{\infty} f(x)\left(\int_{-\infty}^{\infty} e^{-2\pi isx}\varphi(s)\,ds\right)dx = \int_{-\infty}^{\infty} f(x)\mathcal{F}\varphi(x)\,dx = \langle T_f, \mathcal{F}\varphi\rangle\,.
\end{aligned}$$
But now, by our definition of the generalized Fourier transform
$$\langle T_f, \mathcal{F}\varphi\rangle = \langle \mathcal{F}T_f, \varphi\rangle\,.$$
Putting this together with the start of the calculation, we obtain
$$\langle T_{\mathcal{F}f}, \varphi\rangle = \langle \mathcal{F}T_f, \varphi\rangle\,,$$
from which
$$T_{\mathcal{F}f} = \mathcal{F}T_f\,.$$
In words, if the classical Fourier transform of a function defines a distribution, then that distribution is the Fourier transform of the distribution that the function

defines. A marvelous sentence. This is a precise way of saying that the generalized Fourier transform includes the classical Fourier transform.

It's been gratifying, I really hope you agree, to see the familiar formulas from earlier work carry over to the more general setting of distributions. Now for another advance that distributions provide.

4.6. Fluxions Finis: The End of Differential Calculus

Act IV, I believe. Here let's show how introducing distributions completes differential calculus: how we can define the derivative of a distribution and, consequently, how we can differentiate functions your calculus teacher never would have let you differentiate. We'll make use of this for Fourier transforms, too, through the corresponding derivative theorems.

The motivation for how to bring about this remarkable state of affairs goes back to integration by parts, a technique we've used often in our calculations with the Fourier transform. If φ is a test function and f is a function for which $f(x)\varphi(x) \to 0$ as $x \to \pm\infty$ (as usual, not demanding much) and *if* f is differentiable (demanding a little more), *then* we can use integration by parts to write

$$\begin{aligned}\int_{-\infty}^{\infty} f'(x)\varphi(x)\,dx &= \Big[f(x)\varphi(x)\Big]_{-\infty}^{\infty} - \int_{-\infty}^{\infty} f(x)\varphi'(x)\,dx \quad (u=\varphi\,,\ dv = f'(x)\,dx)\\ &= -\int_{-\infty}^{\infty} f(x)\varphi'(x)\,dx.\end{aligned}$$

The derivative has shifted from f to φ.

We can find similar formulas for higher derivatives by repeated integrations by parts, shifting the action from higher derivatives of f to higher derivatives of φ. For example, supposing that the boundary terms in the integrations by parts of $f''(x)\varphi(x)$ tend to 0 as $x \to \pm\infty$, we then find that

$$\begin{aligned}\int_{-\infty}^{\infty} f''(x)\varphi(x)\,dx &= \Big[f'(x)\varphi(x)\Big]_{-\infty}^{\infty} - \int_{-\infty}^{\infty} f'(x)\varphi'(x)\,dx\\ &\qquad (u=\varphi(x)\,,\ dv = f''(x)\,dx)\\ &= -\int_{-\infty}^{\infty} f'(x)\varphi'(x)\,dx\\ &= -\left(\Big[f(x)\varphi'(x)\Big]_{-\infty}^{\infty} - \int_{-\infty}^{\infty} f(x)\varphi''(x)\,dx\right)\\ &\qquad (u=\varphi'(x)\,,\ dv = f'(x)\,dx)\\ &= \int_{-\infty}^{\infty} f(x)\varphi''(x)\,dx\,.\end{aligned}$$

Watch out — there's no minus sign out front when we've shifted the *second* derivative from f to φ.

We'll concentrate just on the formula for the first derivative. Let's write it again:

$$\int_{-\infty}^{\infty} f'(x)\varphi(x)\,dx = -\int_{-\infty}^{\infty} f(x)\varphi'(x)\,dx\,.$$

The right-hand side may make sense even if the left-hand side does not, and we can view the right-hand side as a way of saying how the derivative of $f(x)$ *would* act if $f(x)$ had a derivative. Put in terms of our "try a function first" procedure, if a distribution comes from a function $f(x)$, then this formula tells us how the derivative $f'(x)$ *as a distribution* should be paired with a test function $\varphi(x)$. It should be paired according to the equation above:

$$\langle f', \varphi\rangle = -\langle f, \varphi'\rangle\,.$$

Turning this outcome into a definition, as our general procedure tells us we should do when passing from functions to distributions, we define the derivative of a distribution as another distribution according to:

- If T is a distribution, then its derivative T' is the distribution defined by

$$\langle T', \varphi\rangle = -\langle T, \varphi'\rangle\,.$$

This definition makes sense provided that φ' is again a test function. That's certainly the case for $\mathcal{S}$ and $\mathcal{C}$. Thus derivatives are defined for tempered distributions and for distributions on $\mathcal{C}$.

Naturally, $(T_1+T_2)' = T_1'+T_2'$ and $(\alpha T)' = \alpha T'$. However, there is *no* product rule in general because there's no way to multiply two distributions. I'll discuss this later in connection with convolution.

You can go on to define derivatives of higher orders in a similar way, and I'll let you write down what the general formula for the pairing should be. Watch out for how the minus sign does or does not appear. The striking thing is that you don't have to stop: *distributions are infinitely differentiable*!

Let's see how differentiating a distribution works in practice.

Derivative of the unit step function. Remember the Heaviside step function

$$H(x) = \begin{cases} 0, & x \le 0\,, \\ 1, & x > 0\,. \end{cases}$$

Early on, we mentioned that $H(x)$ determines a tempered distribution because for any Schwartz function φ the paring

$$\langle H, \varphi\rangle = \int_{-\infty}^{\infty} H(x)\varphi(x)\,dx = \int_{0}^{\infty} \varphi(x)\,dx$$

makes sense.

From the definition of the derivative of a distribution, if $\varphi(x)$ is any test function, then

$$\begin{aligned}\langle H',\varphi\rangle &= -\langle H,\varphi'\rangle\\ &= -\int_{-\infty}^{\infty} H(x)\varphi'(x)\,dx\\ &= -\int_{0}^{\infty} 1\cdot\varphi'(x)\,dx\\ &= -(\varphi(\infty)-\varphi(0)) = \varphi(0)\,.\end{aligned}$$

We see that pairing H' with a test function produces the same result as if we had paired δ with a test function:

$$\langle H',\varphi\rangle = \varphi(0) = \langle \delta,\varphi\rangle\,.$$

We conclude that

$$H' = \delta\,.$$

It's also common to instead define $H(0) = 1/2$. There's no effect on the formula $H' = \delta$ nor on subsequent results in any way.

You may have encountered the formula $H' = \delta$ in another class, based on quickly articulated statements like: "$H'(x) = 0$ everywhere for $x \neq 0$ because $H(x)$ is constant and $H(x)$ takes a jump at $x = 0$ so that's an infinite slope and it only jumps by 1 at $x = 0$ so $H'(0) = 1\cdot\infty = \infty$ so $H'(x)$ has the properties of δ except the integral property but never mind that so $H'(x) = \delta$." Right? By contrast, the argument given here is *airtight*!

We can step down at the origin rather than step up via

$$H(-x) = \begin{cases} 1, & x < 0\,,\\ 0, & x \geq 0\,,\end{cases}$$

and for a unit step down the derivative is minus the δ-function:

$$\frac{d}{dx}H(-x) = -\delta(-x) = -\delta(x)\,,$$

the final equation holding because δ is even. The formula for the derivative of $H(-x)$ can be proved directly.[24]

In brief, the derivative of a unit step *up* is a δ-function and the derivative of a unit step *down* is minus a δ-function. If we shift the discontinuity to x_0, then

$$\frac{d}{dx}H(x-x_0) = \delta(x-x_0), \quad \frac{d}{dx}H(-x+x_0) = -\delta(-x+x_0) = -\delta(x-x_0).$$

These statements apply locally; i.e., if a function is equal to a (shifted, up or down) unit step function on a neighborhood of a point of discontinuity, then its derivative on the neighborhood will be plus or minus a δ-function at that point.

[24] Or by appeal to the chain rule applied to distributions. Requires some further definitions.

Derivative of the unit ramp. The *unit ramp* function is defined by

$$u(x) = \begin{cases} 0, & x \le 0, \\ x, & x > 0. \end{cases}$$

In an introductory calculus class, if you were asked, "What is the derivative of $u(x)$?" you might have said, "It's 0 if $x \le 0$ and 1 if $x > 0$, so it looks like the unit step $H(x)$ to me." You'd be right, but your teacher would probably rain shame down upon you because $u(x)$ is not differentiable at $x = 0$. Hah, a temporary inconvenience. Now that you know about distributions, here's why you were right.

For a test function $\varphi(x)$,

$$\begin{aligned}
\langle u', \varphi \rangle &= -\langle u, \varphi' \rangle = -\int_{-\infty}^{\infty} u(x)\varphi'(x)\,dx = -\int_0^{\infty} x\varphi'(x)\,dx \\
&= -\left(\Big[x\varphi(x)\Big]_0^{\infty} - \int_0^{\infty} \varphi(x)\,dx \right) \\
&\qquad (x\varphi(x) \to 0 \text{ as } x \to \infty \text{ because } \varphi(x) \text{ decays faster than any power of } x) \\
&= \int_0^{\infty} \varphi(x)\,dx = \int_{-\infty}^{\infty} H(x)\varphi(x)\,dx = \langle H, \varphi \rangle .
\end{aligned}$$

Since $\langle u', \varphi \rangle = \langle H, \varphi \rangle$, we conclude that $u' = H$ as distributions. *Airtight!* Then of course $u'' = \delta$.

Derivative of the signum function. The *signum* (or sign) function is defined by

$$\operatorname{sgn}(x) = \begin{cases} +1, & x > 0, \\ -1, & x < 0. \end{cases}$$

We didn't define sgn at 0, but it's also common to set $\operatorname{sgn}(0) = 0$. As with the unit step function the value at 0 isn't an issue. Ever.

Let $\varphi(x)$ be any test function. Then

$$\begin{aligned}
\langle \operatorname{sgn}', \varphi \rangle &= -\langle \operatorname{sgn}, \varphi' \rangle = -\int_{-\infty}^{\infty} \operatorname{sgn}(x)\varphi'(x)\,dx \\
&= -\left(\int_{-\infty}^{0} (-1)\varphi'(x)\,dx + \int_0^{\infty} (+1)\varphi'(x)\,dx \right) \\
&= (\varphi(0) - \varphi(-\infty)) - (\varphi(\infty) - \varphi(0)) = 2\varphi(0) .
\end{aligned}$$

The result of pairing sgn' with φ is the same as if we had paired φ with 2δ;

$$\langle \operatorname{sgn}', \varphi \rangle = 2\varphi(0) = \langle 2\delta, \varphi \rangle .$$

Hence

$$\operatorname{sgn}' = 2\delta .$$

Observe that $H(x)$ jumps up by 1 at 0, and its derivative is δ, whereas sgn jumps up by 2 at 0, and its derivative is 2δ.[25]

[25] I suppose that if you define $\operatorname{sgn}(0) = 0$, then at the origin sgn jumps by 1, but it jumps twice, making it a total jump of 2. Likewise if $H(0) = 1/2$, then H jumps by $1/2$ at the origin, but it jumps twice, making it a total jump of 1. I think this is what interpreting the Talmud is like.

Derivatives of δ. To find the derivative of the δ-function we have, for any test function φ,

$$\langle \delta', \varphi \rangle = -\langle \delta, \varphi' \rangle = -\varphi'(0)\,.$$

That's really as much of a formula as we can write. δ itself acts by popping out the value of a test function at 0, and δ' acts by popping out *minus* the value of the derivative of the test function at 0.

Higher derivatives are handled recursively, the only caution being what happens to the minus sign. For example,

$$\langle \delta'', \varphi \rangle = -\langle \delta', \varphi' \rangle = \langle \delta, \varphi'' \rangle = \varphi''(0)\,.$$

In general,

$$\langle \delta^{(n)}, \varphi \rangle = (-1)^n \varphi^{(n)}(0)\,.$$

For the shifted δ,

$$\langle \delta_a^{(n)}, \varphi \rangle = (-1)^n \varphi^{(n)}(a)\,.$$

Derivative of $\ln|x|$. Remember that famous formula from calculus:

$$\frac{d}{dx}\ln|x| = \frac{1}{x}\,.$$

Any chance of something like that being true for distributions? Sure, with the proper interpretation. On the right-hand side, $1/x$ is the principal value distribution discussed in Section 4.4.5.

This is important because of the connection of $1/x$ to the Hilbert transform mentioned as well in Section 4.4.5, a tool that communications engineers use every day. Anticipating its grand entrance later, the Hilbert transform is given by convolution of a signal with $1/\pi x$. Once we learn how to take the Fourier transform of $1/x$, which is coming up, we'll see in Section 8.9 that the Hilbert transform is a filter with the interesting property that magnitudes of the spectral components are unchanged while their phases are shifted by $\pm\pi/2$.

The function $\ln|x|$ becomes negatively infinite at 0 so the standard pairing

$$\langle \ln|x|, \varphi \rangle = \int_{-\infty}^{\infty} \ln|x|\varphi(x)\,dx$$

appears questionable. In fact, it's OK; the integral converges. Because you may see integrals like this elsewhere, I think it's worth going through the argument, which is similar to what we did in Section 4.4.5.

We'll need three facts:

(1) An antiderivative of $\ln x$ is $x\ln x - x$.

(2) $\lim_{|x|\to\infty} |x|^k \ln|x| = 0$ for any $k < 0$.

(3) $\lim_{|x|\to 0} |x|^k \ln|x| = 0$ for any $k > 0$.

Facts (2) and (3) are about the rate of growth of the log — pretty slow — and you can check them with l'Hôpital's rule, for example.

From the second point, because $\varphi(x)$ is decaying so rapidly at $\pm\infty$ there's no problem with the tails of the integrals. Near 0 we have to show that the separate limits

$$\lim_{\epsilon_1 \to 0} \int_{-\infty}^{-\epsilon_1} \ln(-x)\varphi(x)\,dx + \lim_{\epsilon_2 \to 0} \int_{\epsilon_2}^{\infty} \ln x\,\varphi(x)\,dx$$

exist. For this, write

$$\begin{aligned}
&\int_{-\infty}^{-\epsilon_1} \ln(-x)\varphi(x)\,dx + \int_{\epsilon_2}^{\infty} \ln x\,\varphi(x)\,dx \\
&= \int_{-\infty}^{-1} \ln(-x)\varphi(x)\,dx + \int_{-1}^{-\epsilon_1} \ln(-x)\varphi(x)\,dx + \int_{\epsilon_2}^{1} \ln|x|\varphi(x)\,dx + \int_{1}^{\infty} \ln|x|\varphi(x)\,dx\,.
\end{aligned}$$

As I said, the first and last integrals, the ones going off to $\pm\infty$, aren't a problem and only the second and third integrals need work. For these, use a Taylor approximation to $\varphi(x)$, writing $\varphi(x) = \varphi(0) + O(x)$. Then

$$\begin{aligned}
&\int_{-1}^{-\epsilon_1} \ln(-x)(\varphi(0) + O(x))\,dx + \int_{\epsilon_2}^{1} \ln x(\varphi(0) + O(x))\,dx \\
&= \varphi(0)\left(\int_{-1}^{-\epsilon_1} \ln(-x)\,dx + \int_{\epsilon_2}^{1} \ln x\,dx\right) + \int_{-1}^{-\epsilon_1} O(x)\ln(-x)\,dx + \int_{\epsilon_2}^{1} O(x)\ln x\,dx \\
&= \varphi(0)\left(\int_{\epsilon_1}^{1} \ln x\,dx + \int_{\epsilon_2}^{1} \ln x\,dx\right) + \int_{-1}^{-\epsilon_1} O(x)\ln(-x)\,dx + \int_{\epsilon_2}^{1} O(x)\ln x\,dx\,.
\end{aligned}$$

We want to let $\epsilon_1 \to 0$ and $\epsilon_2 \to 0$. You can now use points (1) and (3) above to check that the limits of the first pair of integrals exist, and by point (3) the second pair of integrals are also fine. We've shown that

$$\int_{-\infty}^{\infty} \ln|x|\,\varphi(x)\,dx$$

exists; hence $\ln|x|$ is a distribution. (The pairing by integration is obviously linear. We haven't checked continuity, but we never check continuity.)

Since $\ln|x|$ determines a distribution, it has a derivative, and

$$\langle(\ln|x|)', \varphi(x)\rangle = -\langle\ln|x|, \varphi'\rangle = -\int_{-\infty}^{\infty} \ln|x|\varphi'(x)\,dx\,.$$

The integral on the right also converges (same analysis as above). One can then further show, and here I won't give the details, that an integration by parts produces

$$-\int_{-\infty}^{\infty} \ln|x|\varphi'(x)\,dx = \text{pr.v.} \int_{-\infty}^{\infty} \frac{1}{x}\,\varphi(x)\,dx = \left\langle \frac{1}{x}, \varphi \right\rangle,$$

where pr.v. is the Cauchy principal value of the integral.[26] Thus the derivative $(\ln|x|)'$ is the principal value distribution $1/x$, and all's right with calculus.

[26] Remember, the definition of $1/x$ as a distribution is via the pairing using the Cauchy principal value integral.

4.6.1. The derivative theorems. Another basic property of the Fourier transform is what it does in relation to differentiation. As you'll recall, "differentiation becomes multiplication" is the shorthand way of describing the situation. We know how to differentiate a distribution, and it's an easy step to bring the Fourier transform into the picture, leading to the derivative theorems. We'll then use these to *find* the Fourier transforms for some common functions that we haven't treated.

Let's recall the formulas for functions, best written as

$$f'(t) \rightleftharpoons 2\pi i s F(s) \quad \text{and} \quad -2\pi i t f(t) \rightleftharpoons F'(s)$$

where $f(t) \rightleftharpoons F(s)$.

We first want to find $\mathcal{F}T'$ for a distribution T. For any test function φ,

$$\begin{aligned}
\langle \mathcal{F}T', \varphi\rangle &= \langle T', \mathcal{F}\varphi\rangle = -\langle T, (\mathcal{F}\varphi)'\rangle \\
&= -\langle T, \mathcal{F}(-2\pi i s\varphi)\rangle \quad \text{(from the second formula for functions)} \\
&= -\langle \mathcal{F}T, -2\pi i s\varphi\rangle \quad \text{(moving } \mathcal{F} \text{ back over to } T) \\
&= \langle 2\pi i s\mathcal{F}T, \varphi\rangle
\end{aligned}$$

(canceling minus signs and moving the smooth function $2\pi i s$ back onto $\mathcal{F}T$).

So the second formula for functions has helped us to derive the version of the first formula for distributions:

$$\mathcal{F}T' = 2\pi i s\mathcal{F}T\,.$$

On the right-hand side, that's the smooth function $2\pi i s$ times the distribution $\mathcal{F}T$.

Now let's work with $(\mathcal{F}T)'$:

$$\begin{aligned}
\langle (\mathcal{F}T)', \varphi\rangle &= -\langle \mathcal{F}T, \varphi'\rangle = -\langle T, \mathcal{F}(\varphi')\rangle \\
&= -\langle T, 2\pi i s\mathcal{F}\varphi\rangle \quad \text{(from the first formula for functions)} \\
&= \langle -2\pi i s T, \mathcal{F}\varphi\rangle \\
&= \langle \mathcal{F}(-2\pi i s T), \varphi\rangle\,.
\end{aligned}$$

Therefore

$$(\mathcal{F}T)' = \mathcal{F}(-2\pi i s T)\,.$$

4.6.2. A Fourier transform hit parade, Part 2. Time for a few more Fourier transforms. We can put the derivative formula to use to find the Fourier transform of the sgn function, and from that several others.

Fourier transform of sgn. On the one hand, $\operatorname{sgn}' = 2\delta$, from an earlier calculation, so $\mathcal{F}\operatorname{sgn}' = 2\mathcal{F}\delta = 2$. On the other hand, using the derivative theorem,

$$\mathcal{F}\operatorname{sgn}' = 2\pi i s\,\mathcal{F}\operatorname{sgn}\,.$$

Hence

$$2\pi i s\,\mathcal{F}\operatorname{sgn} = 2\,.$$

We'd like to say that

$$\mathcal{F}\operatorname{sgn} = \frac{1}{\pi i s}\,,$$

where $1/s$ is the principal value distribution from Section 4.4.5. In fact, this is the case, but it requires a little more of an argument.

Yes, $2\pi is\,\mathcal{F}\text{sgn} = 2$, but also $2\pi is(\mathcal{F}\text{sgn} + c\delta) = 2$ for a constant c, since $s\delta = 0$. In fact, if a distribution T is such that $sT = 0$, then $2\pi is(\mathcal{F}\text{sgn} + T) = 2$, and we showed earlier that such a T *must* be $c\delta$ for some constant c. So, for generality, $2\pi is(\mathcal{F}\text{sgn} + c\delta) = 2$ and we have

$$\mathcal{F}\text{sgn} = \frac{1}{\pi is} - c\delta\,.$$

However, sgn is odd and so is its Fourier transform, and so, too, is $1/2\pi is$, while δ is even. The only way $1/\pi is + c\delta$ can be odd is to have $c = 0$.

To repeat, we have now found

$$\mathcal{F}\text{sgn} = \frac{1}{\pi is}\,.$$

By duality we also now know the Fourier transform of $1/x$:

$$\mathcal{F}\left(\frac{1}{x}\right) = \pi i\,\mathcal{F}\mathcal{F}\text{sgn} = \pi i\,\text{sgn}^{-}\,,$$

and since sgn is odd,

$$\mathcal{F}\left(\frac{1}{x}\right) = -\pi i\,\text{sgn}\,s\,.$$

Fourier transform of the unit step. Having found $\mathcal{F}\text{sgn}$ it's also easy to find the Fourier transform of the unit step H. Indeed,

$$H(t) = \tfrac{1}{2}(1 + \text{sgn}\,t)$$

and from this

$$\mathcal{F}H = \tfrac{1}{2}\left(\delta + \frac{1}{\pi is}\right).$$

You might wonder if it's possible to use $H' = \delta$ to derive the formula for $\mathcal{F}H$ much as we did for $\mathcal{F}\text{sgn}$. Yes, it's possible, but somewhat more involved; see the problems.

4.7. Convolutions and the Convolution Theorem

Convolution, that fundamental operation, has been conspicuously absent. It gets its own section, Act V. Convolution of distributions presents some special problems and we're not going to take this as far as it can go. It's not so hard figuring out formally how to define $S*T$ for distributions S and T; it's setting up conditions under which the convolution exists that's somewhat tricky. This is related to the fact of nature that it's impossible to define (in general) the product of two distributions, for we also want to have a convolution theorem that says $\mathcal{F}(S*T) = (\mathcal{F}S)(\mathcal{F}T)$, and both sides of the formula have to make sense.

What works easily is the convolution of a distribution with a test function. We approach this as in our past work defining distributions, with a little twist. Looking ahead, however much you are concerned with the details, I am pleased to report right away that the convolution theorem on Fourier transforms continues to hold: if ψ is a Schwartz function and T is a tempered distribution, then

$$\mathcal{F}(\psi * T) = (\mathcal{F}\psi)(\mathcal{F}T)\,.$$

The right-hand side is the product of a Schwartz function and a tempered distribution, which *is* defined.

Here's the discussion that supports the development of convolution in this setting. Before trying to convolve two distributions, we first consider how to define the convolution of ψ and T. As in every other case of extending operations from functions to distributions, we begin by supposing that T comes from a function f, and we work with the pairing of $\psi * f$ with a test function φ. This is

$$\begin{aligned}\langle \psi * f, \varphi\rangle &= \int_{-\infty}^{\infty} (\psi * f)(x)\varphi(x)\,dx \\ &= \int_{-\infty}^{\infty}\left(\int_{-\infty}^{\infty}\psi(x-y)f(y)\,dy\right)\varphi(x)\,dx \\ &= \int_{-\infty}^{\infty}\int_{-\infty}^{\infty}\psi(x-y)\varphi(x)f(y)\,dy\,dx \\ &= \int_{-\infty}^{\infty}\left(\int_{-\infty}^{\infty}\psi(x-y)\varphi(x)\,dx\right)f(y)\,dy\,.\end{aligned}$$

The interchange of integration in the last line is justified because every function in sight is as nice as can be.

We *almost* see a convolution $\psi*\varphi$ in the inner integral but the sign is wrong; it's $\psi(x-y)$ instead of $\psi(y-x)$. However, bringing back our notation $\psi^-(x) = \psi(-x)$, we can write the inner integral as the convolution $\psi^- * \varphi$, and then

$$\langle \psi * f, \varphi\rangle = \int_{-\infty}^{\infty}(\psi * f)(x)\varphi(x)\,dx = \int_{-\infty}^{\infty}(\psi^- * \varphi)(x)f(x)\,dx = \langle f, \psi^- * \varphi\rangle\,.$$

This tells us what to do in general:

- If T is a distribution and ψ is a test function, then $\psi * T$ is defined by

$$\langle \psi * T, \varphi\rangle = \langle T, \psi^- * \varphi\rangle\,.$$

Unspoken in this definition is the fact that if ψ and φ are test functions, then so is $\psi^- * \varphi$. Unspoken, but true; see Section 4.8.

4.7.1. Convolving with δ. We considered convolution with δ when δ was defined via a limit. Let's see how the distributional definition works to establish this fundamental property:

$$\psi * \delta = \psi\,.$$

On the right-hand side we regard ψ as a distribution. To check:

$$\begin{aligned}\langle \psi * \delta, \varphi\rangle &= \langle \delta, \psi^- * \varphi\rangle = (\psi^- * \varphi)(0) \\ &= \int_{-\infty}^{\infty}\psi^-(-y)\varphi(y)\,dy = \int_{-\infty}^{\infty}\psi(y)\varphi(y)\,dy = \langle \psi, \varphi\rangle\,.\end{aligned}$$

This says that $\psi * \delta$ has the same outcome when paired with φ as ψ does. Thus $\psi * \delta = \psi$. Works like a charm. Airtight.

As pointed out earlier, it's common practice to write the expressions with the variables and with an integral,

$$\psi(x) = (\psi * \delta)(x) = \int_{-\infty}^{\infty} \delta(x-y)\psi(y)\, dy\,.$$

Generations of distinguished engineers and scientists have written this identity in this way, and no harm seems to have befallen them.

Convolving with a shifted δ. Let's also establish the corresponding result for the shifted δ,

$$\psi * \delta_a = \tau_a \psi\,,$$

where τ_a is the shift or delay operator. It's probably more memorable to write this with the variable as

$$\psi(x) * \delta(x-a) = \psi(x-a)\,.$$

In words, convolving a signal with a shifted δ shifts the signal.

For the derivation,

$$\begin{aligned}\langle \psi * \delta_a, \varphi\rangle &= \langle \delta_a, \psi^- * \varphi\rangle = (\psi^- * \varphi)(a)\\ &= \int_{-\infty}^{\infty} \psi^-(a-y)\varphi(y)\, dy\\ &= \int_{-\infty}^{\infty} \psi(y-a)\varphi(y)\, dy\\ &= \langle \tau_a\psi, \varphi\rangle\,.\end{aligned}$$

We can even think of Fourier inversion as a kind of convolution identity. The inversion theorem is sometimes presented in this way (proved, according to some people, though it's circular reasoning). We need to write (formally)

$$\int_{-\infty}^{\infty} e^{2\pi i s x}\, ds = \delta(x)$$

viewing the left-hand side as the inverse Fourier transform of 1, and then, shifting,

$$\int_{-\infty}^{\infty} e^{2\pi i s x} e^{-2\pi i s t}\, ds = \delta(x-t)\,.$$

Now, shamelessly,

$$\begin{aligned}\mathcal{F}^{-1}\mathcal{F}\varphi(x) &= \int_{-\infty}^{\infty} e^{2\pi i s x}\left(\int_{-\infty}^{\infty} e^{-2\pi i s t}\varphi(t)\, dt\right) ds\\ &= \int_{-\infty}^{\infty}\int_{-\infty}^{\infty} e^{2\pi i s x} e^{-2\pi i s t}\varphi(t)\, dt\, dt\\ &= \int_{-\infty}^{\infty}\left(\int_{-\infty}^{\infty} e^{2\pi i s x} e^{-2\pi i s t}\, ds\right)\varphi(t)\, dt = \int_{-\infty}^{\infty} \delta(x-t)\varphi(t)\, dt = \varphi(x)\,.\end{aligned}$$

At least these manipulations didn't lead to a contradiction! I don't mind if you think of the inversion theorem in this way, as long as you know what's behind it and as long as you don't tell anyone where you saw it.

4.7.2. The convolution theorem. Having come this far, we can now derive the convolution theorem for the Fourier transform:

$$\begin{aligned}
\langle \mathcal{F}(\psi * T), \varphi\rangle &= \langle \psi * T, \mathcal{F}\varphi\rangle = \langle T, \psi^- * \mathcal{F}\varphi\rangle \\
&= \langle T, \mathcal{F}\mathcal{F}\psi * \mathcal{F}\varphi\rangle \quad \text{(using the identity } \mathcal{F}\mathcal{F}\psi = \psi^-\text{)} \\
&= \langle T, \mathcal{F}(\mathcal{F}\psi \cdot \varphi)\rangle \\
&\qquad \text{(for functions the convolution of the Fourier transforms is the} \\
&\qquad \text{Fourier transform of the product)} \\
&= \langle \mathcal{F}T, \mathcal{F}\psi \cdot \varphi\rangle \quad \text{(bringing } \mathcal{F} \text{ back to } T\text{)} \\
&= \langle (\mathcal{F}\psi)(\mathcal{F}T), \varphi\rangle \quad \text{(how multiplication by a function is defined).}
\end{aligned}$$

Comparing where we started and where we ended up,

$$\langle \mathcal{F}(\psi * T), \varphi\rangle = \langle (\mathcal{F}\psi)(\mathcal{F}T), \varphi\rangle\,.$$

That is,

$$\mathcal{F}(\psi * T) = (\mathcal{F}\psi)(\mathcal{F}T)\,.$$

Done. Perfect.

One can also show the dual identity:

$$\mathcal{F}(\psi T) = \mathcal{F}\psi * \mathcal{F}T.$$

Pay attention to how everything makes sense here and has been previously defined. The product of the Schwartz function ψ and the distribution T is defined, and as a tempered distribution ψT has a Fourier transform. Since ψ is a Schwartz function, so is its Fourier transform $\mathcal{F}\psi$, and hence $\mathcal{F}\psi * \mathcal{F}T$ is defined.

I'll leave it to you to check that the algebraic properties of the convolution continue to hold for distributions, whenever all the quantities are defined. However, be warned that associativity can be an issue, as it can be for convolution of functions.[27]

Note that the convolution identities are consistent with $\psi * \delta = \psi$ and with $\psi\delta = \psi(0)\delta$. The first of these convolution identities says that

$$\mathcal{F}(\psi * \delta) = \mathcal{F}\psi\mathcal{F}\delta = \mathcal{F}\psi\,,$$

since $\mathcal{F}\delta = 1$, and that jibes with $\psi * \delta = \psi$. The other identity is a little more interesting. We have

$$\mathcal{F}(\psi\delta) = \mathcal{F}\psi * \mathcal{F}\delta = \mathcal{F}\psi * 1 = \int_{-\infty}^{\infty} 1 \cdot \mathcal{F}\psi(x)\,dx = \mathcal{F}^{-1}\mathcal{F}\psi(0) = \psi(0)\,.$$

This is consistent with $\mathcal{F}(\psi\delta) = \mathcal{F}(\psi(0)\delta) = \psi(0)\mathcal{F}\delta = \psi(0)$.

[27] Most easily illustrated by associativity of multiplication of distributions being an issue. For example, working with the principal value distribution $1/x$ together with δ, we can say $(1/x)\cdot(x\cdot\delta) = 0$ since $x\delta = 0$. But $((1/x)\cdot x)\cdot\delta = \delta$ since, as one can show, $(1/x)\cdot x = 1$ as distributions. So be careful out there.

4.7.3. Convolution in general. I said earlier that convolution can't be defined for every pair of distributions. I want to say a little more about this, but only a little, and give a few examples of cases when it works out.

At the beginning of this section we considered, as we always do, what convolution looks like for distributions in the case when the distribution comes from a function. With f playing the role of the distribution and ψ a Schwartz function we wrote

$$\begin{aligned}\langle \psi * f, \varphi \rangle &= \int_{-\infty}^{\infty} (\psi * f)(x)\varphi(x)\, dx \\ &= \int_{-\infty}^{\infty} \left(\int_{-\infty}^{\infty} \psi(x-y)f(y)\, dy \right) \varphi(x)\, dx \\ &= \int_{-\infty}^{\infty} \int_{-\infty}^{\infty} \psi(x-y)\varphi(x)f(y)\, dy\, dy \\ &= \int_{-\infty}^{\infty} \left(\int_{-\infty}^{\infty} \psi(x-y)\varphi(x)\, dx \right) f(y)\, dy\,.\end{aligned}$$

At this point we stopped and expressed this calculation as the pairing

$$\langle \psi * f, \varphi \rangle = \langle f, \psi^- * \varphi \rangle$$

so that we could see how to define $\psi * T$ when T is a distribution.

This time, and for a different reason, I want to take the last inner integral above one step further and write

$$\int_{-\infty}^{\infty} \psi(x-y)\varphi(x)\, dx = \int_{-\infty}^{\infty} \psi(u)\varphi(u+y)\, du \quad \text{(using the substitution } u = x - y\text{)}.$$

This latter integral is the pairing $\langle \psi(x), \varphi(x+y) \rangle$, where I wrote the variable of the pairing (the integration variable) as x and included it in the notation for pairing to indicate that what results from the pairing is a function of y. In fact, what we see from this is that $\langle \psi * f, \varphi \rangle$ can be written as a nested pairing; namely,

$$\langle \psi * f, \varphi \rangle = \langle f(y), \langle \psi(x), \varphi(x+y) \rangle \rangle\,,$$

where I included the variable y in the outside pairing to keep things straight and to help recall that in the end everything gets integrated away and the result of the nested pairing is a number.

Now, this nested pairing tells us how we might define the convolution $S * T$ of two distributions S and T. With a strong proviso:

Convolution of two distributions. If S and T are two distributions, then their convolution is the distribution $S * T$ defined by

$$\langle S * T, \varphi \rangle = \langle S(y), \langle T(x), \varphi(x+y) \rangle \rangle$$

provided the right-hand side exists.

We've written $S(y)$ and $T(x)$ "at points" to keep straight what gets paired with what; $\varphi(x+y)$ makes sense, it's a function of x and y, and it's necessary to indicate which variable x or y is getting hooked up with T in the inner pairing and then with S in the outer pairing.

Why the proviso? Because the inner paring $\langle T(x), \varphi(x+y)\rangle$ produces a function of y which *might not be a test function.* Sad, but true. One can state some general conditions under which $S * T$ exists, but this requires a few more definitions and a little more discussion.[28] Enough is enough. It can be dicey, but, surprise, we'll be open-minded about the existence of convolution and applications of the convolution theorem. Tell the rigor police to take the day off.

Convolving δ with itself. For various applications you may find yourself wanting to use the identity

$$\delta * \delta = \delta\,.$$

By all means use it. In this case the convolution makes sense and the formula follows:

$$\begin{aligned}\langle \delta * \delta, \varphi\rangle &= \langle \delta(y), \langle \delta(x), \varphi(x+y)\rangle\rangle \\ &= \langle \delta(y), \varphi(y)\rangle = \varphi(0) = \langle \delta, \varphi\rangle\,.\end{aligned}$$

A little more generally, we have

$$\delta_a * \delta_b = \delta_{a+b}\,,$$

a nice formula! We can derive this easily from the definition

$$\begin{aligned}\langle \delta_a * \delta_b, \varphi\rangle &= \langle \delta_a(y), \langle \delta_b(x), \varphi(x+y)\rangle\rangle \\ &= \langle \delta_a(y), \varphi(b+y)\rangle = \varphi(b+a) = \langle \delta_{a+b}, \varphi\rangle\,.\end{aligned}$$

It would be more common to write this identity as

$$\delta(x-a) * \delta(x-b) = \delta(x-a-b)\,.$$

In this notation, here's the down-and-dirty version of what we just did (so you know how it looks):

$$\begin{aligned}\delta(x-a) * \delta(x-b) &= \int_{-\infty}^{\infty} \delta(y-a)\delta(x-b-y)\,dy \\ &= \int_{-\infty}^{\infty} \delta(u-b-a)\delta(x-u)\,du \quad (\text{using } u = b+y) \\ &= \delta(x-b-a)\,.\end{aligned}$$

By the way, look at this formula from the Fourier transform perspective:

$$\mathcal{F}(\delta_{a+b}) = e^{-2\pi i s(a+b)} = e^{-2\pi i s a}e^{-2\pi i s b} = (\mathcal{F}\delta_a)(\mathcal{F}\delta_b).$$

Now take the inverse Fourier transform and turn multiplication into convolution:

$$\delta_{a+b} = \mathcal{F}^{-1}((\mathcal{F}\delta_a)(\mathcal{F}\delta_b)) = \delta_a * \delta_b.$$

One more consistency check. Always satisfying.

[28] It inevitably brings in questions about associativity of convolution, which might not hold in general, as it turns out, and a more detailed treatment of the convolution theorem.

4.8. Appendix: Windowing, Convolution, and Smoothing

The main goal of this section is to learn a little more about convolution as a smoothing operation. I'd call this Act VI, but:

- The essential work of this chapter is complete. From here on out it's extra-curricular.

I bulleted that out of common courtesy. You, the audience, may already be ready for the final curtain and you can head for the exit. At least read the epilog.

We'll be revisiting the maxim we had in Chapter 3:

- The convolution $f * g$ is at least as good a function as f and g are separately, maybe better.

When at least one of f and g is continuously differentiable, the maxim is often used in connection with the derivative theorem for convolutions:

$$(f * g)'(x) = (f * g')(x) = (f' * g)(x).$$

On the left-hand side is the derivative of the convolution, while on the right-hand side we put the derivative on whichever factor *has* a derivative.

There's a raft of results making the maxim precise and using it to deduce theorems on approximations of functions and distributions. Though that raft may be carrying more pure mathematicians than applied consumers of convolutions, it's good for you to have an idea of what's going on and to have some quotable results at the ready. I keep getting asked about these things in class — by your fellow engineers, mind you — so I thought you'd appreciate such a discussion, however abbreviated. The emphasis here is as much on *technique* as it is on specific results and in no case are we going for the utmost generality. I hope you come away with an appreciation for the style of the arguments and for how the ideas are deployed.

4.8.1. Functions of compact support and windows.

We'll often be working with functions of compact support, either as a hypothesis in the result under discussion or as an approximation to make in the course of a derivation. Recall that "compact support" means that the function is identically zero for $|x| \geq A$ for some A. Recall also that $\mathcal{C}$ denotes the set of all functions that are infinitely differentiable and have compact support.

How does that property come up? In proofs, primarily, and there are often three ways it's used, not necessarily all at once:

(1) Integrals from $-\infty$ to ∞ become integrals from $-A$ to A, as in

$$(f * g)(x) = \int_{-\infty}^{\infty} f(y)g(x-y)\,dy = \int_{-A}^{A} f(y)g(x-y)\,dy$$

when, say, f has compact support in $|x| \leq A$.

You are happier integrating over a finite interval.

(2) If f and g have compact support, then so does $f * g$.

You should verify this (as a problem) and in doing so try to determine the relationship between the supports of f and g and that of $f * g$. Look back at the problems for Chapter 3.

(3) A continuous function on a compact interval $|x| \leq A$ assumes a finite maximum (and minimum) on the interval. Thus a continuous function of compact support assumes a maximum (and a minimum).

This is a math fact mentioned back in Section 4.2.2. It's fair to say that this is considered a substantial theorem as theorems on continuous functions go.

Watch how these come in as we proceed. I won't always flag them.

Windowing and windowing smoothly. We've had practical reasons to use functions of compact support and to produce functions of compact support via *windowing*. In Chapter 3, an ideal low-pass filter zeros out the spectrum $\mathcal{F}f(s)$ of a signal $f(x)$ beyond a cutoff frequency $\pm\nu_c$ and passes all the frequencies in between. This is accomplished by multiplying the Fourier transform by a scaled rectangle function, forming $\Pi_{2\nu_c}(s)\mathcal{F}f(s) = \Pi(s/2\nu_c)\mathcal{F}f(s)$. That simple process is called windowing and $\Pi_{2\nu_c}(s)$ is called the window — the only part of $\mathcal{F}f(s)$ that you see is what the window lets through. A problem with windowing by a scaled Π is that it may introduce discontinuities in the cutoff at the edges, and there can be practical consequences of this for the filtered signal.[29]

Here's another way windowing is applied. Physical signals don't go on forever, but a mathematical model might well be a function $f(x)$ defined for $-\infty < x < \infty$. Forming the windowed signal $f_n(x) = \Pi(x/n)f(x)$ cuts off $f(x)$ outside the interval $[-n/2, +n/2]$, and so letting $n \to \infty$ approximates $f(x)$ by a sequence $f_n(x)$ of signals of compact support. One can hope that properties of the windowed signals $f_n(x)$, which might be more accessible, for example, to numerical computation, give insight into properties of $f(x)$. Though again there can be problems with discontinuities at sharp cutoffs.

How can we bring a function *smoothly* down to zero? Here's a model for doing this, sort of a smoothed version of the rectangle function. It's amazing that you can write it down, and if you are ever looking for smooth windows, here's one way to get them.[30]

First, the function

$$g(x) = \begin{cases} 0\,, & x \leq 0\,, \\ \exp\left(-\left(\frac{1}{2x}\right)\exp\left(\frac{1}{2x-1}\right)\right), & 0 < x < \frac{1}{2}\,, \\ 1\,, & x \geq \frac{1}{2}, \end{cases}$$

is smooth, i.e., infinitely differentiable! It goes from the constant value 0 to the constant value 1 smoothly on the interval from 0 to 1/2. Here's the plot.

---

[29] Search "ringing".

[30] Amazing, yes, but I have to admit that I've never seen this particular window used in practice. Be the first?

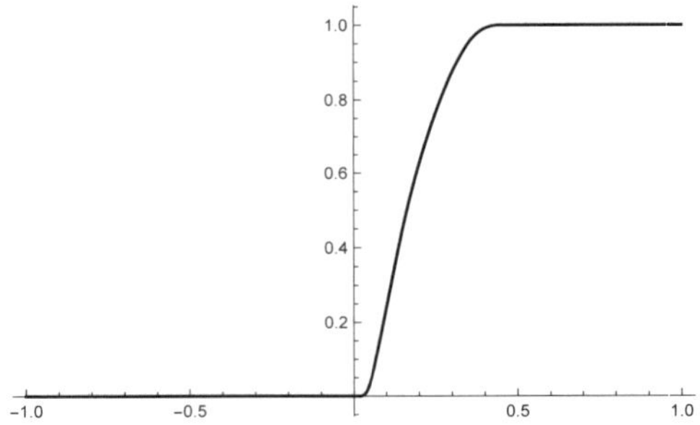

Next, the function $g(1+x)$ goes up smoothly from 0 to 1 over the interval from -1 to $-1/2$ and the function $g(1-x)$ goes down smoothly from 1 to 0 over the interval from $1/2$ to 1. Finally, their product

$$c(x) = g(1+x)g(1-x)$$

is 1 on the interval from $-1/2$ to $1/2$, goes down smoothly to 0 between $\pm 1/2$ and ± 1, and is zero for $x \leq -1$ and for $x \geq 1$. Here's the graph of $c(x)$, which is the one we had earlier in the chapter. Mathematicians refer to it as a *bump function.*

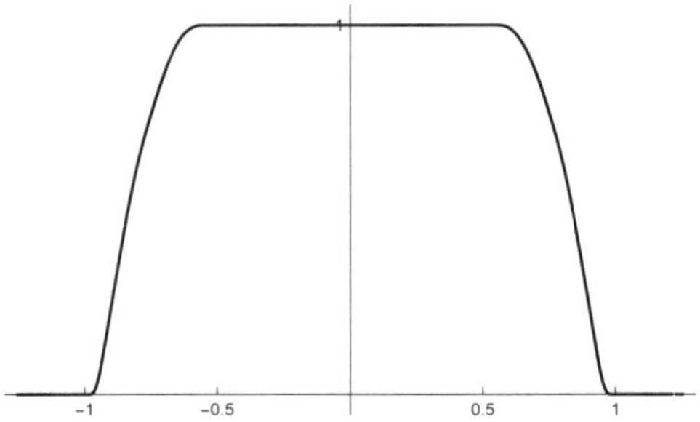

Scaling to $c(x/\alpha)$ and forming $c(x/\alpha)f(x)$ smoothly cuts off a function to be zero outside the interval $[-\alpha, \alpha]$. Letting α increase serves to approximate a general (smooth) function by a smooth function with compact support.

A picture might help. Here's an interesting looking function, $f(x)$, that I cooked up (not periodic, by the way) together with its smooth windowed version $c(x/3)f(x)$ set to be identically 0 after ± 3:

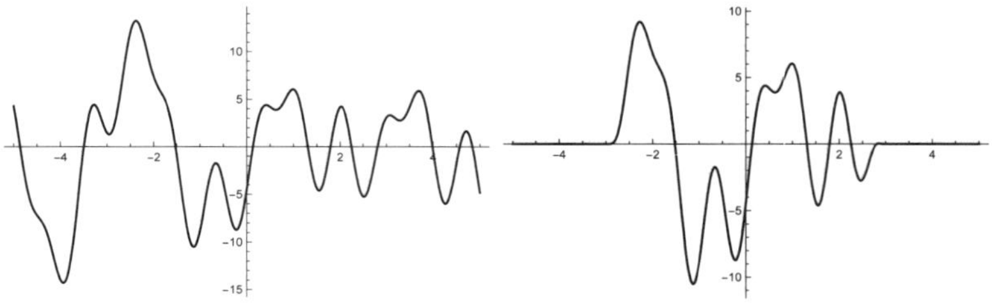

4.8.2. Smoothing and approximation via convolution. Convolution provides a sophisticated and powerful approach to smoothing and approximation. We'll use the smooth window $c(x)$ and then a scaled version to give some model results.

Here's the first one:

- If $f(x)$ has compact support and is k-times continuously differentiable for some $k \geq 0$, then $(c * f)(x)$ is in $\mathcal{C}$.

So a differentiable function (or even just a continuous function if $k = 0$) becomes infinitely differentiable when convolved with $c(x)$, as smooth as you'd want. A common notation for the functions $f(x)$ named in the hypothesis is $C_c^k(\mathbb{R})$. When $k = 0$, the notation is usually simplified to $C_c(\mathbb{R})$.

As remarked earlier, the convolution of two functions with compact support has again compact support, so that aspect (which is one requirement of being in $\mathcal{C}$) is taken care of. I should also note that the integral defining the convolution exists in the first place. For suppose the support of $f(x)$ is $|x| \leq A$. Then

$$(c * f)(x) = \int_{-\infty}^{\infty} f(y)c(x-y)\,dy = \int_{-A}^{A} f(y)c(x-y)\,dy.$$

The integrand is continuous and bounded on the interval and all is well with the integral.

Now we want to differentiate under the integral sign, essentially verifying the derivative formula:

$$\begin{aligned}(f * c)'(x) &= \frac{d}{dx}\int_{-A}^{A} f(y)c(x-y)\,dy \\ &= \int_{-A}^{A} f(y)\frac{d}{dx}c(x-y)\,dy = \int_{-A}^{A} f(y)c'(x-y)\,dy = (f * c')(x)\,.\end{aligned}$$

We did this in the last chapter but without justification (and without stating hypotheses). In the present situation it is justified because $f(x)$ is continuous and $c(x)$ is differentiable with a continuous derivative. The integrand $f(x)c(x-y)$ then satisfies the hypotheses of the Lebesgue dominated convergence theorem, which we encountered briefly in Section 4.3, and that's what allows us to pull the derivative inside the integral. That's what to say if you are confronted by a mathematician.[31] In fact, since $c \in \mathcal{C}$, one can continue justifying and differentiating, and

$$(f * g)^{(n)}(x) = (f * g^{(n)})(x)$$

for any n.

Looking back on the argument, we only used that $c(x)$ is in $\mathcal{C}$. Convolving $f(x)$ with any function in $\mathcal{C}$ will do to smooth it, but using $c(x)$ ties in with the approximation result to follow.

You'll appreciate the next result on approximation more if you go back and read the discussion of approximating δ in Section 4.4.1. To make the tie-in, first scale $c(x)$ to have area 1 by putting

$$\tilde{c}(x) = \frac{c(x)}{\int_{-\infty}^{\infty} c(y)\,dy}\,.$$

[31] The derivative is the limit of a difference quotient, and the Lebesgue dominated convergence is about taking limits inside and outside the integral. One technique to learn is the fine art of quoting theorems to justify an argument. Takes some care.

Now set

$$\tilde{c}_\alpha(x) = \frac{1}{\alpha}\tilde{c}\left(\frac{x}{\alpha}\right).$$

From our work with δ we know

$$\lim_{\alpha\to 0}\int_{-\infty}^{\infty} f(x)\tilde{c}_\alpha(x)\,dx \to f(0).$$

A more complete result is:

- If $f(x)$ has compact support and is k-times continuously differentiable for some $k \geq 0$, then for $0 \leq l \leq k$ the functions $(\tilde{c}_\alpha * f)^{(l)}(x)$ converge to $f^{(l)}(x)$ as $\alpha \to 0$.

The first result on smoothing, above, holds with $\tilde{c}_\alpha(x)$ replacing $c(x)$, and it follows that $(\tilde{c}_\alpha * f)(x)$ is in $\mathcal{C}$ with

$$(\tilde{c}_\alpha * f)^{(l)}(x) = \int_{-\infty}^{\infty} f(y)\tilde{c}_\alpha^{(l)}(x-y)\,dy.$$

The claim is that this converges to $f^{(l)}(x)$ as $\alpha \to 0$. In fact, the convergence is *uniform*, as a consequence of compact support. We encountered uniform convergence in studying (briefly!) convergence of Fourier series in Section 1.8. We won't make a show of it here, but, invoking the word, the result says that via convolution the function $f(x)$ and its derivatives can be *uniformly approximated* by functions in $\mathcal{C}$.[32] That's what you would say to a mathematician. And the mathematician would nod and say to you, in mathspeak:

- $\mathcal{C}$ is *dense* in $C_c^k(\mathbb{R})$.

Here's a sketch of the proof. It's very much like the argument in Section 4.4.1, but simplified somewhat because of the hypotheses of compact support. Start with

$$(f*\tilde{c}_\alpha)(x) = \int_{-\infty}^{\infty} f(y)\tilde{c}_\alpha(x-y)\,dy = \int_{-\infty}^{\infty} f(x-\alpha u)\tilde{c}(u)\,du = \int_{-1}^{1} f(x-\alpha u)\tilde{c}(u)\,du,$$

making the substitution $u = (x-y)/\alpha$ and using that the support of $\tilde{c}(x)$ is $|x| \leq 1$. With $\int_{-\infty}^{\infty}\tilde{c}(u)\,du = 1$ we can further say

$$\begin{aligned}|(f*\tilde{c}_\alpha)(x) - f(x)| &= \left|\int_{-1}^{1}(f(x-\alpha u) - f(x))\tilde{c}(u)\,du\right| \\ &\leq \int_{-1}^{1}|f(x-\alpha u) - f(x)||\tilde{c}(u)|\,du.\end{aligned}$$

And even more, we're justified in differentiating under the integral to say

$$\frac{d^l}{dx^l}(f*\tilde{c}_\alpha)(x) = \int_{-1}^{1}\frac{d^l}{dx^l}f(x-\alpha u)\tilde{c}(u)\,du,$$

the change of variables allowing us to put the derivative on f. Then

$$\left|\frac{d^l}{dx^l}(f*\tilde{c}_\alpha)(x) - \frac{d^l}{dx^l}f(x)\right| = \int_{-1}^{1}\left|\frac{d^l}{dx^l}f(x-\alpha u) - \frac{d^l}{dx^l}f(x)\right||\tilde{c}(u)|\,du.$$

---

[32]The graphs of $\tilde{c}_\alpha * f$ and all its derivatives stay close to the graphs of f and all its derivatives.

Without pressing the details, the hypothesis of compact support on $f(x)$ allows one to be certain that the integrand tends uniformly to 0 as $\alpha \to 0$; boundedness and uniform continuity of $f(x)$ and its derivatives do the trick. So

$$\lim_{\alpha \to 0} \left| \frac{d^l}{dx^l}(f * \tilde{c}_\alpha)(x) - \frac{d^l}{dx^l} f(x) \right| = 0 ,$$

uniformly in x.

Approximating integrable functions. It may be surprising, but it's comforting and useful to know that such an approximation result can be strengthened to approximating just *integrable* functions, and integrable functions can be pretty rough. In fact, let's jump right to $L^p(\mathbb{R})$, $1 \leq p < \infty$, the functions for which

$$\int_{-\infty}^{\infty} |f(x)|^p \, dx < \infty .$$

The L^p-norm is

$$\|f\|_p = \left\{ \int_{-\infty}^{\infty} |f(x)|^p \, dx \right\}^{\frac{1}{p}} .$$

The result is:

- $\mathcal{C}$ is dense in $L^p(\mathbb{R})$ for any $1 \leq p < \infty$.
  Meaning, if $f(x)$ is in $L^p(\mathbb{R})$ and $\epsilon > 0$, then there is a function $g(x)$ in $\mathcal{C}$ with $\|f - g\|_p < \epsilon$.

I'm not going to give the details, but think in terms of two steps. First, a function in $L^p(\mathbb{R})$ can be arbitrarily well approximated by a step function with compact support. Like so:

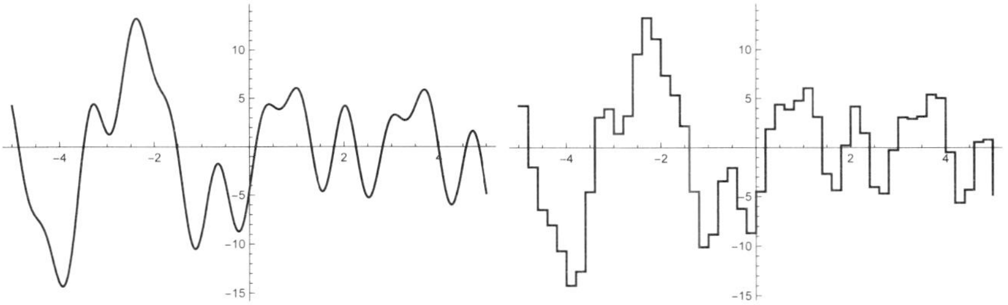

As you and your friends would say, it's approximation via zero-order hold. You were prescient enough to work with step functions back in Section 4.3 on integrals. Second, such a step function, which typically has jump discontinuities, can be smoothed out, via a convolution, and you're on your way.

The proofs for approximating integrable functions are in the same spirit as the arguments for approximating C^k-functions (convolution again), but there are differences. Here "approximated by ..." means we have to estimate the L^p-norm of the difference between the function and its approximation. In the C^k-approximation result we estimated the difference in the values of the function and its approximation.[33]

Finally, note that since $\mathcal{C} \subset \mathcal{S} \subset L^p(\mathbb{R})$ and $\mathcal{C}$ is already dense in $L^p(\mathbb{R})$, so, too, is the bigger set of functions:

- $\mathcal{S}$ is dense in $L^p(\mathbb{R})$ for any $1 \leq p < \infty$.

The less precise headline: Integrable functions can be approximated by Schwarz functions.

4.8.3. The Riemann-Lebesgue lemma. Way back at the beginning of this chapter, in Section 4.1.1, I stated a famous result called the Riemann-Lebesgue lemma on integrability of a function and decay of its Fourier transform:

- If $\displaystyle\int_{-\infty}^{\infty} |f(t)|\, dt < \infty$, then $|\mathcal{F}f(s)| \to 0$ as $s \to \pm\infty$.

I showed at the time that $\mathcal{F}f$ is continuous given that f is integrable, and that was pretty easy. It's a much stronger statement to say that $\mathcal{F}f$ tends to zero at infinity. We'll derive that result from the approximation results above.

Let f be in $L^1(\mathbb{R})$ and choose a sequence of functions f_n in $\mathcal{S}$ so that

$$\|f - f_n\|_1 = \int_{-\infty}^{\infty} |f(t) - f_n(t)|\, dt < \tfrac{1}{n}\,.$$

We can do that because $\mathcal{S}$ is dense in $L^1(\mathbb{R})$. Then use the earlier result that the Fourier transform of a function is bounded by the $L^1(\mathbb{R})$-norm of the function, so that

$$|\mathcal{F}f(s) - \mathcal{F}f_n(s)| \leq \|f - f_n\|_1 < \tfrac{1}{n}\,.$$

Therefore

$$|\mathcal{F}f(s)| \leq |\mathcal{F}f_n(s)| + \tfrac{1}{n}\,.$$

But since f_n is in $\mathcal{S}$, so is $\mathcal{F}f_n$, and hence $\mathcal{F}f_n(s)$ tends to zero as $s \to \pm\infty$. Thus

$$\lim_{s\to\infty} |\mathcal{F}f(s)| < \tfrac{1}{n}$$

for all $n \geq 1$. Now let $n \to \infty$. Pretty slick.

Incidentally, compare this to the situation for the Fourier series

$$\sum_{n=-\infty}^{\infty} \hat{f}(n) e^{2\pi i n t}$$

of a function $f(x)$ in $L^2([0,1])$. As a consequence of Bessel's inequality (see the paragraph on it in Section 1.7) we observed that $|\hat{f}(n)|^2 \to 0$ as $n \to \infty$. That's a much easier result than the Riemann-Lebesgue lemma.

[33] The different notions of convergence, or approximation, for different spaces of functions is one of the things you learn about in a course on mathematical analysis, which is not this course. A mathematician would say that different notions of convergence are different *topologies* on the spaces. Here it's the difference between the L^p-norm and the supremum norm.

4.8.4. Convolution and Schwartz functions. What about those best functions for the Fourier transform? Here's an example of the maxim involving an OK class for Fourier transforms and the best class.

- If $f(x)$ is in $L^1(\mathbb{R})$ and has compact support and if $g(x)$ is a Schwartz function, then $f * g$ is a Schwartz function.

A function that is only assumed to be integrable (about the minimum requirement for the Fourier transform) can be pretty far from continuous, never mind infinitely differentiable, and Schwartz functions are pretty great, so convolution has done a lot for you here. Note, however, that we do assume that f has compact support. The result is not true if we drop that assumption.

Let's show this as a further exercise in estimating and using the definition(s) of Schwartz functions. As above, I'd like this to serve as a helpful guide. You might just need this kind of know-how sometime.

There are two aspects, first that $f * g$ is infinitely differentiable and second that it decays at $\pm\infty$ in the way required of a Schwartz function. The first point is just as before: it's justifiable to differentiate under the integral sign,

$$(f * g)^{(n)}(x) = \int_{-\infty}^{\infty} f(y) g^{(n)}(x-y)\, dy = (f * g^{(n)})(x)\,.$$

Just say: "Lebesgue dominated convergence theorem."

What about the decay at $\pm\infty$? We're going to use one of the alternate characterizations of Schwartz functions from Section 4.2.2. Because g is a Schwartz function, for any nonnegative integers m, n there is a constant C_{mn} such that for all x,

$$(1 + |x|)^m \left| \frac{d^n}{dx^n} g(x) \right| \le C_{mn}\,.$$

Fix an m and n in the rest of the argument.

We've assumed that f has compact support, so say that it's zero for $|x| > A$. Then

$$\begin{aligned}
\left| \frac{d^n}{dx^n}(f * g)(x) \right| &= \left| \int_{-\infty}^{\infty} f(y) \frac{d^n}{dx^n} g(x-y)\, dx \right| \\
&= \left| \int_{-A}^{A} f(y) \frac{d^n}{dx^n} g(x-y)\, dx \right| \\
&\le \int_{-A}^{A} |f(y)| \left| \frac{d^n}{dx^n} g(x-y) \right| dx \\
&\le \int_{-A}^{A} |f(y)| \frac{C_{mn}}{(1 + |x-y|)^m)}\, dx\,.
\end{aligned}$$

In the last integral y is between $-A$ and A, and we want to know first (and foremost) what happens to $(d/dx)^n (f * g)(x)$ as $|x| \to \infty$. So suppose $|x| \ge 2A$. Then using[34] $|x - y| \ge |x| - |y|$,

$$1 + |x-y| \ge 1 + |x| - |y| \ge 1 + |x| - \frac{|x|}{2} = 1 + \frac{|x|}{2}\,.$$

[34]If it's been awhile: $|x| = |y + (x-y)| \le |y| + |x-y|$.

Plug this into the estimate:

$$\begin{aligned}\left|\frac{d^n}{dx^n}(f*g)(x)\right| &\leq \int_{-A}^{A} |f(y)| \frac{C_{mn}}{(1+|x-y|)^m}\, dx \\ &\leq \frac{C_{mn}}{(1+\frac{|x|}{2})^m} \int_{-A}^{A} |f(y)|\, dy \\ &= \frac{2^m C_{mn}}{(2+|x|)^m} \int_{-A}^{A} |f(y)|\, dy \\ &\leq \frac{2^m C_{mn}}{(1+|x|)^m} \int_{-A}^{A} |f(y)|\, dy = \frac{2^m C_{mn}\|f\|_1}{(1+|x|)^m}\,.\end{aligned}$$

So for $|x| \geq 2A$,

$$(1+|x|)^m \left|\frac{d^n}{dx^n}(f*g)(x)\right| \leq 2^m C_{mn}\|f\|_1 = C'_{mn}\,.$$

And then for $|x| \leq 2A$ the continuous function $(1+|x|)^m \left|\frac{d^n}{dx^n}(f*g)(x)\right|$ is bounded by some other constant, say C''_{mn}. We conclude that *for all* x

$$(1+|x|)^m \left|\frac{d^n}{dx^n}(f*g)(x)\right| \leq \max\{C'_{mn}, C''_{mn}\}\,.$$

This shows that $(f*g)(x)$ is a Schwartz function. Cool.

The meta-lesson here is to split the estimates: for one range of values of x ($|x| \geq 2A$) you use one way of estimating and for the other range ($|x| < 2A$) you use another. Mathematical analysts have internalized this kind of argument, and it comes up frequently, but they forget to share their internal life with others.

Convolution of two Schwartz functions. It doesn't follow from what we've done so far that the convolution of two Schwartz functions is again a Schwartz function! That's a true statement, but an appeal to any of the previous results assumes that one of the factors in the convolution has compact support and that's not necessarily true of Schwartz functions. However, we can deduce what we want by invoking the best-functions-for-the-Fourier-transform property of $\mathcal{S}$. Here we use that if f and g are Schwartz functions, then the classical convolution theorem holds. I didn't show this, but every step in the derivation in the classical case is valid (the derivation that motivated the definition of convolution at the beginning of the last chapter), and $\mathcal{F}(f*g) = \mathcal{F}f\,\mathcal{F}g$. Both $\mathcal{F}f$ and $\mathcal{F}g$ are Schwartz functions and so is their product. Then Fourier inversion implies that $f*g = \mathcal{F}^{-1}(\mathcal{F}f\,\mathcal{F}g)$ is again a Schwartz function.

That's fine, but if you'll indulge my wish for you to learn some techniques in making estimates, there's a more direct way of getting this. It's elementary but tricky. Proceed wth caution.[35]

[35] I can't remember where I saw this proof — it's not mine — but I liked its directness and cleverness.

Certainly the convolution of two Schwartz functions is infinitely differentiable, and

$$\frac{d^n}{dx^n}(f*g)(x) = \int_{-\infty}^{\infty} f(y)\frac{d^n}{dx^n}g(x-y)\,dy\,.$$

The issue is the decay. Let m and n be nonnegative integers, fixed in the following. We'll use the same alternate characterization of Schwartz functions as in the preceding argument to estimate the integral to show that, with

$$(1+|x|)^m\left|\frac{d^n}{dx^n}(f*g)(x)\right| \le (1+|x|)^m\int_{-\infty}^{\infty}\left|f(y)\frac{d^n}{dx^n}g(x-y)\right|\,dy\,,$$

the right-hand integral remains bounded for all x by a constant depending on m and n.

The characterization of Schwartz functions allows us to introduce some free parameters, so to speak, that we'll adjust at the end. This is the meta-lesson of this derivation, the ability to twiddle parameters. Learning to take advantage of such a possibility is a slowly acquired skill.

For any k there is a C_k such that

$$|f(y)| \le C_k(1+|y|)^{-k}$$

and for any l there is a C'_{ln} such that

$$\left|\frac{d^n}{dx^n}g(x-y)\right| \le C'_{ln}(1+|x-y|)^{-l}\,.$$

The "free parameters" are k and l. For the integrand,

$$|f(y)|\left|\frac{d^n}{dx^n}g(x-y)\right| \le (1+|y|)^{-k}(1+|x-y|)^{-l}C_kC'_{ln}\,,$$

and then

$$(1+|x|)^m|f(y)|\left|\frac{d^n}{dx^n}g(x-y)\right| \le (1+|x|)^m(1+|y|)^{-k}(1+|x-y|)^{-l}C_kC'_{ln}\,.$$

Now we need a little magic. We need to turn that $(1+|x-y|)$ factor into products of $1+|x|$ with $1+|y|$. You wouldn't think that anything magical could be done with the absolute value, but watch. The triangle inequality is $|x+y| \le |x|+|y|$, and from this

$$1+|x+y| \le 1+|x|+|y| \le 1+|x|+|y|+|x|\,|y| = (1+|x|)(1+|y|)\,.$$

Change y to $-y$ to write instead $1+|x-y| \le (1+|x|)(1+|y|)$, and apply this with x replaced by $x+y$:

$$1+|x| = 1+|(x+y)-y| \le (1+|x+y|)(1+|y|) \quad \text{or} \quad (1+|x|)(1+|y|)^{-1} \le 1+|x+y|\,.$$

Putting the upper and lower bounds together,

$$(1+|x|)(1+|y|)^{-1} \le 1+|x+y| \le (1+|x|)(1+|y|)\,,$$

and again replacing y by $-y$ finally produces

$$(1+|x|)(1+|y|)^{-1} \le 1+|x-y| \le (1+|x|)(1+|y|)\,.$$

Nothing but the triangle inequality!

We now use the left inequality to say

$$(1+|x-y|)^{-l} \le (1+|x|)^{-l}(1+|y|)^{l},$$

from which

$$\begin{aligned}(1+|x|)^m(1+|y|)^{-k}&(1+|x-y|)^{-l}C_kC'_{ln}\\ &\le (1+|x|)^m(1+|y|)^{-k}(1+|x|)^{-l}(1+|y|)^{l}C_kC'_{ln}\\ &= (1+|x|)^{m-l}(1+|y|)^{l-k}C_kC'_{ln}.\end{aligned}$$

Now choose $l = m$ and then choose k so that $l-k < -1$, i.e., so that $k > l+1 = m+1$. We then have

$$(1+|x|)^m \int_{-\infty}^{\infty} \left| f(y)\frac{d^n}{dx^n}g(x-y)\right| dy \le C_kC'_{ln}\int_{-\infty}^{\infty}(1+|y|)^{l-k}\,dy,$$

and that integral is finite because $l-k<-1$. The bound depends only on m and n. Done!

Regularizing a distribution. The technique of smoothing by convolution also applies to distributions, to $\psi * T$, for a test function ψ and a distribution T. The key observation is that, working with the definition of convolution, $\psi * T$ can be regarded as a *function itself* (not just a distribution) and it is smooth. The basic quotable result, which I won't prove, is:

- If $\psi \in \mathcal{C}$ and T is a distribution, then $\psi * T$ is an infinitely differentiable function.

The function $\psi * T$ is called a *regularization* of T, another overused term but it's the word that's used. Here's the function that $\psi * T$ is:

$$(\psi * T)(x) = \langle T(y), \psi(x-y)\rangle.$$

The use of the variables in the pairing means, informally, that T gets paired in the y-variable with $\psi(x-y)$, leaving a function of the variable x. I will leave it to you to fit this to how we defined $\psi * T$ earlier.

The catch is that $\psi * T$ may not have compact support, and having compact support is a desirable property in many cases. Nevertheless, the result as it stands is the basis for the important approximation theorem from Section 4.4.4.

- If T is any distribution, then there are test functions ψ_n such that T_{ψ_n} converges to T.

Here's a sketch of the proof when the test functions are $\mathcal{C}$. The result is also true for Schwartz functions.

Take a test function φ with $\int_{-\infty}^{\infty}\varphi(x)\,dx = 1$. Then the functions $\varphi_n(x) = n\varphi(nx)$, $n = 1, 2, \ldots$, induce distributions T_{φ_n} that converge to δ; refer back to Section 4.4.1 on approximating δ. So we have

$$\varphi_n * T \to \delta * T = T \quad \text{as } n \to \infty, \text{ as distributions.}$$

By the preceding result $\varphi_n * T$ is a smooth function. It doesn't necessarily have compact support, so a modification is needed to get a sequence of distributions on $\mathcal{C}$ that converge to T.

Choose a smooth window $\eta(x)$ which is identically 1 for $|x| < 1$, goes down smoothly to 0, and set

$$\Phi_n(x) = \eta(x/n)(\varphi_n * T)(x), \quad n = 1, 2, \dots .$$

Then the Φ_n are in $\mathcal{C}$ and induce distributions T_{Φ_n}. Let ψ be a test function in $\mathcal{C}$. Now $\eta(x/n) = 1$ for $|x| < n$, and so from some n onward $\eta(x/n)\psi(x) = \psi(x)$ (ψ has compact support). Thus

$$\begin{aligned}\langle T_{\Phi_n}, \psi\rangle &= \int_{-\infty}^{\infty} \Phi_n(x)\psi(x)\,dx \\ &= \int_{-\infty}^{\infty} \eta(x/n)(\varphi_n * T)(x)\psi(x)\,dx = \int_{-\infty}^{\infty} (\varphi_n * T)(x)\psi(x)\,dx\,.\end{aligned}$$

Because $\varphi_n * T \to T$ as distributions, as we saw above, we conclude that $T_{\Phi_n} \to T$.

One more comment on δ. OK, I can believe that you've had enough of these arguments, though there's something invigorating in seeing in play so many of the ideas that we've talked about. One final comment here. The distribution δ is the break-even point for smoothing by convolution — it doesn't do any smoothing; it leaves the function alone, as in

$$f * \delta = f\,.$$

Going further, convolving a differentiable function with derivatives of δ produces derivatives of the function, for example,

$$f * \delta' = f'\,.$$

You can derive this from scratch using the definition of the derivative of a distribution and the definition of convolution or you can also think of

$$f * \delta' = (f * \delta)' = f' * \delta = f'\,.$$

A similar result holds for higher derivatives:

$$f * \delta^{(n)} = f^{(n)}\,.$$

Sometimes one thinks of taking a derivative as making a function less smooth, so counterbalancing the maxim that convolution is a smoothing operation, one should add that convolving with derivatives of δ may roughen a function up.

4.8.5. Whatever happened to L^2? Students in the class also asked about this. Remember the central, organizing role for Fourier series played by the square integrable functions? The key ideas were orthogonality and orthogonal projection. Do those ideas also play a role for the Fourier transform? Sort of, but not really in this class. It's more a math class thing.

To recall, the inner product of two functions f and g in $L^2([0, 1])$, upon which everything is based, is

$$(f, g) = \int_0^1 f(t)\overline{g(t)}\,dt\,.$$

By analogy to geometric vectors, orthogonality of functions is *defined* to be $(f, g) = 0$, and the key calculation is the orthogonality (actually orthonormality) of the complex exponentials: for $n, m = 0, \pm 1, \pm 2, \ldots$,

$$(e^{2\pi int}, e^{2\pi imt}) = \int_0^1 e^{2\pi int} e^{-2\pi imt}\, dt = \delta_{ij} = \begin{cases} 1, & n = m\,, \\ 0, & n \neq m\,. \end{cases}$$

This is a straightforward calculation, nothing fancy. Things get fancy when Lebesgue integration makes a triumphant entrance. The theorem is that

$$e_n(t) = e^{2\pi int}, \quad n = 0, \pm 1, \pm 2, \ldots,$$

are an orthonormal basis of $L^2([0,1])$ and that the Fourier series of $f(t)$ in $L^2([0,1])$,

$$f(t) = \sum_{n=-\infty}^{\infty} \hat{f}(n) e^{2\pi int} = \sum_{n=-\infty}^{\infty} (f, e_n) e^{2\pi int}, \quad \hat{f}(n) = (f, e_n) = \int_0^1 f(t) e^{-2\pi int}\, dt\,,$$

is the expansion of $f(t)$ in that basis.[36] Pretty much none of this goes through for $L^2(\mathbb{R})$.

Now, the definition of the inner product does go through: if $f(x)$ and $g(x)$ are in $L^2(\mathbb{R})$, then $f(x)g(x)$ and $f(x)\overline{g(x)}$ are in $L^1(\mathbb{R})$ and

$$\|fg\|_1 \leq \|f\|\|_2 \|g\|_2\,,$$

a result still known as the Cauchy-Schwarz inequality.[37] That means you can talk about orthogonality, orthonormal bases, etc. But the complex exponentials are not square integrable and the Fourier integral

$$\int_{-\infty}^{\infty} f(x) e^{-2\pi isx}\, dx$$

does not exist for your typical L^2-function. We do allow ourselves to write

$$\int_{-\infty}^{\infty} e^{2\pi ixt} e^{-2\pi iyt}\, dt = \delta(x - y),$$

with the Dirac δ, and I guess you can cling to this. But I wouldn't cling too tightly.

It is nonetheless possible to develop a Fourier transform for L^2-functions. Mathematicians love this, while engineers are less moved. Physicists have more than a passing interest through connections to quantum mechanics. Briefly, the easiest approach is to use the fact that the Schwartz functions are dense in $L^2(\mathbb{R})$. Then if $f(x)$ is in $L^2(\mathbb{R})$, we can find a sequence $f_n(x)$ of Schwartz functions with $f_n \to f$ in L^2, meaning that $\|f - f_n\|_2 \to 0$ as $n \to \infty$. Take m and n large, and apply Parseval's identity to write

$$\|\mathcal{F}f_n - \mathcal{F}f_m\|_2 = \|\mathcal{F}(f_n - f_m)\|_2 = \|f_n - f_m\|_2 \leq \|f_n - f\|_2 + \|f - f_m\|_2\,.$$

---

[36] I feel obliged to remind you that "=" means in L^2, that is

$$\lim_{N \to \infty} \left\| f(t) = \sum_{n=-N}^{N} \hat{f}(n) e^{2\pi int} \right\|_2 = 0\,.$$

[37] A more general statement is Hölder's inequality: if $f(x)$ is in $L^p(\mathbb{R})$ and $g(x)$ is in $L^q(\mathbb{R})$ where $\frac{1}{p} + \frac{1}{q} = 1$, then $f(x)g(x)$ is in $L^1(\mathbb{R})$ and $\|fg\|_1 \leq \|f\|_p \|g\|_q$.

The right-hand side tends to 0 as $m, n \to \infty$. This implies that the sequence of Fourier transforms $\mathcal{F}f_n$ converges to a function F in $L^2(\mathbb{R})$.[38] Now use this procedure to *define*

$$\mathcal{F}f = F\,.$$

That's the Fourier transform on L^2. What do you know — the solution to a problem became a definition! One still has Parseval's identity,

$$\|\mathcal{F}f\|_2 = \|f\|_2\,,$$

sometimes, in this context, called *Plancherel's theorem.*

4.9. Epilog and Some References

Well, I hope you enjoyed the show. Further commentary is available from many sources. Reading the expository chapter on distributions in the book *Some Points of Analysis and Their History* by the mathematician L. Gårding is like turning to a knowledgeable critic. Let me offer a few quotes:

> The book *Theorie des distributions* by Laurent Schwartz (1951), now one of the nonread classics of mathematics, has transformed many branches of analysis, and the theory is familiar to every student who ever took an advanced mathematics course in analysis.

...

> But the most important legacy of the theory of distributions is that it freed Fourier analysis from the prison of absolute and square integrability where it was kept before the war by the classical treatments, for instance Titchmarsh's book (1937) on Fourier integrals and Zygmund's *Trigonometrical Series* (1935).

...

> At the time the theory of distributions got a rather lukewarm and sometimes even hostile reception among mathematicians. Analysts of an older school could joke that "Your distributions may be all right, but you are only really happy when you find a function."

Two short books that I think are particularly helpful are:

F. G. Friedlander and M. Joshi, *Introduction to the Theory of Distributions*,

R. Strichartz, *A Guide to Distribution Theory and Fourier Transforms.*

Schwartz's book *Mathematics for the Physical Sciences* has an enormous amount of material maybe intended more for (French) physicists. A. Zemanian's *Distribution Theory and Transform Analysis* is also quite extensive and is pitched more toward mathematically oriented engineers. Lots more out there.

Exeunt all. Curtain.

---

[38]The result we're using, an important one, is that the sequence of functions $\mathcal{F}f_n$ is a *Cauchy sequence* and that a Cauchy sequence in $L^2(\mathbb{R})$ converges to a function in $L^2(\mathbb{R})$. Once again, it's a math-class thing.

Problems and Further Results

4.1. *The Fourier transform of the* sinc *function*

We claimed the result

$$\mathcal{F}\operatorname{sinc}(s) = \Pi(s)$$

on the basis of an uncritical use of duality, namely

$$\mathcal{F}\operatorname{sinc} = \mathcal{F}(\mathcal{F}\Pi) = \Pi^- = \Pi.$$

We say "uncritical" because, as has been our custom, we proceeded without attention to convergence of the Fourier integrals, in this case

$$\mathcal{F}\operatorname{sinc}(s) = \int_{-\infty}^{\infty} e^{-2\pi i s t}\operatorname{sinc} t\, dt = \int_{-\infty}^{\infty} e^{-2\pi i s t}\frac{\sin \pi t}{\pi t}\, dt.$$

In fact,

$$\int_{-\infty}^{\infty} \left|\frac{\sin \pi t}{\pi t}\right| dt = \infty,$$

so there's certainly an issue in defining $\mathcal{F}\operatorname{sinc}$. But not to worry! The result is correct, the Fourier integral *can* be evaluated, and in this problem we'll see how to do it. In fact, we will calculate that

$$\int_{-\infty}^{\infty} e^{-2\pi i s t}\frac{\sin \pi t}{\pi t}\, dt = \begin{cases} 1, & |s| < \frac{1}{2}, \\ \frac{1}{2}, & |s| = \frac{1}{2}, \\ 0, & |s| > \frac{1}{2}. \end{cases}$$

Aha! This reignites the battle over how $\Pi(t)$ should be defined at $t = \pm 1/2$. I concede.

The subject of Fourier analysis is full of tricks to evaluate particular integrals that come up, the result of mathematicians and other Zen masters of calculus devising ad hoc arguments to justify the formulas that engineers and scientists were always happy to use. Probably the most common approach to the sinc integral is via complex functions, contour integration, and the residue theorem. If you know this material, then a reference is the book *Complex Analysis* by Lars Ahlfors. However, another approach is also possible, and the particular arrangement of the method presented here is from D. Kammler's *A First Course in Fourier Analysis*.

We're actually going to pass over one point and assume that the Fourier integral can be taken as a symmetric limit, as in

$$\int_{-\infty}^{\infty} e^{-2\pi i s t}\frac{\sin \pi t}{\pi t}\, dt = \lim_{a\to\infty}\int_{-a}^{a} e^{-2\pi i s t}\frac{\sin \pi t}{\pi t}\, dt.$$

Strictly speaking, such an improper integral has to be defined by

$$\int_{-\infty}^{\infty} e^{-2\pi i s t}\frac{\sin \pi t}{\pi t}\, dt = \lim_{a,b\to\infty}\int_{-a}^{b} e^{-2\pi i s t}\frac{\sin \pi t}{\pi t}\, dt,$$

where a and b tend to ∞ independently. Onward.

(a) Assuming first that the integrals exist, show that

$$\int_{-\infty}^{\infty} e^{-2\pi ist} \frac{\sin \pi t}{\pi t}\, dt = 2\int_{0}^{\infty} \frac{\sin \pi t}{\pi t} \cos 2\pi st\, dt.$$

We will proceed to show that the integral on the right-hand side does exist, and we will evaluate it.

(b) Show for $p > 0$ and any q that

$$\int_{0}^{\infty} e^{-pt} \cos \pi qt\, dt = \frac{p}{p^2 + \pi^2 q^2}.$$

Use integration by parts, twice. Here there's no question that the integral converges, since $p > 0$.

(c) Integrate the expressions in the previous identity with respect to q from $q = 0$ to $q = a$ to obtain

$$\int_{0}^{\infty} e^{-pt} \frac{\sin \pi at}{\pi t}\, dt = \frac{1}{\pi} \tan^{-1} \frac{\pi a}{p}.$$

(d) Let $p \to 0$ and conclude from the previous identity that

$$(*) \qquad \int_{0}^{\infty} \frac{\sin \pi at}{\pi t}\, dt = \begin{cases} -\frac{1}{2}, & a < 0, \\ 0, & a = 0, \\ \frac{1}{2}, & a > 0. \end{cases}$$

(e) Use the addition formula for the sine function to show that

$$2\int_{0}^{\infty} \frac{\sin \pi t}{\pi s} \cos(2\pi st)\, dt = \int_{0}^{\infty} \frac{\sin(\pi t + 2\pi st)}{\pi t}\, dt + \int_{0}^{\infty} \frac{\sin(\pi t - 2\pi st)}{\pi t}\, dt.$$

(f) In particular, the integral on the left exists. Now figure out which particular values of a you can use in equation (*) to get the final result,

$$2\int_{0}^{\infty} \frac{\sin \pi t}{\pi s} \cos(2\pi st)\, dt = \begin{cases} 1, & |s| < \frac{1}{2}, \\ \frac{1}{2}, & |s| = \frac{1}{2}, \\ 0, & |s| > \frac{1}{2}. \end{cases}$$

4.2. *Decreasing fast, but not fast enough*

For a few increasing, positive integer values of k plot

$$f(x) = \frac{1}{(1 + x^2)^k}.$$

The function $f(x)$ is infinitely differentiable, certainly decreases to 0 at $\pm\infty$, and looks pretty Gaussian. But $f(x)$ is not a Schwartz function. Why not?

4.3. *Even and odd distributions and the derivative*

(a) First for functions: Show that if $f(x)$ is an even (odd) function, then $f'(x)$ is odd (even).

(b) Now for distributions: Show that if T is an even (odd) distribution, then its derivative T' is odd (even).

4.4. *Derivatives of a shifted unit step and of* Π_p
Let $h(x)$ be a shifted Heaviside unit step function,

$$h_b(x) = H(x-b) = \begin{cases} 1, & x \geq b\,, \\ 0, & x < b\,. \end{cases}$$

(a) Regarding $h_b(x)$ as defining a distribution, show that $h_b' = \delta_b$ using the definition of the derivative of a distribution.
(b) Using part (a) show that

$$\Pi_p' = \delta_{-p/2} - \delta_{p/2},$$

where we regard Π_p as defining a distribution.

4.5. *Generalized Fourier transforms*
(a) Find the Fourier transform of the signal

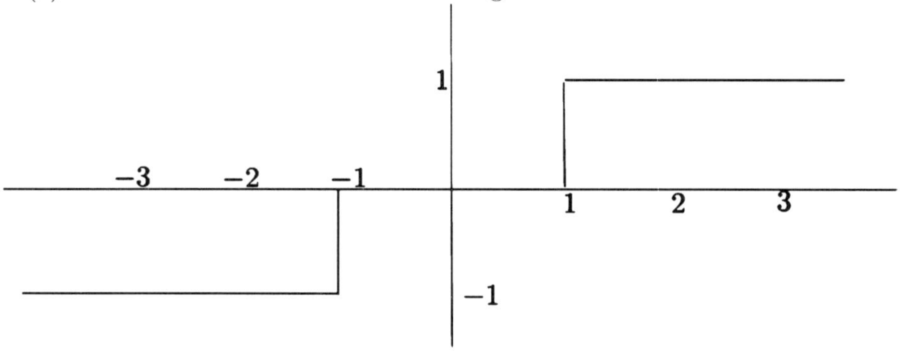

(b) Find the Fourier transform of the function $f(t) = \sin(2\pi|t|)$ plotted below for $-2 \leq t \leq 2$. Simplify your expression as much as possible.

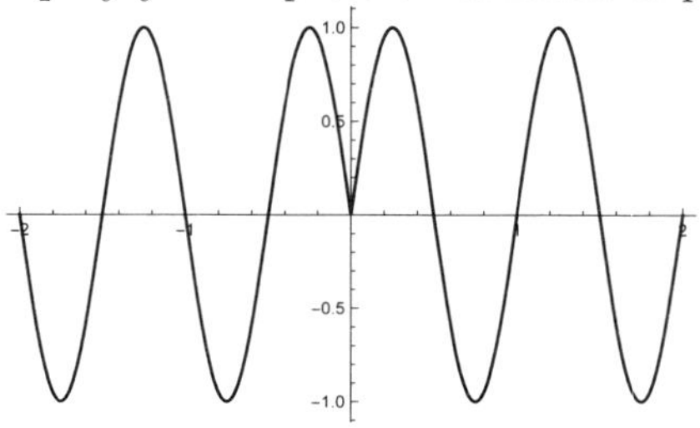

(c) Find the Fourier transform of the absolute value function $f(t) = |t|$. *Hint*: For parts (b) and (c), think about how to use the signum function to express the absolute value.
(d) Find the Fourier transform of a polynomial $f(t) = a_0 + a_1 t + a_2 t^2 + \cdots + a_n t^n$.

4.6. *δ' as a limit*[39]

Let

$$T_n = \frac{n}{2}\delta\Big(x - \frac{1}{n}\Big) - \frac{n}{2}\delta\Big(x + \frac{1}{n}\Big).$$

Show that $\lim_{n\to\infty} T_n = -\delta'$ by finding $\lim_{n\to\infty}\langle T_n, \varphi\rangle$ for a test function φ.

4.7. Express $f(x)\delta''(x)$ in terms of $f(0)$, $f'(0)$, $f''(0)$, $\delta(x)$, $\delta'(x)$, $\delta''(x)$.

4.8. *Some short problems on distributions*

(a) Let $g(t)$ be a Schwartz function. Show that

$$g(t)\delta'(t) = g(0)\delta'(t) - g'(0)\delta(t).$$

(b) Derive the product rule for differentiating a smooth function times a distribution: if T is a distribution and ψ is a smooth function, then $(\psi T)' = \psi' T + \psi T'$. Use this to show

$$\frac{d}{dt}|t| = \operatorname{sgn} t\,.$$

4.9. *Fourier transform of a composition of signals*

Let $h(t) = g(f(t))$ and denote $G = \mathcal{F}g$. Show that the Fourier transform of $h(t)$ is given by

$$H(s) = \int_{-\infty}^{\infty} G(y)P(s,y)\,dy$$

where

$$P(s,y) = \int_{-\infty}^{\infty} e^{2\pi i y f(t)} e^{-2\pi i s t}\,dt.$$

Hint: Work *backwards* from the formula for H and show that it is $\mathcal{F}g(f(t))$. You may freely change the order of integration, etc.

As an application, let's find a particular transform. The hyperbolic cosine function is defined by

$$\cosh x = \frac{e^x + e^{-x}}{2}.$$

It comes up in a wide variety of applications. There is no i in the exponent, but using

$$\cos x = \frac{e^{ix} + e^{-ix}}{2}$$

we can evidently write

$$\cosh x = \cos(ix).$$

Use the first part of the problem to find the Fourier transform of $h(t) = \cos 2\pi i t^2 = \cosh 2\pi t^2$. Your answer should again be in terms of cosh.

[39]From John Gill.

4.10. Derive the formula

$$\mathcal{F}\left\{\int_{-\infty}^{x} y(\tau)\, d\tau\right\} = \frac{1}{2}Y(0)\delta(s) + \frac{Y(s)}{2\pi i s},$$

for the Fourier transform of the accumulator. Here $Y(s)$ is the Fourier transform of $y(t)$.

4.11. Let $X(s)$ be the Fourier transform of the signal $x(t)$. The real part of $X(s)$ is $\left[4/(4+4\pi^2 s^2)\right]$, and the imaginary part of $X(s)$ is $[2/(\pi s)]$. Find $x(t)$ (your answer will be defined in a piecewise manner for the regions $t = 0$, $t > 0$, and $t < 0$).

4.12. *Derivation of the Fourier transform of the Heaviside (unit) step function*[40]

Recall that the Heaviside step function (also often called the unit step function) is defined by

$$H(x) = \begin{cases} 0, & x < 0\,, \\ \frac{1}{2}, & x = 0\,, \\ 1, & x > 0\,. \end{cases}$$

You'll notice that I've changed the definition at $x = 0$ from what we've used before. You'll see why, and though it really does not make a difference, I'm about willing to concede that this form of the definition is better (likewise for other functions with such jump discontinuities).

We had the formula

$$\mathcal{F}H(s) = \frac{1}{2}\left(\delta + \frac{1}{\pi i s}\right)$$

for the Fourier transform, which we derived writing $H(x)$ in terms of the sgn function.[41] With $H(x) = \frac{1}{2}(1 + \operatorname{sgn} x)$ and

$$\mathcal{F}\mathrm{sgn}\,(s) = \frac{1}{\pi i s}\,,$$

we get immediately the formula for $\mathcal{F}H(s)$.

In turn, we derived the formula for $\mathcal{F}\mathrm{sgn}$ using the derivative formula for the Fourier transform together with $\mathrm{sgn}' = 2\delta$. Since $H' = \delta$, you might wonder if we can also find $\mathcal{F}H$ the same way. The answer is yes, and that's what this problem is about.

First, indeed we have

$$\mathcal{F}H' = \mathcal{F}\delta = 1,$$

while from the derivative formula

$$\mathcal{F}H' = 2\pi i s \mathcal{F}H,$$

[40] The setup for the problem is longer than the problem.

[41] Also redefined here as

$$\mathrm{sgn}\,(x) = \begin{cases} 1, & x > 0\,, \\ 0, & x = 0\,, \\ -1, & x < 0\,; \end{cases}$$

ditto to the comment above about the function at $x = 0$.

and thus

$$2\pi is\mathcal{F}H = 1.$$

Now, we cannot say

$$\mathcal{F}H = \frac{1}{2\pi is},$$

but we *can* say that

$$\mathcal{F}H = \frac{1}{2\pi is} + c\delta,$$

for a constant c. The reason is that if $2\pi is\mathcal{F}H = 1$, upon which we're basing the derivation, the same is true of $2\pi is(\mathcal{F}H + c\delta)$. So that extra $c\delta$ term has to be there to get the most general solution. This issue comes up in deriving the formula for $\mathcal{F}\mathrm{sgn}$. In the case of $\mathcal{F}\mathrm{sgn}$ we can conclude that there was no δ term by appealing to the oddness of the function sgn and its transform $\mathcal{F}\mathrm{sgn}$, but for $\mathcal{F}H$ we don't have any such symmetry considerations and we have to figure out what c is.

Finally the problem: By considering also $\mathcal{F}H^-$ and using $\mathcal{F}H = \frac{1}{2\pi is} + c\delta$, show that $c = 1/2$, thus recovering the formula for $\mathcal{F}H(s)$. Would your argument, and the result, be any different if H were defined differently at 0?

4.13. *One more problem on the Fourier transform of the Heaviside step function: Practice using principal value distributions*

Under the heading of beating a dead horse, let's look one more time at the result

$$\mathcal{F}H(s) = \frac{1}{2\pi is} + \frac{1}{2}\delta.$$

The argument using the derivative theorem brings us to

$$\mathcal{F}H(s) = \frac{1}{2\pi is} + c\delta,$$

and we want to show that $c = 1/2$. We'll do this by pairing our distributions with the Gaussian $e^{-\pi s^2}$.

(a) Using the definition of the Fourier transform for distributions (not the formula above for $\mathcal{F}H$), show that

$$\langle \mathcal{F}H, e^{-\pi s^2} \rangle = \frac{1}{2}.$$

(b) With reference to Section 4.4.5, show that

$$\left\langle \frac{1}{s}, e^{-\pi s^2} \right\rangle = 0.$$

(c) Conclude, once again, that $\mathcal{F}H(s) = \frac{1}{2\pi is} + \frac{1}{2}\delta$.

4.14. *Windowing functions*

In signal analysis, it is not realistic to consider a signal $f(t)$ from $-\infty < t < \infty$. Instead, one considers a modified section of the signal, say from $t = -1/2$ to $t = 1/2$. The process is referred to as *windowing* (in case you didn't read the sections on this) and is achieved by multiplying the signal by a *windowing function* $w(t)$. By the convolution theorem, the Fourier transform of the windowed signal, $w(t)f(t)$ is the convolution of the Fourier transform of the original signal with the Fourier transform of the window function, $\mathcal{F}w * \mathcal{F}f$. The ideal window is no window at all; i.e., it's just the constant function 1 from $-\infty$ to ∞, which has no effect in either the time domain (multiplying) or the frequency domain (convolving).

Below are three possible windowing functions. Find the Fourier transform of each (you know the first two) and plot the *log magnitude* of each from $-100 < s < 100$. Which window is closest to the ideal window?

Rectangle window: $w(t) = \Pi(t)$.

Triangular window: $w(t) = 2\Lambda(2t)$.

Hann window:

$$w(t) = \begin{cases} 2\cos^2(\pi t)\,, & |t| < \frac{1}{2}, \\ 0\,, & \text{otherwise.} \end{cases}$$

4.15. *The consistency of math and the world's most sophisticated derivation of an ancient identity*

Use the convolution theorem to show that

$$\cos(a+b) = \cos a\,\cos b - \sin a\,\sin b.$$

Hint: Consider the Fourier transform of

$$f(t) = \cos(2\pi a t)\,\cos(2\pi b t) - \sin(2\pi a t)\,\sin(2\pi b t).$$

4.16. *Another elementary identity from an advanced standpoint*

Use the convolution theorem for Fourier transforms to show that

$$\cos^2\theta = \frac{1}{2}(1 + \cos 2\theta).$$

Hint: Let $f(t) = \cos 2\pi a t$ and take the Fourier transform of $f(t)^2 = f(t)\cdot f(t)$.

4.17. *One more old trig identity via convolution*

The product of two trigonometric functions

$$f(t) = \cos(2\pi s_1 t)\sin(2\pi s_2 t)$$

can be written as the sum of two trigonometric functions. Find the trig identity using convolution. You may assume $s_1 > s_2 > 0$.

4.18. *An alternate way to find the Fourier transform of sinc*

Among our earliest examples of Fourier transform pairs was

$$\mathcal{F}\Pi = \text{sinc} \quad \text{and} \quad \mathcal{F}\,\text{sinc} = \Pi.$$

That $\mathcal{F}\Pi = \text{sinc}$ was a calculation using the definition of the Fourier transform. We derived $\mathcal{F}\,\text{sinc} = \Pi$ via duality (and later gave a derivation via integration in a supplementary problem).

If we regard sinc as a product,

$$\text{sinc}\,x = \frac{\sin \pi x}{\pi x} = \sin \pi x \cdot \frac{1}{\pi x},$$

we can derive $\mathcal{F}\,\text{sinc} = \Pi$ via convolution (using distributions).

Obtain the Fourier transform of sinc this way, confirming yet again that $\mathcal{F}\,\text{sinc} = \Pi$.

4.19. Let $f(x)$ be a signal, and for $h > 0$ let $A_h f(x)$ be the averaging operator,

$$A_h f(x) = \frac{1}{2h} \int_{x-h}^{x+h} f(y)\, dy\,.$$

(a) Show that we have the alternate expressions for $A_h f(x)$:

$$A_h f(x) = \frac{1}{2h} \int_{-h}^{h} f(x+y)\, dy = \frac{1}{2h} \int_{-h}^{h} f(x-y)\, dy\,.$$

(This part is from Problem 3.13. Look it over.)

(b) Show that we should define $A_h T$ for a distribution T by

$$\langle A_h T\,,\, \varphi \rangle = \langle T\,,\, A_h \varphi \rangle\,.$$

(c) What is $A_h \delta$?

4.20. *Averaging operator and periodicity*

(a) Show that

$$(A_h f)'(x) = \frac{1}{2h}(f(x+h) - f(x-h))\,.$$

For this, write $A_h f$ as a convolution and freely use $(g * f)' = g' * f$.

(b) Conclude that $A_h f(x)$ is a constant if and only if $f(x)$ is periodic of period $2h$. *Hint*: Use the theorem from calculus that a function is constant if and only if its derivative is 0. The moral of the story: If the values repeat, the average stays constant.

4.21. *Some short problems on differential equations*[42]

(a) Show the solution to the *one-dimensional Poisson equation*

$$u''(x) = f(x)$$

is given by

$$u(x) = f(x) * \frac{|x|}{2}.$$

(b) Show the solution to the *driven harmonic oscillator*

$$u''(x) + (2\pi\nu)^2 u(x) = f(x)$$

is given by

$$u(x) = \frac{1}{4\pi\nu}\, f(x) * \sin(2\pi\nu|x|).$$

Assume ν is a real positive constant.

---

[42]From Raj Bhatnagar.

Chapter 5

δ Hard at Work

We've put a lot of effort into general theory and it's time to see a few applications. They range from finishing some work on filters to optics and diffraction to X-ray crystallography. Via the Ш-function, which we'll meet in this chapter (very important!), the latter will even lead us toward the sampling theorem in the next chapter. What all these examples have in common is their use of δ's. This chapter is short and selective, but the topics are all interesting and illustrative.[1]

The main properties of δ that we'll need are what happens with convolution and multiplication. Recall that for a function f,

$$(f * \delta)(x) = f(x) \quad \text{and} \quad f(x)\delta(x) = f(0)\delta(x)\,.$$

We'll tend to write δ with variables in this chapter, and we'll also feel free to integrate against a δ, writing

$$(f * \delta)(x) = \int_{-\infty}^{\infty} f(y)\delta(x-y)\,dy = f(x)\,.$$

For a shifted δ,

$$\begin{aligned} f(x) * \delta(x-a) &= (f * \delta)(x-a) = f(x-a)\,, \\ f(x)\delta(x-a) &= f(a)\delta(x-a)\,. \end{aligned}$$

We'll also use the scaling property

$$\delta(ax) = \frac{1}{|a|}\delta(x)\,,$$

and of course we'll need the Fourier transform,

$$\mathcal{F}\delta(x-a) = e^{-2\pi i s a}\,.$$

[1] A popular collection of problems that we won't dip into are ones on solving ordinary differential equations with a δ on the right-hand side. If you took a signals and systems class, you know what I mean. You probably did scads of these, typically using the Laplace transform (not the Fourier transform) and involving lots of partial fractions. Not here.

5.1. Filters, Redux

One of our first applications of convolution was to set up and study a few simple filters. Let's recall the terminology and some work left undone from Section 3.4. The unfiltered input $v(t)$ and the filtered output $w(t)$ are related through convolution with the *impulse response* $h(t)$:

$$w(t) = (h * v)(t)\,.$$

(We're not quite ready to explain why h is called the impulse response.) The action of the filter is easier to understand in the frequency domain, for there, by the convolution theorem, it acts by multiplication:

$$W(s) = H(s)V(s)\,.$$

Here

$$W = \mathcal{F}w, \quad H = \mathcal{F}h, \quad \text{and} \quad V = \mathcal{F}v\,.$$

$H(s)$ is called the *transfer function.*

The simplest example, out of which the others can be built, is the low-pass filter with transfer function:

$$\text{Low}(s) = \Pi_{2\nu_c}(s) = \Pi\Big(\frac{s}{2\nu_c}\Big) = \begin{cases} 1, & |s| < \nu_c\,, \\ 0, & |s| \geq \nu_c\,. \end{cases}$$

The impulse response is

$$\text{low}(t) = 2\nu_c \operatorname{sinc}(2\nu_c t)\,,$$

a scaled sinc function. The output $w(t)$ coming from an input $v(t)$ is the convolution

$$w(t) = v(t) * 2\nu_c \operatorname{sinc}(2\nu_c t) = 2\nu_c \int_{-\infty}^{\infty} \operatorname{sinc}(2\nu_c(t-\tau))v(\tau)\,d\tau\,.$$

High-pass filter. Back in Chapter 3 we displayed the graph of the transfer function for an ideal high-pass filter:

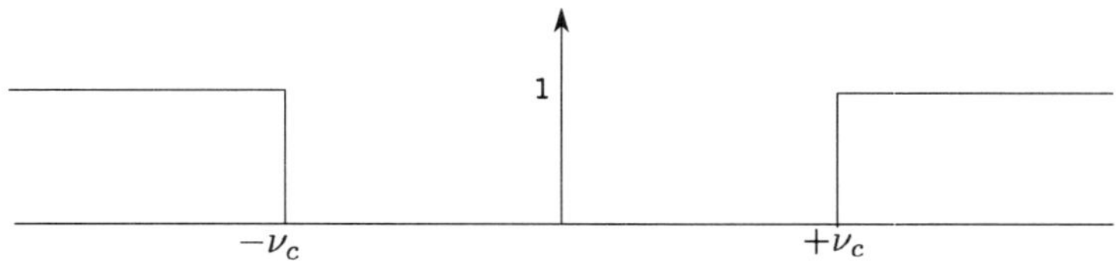

A formula for the transfer function is

$$\text{High}(s) = 1 - \text{Low}(s) = 1 - \Pi_{2\nu_c}(s)\,,$$

where ν_c is the cutoff frequency. At the time, we couldn't finish the analysis simply by taking the inverse Fourier transform because we didn't have δ. Now we do. The impulse response is

$$\text{high}(t) = \delta(t) - 2\nu_c \operatorname{sinc}(2\nu_c t)\,.$$

For an input $v(t)$ the output is then

$$\begin{aligned} w(t) &= (\text{high} * v)(t) \\ &= \big(\delta(t) - 2\nu_c \operatorname{sinc}(2\nu_c t)\big) * v(t) \\ &= v(t) - 2\nu_c \int_{-\infty}^{\infty} \operatorname{sinc}(2\nu_c(t-\tau))v(\tau)\, d\tau\,. \end{aligned}$$

The role of the convolution property of δ in this formula shows us again that the high-pass filter literally subtracts part of the signal away.

Notch filter. The transfer function for the notch filter is just

$$1 - (\text{transfer function for band-pass filter})$$

and it looks like this:

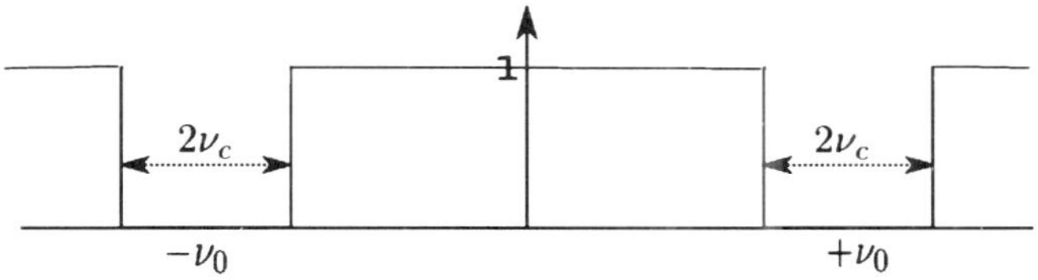

Frequencies in the notch are filtered out and all others are passed through unchanged. Suppose that the notches are centered at $\pm\nu_0$ and that they are $2\nu_c$ wide. The formula for the transfer function, in terms of the transfer function for the low-pass filter with cutoff frequency ν_c, is

$$\text{Notch}(s) = 1 - \big(\text{Low}(s-\nu_0) + \text{Low}(s+\nu_0)\big)\,.$$

For the impulse response we obtain

$$\begin{aligned} \text{notch}(t) &= \delta(t) - (e^{-2\pi i\nu_0 t}\text{low}(t) + e^{2\pi i\nu_0 t}\text{low}(t)) \\ &= \delta(t) - 4\nu_c \cos(2\pi\nu_0 t)\operatorname{sinc}(2\nu_c t)\,. \end{aligned}$$

Thus

$$\begin{aligned} w(t) &= (\delta(t) - 4\nu_c \cos(2\pi\nu_0 t)\operatorname{sinc}(2\nu_c t)) * v(t) \\ &= v(t) - 4\nu_c \int_{-\infty}^{\infty} \cos(2\pi\nu_0(t-\tau))\, \operatorname{sinc}(2\nu_c(t-\tau))\, v(\tau)\, d\tau\,, \end{aligned}$$

and again we see the notch filter subtracting away part of the signal.

We'll return to these filters, one more time, in Chapter 8.

5.2. Diffraction: Sincs Live and in Pure Color

Some of the most interesting applications of the Fourier transform are in the field of optics, understood broadly to include the study of most of the electromagnetic spectrum. An authoritative book on the subject is *Fourier Optics*, by my Stanford colleague J. W. Goodman.

The fundamental phenomenon associated with the wave theory of light is *diffraction*, or *interference*. I'm certain that this phenomenon is familiar to you. In a stab at a definition, A. Sommerfeld says that diffraction is "any deviation of light rays from rectilinear paths which cannot be interpreted as reflection or refraction."

I suppose you have to start somewhere. Is there a difference between diffraction and interference? In his *Lectures on Physics*, R. Feynman says:

> No one has ever been able to define the difference between interference and diffraction satisfactorily. It is just a question of usage, and there is no specific, important physical difference between them.

He does go on to say that interference is usually associated with patterns caused by a few radiating sources, like two, while diffraction is due to many sources.

Whatever the definition, or nondefinition, you've probably seen pictures like this one. It's the diffraction pattern from a single, rectangular aperture, and it's found anywhere diffraction is pictured. The question: Why are there bands of light and dark and not just sharp edges caused by light passing through the sharp-edged aperture?

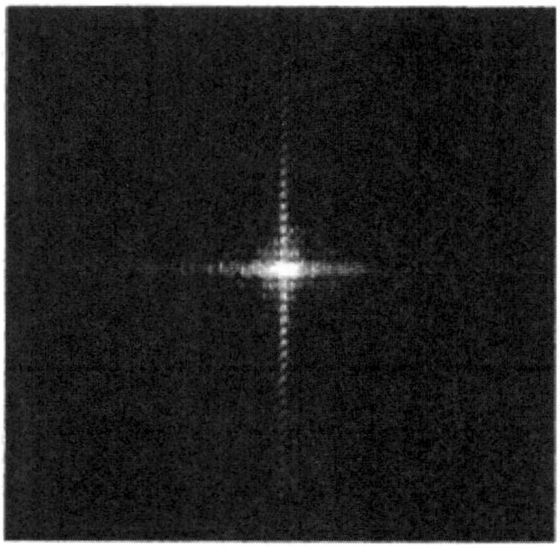

The experiments were easy to perform. Thus spoke Thomas Young[2] in 1803 to the Royal Society:

> The experiments I am about to relate ... may be repeated with great ease, whenever the sun shines, and without any other apparatus than is at hand to everyone.

The shock of the diffraction patterns when first seen was that light + light could be dark. Such images of diffraction/interference patterns, most notably Young's "two slits experiments," which we'll analyze below, were crucial in tipping the balance away from the corpuscular theory of light favored by Newton to the wave theory propounded by Christiaan Huygens. We are thus taking sides in the grand battle between the armies of "light is a wave" and the armies of "light is a particle." It may be that light is truly like nothing you've ever seen before, but for this discussion, it's a wave.

[2]Young also did important work on Egyptian hieroglyphics, completely translating a section of the Rosetta Stone.

Jumping ahead to Maxwell, we assume that light is a periodic electromagnetic wave. In this treatment of diffraction we assume further that the light is:

- Monochromatic
  - Meaning that the periodicity in time is due to a single frequency and so is described by a simple sinusoid.
- Linearly polarized
  - Meaning that the electric field vector stays in a plane as the wave moves. (Hence so, too, does the magnetic field vector.)

We'll also suppose that everything takes place in a vacuum, so the velocity of light is c and light travels at that velocity in any direction. There are no special effects from light going through different layers of different stuff.

Putting these assumptions together means that we're going to discuss an example of *scalar diffraction theory*, while more sophisticated treatments handle the vector theory.

The diffraction problem. Given the assumptions above, the diffraction problem can be stated as follows:

> Light, an electromagnetic wave, is incident on an (opaque) screen with one or more apertures (transparent openings) of various shapes. What is the intensity of the light on a screen some distance from the diffracting screen?

We're going to consider only the case of *Fraunhofer diffraction*[3] based on the *Fraunhofer approximation*, where the analysis is fairly straightforward. This involves a number of additional simplifying assumptions, mentioned below, but the results are used widely. Before we embark on the analysis let me point out that reasoning very similar to what we'll do here is used to understand the radiation patterns of antennas. For this take on the subject see Bracewell's *The Fourier Transform and Its Applications*, Chapter 15.

Light waves. A few more comments on light waves, following the discussion in Goodman's book on Fourier optics. In the scalar theory, under the previous assumptions, we can describe light by a real-valued function of time and position:

$$u(x, y, z, t) = a(x, y, z)\cos(2\pi\nu t + \phi(x, y, z)) .$$

Here, $a(x, y, z)$ is the amplitude, depending only on position in space, ν is the (single) frequency, and $\phi(x, y, z)$ is the phase at $t = 0$, also depending only on position.[4]

The equation

$$\phi(x, y, z) = \text{constant}$$

describes a surface in space. At a fixed time, all the points on such a surface have the same phase, by definition. Or we might say equivalently that the traveling wave reaches all points of such a surface $\phi(x, y, z) = \text{constant}$ at the same time. Thus any one of the surfaces $\phi(x, y, z) = \text{constant}$ is called a *wavefront*. In general, the wave propagates through space in a direction normal to the wavefronts.

[3]Named for the Bavarian physicist Joseph von Fraunhofer. He made many contributions to optics and optical instruments.

[4]It's also common to refer to the whole argument of the cosine, $2\pi\nu t + \phi$, simply as "the phase."

The function $u(x, y, z, t)$ satisfies the three-dimensional wave equation

$$\Delta u = \frac{1}{c^2} \frac{\partial^2 u}{\partial t^2},$$

where

$$\Delta = \frac{\partial^2}{\partial x^2} + \frac{\partial^2}{\partial y^2} + \frac{\partial^2}{\partial z^2}$$

is the Laplacian and c is the speed of light in a vacuum. I'm not going to offer a derivation of the wave equation, but you can easily find one. Unlike the heat equation from earlier work, the wave equation has the *second* derivative with respect to time. That's an important difference.

For many problems it's helpful to separate the spatial behavior of the wave from its temporal behavior and to introduce

$$U(x, y, z) = a(x, y, z) e^{i\phi(x,y,z)} .$$

We obtain the time-dependent function $u(x, y, z, t)$ as

$$u(x, y, z, t) = \text{Re}\left(U(x, y, z) e^{2\pi i \nu t} \right),$$

and so if we know $U(x, y, z)$, we can recover $u(x, y, z, t)$. Plug this form for $u(x, y, z, t)$ into the wave equation. The derivative of the real part is the real part of the derivative (check that) and it quickly follows that $U(x, y, z)$ satisfies the differential equation

$$\Delta U(x, y, z) = -\frac{4\pi^2 \nu^2}{c^2} U(x, y, z),$$

or

$$\Delta U(x, y, z) + k^2 U(x, y, z) = 0,$$

where $k = 2\pi\nu/c = 2\pi/\lambda$; recall that $c = \lambda\nu$. This is called the *Helmholtz equation*, and the fact that it is time independent makes it simpler than the wave equation.

The constant k is called the *wave number*. With dimension 1/length, or radians per length with the 2π factor, it's a spatial frequency, analogous to cycles/second for the usual time frequency ν. For example, if the wavelength is $\lambda = 5$ meters, then the wave has $1/5 = 1/\lambda$ of a wavelength in 1 meter. Throw in the 2π to make that correspond to radian measure, i.e., the length of the corresponding arc of the unit circle — more convenient in some formulas.

5.2.1. Fraunhofer diffraction. We take a sideways view of the situation. Light is coming from a source at a point O and hits a plane S. We assume that the source is so far away from S that the magnitude of the electric field associated with the light is constant on S and has constant phase; i.e., S is a wavefront and we have what is called a *plane wave field*. Let's say the frequency is ν and the wavelength is λ.

Set up coordinates so that the z-axis is perpendicular to S and the x-axis lies in S, perpendicular to the z-axis. (In most diagrams in most optics books it is traditional to have the z-axis be horizontal and the x-axis be vertical.)

In S we have one or more rectangular apertures. We allow the length of the side of the aperture along the x-axis to vary, but we assume that the other side (perpendicular to the plane of the diagram) has length 1. A large distance from S is another parallel plane on which the light shines in all its diffracted glory. Call this the image plane.

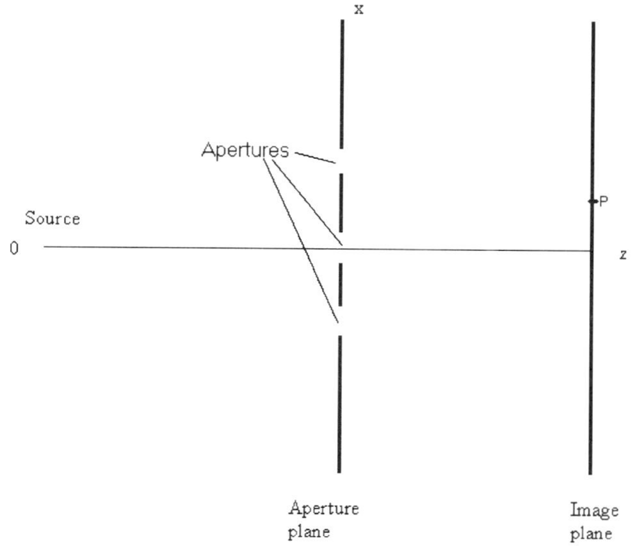

We restate the diffraction problem as follows:

- What is the electric field at a point P in the image plane?

The derivation I'm going to give to answer this question is not as detailed as is possible, but we'll get the correct form of the answer, and the point is to see how the Fourier transform enters. Once again I refer to Goodman's book and also to a very good, short treatment in *A Student's Guide to Fourier Transforms* by J. F. James.

The basis for analyzing diffraction is *Huygens's principle*, which states, roughly, that each point in an aperture on S (and S is a wavefront of the original source) may itself be regarded as a (secondary) source, and the field at P is the sum (integral) of the fields coming from these sources on S.[5]

The wave leaves a point on an aperture in S, a secondary source, and arrives at a point P on the image plane sometime later. Waves from different points on S will arrive at P at different times, and hence there will be a phase difference between the arriving waves. The waves also drop off in amplitude like one over the distance to P, and so by different amounts, but if, as we'll later assume, the size of the apertures on S is small compared to the distance between S and the image plane, then this is not as significant as the phase differences. Light is moving so fast that even small differences between locations of the secondary point sources on S may lead to significant differences in the phases when the waves reach P.

[5]Everybody does it this way, with a few exceptions. One exception is *Principles of Electrodynamics* by the Nobel Prize winner Melvin Schwartz (another Schwartz). In speaking of Huygens's principle he says: "Physically this makes *no* sense at all. Light does not emit light; only accelerating charges emit light. Thus we will begin by throwing out Huygens's principle completely; later we will see that it actually does give the right answer for the wrong reasons." My colleague David Miller wrote an important paper on the subject as well, titled *Huygens's wave propagation principle corrected.*

To express things a little more geometrically, each point in an aperture (as a point source) is a source of a spherical wavefront. The different wavefronts, from different sources, interfere constructively and destructively, leading to the observed diffraction patterns. Hand drawing some sketches, you can start to get a sense of how the size of the aperture relative to the wavelength influences the diffraction. But good luck on this; I always found it difficult. At least these days there are helpful animated demos on the web.

To express things a little less geometrically, if E_0 is the strength of the electric field on S, then an aperture of area dS is a source of strength $dE = E_0\,dS$.[6] At a distance r from this aperture the field strength is $E_0\,dS/r$, and we get the electric field at this distance by integrating over the apertures the elements $E_0\,dS/r$, adjusting for the phase. Let's look more carefully at the phase.

The phase on S is constant and we might as well assume that it's zero. Then we write the electric field on S in complex form as

$$E = E_0 e^{2\pi i\nu t}\,,$$

where E_0 is constant and ν is the frequency of the light. Suppose P is at a distance r from a point x on S. The phase change from x to P can be assessed by comparing r to the wavelength λ, as in how many wavelengths (or fractions of a wavelength) the wave goes through in going a distance r from x to P. This is $2\pi r/\lambda$; that's r times the wave number. (The usefulness of wave numbers.) Just to be sure: the wave travels a distance r in a time r/c seconds, and in that time it goes through $\nu(r/c)$ cycles. Using $c = \lambda\nu$, that's $\nu r/c = r/\lambda$. This is $2\pi r/\lambda$ radians, and that's the phase shift.

Take a thin slice of width dx at a height x above the origin of an aperture on S. Then the field at P due to this source is, on account of the phase change,

$$dE = E_0 e^{2\pi i\nu t} e^{-2\pi i r/\lambda}\,dx\,,$$

ignoring the drop-off in magnitude of E. (Remember, the side of the aperture perpendicular to the plane of the diagram has length 1, so the area of the slice is $1\cdot dx$.) The total field at P is

$$E = \int_{\text{apertures}} E_0 e^{2\pi i\nu t} e^{-2\pi i r/\lambda}\,dx = E_0 e^{2\pi i\nu t}\int_{\text{apertures}} e^{-2\pi i r/\lambda}\,dx\,.$$

There's a Fourier transform coming, but we're not there yet.

We now suppose that

$$r \gg x\,;$$

that is, the distance between the plane S and the image plane is much greater than any x in any aperture. In particular, r is large compared to any aperture size. This assumption is what makes this *Fraunhofer diffraction*; it's also referred to as

[6]This is part of Huygens's principle. "Strength" is a little hazy, but so are the diffraction patterns.

far-field diffraction.[7] With this assumption, we have, approximately,

$$r = r_0 - x \sin\theta\,,$$

where r_0 is the distance between the origin of S to P and θ is the angle between the z-axis and P.

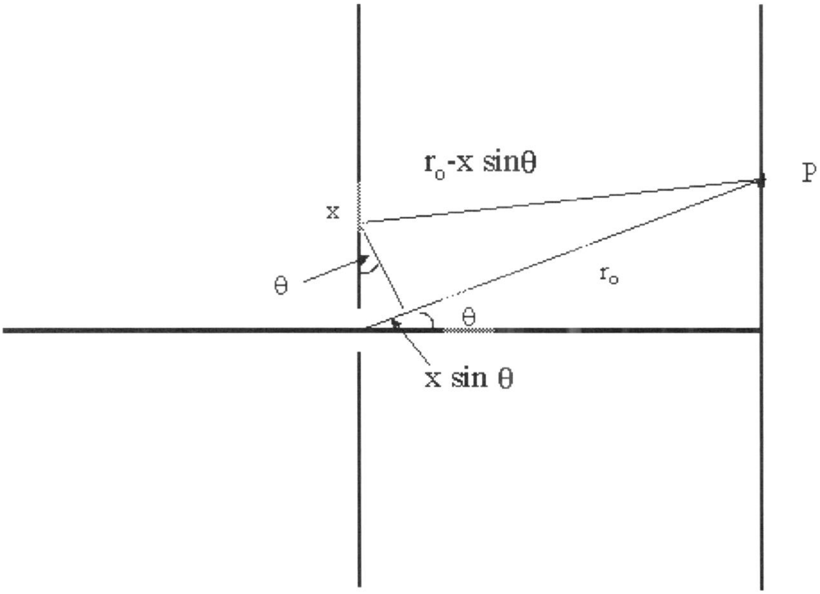

Plug this into the formula for E:

$$E = E_0 e^{-2\pi i\nu t} e^{2\pi i r_0/\lambda} \int_{\text{apertures}} e^{2\pi i x \sin\theta/\lambda}\, dx.$$

Drop those factors out front; as you'll see they won't be important for the rest of our considerations.

We describe the apertures on S by a function $A(x)$, which is zero most of the time (the opaque parts of S) and 1 some of the time (apertures). Thus we can write

$$E \propto \int_{-\infty}^{\infty} A(x) e^{2\pi i x \sin\theta/\lambda}\, dx\,.$$

It's common to introduce the variable

$$p = \frac{\sin\theta}{\lambda}$$

and hence to write

$$E \propto \int_{-\infty}^{\infty} A(x) e^{2\pi i p x}\, dx\,.$$

[7]Near-field diffraction is called Fresnel diffraction. Different assumption, different approximation. See Goodman's book. Augustin-Jean Fresnel was a French civil engineer and physicist. One of his more celebrated contributions was the invention of the Fresnel lens, used in lighthouses to extend the visibility of the light beam. Fascinating story, really. Read about it in *A Short Bright Flash, Augustin Fresnel and the Birth of the Modern Lighthouse* by Theresa Levitt.

There you have it. With these approximations (the *Fraunhofer approximations*) the electric field, up to a multiplicative constant, is the inverse Fourier transform of the aperture!

Note that the variables in the formula are x, a spatial variable, and $p = \sin\theta/\lambda$, in terms of an angle θ. It's the θ that's important, and one always speaks of diffraction "through an angle."

Also note that the *intensity* of the light, which is what we see and what photodetectors register, is proportional to the energy of E, i.e., to $|E|^2$. This is why we dropped the factors $e^{2\pi i\nu t}e^{2\pi i r_0/\lambda}$ multiplying the integral. They have magnitude 1. (We also dropped E_0, the constant field on the aperture plane, which was just going along for the ride.)

An aphorism:

- For Fraunhofer diffraction, the diffraction pattern is determined by the inverse Fourier transform of the aperture function.

But, to repeat, it's the square magnitude that devices register. This result doesn't give any direct information on the phase. Getting the phase is a famous problem.

The inverse Fourier transform ... I knew this unhappy moment was coming. As we've defined it, this wonderful result for this wonderful application involves the *inverse* Fourier transform rather than the Fourier transform. We'd like to say simply that "the diffraction pattern is the Fourier transform of the aperture function," but we can't quite. Not that it makes any substantive difference.

Here you have one of the reasons why some people (primarily physicists) define the Fourier transform to have a + sign in the complex exponential. We all have to live with our choices.

Diffraction by a single slit. Take the case of a single rectangular slit of width a, thus described by $A(x) = \Pi_a(x)$. Then the field at P is

$$E \propto a \operatorname{sinc} ap = a \operatorname{sinc}\left(\frac{a\sin\theta}{\lambda}\right).$$

So the diffraction pattern you see from a single slit, those alternating bright and dark bands, is

$$\text{intensity} = a^2 \operatorname{sinc}^2\left(\frac{a\sin\theta}{\lambda}\right).$$

You don't see a sharp shadow of the aperture. You see bands of light and dark.

Pretty good. The sinc function, or at least its square, live and in black and white. Just as promised. We've seen a plot of sinc^2 before, and you may very well have seen it, without knowing it, as a plot of the intensity from a single slit diffraction experiment. Here's a plot for $a = 2$, $\lambda = 1$, and $-\pi/2 \le \theta \le \pi/2$.

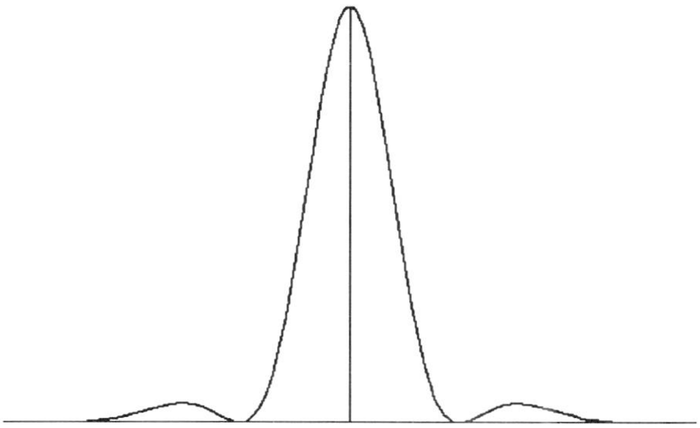

Young's experiment. As mentioned earlier, Thomas Young observed diffraction caused by light passing through two slits. To analyze his experiment using what we've derived, we need an expression for the apertures that's convenient for taking the Fourier transform.

Suppose we have two slits, each of width a, centers separated by a distance b. We can model the aperture function by the sum of two shifted rect functions,

$$A(x) = \Pi_a(x - b/2) + \Pi_a(x + b/2)\,.$$

(Like the transfer function of a band-pass filter.) That's fine, but we can also shift the Π_a's by convolving with shifted δ's, as in

$$\begin{aligned} A(x) &= \delta(x - b/2) * \Pi_a(x) + \delta(x + b/2) * \Pi_a(x) \\ &= (\delta(x - b/2) + \delta(x + b/2)) * \Pi_a(x)\,, \end{aligned}$$

and the advantage of writing $A(x)$ in this way is that the convolution theorem applies to help in computing the Fourier transform. Namely,

$$\begin{aligned} E(p) &\propto (2\cos \pi bp)(a \operatorname{sinc} ap) \\ &= 2a \cos\left(\frac{\pi b \sin\theta}{\lambda}\right) \operatorname{sinc}\left(\frac{a\sin\theta}{\lambda}\right)\,. \end{aligned}$$

Young saw the intensity, and so would we, which is then

$$\text{intensity} = 4a^2\cos^2\left(\frac{\pi b \sin\theta}{\lambda}\right)\operatorname{sinc}^2\left(\frac{a\sin\theta}{\lambda}\right)\,.$$

Here's a plot for $a = 2$, $b = 6$, $\lambda = 1$ for $-\pi/2 \le \theta \le \pi/2$.

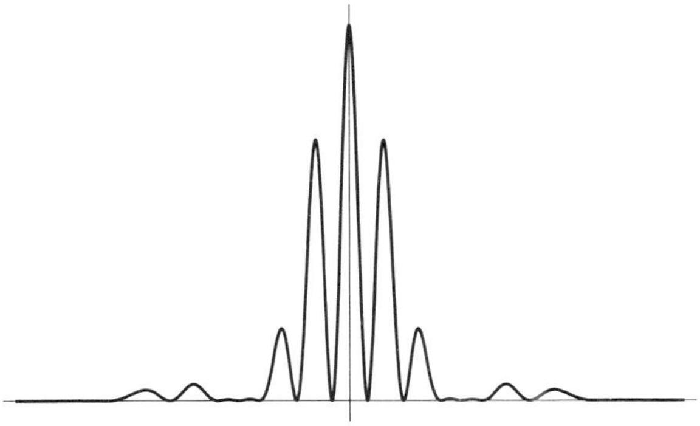

This is quite different from the diffraction pattern for one slit. Again, you don't see something corresponding to two sharp shadows of the two slits; you see a rather complicated pattern of light and dark, with a peak in the middle!

Diffraction by two point sources. Say we have two point sources — the apertures — and that they are at a distance b apart. In this case we can model the apertures by a pair of δ-functions:

$$A(x) = \delta(x - b/2) + \delta(x + b/2)\,.$$

Taking the Fourier transform then gives

$$E(p) \propto 2\cos \pi b p = 2\cos\left(\frac{\pi b \sin\theta}{\lambda}\right)$$

and the intensity as the square magnitude.

$$\text{intensity} = 4\cos^2\left(\frac{\pi b \sin\theta}{\lambda}\right).$$

Here's a plot of this for $b = 6$, $\lambda = 1$ for $-\pi/2 \le \theta \le \pi/2$.

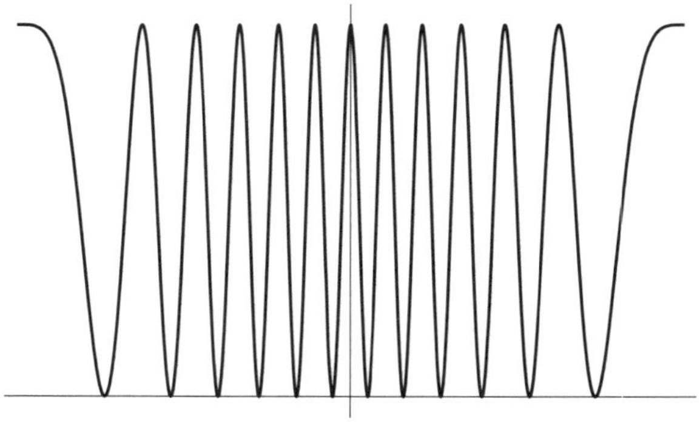

Incidentally, the case of two radiating point sources covers the case of two antennas transmitting in phase from a single oscillator; again, see Bracewell.

An optical interpretation of $\mathcal{F}\delta = 1$. What if we had light radiating from a single point source? What would the pattern be on the image plane? For a single point source there is no diffraction — a *point source*, not a circular aperture of some definite radius — and the image plane is illuminated uniformly. Thus the strength of the field is constant on the image plane. On the other hand, if we regard the aperture as δ and plug into it the formula we have the inverse Fourier transform of δ,

$$E \propto \int_{-\infty}^{\infty} \delta(x) e^{2\pi i p x}\, dx.$$

This gives a physical reason why the Fourier transform of δ should be constant (if not 1).

Also note what happens to the intensity as $b \to 0$ of the diffraction due to two point sources at a distance b. Physically, we have a single point source (of strength 2) and the formula gives

$$\text{intensity} = 4\cos^2\left(\frac{\pi b \sin\theta}{\lambda}\right) \to 4\,.$$

There's more. The lesson of this treatment of diffraction is that nature takes a Fourier transform![8] I refer to Goodman's book and to James's book for more examples. In particular, read the interesting discussions of diffraction gratings (noted briefly, below).

5.3. X-Ray Diffraction

Through a glass darkly [9]. Let's start with a little history. X-rays were discovered by William Roentgen in 1895. It was not known whether they were particles or waves or what, but the wave hypothesis put their wavelength at about 10^{-8} cm. Experiments with visible light, where the wavelengths are in the range between 400 and 700 nanometers (10^{-7} cm), relied on diffraction gratings as a means to separate light of different wavelengths and to measure the wavelengths.[10] A diffraction grating is an aperture plane with a large number of parallel slits, closely spaced, and diffraction effects are only seen if the width and spacing of the slits are comparable to the wavelength. Using diffraction gratings was out of the question for experiments on X-rays because of that extra order of magnitude. Decreasing the width and spacing of the slits from 10^{-7} to 10^{-8} couldn't be done.

A related set of mysteries had to do with the structure of crystals. It was thought that the macroscopic structure of crystals, with their symmetries visibly apparent, could be explained by a periodic arrangement of atoms. But there was no way to test this. In 1912 Max von Laue proposed that the purported periodic structure of crystals could be used to diffract X-rays, just as gratings diffracted visible light. He thus had three hypotheses:

(1) X-rays are waves.

(2) Crystals are periodic at the atomic level.

(3) The spacing between atoms is of the order 10^{-8} cm.

Walter Friedrich and Paul Kniping carried out experiments that confirmed von Laue's hypotheses and the subject of X-ray crystallography was born.[11] But you need to know some math (though it's not the math they used).

Electron density distribution. An important quantity to consider in crystallography is how the electrons are distributed among the atoms in the crystal. This is the *electron density distribution* of the crystal, "distribution" more in the probabilistic sense. We'll represent this as a function and consider its role in an X-ray diffraction experiment.

Let's take the one-dimensional case as an illustration; we'll look at the more realistic higher-dimensional case when we learn about higher-dimensional Fourier transforms. We view a one-dimensional crystal as an evenly spaced collection of atoms along a line. In fact, for purposes of approximation, we suppose that an

[8] Recall from Chapter 1 that nature uses Fourier series for hearing.

[9] 1 Corinthians 13: When I was a child, I spake as a child, I understood as a child, I thought as a child: but when I became a man, I put away childish things. For now we see through a glass, darkly, but then face to face: now I know in part; but then shall I know even as also I am known.

[10] There are many sources out there. See, for example `http://hyperphysics.phy-astr.gsu.edu/hbase/phyopt/grating.html`.

[11] Nobel Prize for Max in 1914.

infinite number of atoms are strung out along a line.[12] If we describe the electron density distribution of a *single* atom by a function $\rho(x)$, then the electron density distribution of the crystal with spacing p between atoms is the *periodic* function of period p,

$$\rho_p(x) = \sum_{k=-\infty}^{\infty} \rho(x - kp)\,.$$

As our discussion of diffraction might indicate, with the space between atoms as apertures the Fourier transform of $\rho_p(x)$ is proportional to the *scattered amplitude* of X-rays diffracted by the crystal.[13] Thus we want to write $\rho_p(x)$ in a form that's amenable to taking the Fourier transform. (Incidentally, it's not unreasonable to suppose that ρ is rapidly decreasing, or even of compact support — the electron density of a single atom dies off as we move away from the atom.)

It's convenient to write the periodized density as a convolution with a sum of shifted δ's:

$$\rho_p(x) = \sum_{k=-\infty}^{\infty} \rho(x - pk) = \sum_{k=-\infty}^{\infty} \delta(x - kp) * \rho(x) = \left(\sum_{k=-\infty}^{\infty} \delta(x - kp) \right) * \rho(x)\,.$$

Now introduce

$$Ш_p(x) = \sum_{k=-\infty}^{\infty} \delta(x - kp)\,,$$

so that, simply,

$$\rho_p = Ш_p * \rho\,.$$

$Ш_p$ is the star of the show. It was Bracewell, I think, who called this the *shah function*, after the Cyrillic letter of that name, and this has caught on.[14] It's also referred to as the *Dirac comb*. The picture is of an infinite train of δ's spaced p apart.

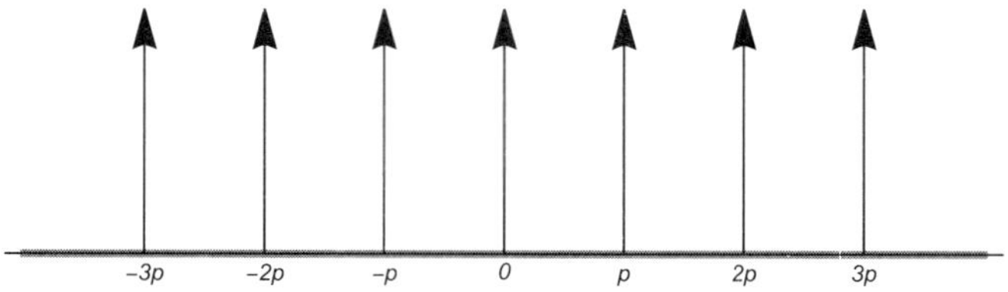

Note that $Ш_p$ is even.

Applying the convolution theorem to $\rho_p = Ш_p * \rho$ we have

$$\mathcal{F}\rho_p = \mathcal{F}\rho \cdot \mathcal{F}Ш_p\,.$$

To make further progress we need to find $\mathcal{F}Ш_p$. That's a really interesting problem.

---

[12] As we consider an infinite rod as an approximation to a really long finite rod when analyzing the heat equation. We'll see later why an infinite array of atoms is a mathematically useful approximation to a finite crystal. It has to do with Fourier transforms.

[13] Well, actually $\mathcal{F}^{-1}$ the way we've defined things, but please just let me use $\mathcal{F}$.

[14] The reason he chose the symbol is because Ш is supposed to be reminiscent of some spikes sticking up, as in the picture.

5.4. The Ш-Function on Its Own

We want to develop the properties of $Ш_p$ and in particular find its Fourier transform. This work is essential in analyzing X-ray diffraction, to which we'll return, and will be just as essential in the next chapter when we study sampling and interpolation.

As a model case we take the spacing p to be 1, so we sum the shifted δ's at the integer points. In this case, for convenience, we drop the subscript and just write

$$Ш(x) = \sum_{k=-\infty}^{\infty} \delta(x-k) \quad \text{or, without the variable,} \quad Ш = \sum_{k=-\infty}^{\infty} \delta_k\,.$$

We'll get back to $Ш_p$ by a scaling result.

To see that the series for Ш makes sense as a distribution, let φ be a test function and pair it with Ш:

$$\langle Ш, \varphi\rangle = \left\langle \sum_{k=-\infty}^{\infty} \delta_k, \varphi \right\rangle = \sum_{k=-\infty}^{\infty} \langle \delta_k, \varphi\rangle = \sum_{k=-\infty}^{\infty} \varphi(k)\,.$$

This sum converges because of the rapid decrease of φ at $\pm\infty$.[15]

In fact, we met Ш earlier, in Chapter 1. Rather, we met its Fourier transform in one form — it's the continuous buzz signal. Taking the Fourier transform of Ш term by term using $\mathcal{F}\delta(x-k) = e^{-2\pi i k s}$ produces

$$\mathcal{F}Ш(s) = \sum_{k=-\infty}^{\infty} e^{-2\pi i k s} = \sum_{k=-\infty}^{\infty} e^{2\pi i k s} \text{ (change } k \text{ to } -k \text{ to get the second sum).}$$

That's the continuous buzz on the right. It sounds like a signal with every harmonic present in equal amounts. It sounds terrible.

The sum of complex exponentials doesn't make sense classically, but it does make sense as a tempered distribution — the sum converges when paired with a Schwartz function — and the result really is the Fourier transform of Ш. So does that do it? No. *It misses the point entirely.* And what is the point? Patience.

5.4.1. Periodizing and sampling with $Ш_p$. There are two facets to Ш's versatility: *periodizing* and *sampling*. We'll consider each in turn, but via the convolution theorem, and what happens with $\mathcal{F}Ш_p$; they turn out to be two sides of the same coin.

Periodizing. Our introduction to $Ш_p$, its first application, was to write as a convolution the periodization of the electron density function ρ of a single atom in a crystal. The periodization of ρ reflects the physical structure of the crystal as a whole.

Periodizing, though, is a general procedure and convolving with $Ш_p$ furnishes a handy way of generating and working with periodic functions and distributions. It is still another example of the usefulness of convolution. One can view $Ш_p$ itself as periodizing δ with period p.

[15]In the middle I used the fact that the pairing of the sum is the sum of the pairings, for an infinite series. If this bothers you, take a limit of partial sums.

You encountered periodization in the problems back in Chapter 1, but not through convolution. Starting with a function $f(x)$ we just formed the sum

$$\sum_{k=-\infty}^{\infty} f(x-pk)$$

to get a periodic function of period p, provided the sum converges (which we'll assume is the case). With the latest technology we can say that

$$\sum_{k=-\infty}^{\infty} f(x-pk) = (f * \text{III}_p)(x)\,,$$

provided convolution with III_p makes sense (ditto). Convolving with III_p thus emerges as the familiar way to produce a periodic function of period p.

We're using the shifting property of δ but do keep in mind that convolving with III_p doesn't just shift f by multiples of p; it shifts and *adds up* the shifted copies. If the shifts overlap, then the values in the sum can pile up, thus causing trouble with the convergence of the series. (Think in terms of graphs of f and its shifts and draw a picture!) A common instance is to form $f * \text{III}_p$ when $f(x)$ is zero for $|x| \geq p/2$. This is one that we'll see in the next chapter. In this case the shifts do not overlap (think of graphs, again), the series is fine, and we naturally say that $f * \text{III}_p$ is the p-periodic extension of f.

Finally, note that

$$f(ax+b) * \text{III}_p(x) = \sum_{k=-\infty}^{\infty} f(ax+b-apk) \quad \text{(check this!)}$$

also has period p, and this can just as well be written in terms of a shifted III:

$$\sum_{k=-\infty}^{\infty} f(at+b-apk) = f(at) * \text{III}_p\Big(t+\frac{b}{a}\Big)\,.$$

Sampling. The flip side of periodizing with III_p is sampling with III_p. If we multiply $\text{III}_p(x)$ by a function $f(x)$, then

$$f(x)\text{III}_p(x) = \sum_{k=-\infty}^{\infty} f(x)\delta(x-kp) = \sum_{k=-\infty}^{\infty} f(kp)\delta(x-kp)\,.$$

We've sampled $f(x)$ at the points 0, $\pm p$, $\pm 2p$, $\dots$, in the sense that we've recorded the values of $f(x)$ at those points as a sum of δ's. This is an important point of view, one whose value we'll soon learn from practice:

- The sampled version of a function is not just a list of values of the function, it's the sum of the values times the corresponding δ-functions.

While one can apply this convention to any collection of sampled values, multiplying with III_p produces *evenly spaced samples.*

To summarize:

- Convolving a function with $Ш_p$ produces a periodic function with period p.
- Multiplying a function by $Ш_p$ samples the function at the points pk, $k = 0, \pm 1, \pm 2, \ldots$.

5.4.2. A scaling identity for $Ш_p$. There's a simple scaling identity that allows us to pass between Ш, with spacing 1 between δ's, and $Ш_p$, with spacing p. Start with

$$Ш(px) = \sum_{k=-\infty}^{\infty} \delta(px - k)\,,$$

and recall the scaling property of δ; for $p > 0$,

$$\delta(px) = \frac{1}{p}\delta(x)\,.$$

Plugging this into the formula for $Ш(px)$ gives

$$\begin{aligned} Ш(px) &= \sum_{k=-\infty}^{\infty} \delta(px - k) \\ &= \sum_{k=-\infty}^{\infty} \delta\left(p\Big(x - \frac{k}{p}\Big)\right) \\ &= \sum_{k=-\infty}^{\infty} \frac{1}{p}\delta\left(x - \frac{k}{p}\right) = \frac{1}{p}Ш_{1/p}(x)\,. \end{aligned}$$

To give it its own display,

$$Ш(px) = \frac{1}{p}Ш_{1/p}(x)\,.$$

Replacing p by $1/p$ in the formula gives as well

$$Ш_p(x) = \frac{1}{p}Ш\left(\frac{1}{p}x\right)\,.$$

It's a good exercise to derive these in a variable-free environment, using the delay operator τ_p and the scaling operator σ_p.

5.4.3. The Fourier transform of Ш, or: The deepest fact about the integers is well known to every electrical engineer and crystallographer. Forgive the hyperbole. I'll explain later where "deepest" came from. The title is meant to encourage engineers to spend a little time with number theory, of all things.

The moment you've been patiently waiting for is here: the Fourier transform $\mathcal{F}Ш_p$. Let's first work with $p = 1$. We go back to the definition of the Fourier transform of a tempered distribution: if φ is a Schwartz function, then

$$\langle \mathcal{F}Ш, \varphi \rangle = \langle Ш, \mathcal{F}\varphi \rangle\,.$$

On the right-hand side,

$$\langle Ш, \mathcal{F}\varphi \rangle = \left\langle \sum_{k=-\infty}^{\infty} \delta_k, \mathcal{F}\varphi \right\rangle = \sum_{k=-\infty}^{\infty} \langle \delta_k, \mathcal{F}\varphi \rangle = \sum_{k=-\infty}^{\infty} \mathcal{F}\varphi(k)\,.$$

And now we have something absolutely remarkable.

The Poisson summation formula: Let φ be a Schwartz function. Then

$$\sum_{k=-\infty}^{\infty} \mathcal{F}\varphi(k) = \sum_{k=-\infty}^{\infty} \varphi(k)\,.$$

This result actually holds for other classes of functions (the Schwartz class was certainly not known to Poisson![16]), but that's not important for us.

Picking up the calculation of $\mathcal{F}\text{Ш}$ where we left off,

$$\begin{aligned}
\langle \mathcal{F}\text{Ш}, \varphi\rangle &= \sum_{k=-\infty}^{\infty} \mathcal{F}\varphi(k) \\
&= \sum_{k=-\infty}^{\infty} \varphi(k) \quad \text{(because of the Poisson summation formula)} \\
&= \sum_{k=-\infty}^{\infty} \langle \delta_k, \varphi\rangle \quad \text{(definition of } \delta_k\text{)} \\
&= \left\langle \sum_{k=-\infty}^{\infty} \delta_k, \varphi \right\rangle \\
&= \langle \text{Ш}, \varphi\rangle\,.
\end{aligned}$$

Comparing where we started to where we finished,

$$\langle \mathcal{F}\text{Ш}, \varphi\rangle = \langle \text{Ш}, \varphi\rangle\,,$$

and we conclude that

$$\mathcal{F}\text{Ш} = \text{Ш}\,.$$

Outstanding. The Ш-function is its own Fourier transform!

From our earlier, simple-minded computation of $\mathcal{F}\text{Ш}$, this also implies the identity (as distributions)

$$\sum_{k=-\infty}^{\infty} \delta(x-k) = \text{Ш}(x) = \mathcal{F}\text{Ш}(x) = \sum_{k=-\infty}^{\infty} e^{2\pi i k x}\,.$$

Here's a plot of $\sum_{k=-20}^{20} e^{2\pi i k x}$ from Section 1.6.2. You can see that it's starting to look like a train of δ's, but we didn't know that at the time. Moreover, it's absolutely true and correct to say that

$$\lim_{N\to\infty} \sum_{k=-N}^{N} e^{2\pi i k x} \to \text{Ш}(x)$$

as distributions.

[16]Poisson was also known for his work on optics, in particular for his opposition to the wave theory as it was used to explain diffraction.

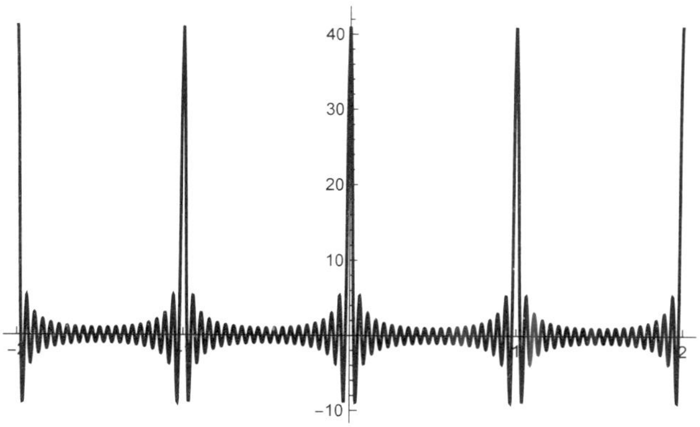

The Fourier transform of $Ш_p$. Using $\mathcal{F}Ш = Ш$, the previous scaling results for Ш, and the stretch theorem for Fourier transforms, we can deduce the formula for $\mathcal{F}Ш_p$. With

$$Ш_p(x) = \frac{1}{p}Ш\left(\frac{1}{p}x\right) \quad \text{and} \quad Ш(px) = \frac{1}{p}Ш_{1/p}(x)\,,$$

we have

$$\begin{aligned}\mathcal{F}Ш_p(s) &= \frac{1}{p}\mathcal{F}\left(Ш\left(\frac{x}{p}\right)\right)\\ &= \frac{1}{p}\,p\,\mathcal{F}Ш(ps) \quad \text{(stretch theorem)}\\ &= Ш(ps)\\ &= \frac{1}{p}Ш_{1/p}(s)\,.\end{aligned}$$

This, too, gets its own display:

$$\mathcal{F}Ш_p = \frac{1}{p}Ш_{1/p}\,.$$

What a great reciprocity result. The spacing of the δ's is p in the time domain and $1/p$ in the frequency domain, with an additional $1/p$ scaling of the strength.

Since $Ш_p$ is even, we also have

$$\mathcal{F}^{-1}Ш_p = \frac{1}{p}Ш_{1/p}\,.$$

Proof of the Poisson summation formula. I sat in on a course on analytic number theory taught by an eminent mathematician and it was he who referred to the Poisson summation formula as the deepest fact we know about the integers.[17] It's well known to every electrical engineer and crystallographer because of what it says about the Fourier transform of Ш, and they all know that $\mathcal{F}Ш = Ш$.

The proof of the Poisson summation formula is an excellent example of the power of having two different representations of the same thing, an idea certainly at the heart of Fourier analysis. Remember the maxim: if you can evaluate an expression in two different ways, it's likely you've done something significant.

[17] Other candidates? This was also prior to the proof of Fermat's Last Theorem. The Riemann hypothesis is still out there.

Given a test function $\varphi(t)$ we periodize to $\Phi(t)$ of period 1:

$$\Phi(t) = (\varphi * Ш)(t) = \sum_{k=-\infty}^{\infty} \varphi(t-k)\,.$$

As a periodic function, Φ has a Fourier series:

$$\Phi(t) = \sum_{m=-\infty}^{\infty} \hat{\Phi}(m)e^{2\pi imt}\,.$$

Let's find the Fourier coefficients:

$$\begin{aligned}
\hat{\Phi}(m) &= \int_0^1 e^{-2\pi imt}\Phi(t)\,dt \\
&= \int_0^1 \sum_{k=-\infty}^{\infty} e^{-2\pi imt}\varphi(t-k)\,dt = \sum_{k=-\infty}^{\infty} \int_0^1 e^{-2\pi imt}\varphi(t-k)\,dt \\
&= \sum_{k=-\infty}^{\infty} \int_{-k}^{-k+1} e^{-2\pi im(t+k)}\varphi(t)\,dt \quad \text{(change of variable)} \\
&= \sum_{k=-\infty}^{\infty} \int_{-k}^{-k+1} e^{-2\pi imt}e^{-2\pi imk}\varphi(t)\,dt \quad (\text{use } e^{-2\pi imk} = 1) \\
&= \int_{-\infty}^{\infty} e^{-2\pi imt}\varphi(t)\,dt \\
&= \mathcal{F}\varphi(m)\,.
\end{aligned}$$

Therefore

$$\Phi(t) = \sum_{m=-\infty}^{\infty} \mathcal{F}\varphi(m)e^{2\pi imt}\,.$$

(We've actually seen this calculation before, in a disguised form; look back to Section 3.5.4 on the relationship between the solutions of the heat equation on the line and on the circle. There was also a problem along these lines.)

Since Φ is a smooth function, the Fourier series converges. Now compute $\Phi(0)$ two ways, one way from plugging into its definition as a periodization and the other from plugging into its Fourier series:

$$\Phi(0) = \sum_{k=-\infty}^{\infty} \varphi(-k) = \sum_{k=-\infty}^{\infty} \varphi(k),$$

$$\Phi(0) = \sum_{k=-\infty}^{\infty} \mathcal{F}\varphi(k)e^{2\pi in0} = \sum_{k=-\infty}^{\infty} \mathcal{F}\varphi(k).$$

Done.

The Poisson summation formula is a statement of equality between *infinite sums*. The values of the individual terms $\varphi(k)$ and $\mathcal{F}\varphi(k)$ don't have anything to do with each other, and neither do their finite sums. It is only in the limit, for the infinite series, that we obtain this truly striking result.

5.4.4. Crystal gazing: Periodizing and sampling working together. Let's return to the setup for X-ray diffraction for a one-dimensional crystal, using the result $\mathcal{F}\text{Ш}_p = (1/p)\text{Ш}_{1/p}$ to marry periodizing and sampling via the convolution theorem. We described the electron density distribution of a single atom by a function $\rho(x)$ and the electron density distribution of the crystal with spacing p as

$$\rho_p(x) = \sum_{k=-\infty}^{\infty} \rho(x - kp) = (\rho * \text{Ш}_p)(x) .$$

Then

$$\begin{aligned} \mathcal{F}\rho_p(s) &= \mathcal{F}(\rho * \text{Ш}_p)(s) \\ &= (\mathcal{F}\rho \cdot \mathcal{F}\text{Ш}_p)(s) \quad \text{(convolution theorem)} \\ &= \mathcal{F}\rho(s)\frac{1}{p}\text{Ш}_{1/p}(s) \\ &= \sum_{k=-\infty}^{\infty} \frac{1}{p}\mathcal{F}\rho\left(\frac{k}{p}\right)\delta\left(s - \frac{k}{p}\right) . \end{aligned}$$

Here's the significance of this. In an X-ray diffraction experiment what you see on the X-ray detector is a bunch of spots, corresponding to $\mathcal{F}\rho_p$. The intensity of each spot is proportional to the squared magnitude of the Fourier transform of the electron density ρ and, according to the formula, the spots are spaced a distance $1/p$ apart, *not* p apart. If you were an X-ray crystallographer and didn't know your Fourier transforms, you might assume that there is a relation of direct proportion between the spacing of the spots on the detector and the atoms in the crystal. But there's not; it's a *reciprocal* relation — kiss your Nobel Prize goodbye. Every crystallographer knows this.

You now also see the significance of approximating the crystal by an infinite array of atoms along a line. It is for such an infinite train of δ's that we have the Fourier transform result $\mathcal{F}\text{Ш}_p = (1/p)\text{Ш}_{1/p}$.

We'll see a similar relation when we consider higher-dimensional Fourier transforms and higher-dimensional Ш-functions. A Ш-function will be associated with a *lattice*, and the Fourier transform will be a Ш-function associated with the *reciprocal* or *dual* lattice. This phenomenon has turned out to be important in image processing, too; see, for example, *Digital Video Processing* by A. M. Tekalp.

To conclude, in diffraction, nature takes a Fourier transform. Nature even knows the Poisson summation formula.

There's no shortage of books on X-ray diffraction and crystallography. One that's very complete and very clear is *Crystals, X-rays, and Proteins* by D. Sherwood and J. Cooper, published by Oxford Press. In particular, you can read about how X-ray crystallography experiments are set up to record the diffraction patterns and hence the atomic spacings. It's not just directly finding a p spacing in Ångströms in the crystal through a $1/p$ spacing in the detector, for that would mean that separation of atoms in the crystal would produce spots that are, oh, planetary distances apart. You'll run across *Ewald spheres* in this context. An added complication for real, 3D crystals is that you're working with projections onto a 2D array of detectors of the 3D configurations.

5.5. Periodic Distributions and Fourier Series

I want to collect a few facts about periodic distributions and Fourier series and show how we can use Ш as a convenient tool for classical Fourier series.

More on periodicity via $Ш_p$. First, the notion of periodicity for distributions can be expressed as invariance under the delay operator τ_p. Namely, a distribution T is periodic with period p if

$$\tau_p T = T\,.$$

This is the variable-free definition of periodicity, and it applies as well to a classical function. If we employ the delay operator, the condition for a function to be periodic is then that $\varphi(x-p) = \tau_p\varphi(x) = \varphi(x)$ rather than the more common way of writing the condition as $\varphi(x+p) = \varphi(x)$. It's just how we defined the delay operator and it doesn't make any real difference which way periodicity is defined. That was an exercise in Chapter 1.

For a distribution, we're not supposed to write $\tau_p T(x) = T(x-p)$. Rather, pairing with a test function φ the definition of $\tau_p T$ is

$$\langle \tau_p T, \varphi\rangle = \langle T, \tau_{-p}\varphi\rangle \quad \text{where } \tau_{-p}\varphi(x) = \varphi(x+p).$$

In terms of a pairing the periodicity condition $\tau_p T = T$ is then

$$\langle T, \varphi\rangle = \langle \tau_p T, \varphi\rangle = \langle T, \tau_{-p}\varphi\rangle\,.$$

Using this definition, it is a pleasure to report that $Ш_p$ is periodic with period p:

$$\tau_p Ш_p = Ш_p\,.$$

As we found earlier, $Ш_p$ is the p-periodization of δ. When we periodize φ by forming the convolution,

$$\Phi(x) = (\varphi * Ш_p)(x)\,,$$

it's natural to view the periodicity of Φ as a *consequence* of the periodicity of $Ш_p$. By this I mean we can appeal to the following:

- For distributions, if S or T is periodic of period p, then $S * T$ (when it is defined) is periodic of period p.

You can show that yourself, but don't get caught up worrying about conditions for the convolution of distributions to exist.

So, on the one hand, convolving with $Ш_p$ produces a periodic function, as we know. On the other hand, suppose Φ is periodic of period p and we cut out one period of it by forming $\Pi_p\Phi$, as in a low-pass filter. We get Φ back, *in toto*, by forming the convolution with $Ш_p$:[18]

$$\Phi = (\Pi_p\Phi) * Ш_p\,.$$

[18] Well, this is almost right. The cutoff $\Pi_p\Phi$ is zero at $\pm p/2$ while $\Phi(\pm p/2)$ may not be zero. These possible exceptions at the endpoints won't affect the discussion here in any substantive way.

Put loosely, the upshot of this is that a function, or a distribution, is periodic *if and only if* it is a convolution with III_p. This is a nice point of view and points to III_p as a fundamental object. I'll push this a little further in Section 5.8.

Fourier transforms and Fourier series. When we first started to work with tempered distributions, I said that we would be able to take the Fourier transform of functions that didn't have a classical Fourier transform. We've made good on that promise, including complex exponentials, for which

$$\mathcal{F}e^{2\pi ikt/p} = \delta\left(s - \frac{k}{p}\right).$$

With this we can now find the Fourier transform of a Fourier series. If

$$\varphi(t) = \sum_{k=-\infty}^{\infty} c_k e^{2\pi ikt/p},$$

then

$$\mathcal{F}\varphi(s) = \sum_{k=-\infty}^{\infty} c_k \mathcal{F}e^{2\pi ikt/p} = \sum_{k=-\infty}^{\infty} c_k \delta\left(s - \frac{k}{p}\right).$$

We can turn this around and rederive the formula for Fourier series as a consequence of our work on Fourier transforms. Suppose Φ is periodic of period p and write, as we know we can,

$$\Phi = \varphi * \text{III}_p,$$

where φ is one period of Φ, say $\varphi = \Pi_p \Phi$. Take the Fourier transform of both sides and boldly invoke the convolution theorem:

$$\mathcal{F}\Phi = \mathcal{F}(\varphi * \text{III}_p) = \mathcal{F}\varphi \cdot \mathcal{F}\text{III}_p = \mathcal{F}\varphi \cdot \frac{1}{p}\text{III}_{1/p},$$

or, at points,

$$\mathcal{F}\Phi(s) = \mathcal{F}\varphi(s)\left(\frac{1}{p}\sum_{k=-\infty}^{\infty}\delta\left(s - \frac{k}{p}\right)\right) = \frac{1}{p}\sum_{k=-\infty}^{\infty}\mathcal{F}\varphi\left(\frac{k}{p}\right)\delta\left(s - \frac{k}{p}\right).$$

Now boldly take the inverse Fourier transform:

$$\Phi(t) = \sum_{k=-\infty}^{\infty}\frac{1}{p}\mathcal{F}\varphi\left(\frac{k}{p}\right)e^{2\pi ikt/p} \quad \left(\text{the } \mathcal{F}\varphi\left(\frac{k}{p}\right) \text{ are constants}\right).$$

But

$$\begin{aligned}\frac{1}{p}\mathcal{F}\varphi\left(\frac{k}{p}\right) &= \frac{1}{p}\int_{-\infty}^{\infty} e^{-2\pi i(k/p)t}\varphi(t)\,dt \\ &= \frac{1}{p}\int_{-\infty}^{\infty} e^{-2\pi i(k/p)t}\,\Pi_p(t)\Phi(t)\,dt = \frac{1}{p}\int_{-p/2}^{p/2} e^{-2\pi i(k/p)t}\,\Phi(t)\,dt,\end{aligned}$$

and this is the kth Fourier coefficient c_k of Φ. We've rederived

$$\Phi(t) = \sum_{k=-\infty}^{\infty} c_k e^{2\pi ikt/p}, \quad \text{where } c_k = \frac{1}{p}\int_{-p/2}^{p/2} e^{-2\pi i(k/p)t}\,\Phi(t)\,dt.$$

As always, cool. I'm not saying this gives an easy way of computing Fourier series; it just closes the loop.

5.6. A Formula for δ Applied to a Function, and a Mention of Pullbacks

For some applications it's useful to have an impulse applied to a function, meaning an expression of the form $\delta(g(x))$ where $g(x)$ is a real-valued function. It's a designer δ, where we expect (we want) $\delta(g(x))$ to be impulsive at any points where $g(x) = 0$. We'd like a formula that puts the zeros of $g(x)$ explicitly on display. To find this, the idea is to use systematically the unit step function $H(x)$ and its derivative $H' = \delta$.

We assume that $g(x)$ is differentiable and that the x's where $g'(x) = 0$ constitute a discrete set of points. Then the zeros of $g(x)$ are also a discrete set. (Reason that through.) We further assume that $g'(x)$ is *not* zero at any place where $g(x)$ *is* zero, so $g(x)$ *crosses* the x-axis wherever it hits the x-axis and is not tangent at such points.

Here's one way to analyze the situation. We will proceed without justifying some of the steps — the derivation is simple and will make clear *why* the formula comes out the way it does — but every step can be rigorously established. Consider the function $H(g(x))$. Written out in symbols, the definition is

$$H(g(x)) = \begin{cases} 1, & g(x) > 0\,, \\ 0, & g(x) \le 0\,, \end{cases}$$

though it's probably simpler to think in words and in terms of the graph: $H(g(x))$ is 1 on the intervals where $g(x)$ is positive and it is 0 on the intervals where $g(x)$ is negative or zero. (Put another way, $H(g(x))$ quantizes $g(x)$ to two levels, rounding to 1 when $g(x)$ is positive and rounding to 0 when $g(x)$ is negative or zero.) There are jump discontinuities in $H(g(x))$ at the points where $g(x)$ crosses the x-axis; moving from left to right along the x-axis, $H(g(x))$ takes unit steps up and unit steps down at the points where $g(x) = 0$. Here's a typical picture.

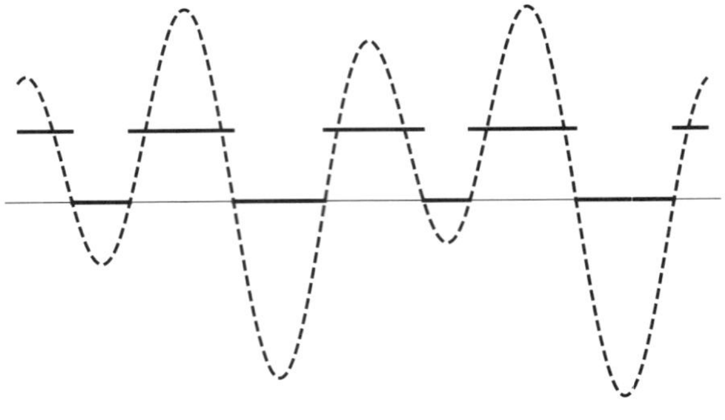

Let $\{x_k\}$ be the set of zeros of $g(x)$. Each x_k is either a "step up" or a "step down" discontinuity of $H(g(x))$, briefly "Up" or "Down" points, and whichever kind of zero x_k is, the next zero x_{k+1} is the other kind.

What about the derivative $(d/dx)H(g(x))$? On the one hand, it follows directly from the definition of $H(g(x))$ and the discussion above that

$$\frac{d}{dx}H(g(x)) = \sum_{x_k \in \mathrm{Up}} \delta(x - x_k) - \sum_{x_k \in \mathrm{Down}} \delta(x - x_k)\,.$$

On the other hand, applying the chain rule (legit) yields

$$\frac{d}{dx}H(g(x)) = H'(g(x))g'(x) = \delta(g(x))g'(x)\,.$$

Thus

$$\delta(g(x))g'(x) = \sum_{x_k \in \mathrm{Up}} \delta(x - x_k) - \sum_{x_k \in \mathrm{Down}} \delta(x - x_k)\,.$$

When $g'(x) \neq 0$, we can divide and write

$$\begin{aligned}\delta(g(x)) &= \sum_{x_k \in \mathrm{Up}} \frac{\delta(x - x_k)}{g'(x)} - \sum_{x_k \in \mathrm{Down}} \frac{\delta(x - x_k)}{g'(x)}\\ &= \sum_{x_k \in \mathrm{Up}} \frac{\delta(x - x_k)}{g'(x_k)} - \sum_{x_k \in \mathrm{Down}} \frac{\delta(x - x_k)}{g'(x_k)}.\end{aligned}$$

Observe that $g'(x_k)$ is positive at $x_k \in$ Up and negative at $x_k \in$ Down. We can then incorporate the minus sign in front of the second sum and combine the two sums to obtain the formula

$$\delta(g(x)) = \sum_k \frac{\delta(x - x_k)}{|g'(x_k)|}\,.$$

There it is! (This is actually a pretty well-known formula.)

And how does $\delta(g(x))$ operate (pair) with a test function $\varphi(x)$? Writing the pairing as an integral, it is

$$\int_{-\infty}^{\infty} \delta(g(x))\varphi(x)\,dx = \sum_k \frac{\varphi(x_k)}{|g'(x_k)|}\,.$$

5.6.1. Pullbacks. In more dignified mathematical treatments, with even more emphasis on distributions and test functions than you've seen here, expressions such as $\delta(g(x))$ are defined via pullbacks. A word about this here, with further discussion in Section 9.9.

For functions, "pullback" is just another way of talking about composition, but the terminology seems more dynamic. Remember the composition notation $(f \circ g)(x) = f(g(x))$. Instead of saying "f composed with g," one says that $f \circ g$ is the *pullback* of f *by* g, and mathematicians write this as

$$f \circ g = g^* f \quad \text{and} \quad f(g(x)) = (g^* f)(x)\,.$$

That's * as a superscript, *not* to be confused with any kind of convolution. The domain of the composition $f \circ g$ is the domain of g, and, in that light, the points where $g^* f$ is to be applied have been "pulled back" from where f is defined to where g is defined. The terminology has caught on among mathematicians. Impress your friends. It's possible (some would say desirable) to look at the delay operator τ_b and the scaling operator σ_a in terms of pullbacks. For this, define

$$\tau_b(x) = x - b\,,$$

$$\sigma_a(x) = ax\,, \quad a \neq 0\,,$$

just as functions defined on $\mathbb{R}$, rather than as operating on signals. Then to delay or to scale a signal $f(x)$ is to form the compositions

$$(f \circ \tau_b)(x) = f(\tau_b(x)) = f(x-b)\,,$$
$$(f \circ \sigma_a)(x) = f(\sigma_a(x)) = f(ax)\,.$$

Using the pullback notation we would write

$$(\tau_b^* f)(x) = f(x-b)\,,$$
$$(\sigma_a^* f)(x) = f(ax)\,.$$

Here's what the shift and stretch theorems look like in pullback notation:

$$\mathcal{F}(\tau_b^* f)(s) = e^{-2\pi i s b}\mathcal{F}f(s)\,,$$
$$\mathcal{F}(\sigma_a^* f)(s) = \frac{1}{|a|}(\sigma_{1/a}^* \mathcal{F}f)(s)\,.$$

We haven't taken this point of view on the various occasions when we've talked about delays and scaling, preferring instead to think of delay and scaling as operators and writing $\tau_b f$ and $\sigma_a f$ (no $*$'s). But the pullback notion and notation are used, and I wanted you to be aware of it.

The preceding discussion is just for functions. In considering $\delta(g(x))$ we'd be looking to define $g^*\delta$ as a distribution and that's more complicated to do because distributions aren't evaluated at points; $g^*\delta$ is not a simple composition $\delta \circ g$. We have to define $g^*\delta$ via a pairing with a test function $\varphi(x)$, and writing this as an integral we'd want to work with

$$\int_{-\infty}^{\infty} \delta(g(x))\varphi(x)\,dx\,.$$

This naturally calls for the substitution $u = g(x)$. This approach can be coaxed into giving the formula we had before, but not easily.

Unnecessary complication and abstraction? Not once you're in the distributional mindframe; see, again, Friedlander's *Introduction to the Theory of Distributions*. The more general viewpoint offers a systematic approach to a construction that comes up often, and that's useful to know. But this is not to say that you should be worried about writing $\delta(g(x))$ and applying the earlier formula as above. We'll come back to pullback of a distribution in Section 9.9 when we talk about line impulses.

One final comment. It's also possible to view the shifted and scaled δ in terms of pullbacks by the functions τ_b and σ_a. We have that $\tau_b^*\delta = \delta_b$ is the shifted δ and $\sigma_a^*\delta$ is the scaled δ. The formula $\delta(ax) = (1/|a|)\delta(x)$ can be written as

$$\sigma_a^*\delta = \frac{1}{|a|}\delta\,.$$

5.7. Cutting Off a δ

In the last chapter we defined the product of a function g and a distribution T by the pairing

$$\langle gT, \varphi \rangle = \langle T, g\varphi \rangle \,,$$

where φ is a test function. This led to the hard-working sampling property $g(x)\delta(x-a) = g(a)\delta(x-a)$, in use every day all around the world.

The catch is that for $\langle T, g\varphi \rangle$ to make sense, $g\varphi$ must also be a test function. Hence we have to assume something about g, typically that it's smooth. There are occasions — there will be some in the next chapter — where one wants to roughen things up. Specifically, we would like to know what to do with the product of a rectangle function and a δ, as in $\Pi_p(x)\delta(x-q)$.

In two situations it's clear what should happen. If either $|q| < p/2$ or $|q| > p/2$, then $\delta(x-q)$ is strictly inside or strictly outside the rectangle, and we should declare

$$\Pi_p(x)\delta(x-q) = \begin{cases} \delta(x-q)\,, & |q| < \frac{p}{2}\,, \\ 0\,, & |q| > \frac{p}{2}\,. \end{cases}$$

The interesting case is on the edge. What to make of $\Pi_p(x)\delta(x \pm p/2)$?

We'll see what to do via the limit of approximating rectangle functions for $\delta(t - p/2)$, cut off by $\Pi_p(x)$. Take the right endpoint, $p/2$, and for $0 < \epsilon < p/2$ let

$$R_\epsilon(x) = \Pi_p(x)\frac{1}{\epsilon}\Pi_\epsilon\Big(x - \frac{p}{2}\Big) = \begin{cases} \frac{1}{\epsilon}, & \frac{p}{2} - \frac{\epsilon}{2} < x < \frac{p}{2}, \\ 0, & \text{otherwise.} \end{cases}$$

Integrate $R_\epsilon(x)$ against a test function φ and use a Taylor approximation to φ at $p/2$:

$$\begin{aligned} \langle R_\epsilon, \varphi \rangle &= \frac{1}{\epsilon}\int_{\frac{p}{2}-\frac{\epsilon}{2}}^{\frac{p}{2}} \varphi(x)\,dx \\ &= \frac{1}{\epsilon}\int_{\frac{p}{2}-\frac{\epsilon}{2}}^{\frac{p}{2}} (\varphi(p/2) + \varphi'(p/2))\Big(x - \frac{p}{2}\Big) + O\Big(\Big(x - \frac{p}{2}\Big)^2\Big)\,dx \\ &= \frac{1}{2}\varphi\,(p/2) + O(\epsilon)\,. \end{aligned}$$

In the limit,

$$\lim_{\epsilon \to 0}\langle R_\epsilon, \varphi \rangle = \frac{1}{2}\varphi\,(p/2) = \Big\langle \frac{1}{2}\delta_{p/2}, \varphi \Big\rangle\,.$$

This gives us the interesting result

$$\Pi_p(x)\delta\Big(x - \frac{p}{2}\Big) = \frac{1}{2}\delta\Big(x - \frac{p}{2}\Big)\,.$$

We'd have the analogous result at the other endpoint, $-p/2$. So:

- Cutting off a δ at the edges of a rectangle function gives half the δ at that edge:

$$\Pi_p(x)\delta\Big(x \pm \frac{p}{2}\Big) = \frac{1}{2}\delta\Big(x \pm \frac{p}{2}\Big)\,.$$

If we define[19] $\Pi(\pm p/2) = 1/2$, then we can state all the cutoff results together consistently with the sampling property of δ:

$$\Pi_p(x)\delta(x-q) = \Pi_p(q)\delta(x-q)\,.$$

5.8. Appendix: How Special Is Ш?

I'd like to elaborate on Ш as a kind of fundamental object in discussing periodicity. To begin with, let's show the following:

- If S is a periodic, tempered distribution of period 1 with $\mathcal{F}S = S$, then $S = c$Ш for a constant $c \in \mathbb{R}$.

The argument goes like this. First we observe that such an S is even, for

$$\mathcal{F}S^- = \mathcal{F}^{-1}S = S = \mathcal{F}S\,,$$

from which

$$S^- = S\,.$$

Now expand S in a Fourier series

$$S = \sum_{n=-\infty}^{\infty} c_n e^{2\pi int}\,.$$

This is perfectly legit for periodic, tempered distributions, though I haven't previously quoted the requisite results justifying it (and I won't). Since S is even, the coefficients c_n are real, and then also $c_{-n} = c_n$. Using $\mathcal{F}S = S$ we may write

$$\sum_{n=-\infty}^{\infty} c_n e^{2\pi int} = S = \mathcal{F}S = \sum_{n=-\infty}^{\infty} c_n\delta_n\,.$$

We'll prove by induction that

$$c_n = c_0 \quad \text{for all } n \geq 0$$

and hence that

$$S = c_0 \sum_{n=-\infty}^{\infty} \delta_n = c_0\text{Ш}\,.$$

The assertion is trivially true for $n = 0$. Suppose it has been proved for $n \leq N-1$. Let $p = 2N+1$ and consider the cutoff $\Pi_p\mathcal{F}S$. Then on the one hand,

$$\Pi_p\mathcal{F}S = \sum_{n=-(N-1)}^{N-1} c_n\delta_n + (c_N\delta_N + c_{-N}\delta_{-N}) = c_0 \sum_{n=-(N-1)}^{N-1} \delta_n + c_N(\delta_N + \delta_{-N})$$

using the induction hypothesis. On the other hand,

$$\Pi_p\mathcal{F}S = \Pi_p S = \sum_{n=-\infty}^{\infty} c_n \Pi_p e^{2\pi ins}.$$

[19] I'm about ready to concede this.

Integrate both sides from $-\infty$ to ∞ (or pair with an appropriate test function — however you'd like to say it). This results in

$$(2N-1)c_0 + 2c_N = \sum_{n=-\infty}^{\infty} c_n \int_{-N-\frac{1}{2}}^{N+\frac{1}{2}} e^{2\pi i n s}\, ds = (2N+1)c_0 = (2N-1)c_0 + 2c_0\,.$$

Thus

$$c_N = c_0\,,$$

completing the induction.

Remark. The hypothesis that the period is 1 is necessary, since $\text{III}(2x)+\frac{1}{2}\text{III}(x/2)$ is periodic of period two and is its own Fourier transform. (In general, if f is even, then $f + \mathcal{F}f$ is its own Fourier transform.)

The preceding uniqueness result coupled with the existence result $\mathcal{F}\text{III} = \text{III}$, and with a normalization on c_0 tacked on, results in:

- There is exactly one periodic, tempered distribution of period 1 that is equal to its Fourier transform and has zeroth Fourier coefficient equal to 1.

In mathematics there are many such "there is exactly one of these" theorems. They often indicate what one might take to be essential building blocks of a subject. It's a little late in the day to decide to base all discussion of periodic functions on III, but there might be some things to do.

Problems and Further Results

5.1. Suppose we have a real and even function $g(t)$ with compact support as shown below.

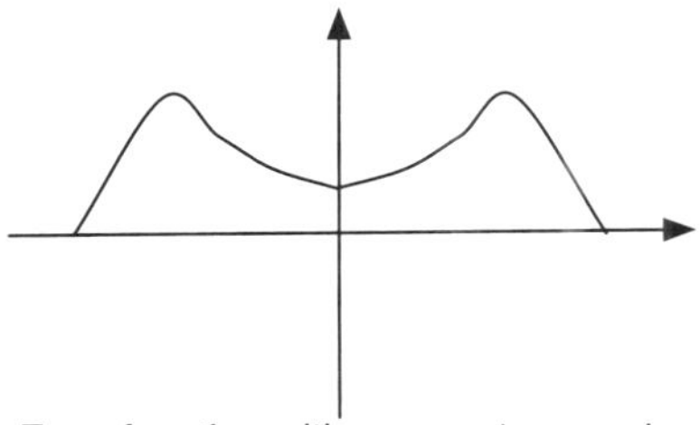

Even function with compact support

Now construct the function $f(t) = g(t) * [\delta(t+t_1) - \delta(t-t_1)]$ where $t_1 > t_0$. Can the phase profile $\angle\mathcal{F}f(s)$ be of the form shown below? Explain your reasoning.

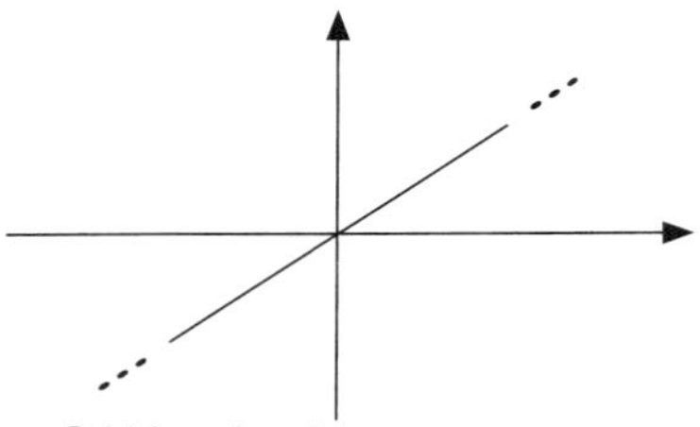

5.2. *Modified cosines*

Recall that the Fourier transform of $\cos 2\pi a t$ is $\frac{1}{2}(\delta(x-a)+\delta(x+a))$. Find the Fourier transform of the following modified cosine functions.

(a)

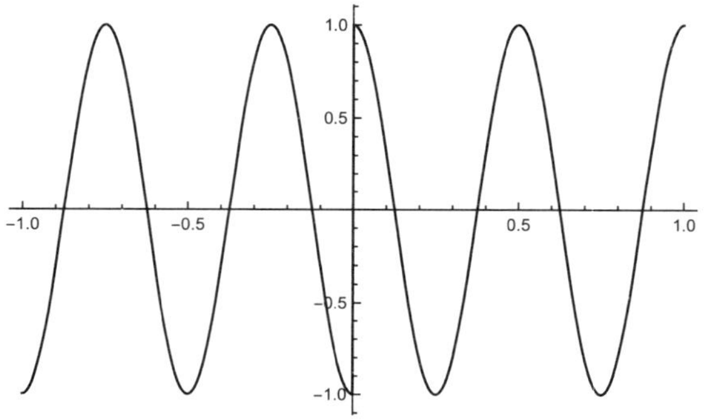

(The signals continue to the right and left.)

(b)

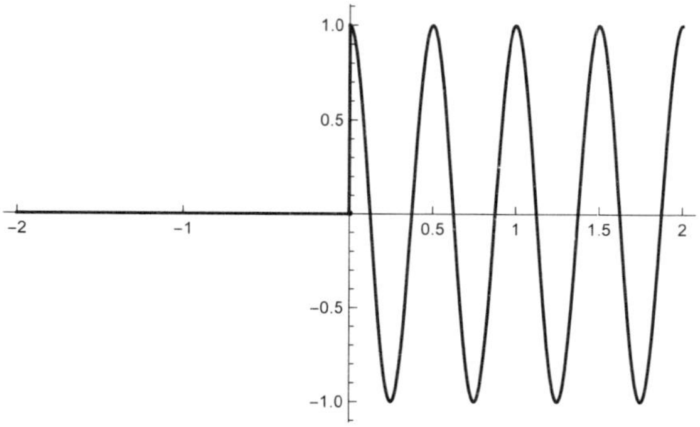

(The signals continue to the right and left.)

(c)

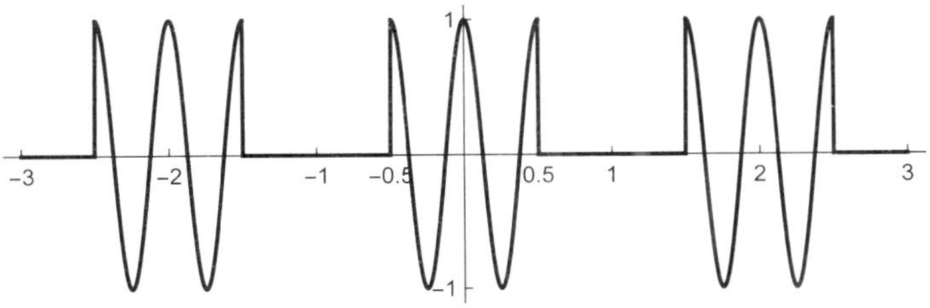

(The signal is periodic.)

5.3. *Short-time Fourier transform of* δ

Recall the short-time Fourier transform (STFT) of a signal $f(t)$, from the problems in Chapter 2:

$$F(\tau, s) = \int_{-\infty}^{\infty} f(t)w(t-\tau)e^{-2\pi i s t}dt.$$

The window function $w(t)$ is real valued, has a Fourier transform $W(s)$, and satisfies the scaling condition:

$$W(0) = \int_{-\infty}^{\infty} w(t)dt = 1.$$

Find the STFT of $f(t) = \delta(t - t_0)$.

5.4. *More on windowing*[20]

Consider a signal of the form

$$f(t) = \sum_{n-=0}^{N-1} A_n e^{2\pi i n t}. \qquad (*)$$

From the spectrum $\mathcal{F}f(s)$, we wish to identify the amplitudes A_n. Note that $f(t)$ is a signal with infinite support (a signal that goes on infinitely) and hence, in practice, we will only be able to observe it on a finite window. We model this limitation with a windowing function $w(t)$ — in other words, we only observe the spectrum of the windowed signal $g(t) = f(t)w(t)$.

We have seen various window functions. Here we take $w(t) = a \operatorname{sinc} at$, $a > 0$. The number $1/a$ is half the width of the window.

We say that the width a is *valid* if for any $f(t)$ of the form (*), we have

$$\mathcal{F}g(n) = A_n, \qquad n = 0, 1, 2, \dots, N-1.$$

Ideally, we would like a to be as large as possible, because that allows us to determine f from as few readings as possible.

(a) Find and sketch the Fourier transform of the windowed function $g(t) = f(t)w(t)$.

(b) What is the largest valid a?

(c) What is the largest valid a when $w(t) = a \operatorname{sinc}^2 at$?

---

[20]From A. Siripuram.

5.5. *Frequency modulation and music*[21]

A frequency modulated (FM) signal is one whose frequency is a function of time:

$$x(t) = A\cos(2\pi f(t)).$$

FM signals are central to many scientific fields. Most notably, they are used in communications where

$$f(t) = f_c t + k\int_{-\infty}^{t} m(t)\,dt.$$

Here f_c is the *carrier frequency* (typically a large value is necessary for the physics of wave propagation), k is a constant known as the *frequency modulation index*, and $m(t)$ is the function with the information and doing the modulating. You set your receiver to f_c to listen to the signal.

Another famous (and profitable) application of FM signals is in the digital synthesis of music, pioneered by John Chowning at Stanford. We'll take a closer look at how this done later in this problem.

Let's start by taking the case where we modulate a pure tone,

$$x(t) = A\cos(2\pi f_c t + k\sin(2\pi f_m t)).$$

Here's a plot for $0 \le t \le 1$ with $A = 1$, $f_c = 20$, $k = 2$, and $f_m = 5$.

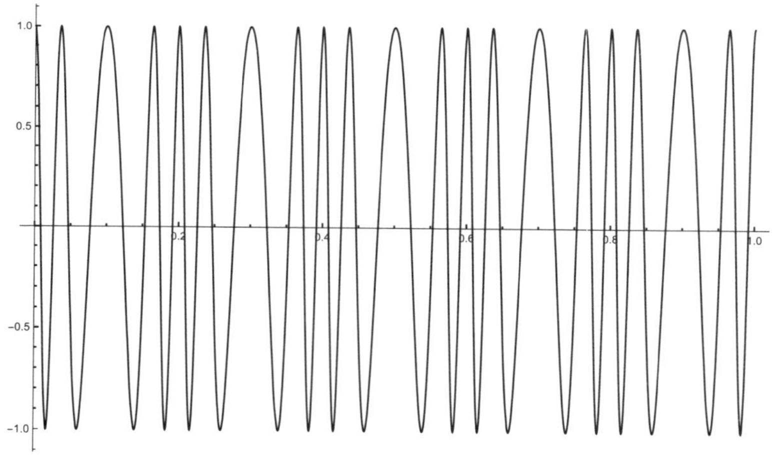

What is the spectrum? Remember Bessel functions, introduced in a problem back in Chapter 1? The answer depends on these. Let's recall: the Bessel equation of order n is

$$x^2y'' + xy' + (x^2 - n^2)y = 0\,.$$

A solution of the equation is the *Bessel function of the first kind of order* n, given by the integral

$$J_n(x) = \frac{1}{2\pi}\int_0^{2\pi} \cos(x\sin\theta - n\theta)\,d\theta\,.$$

[21] From D. Kammler's book via Raj Bhatnagar.

You showed in a problem in Chapter 1 that

$$e^{ix\sin\theta} = \sum_{n=-\infty}^{\infty} J_n(x)e^{in\theta} \qquad \text{and} \qquad e^{ix\cos\theta} = \sum_{n=-\infty}^{\infty} i^n J_n(x)e^{in\theta}.$$

(a) Show the Fourier series relationship

$$\exp(2\pi i f_c t + ik\sin(2\pi f_m t)) = \sum_{n=-\infty^{\infty}} J_n(k)\exp(2\pi i(f_c + nf_m)t).$$

Use this result to show that the Fourier transform of $x(t)$ is

$$\mathcal{F}x(s) = \frac{A}{2}\sum_{n=-\infty}^{\infty} J_n(k)[\delta(s-(f_c+nf_m)) + \delta(s+(f_c+nf_m))].$$

Hint: What is the real part of the Fourier series relationship?

(b) Download the file `fm.mat` from the course site. It contains `g`, a guitar playing the G above middle C — approximately 392 Hz; `t`, the time vector; and `fs`, the sampling frequency. Listen to the note and plot its spectrogram:

```
load fm.mat;
soundsc(g,fs);
spectrogram(g, 256, 64, 256, fs, 'yaxis');
```

Approximately how many frequencies make up this note? Let's say we modeled this note by the function

$$g(t) = \sum_{n=1}^{N} A_n(t)\sin(2\pi f_n t),$$

where the amplitudes $A_n(t)$ are of the form $A_n e^{-t/t_n}$ so that the note dies down after some time. About how many parameters (f_n, A_n, t_n) would we need?

Early pioneers in computer music synthesized notes in exactly this way, a technique now known as *additive synthesis*. The large number of parameters needed to characterize each note of each instrument led musicians to look for other ways to produce musical tones. The rest of this problem will explore an approach devised by John Chowning in the 1970s that uses frequency modulation synthesis to dynamically change the signal's frequency content over time. Chowning's model was easy to implement as he was able borrow many ideas from FM circuits built by communications engineers. The first digital implementation of Chowning's research was realized in 1983 in Yamaha's DX7 synthesizer.

(c) Set $f_c = 392$ Hz and $f_m = 196$ Hz and plot $\mathcal{F}x(s)$ when the modulation index k is an integer between 1 and 4 inclusive (MATLAB's `besselj` command will be useful). Which frequencies are significant and which are negligible? How does this depend on k? You should find that the energy of an FM signal is concentrated in a band of frequencies dependent on k, an observation known as *Chowning's rule*.

(d) Let's take a closer look at that guitar note again:

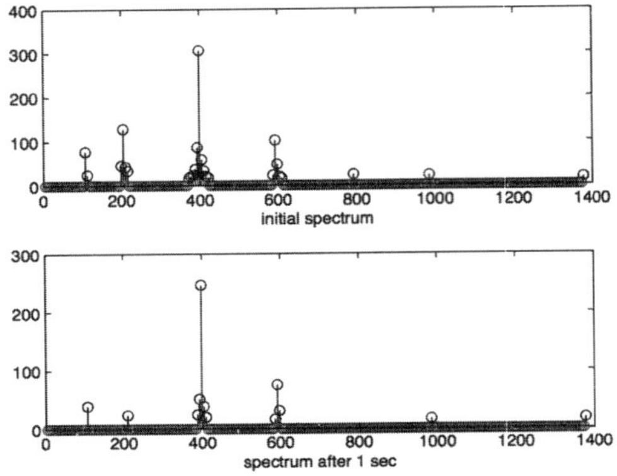

Notice that the spectrum depends on time: initially, there are many frequencies, but after a second or so, we're left with just the fundamental frequency and a few smaller overtones. The FM synthesis model of a note takes this into account by letting the modulation index k be a function of time:

$$\begin{aligned} &A(t)\cos(2\pi f_c t + k(t)\sin(2\pi f_m t)) \\ &\qquad = A(t)\sum_{n=-\infty}^{\infty} J_n(k(t))\cos(2\pi(f_c + nf_m)t). \end{aligned}$$

What values of $A(t)$ and $k(t)$ should we use to mimic a guitar playing the G above middle C using this model? Your note will not sound exactly like a guitar, but it should be somewhat better than a pure sinusoid. *Hint: Use the same values of f_c and f_m from part* (c) *and try using exponential decays for $k(t)$ and $A(t)$.*

(e) Chowning tested the FM model with the parameters $f_c = 80$ Hz, $f_m = 110$ Hz, $k(t) = 4e^{-t/11}$, and $A(t) = e^{-t/16.66}$. Listen to the sound that this produces. Which of these instruments do you think Chowning was trying to model: piano, violin, bell, or drum?

More information about FM synthesis and music can be found in Chowning's paper at `http://users.ece.gatech.edu/~mcclella/2025/labs-s05/Chowning.pdf`.

5.6. *Poisson summation based on period T*

The Poisson summation formula can be formulated more generally. Let φ be a Schwartz function and let $T > 0$. Show that

$$\sum_{n=-\infty}^{\infty} \varphi(nT) = \frac{1}{T}\sum_{n=-\infty}^{\infty} \mathcal{F}\varphi(n/T).$$

Do this in two ways, by following the earlier proof but periodizing φ to have period T (just for practice, to appreciate the argument) and by appealing directly to the original Poisson summation formula but applied to a scaled version of φ.

5.7. *Duality via the Poisson summation formula*

One of our basic duality formulas, from back in Chapter 2, is $\mathcal{F}\mathcal{F}f = f^-$; we used it a lot. This follows from Fourier inversion and is actually equivalent to Fourier inversion. Here we'll use the Poisson summation formula (as given in Problem 5.6) to derive the duality formula when f is a Schwartz function. This means that Fourier inversion is itself a consequence of the Poisson summation formula.[22]

Let f be a Schwartz function.

(a) Show that

$$\mathcal{F}\mathcal{F}(\tau_a f) = \tau_{-a}(\mathcal{F}\mathcal{F}f).$$

Here $\tau_a f(x) = f(x-a)$, as usual.

(b) Using this result, show that

$$\mathcal{F}\mathcal{F}f(0) = f(0) \quad \Rightarrow \quad \mathcal{F}\mathcal{F}f(a) = f(-a) \quad \text{for any } a.$$

Thus if we can prove the duality formula at 0, meaning $\mathcal{F}\mathcal{F}f(0) = f^-(0) = f(0)$, then it holds everywhere.

The argument is completed as follows: from the Poisson summation formula we have for $T \neq 0$

$$\sum_{n=-\infty}^{\infty} f(Tn) = \frac{1}{T} \sum_{n=-\infty}^{\infty} \mathcal{F}f(n/T).$$

Let $T \to \infty$.

(c) The left-hand side tends to $f(0)$. Why?

(d) The right-hand side tends to $\int_{-\infty}^{\infty} \mathcal{F}f(y)\,dy$. Why?

The integral is $\mathcal{F}\mathcal{F}f(0)$, and we're done.

5.8. *Fourier transform of* sech *using the Poisson summation formula*[23]

The *hyperbolic secant*, denoted by sech, is defined by

$$\operatorname{sech} t = \frac{1}{\cosh t} = \frac{2}{e^t + e^{-t}}.$$

[22] A. Robert, A short proof of the Fourier inversion formula, Proc. Amer. Math. Soc. **59** (1976), 287–288.

[23] From Aditya Siripuram.

(An alternate expression, which will come up in the calculations in this problem, is $\operatorname{sech} t = 2e^{-t}/(1+e^{-2t})$.)

The Fourier transform of $\operatorname{sech} \pi t$ is $\operatorname{sech} \pi s$, certainly not an obvious result. The usual approach to proving this seems to be contour integration, but here is a way to do it via the Poisson summation formula.

(a) Use the Poisson summation formula to show that

$$\operatorname{sech}(\pi x) = \frac{1}{\pi} \sum_{k=0}^{\infty} \frac{(-1)^k(2k+1)}{(k+1/2)^2 + x^2}.$$

(b) Deduce that the Fourier transform of $\operatorname{sech} \pi t$ is $\operatorname{sech} \pi s$.

5.9. A diffraction grating consists of N slits, each of width w, with the centers of each of the slits separated from its neighbors by an amount a. Which of the following is a possible expression for this grating (the aperture function)? Explain.

(a) $\Pi_{Na}(x) * \Pi_w(x) * Ш_a(x)$.
(b) $\Pi_{Na}(x)(\Pi_w(x) * Ш_a(x))$.
(c) $\Pi_{Na}(x) * (\Pi_w(x) Ш_a(x))$.
(d) $(\Pi_{Na}(x) * \Pi_w(x)) Ш_a(x)$.

5.10. *Sampling patterns*

Consider sampling a real, even signal f whose spectrum is shown below.

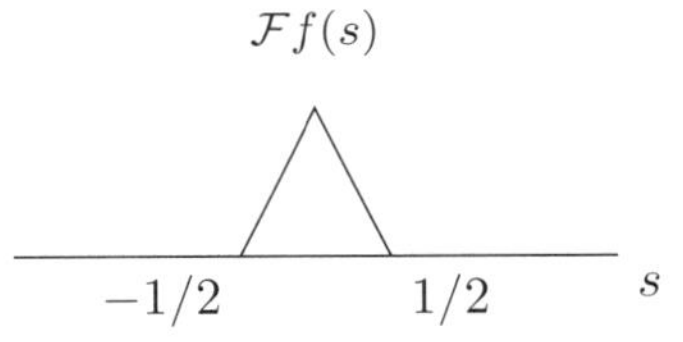

The signal f is sampled at uniformly spaced set of points $\{\dots, t_{-1}, t_0, t_1, t_2, \dots\}$ (we will refer to this set of points as the *sampling pattern*) to obtain the sampled signal $g(t)$:

$$g(t) = f(t) \sum_{k=-\infty}^{\infty} \delta(t - t_k), \quad t_k \text{ uniformly spaced.}$$

Note that 0 *need not* be one of the sample points.

Below are the plots for the sampled spectrum, $\mathcal{F}g(s)$ for various sampling patterns. For each plot (a)–(e), identify the corresponding sampling pattern $\{\dots, t_{-1}, t_0, t_1, t_2, \dots\}$.

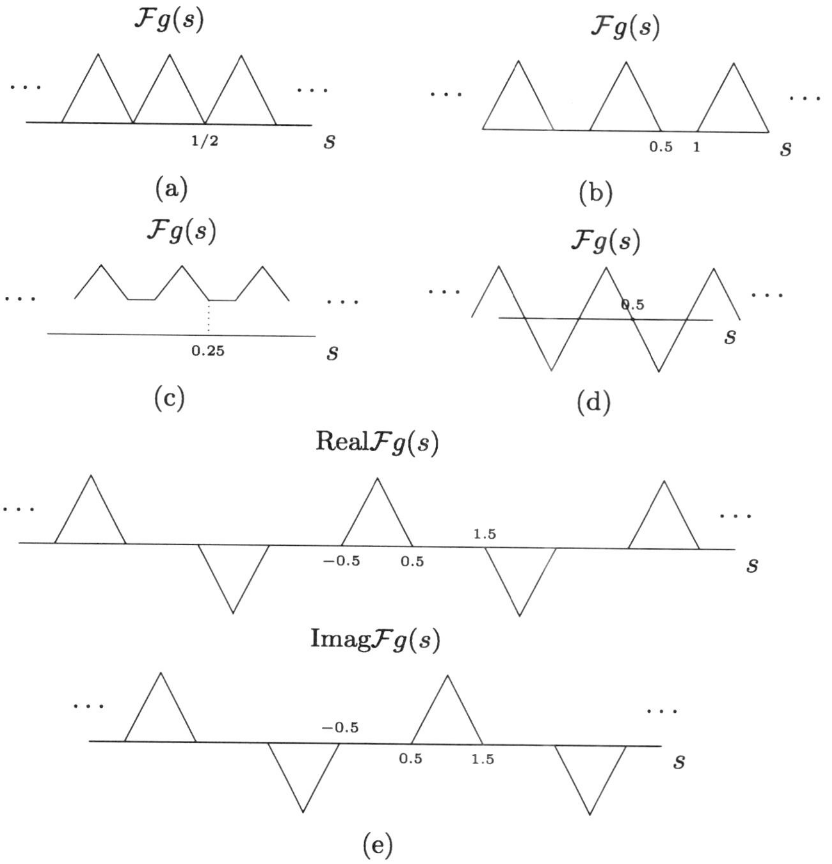

5.11. *The Dirichlet problem for the upper half-plane*

Recall that a function $u(x, y)$ is called *harmonic* if it satisfies Laplace's equation:

$$\Delta u(x, y) = \left(\frac{\partial^2}{\partial x^2} + \frac{\partial^2}{\partial y^2}\right) u(x, y) = 0.$$

The *Dirichlet problem* for the upper half-plane is as follows:

Given a function $f(x)$, $-\infty < x < \infty$, find a function $u(x, y)$ that is harmonic on the upper half-plane

$$H = \{(x, y) : -\infty < x < \infty, y \geq 0\}$$

with boundary values $f(x)$, i.e., with $u(x, 0) = f(x)$.

We will solve this via Fourier transforms and convolution. In Problem 1.32 you used Fourier series to solve the Dirichlet problem for the unit disk, and the solution is convolution of the boundary values with the Poisson kernel. Here you'll find the Poisson kernel for the upper half-plane.

(a) Suppose we can find a function $g(x, y)$ which is harmonic in the upper half-plane and satisfies $g(x, 0) = \delta(x)$. Show that $u(x, y) = f(x)*g(x, y)$ is a solution of the Dirichlet problem for the upper half-plane with boundary values $f(x)$:

$$\Delta u(x, y) = 0, \quad u(x, 0) = f(x).$$

[Use the result: the derivative of the convolution is the convolution with the derivative.]

(b) We use the Fourier transform to find $g(x, y)$, supposing that it satisfies the conditions above. Let $G(s, y)$ be the Fourier transform of $g(x, y)$ with respect to x. Show that $G(s, y)$ satisfies

$$\frac{\partial^2}{\partial y^2} G(s, y) - 4\pi^2 s^2 G(s, y) = 0, \quad G(s, 0) = 1.$$

(c) Find $G(s, y)$. Your answer should involve one arbitrary constant. [Recall that the differential equation $u'' = \lambda u$, $\lambda > 0$, has solutions of the form $u = Ae^{\sqrt{\lambda}t} + Be^{-\sqrt{\lambda}t}$. Also, you should use $\sqrt{s^2} = |s|$.]

(d) Because $G(s, y)$ comes from taking the Fourier transform of $g(x, y)$ we should be able to take the inverse Fourier transform of $G(s, y)$ to get $g(x, y)$. Argue that this rules out one of the exponentials in the solution for $G(s, y)$ and then show that $g(x, y)$ is

$$g(x, y) = \frac{1}{\pi} \frac{y}{x^2 + y^2}.$$

From part (a) deduce that

$$u(x, y) = \frac{1}{\pi} \int_{-\infty}^{\infty} \frac{y}{(x - t)^2 + y^2} f(t)\, dt.$$

5.12. *An identity for III*

Derive the identity

$$\sum_{n=-\infty}^{\infty} \cos 2\pi n t = Ш(t) = \sum_{n=-\infty}^{\infty} \delta(t - n).$$

Of course the sum of cosines doesn't make sense classically, but it does define a tempered distribution. As such it has a Fourier transform. Use the fact that the Fourier transform of the sum is the sum of the Fourier transforms, even for this infinite (distributional) sum. This requires a further argument, but you can assume it's OK.

5.13. *Convolving III's?*

Convolving with Ш periodizes. If we form $Ш_1 * Ш_2$, we get a distribution that has period 1 *and* period 2. Yes? No?

Chapter 6

Sampling and Interpolation

In the previous chapter we studied three properties of Ш that make it so useful in many applications. They are:

- Periodizing.
  - Convolving with Ш periodizes a function.
- Sampling.
  - Multiplying by Ш samples a function.
- The Fourier transform of Ш is Ш.
  - Through this and the convolution theorem for Fourier transforms, periodizing and sampling are flip sides of the same coin.

We are about to combine these ideas in a spectacular way to treat the problem of sampling and interpolation.

Let me state the problem this way:

- Given a signal $f(t)$ and a collection of *samples* of the signal, i.e., values of the signal at a set of points $f(t_0)$, $f(t_1)$, $f(t_2)$, ..., how can one interpolate the values $f(t)$ at other points?

This is an old question and a broad one, and it would appear on the surface to have nothing to do with Ш's or Fourier transforms or any of that. But we've already seen some clues, and in at least one important case the full solution is set to unfold.

6.1. Sampling sines and the Idea of a Bandlimited Signal

Imagine putting down a bunch of dots — theoretically, maybe infinitely many — and asking someone to pass a curve through them that *agrees everywhere exactly* with a predetermined mystery function passing through those dots. Plainly a ridiculous request. Isn't it?

Try this with a sum of two sinusoids. The curve and the sample points are shown. The sample points in this case are evenly spaced though in general they need not be. And I don't mean to be mysterious about this: I played around and settled on the signal $f(t) = 4\cos(2\pi(.2)t) + 1.5\sin(2\pi(1.414)t)$ for $0 \leq t \leq 12$. The sample points are spaced 0.5 apart.

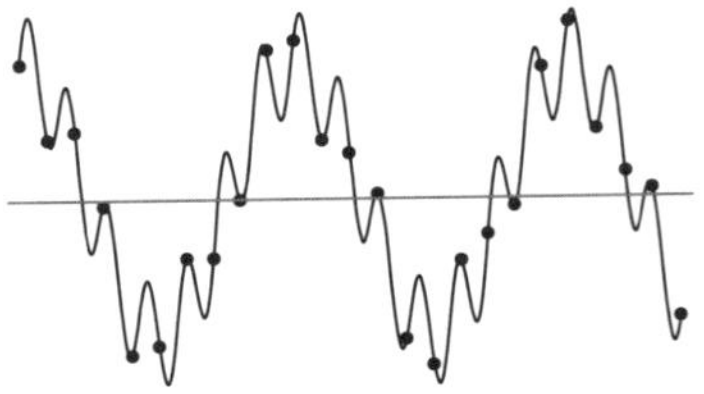

Two simple methods for drawing an approximating curve are *linear interpolation*[1] and *sample-and-hold*.

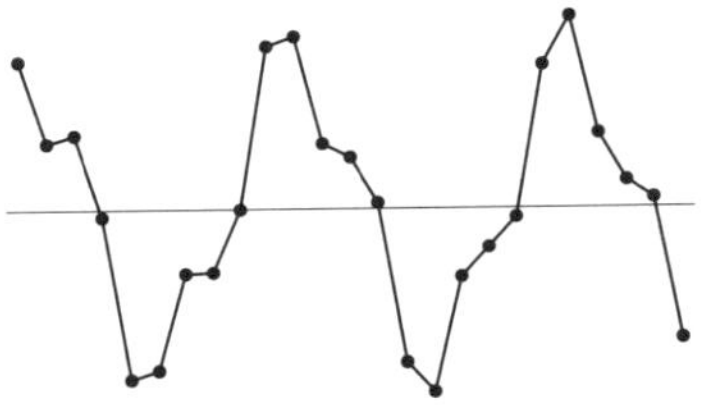

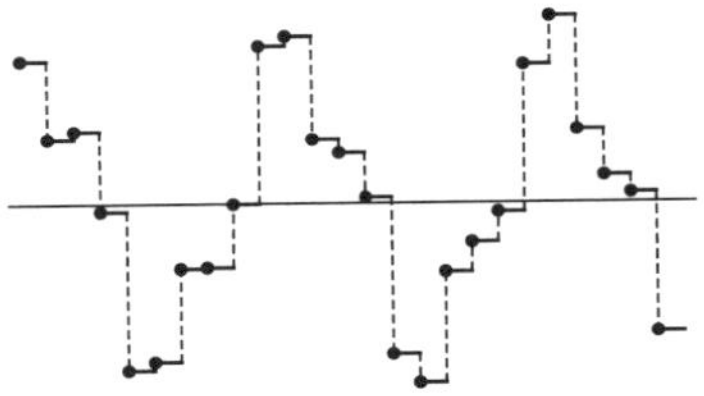

Comparing with the actual curve, the interpolated curves clearly miss the more rapid oscillations that take place between sample values. No surprise that this might happen. We could add more sample points but, to repeat, is it just ridiculous to expect to be able to do exact interpolation? There will always be some uncertainty in the interpolated values. Won't there?

It's *not* ridiculous. If a relatively simple hypothesis is satisfied, then exact interpolation can be done! Here's one way of getting some intuitive sense of the problem and what that hypothesis should be.

Suppose we *know* that the mystery signal is a single sinusoid, say of the form $A\sin(2\pi\nu t + \phi)$. A sinusoid repeats, so if we have enough information to pin it down over one cycle, then we know the whole thing. How many samples — how many values of the function — within one cycle do we need to know which sinusoid we have? We need three samples *strictly within* one cycle. You can think of the graph or you can think of the equation. There are three unknowns, the amplitude A, the frequency ν, and the phase ϕ. We would expect to need three equations to find the unknowns; hence we need values of the function at three points, three samples. I'm not saying that it's so easy to solve for the unknowns, only that it's what we might expect.

[1]Otherwise known as *connect the dots*. There used to be connect the dots books for little kids, not featuring sinusoids.

What if the signal is a sum of sinusoids, say

$$\sum_{n=1}^{N} A_n \sin(2\pi n\nu_n t + \phi_n)\,.$$

Sample points for the sum are morally sample points for the individual harmonics, though not explicitly. We need to take enough samples to get sufficient information to determine all of the unknowns for all of the harmonics. Now, in the time it takes for the combined signal to go through one cycle, the individual harmonics will have gone through several cycles, the lowest frequency harmonic through one cycle, the lower frequency harmonics through a few cycles, say, and the higher frequency harmonics through many. We have to take enough samples of the combined signal so that as the individual harmonics go rolling along we'll be sure to have at least three samples in *some* cycle of *every* harmonic.

To simplify and standardize, we assume that we take evenly spaced samples. That's what you'd typically record with a measuring instrument. Since we've phrased things in terms of cycles per second, it's then also better to think in terms of *sampling rate*, i.e., samples/sec instead of number of samples. If we are to have at least three samples strictly within a cycle, then the sample points must be strictly less than a half-cycle apart. A sinusoid of frequency ν goes through a half-cycle in $1/2\nu$ seconds so we want

$$\text{spacing between samples} = \frac{\text{number of seconds}}{\text{number of samples}} < \frac{1}{2\nu}\,.$$

The more usual way of putting this is

$$\text{sampling rate} = \text{samples/sec} > 2\nu\,.$$

This is the rate at which we should evenly sample a given sinusoid of frequency ν to guarantee that a single cycle will contain at least three sample points. Furthermore, if we sample at this rate for a given frequency, we will certainly have more than three sample points in some cycle of any harmonic at a *lower* frequency. Note again that the sampling rate has units 1/second and that sample points are 1/(sampling rate) seconds apart.

For the combined signal — a sum of harmonics — the higher frequencies are driving up the sampling rate; specifically, the *highest* frequency is driving up the rate. To think of the interpolation problem geometrically, high frequencies cause more rapid oscillations, i.e., rapid changes in the function over small intervals. To hope to interpolate such fluctuations accurately we'll need a high sampling rate. If we sample at too low a rate, we might miss the wiggles entirely. We might mistakenly think we had only a low frequency sinusoid, and, moreover, since all we have to go on are the samples, we wouldn't even know we'd made a mistake! We'll come back to just this problem (called *aliasing*) a little later.

If we sample at a rate greater than twice the highest frequency, our sense is that we will be sampling often enough for all the lower harmonics as well, and we should be able to determine everything. Well, maybe. I'm not saying that we can disentangle the sample values in any easy way to determine the individual harmonics, but in principle it looks like we have enough information.

The problem is if the spectrum is *unbounded*. If we have a full Fourier series and not just a finite sum of sinusoids, then, in this argument at least, we can't expect to sample frequently enough to determine the combined signal from the samples; there is *no* "highest frequency."

Bandlimited signals. It's time to define ourselves out of trouble. From the point of view above, whatever the caveats, the problem for interpolation is high frequencies. The best thing a signal can be is a finite Fourier series, in which case the signal has a discrete, finite set of frequencies that stop at some point. This is much too restrictive for applications, of course, so what's the next best thing a signal can be? It's one for which the spectrum is zero from some point on. These are the *bandlimited* signals — signals whose Fourier transforms are identically zero outside of a bounded interval, outside of a bounded band of frequencies. More formally:

- A signal $f(t)$ is bandlimited if there is a finite number p such that $\mathcal{F}f(s) = 0$ for all $|s| \geq p/2$.

Interesting that an adjective attached to the signal in the time domain is defined by a property in the frequency domain. The Fourier transform $\mathcal{F}f(s)$ may well have zeros at points s with $|s| \leq p/2$, but for sure it's *identically zero* for $|s| \geq p/2$. Recalling terminology from Chapter 4, you will recognize the definition as saying that $f(t)$ is bandlimited if $\mathcal{F}f(s)$ has compact support.[2]

- The smallest number B for which the condition is satisfied is called the *bandwidth*. I'm sure you're familiar with that term, but there are some subtleties. We'll come back to this. In the meantime note carefully that the bandwidth is B, not $B/2$.

Singling out the collection of bandlimited signals is an example of one of the working principles laid down by the eminent British mathematician J. E. Littlewood, one of my mathematical forebearers. To wit: "Make your worst enemy your best friend." The idea is to identify the stumbling block and make an assumption, a definition, that will eliminate or weaken the opposition. Not always easy to do, but an important strategy to keep in mind. The enemy is high frequencies. We wish them away and see what happens.

6.2. Sampling and Interpolation for Bandlimited Signals

We're about to solve the interpolation problem for bandlimited signals. We'll show that interpolation is possible by finding an explicit formula that does the job. This uses all the important properties of III_p, but it goes so fast that you might miss the fun entirely if you read too quickly. I'll make an effort to slow things down.

Suppose $f(t)$ is a bandlimited signal with $\mathcal{F}f(s)$ identically zero for $|s| \geq p/2$. We periodize $\mathcal{F}f$ using III_p and then cut off to get $\mathcal{F}f$ back again:

$$\mathcal{F}f = \Pi_p(\mathcal{F}f * \text{III}_p)\,.$$

[2]To repeat from Chapter 4, the support is the smallest closed set containing the set $\{x\colon f(x) \neq 0\}$. To say that $\mathcal{F}f(s)$ has compact support is to say that it is identically zero outside a bounded set.

This is the crucial equation. The condition of being bandlimited to $|s| \le p/2$ means that convolving $\mathcal{F}f(s)$ with III_p shifts the spectrum off itself — no overlaps, not even any (nonzero) touching of the shifted copies since $\mathcal{F}f(\pm p/2) = 0$. We recover the original Fourier transform $\mathcal{F}f(s)$ by multiplying the periodized function $\mathcal{F}f * \text{III}_p$ by a rectangle function of width p. (Low-pass filtering.)

Here's the picture to keep in mind. Take a real, bandlimited function $f(t)$ and plot the *magnitude* $|\mathcal{F}f(s)|$ (the best we can do), which is even (because $f(t)$ is real):

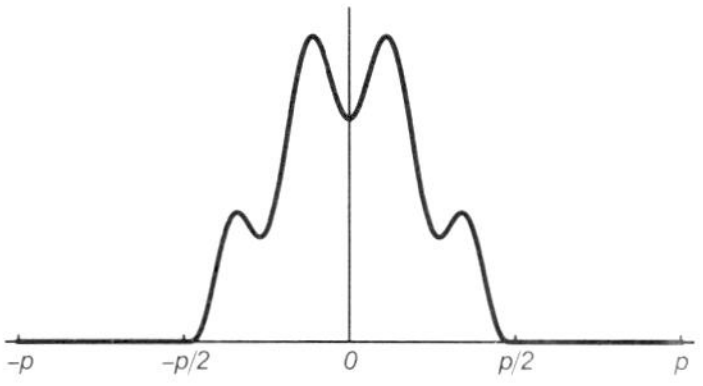

The periodized $\mathcal{F}f(s)$, or rather its magnitude, looks like

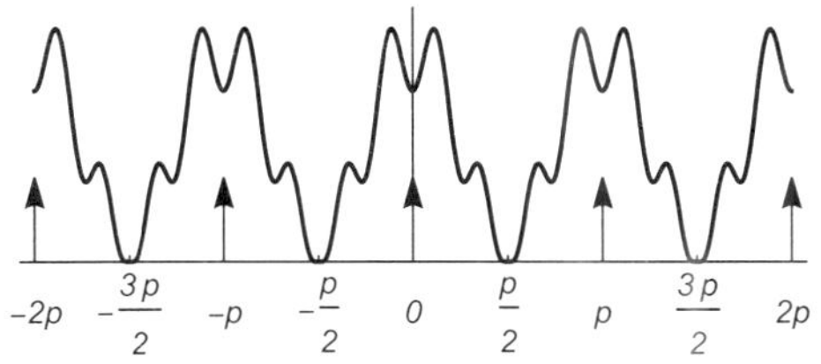

I put a plot of III_p in there, too, so you could see how convolution with III_p shifts the spectrum (and adds the shifts together). Then what remains after multiplying by Π_p is the original $\mathcal{F}f(s)$:

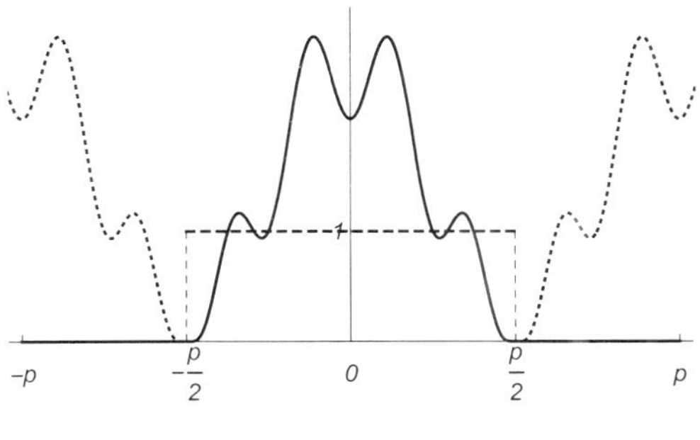

Brilliant. We did something (periodize) and then undid it (cut off), and we're back to where we started, with $\mathcal{F}f(s)$. Have we done nothing? To the contrary, watch! Take the inverse Fourier transform of $\mathcal{F}f = \Pi_p(\mathcal{F}f * Ш_p)$:

$$\begin{aligned}
f(t) &= \mathcal{F}^{-1}\mathcal{F}f(t) = \mathcal{F}^{-1}(\Pi_p(\mathcal{F}f * Ш_p))(t)\\
&= \mathcal{F}^{-1}\Pi_p(t) * \mathcal{F}^{-1}(\mathcal{F}f * Ш_p)(t)\\
&\qquad \text{(taking } \mathcal{F}^{-1} \text{ turns multiplication into convolution)}\\
&= \mathcal{F}^{-1}\Pi_p(t) * (\mathcal{F}^{-1}\mathcal{F}f(t) \cdot \mathcal{F}^{-1}Ш_p(t))\\
&\qquad \text{(on this line it's convolution turning into multiplication)}\\
&= p \operatorname{sinc} pt * \left(f(t) \cdot \frac{1}{p} Ш_{1/p}(t)\right) \quad \text{(note that the } p\text{'s cancel)}\\
&= \operatorname{sinc} pt * \sum_{k=-\infty}^{\infty} f\left(\frac{k}{p}\right) \delta\left(t - \frac{k}{p}\right) \quad \text{(the sampling property of } Ш_p)\\
&= \sum_{k=-\infty}^{\infty} f\left(\frac{k}{p}\right) \operatorname{sinc} pt * \delta\left(t - \frac{k}{p}\right)\\
&= \sum_{k=-\infty}^{\infty} f\left(\frac{k}{p}\right) \operatorname{sinc} p\left(t - \frac{k}{p}\right) \quad \text{(the convolution property of } \delta).
\end{aligned}$$

6.2.1. The Nyquist-Shannon sampling theorem. We've just established the classic *Nyquist-Shannon sampling theorem*, though it might be better to call it the interpolation theorem. Here it is as a single statement:

- If $f(t)$ is a signal with $\mathcal{F}f(s)$ identically zero for $|s| \geq p/2$, then

$$f(t) = \sum_{k=-\infty}^{\infty} f\left(\frac{k}{p}\right) \operatorname{sinc} p\left(t - \frac{k}{p}\right).$$

The attribution is in honor of Harry Nyquist, God of Sampling, who was the first engineer to consider these problems for the purpose of communications, and Claude Shannon, overall genius and founder of information theory, who in 1949 presented the result as it appears here. There are other names associated with sinc interpolation, most notably the mathematician E. Whittaker whose paper was published in 1915.[3]

People usually refer to this expression as the sampling formula, or the interpolation formula, and it's often written as

$$f(t) = \sum_{k=-\infty}^{\infty} f\left(\frac{k}{p}\right) \operatorname{sinc}(pt - k).$$

I generally prefer to emphasize the sample points k/p within the sinc functions. If we further write

$$t_k = \frac{k}{p},$$

[3]Whittaker called the sum of sincs the *cardinal series*. (I'm sure he didn't write "sinc.") Impress your friends with your erudition by using this term, but look up some of the history first.

then the formula is

$$f(t) = \sum_{k=-\infty}^{\infty} f(t_k)\,\mathrm{sinc}\,p(t - t_k)\,.$$

As a quick reality check, let's verify that the formula when evaluated at a sample point returns the value of the signal at that point. Plug in $t_\ell = \ell/p$:

$$\sum_{k=-\infty}^{\infty} f\left(\frac{k}{p}\right)\mathrm{sinc}\,p\left(\frac{\ell}{p} - \frac{k}{p}\right) = \sum_{k=-\infty}^{\infty} f\left(\frac{k}{p}\right)\mathrm{sinc}(\ell - k)\,.$$

Now remember that the sinc function is zero at the integers, except that $\mathrm{sinc}\,0 = 1$, so

$$\mathrm{sinc}(\ell - k) = \begin{cases} 1, & \ell = k, \\ 0, & \ell \neq k\,, \end{cases}$$

and hence

$$\sum_{k=-\infty}^{\infty} f\left(\frac{k}{p}\right)\mathrm{sinc}(\ell - k) = f\left(\frac{\ell}{p}\right)\,.$$

Good thing.

For me, the sampling theorem is identical to the proof of the sampling theorem; it's so automatic. It all depends on having

$$\mathcal{F}f = \Pi_p(\mathcal{F}f * \text{III}_p)\,.$$

Say it with me: periodize and cut off to get back the original Fourier transform, and then interchange periodizing and sampling via the convolution theorem. It's almost indecent the way this works. There are variations on the theme, if there are perhaps special aspects of the spectrum. You'll see examples in the problems, e.g., "islands" where the spectrum is nonzero, separated by intervals where the spectrum vanishes. But the idea of periodize-and-cut-off is *always there*.

Bandwidth, sampling rates, and the Nyquist frequency. There is a series of observations on this wondrous result. The first is yet another reciprocal relationship to keep in mind:

- The sample points are spaced $1/p$ apart.
- The *sampling rate* is p, in units of Hz. This is the setting on your sampling machine. The higher the sampling rate, the more closely spaced the sample points.

I point this out because there are higher-dimensional versions of the sampling theorem and more subtle reciprocal relationships. This will be a topic in Chapter 9.

Next some matters of definition. If you look up the sampling theorem elsewhere, you may find other conventions and other notations. Often the assumption is written as $\mathcal{F}f(s) = 0$ for $|s| > \nu_{\max}$, and the conclusion is then that the signal can be interpolated, with the formula as above, using any sampling rate $> 2\nu_{\max}$. Note the strict inequalities here. (This way of writing the sampling rate does jibe with our initial discussion of sampling a sinusoid.)

The number $2\nu_{\max}$ is often called the *Nyquist frequency*, as well as being called the bandwidth. You'll hear people say things like: "For interpolation, sample at a rate greater than twice the highest frequency." OK, but those same people are not always clear what they mean by "highest frequency." Is it the largest frequency for which $\mathcal{F}f(s) \neq 0$ or is it the smallest frequency for which $\mathcal{F}f(s)$ is identically zero beyond that point?

Mathematically, the way out is to use the *infimum* of a set of numbers, abbreviated inf, to define $\nu_{\max}$ and the bandwidth. The infimum of a set of numbers is the greatest lower bound of the set. For example,

$$\sqrt{2} = \inf\{x : x \text{ is a positive rational number and } x^2 \geq 2\}\,.$$

The picture, with the set in the heavier line:

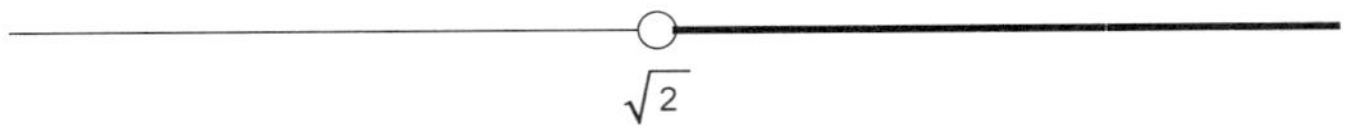

Note that $\sqrt{2}$ is not itself in the set since it's irrational. The various approximations that you know, $1.4, 1.41, 1.414, \ldots$, are all lower bounds for any number in the set, and $\sqrt{2}$ is the *greatest* lower bound.[4]

For a signal $f(t)$ we consider the set

$$\{p : \mathcal{F}f(s) = 0 \text{ for all } |s| \geq p/2\}$$

and we find the infimum

$$\inf\{p : \mathcal{F}f(s) = 0 \text{ for all } |s| \geq p/2\}\,.$$

The infimum is what people mean by $\nu_{\max}$, and the bandwidth is twice this:

$$B = 2\inf\{p : \mathcal{F}f(s) = 0 \text{ for all } |s| > p/2\} = 2\nu_{\max}\,.$$

As with the $\sqrt{2}$ example, the number $\nu_{\max}$ may not be in the set whose infimum defines it; i.e., we may not have $\mathcal{F}f(\pm\nu_{\max}) = 0$. (However if $f(t)$ is an integrable function, then $\mathcal{F}f(s)$ is continuous, so the fact that $\mathcal{F}f(s) = 0$ for all $|s| > B/2$ implies that $\mathcal{F}f(\pm B/2) = 0$ as well.) In any event, if $p/2 > \nu_{\max}$, we do have $\mathcal{F}f(s) = 0$ for $|s| \geq p/2$, so including the endpoints $\pm p/2$, and this is what we need for the derivation of the sampling theorem.

Why these niggling remarks? Take, for example, a simple sinusoid, say $f(t) = \sin 2\pi(B/2)t$ (not integrable!), for which $\mathcal{F}f(s) = (1/2i)(\delta(s - B/2) - \delta(s + B/2))$. What is the bandwidth? Everyone in the world would say it's B, and so would I. Sure enough, this conforms to the definition, if you're willing to grant that the δ's are zero away from $\pm B/2$.[5] To apply the sampling formula to $\sin 2\pi(B/2)t$ (which, for me, is to apply the derivation of the sampling formula) we'd sample at any rate $> B$.

---

[4] It is a deep property of the real numbers, taken in some form as an axiom, that any set of numbers that doesn't stretch down to $-\infty$ *has* an infimum. This is called the *completeness property* of the real numbers and it underlies many of the existence theorems in calculus.

[5] We know that we don't really talk about the values of δ's at points, but here we are around half-way through the book and I'm willing to let that slide.

In fact, to conclude, let's restate the sampling theorem using the bandwidth.

- If $f(t)$ is a bandlimited signal with bandwidth B, then

$$f(t) = \sum_{k=-\infty}^{\infty} f\left(\frac{k}{p}\right) \operatorname{sinc} p\left(t - \frac{k}{p}\right)$$

  for any $p > B$.

Got it? The formula is such a nice thing. We're saving the nasty things for later.[6]

After this abstraction here's one real-world example, the sampling rate for the audio on compact discs. It's 44.1 kHz. Where does this number come from? We can hear frequencies up to about 20 kHz; that's ν_{max}, so the sampling rate should be upwards of $2\nu_{\text{max}} = 40$ kHz.[7] The precise number 44.1 kHz allows for some slack but really comes from the legacy equipment used for analog recordings.

Bandlimited and timelimited signals. You may have noted, with some sadness, that the sampling theorem involves an infinite sum and infinitely many sample points. Two questions come to mind:

- Do I have to worry about convergence of the series?
- Do I really need infinitely many sample points?

On the first point, see the problems for a theorem that guarantees convergence of a sum of the form

$$\sum_{k=-\infty}^{\infty} a_k \operatorname{sinc} p\left(t - \frac{k}{p}\right).$$

That result suffices for almost everything. We'll be meeting various infinite series, and we won't be concerned with questions of convergence. The rigor police are back to hanging around outside the room.

The second point is really the more interesting one. It points to a phenomenon that has a consequence for the sampling theorem but is independent of it.

It is unphysical to consider a signal as lasting forever in time. A physical signal $f(t)$ is naturally *timelimited*, meaning that $f(t)$ is identically zero for $|t| \geq q/2$ for some q. There just isn't any signal beyond some point. (In mathematical language, a physical signal has compact support.) On the other hand, it is very physical to consider a bandlimited signal, one with no frequencies beyond a certain point, or at least no frequencies that our instruments can register. Well, we can't have both, at least not in the ideal world of mathematics. Here is where mathematical description meets physical expectation, and they disagree. The fact is:

- A signal cannot be both timelimited and bandlimited unless it is identically zero.

What this means in practice is that there must be inaccuracies in a mathematical model of a phenomenon that assumes a signal is both timelimited and bandlimited.

[6]See *Lucky Jim* by Kingsley Amis for an appreciation of nice things versus nasty things. Mull over his description of much academic work as "throwing pseudo-light on nonproblems."

[7]As a trombonist (see Chapter 1), I have spent years sitting in front of a trumpet section. My hearing probably no longer goes up to 20 kHz.

Such a model can be at best an approximation, and one has to be prepared to estimate the errors as they may affect measurements and conclusions.

Here's one argument why the statement is true: I'll give another, more refined, statement and proof in Section 6.5. Suppose $f(t)$ is bandlimited, say, $\mathcal{F}f(s)$ is zero for $|s| \geq p/2$. Then

$$\mathcal{F}f = \Pi_p \cdot \mathcal{F}f\,.$$

No periodization here, just cutting off the already limited Fourier transform. Take the inverse Fourier transform of both sides to obtain

$$f(t) = p \operatorname{sinc} pt * f(t)\,.$$

Now $\operatorname{sinc} pt$ goes on forever; it decays but it has nonzero values all the way out to $\pm\infty$. Hence the convolution with f also goes on forever, so $f(t)$ is not timelimited.

sinc *as an identity for convolution.* There's an interesting observation that goes along with the argument I just gave. We're familiar with δ acting as an identity element for convolution, meaning

$$f * \delta = f\,.$$

This important property of δ holds for *all* signals for which the convolution is defined. We've just seen for the more restricted class of bandlimited functions, with spectrum from $-p/2$ to $+p/2$, that the sinc function also has this property:

$$p \operatorname{sinc} pt * f(t) = f(t)\,.$$

We can also shift:

$$p \operatorname{sinc} p(t-a) * f(t) = f(t-a)\,.$$

What is the consequence of the bandlimited vs. timelimited phenomenon for the sampling theorem? If the signal $f(t)$ is bandlimited, then it cannot be timelimited; $f(t)$ has nonzero values all the way out to $\pm\infty$. Consequently, we should expect to need sample points out to $\pm\infty$. It's not reasonable to take samples only up to a finite point and still interpolate values of the function way beyond (infinitely beyond) that point. This means, of course, that any practical application of the sampling theorem, which must use a finite sum, will have to be an approximation.[8] These problems are absolutely inevitable. The approaches are via filters, first low-pass filters done *before* sampling to force a signal to be bandlimited and then other kinds of filters (smoothing) following whatever reconstruction is made from the samples. Particular kinds of filters are designed for particular kinds of signals, e.g., sound or images.

[8] Later we'll walk this back in the case of sampling and interpolating periodic signals. See also J. R. Higgins, *Sampling Theory in Fourier and Signal Analysis*. A nice reference with all sorts of nice topics.

6.2.2. An example. Before we consider how the sampling formula might go rogue, let's see how it works nicely with the curve we looked at first, back in Section 6.1. The signal is

$$f(t) = 4\cos(2\pi(.2)t) + 1.5\sin(2\pi(1.414)t)$$

on the interval $0 \le t \le 12$. The bandwidth is 2.828 so we need a sampling rate greater than this to apply the formula and expect to get the signal back.

Jumping all the way up to $p = 3$, giving 37 sample points, here's a plot of the original curve, the sample points, and the finite sum:

$$\sum_{k=0}^{36} f\left(\frac{k}{3}\right) \operatorname{sinc} 3\left(t - \frac{k}{3}\right).$$

The original curve is solid and the approximating curve is dashed. You can spot some differences, particularly where $f(t)$ is bending the most, naturally, but even with the finite approximation there's very little daylight between the two curves.

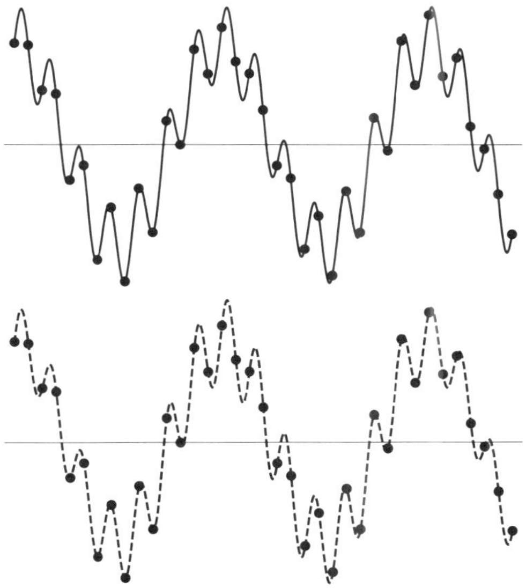

6.2.3. Interpolation and orthogonality. Here's a brief take on orthogonality playing a role in the sinc interpolation formula, very much analogous to orthogonality and Fourier series.

Recall that the inner product on $L^2(\mathbb{R})$, the square integrable functions defined on $\mathbb{R}$, is

$$(f, g) = \int_{-\infty}^{\infty} f(t)\overline{g(t)}\, dt\,.$$

Two functions are orthogonal if their inner product is zero.

Another amazing property of sinc functions is that the family of shifted sincs, $\operatorname{sinc}(t - n)$, $n = 0, \pm 1, \pm 2, \ldots$, of bandwidth 1, is orthonormal. The calculation to establish this is the general Parseval identity, which you'll recall says

$$\int_{-\infty}^{\infty} f(t)\overline{g(t)}\,dt = \int_{-\infty}^{\infty} \mathcal{F}f(s)\overline{\mathcal{F}g(s)}\,ds\,.$$

For the shifted sincs we have

$$\begin{aligned}\int_{-\infty}^{\infty} \operatorname{sinc}(t-n)\,\operatorname{sinc}(t-m)\,dt &= \int_{-\infty}^{\infty} (e^{-2\pi i s n}\Pi(s))\,\overline{(e^{-2\pi i s m}\Pi(s))}\,ds \\ &= \int_{-\infty}^{\infty} e^{2\pi i s(m-n)}\Pi(s)\Pi(s)\,ds \\ &= \int_{-1/2}^{1/2} e^{2\pi i s(m-n)}\,ds\,.\end{aligned}$$

Welcome back to that celebrated integral, last seen playing a fundamental role for Fourier series, in Chapter 1. Then, as now, direct integration gives you 1 when $n = m$ and 0 when $n \neq m$. And in case you're fretting over it, the sinc function *is* in $L^2(\mathbb{R})$ and the product of two sinc functions is integrable. Parseval's identity holds for functions in $L^2(\mathbb{R})$.

Sticking with bandwidth 1 for the moment, coupled with the result on orthogonality, the formula

$$f(t) = \sum_{n=-\infty}^{\infty} f(n)\operatorname{sinc}(t-n)$$

suggests that the family of sinc functions forms an orthonormal *basis* for the space of bandlimited signals with spectrum in $-1/2 \leq s \leq 1/2$ and that we're expressing $f(t)$ in terms of this basis. That's exactly what's going on. The coefficients (the sample values $f(n)$) are obtained as the inner product of $f(t)$ with $\operatorname{sinc}(t-n)$. We have, again using Parseval,

$$\begin{aligned}(f(t), \operatorname{sinc}(t-n)) &= \int_{-\infty}^{\infty} f(t)\,\operatorname{sinc}(t-n)\,dt \\ &= \int_{-\infty}^{\infty} \mathcal{F}f(s)\mathcal{F}(\operatorname{sinc}(t-n))\,ds \quad \text{(by Parseval)} \\ &= \int_{-\infty}^{\infty} \mathcal{F}f(s)\overline{(e^{-2\pi i s n}\Pi(s))}\,ds \\ &= \int_{-1/2}^{1/2} \mathcal{F}f(s)e^{2\pi i n s}\,ds \\ &= \int_{-\infty}^{\infty} \mathcal{F}f(s)e^{2\pi i n s}\,ds \quad \text{(because } f \text{ is bandlimited)} \\ &= f(n) \quad \text{(by Fourier inversion).}\end{aligned}$$

It's perfect! The interpolation formula says that $f(t)$ is written in terms of an orthonormal basis, and the coefficient $f(n)$, the nth sampled value of $f(t)$, is exactly the projection of $f(t)$ onto the nth basis element:

$$f(t) = \sum_{n=-\infty}^{\infty} f(n)\operatorname{sinc}(t-n) = \sum_{n=-\infty}^{\infty} \big(f(t), \operatorname{sinc}(t-n)\big)\operatorname{sinc}(t-n)\,.$$

Changing to bandwidth p is analogous to changing the period for Fourier series. For the shifted sincs we again use Parseval to calculate the inner product:

$$\begin{aligned}\left(\operatorname{sinc} p\left(t-\frac{n}{p}\right), \operatorname{sinc} p\left(t-\frac{m}{p}\right)\right) &= \int_{-\infty}^{\infty} \operatorname{sinc} p\left(t-\frac{n}{p}\right) \overline{\operatorname{sinc} p\left(t-\frac{m}{p}\right)}\, dt \\ &= \int_{-\infty}^{\infty} \frac{1}{p} e^{-2\pi i s(n/p)} \Pi_p(s)\, \overline{\frac{1}{p} e^{-2\pi i s(m/p)} \Pi_p(s)}\, ds \\ &= \frac{1}{p^2} \int_{-p/2}^{p/2} e^{2\pi i (s/p)(m-n)}\, ds \\ &= \begin{cases} 1/p, & m=n, \\ 0, & m \neq n\,. \end{cases}\end{aligned}$$

Orthogonal but not orthonormal. This is telling us that the *orthonormal* family to consider is

$$\sqrt{p}\operatorname{sinc} p\left(t-\frac{n}{p}\right), \quad n = 0, \pm 1, \pm 2, \dots .$$

And the sample values? For the projection of $f(t)$, of bandwidth p, onto the sincs we have, with Parseval (skipping a few steps),

$$\begin{aligned}\left(f(t), \sqrt{p}\operatorname{sinc} p\left(t-\frac{n}{p}\right)\right) &= \int_{-\infty}^{\infty} \mathcal{F}f(s)\, \sqrt{p}\, \frac{1}{p}\, e^{2\pi i s(n/p)} \Pi_p(s)\, ds \\ &= \frac{1}{\sqrt{p}} \int_{-\infty}^{\infty} \mathcal{F}f(s) e^{2\pi i s(n/p)}\, ds = \frac{1}{\sqrt{p}} f\left(\frac{n}{p}\right).\end{aligned}$$

Then expanding in terms of the orthonormal basis cancels the $\sqrt{p}$'s:

$$\begin{aligned}&\sum_{k=-\infty}^{\infty} \left(f(t), \sqrt{p}\operatorname{sinc} p\left(t-\frac{k}{p}\right)\right) \sqrt{p}\operatorname{sinc} p\left(t-\frac{k}{p}\right) \\ &\qquad = \sum_{k=-\infty}^{\infty} \frac{1}{\sqrt{p}} f\left(\frac{k}{p}\right) \sqrt{p}\operatorname{sinc} p\left(t-\frac{k}{p}\right) = \sum_{k=-\infty}^{\infty} f\left(\frac{k}{p}\right) \operatorname{sinc} p\left(t-\frac{k}{p}\right).\end{aligned}$$

It's perfect, again.

6.3. Undersampling and Aliasing

The troubles. What if we work a little less hard than as dictated by the bandwidth. What if we undersample a bit and try to apply the interpolation formula with a little lower sampling rate and with the sample points spaced a little farther apart. Will the interpolation formula produce almost a good fit, good enough to listen to or to see? Maybe yes, maybe no. A disaster is a definite possibility.

6.3.1. Sampling sinusoids, still. Let's revisit the question of sampling and interpolation for a simple sine function and let's work with an explicit example. Take the signal given by

$$f(t) = \cos \tfrac{9\pi}{2} t\,.$$

The frequency of this signal is 9/4 Hz. If we want to apply the sampling formula, we can take the sampling rate to be anything $> 9/2 = 4.5$. Suppose our sampler is stuck in low and we can only take one sample every second. Then our samples have values

$$\cos \tfrac{9\pi}{2} n, \quad n = 0, 1, 2, 3, \ldots .$$

There is another, lower frequency signal that has the same samples. To find it, take away from $9\pi/2$ the largest multiple of 2π that leaves a remainder of less than π in absolute value, so there's a spread of less than 2π, one full period, to the left and right. You'll see what I mean as the example proceeds. Here we have

$$\tfrac{9\pi}{2} = 4\pi + \tfrac{\pi}{2} .$$

Then

$$\cos \tfrac{9\pi}{2} n = \cos\left(\left(4\pi + \tfrac{\pi}{2}\right) n\right) = \cos \tfrac{\pi}{2} n .$$

The signal $f(t)$ has the same samples at $0, \pm 1, \pm 2, \ldots$ as the signal

$$g(t) = \cos \tfrac{\pi}{2} t ,$$

whose frequency is only $1/4$. The two functions *are not* the same everywhere, but their samples at the integers are equal.

Here are plots of the original signal $f(t)$ and of $f(t)$ and $g(t)$ plotted together, showing how the curves match up at the sample points.[9]

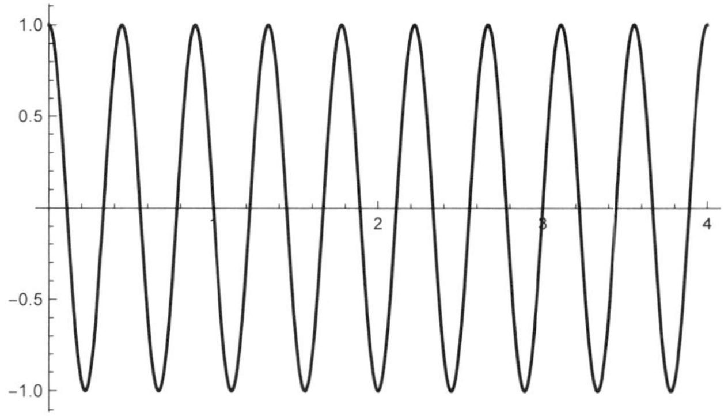

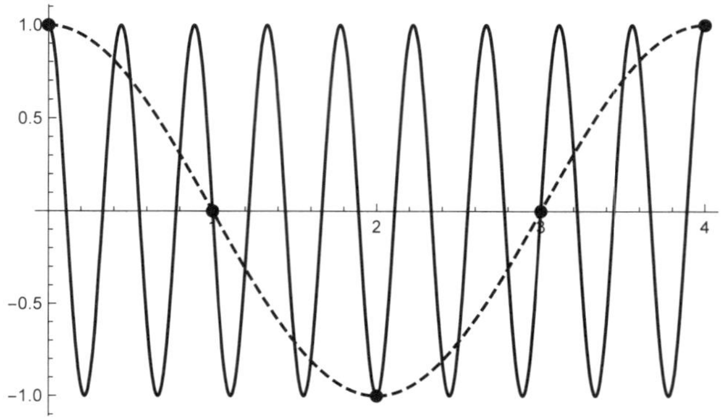

If we're taking measurements of the mystery signal, in this case $f(t)$, and all we know are the sample values, there's no reason to prefer $f(t)$ over $g(t)$. They are indistinguishable as far as their sample values go. The terminology is to say that

[9]They match up wherever they intersect, but we're only aware of the measurements we make.

$g(t)$ is an *alias* of $f(t)$. The lower frequency signal is masquerading as the higher frequency signal via their identical sample values.

You have probably seen physical manifestations of this phenomenon with a strobe light flashing on and off on a moving fan, for example. There are more dramatic examples; see the problems.

Now let's analyze this example in the frequency domain, essentially repeating the derivation of the sampling formula for this particular function at the particular sampling rate of 1 Hz. The Fourier transform of $f(t) = \cos 9\pi t/2$ is

$$\mathcal{F}f(s) = \tfrac{1}{2}\left(\delta\left(s - \tfrac{9}{4}\right) + \delta\left(s + \tfrac{9}{4}\right)\right) .$$

To sample at $p = 1$ Hz means, first, that in the frequency domain we:

- Periodize $\mathcal{F}f$ by $Ш_1$.
- Cut off by Π_1.

After that we take the inverse Fourier transform and, *by definition*, this gives the interpolation to $f(t)$ using the sample points $f(0), f(\pm 1), f(\pm 2), \ldots$. The question is whether this interpolation gives back $f(t)$. We know it doesn't, but what goes wrong?

The Fourier transform of $\cos 9\pi t/2$ looks like

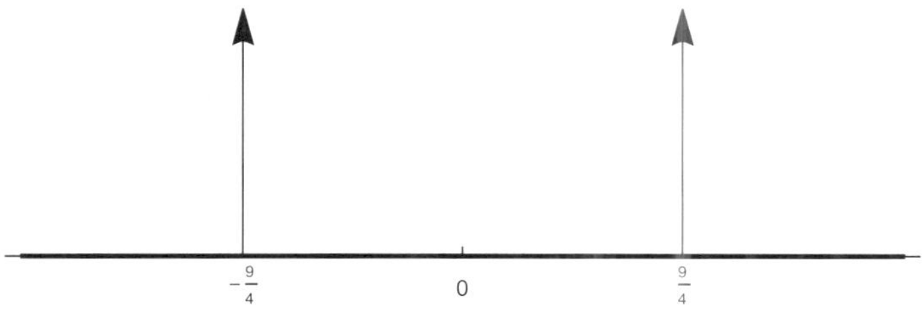

For the periodization step, direct calculation results in

$$\begin{aligned}
\mathcal{F}f(s) * Ш_1(s) &= \tfrac{1}{2}\left[\delta\left(s - \tfrac{9}{4}\right) + \delta\left(s + \tfrac{9}{4}\right)\right] * \sum_{k=-\infty}^{\infty} \delta(s-k) \\
&= \tfrac{1}{2} \sum_{k=-\infty}^{\infty} \left(\delta\left(s - \tfrac{9}{4}\right) * \delta(s-k) + \delta\left(s + \tfrac{9}{4}\right) * \delta(s-k)\right) \\
&= \tfrac{1}{2} \sum_{k=-\infty}^{\infty} \left(\delta\left(s - \tfrac{9}{4} - k\right) + \delta\left(s + \tfrac{9}{4} - k\right)\right)
\end{aligned}$$

(remember the formula $\delta_a * \delta_b = \delta_{a+b}$).

Multiplying by Π_1 cuts off outside $(-1/2, +1/2)$, and we get δ's within $-1/2 < s < 1/2$ if, working separately with $\delta\left(s - \frac{9}{4} - k\right)$ and $\delta\left(s + \frac{9}{4} - k\right)$, we have

$$-\tfrac{1}{2} < -\tfrac{9}{4} - k < \tfrac{1}{2}, \qquad -\tfrac{1}{2} < \tfrac{9}{4} - k < \tfrac{1}{2},$$

$$\tfrac{7}{4} < -k < \tfrac{11}{4}, \qquad -\tfrac{11}{4} < k < -\tfrac{7}{4},$$

$$-\tfrac{11}{4} < k < -\tfrac{7}{4}, \qquad \tfrac{7}{4} < k < \tfrac{11}{4}.$$

Thus we get δ's within $-1/2 < s < 1/2$ for

$$k = -2 \quad \text{and the term} \quad \delta\left(s - \tfrac{9}{4} - (-2)\right) = \delta\left(s - \tfrac{1}{4}\right)$$

and

$$k = 2 \quad \text{and the term} \quad \delta\left(s + \tfrac{9}{4} - 2\right) = \delta\left(s + \tfrac{1}{4}\right).$$

All other δ's in $\mathcal{F}f(s) * Ш_1(s)$ will be outside the range $-1/2 < s < 1/2$, and the final result is

$$\Pi_1(s)(\mathcal{F}f(s) * Ш_1(s)) = \tfrac{1}{2}\left(\delta\left(s + \tfrac{1}{4}\right) + \delta\left(s - \tfrac{1}{4}\right)\right).$$

We *do not have*

$$\Pi_1(\mathcal{F}f * Ш_1) = \mathcal{F}f.$$

So if we take the inverse Fourier transform of $\Pi_1(\mathcal{F}f * Ш_1)$, we *do not get* f back. But we can take the inverse Fourier transform of $\Pi_1(\mathcal{F}f * Ш_1)$ anyway, and this produces

$$\mathcal{F}^{-1}\left(\tfrac{1}{2}\left(\delta\left(s - \tfrac{1}{4}\right) + \delta\left(s + \tfrac{1}{4}\right)\right)\right) = \tfrac{1}{2}(e^{\pi i t/2} + e^{-\pi i t/2}) = \cos \tfrac{\pi}{2} t.$$

There's the aliased signal!

$$g(t) = \cos \tfrac{\pi}{2} t.$$

It's still quite right to think of $\mathcal{F}^{-1}(\Pi_1(\mathcal{F}f * Ш_1))$ as an interpolation based on sampling $f(t)$ at 1 Hz. That's exactly what it is; it's just not a good one. The sampling formula is

$$\begin{aligned} \mathcal{F}^{-1}(\Pi_1(\mathcal{F}f * Ш_1))(t) &= \operatorname{sinc} t * (f(t) \cdot Ш_1(t)) \\ &= \operatorname{sinc} t * \sum_{k=-\infty}^{\infty} f(k)\delta(t-k) \\ &= \sum_{k=-\infty}^{\infty} f(k) \operatorname{sinc}(t-k) = \sum_{k=-\infty}^{\infty} \cos \frac{9\pi k}{2} \operatorname{sinc}(t-k). \end{aligned}$$

This sum of sincs provided by the sampling formula isn't $f(t) = \cos \frac{9\pi}{2} t$; it's $g(t) = \cos \frac{\pi}{2} t$ (though you'd never know that just from the formula). To say it again, interpolating the samples of f according to the formula at the sampling rate of 1 Hz — *too low a sampling rate* — has not produced $f(t)$; it has produced $g(t)$, an alias of f. Cool.

Before we leave this example let's take one more look at

$$\mathcal{F}f(s) * Ш_1(s) = \tfrac{1}{2} \sum_{k=-\infty}^{\infty} \left(\delta\left(s - \tfrac{9}{4} - k\right) + \delta\left(s + \tfrac{9}{4} - k\right)\right).$$

Being a convolution with III_1, this is periodic of period 1, but, actually, it has a *smaller* period. To find it, write

$$\begin{aligned}\tfrac{1}{2}\sum_{k=-\infty}^{\infty}\left(\delta\left(s-\tfrac{9}{4}-k\right)+\delta\left(s+\tfrac{9}{4}-k\right)\right) &= \tfrac{1}{2}\left(\text{III}_1\left(s-\tfrac{9}{4}\right)+\text{III}_1\left(s+\tfrac{9}{4}\right)\right)\\ &= \tfrac{1}{2}\left(\text{III}_1\left(s-2-\tfrac{1}{4}\right)+\text{III}_1\left(s+2+\tfrac{1}{4}\right)\right)\end{aligned}$$

(this is pretty much what we did at the top of the section)

$$= \tfrac{1}{2}\left(\text{III}_1\left(s-\tfrac{1}{4}\right)+\text{III}_1\left(s+\tfrac{1}{4}\right)\right)$$

(because III_1 is periodic of period 1).

This sum of those two III's is periodic of period 1/2, for

$$\begin{aligned}\tfrac{1}{2}\left(\text{III}_1\left(s-\tfrac{1}{4}+\tfrac{1}{2}\right)+\text{III}_1\left(s+\tfrac{1}{4}+\tfrac{1}{2}\right)\right) &= \tfrac{1}{2}\left(\text{III}_1\left(s+\tfrac{1}{4}\right)+\text{III}_1\left(s+\tfrac{3}{4}\right)\right)\\ &= \tfrac{1}{2}\left(\text{III}_1\left(s+\tfrac{1}{4}\right)+\text{III}_1\left(s+1-\tfrac{1}{4}\right)\right)\\ &= \tfrac{1}{2}\left(\text{III}_1\left(s+\tfrac{1}{4}\right)+\text{III}_1\left(s-\tfrac{1}{4}\right)\right)\end{aligned}$$

(using the periodicity of III_1).

You can also see the reduced periodicity of $(1/2)(\text{III}(s-9/4)+\text{III}(s+9/4))$ graphically from the way $\text{III}_1(s-9/4)$ and $\text{III}_1(s+9/4)$ line up. Here's a plot of part of

$$\frac{1}{2}\text{III}\left(s-\frac{9}{4}\right)=\tfrac{1}{2}\sum_{k=-\infty}^{\infty}\delta\left(s-k-\tfrac{9}{4}\right).$$

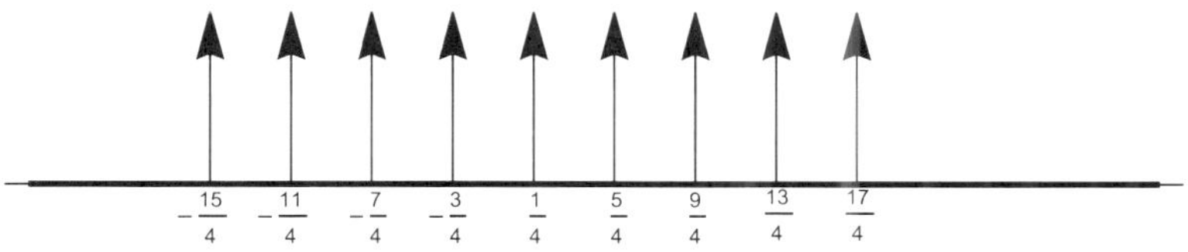

Here's a plot of part of

$$\frac{1}{2}\text{III}\left(s+\frac{9}{4}\right)=\tfrac{1}{2}\sum_{k=-\infty}^{\infty}\delta\left(s-k+\tfrac{9}{4}\right).$$

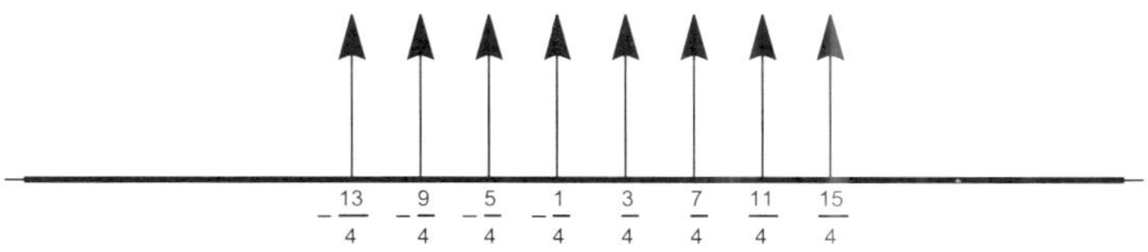

Here's a plot of the sum of the two.

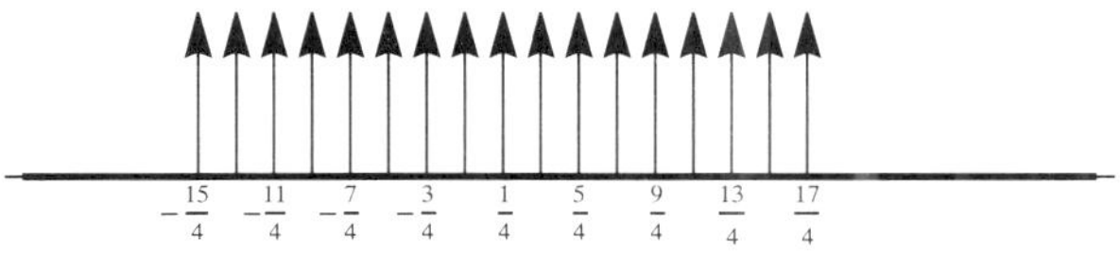

You can see that the δ's in the sum are spaced 1/2 apart. Cool? You might try working with $\cos 9\pi/2$ in the frequency domain using other sampling rates. See what periodizations look like and what happens when you cut off.

6.3.2. Sampling at the bandwidth. When we talked about the definition of bandwidth B of a signal, we raised the issue of what happens at the endpoints of the spectrum, $\pm B$, whether or not the Fourier transform is zero there, and how this sometimes requires special consideration. Here's an example of what I had in mind. Take two very simple signals,

$$f(t) = \sin 2\pi t \quad \text{and} \quad g(t) = \cos 2\pi t\,,$$

each of period 1. The Fourier transforms are

$$\mathcal{F}f(s) = \frac{1}{2i}\left(\delta(s-1) - \delta(s+1)\right) \quad \text{and} \quad \mathcal{F}g(s) = \frac{1}{2}\left(\delta(s-1) + \delta(s+1)\right)\,.$$

The bandwidth is 2 for each signal and the Fourier transforms are not zero at the endpoints $\pm p/2 = \pm 1$.

If we apply the sampling formula with $p = 2$ to $\sin 2\pi t$, we get the upsetting news that

$$\sin 2\pi t = \sum_{k=-\infty}^{\infty} \sin \tfrac{2\pi k}{2} \operatorname{sinc}\left(2\left(t - \frac{k}{2}\right)\right) = \sum_{k=-\infty}^{\infty} \sin k\pi \operatorname{sinc}(2t-k) = 0\,.$$

On the other hand, for $\cos 2\pi t$ the formula gives, again with $p = 2$,

$$\begin{aligned}
\cos 2\pi t &= \sum_{k=-\infty}^{\infty} \cos \tfrac{2\pi k}{2} \operatorname{sinc}\left(2\left(t - \frac{k}{2}\right)\right) \\
&= \sum_{k=-\infty}^{\infty} \cos k\pi \operatorname{sinc}(2t-k) = \sum_{k=-\infty}^{\infty} \operatorname{sinc}(2t-2k) - \sum_{k=-\infty}^{\infty} \operatorname{sinc}(2t-2k-1)\,,
\end{aligned}$$

which we might like to believe — at least both the series of sinc functions have period 1. Here's a plot (really) of

$$\sum_{k=-50}^{50} \cos k\pi \operatorname{sinc}(2t-k)$$

for some further encouragement.

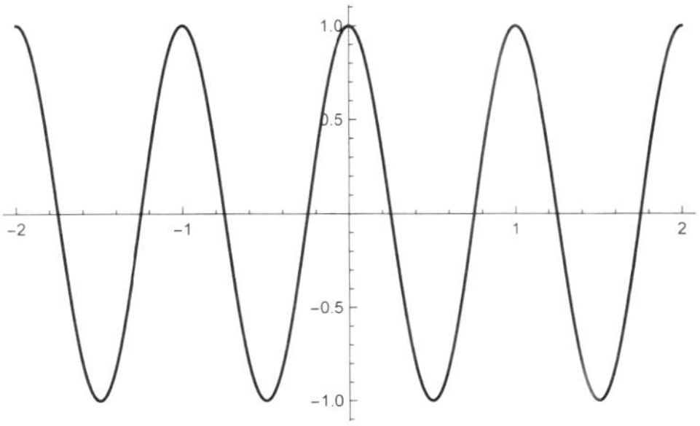

Pretty impressive.

It's easy to see what goes wrong with the sampling formula in the example of $\sin 2\pi t$. The first step in the derivation is to periodize the Fourier transform, and for $\sin 2\pi t$ this results in

$$\begin{aligned}\mathcal{F}f * \text{III}_2 &= \frac{1}{2i}(\delta_1 - \delta_{-1}) * \sum_{k=-\infty}^{\infty} \delta_{2k} \quad \text{(using the notation } \delta_a \text{ for } \delta(t-a)\text{)} \\ &= \frac{1}{2i} \sum_{k=-\infty}^{\infty} (\delta_{2k+1} - \delta_{2k-1}) = 0.\end{aligned}$$

The series telescopes and the terms cancel, so the sum is zero.

On the other hand, for $\sin 2\pi t$, taking the sampling rate to be 2.1, just 0.1 beyond the bandwidth, here's a plot of

$$\sum_{k=-50}^{50} \sin\left(\frac{2\pi k}{2.1}\right) \operatorname{sinc} 2.1\left(t - \frac{k}{2.1}\right).$$

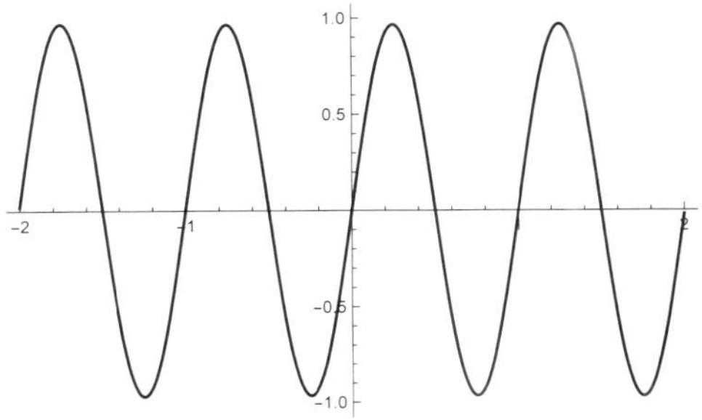

Some comfort in that, at least.

For $\cos 2\pi t$ we find something different in applying the derivation of the sampling formula:

$$\begin{aligned}\mathcal{F}g * Ш_2 &= \frac{1}{2}(\delta_1 + \delta_{-1}) * \sum_{k=-\infty}^{\infty} \delta_{2k} \\ &= \frac{1}{2} \sum_{k=-\infty}^{\infty} (\delta_{2k+1} + \delta_{2k-1}) = \sum_{k=-\infty}^{\infty} \delta_{2k+1}\,.\end{aligned}$$

The series telescopes and this time the terms add. So far so good, but is it true that $\mathcal{F}g = \Pi_2(\mathcal{F}g * Ш_2)$, as needed in the second step of the derivation of the sampling formula? We are asking whether

$$\Pi_2 \cdot Ш_{2n+1} = \Pi_2 \cdot \sum_{n=-\infty}^{\infty} \delta_{2n+1} = \tfrac{1}{2}(\delta_1 + \delta_{-1})\,.$$

This is correct — cutting off a δ at the edge by a rectangle function results in half the δ, as in the above, and cutting off the δ's outside the support of Π_2 gives zero. See Section 5.7; here's one place where we needed the result from that section.

I chose $\sin 2\pi t$ and $\cos 2\pi t$ as simple examples illustrating the extreme cases in setting the sampling rate right at the bandwidth. The signal $\sin 2\pi t$ is aliased to zero while $\cos 2\pi t$ is reconstructed without a problem. I'm hesitant to attempt to formulate a general principle here. I think it's best to say, as I did earlier, that any particular endpoint problem should call for special considerations.

6.3.3. Aliasing in general.

I sometimes think of aliasing as the natural phenomenon associated with the sampling formula. The sampling formula is a machine; you pass it data a_k, $k = 0, \pm 1, \pm 2, \ldots$, that you've collected, measured sample values determined by a sampling rate q, and it returns a signal

$$g(t) = \sum_{k=-\infty}^{\infty} a_k \operatorname{sinc} q\left(t - \frac{k}{q}\right)$$

that has the sample values you gave it,

$$g(n/q) = a_n\,.$$

Remember, this is true for any sampling rate q, because, to remind you,

$$\operatorname{sinc} q\left(\frac{n}{q} - \frac{k}{q}\right) = \operatorname{sinc}(n-k) = \begin{cases} 1, & n = k\,, \\ 0, & n \neq k\,. \end{cases}$$

It's not the formula's fault if $g(t)$ isn't the signal you wanted, or thought you wanted. If you used too low a sampling rate, you got an alias — the right sample values, but that's all you can say.

By the way, it's also true that the sampling formula returns a bandlmited signal of bandwidth q. To see this, we have

$$\begin{aligned}\mathcal{F}g(s) &= \sum_{k=-\infty}^{\infty} a_k \mathcal{F}\operatorname{sinc} q\left(t - \frac{k}{q}\right)\\ &= \sum_{k=-\infty}^{\infty} a_k e^{-2\pi i(k/q)s}\frac{1}{q}\Pi_q(s) = \frac{1}{q}\Pi_q(s)\sum_{k=-\infty}^{\infty} a_k e^{-2\pi i(k/q)s},\end{aligned}$$

and you're cutting off the series outside the interval $-q/2 < s < q/2$.

As in the examples we had above, aliasing occurs because of a breakdown in the derivation of the sampling formula caused by too low a sampling rate. To close out this discussion let's look back to the signal $f(t)$ we had in Section 6.2 to get a picture of this breakdown for a generic bandlimited signal.

Here's the plot of the magnitude of its Fourier transform (remember, it's the magnitude that we typically have to plot).

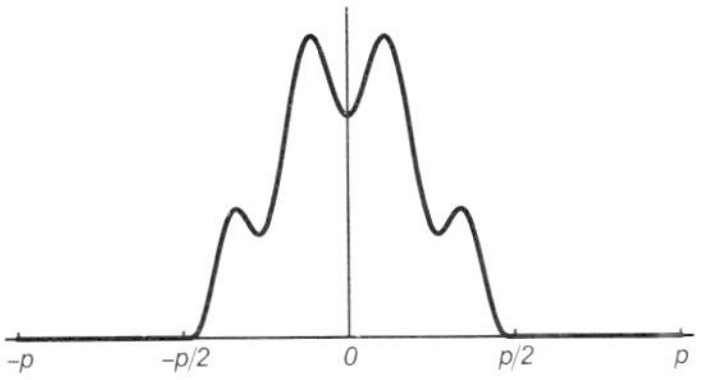

The bandwidth is p. For a sampling rate $q < p$ here's what happens when you periodize via $\mathcal{F}f * \text{III}_q$. Think of this in two steps. First the spectrum is shifted by integer multiples of q:

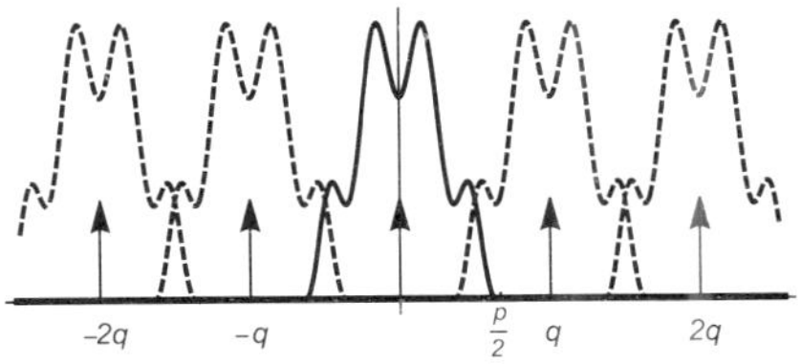

We see that the spectrum is not shifted off itself. There are overlaps. Then the shifts are added:

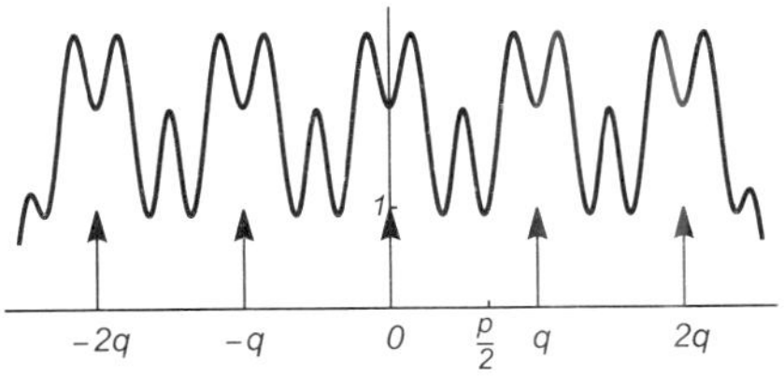

Then we cut off: $\Pi_q(\mathcal{F}f * Ш_q)$. Here's a plot of the result, with the cutoff boxed together with the original $|\mathcal{F}f(s)|$; the latter is shown dashed.

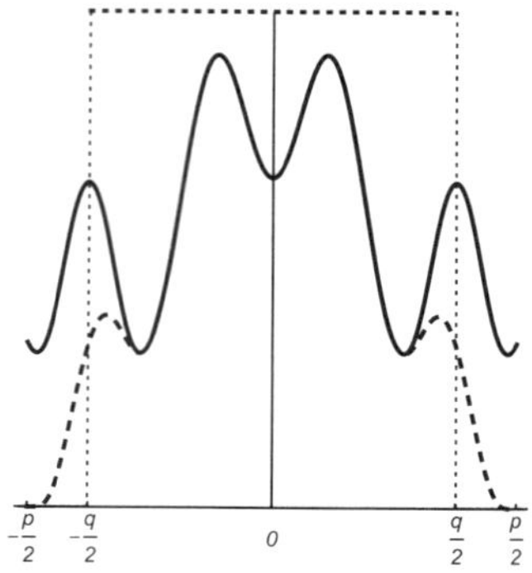

We *don't* have $\mathcal{F}f = \Pi_q(\mathcal{F}f * Ш_q)$. That's the breakdown. Taking the inverse Fourier transform will *not* give us back the signal f. It will give us an aliased signal. From the pictured spectrum, $\Pi_q(\mathcal{F}f * Ш_q)$, the aliased signal will have more energy in the frequencies near $\pm q/2$. The frequencies toward the middle of the spectrum are unaffected.

To avoid aliasing, one might pass the signal through an initial low-pass filter, in effect setting the bandwidth. That has problems of its own, depending on how sharp the filter is at the edges, but it's done. Also, despite what may sound like dire warnings, aliasing is not always to be avoided. See the problems.

6.4. Finite Sampling for a Bandlimited Periodic Signal

We started the whole discussion of sampling and interpolation by arguing that one ought to be able to interpolate the values of a finite sum of sinusoids from knowledge of a finite number of samples. Let's see how this works out, but rather than starting from scratch, let's use what we've learned about sampling for general bandlimited signals.

As always, it's best to work with the complex form of a sum of sinusoids, so we consider a real signal given by

$$f(t) = \sum_{k=-N}^{N} c_k e^{2\pi i k t/q}, \quad c_{-k} = \overline{c_k}\,.$$

$f(t)$ is periodic of period q. Some of the coefficients may be zero, but we assume that $c_N \neq 0$.

There are $2N+1$ terms in the sum (don't forget $k = 0$) and it should take $2N+1$ sampled values over one period to determine $f(t)$ completely. You might think it would take twice that many sampled values because the values of $f(t)$ are real and we have to determine *complex* coefficients. However, since $c_{-k} = \overline{c_k}$, if we know c_k, we know c_{-k}. Think of the $2N+1$ sample values as enough information to determine the real number c_0 and the N complex numbers $c_1, c_2, \ldots, c_N$.

The Fourier transform of f is

$$\mathcal{F}f(s) = \sum_{k=-N}^{N} c_k \delta\left(s - \frac{k}{q}\right)$$

and the spectrum goes from $-N/q$ to N/q. The sampling formula applies to $f(t)$, and we can write an equation of the form

$$f(t) = \sum_{k=-\infty}^{\infty} f(t_k) \operatorname{sinc} p(t - t_k)\,.$$

It's a question of what to take for the sampling rate and hence how to space the sample points.

We want to make use of the known periodicity of $f(t)$. If the sample points t_k are a fraction of a period apart, say q/M for an M to be determined, then the values $f(t_k)$ with $t_k = kq/M$, $k = 0, \pm 1, \pm 2, \dots$, will repeat after M samples. We'll see how this collapses the interpolation formula.

To find the right sampling rate, p, think about the derivation of the sampling formula, the first step being: "periodize $\mathcal{F}f$." The Fourier transform $\mathcal{F}f$ is a train of δ's spaced $1/q$ apart and scaled by the coefficients c_k. The natural periodization of $\mathcal{F}f$ is to keep the spacing $1/q$ in the periodized version, essentially making the periodized $\mathcal{F}f$ a version of $\text{Ш}_{1/q}$ with the coefficients scaled by the c_k. We do this by convolving $\mathcal{F}f$ with Ш_p where $p/2$ is the midpoint between N/q, the last point in the spectrum of $\mathcal{F}f$, and the point $(N+1)/q$, which is the next point $1/q$ away. Here's a picture (unscaled δ's).

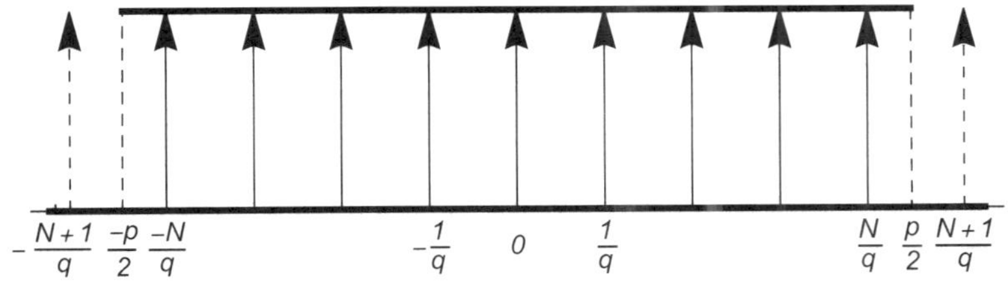

Thus we find p from

$$\frac{p}{2} = \frac{1}{2}\left(\frac{N}{q} + \frac{N+1}{q}\right) = \frac{(2N+1)}{2q}, \quad \text{or} \quad p = \frac{2N+1}{q}\,.$$

We periodize $\mathcal{F}f$ by Ш_p (draw yourself a picture of this!), cut off by Π_p, and then take the inverse Fourier transform. The sampling formula back in the time domain is

$$f(t) = \sum_{k=-\infty}^{\infty} f(t_k) \operatorname{sinc} p(t - t_k)$$

with

$$t_k = \frac{k}{p}\,.$$

With our particular choice of p let's now see how the q-periodicity of $f(t)$ comes into play. Write

$$M = 2N + 1\,,$$

so that

$$t_k = \frac{k}{p} = \frac{kq}{M}.$$

Then, to repeat what we said earlier, the sample points are spaced a fraction of a period apart, q/M, and after $f(t_0)$, $f(t_1)$, …, $f(t_{M-1})$ the sample values repeat; e.g., $f(t_M) = f(t_0)$, $f(t_{M+1}) = f(t_1)$, and so on. More succinctly,

$$t_{k+k'M} = t_k + k'q,$$

and so

$$f(t_{k+k'M}) = f(t_k + k'q) = f(t_k),$$

for any k and k'. Using this periodicity of the coefficients in the sampling formula, the single sampling sum splits into M sums as

$$\begin{aligned}
&\sum_{k=-\infty}^{\infty} f(t_k)\operatorname{sinc} p(t-t_k) \\
&= f(t_0)\sum_{m=-\infty}^{\infty} \operatorname{sinc}(pt - mM) + f(t_1)\sum_{m=-\infty}^{\infty} \operatorname{sinc}(pt-(1+mM)) \\
&\quad + f(t_2)\sum_{m=-\infty}^{\infty} \operatorname{sinc}(pt-(2+mM)) + \cdots + f(t_{M-1})\sum_{m=-\infty}^{\infty} \operatorname{sinc}(pt-(M-1+mM)).
\end{aligned}$$

Those sums of sincs on the right are periodizations of $\operatorname{sinc} pt$ and, remarkably, they have a simple closed form expression. The kth sum is

$$\begin{aligned}
\sum_{m=-\infty}^{\infty} \operatorname{sinc}(pt-k-mM) &= \operatorname{sinc}(pt-k) * Ш_{M/p}(t) \\
&= \frac{\operatorname{sinc}(pt-k)}{\operatorname{sinc}(\frac{1}{M}(pt-k))} = \frac{\operatorname{sinc}(p(t-t_k))}{\operatorname{sinc}(\frac{1}{q}(t-t_k))}.
\end{aligned}$$

I'll give a derivation of this at the end of this section. Using these identities, we find that the sampling formula to interpolate

$$f(t) = \sum_{k=-N}^{N} c_k e^{2\pi i k t/q}$$

from $2N+1 = M$ sampled values is

$$f(t) = \sum_{k=0}^{2N} f(t_k)\frac{\operatorname{sinc}(p(t-t_k))}{\operatorname{sinc}(\frac{1}{q}(t-t_k))}, \quad \text{where } p = \frac{2N+1}{q},\ t_k = \frac{k}{p} = \frac{kq}{2N+1}.$$

This is the *finite sampling theorem* for periodic functions.

Notice, by the way, that the zeros of the denominator of the ratio of sincs are exactly those of the numerator, so we're spared any embarrassment there. In fact, just as it should be,

$$\frac{\operatorname{sinc}(p(t_j-t_k))}{\operatorname{sinc}(\frac{1}{q}(t_j-t_k))} = \delta_{jk} \quad \text{(Kronecker delta)},$$

for

$$\frac{\operatorname{sinc}(p(t_j - t_k))}{\operatorname{sinc}(\frac{1}{q}(t_j - t_k))} = \frac{\operatorname{sinc}(p(\frac{j}{p} - \frac{k}{p}))}{\operatorname{sinc}(\frac{1}{q}(\frac{j}{p} - \frac{k}{p}))} = \frac{\operatorname{sinc}(j-k)}{\operatorname{sinc}(\frac{1}{2N+1}(j-k))}.$$

It might also be helpful to write the sampling formula in terms of frequencies. Thus, if the lowest frequency is $\nu_{\min} = 1/q$ and the highest frequency is $\nu_{\max} = N\nu_{\min}$, then

$$f(t) = \sum_{k=0}^{2N} f(t_k) \frac{\operatorname{sinc}((2\nu_{max} + \nu_{min})(t - t_k))}{\operatorname{sinc}(\nu_{\min}(t - t_k))}, \quad \text{where } t_k = \frac{kq}{2N+1}.$$

The sampling rate is $2\nu_{\max} + \nu_{min}$. Compare this to the sampling rate $> 2\nu_{\max}$ for a general bandlimited function.

Here's a simple example of the formula. Take $f(t) = \cos 2\pi t$. There's only one frequency, and $\nu_{\min} = \nu_{\max} = 1$. Then $N = 1$, the sampling rate is 3, and the sample points are $t_0 = 0$, $t_1 = 1/3$, and $t_2 = 2/3$. The formula says

$$\cos 2\pi t = \frac{\operatorname{sinc} 3t}{\operatorname{sinc} t} + \cos\left(\frac{2\pi}{3}\right)\frac{\operatorname{sinc}(3(t - \frac{1}{3}))}{\operatorname{sinc}(t - \frac{1}{3})} + \cos\left(\frac{4\pi}{3}\right)\frac{\operatorname{sinc}(3(t - \frac{2}{3}))}{\operatorname{sinc}(t - \frac{2}{3})}.$$

Does this really work? I'm certainly not going to plow through the trig identities needed to check it! However, here's a plot of the right-hand side. (Trust me; it's really the right-hand side.)

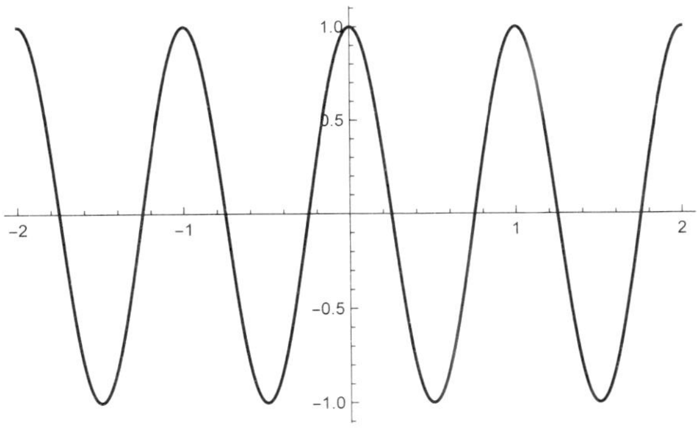

Any questions? Ever thought you'd see such a complicated way of writing $\cos 2\pi t$?

Periodizing sinc *functions.* In applying the general sampling theorem to the special case of a periodic signal, we wound up with sums of sinc functions that we recognized (sharp-eyed observers that we are) to be periodizations. Then, out of nowhere, came a closed form expression for such periodizations as a ratio of sinc functions. Here's where this comes from, and here's a fairly general result that covers it:

Lemma. Let p, $q > 0$ and let $N = \lfloor \frac{pq}{2} \rfloor$, the largest integer $\leq pq/2$. Then

$$\operatorname{sinc}(pt) * \text{III}_q(t) = \begin{cases} \dfrac{1}{pq}\dfrac{\sin((2N+1)\pi t/q)}{\sin(\pi t/q)} & \text{if } N < \frac{pq}{2}, \\[2ex] \dfrac{1}{pq}\left(\dfrac{\sin((2N-1)\pi t/q)}{\sin(\pi t/q)} + \cos(2\pi N t/q)\right) & \text{if } N = \frac{pq}{2}. \end{cases}$$

Using the identity

$$\sin((2N+1)\alpha) = \sin((2N-1)\alpha) + 2\sin\alpha\cos 2\alpha$$

we can write

$$\frac{1}{pq}\frac{\sin((2N+1)\pi t/q)}{\sin(\pi t/q)} = \frac{1}{pq}\left(\frac{\sin((2N-1)\pi t/q)}{\sin(\pi t/q)} + 2\cos(2\pi N t/q)\right),$$

so, if pressed, we could combine the two cases into a single formula:

$$\operatorname{sinc}(pt) * \text{III}_q(t) = \frac{1}{pq}\left(\frac{\sin((2N-1)\pi t/q)}{\sin(\pi t/q)} + \left(1 + \left\lceil \frac{pq}{2} \right\rceil - \left\lfloor \frac{pq}{2} \right\rfloor\right)\cos(2\pi N t/q)\right).$$

Here, in standard notation, $\lceil pq/2 \rceil$ is the smallest integer $\geq pq/2$ and $\lfloor pq/2 \rfloor$ is the largest integer $\leq pq/2$.

Having written this lemma down so grandly I now have to admit that it's really just a special case of the sampling theorem as we've already developed it, though I think it's fair to say that this is only obvious in retrospect. The functions on the right-hand side of the equation are each bandlimited — obvious in retrospect — and $\operatorname{sinc}(pt) * \text{III}_q(t)$ is the sampling series. One usually thinks of the sampling theorem as going *from* a function *to* a series of sampled values, but it can also go the other way.

This admission notwithstanding, I still want to go through the proof. The distinction between the two cases comes from cutting off just beyond a δ versus cutting off right at a δ. See Section 5.7. In practice, if q is given, it seems most natural to then choose p to make use of the first formula rather than the second.

In terms of sinc functions the first formula is

$$\frac{2N+1}{pq}\frac{\operatorname{sinc}((2N+1)t/q)}{\operatorname{sinc}(t/q)}.$$

The factor $2N+1$ can be expressed in terms of p and q as

$$2N + 1 = \left\lfloor \frac{pq}{2} \right\rfloor + \left\lceil \frac{pq}{2} \right\rceil.$$

It's also easy to extend the lemma slightly to include periodizing a shifted sinc function $\operatorname{sinc}(pt+b)$. We only write the formula in the first case, for $N < pq/2$, which is what we used for the finite sampling formula:

$$\operatorname{sinc}(pt+b) * \text{III}_q(t) = \frac{2N+1}{pq}\frac{\operatorname{sinc}(\frac{2N+1}{pq}(pt+b))}{\operatorname{sinc}(\frac{1}{pq}(pt+b))}.$$

One more thing. If $p = q = 1$, so that $N = 0$, the formula in the lemma gives

$$\sum_{n=-\infty}^{\infty} \operatorname{sinc}(t-n) = \operatorname{sinc} t * \text{Ш}_1(t) = 1\,.$$

Striking. Still don't believe it? Here's a plot of

$$\sum_{n=-100}^{100} \operatorname{sinc}(t-n)\,.$$

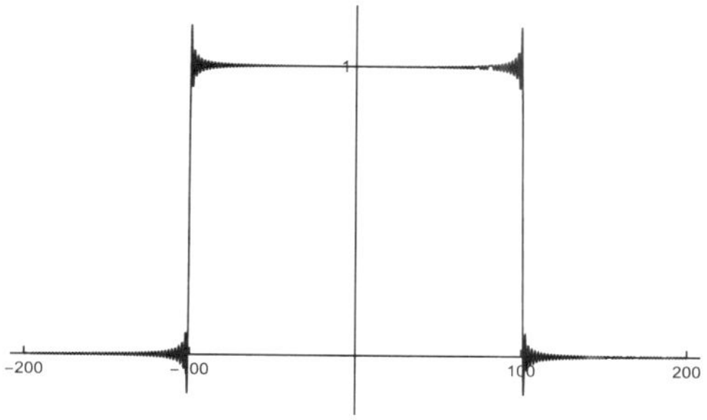

Note the scale on the axes — how else was I supposed to display it. There's a Gibbs-like phenomenon at the edges. This means there's some issue with what kind of convergence is involved, which is the last thing you and I want to worry about.

We proceed with the proof, which will look awfully familiar. Take the case when $N < \lfloor pq/2 \rfloor$ and take the Fourier transform of the convolution:

$$\begin{aligned}\mathcal{F}(\operatorname{sinc}(pt) * \text{Ш}_q(t)) &= \mathcal{F}(\operatorname{sinc}(pt)) \cdot \mathcal{F}\text{Ш}_q(t) \\ &= \frac{1}{p}\Pi_p(s) \cdot \frac{1}{q}\text{Ш}_{1/q}(s) \\ &= \frac{1}{pq}\sum_{n=-N}^{N} \delta\Big(s - \frac{n}{q}\Big)\,.\end{aligned}$$

See the figure below.

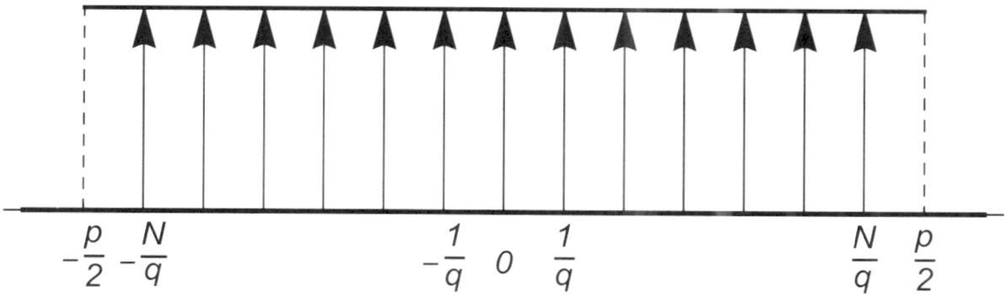

And now take the inverse Fourier transform:

$$\mathcal{F}^{-1}\bigg(\frac{1}{pq}\sum_{n=-N}^{N} \delta\Big(s - \frac{n}{q}\Big)\bigg) = \frac{1}{pq}\sum_{n=-N}^{N} e^{2\pi i n t/q} = \frac{1}{pq}\,\frac{\sin(\pi(2N+1)t/q)}{\sin(\pi t/q)}\,.$$

There it is. One reason I wanted to go through this is because it is another occurrence of the sum of exponentials and the identity

$$\sum_{n=-N}^{N} e^{2\pi int/q} = \frac{\sin(\pi(2N+1)t/q)}{\sin(\pi t/q)},$$

which we've seen on other occasions. Reading the equalities backwards we have

$$\mathcal{F}\left(\frac{\sin(\pi(2N+1)t/q)}{\sin(\pi t/q)}\right) = \mathcal{F}\left(\sum_{n=-N}^{N} e^{2\pi int/q}\right) = \sum_{n=-N}^{N} \delta\left(s - \frac{n}{q}\right).$$

This substantiates the earlier claim that the ratio of sines is bandlimited, and hence we could have appealed to the sampling formula directly instead of going through the argument we just did. But who would have guessed it?

The second case is when $N/q = p/2$. This time cutting off with Π_p gives (à la Section 5.7)

$$\frac{1}{pq}\left(\sum_{n=-(N-1)}^{N-1} \delta\left(s - \frac{n}{q}\right) + \frac{1}{2}(\delta_{N/q} + \delta_{-N/q})\right).$$

The half δ's account for the cosine.

6.5. Appendix: Timelimited vs. Bandlimited Signals

Here's a more careful treatment of the result that a bandlimited signal cannot be timelimited. We'll actually prove a more general statement and perhaps I should have said that no *interesting* signal can be both timelimited and bandlimited because, precisely:

- Suppose $f(t)$ is a bandlimited signal. If there is some interval $a < t < b$ on which $f(t)$ is identically zero, then $f(t)$ is identically zero for all t.

There's a very cunning argument for this, due as far as I know to Dym and McKean from their book *Fourier Series and Integrals* mentioned back in Chapter 1. Here we go.

The signal $f(t)$ is bandlimited so $\mathcal{F}f(s)$ is identically zero, say, for $|s| \geq p/2$. The Fourier inversion formula says[10]

$$f(t) = \int_{-\infty}^{\infty} \mathcal{F}f(s)e^{2\pi ist}\, ds = \int_{-p/2}^{p/2} \mathcal{F}f(s)e^{2\pi ist}\, ds.$$

Suppose $f(t)$ is zero for $a < t < b$. Then for t in this range,

$$\int_{-p/2}^{p/2} \mathcal{F}f(s)e^{2\pi ist}\, ds = 0.$$

Differentiate with respect to t under the integral. If we do this n times, we get

$$0 = \int_{-p/2}^{p/2} \mathcal{F}f(s)(2\pi is)^n e^{2\pi ist}\, ds = (2\pi i)^n \int_{-p/2}^{p/2} \mathcal{F}f(s)s^n e^{2\pi ist}\, ds,$$

[10] We assume the signal is such that Fourier inversion holds. You can take $f(t)$ to be a Schwartz function, but some more general signals will do.

so that

$$\int_{-p/2}^{p/2} \mathcal{F}f(s)s^n e^{2\pi ist}\,ds = 0\,.$$

Again, this holds for all t with $a < t < b$; pick one, say t_0. Then

$$\int_{-p/2}^{p/2} \mathcal{F}f(s)s^n e^{2\pi ist_0}\,ds = 0\,.$$

But now for *any* t (anywhere, not just between a and b) we can write

$$\begin{aligned} f(t) &= \int_{-p/2}^{p/2} \mathcal{F}f(s)e^{2\pi ist}\,ds = \int_{-p/2}^{p/2} \mathcal{F}f(s)e^{2\pi is(t-t_0)}e^{2\pi ist_0}\,ds \\ &= \int_{-p/2}^{p/2} \sum_{n=0}^{\infty} \frac{(2\pi i(t-t_0))^n}{n!} s^n e^{2\pi ist_0}\mathcal{F}f(s)\,ds \\ &\qquad \text{(using the Taylor series expansion for } e^{2\pi is(t-t_0)}\text{)} \\ &= \sum_{n=0}^{\infty} \frac{(2\pi i(t-t_0))^n}{n!} \int_{-p/2}^{p/2} s^n e^{2\pi ist_0}\mathcal{F}f(s)\,ds = \sum_{n=0}^{\infty} \frac{(2\pi i(t-t_0))^n}{n!} 0 = 0\,. \end{aligned}$$

Hence $f(t)$ is zero for all t.

The same argument *mutatis mutandis* will show:

- If $f(t)$ is timelimited and if $\mathcal{F}f(s)$ is identically zero on any interval $a < s < b$, then $\mathcal{F}f(s)$ is identically zero for all s.

Then $f(t)$ is identically zero, too, by Fourier inversions.

Remark 1, for eager seekers of knowledge. This bandlimited vs. timelimited result is often proved by establishing a relationship between timelimited signals and analytic functions (of a complex variable) and then appealing to results from the theory of analytic functions. That connection opens up an important direction for applications of the Fourier transform, but it involves a considerable amount of background and the direct argument makes this approach unnecessary.

Remark 2, for overwrought math students and careful engineers. Where in the preceding argument did we use that $p < \infty$? It's needed in switching integration and summation, in the line

$$\int_{-p/2}^{p/2} \sum_{n=0}^{\infty} \frac{(2\pi i(t-t_0))^n}{n!} s^n e^{2\pi ist_0}\mathcal{F}f(s)\,ds$$
$$= \sum_{n=0}^{\infty} \frac{(2\pi i(t-t_0))^n}{n!} \int_{-p/2}^{p/2} s^n e^{2\pi ist_0}\mathcal{F}f(s)\,ds.$$

The theorems that tell us "the integral of the sum is the sum of the integral" require as an essential hypothesis that the series converges *uniformly*.[11] In the sum-and-integral expression, above, the variable s ranges over a finite interval, from $-p/2$ to $+p/2$. Over such a finite interval the series for the exponential converges uniformly,

[11] Recall that "uniformly" means, loosely, that if we plug a particular value into the converging series, we can estimate the rate at which the series converges *independent* of that particular value. We can make "uniform" estimates, in other words. We saw this sort of thing in the notes on convergence of Fourier series.

essentially because the terms can only get so big — so they can be estimated uniformly — when s can only get so big. We can switch integration and summation in this case. If, however, we had to work with

$$\int_{-\infty}^{\infty} \sum_{n=0}^{\infty} \frac{(2\pi i(t-t_0))^n}{n!} s^n e^{2\pi i s t_0} \mathcal{F}f(s)\, ds\,,$$

i.e., if we did not have the assumption of bandlimitedness, then we could not make uniform estimates for the convergence of the series, and switching integration and summation would not be justified.

It's not only unjustified, it's really wrong. If we could drop the assumption that the signal is bandlimited, we'd be buying into the statement: if $f(t)$ is identically zero on an interval, then it's identically zero. Think of the implications of such a dramatic statement. In a phone conversation, if you paused for a few seconds to collect your thoughts, your signal would be identically zero on that interval of time, and therefore you would have nothing to say at all, ever again. Be careful.

6.6. Appendix: Linear Interpolation via Convolution

Remember linear interpolation from Section 6.1 as a grown-up version of connect-the-dots? There's a nice way to describe how to do it and to get a formula using the technology of Ш's and convolution. No Fourier transforms, and everything happens in the time domain. You'll enjoy this.

Suppose we sample a function $f(t)$ at a sampling rate p. The sample points, in time, are $t_k = k/p$, spaced $1/p$ apart. The sampled function is

$$f(t)Ш_{1/p}(t) = \sum_{k=-\infty}^{\infty} f(t_k)\delta(t-k/p) = \sum_{k=-\infty}^{\infty} f(t_k)\delta(t-t_k)\,.$$

This is meant to include the possibility of a finite number of samples, where the $f(t_k)$ are zero beyond a certain point. Now bring in the triangle function

$$\Lambda(t) = \begin{cases} 1-|t|, & |t| \le 1, \\ 0, & |t| \ge 1\,, \end{cases}$$

and its scaled version

$$\Lambda_{1/p}(t) = \Lambda(pt) = \begin{cases} 1-|pt|, & |t| \le 1/p, \\ 0, & |t| \ge 1/p\,, \end{cases}$$

languishing since Chapter 3:

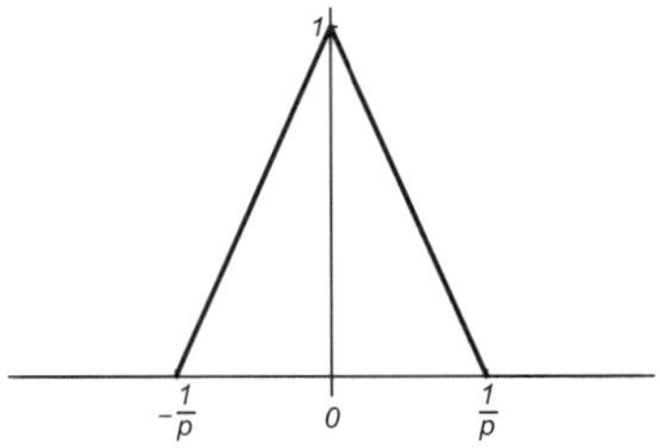

Convolve this with $f(t)Ш_{1/p}(t)$, forming

$$L(t) = \Lambda_{1/p}(t) * (f(t)Ш_{1/p}(t)) = \sum_{k=-\infty}^{\infty} f(t_k)\Lambda_{1/p}(t - t_k)$$
$$= \sum_{k=-\infty}^{\infty} f\left(\frac{k}{p}\right) \Lambda\left(p\left(t - \frac{k}{p}\right)\right).$$

That's the linear interpolation! $\Lambda_{1/p}$ is playing the role of the sinc.

Below is a picture of

$$L(t) = 3\Lambda_{1/p}(t - 1/p) + 2.5\Lambda_{1/p}(t - 2/p) - 1\Lambda_{1/p}(t - 3/p) + 1.5\Lambda_{1/p}(t - 4/p)$$

showing how the triangles combine. Because Λ's vanish at their endpoints, the first and last interpolated points are on the t-axis (sample value 0), so you have to take account of that.

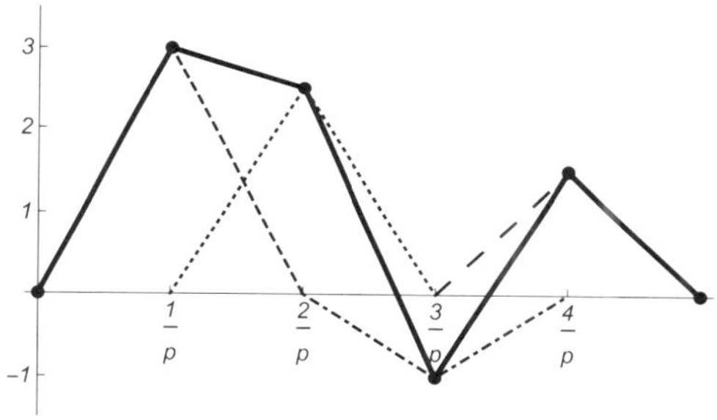

To see that the formula is producing what you think it is, first check the sample points. Λ has the magical property (like sinc)

$$\Lambda(n-k) = \begin{cases} 1, & n = k, \\ 0, & n \neq k, \end{cases}$$

so

$$L(t_n) = \sum_{k=-\infty}^{\infty} f(t_k)\Lambda\left(p\left(\frac{n}{p} - \frac{k}{p}\right)\right) = \sum_{k=-\infty}^{\infty} f(t_k)\Lambda(n-k) = \begin{cases} f(t_n), & n = k, \\ 0, & n \neq k. \end{cases}$$

Next, as a sum of piecewise linear functions $L(t)$ is also piecewise linear. The only question is whether, in the sum defining $L(t)$, two adjacent Λ's for two adjacent sample points combine to give you a single segment joining the two sample points. Yes, that's what happens and I'll leave the verification to you. This is enough to be sure that $L(t)$ is the very same connect-the-dots linear interpolation that you drew when you were young.

What about the even simpler sample-and-hold interpolation? That's a sum of shifted rectangles — also a convolution. I'll let you write that down, giving a derivation modeled on the one we just did.

6.7. Appendix: Lagrange Interpolation

Finally, a nod to polynomial interpolation, a separate topic and a large one. Going way back, it was desirable to find readily computable approximations of complicated functions by simple functions, particularly in applications arising from solutions to differential equations. Even as computational power has increased, this is still of some interest, though much has changed to say the least. The next step up from linear interpolation is to use polynomials of higher degree, and this has been the classic way to interpolate and approximate. One old method, presented here for your general background and know-how, is due to J.-L. Lagrange.[12]

Suppose we have n points $t_1, t_2, \dots, t_n$ at which we have made n measurements. We want a polynomial of degree $n-1$ that assumes the measured values at the respective t's (2 points, a line, degree 1; 3 points, a quadratic, degree 2; etc.). For this, start with an nth degree polynomial that vanishes exactly at the t_k. This is

$$p(t) = (t-t_1)(t-t_2)\cdots(t-t_n)\,.$$

Next put

$$p_k(t) = \frac{p(t)}{t-t_k}\,.$$

Then $p_k(t)$ is a polynomial of degree $n-1$; we divide out the factor $(t-t_k)$ and so $p_k(t)$ vanishes at the same points as $p(t)$ except at t_k. Next consider the quotient

$$\frac{p_k(t)}{p_k(t_k)}\,.$$

This is again a polynomial of degree $n-1$. The key property is that $p_k(t)/p_k(t_k)$ vanishes at the sample points t_j *except* at the point t_k where its value is 1; i.e.,

$$\frac{p_k(t_j)}{p_k(t_k)} = \begin{cases} 1, & j = k\,, \\ 0, & j \neq k\,. \end{cases}$$

There's the magical property again, just like sinc and just like Λ.

To interpolate/approximate a function by a polynomial (to fit a curve through a given set of points) we just scale and add. Suppose we have a function $f(t)$ and we want a polynomial that has values $f(t_1), f(t_2), \dots, f(t_n)$ at the points $t_1, t_2, \dots, t_n$. We get this by forming the sum

$$p(t) = \sum_{k=1}^{n} f(t_k)\frac{p_k(t)}{p_k(t_k)}\,.$$

This does the trick. It is known as the *Lagrange interpolation polynomial.*

The shifted, scaled sinc functions $\operatorname{sinc} p(t-t_k)$ are the analogs for Fourier interpolation of the $p_k(t)/p_k(t_k)$. As with linear interpolation and sample-and-hold, and unlike the sinc interpolation formula, we're not aiming to reconstruct exactly all the values of $f(t)$ from a set of sample values. On the other hand, there are no restrictions, such as being bandlimited. The aim is to approximate $f(t)$ by a

[12] Joseph-Louis Lagrange made many important contributions to mathematics and its applications. Engineers often first run into his work through the method of Lagrange multipliers in studying optimization problems and are then later tormented in a class on ODEs by his method of variation of parameters. If you took an advanced mechanics course, you probably also learned about Lagrangians and his formulation of mechanics through variational principles. All told, quite remarkable.

polynomial that has the same values as $f(t)$ at a prescribed set of points. Also, for Lagrange interpolation we're not assuming that the t_k are evenly spaced, as we do in the sampling theorem. Moreover, unlike in the sinc interpolation formula, there are only a finite number of sample points.

As an aside, let me mention that the general topic of orthogonal polynomials (which the Lagrange polynomials are not, actually) is a vast masterpiece of classical mathematics. ("Orthogonal" means with respect to an inner product defined via an integral.) Sometimes such polynomials appear in the service of applications, including approximations, and sometimes they have their own intrinsic interest. You saw Legendre polynomials in Problem 1.19. Various families of polynomials and related functions are built into the standard mathematical software packages. Experiment.

Problems and Further Results

6.1. *Convergence of* sinc *series*[13]

There are many theorems, of various degrees of generality, on the convergence of a series of sinc functions, say a series of the form

$$\sum_{n=-\infty}^{\infty} a_n \operatorname{sinc}\left(p\left(t-\frac{n}{p}\right)\right).$$

A basic result is on *absolute convergence*, meaning

$$\sum_{n=-\infty}^{\infty} \left|a_n \operatorname{sinc}\left(p\left(t-\frac{n}{p}\right)\right)\right| < \infty \quad \text{for any } t.$$

(Look up why this is a property you'd like to have.) In this problem you'll show that this property holds for every real number t if and only if

$$\sum_{n\neq 0} \left|\frac{a_n}{n}\right| < \infty.$$

Fix a number t. If $t = m/p$ for some integer m, then the series reduces to a single term and there's nothing more to say about its convergence. So we can assume that t is not of this form, and that's important for the argument.

(a) Derive the estimate

$$\left|a_n \operatorname{sinc}\left(p\left(t-\frac{n}{p}\right)\right)\right| \leq \left|\frac{a_n}{n}\right| \frac{1}{\pi p\left|\frac{t}{n}-\frac{1}{p}\right|},$$

---

[13]From J. R. Higgins, *Sampling Theory in Fourier and Signal Analysis*, Oxford University Press, 1996.

and deduce that if $|n|$ is sufficiently large, then

$$\left|a_n \operatorname{sinc}\left(p\left(t - \frac{n}{p}\right)\right)\right| \leq \frac{2}{\pi}\left|\frac{a_n}{n}\right|.$$

(The constant $2/\pi$ isn't important, just that there's *some* constant for which the inequality holds.) Conclude that the condition on the sum of the a_n's implies the absolute convergence of the sinc series.

(b) Next suppose that the sinc series converges absolutely. Why does it follow that

$$\sum_{n \neq 0} \frac{|a_n|}{\pi p|t - \frac{n}{p}|} < \infty?$$

(c) Show that by taking $|n|$ large enough we have

$$\frac{|a_n|}{\pi|n|p|\frac{t}{n} - \frac{1}{p}|} \geq \frac{1}{2\pi}\left|\frac{a_n}{n}\right|.$$

Conclude that the absolute convergence implies the condition on the sum of the a_n.

6.2. *A surprising identity*

The sinc function satisfies

$$\sum_{n=-\infty}^{\infty} \operatorname{sinc}^2(t - n) = 1.$$

Show this in three different ways:

(a) Periodize $\operatorname{sinc}^2 t$ to have period 1 and find its Fourier series. (Recall the relationship between the Fourier coefficients and the Fourier transform.)

(b) Use the sampling formula applied to $\operatorname{sinc}(t - x)$ regarded as a function of x with t fixed. Show that this leads to the formula

$$\operatorname{sinc}(t - x) = \sum_{n=-\infty}^{\infty} \operatorname{sinc}(t - n)\operatorname{sinc}(x - n).$$

This is itself an interesting formula.

(c) Use the Poisson summation formula applied to $\operatorname{sinc}^2(t - x)$ regarded as a function of x with t fixed.

6.3. *Energy of a bandlimited signal*

The energy of a signal $g(t)$ is the integral

$$\int_{-\infty}^{\infty} |g(t)|^2\, dt.$$

Suppose that $g(t)$ is bandlimited with

$$\mathcal{F}g(s) = 0, \quad |s| \geq \frac{1}{2}.$$

Express the energy of $g(t)$, in terms of the sample values of $g(t)$ at the integers, $g(n)$, $n = 0, \pm 1, \pm 2, \ldots$. *Hint*: Use the Fourier series of the periodization of $\mathcal{F}g(s)$ of period 1.

6.4. *Filtering for interpolation*

Suppose you have sampled a signal $f(t)$ at intervals of one unit to obtain $f_{\text{sampled}}(t)$, shown below:

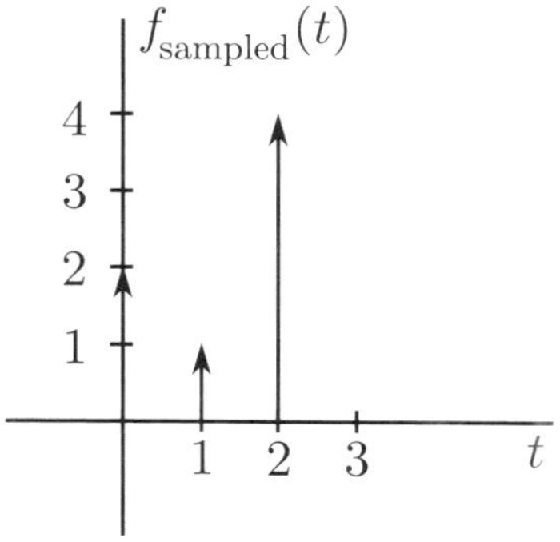

The arrows represent δ-functions of different strengths at 0, 1, and 2.

Sketch the following interpolations of $f(t)$:

(a) $f_1(t) = \mathcal{F}^{-1}\{\text{sinc}(s)\, \mathcal{F}f_{\text{sampled}}(s)\}$.
(b) $f_2(t) = \mathcal{F}^{-1}\{e^{-i\pi s}\, \text{sinc}(s)\, \mathcal{F}f_{\text{sampled}}(s)\}$.
(c) $f_3(t) = \mathcal{F}^{-1}\{\text{sinc}^2(s)\, \mathcal{F}f_{\text{sampled}}(s)\}$.

These filters correspond to common interpolation methods: nearest neighbor, zero-order hold, and linear interpolation.

6.5. *And yet it flies*[14]

Watch the video "chopper" available at

`http://www.youtube.com/watch?v=bZCUB_BiY_4.`

The phenomenon you are observing can be attributed to aliasing. Suppose the frame rate of the video camera is R_1; i.e., the camera is taking R_1 still shots per second; and the rotation rate of the main rotor is R_2 rotations per second.

(a) Suppose R_1 is fixed and the chopper has 5 rotor blades. What values of R_2 (expressed in terms of R_1) cause the rotor to appear stationary as in the video?
(b) In part (a), we assumed that the chopper has 5 rotor blades. Is this assumption valid? If you had seen 6 blades in the video, how many blades do you think the chopper has?

6.6. *Handel's Hallelujah*[15]

In this problem we will explore the effects of sampling with or without anti-aliasing filters. There can be a significant distortion of music due to aliasing if we sample slower than twice the highest frequency. However, if we can suppress the high frequency components before sampling, we can possibly

[14] From T. John.
[15] From Logi Vidarsson.

avoid this distortion. In this problem we will use an anti-aliasing filter $H(s)$ whose Fourier transform is shown below. $H(s)$ is available to download in the MATLAB file `anti-aliasing.mat`, which contains $H(s)$ in the vector `Hs`.

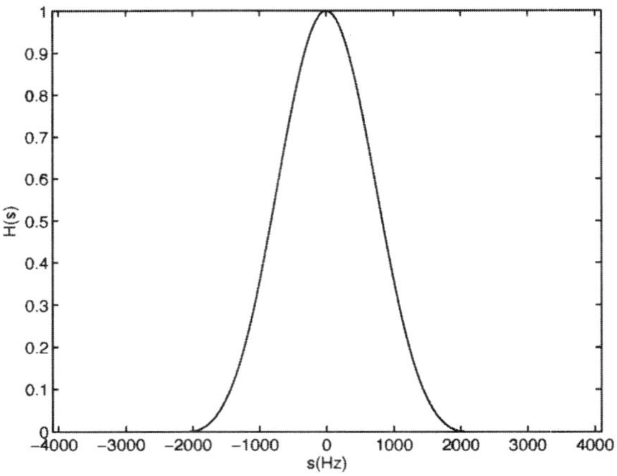

Figure 6.1. Anti-aliasing filter.

Built into MATLAB is a snippet of Handel's "Hallelujah Chorus". You load it into the workspace by typing

```
load handel
```

This loads two variables into the workspace: `y`, which contains about 8 seconds of Handel's "Hallelujah Chorus", and `Fs`, which is the sampling frequency used.

Finally, here is the problem. Resample the snippet of Handel's "Hallelujah Chorus" down to a sampling frequency of $f_s = 4{,}096$ Hz that should be half of the original sampling frequency.

Now apply the anti-aliasing filter to Handel's "Hallelujah Chorus" so that you cut off all frequencies higher than 2,048 Hz, and then resample down to $f_s = 4{,}096$ Hz. Is there any audible difference between the two versions? Why or why not. Discuss audible differences you heard or did not hear.

Hints: To resample at half the sampling rate, you can use

```
xhalf = x(1:2:length(x));
```

Remember to adjust the sampling rate correctly when you use `sound` or `wavwrite`.

Recall that you can use `fft` to take the Fourier transform, and `ifft` to take the inverse Fourier transform. `Hs` has been arranged in the same way MATLAB's `fft` returns Fourier transforms.

To evaluate $H(s)X(s)$ tryusing the `.*` operator.

6.7. *Nyquist rate and spectral islands*

The signal $f(t)$ has the Fourier transform $F(s)$ as shown below.

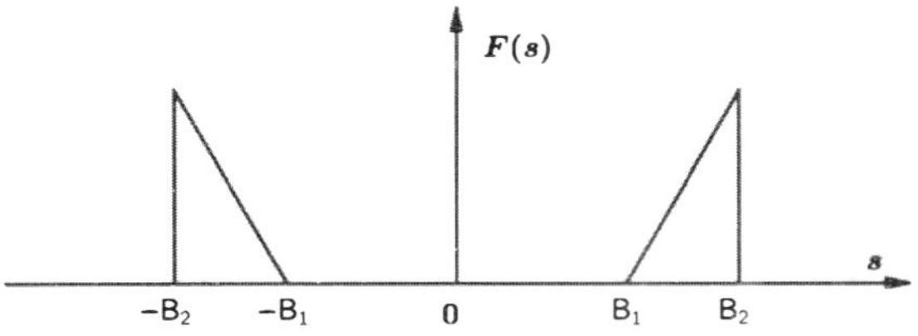

The Nyquist frequency is $2B_2$ since the highest frequency in the signal is B_2. The sampling theorem tells us that if we sample above the Nyquist rate, no aliasing will occur. Is it possible, however, to sample at a lower frequency in this case and not get aliasing effects? If it is possible, then explain how it can be done and specify at least one range of valid sampling frequencies below the Nyquist rate that will not result in aliasing. If it is not possible, explain why not.

6.8. *Natural sampling*

Suppose the signal $f(t)$ is bandlimited with $\mathcal{F}f(s) = 0$ for $|s| \geq B$. Instead of sampling with a train of δ's we sample $f(t)$ with a train of very narrow pulses. The pulse is given by a function $p(t)$, we sample at a rate T, and the sampled signal then has the form

$$g(t) = f(t)\left(\sum_{k=-\infty}^{\infty} Tp(t-kT)\right).$$

(a) Is it possible to recover the original signal f from the signal g?
(b) If not, why not. If it is possible, what conditions on the parameters T and B and on the pulse $p(t)$ make it possible?

6.9. *Nonuniform sampling: A variation on some earlier problems*

We wish to sample and reconstruct signals whose spectrum is known to be nonzero within the frequency bands $0 \leq s \leq 1$ and $2 \leq s \leq 3$. (Suppose the signals have been phase shifted so that there are no negative frequencies.) For example, a signal $f(t)$ whose spectrum $F(s) = \mathcal{F}f(s)$ is as shown below.

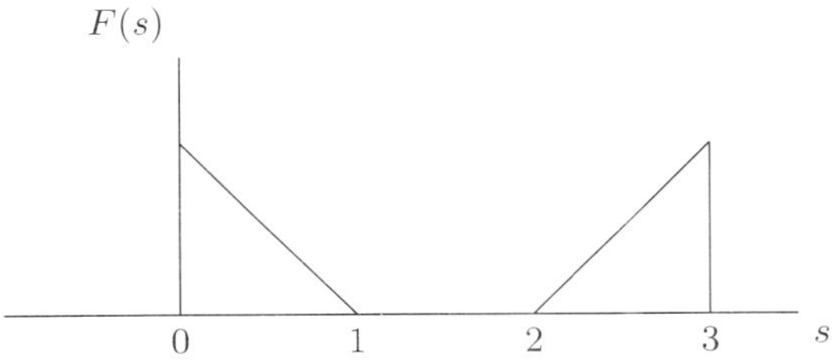

If we sum shifted copies of $F(s)$, as in forming the periodization

$$\sum_{n=-\infty}^{\infty} F(s-n),$$

we may get some overlapping, and the way the shifts overlap depends on if we shift by even or odd integers.

(a) On one set of axes sketch

$$\sum_{n\,\mathrm{odd}} F(s-n) = F(s-1) + F(s+1) + F(s-3) + F(s+3) + \cdots .$$

On a second set of axes sketch

$$\sum_{n\,\mathrm{even}} F(s-n) = F(s) + F(s-2) + F(s+2) + F(s-4) + F(s+4) + \cdots .$$

On a third set of axes sketch the full sum

$$\sum_{n=-\infty}^{\infty} F(s-n).$$

If we wanted to reconstruct the signal $f(t)$ from sampled values and use uniform sampling (i.e., applying the usual sampling theorem, where the sampling points are evenly spaced), we would have to use a sampling rate of at least 3. Instead, we consider a modified sampling scheme that uses a different, nonuniform sampling pattern. The spacing of the sample points is determined by choosing a value t_0 from among $\{1/4, 1/2, 3/4\}$ — you will be asked which one(s) work. Let

$$p(t) = \delta(t) + \delta(t - t_0) \quad \text{and} \quad f_{\text{sampled}}(t) = f(t)\left(\sum_{n=-\infty}^{\infty} p(t-n)\right).$$

The train of δ's for $t_0 = 1/4$ is shown below. The other choices of t_0 have a similar pattern. There are two samples per second, though not evenly spaced within a one-second interval, so we might still say that the sampling rate is 2.

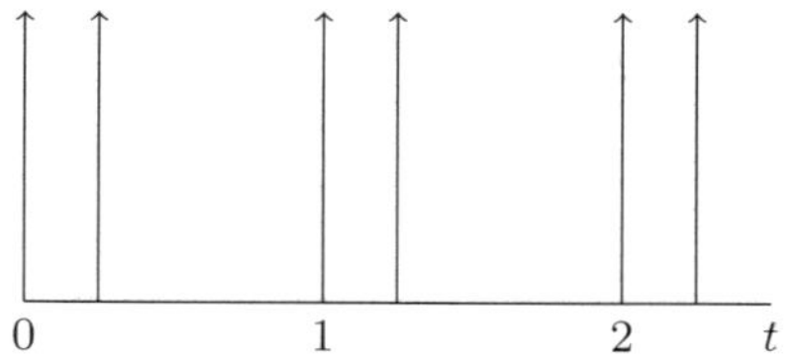

(b) Show that

$$\mathcal{F} f_{\text{sampled}}(s) = \sum_{n=-\infty}^{\infty} \mathcal{F}p(n)F(s-n).$$

Hint: You did this in the problem on natural sampling: write $\sum_n p(t-n) = (p * Ш)(t)$.

So the Fourier transform of $f_{\text{sampled}}(t)$ is a sum of shifted copies of $F(s)$, *and each copy is scaled by $\mathcal{F}p(n)$*, the Fourier transform of the pulse at integer values.

The possibility of reconstructing $f(t)$ from its samples depends on whether the scaled shifts interfere with each other, i.e., whether you can isolate the original spectrum $F(s)$, and this depends on the value you take for t_0 as it affects $\mathcal{F}p(n)$.

(c) Which values t_0 chosen from $\{1/4, 1/2, 3/4\}$ will allow you to recover the signal?

6.10. *Oversampling*

Let $f(t)$ be a bandlimited signal with spectrum contained in the interval $-1/2 < s < 1/2$. Suppose you sample $f(t)$ at intervals of $1/2$ (that is, at *twice* the Nyquist rate), to obtain

$$f_{\text{sampled}}(t) = \frac{1}{2} Ш_{1/2}(t)\, f(t).$$

(a) Qualitatively explain why the following equation is correct:

$$f(t) = \mathcal{F}^{-1}\{K(s)\ \mathcal{F}f_{\text{sampled}}(s)\}$$

where $K(s)$ is defined by

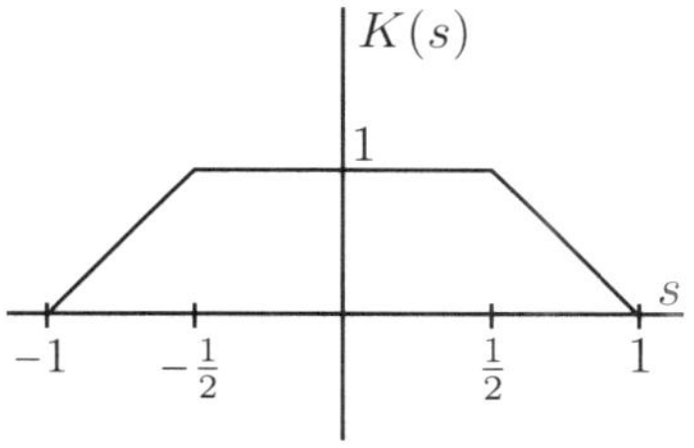

(b) Show that you can reconstruct $f(t)$ by

$$f(t) = \frac{1}{2} \sum_{n=-\infty}^{\infty} f\left(\frac{n}{2}\right) k\left(t - \frac{n}{2}\right) \qquad \text{where } k(y) = \frac{\cos(\pi y) - \cos(2\pi y)}{\pi^2 y^2}.$$

You may use the fact that

$$K(s) = \mathcal{F}\left\{ \frac{\cos(\pi t) - \cos(2\pi t)}{\pi^2 t^2} \right\}.$$

(c) Describe an advantage the reconstruction formula in part (b) has over the usual sinc interpolation formula, say in terms of the accuracy of the series if one only uses a finite number of terms.

6.11. *Can undersampling be overcome?*[16]

Let $f(t)$ be a bandlimited signal whose Fourier transform satisfies $|\mathcal{F}f(s)| = 0$ for $|s| \geq 1$. According to the sampling theorem, one has to sample $f(t)$ with $Ш_{1/2}(t)$ to reproduce the signal without aliasing.

[16] From T. John.

You try to test this out in the lab, but there is something wrong with the ideal sampler — it can *only* sample using $Ш_1(t)$. This will inevitably cause aliasing, but you think you can devise a scheme to somehow reconstruct $f(t)$ without aliasing.

This problem explores how this can be done by using this faulty sampler and some filter $h(t)$.

(a) Let $h(t) = -1/(\pi i t)$. If $g(t) = (f * h)(t)$, express $G(s)$ in terms of $F(s)$ *only*. *Hint*: Consider the cases $s = 0$, $s > 0$, and $s < 0$.

(b) Suppose you sample $f(t)$ with $Ш_1(t)$ to yield $y(t)$. For the case of $0 < s < 1$, express $Y(s)$ in terms of $F(s)$ and $F(s-1)$ *only*.

(c) Suppose you sample $g(t)$ with $Ш_1(t)$ to yield $x(t)$. For the case of $0 < s < 1$, express $X(s)$ in terms of $F(s)$ and $F(s-1)$ *only*.

(d) Using the two equations you have from parts (b) and (c), show how you might reconstruct $f(t)$ without aliasing.

6.12. *Downconversion*

A common problem in radio engineering is "downconversion to baseband." Consider a signal $f(t)$ whose spectrum $\mathcal{F}f(s)$ satisfies

$$\mathcal{F}f(s) = 0, \quad |s - s_0| \geq B.$$

To downconvert $\mathcal{F}f(s)$ to baseband means to move the spectrum so that it is centered around 0. Devise a strategy to downconvert using convolution with an appropriate Ш and a single ideal low-pass filter. What is the new signal in terms of the old? (Note that you can assume $s_0 > 2B$.)

6.13. *Sampling and (single-sided) modulation*

Communication channels are limited, and everyone wants to get his or her message through. In this problem you'll see how two signals can be combined, the combination sampled, and then the individual signals reconstructed from the samples of the combined signal.

Suppose $g(t)$ and $h(t)$ are bandlimited signals with bandwidth p; i.e. (using uppercase letters to denote Fourier transforms),

$$G(s) \equiv 0, \quad H(s) \equiv 0 \quad \text{for } |s| \geq \frac{p}{2}.$$

Form the signal

$$\begin{aligned} f(t) &= e^{-2\pi i (p/2) t} g(t) + e^{2\pi i (p/2) t} h(t) \\ &= e^{-\pi i p t} g(t) + e^{\pi i p t} h(t). \end{aligned}$$

Note the minus sign in the first complex exponential and the plus sign in the second.

We're going to recover $g(t)$ and $h(t)$ in terms of samples of $f(t)$.

(a) What is $F(s) = \mathcal{F}f(s)$ in terms of $G(s)$ and $H(s)$? Is $f(t)$ bandlimited? What is its bandwidth?

Suppose, for simplicity, that plots of $|G(s)|$ and $|H(s)|$ look like

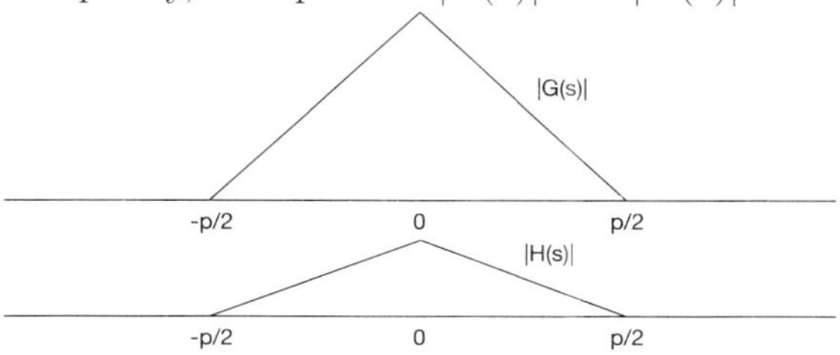

Sketch a plot of $|F(s)|$.

(b) Explain why

$$\begin{aligned}H(s) &= \Pi_p(s) \cdot (\text{III}_{2p} * F)\Big(s + \frac{p}{2}\Big) \\ &= \Pi_p(s) \cdot \Big(\text{III}_{2p}(s) * F\Big(s + \frac{p}{2}\Big)\Big),\end{aligned}$$

and use this to derive the interpolation formula

$$h(t) = \frac{1}{2} \sum_{k=-\infty}^{\infty} (-i)^k f\Big(\frac{k}{2p}\Big) \operatorname{sinc}\Big(p\Big(t - \frac{k}{2p}\Big)\Big).$$

Thus the signal $h(t)$ is reconstructed from the samples of $f(t)$.

You are not asked to show this but, likewise,

$$G(s) = \Pi_p(s) \cdot (\text{III}_{2p} * F)\Big(s - \frac{p}{2}\Big),$$

and using this, one obtains in the same manner the result

$$g(t) = \frac{1}{2} \sum_{k=-\infty}^{\infty} i^k f\Big(\frac{k}{2p}\Big) \operatorname{sinc}\Big(p\Big(t - \frac{k}{2p}\Big)\Big).$$

We see that both $g(t)$ and $h(t)$ can be interpolated from the *same samples* of $f(t)$!

6.14. *Demodulation by sampling*[17]

(a) Suppose a signal $x_0(t)$ is bandlimited so that $X_0(s) = 0$ for $|s| \geq W$. It is then modulated to carrier frequency s_0: $x(t) = x_0(t)e^{2\pi i s_0 t}$. Assume $W > 0$ and $s_0 \gg W$ are known. Show that the original signal $x_0(t)$ can be recovered from an ideal sampler: $y(t) = x(t) \cdot \text{III}_T(t)$. What is the largest T that can be used?

(b) Ideal samplers cannot be built in practice. For instance, a more realistic sampling device might provide the *average* value of the signal over each sample period. That is, for signal $x_0(t)$ and a sampling period of T, rather than the sample values being $v_k = x_0(kT)$, we get

$$v_k = \frac{1}{T} \int_{kT - \frac{T}{2}}^{kT + \frac{T}{2}} x_0(t)dt.$$

[17] From Adam Wang.

From the samples $w(t) = \sum_{k=-\infty}^{\infty} v_k \delta(t - kT)$, how can we recover the original signal $x_0(t)$? State any assumptions you make about $x_0(t)$ and T.

(c) Will the sampling device in (b) work for your demodulating scheme in (a)? Why might this be a bad idea?

6.15. *Sampling oscilloscope, warmup*

Let $f(t) = \cos 2\pi t$. Suppose we sample $f(t)$ at a rate of $2/3$ Hz and then interpolate using a low-pass filter with cutoff frequency $2/3$ Hz. What signal, $g(t)$, is the result? Sketch $f(t)$ and $g(t)$ on the same axes and comment on what you see. Is $g(t)$ an alias of $f(t)$ for this sampling rate?

The process illustrated in this problem is the basis of the *sampling oscilloscope*, below.

6.16. *Sampling oscilloscope*[18]

Sometimes signals change much faster than electronic devices can sample in order to reconstruct or display the signal on an oscilloscope. However, if a signal is bandlimited and periodic and if we take regular samples spaced somewhat more than the period (a sampling rate that is lower than the Nyquist rate), we can recover a *stretched version* of the signal. This is the principle of the *sampling oscilloscope*.

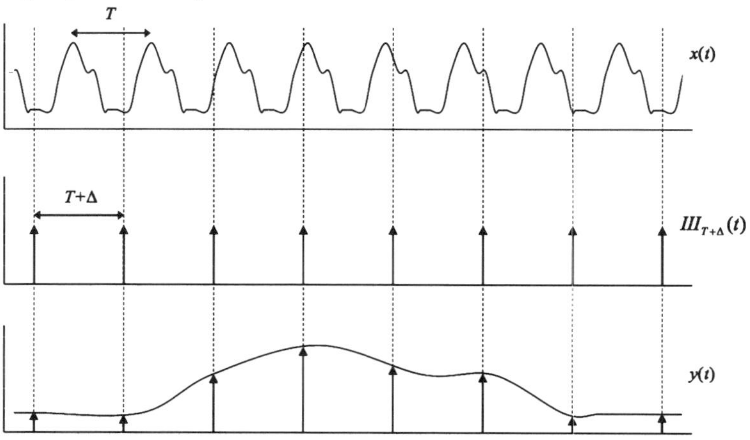

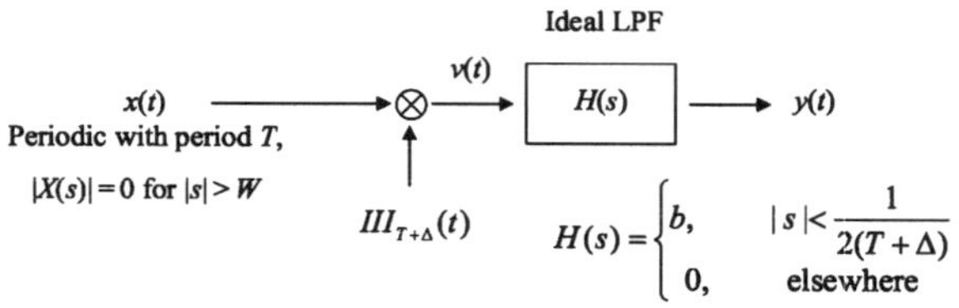

[18]From A. Oppenheim and A. Willsky via Aakanksha Chowdhery.

In the figures above, drawn in the time domain, the periodic signal $x(t)$, with period T, is sampled every $T+\Delta$ producing the sampled function $v(t) = x(t)\text{III}_{T+\Delta}(t)$. The samples are then interpolated by convolving with a scaled sinc function (an ideal low-pass filter), $(b/(T+\Delta))\ \text{sinc}(t/(T+\Delta))$, to get $y(t)$, a stretched version of the original signal $x(t)$. You will see how to choose the parameters Δ and b to obtain $y(t)$ by examining the process in the frequency domain.

(a) Consider a periodic signal $x(t) = \alpha + \beta e^{-2\pi it/T} + \gamma e^{2\pi it/T}$, which is periodic with period T.

(i) Find the Fourier transform $X(s)$ of $x(t)$. Sketch it for $\alpha = 2$, $\beta = \gamma = 1$.

(ii) Find the Fourier transform $V(s)$ of $v(t)$. Sketch it for $\alpha = 2$, $\beta = \gamma = 1$.

(iii) How should Δ be chosen so that $Y(s)$ looks like a compressed version of $X(s)$ (sketch $Y(s)$)? How should b be chosen so that $y(t)$ is a stretched version of $x(t)$, i.e., $y(t) = x(at)$ for some a (which you will find)?

(b) Consider a signal $x_0(t)$ with spectrum $X_0(s)$, drawn below, where $X_0(s) = 0$ for $|s| > W$. Define $x(t)$ as the periodization of $x_0(t)$; i.e., $x(t) = \sum_{n=-\infty}^{\infty} x_0(t - nT) = (x_0 * \text{III}_T)(t)$.

(i) Sketch each of the following, using $T = 1$, $\Delta = 1/9$, and $W = 4$.

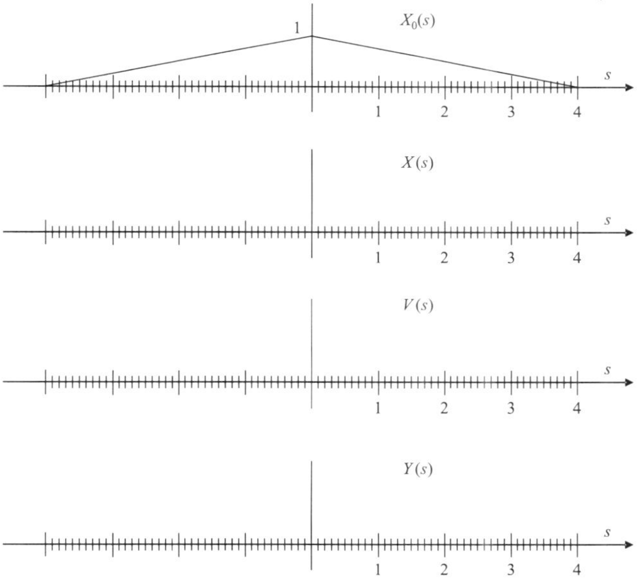

(ii) Determine what constraint applies to Δ so that $y(t)$ is a perfect reconstruction of a scaled version of $x(t)$. (*Hint*: Surprisingly, this method does not depend on the period of the signal, only the bandwidth W.)

6.17. *Compensating for distortions by adjusting the sampling rate*[19]

You are given a bandlimited, real signal $f(t)$ with bandwidth B. We will also assume that $f(t)$ is nonnegative.

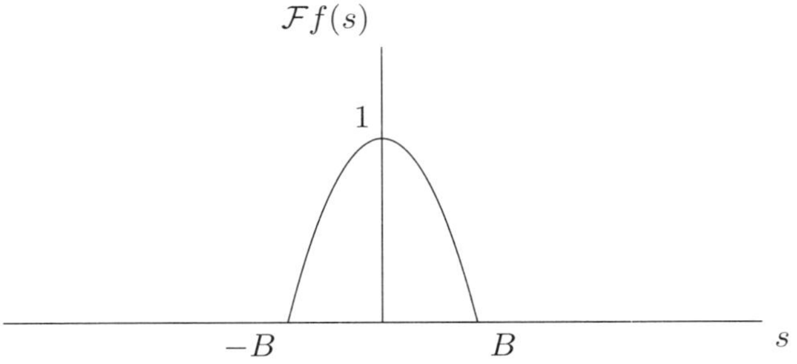

The signal is sampled at a rate $R = 1/T$ to obtain the samples $f(nT)$, $n = 0, \pm 1, \pm 2, \ldots$. Some discrete time processing is subsequently done on these sample values, which unintentionally ends up squaring the sample values. Beyond your control, your interpolator has access to $|f(nT)|^2$ instead of $f(nT)$, but no information is lost since f is nonnegative. The whole system operates the usual way: it constructs

$$f_s(t) = \sum_n |f(nT)|^2 \delta(t - nT)$$

and filters $f_s(t)$ using an ideal low-pass filter of bandwidth $1/2T$ to obtain $g(t)$.

(a) Suppose the sampling rate R is equal to the Nyquist rate for f (i.e., $R = 2B$). Sketch the spectrum of the output $g(t)$.

(b) How will you pick the sampling rate R so that $f(t)$ can be recovered from $g(t)$?

6.18. *An infinite product for sinc*[20]

Let $f(x)$ be the infinite product of cosines defined as

$$f(x) = \cos\frac{\pi x}{2}\cos\frac{\pi x}{4}\cos\frac{\pi x}{8}\cdots = \prod_{k=1}^{\infty}\cos\frac{\pi x}{2^k}.$$

(a) Find $f(0)$.

(b) Find $f(\pm 1), f(\pm 2), \ldots$. Generalize the pattern to find $f(n)$ when n is a nonzero integer.

(c) Argue that f has bandwidth $1/2$. Recall that the largest frequency in the product of cosines $\cos(2\pi\nu_1 t)\cos(2\pi\nu_2 t)$ is $\nu_1 + \nu_2$.

(d) Conclude that $f(x) = \operatorname{sinc} x$.

[19] From A. Siripuram.
[20] From J. Gill.

6.19. *Nyquist-Shannon interpolation: Stability*[21]

Consider an interpolation of a signal $x(t)$ from its samples $x(n)$ in the form

$$x(t) = \sum_{n=-\infty}^{\infty} x(n)s_n(t).$$

In the case of Nyquist-Shannon interpolation we have $s_n(t) = \operatorname{sinc}(t-n)$. The interpolation process is called *stable in the energy sense* if small errors in the sample values do not lead to large errors in the interpolated signal. To make this precise, suppose we have an error e_n in recording the nth sample $x(n)$. Then the overall error is given by

$$e(t) = \sum_{n=-\infty}^{\infty} e_n s_n(t).$$

The interpolation process is stable in the energy sense if

$$\int_{-\infty}^{\infty} |e(t)|^2 dt \le C \sum_{n=-\infty}^{\infty} |e_n|^2,$$

for some constant C.

(a) Show that the Nyquist-Shannon interpolation is stable in the energy sense. (For this, use Parseval's theorem and the fact that the functions $\{\operatorname{sinc}(t-n)\}$ for $n = 0, \pm 1, \pm 2, \dots$ are orthonormal.)

Note that for energy-stable interpolation, even though the overall energy in the error $e(t)$ is not too large, it is possible that the value of the error $e(t)$ is very large at some particular t. For this reason, we can define *stability in the pointwise sense* if

$$|e(t)|^2 \le C \sum_{n=-\infty}^{\infty} |e_n|^2 \quad \text{for all } t,$$

where C is a constant.

(b) Use the Cauchy-Schwarz inequality (see Chapter 1) to show that the Nyquist-Shannon interpolation is stable in the pointwise sense. You'll need the result from Problem 6.2.

6.20. *Sampling using the derivative*[22]

Suppose that $f(t)$ is a bandlimited signal with $\mathcal{F}f(s) = 0$ for $|s| \ge 1$ (bandwidth 2). According to the sampling theorem, knowing the values $f(n)$ for all integers n (sampling rate of 1) is not sufficient to interpolate the values $f(t)$ for all t. However, if *in addition* one knows the values of the derivative $f'(n)$ at the integers, then there is an interpolation formula with a sampling rate of 1. In this problem you will derive that result.

[21] From A. Siripuram.
[22] From D. Kammler.

Let $F(s) = \mathcal{F}f(s)$ and let $G(s) = \frac{1}{2\pi i}(\mathcal{F}f')(s) = sF(s)$.

(a) For $0 \le s \le 1$ show that

$$(\text{III} * F)(s) = F(s) + F(s-1),$$
$$(\text{III} * G)(s) = sF(s) + (s-1)F(s-1),$$

and then show that

$$F(s) = (1-s)(\text{III} * F)(s) + (\text{III} * G)(s)\,.$$

(b) For $-1 \le s \le 0$ show that

$$(\text{III} * F)(s) = F(s) + F(s+1),$$
$$(\text{III} * G)(s) = sF(s) + (s+1)F(s+1),$$

and then show that

$$F(s) = (1+s)(\text{III} * F)(s) - (\text{III} * G)(s)\,.$$

(c) Using parts (a) and (b) show that for all s, $-\infty < s < \infty$,

$$F(s) = \Lambda(s)(\text{III} * F)(s) - \Lambda'(s)(\text{III} * G)(s)\,,$$

where $\Lambda(s)$ is the triangle function

$$\Lambda(s) = \begin{cases} 1 - |s|, & |s| \le 1\,, \\ 0, & |s| \ge 1\,. \end{cases}$$

(d) From part (c) derive the interpolation formula

$$f(t) = \sum_{n=-\infty}^{\infty} f(n)\text{sinc}^2(t-n) + \sum_{n=-\infty}^{\infty} f'(n)(t-n)\text{sinc}^2(t-n)\,.$$

6.21. In this problem[23] we will show that for $0 \le t, \tau \le 1/2$,

$$\min(t,\tau) = \sum_{\substack{n=-\infty \\ n \text{ odd}}}^{\infty} \frac{4\sin n\pi\tau \sin n\pi t}{n^2\pi^2}.$$

(This relationship is useful, for example, in identifying the Fourier series expansion of a Wiener process.)

For a fixed τ, consider the function f given in the figure.

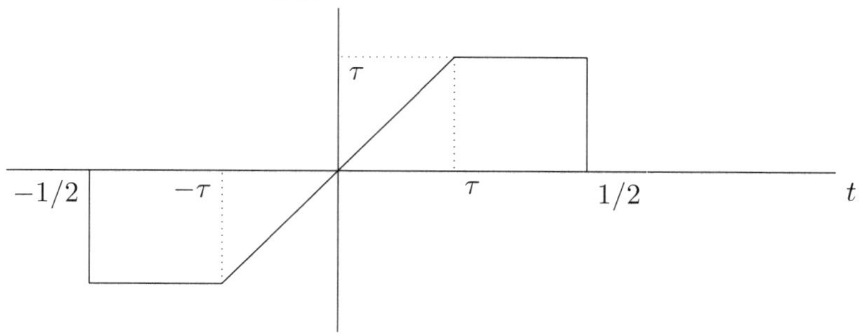

[23] From A. Siripuram.

(a) Since f has a *time width* of 1, we should be able to sample $\mathcal{F}f(s)$ at a sampling rate of 1 and still recover f. Consider sampling $\mathcal{F}f(s)$ by multiplying it with $\text{III}(s - 1/2)$:

$$G(s) = \mathcal{F}f(s)\text{III}(s - 1/2).$$

Let $g(t)$ be the Fourier inverse of $G(s)$. Show that

$$g(t) = \sum_{n=-\infty}^{\infty} (-1)^n f(t - n).$$

(b) Note that g is periodic. What is its period? Find the Fourier series for g (you can leave your answer in terms of $\mathcal{F}f$).

(c) Argue why $\mathcal{F}f(s)$ satisfies

$$\pi i s \mathcal{F}f(s) = \tau \operatorname{sinc} 2\tau s - \tau \cos \pi s.$$

From this, it follows that

$$\mathcal{F}f(s) = \frac{1}{\pi i s}\left(\tau \operatorname{sinc} 2\tau s - \tau \cos \pi s\right).$$

(You do not have to prove this).

(d) Using the result of (c) in (b), deduce that

$$g(t) = \sum_{\substack{n=-\infty \\ n \text{ odd}}}^{\infty} \frac{4}{n^2\pi^2} \sin n\pi\tau \sin n\pi t.$$

In particular, for $0 \leq t, \tau \leq 1/2$,

$$\min(t, \tau) = \sum_{\substack{n=-\infty \\ n \text{ odd}}}^{\infty} \frac{4 \sin n\pi\tau \sin n\pi t}{n^2\pi^2}.$$

6.22. *Sampling with half a shah*

We define a new sampling function $\text{II}(t)$, which is a train of evenly spaced δ-function *pairs*, as shown in the figure. This sampling function is parameterized by two parameters: q, the spacing between the δ-function pairs, and b, the spacing within the pairs of δ-functions.

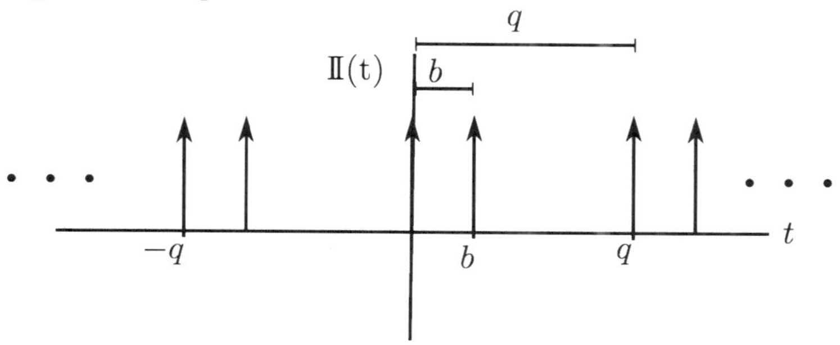

We will apply this sampling scheme to a function $f(t)$, which has a Fourier transform $\mathcal{F}f(s)$ and is bandlimited with bandwidth p (i.e.,, $\mathcal{F}f(s) = 0$ for $|s| \geq p/2$).

(a) Write $\text{Ш}(t)$ as the sum of two Shah functions (in terms of q and b).
(b) Define $g(t)$ to be the sampled version of $f(t)$ using our new sampling scheme, so that $g(t) = \text{Ш}(t)f(t)$. Write the Fourier transform of $g(t)$ in terms of $\mathcal{F}f(s)$, q, and b.
(c) Consider the case where $b = q/2$. What is the maximum bandwidth p for which we can guarantee reconstruction of $f(t)$ from its sampled version $g(t)$? What is a possible reconstruction scheme?
(d) Now consider the case when $b = q/4$. We further assume that $f(t)$ is real and even. What is the maximum bandwidth p for which we can guarantee reconstruction of $f(t)$ from its sampled version $g(t)$? For a signal of maximum bandwidth, what is a possible reconstruction scheme? *Hint*: Consider reconstruction based only on the real or the imaginary part of the Fourier transform of $g(t)$.

6.23. *Papoulis Generalized Sampling Theorem*[24]
Let $f(t)$ be a function with bandwidth $2B$ and Fourier transform $F(s)$, so that $F(s) = 0$ for $|s| \geq B$. If we sample $f(t)$ at the Nyquist rate $2B$, we know we can reconstruct the function through sinc interpolation. This problem will explore a more general sampling and reconstruction scheme due to A. Papoulis.[25]

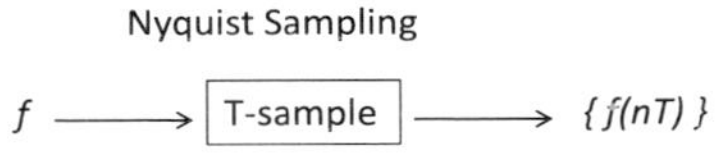

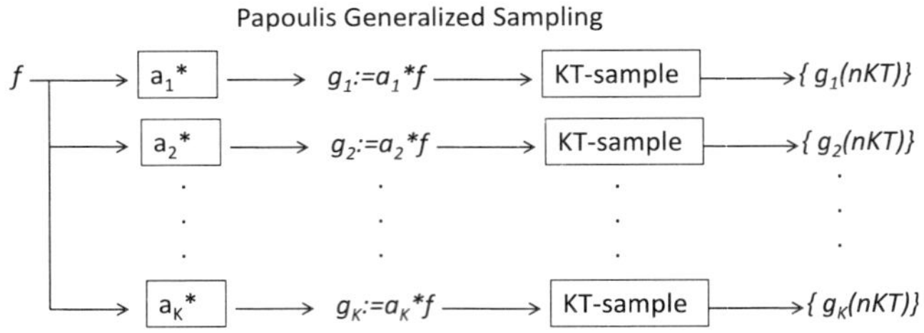

(a) First input the signal $f(t)$ into K linear systems with known impulse responses $a_k(t)$. The outputs of the systems are denoted g_k, so

$$g_k(t) = (a_k * f)(t) \qquad \text{for } k = 0, \ldots, K-1,$$

or in the frequency domain

$$G_k(s) = A_k(s)F(s) \qquad \text{for } k = 0, \ldots, K-1,$$

where the $A_k(s)$ are the transfer functions of the systems. Note that since $f(t)$ has bandwidth $2B$, so do all of the $g_k(t)$.

[24]From Raj Bhatnagar.
[25]Athanasios Papoulis was a Greek-American mathematician who made many contributions to signal processing, of which this sampling theorem may be the best known. He wrote a book on Fourier transforms, too, like everybody has.

(b) Next, sample each of the outputs $g_k(t)$ for $k = 0, \ldots, K-1$ at the sub-Nyquist rate

$$B_K := 2B/K$$

to obtain the samples $g_k(nT_K)$ where $T_K = 1/B_K$.

(c) Then, even though we are sampling below the Nyquist rate, we can reconstruct our signal $f(t)$ from the interpolation formula

$$f(t) = \sum_{k=0}^{K-1} \sum_{n=-\infty}^{\infty} g_k(nT_K) p_k(t - nT_K),$$

where the functions p_k are determined only from A_k and B_K.

Background. Before we discuss how to prove this result (and specify p_k), let's discuss why this result is true. In the interval $|s| < B_K$, we have the original spectrum $A_0(s)F(s)$ as well as $K-1$ aliases, $A_k(s)F(s-kB_K)$. Think of the $F(s-kB_K)$ for $k = 0, \ldots, K-1$ as K unknowns. If we can find a linear system of K equations relating $F(s-kB_K)$ to known quantities, then we can recover the original spectrum and find $f(t)$.

Details. Here's how we find $p_k(t)$. First, for $-B \leq s < -B + B_K$, define the matrices

$$A = \begin{bmatrix} A_0(s) & A_1(s) & \cdots & A_{K-1}(s) \\ A_0(s+B_K) & A_1(s+B_K) & \cdots & A_{K-1}(s+B_K) \\ \vdots & \vdots & \ddots & \vdots \\ A_0(s+(K-1)B_K) & A_1(s+(K-1)B_K) & \cdots & A_{K-1}(s+(K-1)B_K) \end{bmatrix},$$

$$P = \begin{bmatrix} P_0(s,t) \\ P_1(s,t) \\ \vdots \\ P_{K-1}(s,t) \end{bmatrix}, \quad \text{and} \quad E = \begin{bmatrix} 1 \\ e^{2\pi i B_K t} \\ \vdots \\ e^{2\pi i (K-1) B_K t} \end{bmatrix}.$$

Then solve the linear system $P = A^{-1}E$ (assume that $\det A \neq 0$ so that the system has a solution). Notice that A is a function of s while E is a function of t. P_k is a function of both: in s, it is bandlimited to $(-B, -B+B_K)$, and in t, it has period $1/kB_K = T_K/k$. Finally, we define

$$p_k(t) := \frac{1}{B_K} \int_{-B}^{-B+B_K} P_k(s,t) e^{2\pi i s t} ds.$$

The Problem.

(a) Denote the periodization of $P_k(s,t)e^{2\pi i s t}$ in the s-domain by

$$\tilde{P}_k(s,t) := \sum_{m=-\infty}^{\infty} P_k(s + mB_K, t) e^{2\pi i (s+mB_K)t}.$$

Show that the $p_k(t - nT_K)$ are the Fourier series coefficients of $\tilde{P}_k(s,t)$.

(b) Explain why

$$\sum_{k=0}^{K-1} A_k(s) \tilde{P}_k(s,t) = e^{2\pi i s t} \quad \text{for } -B \leq s < B.$$

Hint: Consider multiplying the first equation in the linear system $A = PE$ by $e^{2\pi ist}$. What happens if you do the same with the second equation in the system?

(c) Using part (b), we have

$$f(t) = \int_{-B}^{B} F(s)e^{2\pi ist}\,ds = \int_{-B}^{B} F(s)\sum_{k=0}^{K-1} A_k(s)\tilde{P}_k(s,t)\,ds\,.$$

Use this to show the interpolation formula

$$f(t) = \sum_{k=0}^{K-1}\sum_{n=-\infty}^{\infty} g_k(nT_K)p_k(t - nT_K).$$

6.24. *Chopper amplifiers*[26]

It is difficult to build an amplifier for DC signals (or, more generally, low frequency and bandlimited signals) that provides high gain and low noise. A device that overcomes these problems is the *chopper amplifier*, and its design is a nice example of passing between the time and frequency domains. A figure illustrating the entire process is below, following the statements of the problem.

The basic component, the *chopper*, is essentially a rapid on/off switch, acting on a signal in the time domain by multiplication with a periodized rect function,

$$c(t) = \sum_{k=-\infty}^{\infty} \Pi_T(t - 2kT)\,.$$

Start with a signal $x(t)$, let $X(s) = \mathcal{F}x(s)$, and suppose

$$X(s) \equiv 0 \quad \text{for } |s| \geq W.$$

(a) Form the chopped signal

$$x_1(t) = c(t)x(t).$$

Show that the Fourier transform of $x_1(t)$ is

$$X_1(s) = \frac{1}{2}\sum_{k=-\infty}^{\infty} \operatorname{sinc}(k/2)X\Big(s - \frac{k}{2T}\Big).$$

(b) What relationship must W and T satisfy for there to be no overlaps in the shifts of $X(s)$?

(c) Assuming the conditions of part (b), we next produce two spectral islands of $X_1(s)$, with a gain of A, by forming

$$\begin{aligned} X_2(s) &= A\Big(\Pi_{\frac{1}{2T}}\Big(s + \frac{1}{2T}\Big) + \Pi_{\frac{1}{2T}}\Big(s - \frac{1}{2T}\Big)\Big)X_1(s) \\ &= \frac{A}{2}\operatorname{sinc}(1/2)\Big(X\Big(s - \frac{1}{2T}\Big) + X\Big(s + \frac{1}{2T}\Big)\Big)\,. \end{aligned}$$

(This is a band-pass filter.)

[26] From Oppenheim and Willsky via Adam Wang.

Let $x_2(t)$ be the corresponding signal in the time domain, chop $x_2(t)$ to form

$$x_3(t) = c(t)x_2(t),$$

and, writing $X_3(s) = \mathcal{F}x_3(s)$, show that

$$X_3(s) = \frac{A}{4}\,\text{sinc}(1/2)\sum_{k=-\infty}^{\infty}\left(\text{sinc}\left(\frac{k-1}{2}\right) + \text{sinc}\left(\frac{k+1}{2}\right)\right) X\left(s - \frac{k}{2T}\right).$$

(d) Finally, multiply $X_3(s)$ by $\Pi_{1/2T}(s)$ (a low-pass filter) to obtain $Y(s)$, and let $y(t)$ be the corresponding signal in the time domain. Find $y(t)$. How does $y(t)$ compare to the original signal $x(t)$? In particular, what is the overall gain, B, of the system?

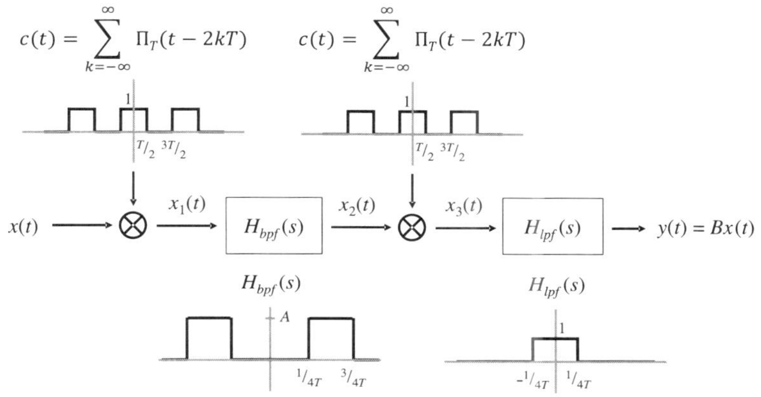

Chapter 7

Discrete Fourier Transform

7.1. The Modern World

Many would say that the modern world began in 1965 when J. Cooley and J. Tukey published their account of an efficient method for numerical computation of Fourier series.[1] By that measure the modern world might actually have started rather earlier, for the idea at the heart of the algorithm is clearly present in an unpublished paper of Gauss that appeared posthumously in 1866.[2] Take your pick.

Whatever the past, the demands of the present and future are that we process continuous signals and their spectra by means of discrete methods; computers in all their forms can work with finite sums only. To turn the continuous into the discrete and finite requires, for our purposes, that a signal be both timelimited and bandlimited, something we know cannot be true, and that we take a finite number of samples, something we know cannot suffice. But it works. It works to the extent that a large fraction of the world's economy depends upon it working, and that's not a bad measure of success. Consider this a proof by economics.

Some would argue that one shouldn't think in terms of "turning the continuous into the discrete" at all, but rather that measurements and data in the real world come to us in discrete forms, and that's how we should understand the world and work with it. Period. Though things seem to have settled down, battles over "discrete" versus "continuous" can rise to a fever pitch. One such battle can be fought over the different approaches to the discrete Fourier transform, abbreviated DFT. As with everything else in this book, there are choices to make. My choice is to make the discrete look like the continuous as much as we can. We're thus taking sides, at least initially, in favor of "*from* continuous *to* discrete" as a way of motivating the definition of the DFT. For one thing, we have built up a lot of

[1] The title of the paper is "An algorithm for the machine calculation of complex Fourier series." It appeared in Math. Comput. **19** (1965), 297–301. Incidentally, Tukey is also credited with coining the term "bit" as an abbreviation for "binary digit" — how about that for immortality!

[2] Much historical scholarship has gone into this. Google the paper "Gauss and the history of the Fast Fourier Transform" by M. T. Heideman, D. H. Johnson, and C. S. Burrus for a thorough discussion.

intuition and understanding of the Fourier transform, its properties, and its uses, and a reasonable goal is to leverage that intuition as we now work with the discrete Fourier transform. For a more discrete-centric approach, and more detail on many of the topics we'll cover here, I highly recommend the book *The DFT: An Owner's Manual* by W. Briggs and V. Henson.

7.2. From Continuous to Discrete

Start with a signal $f(t)$ and its Fourier transform $\mathcal{F}f(s)$, each a function of a continuous variable. We want to:

- Find a discrete version of $f(t)$ that's a reasonable approximation to $f(t)$.
- Find a discrete version of $\mathcal{F}f(s)$ that's a reasonable approximation to $\mathcal{F}f(s)$.
- Find a way to relate the discrete version of $\mathcal{F}f(s)$ to the discrete version of $f(t)$ that's a reasonable approximation to the way $\mathcal{F}f(s)$ is related to $f(t)$.

Good things to try for, but it's not quite straightforward. Here's the setup.

We suppose that $f(t)$ is zero outside of $0 \leq t \leq L$. We also suppose that the Fourier transform $\mathcal{F}f(s)$ is zero, or effectively zero (beyond our ability to measure, negligible energy, whatever) outside of $0 < s < 2B$ (B for bandwidth). We are taking the support of $\mathcal{F}f$ to be the interval from 0 to $2B$ instead of $-B$ to B only because it will make the *initial indexing* of sample points easier; this will not be an issue in the end. We'll also take L and B to both be integers so we don't have to round up or down in any of the considerations that follow; you can think of that as our first concession to the discrete.

Thus we are regarding $f(t)$ as both timelimited and bandlimited, with the knowledge that this can only be approximately true. Remember, however, that we're ultimately going to come up with a definition of a discrete Fourier transform that will make sense in and of itself regardless of shaky initial assumptions. After the definition is written down we could (and some do) erase all that came before it, perhaps casting only a brief, wistful glance back from the discrete to the continuous with a few comments on how the former approximates the latter. Many treatments of the discrete Fourier transform that start with the discrete and stay with the discrete do just that. We're trying not to do that.

According to the sampling theorem (misapplied here, yes, but play along), we can reconstruct $f(t)$ perfectly from its samples if we sample at the rate of $2B$ samples per second. Since $f(t)$ is defined on an interval of length L and the samples are $1/2B$ apart, that means that we want a total of

$$N = \frac{L}{1/2B} = 2BL \quad \text{(note that } N \text{ is therefore even)}$$

evenly spaced samples. Starting at 0 these are at the points

$$t_0 = 0,\, t_1 = \frac{1}{2B},\, t_2 = \frac{2}{2B}, \dots,\, t_{N-1} = \frac{N-1}{2B}\,.$$

Draw yourself a picture. To know the values $f(t_k)$ is to know $f(t)$ reasonably well.

Thus we state:

- The discrete version of $f(t)$ is the list of sampled values $f(t_0),\ldots,f(t_{N-1})$.

Next, represent the discrete version of $f(t)$ (the list of sampled values) continuously with the aid of a finite impulse train (a finite Ш-function) at the sample points. Namely, using

$$\sum_{n=0}^{N-1} \delta(t-t_n)\,,$$

let

$$f_{\text{sampled}}(t) = f(t)\sum_{n=0}^{N-1} \delta(t-t_n) = \sum_{n=0}^{N-1} f(t_n)\delta(t-t_n)\,.$$

This is what we have considered previously as the sampled form of $f(t)$, hence the subscript "sampled." The Fourier transform of f_{sampled} is

$$\mathcal{F}f_{\text{sampled}}(s) = \sum_{n=0}^{N-1} f(t_n)\mathcal{F}\delta(t-t_n) = \sum_{n=0}^{N-1} f(t_n)e^{-2\pi i s t_n}\,.$$

This is close to what we want — it's the continuous Fourier transform of the sampled form of $f(t)$.

Now let's change perspective and look at sampling in the frequency domain. The function $f(t)$ is limited to $0 \le t \le L$, and this determines a sampling rate for reconstructing $\mathcal{F}f(s)$ from *its* samples in the *frequency* domain. The sampling rate is L and the spacing of the sample points is $1/L$. We sample $\mathcal{F}f(s)$ over the interval from 0 to $2B$ in the frequency domain at points spaced $1/L$ apart. The number of sample points is

$$\frac{2B}{1/L} = 2BL = N\,,$$

the same number of sample points as for $f(t)$. The sample points for $\mathcal{F}f(s)$ are of the form m/L, and there are N of them. Starting at 0,

$$s_0 = 0,\ s_1 = \frac{1}{L},\ \ldots,\ s_{N-1} = \frac{N-1}{L}\,.$$

The discrete version of $\mathcal{F}f(s)$ that we take is *not* $\mathcal{F}f(s)$ evaluated at these sample points s_m. Rather, it is $\mathcal{F}f_{\text{sampled}}(s)$ evaluated at the sample points. We base the discrete approximation of $\mathcal{F}f(s)$ on the sampled version of $f(t)$. To ease the notation write $F(s)$ for $\mathcal{F}f_{\text{sampled}}(s)$. Then:

- The discrete version of $\mathcal{F}f(s)$ is the list of values

$$F(s_0) = \sum_{n=0}^{N-1} f(t_n)e^{-2\pi i s_0 t_n},\ F(s_1) = \sum_{n=0}^{N-1} f(t_n)e^{-2\pi i s_1 t_n},\ \ldots,$$
$$F(s_{N-1}) = \sum_{n=0}^{N-1} f(t_n)e^{-2\pi i s_{N-1} t_n}\,.$$

By this definition, we now have a way of going from the discrete version of $f(t)$ to the discrete version of $\mathcal{F}f(s)$, namely the expressions

$$F(s_m) = \sum_{n=0}^{N-1} f(t_n)e^{-2\pi i s_m t_n} .$$

These sums, one for each m from $m = 0$ to $m = N - 1$, are supposed to be an approximation to the Fourier transform going from $f(t)$ to $\mathcal{F}f(s)$.

In what sense is this a discrete approximation to the Fourier transform? Since $f(t)$ is timelimited to $0 \leq t \leq L$, we have

$$\mathcal{F}f(s) = \int_0^L e^{-2\pi i s t} f(t)\, dt .$$

Thus at the sample points s_m,

$$\mathcal{F}f(s_m) = \int_0^L e^{-2\pi i s_m t} f(t)\, dt ,$$

and to know the values $\mathcal{F}f(s_m)$ is to know $\mathcal{F}f(s)$ reasonably well. Now use the sample points t_k for $f(t)$ to write a Riemann sum approximation for the integral. The spacing Δt of the points is $1/2B$, so

$$\begin{aligned}\mathcal{F}f(s_m) = \int_0^L f(t)e^{-2\pi i s_n t}\, dt &\approx \sum_{n=0}^{N-1} f(t_n)e^{-2\pi i s_m t_n}\Delta t \\ &= \frac{1}{2B}\sum_{n=0}^{N-1} f(t_n)e^{-2\pi i s_m t_n} = \frac{1}{2B}F(s_m) .\end{aligned}$$

This is the final point:

- Up to the factor $1/2B$, the values $F(s_m)$ provide an approximation to the values $\mathcal{F}f(s_m)$.

Writing a Riemann sum as an approximation to the integral defining $\mathcal{F}f(s_m)$ essentially discretizes the integral, and this is an alternate way of getting to the expression for $F(s_n)$, up to the factor $2B$. We short-circuited this route by working directly with $\mathcal{F}f_{\text{sampled}}(s)$.

You may find the "up to the factor $1/2B$" unfortunate in this part of the discussion, but it's in the nature of the subject. In fact, back in Chapter 2 we encountered a similar kind of "up to the factor ..." phenomenon when we obtained the Fourier transform as a limit of the Fourier coefficients for a Fourier series.

7.3. The Discrete Fourier Transform

We are almost ready for a definition, with one final observation remaining to clear the way. The sample points are

$$t_n = \frac{n}{2B}, \quad s_m = \frac{m}{L},$$

and so

$$s_m t_n = \frac{mn}{2BL} = \frac{mn}{N} .$$

Hence

$$F(s_m) = \sum_{n=0}^{N-1} f(t_n)e^{-2\pi i s_m t_n} = \sum_{n=0}^{N-1} f(t_n)e^{-2\pi inm/2BL} = \sum_{n=0}^{N-1} f(t_n)e^{-2\pi inm/N}.$$

The last form of the exponential, $e^{-2\pi inm/N}$, highlights the *indices* of the inputs (index n) and outputs (index m) and the number of sample points (N). It hides the sample points themselves. We thus take the needed last step in moving from the continuous to the discrete.

We identify the point t_n with its *index* n, and the list of N values $f(t_0), f(t_1), \dots, f(t_{N-1})$ with the list of values of the discrete signal $\underline{f}[n]$, $n = 0, 1, \dots, N-1$, by defining

$$\underline{f}[n] = f(t_n).$$

Here we use the bracket notation (common practice) for functions of a discrete variable, as well as using the underline notation to distinguish $\underline{f}$, as a function, from f. So, again, as a discrete function

$$\underline{f}[0] = f(t_0), \underline{f}[1] = f(t_1), \dots, \underline{f}[N-1] = f(t_{N-1}).$$

Likewise we identify the point s_m with its index m, and we identify the list of N values $F(s_0), F(s_1), \dots, F(s_{N-1})$ with values of the discrete signal $\underline{F}[m]$, $m = 0, 1, \dots, N-1$, by

$$\underline{F}[m] = F(s_m).$$

Written out,

$$\underline{F}[0] = F(s_0), \underline{F}[1] = F(s_1), \dots, \underline{F}[N-1] = F(s_{N-1}).$$

Purely in terms of the discrete variables n and m,

$$\underline{F}[m] = \sum_{n=0}^{N-1} e^{-2\pi inm/N}\underline{f}[n].$$

It remains to turn this around and make a definition solely in terms of discrete variables and discrete signals, pretending that we never heard of continuous signals and their sampled values.

Officially:

- Let $\underline{f}[n]$, $n = 0, \dots, N-1$, be a discrete signal of length N. Its *discrete Fourier transform* (DFT) is the discrete signal $\underline{\mathcal{F}}\underline{f}$ of length N, defined by

$$\underline{\mathcal{F}}\underline{f}[m] = \sum_{n=0}^{N-1} \underline{f}[n]e^{-2\pi imn/N}, \quad m = 0, 1, \dots, N-1.$$

  We'll write $\underline{\mathcal{F}}_N\underline{f}$, with the subscript, if we need to call attention to the N.

It's perfectly legitimate to let the inputs $\underline{f}[n]$ be complex numbers, though for applications they'll typically be real. The computed values $\underline{\mathcal{F}}\underline{f}[m]$ are generally complex, being sums of complex exponentials.

Giving software its due, MATLAB thinks the DFT is:

- An operation that accepts as input a list of N numbers and returns as output a list of N numbers, as specified above.

There are actually a number of things to say about the inputs and outputs of this operation, and I'll try not to say them all at once. One can think of the inputs and outputs as N-vectors, but note that (for now) the indexing goes from 0 to $N-1$ rather than from 1 to N as usual. MATLAB gags on that, it's not alone, and indexing is one of the things we'll talk about.

There are some instances when the vector point of view is more natural and others where it's more natural to think in terms of discrete signals and their values. I'll use the underline notation for both. You have to keep both interpretations in your head. It's not hard.

7.4. Notations and Conventions 1

I said that I wanted to set things up to look as much like the continuous case as possible, and the usual operations on functions are part of that. For example, for

$$\underline{x} = (\underline{x}[0], \underline{x}[1], \dots, \underline{x}[N-1]) \quad \text{and} \quad \underline{y} = (\underline{y}[0], \underline{y}[1], \dots, \underline{y}[N-1]),$$

we have

$$\underline{x}\,\underline{y} = (\underline{x}[0]\underline{y}[0], \underline{x}[1]\underline{y}[1], \dots, \underline{x}[N-1]\underline{y}[N-1])$$

(componentwise product, not to be confused with the dot product of vectors — that comes later),

$$\frac{\underline{x}}{\underline{y}} = \left(\frac{\underline{x}[0]}{\underline{y}[0]}, \frac{\underline{x}[1]}{\underline{y}[1]}, \dots, \frac{\underline{x}[N-1]}{\underline{y}[N-1]}\right) \quad \text{(when the individual quotients make sense)},$$

$$\underline{x}^p = (\underline{x}[0]^p, \underline{x}[1]^p, \dots, \underline{x}[N-1]^p) \quad \text{(when the individual powers make sense)},$$

and so on. We even allow a function of one variable (think sine or cosine, for example) to operate componentwise via

$$f\big((\underline{x}[0], \underline{x}[1], \dots, \underline{x}[N-1])\big) = \big(f(\underline{x}[0]), f(\underline{x}[1]), \dots, f(\underline{x}[N-1])\big).$$

You wouldn't usually write expressions like this if you were thinking of $\underline{x}$ and $\underline{y}$ as vectors, but standard mathematical software packages (MATLAB, Mathematica) incorporate such componentwise operations on lists. Be careful, but keep an open mind.

We'll also use the notation $[r:s]$ for the tuple of numbers $(r, r+1, r+2, \dots, s)$. Two special signals. We'll write

$$\underline{0} = (0, 0, \dots, 0)$$

for the zero signal and

$$\underline{1} = (1, 1, \dots, 1)$$

for the signal of all 1's. A little notation can go a long way.

Finally, I'm aware of the fact that for various reasons it's better to write an N-vector as a column (an $N \times 1$ matrix). This may have been drummed into you in a linear algebra class. Nevertheless, most of the time we'll write vectors as horizontal N-tuples, as above. It's typography. Sue me.

7.4.1. Discrete complex exponentials. The definition of the discrete Fourier transform, like that of the continuous Fourier transform, involves a complex exponential. Much of the theory and practice of the DFT depends on the properties of the complex exponential, and it's helpful to have some special notations to save writing.

We let

$$\omega = e^{2\pi i/N}.$$

Occasionally we'll decorate this with a subscript to

$$\omega_N = e^{2\pi i/N}$$

when we want to emphasize the N. From Euler's formula,

$$\operatorname{Re}\omega_N = \cos 2\pi/N\,, \quad \operatorname{Im}\omega_N = \sin 2\pi/N\,.$$

ω_N is an Nth *root of unity*, meaning

$$\omega_N^N = e^{2\pi i N/N} = e^{2\pi i} = 1\,.$$

There are N distinct numbers in the list of the powers ω_N^n as n goes from 0 to $N-1$:

$$1 = \omega_N^0\,, \omega_N^1\,, \omega_N^2\,, \dots, \omega_N^{N-1}\,,$$

and each of the ω_N^n is itself an Nth root of unity; i.e.,

$$\omega_N^{nN} = 1\,.$$

Then, in general for any integers n and k,

$$\omega_N^{k+nN} = \omega_N^k\,.$$

All obvious, but these properties come up often enough that it's worth pointing them out.

Also note that

$$\omega_N^{N/2} = e^{2\pi i N/2N} = e^{i\pi} = -1 \quad \text{and hence} \quad \omega_N^{kN/2} = (-1)^k\,.$$

These come up, too.

On the notation, some people write $\omega_N = e^{-2\pi i/N}$ (minus instead of plus in the exponential) and others write $W = e^{2\pi in/N}$. I've seen all sorts of things, so be aware of different conventions as you peruse the literature.

For the discrete Fourier transform it's helpful to bundle the powers of ω together as a discrete signal of length N. We define the discrete signal $\underline{\omega}$ (or $\underline{\omega}_N$ if the N is wanted for emphasis) by

$$\underline{\omega}[m] = \omega^m, \quad m = 0, \dots, N-1\,.$$

Writing the values as a vector,

$$\underline{\omega} = (1, \omega, \omega^2, \dots, \omega^{N-1})\,.$$

This is the *discrete complex exponential*.[3] Recasting what we said above, the real and imaginary parts of $\underline{\omega}[m]$ are

$$\operatorname{Re}\underline{\omega}[m] = \cos\left(\frac{2\pi m}{N}\right), \quad \operatorname{Im}\underline{\omega}[m] = \sin\left(\frac{2\pi m}{N}\right), \quad m = 0, 1, \ldots, N-1.$$

The integer powers of $\underline{\omega}$ are

$$\underline{\omega}^k[m] = \omega^{km} \quad \text{or} \quad \underline{\omega}^k = (1, \omega^k, \omega^{2k}, \ldots, \omega^{(N-1)k}).$$

There's a symmetry here:

$$\underline{\omega}^k[m] = \omega^{km} = \underline{\omega}^m[k]\,.$$

The power k can be positive or negative, and of course

$$\underline{\omega}^{-k}[m] = \omega^{-km}, \quad \text{or} \quad \underline{\omega}^{-k} = (1, \omega^{-k}, \omega^{-2k}, \ldots, \omega^{-(N-1)k})\,.$$

In terms of complex conjugates,

$$\underline{\omega}^{-k} = \overline{\underline{\omega}^k}\,.$$

Note also that

$$\underline{\omega}^0 = \underline{1}\,.$$

In all of this you can see why it's important to use notations that distinguish a discrete signal or a vector from a scalar. Don't let your guard down.

Finally,

$$\underline{\omega}^k[m + nN] = \omega^{km+knN} = \omega^{km} = \underline{\omega}^k[m].$$

This is the *periodicity property* of the discrete complex exponential:

- $\underline{\omega}^k$ is periodic of period N for any integer k.

As simple as it is, this is a crucial fact in working with the DFT.

The DFT rewritten. Introducing the discrete complex exponential allows us to write the formula defining the discrete Fourier transform with a notation that, I think, really looks like a discrete version of the continuous transform. The DFT of a discrete signal $\underline{f}$ is

$$\underline{\mathcal{F}}\underline{f} = \sum_{k=0}^{N-1} \underline{f}[k]\underline{\omega}^{-k}\,.$$

At an index m,

$$\underline{\mathcal{F}}\underline{f}[m] = \sum_{k=0}^{N-1} \underline{f}[k]\underline{\omega}^{-k}[m] = \sum_{k=0}^{N-1} \underline{f}[k]\omega^{-km} = \sum_{k=0}^{N-1} \underline{f}[k]e^{-2\pi ikm/N}\,.$$

We still think of the indices m as frequencies, but the definition has $m = 0, 1, \ldots, N-1$. Where are the negative frequencies, you may well ask? Soon, very soon, you will know.

The frequencies for which $\underline{\mathcal{F}}\underline{f}[m] \neq 0$ constitute the *spectrum* of $\underline{f}$, though, as with the continuous Fourier transform, we sometimes also refer to the values $\underline{\mathcal{F}}\underline{f}[m]$ as the spectrum.

[3]It's also referred to as the *vector complex exponential* (keeping an open mind).

We note one special value:

$$\underline{\mathcal{F}}\,\underline{f}[0] = \sum_{k=0}^{N-1} \underline{f}[k]\underline{\omega}^{-k}[0] = \sum_{k=0}^{N-1} \underline{f}[k]\,,$$

the sum of the values of the input $\underline{f}$. Some people define the DFT with a $1/N$ in front so that the zeroth component of the output is the average of the components of the input, just as the zeroth Fourier coefficient of a periodic function is the average of the function over one period. We're not doing this, but the choice of where and whether to put a factor of $1/N$ in a formula haunts the subject. Wait and see.

We also note that the DFT is linear. To state this formally as a property,

$$\underline{\mathcal{F}}(\underline{f}_1 + \underline{f}_2) = \underline{\mathcal{F}}\,\underline{f}_1 + \underline{\mathcal{F}}\,\underline{f}_2 \quad \text{and} \quad \underline{\mathcal{F}}(\alpha \underline{f}) = \alpha \underline{\mathcal{F}}\,\underline{f}\,.$$

Showing this is easy:

$$\begin{aligned} \underline{\mathcal{F}}(\underline{f}_1 + \underline{f}_2) &= \sum_{k=0}^{N-1} (\underline{f}_1 + \underline{f}_2)[k]\underline{\omega}^{-k} = \sum_{k=0}^{N-1} (\underline{f}_1[k] + \underline{f}_2[k])\underline{\omega}^{-k} \\ &= \sum_{k=0}^{N-1} \underline{f}_1[k]\underline{\omega}^{-k} + \sum_{k=0}^{N-1} \underline{f}_2[k]\underline{\omega}^{-k} = \underline{\mathcal{F}}\,\underline{f}_1 + \underline{\mathcal{F}}\,\underline{f}_2\,. \end{aligned}$$

The DFT in matrix form. Switching the point of view to vectors and matrices, as a linear transformation taking vectors in $\mathbb{C}^N$ to vectors in $\mathbb{C}^N$, computing the DFT is exactly multiplying by a matrix. With $\underline{F} = \underline{\mathcal{F}}\,\underline{f}$,

$$\begin{pmatrix} \underline{F}[0] \\ \underline{F}[1] \\ \underline{F}[2] \\ \vdots \\ \underline{F}[N-1] \end{pmatrix} = \begin{pmatrix} 1 & 1 & 1 & \cdots & 1 \\ 1 & \omega^{-1\cdot 1} & \omega^{-1\cdot 2} & \cdots & \omega^{-1(N-1)} \\ 1 & \omega^{-2\cdot 1} & \omega^{-2\cdot 2} & \cdots & \omega^{-2(N-1)} \\ \vdots & \vdots & \vdots & \ddots & \vdots \\ 1 & \omega^{-(N-1)\cdot 1} & \omega^{-(N-1)\cdot 2} & \cdots & \omega^{-(N-1)^2} \end{pmatrix} \begin{pmatrix} \underline{f}[0] \\ \underline{f}[1] \\ \underline{f}[2] \\ \vdots \\ \underline{f}[N-1] \end{pmatrix}.$$

The discrete Fourier transform is the big old $N \times N$ matrix

$$\underline{\mathcal{F}} = \begin{pmatrix} 1 & 1 & 1 & \cdots & 1 \\ 1 & \omega^{-1} & \omega^{-2} & \cdots & \omega^{-(N-1)} \\ 1 & \omega^{-2} & \omega^{-4} & \cdots & \omega^{-2(N-1)} \\ \vdots & \vdots & \vdots & \ddots & \vdots \\ 1 & \omega^{-(N-1)} & \omega^{-2(N-1)} & \cdots & \omega^{-(N-1)^2} \end{pmatrix}.$$

Of course, if someone gave you the matrix for the DFT as a starting point, you would write down the corresponding transform. You would write

$$\underline{\mathcal{F}}\,\underline{f}[m] = \sum_{k=0}^{N-1} \underline{f}[k]\underline{\omega}^{-km} = \sum_{k=0}^{N-1} \underline{f}[k]\underline{\omega}^{-k}[m].$$

You would do that.

In the usual compact matrix notation,

$$(\underline{\mathcal{F}})_{mn} = \omega^{-mn} = e^{-2\pi imn/N}.$$

The columns of $\underline{\mathcal{F}}$ are just the negative powers of the discrete complex exponentials (as are the rows). Again, we take the indices for $\underline{f}$ and $\underline{F}$ to go from 0 to $N-1$ instead of from 1 to N, and the same for the rows and columns of $\underline{\mathcal{F}}$. Again I issue the warning, be careful how your favorite software package indexes in computing the DFT.

Right away we can see that the matrix for the DFT is *symmetric* (equal to its transpose):

$$(\underline{\mathcal{F}})_{mn} = (\underline{\mathcal{F}})_{nm}.$$

In fact, it has the additional property of being (almost) unitary, but that comes a little later.

7.5. Two Grids, Reciprocally Related

Before surrendering completely to the discrete world, we look back fondly one more time to the continuous. Invoke our understanding that the DFT finds the sampled Fourier transform of a sampled signal. We have a grid of points in the time domain and a grid of points in the frequency domain where the discrete version of the signal and the discrete version of its Fourier transform are known. From the discrete point of view, the values of the signal at the points in the time domain are all we know about the signal, and the values we compute according to the DFT formula are all we know about its transform.

A quick recap: In the time domain the signal is limited to an interval of length L. In the frequency domain the transform is limited to an interval of length $2B$. The grid points in the time domain are spaced $1/2B$ apart. The grid points in the frequency domain are spaced $1/L$ apart, so note (again) that the spacing in one domain is determined by properties of the signal in the other domain. The two grid spacings are related to the third quantity in the setup, the number of sample points, N. The equation is

$$N = 2BL.$$

Now, any two of the quantities B, L, or N determine the third via this relationship. The equation is often written another way, in terms of the grid spacings. If $\Delta t = 1/2B$ is the grid spacing in the time domain and $\Delta\nu = 1/L$ is the grid spacing in the frequency domain, then

$$N = \frac{1}{\Delta t \Delta\nu} \quad \text{or} \quad \frac{1}{N} = \Delta t \Delta\nu.$$

These two equivalent equations are referred to as the *reciprocity relations*. For a fixed number of sample points N, making Δt small means making $\Delta\nu$ large, and vice versa (relatively — depends on the size of N). Put that in your head.

Here's why all of this is important. For a given problem you want to solve, for a given signal you want to analyze by taking the Fourier transform, you typically either *know* or *choose* two of the following items:

- How long the signal lasts, i.e., how long you're willing to sit there taking measurements — that's L.

- How many measurements you make — that's N.
- How often you make a measurement — that's Δt.
- How fine a resolution you want in the spectrum — that's $\Delta\nu$.

Once two of these are determined, everything else is set. See the book by Briggs and Henson, mentioned earlier, for further discussion and examples.

7.6. Getting to Know Your Discrete Fourier Transform

We want to develop some general properties of the DFT as well as to compute the DFT of some special signals, much as we did when we first introduced the continuous Fourier transform. For the DFT there's less of an emphasis on finding explicit transforms, though a few examples are important, the discrete δ for example.

Many properties of the DFT correspond pretty directly to properties of the continuous Fourier transform, but there are differences. You should try to make the most of these correspondences, if only to decide for yourself when they are close and when they are not. Use what you know from your work in the continuous setting. We'll find formulas that are interesting and that are very important in applications.

Derivations using the DFT are often easier, technically, than those for the Fourier transform in that there are no worries about integrals converging, etc., but discrete derivations can have their own complications. The skills you need are in manipulating sums, particularly sums of complex exponentials, and in manipulating matrices. In both instances, if you're troubled, it's often a good idea to first calculate a few examples, say for DFTs of size three or four, or to write out the first several terms in a sum to see what the patterns are.

One thing to be mindful of in deriving formulas, in particular when working with sums, is how the index of summation (analogous to the variable of integration) enters and not to get it confused or in conflict with other indices that are in use; see Appendix C, on geometric sums. Derivations might also involve changing the index of summation, analogous to changing the variable of integration. This procedure (technique) seems easier in the case of integrals than in sums, maybe because of all the practice you've had making substitutions to find integrals. At any rate, you'll see the sorts of things that come up in the derivations in the chapter, and in the problems.

Two important facts. Before embarking, I am pleased to announce at once that all interesting properties of the DFT derive essentially from two facts. They are:

- Orthogonality of the discrete complex exponentials.
- Periodicity of the discrete complex exponentials, and hence periodicity of the inputs and outputs.

After treating ourselves to the DFT of one special signal, we'll discuss each of these in turn.

7.6.1. The discrete δ. Let's begin with an explicit calculation rather than a general property, the DFT of the discrete δ.[4] And good news! No need to go through the theory of distributions to define a δ in the discrete case, no pairings, no "where did we start and where did we finish." We can define it directly and easily by setting

$$\underline{\delta}_0 = (1, 0, \ldots, 0)\,.$$

In words, there's a 1 in the zeroth slot and 0's in the remaining $N-1$ slots. We didn't specify N, and so, strictly speaking, there's a $\underline{\delta}_0$ for each N, but since $\underline{\delta}_0$ will always arise in a context where the N is otherwise specified, we'll set aside that detail and we won't modify the notation.

To find the DFT of $\underline{\delta}_0$, we have, for any $m = 0, 1, \ldots, N-1$,

$$\begin{aligned}\underline{\mathcal{F}\,\delta}[m] &= \sum_{k=0}^{N-1} \underline{\delta}_0[k]\underline{\omega}^{-k}[m] \\ &= 1\,\underline{\omega}^0[m] \quad \text{(only the term } k=0 \text{ survives)} \\ &= 1.\end{aligned}$$

The same calculation done all at once is

$$\underline{\mathcal{F}\,\delta}_0 = \sum_{k=0}^{N-1} \underline{\delta}_0[k]\underline{\omega}^{-k} = \underline{\omega}^0 = \underline{1}\,.$$

Great — it looks just like $\mathcal{F}\delta = 1$ in the continuous case, and no tempered distributions in sight!

The shifted discrete $\underline{\delta}$ is just what you think it is,

$$\underline{\delta}_n = (0, \ldots, 0, 1, 0, \ldots, 0)$$

with a 1 in the nth slot and zeros elsewhere. This is another name for the Kronecker δ, written for indices m and n as

$$\underline{\delta}_n[m] = \delta_{mn} = \begin{cases} 1, & m = n\,, \\ 0, & m \neq n\,. \end{cases}$$

For the DFT of $\underline{\delta}_n$ we have

$$\underline{\mathcal{F}\,\delta}_n = \sum_{k=0}^{N-1} \underline{\delta}_n[k]\underline{\omega}^{-k} = \underline{\omega}^{-n}\,.$$

These are our first explicit transforms, and if we believe that the discrete case can be made to look like the continuous case, the results are encouraging. We state them again.

[4] If you're waiting for a discrete version of the rectangle function Π and its Fourier transform, which was our first example in the continuous setting, that comes later.

- The DFTs of the discrete $\underline{\delta}_0$ and shifted $\underline{\delta}_n$ are

$$\underline{\mathcal{F}}\,\underline{\delta}_0 = \underline{1} \quad \text{and} \quad \underline{\mathcal{F}}\,\underline{\delta}_n = \underline{\omega}^{-n}.$$

We'll establish other properties of discrete $\underline{\delta}_n$'s (convolution, sampling) later.

The vector view. There's a vector/matrix point of view that you should keep in mind. Regarded as vectors, the $\underline{\delta}$'s are the natural basis vectors $\underline{\delta}_0, \underline{\delta}_1, \ldots, \underline{\delta}_{N-1}$ of $\mathbb{C}^N$. True, we're indexing from 0 to $N-1$ instead of from 1 to N, but get over it. Thinking of the DFT as a matrix, $\underline{\mathcal{F}}\,\underline{\delta}_n$ is the nth column of $\underline{\mathcal{F}}$, and this is $\underline{\omega}^{-n}$. Good.

Or think in terms of vectors and $\underline{\mathcal{F}}$ as a linear transformation from $\mathbb{C}^N$ to $\mathbb{C}^N$. For an arbitrary discrete signal (vector) $\underline{f}$ we can write

$$\underline{f} = \sum_{k=0}^{N-1} \underline{f}[k]\underline{\delta}_k\,,$$

and if we had defined the DFT by what it does to a basis, we'd say it's the linear transformation for which $\underline{\mathcal{F}}\,\underline{\delta}_k = \underline{\omega}^{-k}$. We would find from this

$$\underline{\mathcal{F}}\,\underline{f} = \sum_{k=0}^{N-1} \underline{f}[k]\underline{\mathcal{F}}\,\underline{\delta}_k = \sum_{k=0}^{N-1} \underline{f}[k]\underline{\omega}^{-k}.$$

Good.

7.6.2. Orthogonality of the discrete complex exponentials.

To state this crucial property of the discrete complex exponentials we again regard

$$\underline{\omega} = (1, \omega, \omega^2, \ldots, \omega^{N-1})$$

and its kth power

$$\underline{\omega}^k = (1, \omega^k, \omega^{2k}, \ldots, \omega^{(N-1)k})$$

as vectors. Then the inner products satisfy:

- For k and ℓ in $[0 : N-1]$,

$$\underline{\omega}^k \cdot \underline{\omega}^\ell = \begin{cases} 0, & k \neq \ell\,, \\ N, & k = \ell\,. \end{cases}$$

For the norms, for each k,

$$\|\underline{\omega}^k\| = \sqrt{N}\,.$$

Thus the powers of the discrete complex exponentials are orthogonal and *almost* orthonormal. That's the result in the title of this section. Later we'll give another version of orthogonality that incorporates periodicity.

We could make the powers orthonormal by considering instead

$$\frac{1}{\sqrt{N}}(1, \omega^k, \omega^{2k}, \ldots, \omega^{(N-1)k})\,.$$

We won't do this, and neither does anyone else, pretty much. But we'll pay for it in the way a factor of N comes up in various formulas.

To remind you, in the continuous case the analogous result is that the family of functions $(1/\sqrt{T})e^{2\pi int/T}$ (periodic of period T) is orthonormal with respect to the inner product on $L^2([0,T])$:

$$\int_0^T \frac{1}{\sqrt{T}}e^{2\pi imt/T}\frac{1}{\sqrt{T}}e^{-2\pi int/T}\,dt = \frac{1}{T}\int_0^T e^{2\pi i(m-n)t/T}\,dt = \begin{cases} 1, & m=n\,, \\ 0, & m\neq n\,. \end{cases}$$

Orthogonality in the continuous case may seem easier to establish than in the discrete case because, sometimes, integration may seem easier than summation. Depends on what you're used to. There are two things we'll need for the derivation. The first is that $\omega^{kN}=1$ when k is any integer. The second is a geometric sum:

$$1+z+z^2+\cdots+z^{N-1} = \begin{cases} \dfrac{1-z^N}{1-z}, & z\neq 1\,, \\ N, & z=1\,. \end{cases}$$

See Appendix C (again) if you need some review.

We'll use this formula for z a power of ω. Observe that if $k\neq \ell$, then, as both k and ℓ are in $[0:N-1]$,

$$\omega^{k-\ell}=e^{2\pi i(k-\ell)/N}\neq 1 \quad\text{while}\quad \omega^{(k-\ell)N}=e^{2\pi i(k-\ell)}=1.$$

With this, let's compute the inner product $\underline{\omega}^k\cdot\underline{\omega}^\ell$. In full detail,

$$\begin{aligned}\underline{\omega}^k\cdot\underline{\omega}^\ell &= \sum_{n=0}^{N-1}\underline{\omega}^k[n]\overline{\underline{\omega}^\ell[n]} = \sum_{n=0}^{N-1}\underline{\omega}^k[n]\underline{\omega}^{-\ell}[n] \\ &= \sum_{n=0}^{N-1}\underline{\omega}^{k-\ell}[n] = \sum_{n=0}^{N-1}\omega^{(k-\ell)n} = \sum_{n=0}^{N-1}(\omega^{(k-\ell)})^n \\ &= \frac{1-\omega^{(k-\ell)N}}{1-\omega^{k-\ell}} \\ &= \begin{cases} 0, & k\neq\ell, \\ N, & k=\ell. \end{cases}\end{aligned}$$

Done. By taking complex conjugates we also deduce that

$$\underline{\omega}^{-k}\cdot\underline{\omega}^{-\ell} = \begin{cases} 0, & k\neq\ell, \\ N, & k=\ell. \end{cases}$$

$\underline{\mathcal{F}}$ is invertible. From this result we conclude that the N distinct vectors $\underline{1}$, $\underline{\omega}^1$, $\underline{\omega}^2, \dots, \underline{\omega}^{(N-1)}$ are a basis of $\mathbb{C}^N$; they are orthogonal, so linearly independent, and there are N of them. Likewise $\underline{1}$, $\underline{\omega}^{-1}$, $\underline{\omega}^{-2}, \dots, \underline{\omega}^{-(N-1)}$ are also a basis of $\mathbb{C}^N$. From the earlier result that $\underline{\mathcal{F}}\,\underline{\delta}_k=\underline{\omega}^{-k}$ we can then conclude that $\underline{\mathcal{F}}$ is invertible because it's a linear transformation that takes a basis to a basis. This doesn't tell us what the inverse is, however. We have to work a little harder for that.

The DFT of the discrete complex exponential. With the orthogonality of the discrete complex exponentials established, a number of other important results are within easy reach. For example, we can now find $\underline{\mathcal{F}}\,\underline{\omega}^k$.

By definition,

$$\underline{\mathcal{F}\omega}^k = \sum_{n=0}^{N-1} \underline{\omega}^k[n]\underline{\omega}^{-n},$$

and its ℓth component is then

$$\begin{aligned}\underline{\mathcal{F}\omega}^k[\ell] &= \sum_{n=0}^{N-1} \underline{\omega}^k[n]\underline{\omega}^{-n}[\ell] \\ &= \sum_{n=0}^{N-1} \omega^{kn}\omega^{-n\ell} = \underline{\omega}^k \cdot \underline{\omega}^\ell = \begin{cases} 0, & \ell \neq k, \\ N, & \ell = k. \end{cases}\end{aligned}$$

That is, $\underline{\mathcal{F}\omega}^k = (0, \dots, N, \dots, 0)$ with an N in the kth slot. We recognize this, and we are pleased.

- The discrete Fourier transform of $\underline{\omega}^k$ is

$$\underline{\mathcal{F}\omega}^k = N\underline{\delta}_k.$$

- In particular, taking $k = 0$ we have

$$\underline{\mathcal{F}1} = N\underline{\delta}_0.$$

Perhaps we are *almost* pleased. There's a factor of N that comes in that we don't see, in any way, in the continuous case. Here it traces back, ultimately, to $||\underline{\omega}||^2 = N$.

The appearance of a factor N or $1/N$ in various formulas is *always* wired somehow to $||\underline{\omega}||^2 = N$. It's one thing that makes the discrete case appear different from the continuous case, and it's a pain in the neck to keep straight. This is what you heard me warning and complaining about.

7.6.3. Inverting the DFT.

By now it should be second nature to you to expect that any useful transform ought to be invertible. The DFT is no exception; we commented on that just above. The DFT does have an inverse and the key to finding the formula is the orthogonality of the discrete complex exponentials. With a nod to the matrix form of the DFT it's actually pretty easy to realize what $\underline{\mathcal{F}}^{-1}$ is, if somewhat unmotivated. (I'll give an alternate way to identify $\underline{\mathcal{F}}^{-1}$ a little later.)

All the inner products $\underline{\omega}^k \cdot \underline{\omega}^\ell$ can be bundled together in the product of two matrices. Consider the conjugate transpose (the *adjoint*), $\underline{\mathcal{F}}^*$, of the DFT matrix. Since the DFT matrix is already symmetric, this just takes the conjugate of all the entries. The kth row of $\underline{\mathcal{F}}^*$ is

$$\begin{pmatrix} 1 & \omega^{1k} & \omega^{2k} & \cdots & \omega^{(N-1)k} \end{pmatrix}.$$

The ℓth column of $\underline{\mathcal{F}}$ is

$$\begin{pmatrix} 1 \\ \omega^{-1\cdot\ell} \\ \omega^{-2\cdot\ell} \\ \vdots \\ \omega^{-(N-1)\ell} \end{pmatrix}.$$

The $k\ell$-entry of $\underline{\mathcal{F}}^*\,\underline{\mathcal{F}}$ is the product of the kth row of $\underline{\mathcal{F}}^*$ and the ℓth column of $\underline{\mathcal{F}}$, which is, as highlighted in bold,

$$\begin{aligned}
&(\underline{\mathcal{F}}^*\,\underline{\mathcal{F}})_{k\ell} \\
&= \begin{pmatrix} 1 & 1 & 1 & \cdots & 1 \\ 1 & \omega^{1\cdot 1} & \omega^{2\cdot 1} & \cdots & \omega^{(N-1)1} \\ 1 & \omega^{2\cdot 1} & \omega^{2\cdot 2} & \cdots & \omega^{(N-1)2} \\ \vdots & \vdots & \vdots & \ddots & \vdots \\ \mathbf{1} & \boldsymbol{\omega^{k\cdot 1}} & \boldsymbol{\omega^{k\cdot 2}} & \cdots & \boldsymbol{\omega^{k(N-1)}} \\ \vdots & \vdots & \vdots & \ddots & \vdots \\ 1 & \omega^{(N-1)\cdot 1} & \omega^{(N-1)\cdot 2} & \cdots & \omega^{(N-1)^2} \end{pmatrix} \begin{pmatrix} 1 & 1 & \cdots & \mathbf{1} & \cdots \\ 1 & \omega^{-1\cdot 1} & \cdots & \boldsymbol{\omega^{-1\cdot\ell}} & \cdots \\ 1 & \omega^{-1\cdot 2} & \cdots & \boldsymbol{\omega^{-2\cdot\ell}} & \cdots \\ \vdots & \vdots & \ddots & \vdots & \cdots \\ \vdots & \vdots & \ddots & \vdots & \cdots \\ \vdots & \vdots & \ddots & \vdots & \cdots \\ 1 & \omega^{-1(N-1)} & \cdots & \boldsymbol{\omega^{-(N-1)\ell}} & \cdots \end{pmatrix} \\
&= \sum_{n=0}^{N-1} \omega^{kn}\omega^{-n\ell}.
\end{aligned}$$

This sum is the inner product,

$$\sum_{n=0}^{N-1} \omega^{kn}\omega^{-n\ell} = \underline{\omega}^k \cdot \underline{\omega}^\ell = \begin{cases} 0, & \ell \neq k, \\ N, & \ell = k. \end{cases}$$

All the off-diagonal elements of $\underline{\mathcal{F}}^*\underline{\mathcal{F}}$ are zero, and the elements along the main diagonal are all equal to N. Combining the two statements,

$$\underline{\mathcal{F}}^*\underline{\mathcal{F}} = NI,$$

where I is the $N \times N$ identity matrix. A similar calculation would also give $\underline{\mathcal{F}}\,\underline{\mathcal{F}}^* = NI$, multiplying the matrices in the other order.

And now the inverse of the DFT appears before us. Rewrite $\underline{\mathcal{F}}^*\underline{\mathcal{F}} = NI$ as

$$\left(\frac{1}{N}\underline{\mathcal{F}}^*\right)\underline{\mathcal{F}} = I$$

and, behold, the inverse of the DFT is

$$\underline{\mathcal{F}}^{-1} = \frac{1}{N}\underline{\mathcal{F}}^*.$$

You would also write this as

$$\underline{\mathcal{F}}^{-1}\underline{f}[m] = \frac{1}{N}\sum_{k=0}^{N-1} \underline{f}[k]\omega^{km} = \frac{1}{N}\sum_{k=0}^{N-1} \underline{f}[k]\underline{\omega}^k[m],$$

for $m = 0, 1, \ldots, N-1$. You would do that.

Viewed together, the two transforms

$$\underline{\mathcal{F}}\,\underline{f} = \sum_{k=0}^{N-1} \underline{f}[k]\underline{\omega}^{-k} \qquad \text{and} \qquad \underline{\mathcal{F}}^{-1}\underline{f} = \frac{1}{N}\sum_{k=0}^{N-1} \underline{f}[k]\underline{\omega}^{k}$$

are imitating the relationship between the continuous-time Fourier transform and its inverse in that there's just a change in the sign of the complex exponential to go from one to the other — except, of course, for the ever irritating factor of $1/N$.

A brief digression on matrices. This material is probably familiar to you from a course in linear algebra, but I thought it would be helpful to include. We've already used some of the terminology, and we'll see these things again when we study linear systems themselves in Chapter 8.

For the definitions it's first necessary to remember that the transpose of a matrix A, denoted by A^{T}, is obtained by interchanging the rows and columns of A. If A is an $M \times N$ matrix, then A^{T} is an $N \times M$ matrix, and in the case of a square matrix ($M = N$) taking the transpose amounts to reflecting the entries across the main diagonal. As a linear transformation, if $A\colon \mathbb{R}^N \to \mathbb{R}^M$, then $A^{\mathsf{T}}\colon \mathbb{R}^M \to \mathbb{R}^N$. If, for shorthand, we write A generically in terms of its entries, as in $A = (a_{ij})$, then we write $A^{\mathsf{T}} = (a_{ji})$; note that the diagonal entries a_{ii}, where $i = j$, are unaffected by taking the transpose.

Square matrices can have a special property with respect to taking the transpose — they get to be symmetric: a square matrix A is *symmetric* if

$$A^{\mathsf{T}} = A\,.$$

In words, interchanging the rows and columns gives back the same matrix — it's symmetric across the main diagonal. (The diagonal entries need not be equal to each other!) The DFT, though a complex matrix, is symmetric.

A different notion also involving a matrix and its transpose is orthogonality. A square matrix is *orthogonal* if

$$A^{\mathsf{T}}A = I\,,$$

where I is the identity matrix. A matrix is orthogonal if and only if its columns (or rows) are orthonormal with respect to the inner product and hence form an orthonormal basis of $\mathbb{R}^N$. Now be careful, "symmetric" and "orthogonal" are *independent notions* for matrices. A matrix can be one and not the other.

For matrices with complex entries (operating on real or complex vectors) the appropriate notion corresponding to simple symmetry in the real case is *Hermitian symmetry*. For this we form the transpose and take the complex conjugate of the entries. If A is a complex matrix, then we use A^* to denote the conjugate transpose, known as the *adjoint* of A. A square matrix A is *Hermitian* if

$$A^* = A\,.$$

The DFT is *not* Hermitian.

Finally, a square matrix is *unitary* if

$$A^*A = I\,.$$

Once again, "Hermitian" and "unitary" are independent notions for complex matrices. And once again, a matrix is unitary if and only if its columns (or rows) are orthonormal with respect to the complex inner product and hence form an orthonormal basis of $\mathbb{C}^N$.

The fundamental result above on the DFT is

$$\underline{\mathcal{F}}^*\underline{\mathcal{F}} = \underline{\mathcal{F}}\,\underline{\mathcal{F}}^* = NI\,,$$

and one could fairly say that $\underline{\mathcal{F}}$ is "almost" unitary. It misses by the factor N, which is there because the discrete complex exponentials are orthogonal but not orthonormal. Oh, the agony.

7.6.4. Periodicity. Here comes the second important fact about the DFT, the periodicity of the inputs and outputs. This is a *difference* between the discrete and continuous cases rather than a similarity. The definition of the DFT, specifically the periodicity of the discrete complex exponentials suggests, even compels, some additional structure to the outputs and inputs.

As with the input values, the output values $\underline{F}[m]$ are defined initially only for $m = 0$ to $m = N - 1$, but their definition as

$$\underline{F}[m] = \sum_{k=0}^{N-1} \underline{f}[k]\omega^{-km}$$

implies a periodicity property. Since

$$\omega^{-k(m+N)} = \omega^{-km},$$

we have

$$\sum_{k=0}^{N-1} \underline{f}[k]\omega^{-k(m+N)} = \sum_{k=0}^{N-1} \underline{f}[k]\omega^{-km} = \underline{F}[m]\,.$$

If we consider the left-hand side as the DFT formula producing an output $\underline{F}$, then that output would be $\underline{F}[m+N]$. The equations together then say that $\underline{F}[m+N] = \underline{F}[m]$, a periodicity property. More generally, and by the same kind of calculation, we would have

$$\underline{F}[m + nN] = \underline{F}[m]$$

for any integer n. Thus, instead of just working with $\underline{F}$ as defined for $m = 0, 1, \ldots, N - 1$ it's natural to extend it to be a periodic signal of period N, defined on all the integers.

For example, if we start off with $N = 4$ and the values $\underline{F}[0], \underline{F}[1], \underline{F}[2], \underline{F}[3]$, then, by definition, the periodic extension of $\underline{F}$ has $\underline{F}[4] = \underline{F}[0]$, $\underline{F}[5] = \underline{F}[1]$, $\underline{F}[6] = \underline{F}[2]$, and so on, and going in the other direction, $\underline{F}[-1] = \underline{F}[3]$, $\underline{F}[-2] = \underline{F}[2]$, and so on. In general,

$$\underline{F}[p] = \underline{F}[q] \quad \text{if } p - q \text{ is a multiple of } N\text{, positive or negative.}$$

Or put another way

$$\underline{F}[p] = \underline{F}[q] \quad \text{if } p \equiv q \bmod N.$$

(I'm assuming that you're familiar with the "≡" notation as used in modular arithmetic.) We then have the formula

$$\underline{F}[m] = \sum_{k=0}^{N-1} \underline{f}[k]\omega^{-km}$$

for *all* integers m.

Because of these observations and unless instructed otherwise:

- We will always assume that the output of the DFT is a periodic signal of period N.

Furthermore, the form of the inverse DFT means that an input $\underline{f}$ to the DFT also extends naturally to be a periodic signal of period N. It's the same argument, for if we start with $\underline{F} = \underline{\mathcal{F}}\,\underline{f}$, input $\underline{f}$ and output $\underline{F}$, then

$$\underline{f}[m] = \underline{\mathcal{F}}^{-1}\underline{F}[m] = \frac{1}{N}\sum_{k=0}^{N-1} \underline{F}[k]\underline{\omega}^k[m],$$

and the right-hand side defines a periodic signal of period N. Thus:

- We will always assume that the input to the DFT is a periodic signal of period N.

So, briefly, when the DFT is involved, we assume that *all* our discrete signals are periodic.

As for how this differs from the continuous case, we certainly *can* consider periodicity — that's what the subject of Fourier series is all about, after all — but when working with the Fourier transform, we don't *have* to consider periodicity. In the discrete case we really do.

The discrete complex exponential and discrete $\underline{\delta}$ as periodic signals. To take an important example, if, according to the periodicity dictum, we consider the discrete complex exponential as a periodic discrete signal, then we can define it simply by

$$\underline{\omega}[m] = \omega^m = e^{2\pi i m/N}, \quad m \text{ an integer.}$$

The orthogonality property looks a little more general:

$$\underline{\omega}^k \cdot \underline{\omega}^\ell = \begin{cases} 0, & k \not\equiv \ell \bmod N, \\ N, & k \equiv \ell \bmod N. \end{cases}$$

The discrete δ, as a periodic signal, also has a more general looking definition:

$$\underline{\delta}_n[m] = \begin{cases} 0, & m \not\equiv n \bmod N, \\ 1, & m \equiv n \bmod N. \end{cases}$$

On the definition of the DFT. If we were developing the DFT from a purely mathematical point of view, we would probably incorporate periodicity as part of the initial definition, and this is sometimes done. For the setup, we let

$$\mathbb{Z}_N = \{0, 1, \dots, N-1\},$$

with addition understood to be modulo N. The symbol "$\mathbb{Z}$" is for the German word *Zahlen*, which means "numbers." Tell your friends. So $\mathbb{Z} = \{0, \pm 1, \pm 2, \dots\}$.

To say a discrete signal is defined on $\mathbb{Z}_N$ is operationally the same as saying that it's defined on all the integers, $\mathbb{Z}$, and is periodic of period N. The DFT, with the same definition as before, operates on signals that are defined on $\mathbb{Z}_N$ and produces signals that are defined on $\mathbb{Z}_N$. That's what a mathematician would say.

Adopting this point of view would make some parts of the mathematical development a little smoother (though no different in substance), but I think it's extra baggage early on. It doesn't apply to how we motivated the definition originally (sampling), and in general it can make the tie in with physical applications more awkward.

7.7. Notations and Conventions 2

The definition of the DFT that we've given is pretty standard, and it's the one we'll use. One sometimes finds an alternate definition of the DFT (used especially in the two-dimensional setting in applications imaging) where N is assumed to be even, and the index set for *both* the inputs $\underline{f}$ and the outputs $\underline{F}$ is taken to be $[-(N/2)+1 : N/2] = (-(N/2)+1, -(N/2)+2, \dots, -1, 0, 1, \dots, N/2)$. There really are N points in that list, and the reason for starting the bottom at $-(N/2)+1$ and going up to $N/2$ will become clear in a bit. The definition of the DFT is then[5]

$$\underline{\mathcal{F}}\,\underline{f} = \sum_{k=-N/2+1}^{N/2} \underline{f}[k]\underline{\omega}^{-k}, \quad \text{or in components} \quad \underline{F}[m] = \sum_{k=-N/2+1}^{N/2} \underline{f}[k]\omega^{-km}\,.$$

This puts the $k=0$ term in the middle. (In 2D it puts the center of the spectrum of the image at the origin.)

We would be led to this indexing of the inputs and outputs if in the sampling-based derivation we originally gave of the DFT, we sampled on the time interval from $-L/2$ to $L/2$ and on the frequency interval from $-B$ to B. Then, using the index set $[-(N/2)+1 : N/2]$, the sample points in the time domain would be of the form

$$t_{-N/2+1} = -\frac{L}{2} + \frac{1}{2B} = \frac{-N/2+1}{2B}, \quad t_{-N/2+2} = \frac{-N/2+2}{2B}, \dots,$$
$$t_{N/2} = \frac{-N/2+N}{2B} = \frac{N/2}{2B} = \frac{L}{2},$$

and in the frequency domain of the form

$$s_{-N/2+1} = -B + \frac{1}{L} = \frac{-N/2+1}{L}, \quad s_{-N/2+2} = \frac{-N/2+2}{L}, \dots,$$
$$s_{N/2} = \frac{-N/2+N}{L} = \frac{N/2}{L} = B\,.$$

Maybe you can see why I didn't want to set things up this way for our first encounter. A little too much to take in at first glance.

The new definition above of the DFT is completely equivalent to the first definition *because* of periodicity. There's a phrase one sees that's supposed to alleviate

[5]Briggs and Henson use this convention in their book on the DFT. They also put a $1/N$ in front of the DFT and leave it off of the inverse DFT.

the tension over this and other similar sorts of things. It goes:

- The DFT can be defined over any set of N consecutive indices.

What this means most often in practice is that we can write

$$\underline{\mathcal{F}}\,\underline{f} = \sum_{k=p}^{p+N-1} \underline{f}[k]\underline{\omega}^{-k},$$

for any integer p. We'll explain this thoroughly in just a bit. It's a little tedious, but not difficult, and it's important to understand. Thinking of an input (or output) $\underline{f}$ as a periodic discrete signal (as we must, but something your software package can't really do), then you don't have to worry about how it's indexed. It goes on forever, and any block of N consecutive values, $\underline{f}[p], \underline{f}[p+1], \ldots, \underline{f}[p+N-1]$, should be as good as any other because the values of $\underline{f}$ repeat. You still have to establish the quoted remark, however, to be assured that finding the DFT gives the same result on any such block. This is essentially a discrete form of the statement for continuous-time periodic functions that the Fourier coefficients can be calculated by integrating over any period.

Negative frequencies. In case you hadn't noticed, negative frequencies have appeared on the scene! Indeed, once the signal has been extended periodically to be defined on all the integers there's no issue with considering negative frequencies. However, unless and until periodicity comes to the rescue, be aware that many standard treatments of the DFT cling to indexing from 0 to $N-1$ and talk about which frequencies in that range should be thought of as the negative frequencies "by convention." This is also an issue with software packages.

Suppose that N is even. This makes things a little easier and is often assumed.[6] Suppose also that we consider real inputs $\underline{f} = (\underline{f}[0], \underline{f}[1], \ldots, \underline{f}[N-1])$. Something special happens at the midpoint, $N/2$, of the spectrum. We find

$$\begin{aligned}\underline{\mathcal{F}}\,\underline{f}[N/2] &= \sum_{k=0}^{N-1} \underline{f}[k]\underline{\omega}^{-k}[N/2] = \sum_{k=0}^{N-1} \underline{f}[k]\omega^{-kN/2} \\ &= \sum_{k=0}^{N-1} \underline{f}[k]e^{-\pi ik} = \sum_{k=0}^{N-1}(-1)^{-k}\underline{f}[k] \quad (\text{using } \omega^{N/2} = -1).\end{aligned}$$

The value of the transform $\underline{\mathcal{F}}\,\underline{f}[N/2]$ is an alternating sum of the components of the input signal $\underline{f}$. In particular, $\underline{\mathcal{F}}\,\underline{f}[N/2]$ is *real*.

Furthermore, the spectrum splits at $N/2$. I explain. For a start, look at $\underline{\mathcal{F}}\,\underline{f}[(N/2)+1]$ and $\underline{\mathcal{F}}\,\underline{f}[(N/2)-1]$:

$$\underline{\mathcal{F}}\,\underline{f}\Big[\frac{N}{2}+1\Big] = \sum_{k=0}^{N-1} \underline{f}[k]\underline{\omega}^{-k}\Big[\frac{N}{2}+1\Big] = \sum_{k=0}^{N-1} \underline{f}[k]\omega^{-k}\omega^{-Nk/2} = \sum_{k=0}^{N-1} \underline{f}[k]\omega^{-k}(-1)^{-k},$$

$$\underline{\mathcal{F}}\,\underline{f}\Big[\frac{N}{2}-1\Big] = \sum_{k=0}^{N-1} \underline{f}[k]\underline{\omega}^{-k}\Big[\frac{N}{2}-1\Big] = \sum_{k=0}^{N-1} \underline{f}[k]\omega^{k}\omega^{-Nk/2} = \sum_{k=0}^{N-1} \underline{f}[k]\omega^{k}(-1)^{-k}.$$

[6] Without that assumption one has to write some formulas using ceilings and floors of $N/2$, and I'm never confident that I'm doing it right.

Comparing the two calculations, we see that

$$\underline{\mathcal{F}}\underline{f}\Big[\frac{N}{2}+1\Big] = \overline{\underline{\mathcal{F}}\underline{f}\Big[\frac{N}{2}-1\Big]}\,.$$

Similarly, we get

$$\underline{\mathcal{F}}\underline{f}\Big[\frac{N}{2}+2\Big] = \sum_{k=0}^{N-1} \underline{f}[k]\omega^{-2k}(-1)^{-k}\,, \qquad \underline{\mathcal{F}}\underline{f}\Big[\frac{N}{2}-2\Big] = \sum_{k=0}^{N-1} \underline{f}[k]\omega^{2k}(-1)^{-k}\,,$$

so that

$$\underline{\mathcal{F}}\underline{f}\Big[\frac{N}{2}+2\Big] = \overline{\underline{\mathcal{F}}\underline{f}\Big[\frac{N}{2}-2\Big]}\,.$$

This pattern persists down to the pair of values

$$\underline{\mathcal{F}}\underline{f}[1] = \sum_{k=0}^{N-1} \underline{f}[k]\omega^{-k}$$

and

$$\underline{\mathcal{F}}\underline{f}[N-1] = \sum_{k=0}^{N-1} \underline{f}[k]\omega^{-k(N-1)} = \sum_{k=0}^{N-1} \underline{f}[k]\omega^{k}\omega^{-kN} = \sum_{k=0}^{N-1} \underline{f}[k]\omega^{k}\,,$$

i.e., to

$$\underline{\mathcal{F}}\underline{f}[1] = \overline{\underline{\mathcal{F}}\underline{f}[N-1]}\,.$$

This is where it stops; recall that $\underline{\mathcal{F}}\underline{f}[0]$ is the sum of the components of $\underline{f}$.

This result is analogous to the symmetry relation $\mathcal{F}f(-s) = \overline{\mathcal{F}f(s)}$ in the continuous case, which we'll come back to. *Because of it*, when the spectrum is indexed from 0 to $N-1$, the *convention* is to say that the frequencies from $m=1$ to $m = N/2-1$ are the *positive frequencies* and those from $N/2+1$ to $N-1$ are the *negative frequencies*. Whatever adjectives one uses, the important upshot is that for a real input $\underline{f}$ *all the information in the spectrum is in*:

- The first component $\underline{\mathcal{F}}\underline{f}[0]$ (the "DC" component, the sum of components of the input).
- The components $\underline{\mathcal{F}}\underline{f}[1]$, $\underline{\mathcal{F}}\underline{f}[2]$, ..., $\underline{\mathcal{F}}\underline{f}[N/2-1]$.
- The special value $\underline{\mathcal{F}}\underline{f}[N/2]$ (the alternating sum of the components of the input).

The remaining components of $\underline{\mathcal{F}}\underline{f}$ are just the complex conjugates of those from 1 to $N/2-1$.

Fine. But periodicity makes this much more clear. It makes honest negative frequencies correspond to negative frequencies "by convention." Suppose we have a periodic input $\underline{f}$ and periodic output $\underline{F} = \underline{\mathcal{F}}\,\underline{f}$ indexed from $-(N/2)+1$ to $N/2$. We would certainly say in this case that the negative frequencies go from $-(N/2)+1$ to -1, with corresponding outputs $\underline{F}[-(N/2)+1]$, $\underline{F}[-(N/2)+2]$, ..., $\underline{F}[-1]$. Where do these frequencies go if we reindex from 0 to $N-1$? Using periodicity,

$$\begin{aligned}
&\underline{F}\Big[-\frac{N}{2}+1\Big] = \underline{F}\Big[-\frac{N}{2}+1+N\Big] = \underline{F}\Big[\frac{N}{2}+1\Big],\\
&\underline{F}\Big[-\frac{N}{2}+2\Big] = \underline{F}\Big[-\frac{N}{2}+2+N\Big] = \underline{F}\Big[\frac{N}{2}+2\Big],\\
&\qquad\text{and so on up to}\\
&\underline{F}[-1] = \underline{F}[-1+N].
\end{aligned}$$

The "honest" negative frequencies at $-(N/2)+1, \dots, -1$ are by periodicity the "negative frequencies by convention" at $N/2+1, \dots, N-1$. Convince yourself as well why it's more natural to index from $-(N/2)+1$ up to $N/2$, as we have, rather than from $-N/2$ to $(N/2)-1$.

We worked through this in so much detail to help you understand and interpret what you're seeing when you compute and plot DFTs using any of the standard software packages, where indexing (usually) goes from 1 to N. Armed with this new understanding, look at Problem 7.5.

7.7.1. Different definitions for the DFT. The purpose of this section is to get a precise understanding of the following statement:

"The DFT can be defined over any set of N consecutive indices,"

mentioned earlier. Once understood precisely, this statement is behind the different indexing conventions that are in use for the DFT. Sort of mathy, but a nice exercise in periodicity, which is all that's involved. You can skip the derivation and still have a happy life.

For periodic functions $f(t)$ in the continuous setting, say functions of period 1, the nth Fourier coefficient is

$$\hat{f}(n) = \int_0^1 e^{-2\pi int} f(t)\,dt.$$

The periodicity of $f(t)$ implies that $\hat{f}(n)$ can be obtained by integrating over any interval of length 1, meaning

$$\hat{f}(n) = \int_a^{a+1} e^{-2\pi int} f(t)\,dt$$

for any a. Morally, we're looking for the discrete version of this. Morally, we'll give the same argument that we did back in Chapter 1, which goes: take the derivative with respect to a:

$$\begin{aligned}\frac{d}{da}\int_a^{a+1} e^{-2\pi int} f(t)\,dt &= e^{-2\pi i(a+1)n} f(a+1) - e^{-2\pi ian} f(a)\\ &= e^{-2\pi ian}e^{-2\pi in} f(a+1) - e^{-2\pi ian} f(a)\\ &= e^{-2\pi ian} f(a) - e^{-2\pi ian} f(a) = 0\,,\end{aligned}$$

using $e^{2\pi in} = 1$ and the periodicity of $f(t)$. Since the derivative is identically zero, the value of the integral is the same for any value of a. In particular, taking $a = 0$,

$$\int_a^{a+1} e^{-2\pi int} f(t)\,dt = \int_0^1 e^{-2\pi int} f(t)\,dt = \hat{f}(n)\,.$$

Back to the discrete case. Take a good backward glance at what's been done and start the whole thing over, so to speak, by giving a more general definition of the DFT:

> We consider discrete signals $\underline{f}$ that are periodic of period N. Let $\mathcal{P}$ and $\mathcal{Q}$ be index sets of N consecutive integers, say,
>
> $$\mathcal{P} = [p : p+N-1], \quad \mathcal{Q} = [q : q+N-1]\,.$$
>
> The DFT based on $\mathcal{P}$ and $\mathcal{Q}$ is defined by
>
> $$\underline{\mathcal{G}}\underline{f}[m] = \sum_{k\in\mathcal{P}} \underline{f}[k]\omega^{-mk} = \sum_{k=p}^{p+N-1} \underline{f}[k]\omega^{-mk}, \quad m \in \mathcal{Q}\,.$$

I've called the transform $\underline{\mathcal{G}}$ to distinguish it from $\underline{\mathcal{F}}$, which in the present setup corresponds to the special choice of index sets $\mathcal{P} = \mathcal{Q} = [0 : N-1]$.

Since $\underline{f}$ is periodic of period N, knowing $\underline{f}$ on any set of N consecutive numbers determines it everywhere. We hope the same will be true of a transform of $\underline{f}$, but this is what must be shown. Thus one wants to establish that the definition of $\underline{\mathcal{G}}$ is independent of $\mathcal{P}$ and $\mathcal{Q}$. This is a sharper version of the informal statement in the first quote, above, but we have to say what "independent of $\mathcal{P}$ and $\mathcal{Q}$" means.

First $\mathcal{Q}$. Allowing for a general set $\mathcal{Q}$ of N consecutive integers to index the values $\underline{\mathcal{G}}\underline{f}[m]$ doesn't really come up in applications and is included in the definition for the sake of generality. We can dispose of the question quickly. To show that the transform is independent of $\mathcal{Q}$, we do the following:

(a) Extend $\underline{\mathcal{G}}\underline{f}$ to be periodic of period N; thus $\underline{\mathcal{G}}\underline{f}[m]$ is defined, by periodicity, for all integers m.

(b) Use the periodicity of the exponentials in the definition of the transform to show that the extension is again given by the formula for $\underline{\mathcal{G}}\underline{f}$.

To wit, for (b), let $m \in \mathbb{Z}$ and write $m = n + \ell N$ where $n \in \mathcal{Q}$ and ℓ is an integer. Then

$$\begin{aligned}
\sum_{k=p}^{p+N-1} \underline{f}[k]\omega^{-mk} &= \sum_{k=p}^{p+N-1} \underline{f}[k]\omega^{-(n+\ell N)k} \\
&= \sum_{k=p}^{p+N-1} \underline{f}[k]\omega^{-nk}\omega^{-\ell N k} \\
&= \sum_{k=p}^{p+N-1} \underline{f}[k]\omega^{-nk} \quad (\text{since } \omega^{-\ell N k} = 1) \\
&= \underline{\mathcal{G}}\underline{f}[n] \quad (\text{from the original definition of } \underline{\mathcal{G}}, \text{ which needs } n \in \mathcal{Q}) \\
&= \underline{\mathcal{G}}\underline{f}[n + \ell N] \quad (\text{because we've extended } \underline{\mathcal{G}} \text{ to be periodic}) \\
&= \underline{\mathcal{G}}\underline{f}[m]\,.
\end{aligned}$$

Reading from bottom right to top left[7] then shows that

$$\underline{\mathcal{G}}\underline{f}[m] = \sum_{k=p}^{p+N-1} \underline{f}[k]\omega^{-mk}$$

for all $m \in \mathbb{Z}$. That's *all* $m \in \mathbb{Z}$, with a formula for $\underline{\mathcal{G}}$ depending only on $\mathcal{P}$. That's the point, and it shows that the definition of $\underline{\mathcal{G}}$ is independent of the initial choice of the index set $\mathcal{Q}$.

Now for independence from $\mathcal{P}$. Write, more compactly,

$$\underline{\mathcal{G}}_p \underline{f} = \sum_{k=p}^{p+N-1} \underline{f}[k]\, \underline{\omega}^{-k}\,.$$

Analogous to taking the derivative with respect to a of the integral from a to $a+1$ that defines the Fourier coefficient, apply the *difference operator* to $\underline{\mathcal{G}}_p \underline{f}$ as a function of p, meaning $\Delta \underline{\mathcal{G}}_p \underline{f} = \underline{\mathcal{G}}_{p+1} \underline{f} - \underline{\mathcal{G}}_p \underline{f}$. Then

$$\begin{aligned}
\Delta \underline{\mathcal{G}}_p \underline{f} &= \underline{\mathcal{G}}_{p+1} \underline{f} - \underline{\mathcal{G}}_p \underline{f} \\
&= \sum_{k=p+1}^{p+N} \underline{f}\, \underline{\omega}^{-k} - \sum_{k=p}^{p+N-1} \underline{f}\, \underline{\omega}^{-k} \\
&= \underline{f}[p+N]\underline{\omega}^{-(p+N)} - \underline{f}[p]\underline{\omega}^{-p} \quad (\text{all other terms cancel}) \\
&= \underline{f}[p+N]\underline{\omega}^{-p}\underline{\omega}^{-N} - \underline{f}[p]\underline{\omega}^{-p} \\
&= \underline{f}[p]\underline{\omega}^{-p} - \underline{f}[p]\underline{\omega}^{-p} = 0
\end{aligned}$$

using $\underline{\omega}^{-N} = \underline{1}$ and the periodicity of $\underline{f}$. Since $\Delta \underline{\mathcal{G}}_p \underline{f} = 0$, it follows that the values of $\underline{\mathcal{G}}_p \underline{f}$ are independent of p, which is just what we wanted to show. In particular, taking $p = 0$ we have

$$\underline{\mathcal{G}}_p \underline{f} = \sum_{k=p}^{p+N-1} \underline{f}[k]\, \underline{\omega}^{-k} = \sum_{k=0}^{N-1} \underline{f}[k]\, \underline{\omega}^{-k} = \underline{\mathcal{F}}\, \underline{f}\,.$$

[7] As in: Where did we start; where did we finish?

In other words, *any* DFT is *the* DFT as we originally defined it, when considered on periodic signals. And we have made the argument to show this look very much like the one for the continuous case.

Finally, in this same circle of ideas, so, too, do we have: "the inverse DFT can be defined over any set of N consecutive indices." No argument from me, and no argument given here.

7.8. Getting to Know Your DFT, Better

Time to push harder on the similarities and differences between the discrete and continuous settings. Start with the following:

7.8.1. Reversed signals and their DFTs. For a discrete signal, $\underline{f}$, defined on the integers, periodic or not, the corresponding reversed signal, $\underline{f}^-$, is defined by

$$\underline{f}^-[m] = \underline{f}[-m] .$$

If $\underline{f}$ is periodic of period N, as we henceforth again assume, and if we write it as the vector

$$\underline{f} = (\underline{f}[0], \underline{f}[1], \dots, \underline{f}[N-1]), \quad \text{so that} \quad \underline{f}^- = (\underline{f}[0], \underline{f}[-1], \dots, \underline{f}[-N+1]),$$

then by periodicity

$$\underline{f}^- = (\underline{f}[N], \underline{f}[N-1], \dots, \underline{f}[1]) \quad (\text{using } \underline{f}[N] = \underline{f}[0]),$$

which makes the description of $\underline{f}^-$ as "reversed" even more apt (though, as in many irritating instances, the indexing is a little off). Defined directly in terms of its components this is

$$\underline{f}^-[n] = \underline{f}[N-n]$$

and this formula is good for all integers n. This description of $\underline{f}^-$ is often quite convenient.

Note that reversing a signal satisfies the principle of superposition (is linear as an operation on signals):

$$(\underline{f} + \underline{g})^- = \underline{f}^- + \underline{g}^- \quad \text{and} \quad (\alpha \underline{f})^- = \alpha \underline{f}^- .$$

It's even more than that, for we also have

$$(\underline{f}\,\underline{g})^- = (\underline{f}^-)(\underline{g}^-) .$$

Let's consider two special cases of reversed signals. First, clearly

$$\underline{\delta}_0^- = \underline{\delta}_0 ,$$

and though we'll pick up more on evenness and oddness later, this says that $\underline{\delta}_0$ is even. For the shifted $\underline{\delta}$,

$$\underline{\delta}_k^- = \underline{\delta}_{-k} .$$

I'll let you verify that. With this result we can write

$$\underline{f}^- = \left(\sum_{k=0}^{N-1} \underline{f}[k] \underline{\delta}_k \right)^- = \sum_{k=0}^{N-1} \underline{f}[k] \underline{\delta}_{-k} .$$

One might say that the $\underline{\delta}_k$ are a basis for the forward signals and the $\underline{\delta}_{-k}$ are a basis for the reversed signals.

Next let's look at $\underline{\omega}$. First, we have

$$\underline{\omega}^- = (\underline{\omega}[N], \underline{\omega}[N-1], \underline{\omega}[N-2], \dots, \underline{\omega}[1]) = (1, \omega^{N-1}, \omega^{N-2}, \dots, \omega).$$

But now notice (as we could have noticed earlier) that

$$\omega^{N-1}\omega = \omega^N = 1 \Rightarrow \omega^{N-1} = \omega^{-1}.$$

Likewise

$$\omega^{N-2}\omega^2 = \omega^N = 1 \Rightarrow \omega^{N-2} = \omega^{-2}.$$

Continuing in this way we see, very attractively,

$$\underline{\omega}^- = (1, \omega^{-1}, \omega^{-2}, \dots, \omega^{-(N-1)}) = \underline{\omega}^{-1}.$$

In the same way we find, equally attractively,

$$(\underline{\omega}^k)^- = \underline{\omega}^{-k}.$$

Of course then also

$$(\underline{\omega}^{-k})^- = \underline{\omega}^k.$$

Duality. This has an important consequence for the DFT, our first discrete duality result. Let's consider $\underline{\mathcal{F}}\,\underline{f}^-$, the DFT of the reversed signal. To work with the expression for $\underline{\mathcal{F}}\,\underline{f}^-$ we'll need to use periodicity of $\underline{f}$ and do a little fancy footwork changing the variable of summation in the definition of the DFT. Here's how it goes:

$$\begin{aligned}
\underline{\mathcal{F}}\,\underline{f}^- &= \sum_{k=0}^{N-1} \underline{f}^-[k]\underline{\omega}^{-k} \\
&= \sum_{k=0}^{N-1} \underline{f}[N-k]\,\underline{\omega}^{-k} \quad \text{(reversing } \underline{f}\text{)} \\
&= \sum_{\ell=N}^{1} \underline{f}[\ell]\,\underline{\omega}^{\ell-N} \quad \text{(letting } \ell = N-k\text{)} \\
&= \sum_{\ell=N}^{1} \underline{f}[\ell]\,\underline{\omega}^{\ell} \quad \text{(since } \underline{\omega}^{-N} = 1\text{)}.
\end{aligned}$$

But using $\underline{f}[N] = \underline{f}[0]$ and $\underline{\omega}^N = \underline{\omega}^0 = \underline{1}$ we can clearly write

$$\begin{aligned}
\sum_{\ell=N}^{1} \underline{f}[\ell]\underline{\omega}^{\ell} &= \sum_{\ell=0}^{N-1} \underline{f}[\ell]\underline{\omega}^{\ell} \\
&= \left(\sum_{\ell=0}^{N-1} \underline{f}[\ell]\underline{\omega}^{-\ell}\right)^- = (\underline{\mathcal{F}}\,\underline{f})^-.
\end{aligned}$$

We have shown $\underline{\mathcal{F}}\,\underline{f}^- = (\underline{\mathcal{F}}\,\underline{f})^-$. Cool. A little drawn out, but cool.

This then tells us that

$$\underline{\mathcal{F}}\,\underline{\omega}^{-k} = (\underline{\mathcal{F}}\,\underline{\omega}^{k})^{-} = (N\underline{\delta}_k)^{-} = N\underline{\delta}_{-k}\,.$$

In turn, from here we get a second duality result. Start with

$$\underline{\mathcal{F}}\,\underline{f} = \sum_{k=0}^{N-1} \underline{f}[k]\underline{\omega}^{-k}$$

and apply $\underline{\mathcal{F}}$ *again*. This produces

$$\underline{\mathcal{F}}\,\underline{\mathcal{F}}\,\underline{f} = \sum_{k=0}^{N-1} \underline{f}[k]\underline{\mathcal{F}}\,\underline{\omega}^{-k} = N\sum_{k=0} \underline{f}[k]\underline{\delta}_{-k} = N\underline{f}^{-}\,.$$

To give the two results their own display:

- Duality relations for the DFT are

$$\underline{\mathcal{F}}\,\underline{f}^{-} = (\underline{\mathcal{F}}\,\underline{f})^{-} \quad \text{and} \quad \underline{\mathcal{F}}\,\underline{\mathcal{F}}\,\underline{f} = N\underline{f}^{-}\,.$$

We'll do more on reversed signals, evenness and oddness, etc., but it's interesting to note that the duality results also lead to the definition of the inverse DFT. We take our cue from the duality results in the continuous case that say

$$\mathcal{F}^{-1}f = \mathcal{F}f^{-} = (\mathcal{F}f)^{-}\,.$$

This equation tells us how we might try defining the inverse DFT. Because I know what's going to happen, I'll put a factor of $1/N$ where it belongs and claim that

$$\underline{\mathcal{F}}^{-1}\underline{f} = \frac{1}{N}\underline{\mathcal{F}}\,\underline{f}^{-}.$$

So equivalently

$$\underline{\mathcal{F}}^{-1}\underline{f} = \frac{1}{N}(\underline{\mathcal{F}}\,\underline{f})^{-} \quad \text{and also} \quad \underline{\mathcal{F}}^{-1}\underline{f} = \frac{1}{N}\sum_{n=0}^{N-1} \underline{f}[n]\underline{\omega}^{n}\,,$$

the second formula being what we know from our earlier work. Let's see why $\underline{\mathcal{F}}^{-1}\underline{f} = (1/N)\underline{\mathcal{F}}\,\underline{f}^{-}$ really does give us an inverse of $\underline{\mathcal{F}}$.

It's clear that $\underline{\mathcal{F}}^{-1}$ as defined this way is linear. We also need to know

$$\begin{aligned}
\underline{\mathcal{F}}^{-1}\underline{\omega}^{-k} &= \frac{1}{N}\underline{\mathcal{F}}(\underline{\omega}^{-k})^{-} \quad (\text{definition of } \underline{\mathcal{F}}^{-1}) \\
&= \frac{1}{N}\underline{\mathcal{F}}\,\underline{\omega}^{k} \quad (\text{using } (\underline{\omega}^{-k})^{-} = \underline{\omega}^{k}) \\
&= \frac{1}{N}N\underline{\delta}_k = \underline{\delta}_k\,.
\end{aligned}$$

With this,

$$\begin{aligned}
\underline{\mathcal{F}}^{-1}\underline{\mathcal{F}}\,\underline{f} &= \underline{\mathcal{F}}^{-1}\left(\sum_{k=0}^{N-1} \underline{f}[k]\underline{\omega}^{-k}\right) \\
&= \sum_{k=0}^{N-1} \underline{f}[k]\underline{\mathcal{F}}^{-1}\underline{\omega}^{-k} = \sum_{k=0}^{N-1} \underline{f}[k]\underline{\delta}_k = \underline{f}\,.
\end{aligned}$$

A similar calculation shows that

$$\underline{\mathcal{F}}\,\underline{\mathcal{F}}^{-1}\underline{f} = \underline{f}\,.$$

Good show. We have shown that $(1/N)\underline{\mathcal{F}}\,\underline{f}^-$ really, really does give an inverse to $\underline{\mathcal{F}}$. A longer route to the inverse DFT, but it took us past the signposts marking the way that said, "The discrete corresponds to the continuous. Use what you know."

Let's also note that

$$\underline{\mathcal{F}}^{-1}\underline{\delta}_k = \frac{1}{N}\underline{\omega}^k\,.$$

Symmetries. Remember those evenness, oddness results for the Fourier transform? Same sort of thing for the DFT. A discrete signal $\underline{f}$ is even if $\underline{f}^- = \underline{f}$, i.e., if $\underline{f}[-n] = \underline{f}[n]$. It's odd if $\underline{f}^- = -\underline{f}$, i.e., $\underline{f}[-n] = -\underline{f}[n]$. The signal and its DFT have the same symmetry. If, for example, $\underline{f}$ is even, then using duality,

$$(\underline{\mathcal{F}}\,\underline{f})^- = \underline{\mathcal{F}}\,\underline{f}^- = \underline{\mathcal{F}}\,\underline{f}.$$

The same sort of argument shows that if $\underline{f}$ is odd, then so is $\underline{\mathcal{F}}\,\underline{f}$.

Finally, if $\underline{f}$ is a *real* signal, then DFT has the conjugate symmetry that we saw in the continuous setting, namely,

$$\underline{\mathcal{F}}\,\underline{f}[-n] = \overline{\underline{\mathcal{F}}\,\underline{f}[n]}.$$

Let's go through this, if only for some practice in duality, sums, and summing over any consecutive N indices to compute the DFT. Onward!

$$\begin{aligned}
(\underline{\mathcal{F}}\,\underline{f})^- &= \underline{\mathcal{F}}\,\underline{f}^- \quad \text{(duality)}\\
&= \sum_{k=0}^{N-1} \underline{f}^-[k]\underline{\omega}^{-k}\\
&= \sum_{k=0}^{N-1} \underline{f}[-k]\underline{\omega}^{-k}\\
&= \sum_{\ell=0}^{-(N-1)} \underline{f}[\ell]\underline{\omega}^{\ell} \quad (\text{letting } \ell = -k)\\
&= \overline{\left(\sum_{\ell=0}^{-(N-1)} \underline{f}[\ell]\underline{\omega}^{-\ell}\right)} \quad (\text{using } \underline{\omega}^{-\ell} = \overline{\underline{\omega}^{\ell}}, \text{ and } \overline{\underline{f}} = \underline{f} \text{ because } \underline{f} \text{ is real})\\
&= \overline{\underline{\mathcal{F}}\,\underline{f}} \quad (\text{because we can sum over any block of } N \text{ indices}).
\end{aligned}$$

We have a few additional properties and formulas for the DFT that are analogous to those in the continuous case. This will mostly be a listing of results, often without much discussion. Use it as a reference.

7.8.2. Parseval's identity. There's a version of Parseval's identity for the DFT, featuring an extra factor of N that one has to keep track of:

$$\underline{\mathcal{F}}\underline{f} \cdot \underline{\mathcal{F}}\underline{g} = N(\underline{f} \cdot \underline{g}) .$$

That's the dot product. The derivation goes like this, using properties of the complex inner product and the orthogonality of the discrete complex exponentials:

$$\begin{aligned}\underline{\mathcal{F}}\underline{f} \cdot \underline{\mathcal{F}}\underline{g} &= \left(\sum_{k=0}^{N-1} \underline{f}[k]\underline{\omega}^{-k}\right) \cdot \left(\sum_{\ell=0}^{N-1} \underline{g}[\ell]\underline{\omega}^{-\ell}\right)\\ &= \sum_{k=0}^{N-1}\sum_{\ell=0}^{N-1} \underline{f}[k]\,\overline{\underline{g}[\ell]}(\underline{\omega}^{-k} \cdot \underline{\omega}^{-\ell}) = N\sum_{k=0}^{N-1} \underline{f}[k]\overline{\underline{g}[k]} = N(\underline{f} \cdot \underline{g}) .\end{aligned}$$

If $\underline{f} = \underline{g}$, then the identity becomes

$$||\underline{\mathcal{F}}\underline{f}||^2 = N||\underline{f}||^2 .$$

Parseval's identity is still another way of saying that $\underline{\mathcal{F}}$ is almost unitary as a matrix. A unitary matrix A has the property that

$$A\underline{f} \cdot A\underline{g} = \underline{f} \cdot \underline{g} ,$$

so A *preserves* the inner product. $\underline{\mathcal{F}}$ almost does this.

7.8.3. Shifts and the shift theorem. In formulating the shift theorem for the DFT it's helpful to introduce the delay operator for a discrete signal $\underline{f}$. For an integer p we define the delayed signal $\tau_p\underline{f}$ by

$$\tau_p\underline{f}[n] = \underline{f}[n-p] .$$

The version of the shift theorem for the DFT looks just like its continuous cousin:

$$\underline{\mathcal{F}}(\tau_p\underline{f}) = \underline{\omega}^{-p}\underline{\mathcal{F}}\underline{f} .$$

The verification of this is left as a problem. Note that we need $\underline{f}$ to be defined on all integers for shifting and the shift theorem to make sense.

7.8.4. The modulation theorem. Modulation also works as in the continuous case. The modulation of a discrete signal $\underline{f}$ is, by definition, a signal of the form

$$(\underline{\omega}^n\underline{f})[m] = \underline{\omega}^n[m]\underline{f}[m] .$$

We can find $\underline{\mathcal{F}}(\underline{\omega}^n\,\underline{f})$ directly from the definition:

$$\underline{\mathcal{F}}(\underline{\omega}^n\underline{f}) = \sum_{k=0}^{N-1} \underline{\omega}^n[k]\underline{f}[k]\underline{\omega}^{-k} ,$$

and so the mth component is

$$\underline{\mathcal{F}}(\underline{\omega}^n\underline{f})[m] = \sum_{k=0}^{N-1} \underline{f}[k]\omega^{kn}\underline{\omega}^{-km} = \sum_{k=0}^{N-1} \underline{f}[k]\omega^{-k(m-n)} .$$

But if we shift $\underline{\mathcal{F}}\,\underline{f}$ by n, we obtain

$$\tau_n(\underline{\mathcal{F}}\,\underline{f}) = \tau_n\left(\sum_{k=0}^{N-1} \underline{f}[k]\underline{\omega}^{-k}\right) = \sum_{k=0}^{N-1} \underline{f}[k]\,\tau_n\underline{\omega}^{-k},$$

and the mth component of the right-hand side is

$$\left(\sum_{k=0}^{N-1} \underline{f}[k]\,\tau_n\underline{\omega}^{-k}\right)[m] = \sum_{k=0}^{N-1} \underline{f}[k]\,(\tau_n\underline{\omega}^{-k})[m] = \sum_{k=0}^{N-1} \underline{f}[k]\omega^{-k(m-n)},$$

just as we had above. We conclude that

$$\underline{\mathcal{F}}(\underline{\omega}^n \underline{f}) = \tau_n(\underline{\mathcal{F}}\,\underline{f}).$$

This is the modulation theorem.

7.8.5. Convolution. Convolution combined with the DFT is the basis of digital filtering. This may be the greatest employment of computation in the known universe. It's a topic in the next chapter.

In considering how convolution is to be defined, let me ask again the question we asked in the continuous case: How can we use one signal to modify another to result in scaling the frequency components? In the continuous case we discovered convolution in the time domain precisely by looking at multiplication in the frequency domain, and we'll do the same thing now.

Given $\underline{F}$ and $\underline{G}$, we can consider their componentwise product $\underline{F}\,\underline{G}$. The question is:

- If $\underline{F} = \underline{\mathcal{F}}\,\underline{f}$ and $\underline{G} = \underline{\mathcal{F}}\,\underline{g}$, is there an $\underline{h}$ so that $\underline{F}\,\underline{G} = \underline{\mathcal{F}}\,\underline{h}$?

The technique to analyze this is to interchange the order of summation, much as we often interchanged the order of integration (e.g., $dx\,dy$ instead of $dy\,dx$) in deriving formulas for Fourier integrals. We did exactly that in the process of coming up with convolution in the continuous setting.

For the DFT and the question we have posed,

$$\begin{aligned}
(\underline{\mathcal{F}}^{-1}(\underline{F}\,\underline{G}))[m] &= \frac{1}{N}\sum_{n=0}^{N-1} \underline{F}[n]\,\underline{G}[n]\,\omega^{mn} \\
&= \frac{1}{N}\sum_{n=0}^{N-1}\left[\sum_{k=0}^{N-1} \underline{f}[k]\omega^{-kn}\right]\left[\sum_{\ell=0}^{N-1} \underline{g}[\ell]\omega^{-\ell n}\right]\omega^{mn} \\
&\qquad \text{(we collect the powers of } \omega\text{, all of which have an } n\text{)} \\
&= \sum_{k=0}^{N-1} \underline{f}[k] \sum_{\ell=0}^{N-1} \underline{g}[\ell]\left[\frac{1}{N}\sum_{n=0}^{N-1} \omega^{-kn}\omega^{-\ell n}\omega^{mn}\right] \\
&= \sum_{k=0}^{N-1} \underline{f}[k] \sum_{\ell=0}^{N-1} \underline{g}[\ell]\left[\frac{1}{N}\sum_{n=0}^{N-1} \omega^{n(m-k-\ell)}\right].
\end{aligned}$$

Now look at the final sum in brackets. As in earlier calculations, this is a finite geometric series whose sum is N when $m-k-\ell \equiv 0 \bmod N$ and is zero if $m-k-\ell \not\equiv 0 \bmod N$. This takes the periodicity of the inputs and outputs into account, and we really must work modulo N in order to do that because $m-k-\ell$ could be less than 0 or bigger than $N-1$. Thus what survives is when $\ell \equiv m-k \mod N$, and the final line above becomes

$$\sum_{k=0}^{N-1} \underline{f}[k]\, \underline{g}[m-k]\,.$$

Therefore, if

$$\underline{h}[m] = \sum_{k=0}^{N-1} \underline{f}[k]\, \underline{g}[m-k]\,, \quad m = 0, \dots, N-1\,,$$

then $\underline{\mathcal{F}}\,\underline{h} = \underline{F}\,\underline{G}$. Again notice that the periodicity of $\underline{g}$ has to be used in defining $\underline{h}$ because the index on $\underline{g}$ will be negative for $m<k$. Also observe that $\underline{h}$ is periodic.

To summarize:

- **Convolution of discrete signals.** Let $\underline{f}$ and $\underline{g}$ be periodic discrete signals. Define the *convolution* of $\underline{f}$ and $\underline{g}$ to be the periodic discrete signal

$$(\underline{f} * \underline{g})[m] = \sum_{k=0}^{N-1} \underline{f}[k]\, \underline{g}[m-k]\,.$$

Then

$$\underline{\mathcal{F}}(\underline{f} * \underline{g}) = (\underline{\mathcal{F}}\,\underline{f})(\underline{\mathcal{F}}\,\underline{g})\,.$$

The product on the right-hand side is the componentwise product of the DFTs.

You can verify (if you have the patience) the various algebraic properties of convolution, namely linearity, commutativity, and associativity.

- It's also true that the DFT turns a product into a convolution:

$$\underline{\mathcal{F}}(\underline{f}\underline{g}) = \frac{1}{N}(\underline{\mathcal{F}}\,\underline{f} * \underline{\mathcal{F}}\,\underline{g})\,.$$

(An extra factor of $1/N$. Agony.)

This equation can be derived from the first convolution property, using duality. Let $\underline{F} = \underline{\mathcal{F}}^{-1}\underline{f}$ and $\underline{G} = \underline{\mathcal{F}}^{-1}\underline{g}$. Then $\underline{f} = \underline{\mathcal{F}}\,\underline{F}$, $\underline{g} = \underline{\mathcal{F}}\,\underline{G}$, and

$$\underline{f}\underline{g} = (\underline{\mathcal{F}}\,\underline{F})(\underline{\mathcal{F}}\,\underline{G}) = \underline{\mathcal{F}}(\underline{F} * \underline{G})\,.$$

Hence

$$\begin{aligned}\underline{\mathcal{F}}(\underline{f}\underline{g}) &= \underline{\mathcal{F}}\,\underline{\mathcal{F}}(\underline{F} * \underline{G}) = N(\underline{F} * \underline{G})^{-} \\ &= N\Big(\frac{1}{N}(\underline{\mathcal{F}}\,\underline{f})^{-} * \frac{1}{N}(\underline{\mathcal{F}}\,\underline{g})^{-}\Big)^{-} = \frac{1}{N}(\underline{\mathcal{F}}\,\underline{f} * \underline{\mathcal{F}}\,\underline{g})\,.\end{aligned}$$

So you know where the N's go, here are the corresponding results for the inverse DFT. You can derive them via duality:

$$\begin{aligned}\underline{\mathcal{F}}^{-1}(\underline{f} * \underline{g}) &= N(\underline{\mathcal{F}}^{-1}\underline{f}\,\underline{\mathcal{F}}^{-1}\underline{g}), \\ \underline{\mathcal{F}}^{-1}(\underline{f}\,\underline{g}) &= \underline{\mathcal{F}}^{-1}\underline{f} * \underline{\mathcal{F}}^{-1}\underline{g}.\end{aligned}$$

Shifts and convolution. We note one general property combining convolution with delays, namely that the discrete shift works with discrete convolution just as it does in the continuous case:

$$\begin{aligned}((\tau_p \underline{f}) * \underline{g})[n] &= \sum_{k=0}^{N-1} \tau_p \underline{f}[n-k]\underline{g}[k] \\ &= \sum_{k=0}^{N-1} \underline{f}[n-k-p]\underline{g}[k] = (\underline{f} * \underline{g})[n-p] = \tau_p(\underline{f} * \underline{g})[n] .\end{aligned}$$

Thus, since convolution is commutative,

$$(\tau_p \underline{f}) * \underline{g} = \tau_p(\underline{f} * \underline{g}) = \underline{f} * (\tau_p \underline{g}) .$$

7.8.6. More properties of $\underline{\delta}$.

Two of the most useful properties of the continuous δ, if we can get away with the term "continuous" in connection with δ, are what it does when multiplied or convolved with a smooth function. For the discrete $\underline{\delta}$ we have similar results. For multiplication,

$$\underline{f}\underline{\delta}_0 = (\underline{f}[0] \cdot 1,\ \underline{f}[1] \cdot 0,\ \ldots,\ \underline{f}[N-1] \cdot 0) = (\underline{f}[0], 0, \ldots, 0) = \underline{f}[0]\underline{\delta}_0 .$$

For convolution,

$$(\underline{f} * \underline{\delta}_0)[m] = \sum_{n=0}^{N-1} \underline{f}[m-n]\underline{\delta}_0[n] = \underline{f}[m] ;$$

i.e., $\underline{f} * \underline{\delta}_0 = \underline{f}$.

There are analogous properties for the shifted discrete $\underline{\delta}$. For multiplication,

$$\underline{f}\underline{\delta}_k = \underline{f}[k]\underline{\delta}_k .$$

For convolution,

$$\underline{f} * \underline{\delta}_k = \underline{f} * \tau_k \underline{\delta}_0 = \tau_k(\underline{f} * \underline{\delta}_0) = \tau_k \underline{f} ,$$

or in components

$$(\underline{f} * \underline{\delta}_k)[m] = \underline{f}[m-k] ,$$

again in agreement with what we would expect from the continuous case.

Observe, incidentally, that

$$\underline{\delta}_k = \tau_k \underline{\delta}_0, \quad \underline{\delta}_k[m] = \underline{\delta}_0[m-k] ,$$

so a shifted discrete $\underline{\delta}$ really does appear as $\underline{\delta}$ delayed by k.

There's more. Note that

$$\underline{\delta}_p \underline{\delta}_q = \begin{cases} \underline{\delta}_p, & p = q , \\ \underline{0}, & p \neq q , \end{cases}$$

and that

$$\underline{\delta}_p * \underline{\delta}_q = \underline{\delta}_{p+q} .$$

The former operation, multiplying δ's, is against the law in the continuous case, but not in the discrete case, where we're just multiplying honest discrete signals. A cute observation making use of the first result and the convolution theorem is that

$$\underline{\omega}^p * \underline{\omega}^q = \begin{cases} N\underline{\omega}^p, & p = q\,, \\ \underline{0}, & p \neq q\,. \end{cases}$$

Of course you can also see this directly, but it might not occur to you to look.

7.9. The Discrete Rect and Its DFT

Unlike for the Fourier transform, where computing a storehouse of particular transforms is necessary for an understanding and for applications, we've only explicitly computed and needed the DFT for $\underline{\delta}$'s and for discrete complex exponentials. We want to add to that short list the DFT of a discrete rectangle function, a discrete version of the very first example that we computed for the continuous Fourier transform. There are times when such an expression is useful, say in the design of digital filters as we'll discuss in the next chapter, or in a discrete version of the Nyquist-Shannon sampling theorem as we'll discuss in the next section. The formulas work out nicely, by and large, and further advance the notion that the discrete case can be made to look like the continuous case.

In this section we'll index discrete signals from $-\frac{N}{2}+1$ to $\frac{N}{2}$, so in particular we're assuming that N is even. The discrete rectangle is a sum of shifted $\underline{\delta}$'s going from $-p/2$ to $p/2$, $0 < p < N/2$, so we also assume that p is even. We let

$$\underline{\Pi}_{p,\alpha} = \alpha(\underline{\delta}_{p/2} + \underline{\delta}_{-p/2}) + \sum_{k=-p/2+1}^{p/2-1} \underline{\delta}_k\,,$$

where α is any real number. Why the extra parameter α in setting the value at the endpoints $\pm p/2$? In the continuous case one also encounters different normalizations of the rect function at the points of discontinuity, but it hardly makes a difference in any formulas and calculations; most of the time there's an integral involved, and changing the value of a function at a few points has no effect on the value of the integral. Not so in the discrete case, where sums instead of integrals are the operations that one encounters. With an eye toward flexibility in applications, we're allowing an α.[8]

Just as $\underline{\Pi}_{p,\alpha}$ comes in two parts, so, too, does its DFT. The first part is easy:

$$\underline{\mathcal{F}}(\alpha(\underline{\delta}_{p/2} + \underline{\delta}_{-p/2})) = \alpha(\underline{\omega}^{-p/2} + \underline{\omega}^{p/2}) = 2\alpha\,\mathrm{Re}\{\underline{\omega}^{p/2}\}\,;$$

hence

$$\underline{\mathcal{F}}(\alpha(\underline{\delta}_{p/2} + \underline{\delta}_{-p/2}))[m] = 2\alpha\cos(\pi p m/N).$$

The second part takes more work:

$$\underline{\mathcal{F}}\left(\sum_{k=-p/2+1}^{p/2-1} \underline{\delta}_k\right)[m] = \sum_{k=-p/2+1}^{p/2-1} \underline{\omega}^{-k}[m] = \sum_{k=-p/2+1}^{p/2-1} \omega^{-km}\,,$$

[8]Here's a hint: $\alpha = 1/2$ is often the best choice — but not always. This opens up wounds from the endpoint battles in the continuous case, I know.

and the right-hand side is a geometric sum. Before plunging in, we have, directly,

$$\underline{\mathcal{F}\,\Pi}_{p,\alpha}[0] = 2\alpha + p - 1\,.$$

Now, if you look at Appendix C, you will see a formula (in the problems) that helps. When $m \neq 0$,

$$\sum_{k=-p/2+1}^{p/2-1} \omega^{-km} = \frac{\omega^{-m(-\frac{p}{2}+1)} - \omega^{-m\frac{p}{2}}}{1-\omega^{-m}}\,.$$

There's a little hocus pocus that goes into modifying this expression, not unfamiliar to us but I'll provide the details because I want to write it in a particular form. I'm aiming to bring in a ratio of sines once we substitute $\omega = e^{2\pi i/N}$.

For the numerator,

$$\begin{aligned}\omega^{-m(-\frac{p}{2}+1)} - \omega^{-m\frac{p}{2}} &= \omega^{-\frac{m}{2}}(\omega^{\frac{m}{2}(p-1)} - \omega^{-\frac{m}{2}(p-1)})\\ &= \omega^{-\frac{m}{2}}(e^{\pi i m(p-1)} - e^{-\pi i m(p-1)})\\ &= \omega^{-\frac{m}{2}}2i\sin(\pi m(p-1)/N)\,.\end{aligned}$$

Same idea for the denominator:

$$1-\omega^{-m} = \omega^{-m/2}(\omega^{m/2} - \omega^{-m/2}) = \omega^{-m/2}2i\sin(\pi m/N)\,.$$

Putting it all together we get

$$\underline{\mathcal{F}\,\Pi}_{p,\alpha}[m] = \begin{cases} 2\alpha + p - 1, & m = 0,\\ 2\alpha\cos(\pi pm/N) + \dfrac{\sin(\pi m(p-1)/N)}{\sin(\pi m/N)}, & m \neq 0. \end{cases}$$

One additional point before we continue with special cases. As with all of our discrete signals we regard $\underline{\Pi}_{p,\alpha}$ as periodic of period N, so the pattern of $\underline{\delta}$'s repeats. Since p is even and $p-1$ is odd, we observe that $\underline{\mathcal{F}\,\Pi}_{p,\alpha}$ is likewise periodic of period N, which it had better be. Because $p-1$ is odd, both the numerator and the denominator of the second term change sign if m is replaced by $m+N$, while, because p is even, the cosine term is unchanged. Thus the equations above should actually read $m \equiv 0 \mod N$ and $m \not\equiv 0 \mod N$. The DFT is also even, which it had better be.

The most common choices of α are $\alpha = 1$ and $\alpha = 1/2$. (If $\alpha = 0$, we just have a thinner version of a $\underline{\Pi}$ with $\alpha = 1$.) For $\alpha = 1$ we use the addition formula for sines to write

$$2\cos(\pi pm/N)\sin(\pi m/N) + \sin(\pi(p-1)m/N) = \sin(\pi(p+1)m/N)\,.$$

Thus

$$\underline{\mathcal{F}\,\Pi}_{p,1}[m] = \begin{cases} p+1, & m \equiv 0 \mod N,\\ \dfrac{\sin(\pi(p+1)m/N)}{\sin(\pi m/N)}, & m \not\equiv 0 \mod N\,. \end{cases}$$

For $\alpha = 1/2$,

$$\cos(\pi pm/N)\sin(\pi m/N) + \sin(\pi(p-1)m/N) = \cos(\pi m/N)\sin(\pi pm/N)\,,$$

and

$$\underline{\mathcal{F}}\,\underline{\Pi}_{p,1/2}[m] = \begin{cases} p, & m \equiv 0 \mod N, \\ \dfrac{\cos(\pi m/N)\sin(\pi pm/N)}{\sin(\pi m/N)}, & m \not\equiv 0 \mod N\,. \end{cases}$$

We'll see this in the next section.

7.10. Discrete Sampling and Interpolation

For the fun of it, let's derive a discrete version of the sampling theorem.[9] Bolstered by the faith that things should work as in the continuous case, the main ingredients should be

(1) a class of bandlimited discrete signals,
(2) a discrete Ш and its DFT,
(3) a discrete Π and its DFT.

We'll work with periodic signals of length N, and we'll index from $-\frac{N}{2}+1$ to $\frac{N}{2}$.

Bandlimited discrete signals. We say that a discrete signal $\underline{f}$ is bandlimited to a set $\mathcal{I}$ of consecutive indices if $\underline{\mathcal{F}}\,\underline{f}[m] = 0$ for $m \notin \mathcal{I}$. In words, $\underline{\mathcal{F}}\,\underline{f}$ vanishes outside $\mathcal{I}$; there *may* be places $m \in \mathcal{I}$ where $\underline{\mathcal{F}}\,\underline{f}[m] = 0$ as well (we're not ruling that out), but for sure the DFT is always 0 off $\mathcal{I}$.

That's a fairly general definition, but for our purposes, to make things a little simpler and to make them look like the continuous case, let's suppose that $\underline{f}$ has $\underline{\mathcal{F}}\,\underline{f}[m] = 0$ for $|m| \geq \frac{p}{2}$. We thus assume that p is even and, for reasons having to do with the discrete Ш, *we also assume that p divides N*. These are definitely restrictions on the argument to follow, but really not severe ones.

Discrete Ш. Remember that the usual $Ш_p$ is defined as a sum of δ's spaced p apart,

$$Ш_p(x) = \sum_{k=-\infty}^{\infty} \delta(x - kp),$$

and is a periodic distribution of period p. To define a discrete Ш we first of all have to recognize that, like all our discrete signals, it's periodic of period N. So what we want is an evenly spaced set of $\underline{\delta}$'s *within* one period. We can do this if the spacing *p is a divisor of N*, setting

$$\underline{Ш}_p = \sum_{k=0}^{\frac{N}{p}-1} \underline{\delta}_{kp}\,.$$

From this comes the divisibility assumption mentioned above.

At an index m,

$$\underline{Ш}_p[m] = \begin{cases} 1, & m \equiv 0, p, 2p, \dots, (\frac{N}{p}-1)p\,, \quad \mod N, \\ 0, & \text{otherwise.} \end{cases}$$

[9] Thanks to Aditya Siripuram and William Wu for their contributions to this subject, beyond this section.

The $\underline{\delta}$'s are spaced p-apart, and there are N/p of them in one N-period of $\underline{\text{III}}_p$. Phrased differently, we put a $\underline{\delta}$ at each pth sample. Here's a plot of $\underline{\text{III}}_4$ when $N = 20$. Only the $\underline{\delta}$'s at the positive integers are shown, but remember that the pattern repeats periodically.

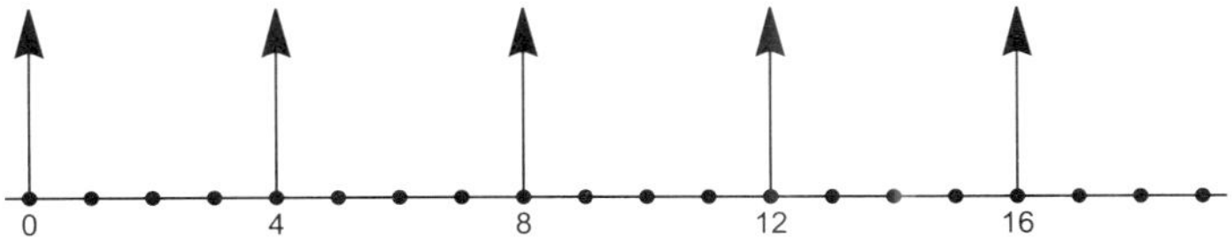

Note that if N is a prime number (the first time a number-theoretic consideration has come up!), then in $[0 : N-1]$ there's only $\underline{\text{III}}_1$, the sum of N shifted $\underline{\delta}$'s, and $\underline{\text{III}}_N$, consisting of the single $\underline{\delta}_0$. Also note that $\underline{\text{III}}_p$ is even.

Just as in the continuous case, the two fundamental properties of III_p are:

- Periodizing.
  For a discrete signal $\underline{f}$ of length N, convolving with $\underline{\text{III}}_p$,

$$(\underline{f} * \underline{\text{III}}_p)[m] = \sum_{k=0}^{\frac{N}{p}-1} (\underline{f} * \underline{\delta}_{kp})[m] = \sum_{k=0}^{\frac{N}{p}-1} \underline{f}[m - kp],$$

  produces a signal that is periodic of period p (as well as periodic of period N).

- Sampling.
  For a discrete signal $\underline{f}$ of length N, multiplying by $\underline{\text{III}}_p$,

$$(\underline{f} \cdot \underline{\text{III}}_p)[m] = \sum_{k=0}^{\frac{N}{p}-1} (\underline{f} \cdot \underline{\delta}_{kp})[m] = \sum_{k=0}^{\frac{N}{p}-1} \underline{f}[kp]\underline{\delta}[m - kp],$$

  samples $\underline{f}$ at the points $0, p, 2p, \dots, (\frac{N}{p} - 1)p$.

What is $\underline{\mathcal{F}\,\text{III}}_p$? Watch how this works out. Recall that $\underline{\mathcal{F}\,\delta}_j = \underline{\omega}^{-j}$, so

$$\underline{\mathcal{F}\,\text{III}}_p = \sum_{k=0}^{\frac{N}{p}-1} \underline{\omega}^{-kp}\,.$$

Then a calculation of the geometric sum gives

$$\begin{aligned}
\underline{\mathcal{F}\,\text{III}}_p[m] &= \sum_{k=0}^{\frac{N}{p}-1} \omega^{-kpm} = \sum_{k=0}^{\frac{N}{p}-1} (\omega^{-pm})^k \\
&= \begin{cases} \frac{N}{p}\,, & pm \equiv 0 \mod N, \\ 0\,, & pm \not\equiv 0 \mod N, \end{cases} \\
&= \begin{cases} \frac{N}{p}\,, & m \equiv 0 \mod \frac{N}{p}, \\ 0\,, & m \not\equiv 0 \mod \frac{N}{p}\,. \end{cases}
\end{aligned}$$

Up to the factor N/p this is another $\underline{\text{III}}$, with spacing N/p. Compactly, and beautifully,

$$\underline{\mathcal{F}}\,\underline{\text{III}}_p = \frac{N}{p}\underline{\text{III}}_{N/p}\,.$$

By duality,

$$\underline{\mathcal{F}}^{-1}\,\underline{\text{III}}_p = \frac{1}{N}(\underline{\mathcal{F}}\,\underline{\text{III}}_p)^- = \frac{1}{p}\underline{\text{III}}_{N/p}.$$

Discrete rectangle function. In the last section I introduced the discrete rectangle with a parameter α, and the question is which α is suited for our problem. The answer is $\alpha = 1/2$. That's not obvious till we actually get to the derivation of the sampling theorem, so you'll have to wait for a few paragraphs. Since we'll only be working with $\alpha = 1/2$, let's simplify the notation for now and just write $\underline{\Pi}_p$ for $\underline{\Pi}_{p,1/2}$. It heightens the drama.

Playing the role of a discrete sinc, let[10]

$$\text{dinc}_p[m] = \begin{cases} 1\,, & m \equiv 0 \mod N\,, \\ \dfrac{1}{p}\dfrac{\cos(\pi m/N)\sin(\pi p m/N)}{\sin(\pi m/N)}\,, & m \not\equiv 0 \mod N\,. \end{cases}$$

For the DFT we thus have

$$\underline{\mathcal{F}}\,\underline{\Pi}_p[m] = p\,\text{dinc}_p[m]\,.$$

This is as close as I could come to making things look like the continuous case, where

$$\mathcal{F}\Pi_p(s) = p\,\text{sinc}\,ps.$$

For the inverse transform, by duality,

$$\underline{\mathcal{F}}^{-1}\underline{\Pi}_p[m] = \frac{1}{N}(\underline{\mathcal{F}}\,\underline{\Pi}_p)^-[m] = \frac{p}{N}\text{dinc}_p[-m] = \frac{p}{N}\text{dinc}_p[m],$$

and

$$\underline{\mathcal{F}}\text{dinc}_p = \frac{N}{p}\underline{\Pi}_p\,.$$

Here's a plot of dinc_p for $p = 6$ and $N = 30$.

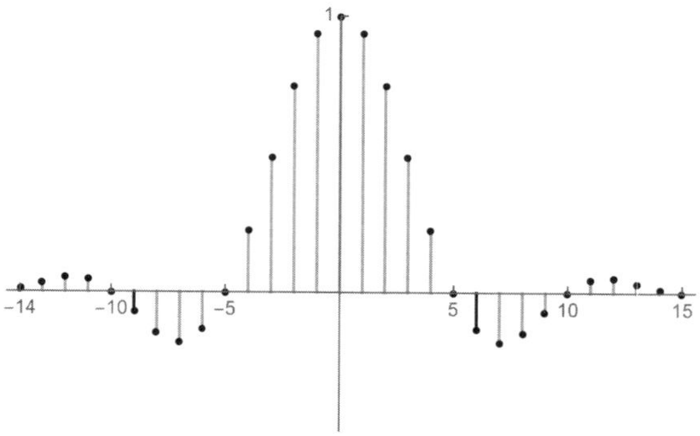

[10]I made up the name "dinc," for discrete sinc. It hasn't caught on. You can help.

A discrete sampling theorem. We're all set. Our assumptions are that $\underline{\mathcal{F}}\,\underline{f}[m] = 0$ for $|m| \geq \frac{p}{2}$ and that p divides N.

Just as in the continuous setting, forming the convolution $\underline{\text{III}}_p * \underline{\mathcal{F}}\,\underline{f}$ shifts the spectrum off itself, and then cutting off by $\underline{\Pi}_p$ recovers $\underline{\mathcal{F}}\,\underline{f}$. We have the fundamental equation

$$\underline{\mathcal{F}}\,\underline{f} = \underline{\Pi}_p(\underline{\text{III}}_p * \underline{\mathcal{F}}\,\underline{f}).$$

Now take the inverse DFT:

$$\begin{aligned}
\underline{f} &= \underline{\mathcal{F}}^{-1}(\underline{\Pi}_p(\underline{\text{III}}_p * \underline{\mathcal{F}}\,\underline{f})) \\
&= (\underline{\mathcal{F}}^{-1}\underline{\Pi}_p) * (\underline{\mathcal{F}}^{-1}(\underline{\text{III}}_p * \underline{\mathcal{F}}\,\underline{f})) \\
&= \frac{p}{N}\text{dinc}_p * N\Big(\frac{1}{p}\underline{\text{III}}_{N/p} \cdot \underline{f}\Big)
\end{aligned}$$

(the factor N comes in because of the convolution theorem for $\underline{\mathcal{F}}^{-1}$, and with $\alpha = 1/2$ we get factors of p cancelling)

$$= \text{dinc}_p * \sum_{k=0}^{p-1} \underline{f}[Nk/p]\underline{\delta}_{Nk/p}\,.$$

Thus at an index m,

$$\underline{f}[m] = \sum_{k=0}^{p-1} \underline{f}\left[\frac{Nk}{p}\right] \text{dinc}_p\left[m - \frac{Nk}{p}\right].$$

Perfect.

As a check, you can verify using the definition of dinc_p that the formula returns the sample values:

$$\sum_{k=0}^{p-1} \underline{f}\left[\frac{Nk}{p}\right] \text{dinc}_p\left[\frac{N\ell}{p} - \frac{Nk}{p}\right] = \underline{f}\left[\frac{N\ell}{p}\right].$$

The next topic along these lines is aliasing, often discussed in the context of upsampling and downsampling. There are some problems on the latter, and a little googling will suggest applications, but a more thorough treatment is for a dedicated course on digital signal processing.

7.11. The FFT Algorithm

I'm sure you've been waiting for this. It's time to consider the practical problem of how the DFT is computed. In this section we'll go through the famous Cooley-Tukey Fast Fourier Transform algorithm.

You can't beat the matrix form

$$(\underline{\mathcal{F}})_{mn} = \omega^{-mn}, \quad m, n = 0, 1, \ldots, N-1.$$

as a compact way of writing the DFT. It contains all you need to know, and finding the DFT is just multiplication by an $N \times N$ matrix. But you can beat it into submission. The FFT is an algorithm for computing the DFT with fewer than the N^2 multiplications that would seem to be required to find $\underline{F} = \underline{\mathcal{F}}_N \underline{f}$ by multiplying $\underline{f}$ by the $N \times N$ matrix. Here's how it works.

Reducing calculations: Merge and sort. To set the stage for a discussion of the FFT algorithm, I thought it first might be useful to see an example of a somewhat simpler but related idea, a way of reducing the total number of steps in a multistep calculation by a clever arranging and rearranging of the individual steps.

Consider the classic (and extremely important) problem of sorting N numbers from smallest to largest. Say the numbers are

5	2	3	1	6	4	8	7

From this list we want to create a new list with the numbers ordered from 1 to 8. The direct assault on this problem is to search the entire list for the smallest number, remove that number from the list and put it in the first slot of the new list, then search the remaining original list for the smallest number, remove that number and put it in the second slot of the new list, and so on:

Zeroth step	5	2	3	1	6	4	8	7				
First step	5	2	3	6	4	8	7		1			
Second step	5	3	6	4	8	7			1	2		
Third step	5	6	4	8	7				1	2	3	
Fourth step	5	6	8	7					1	2	3	4

The successive steps each produce two lists, one that is unsorted and one that is sorted. The sorted list is created one number at a time and the final sorted list emerges in step 8.

In general, how many operations does such an algorithm require to do a complete sort? If we have N numbers, then each step requires roughly N comparisons — true, the size of the list is shrinking, but the number of comparisons is *of the order* N — and we have to repeat this procedure N times. (There are N steps, not counting the zeroth step which just inputs the initial list.) Thus the number of operations used to sort N numbers by this procedure is of the order N^2, or, as it's usually written $O(N^2)$ (read "Big Oh").

The problem with this simple procedure is that the $(n+1)$st step doesn't take into account the comparisons done in the nth step. All that work is wasted, and it's wasted over and over. An alternate approach, one that makes use of intermediate comparisons, is to sort sublists of the original list, merge the results, and sort again. Here's how it goes. We'll assess the efficiency after we see how the method works.[11]

Start by going straight through the list and breaking it up into sublists that have just two elements; say that's step zero.

5	2		3	1		6	4		8	7

[11]This is a pretty standard topic in many introductory algorithm courses in computer science. I wanted an EE source, and I'm following the discussion in K. Steiglitz, *A Digital Signal Processing Primer*. I've referred to this book before; highly recommended.

Step one is to sort each of these (four) 2-lists, producing two sets of 2-lists, called "top" lists and "bottom" lists just to keep then straight (and we'll also use top and bottom in later work on the FFT):

top	2	5
	1	3
bottom	4	6
	7	8

This step requires four comparisons.

Step two merges these 2-lists into two sorted 4-lists (again called top and bottom). Here's the algorithm, applied separately to the top and bottom 2-lists. The numbers in the first slots of each 2-list are the smaller of those two numbers. Compare these two numbers and promote the smaller of the two to the first slot of the top (respectively, bottom) 4-list. That leaves a 1-list and a 2-list. Compare the single element of the 1-list to the first element of the 2-list and promote the smaller to the second slot of the top (resp. bottom) 4-list. We're down to two numbers. Compare them and put them in the three and four slots of the top (resp. bottom) 4-list. For the example we're working with, this results in the two 4-lists:

top	1	2	3	5
bottom	4	6	7	8

With this example this step requires five comparisons.

Following this same sort procedure, step three is to merge the top and bottom sorted 4-lists into a single sorted 8-list:

1	2	3	4	5	6	7	8

With this example this step requires five comparisons.

In this process we haven't cut down (much) the number of comparisons we have to make at each step, but we have cut down the number of steps from 8 to 3.[12] In general how many operations are involved in getting to the final list of sorted numbers? It's not hard to see that the number of comparisons involved in merging two sublists is of the order of the total length of the sublists. Thus with N numbers total (at the start) the number of comparisons in any merge-sort is $O(N)$:

$$\text{number of comparisons in a merge-sort} = O(N)\,.$$

How many merge-sort steps are there? At each stage we halve the number of sublists, or, working the other way, from the final sorted list each step "up" doubles the number of sublists. Thus if there are n doublings (n steps), then $2^n = N$, or

$$\text{number of merge-sort steps} = \log_2 N\,.$$

We conclude that

$$\text{number of steps to sort } N \text{ numbers} = O(N \log N)\,.$$

[12]It wasn't an accident that I took eight numbers here. The procedure is most natural when we have a list of numbers to sort that is a power of 2, something we'll see again when we look at the FFT.

(We can take the log in any base since this only changes the log by a constant factor, and that's thrown into the "big Oh.") If N is large, this is a *huge* savings in steps from the $O(N^2)$ estimate for the simple sort that we did first. For example, if N is one million, then $O(N^2)$ is a million million or 10^{12} steps while $N \log_{10} N = 10^6 \times 6$, a mere six million operations.

That's a correct accounting of the number of operations involved, but *why* is there a savings in using merge-sort rather than a straight comparison? By virtue of the sorting of sublists, we only need to compare first elements of the sublists in the merge part of the algorithm. In this way the $(n+1)$st step takes advantage of the comparisons made in the nth step, the thing that is not done in the straight comparison method.

A sample calculation of the DFT. As we consider how we might calculate the DFT more efficiently than by straight matrix multiplication, let's do a sample calculation with $N = 4$ so we have on record what the answer is and what we're supposed to come up with by other means. The DFT matrix is

$$\underline{\mathcal{F}}_4 = \begin{pmatrix} 1 & 1 & 1 & 1 \\ 1 & \omega_4^{-1} & \omega_4^{-2} & \omega_4^{-3} \\ 1 & \omega_4^{-2} & \omega_4^{-4} & \omega_4^{-6} \\ 1 & \omega_4^{-3} & \omega_4^{-6} & \omega_4^{-9} \end{pmatrix}.$$

We want to reduce this as much as possible, "reduction" being somewhat open to interpretation.

Using $\omega_4 = e^{2\pi i/4}$ and $\omega_4^{-4} = 1$ we have

$$\omega_4^{-6} = \omega_4^{-4}\omega_4^{-2} = \omega_4^{-2} \quad \text{and} \quad \omega_4^{-9} = \omega_4^{-8}\omega_4^{-1} = \omega_4^{-1}.$$

In general, to simplify ω_4 to a power, we take the remainder of the exponent on dividing by 4; that is, we reduce modulo 4. With these reductions $\underline{\mathcal{F}}_4$ becomes

$$\begin{pmatrix} 1 & 1 & 1 & 1 \\ 1 & \omega_4^{-1} & \omega_4^{-2} & \omega_4^{-3} \\ 1 & \omega_4^{-2} & \omega_4^{-4} & \omega_4^{-6} \\ 1 & \omega_4^{-3} & \omega_4^{-6} & \omega_4^{-9} \end{pmatrix} = \begin{pmatrix} 1 & 1 & 1 & 1 \\ 1 & \omega_4^{-1} & \omega_4^{-2} & \omega_4^{-3} \\ 1 & \omega_4^{-2} & 1 & \omega_4^{-2} \\ 1 & \omega_4^{-3} & \omega_4^{-2} & \omega_4^{-1} \end{pmatrix}.$$

But we don't stop there. Note that

$$\omega_4^{-2} = (e^{2\pi i/4})^{-2} = e^{-\pi i} = -1,$$

also a worthy simplification. (We have called attention to this $N/2$th power of ω_N before.) So, finally,

$$\underline{\mathcal{F}}_4 = \begin{pmatrix} 1 & 1 & 1 & 1 \\ 1 & \omega_4^{-1} & \omega_4^{-2} & \omega_4^{-3} \\ 1 & \omega_4^{-2} & \omega_4^{-4} & \omega_4^{-6} \\ 1 & \omega_4^{-3} & \omega_4^{-6} & \omega_4^{-9} \end{pmatrix} = \begin{pmatrix} 1 & 1 & 1 & 1 \\ 1 & \omega_4^{-1} & -1 & -\omega_4^{-1} \\ 1 & -1 & 1 & -1 \\ 1 & -\omega_4^{-1} & -1 & \omega_4^{-1} \end{pmatrix}.$$

Therefore we find

$$\underline{\mathcal{F}}_4\underline{f} = \begin{pmatrix} 1 & 1 & 1 & 1 \\ 1 & \omega_4^{-1} & -1 & -\omega_4^{-1} \\ 1 & -1 & 1 & -1 \\ 1 & -\omega_4^{-1} & -1 & \omega_4^{-1} \end{pmatrix} \begin{pmatrix} \underline{f}[0] \\ \underline{f}[1] \\ \underline{f}[2] \\ \underline{f}[3] \end{pmatrix} = \begin{pmatrix} \underline{f}[0] + \underline{f}[1] + \underline{f}[2] + \underline{f}[3] \\ \underline{f}[0] + \underline{f}[1]\omega_4^{-1} - \underline{f}[2] - \underline{f}[3]\omega_4^{-1} \\ \underline{f}[0] - \underline{f}[1] + \underline{f}[2] - \underline{f}[3] \\ \underline{f}[0] - \underline{f}[1]\omega_4^{-1} - \underline{f}[2] + \underline{f}[3]\omega_4^{-1} \end{pmatrix}.$$

The matrix looks simpler, true, but it still took 16 multiplications to get the final answer. You can see those two special components that we called attention to earlier, the sum of the inputs in the zero slot and the alternating sum of the inputs in the $N/2 = 2$ slot.

This is about as far as we can go without being smart. Fortunately, there have been some smart people on the case.

7.11.1. Half the work is twice the fun: The Fast Fourier Transform. We agree that the DFT has a lot of structure. The trick to a faster computation of a DFT of order N is to use that structure to rearrange the products to bring in DFTs of order $N/2$. Here's where we use that N is even; in fact, to make the algorithm most efficient in being applied repeatedly we'll eventually want to assume that N is a power of 2.

We need a few elementary algebraic preliminaries on the ω_N, all of which we've used before. We also need to introduce some temporary (!) notation or we'll sink in a sea of subscripts and superscripts. Let's write powers of ω with two arguments:

$$\omega[p, q] = \omega_p^q = e^{2\pi i q/p}.$$

I think this will help. It can't hurt. For our uses p will be N, $N/2$, etc.

First, notice that

$$\omega[N/2, -1] = e^{-2\pi i/(N/2)} = e^{-4\pi i/N} = \omega[N, -2].$$

Therefore powers of $\omega[N/2, -1]$ are *even* powers of $\omega[N, -1] = \omega_N^{-1}$:

$$\omega[N/2, -n] = \omega[N, -2n]$$

and, in general,

$$\omega[N, -2nm] = \omega[N/2, -nm].$$

What about odd powers of $\omega_N^{-1} = \omega[N, -1]$? An odd power is of the form $\omega[N, -(2n + 1)]$ and so

$$\omega[N, -(2n + 1)] = \omega[N, -1]\,\omega[N, -2n] = \omega[N, -1]\,\omega[N/2, -n].$$

Thus also

$$\omega[N, -(2n + 1)m] = \omega[N, -m]\,\omega[N/2, -nm].$$

Finally, recall that

$$\omega[N, -N/2] = e^{(-2\pi i/N)(N/2)} = e^{-\pi i} = -1.$$

Splitting the sums. Here's how we'll use this. For each m we want to split the single sum defining $\underline{F}[m]$ into two sums, one over the even indices and one over the odd indices:

$$\begin{aligned} \underline{F}[m] &= \sum_{n=0}^{N-1} \underline{f}[n]\,\omega[N, -nm] \\ &= \sum_{n=0}^{N/2-1} \underline{f}[2n]\,\omega[N, -(2n)m] + \sum_{n=0}^{N/2-1} \underline{f}[2n + 1]\,\omega[N, -(2n + 1)m]. \end{aligned}$$

Everything is accounted for here; all the terms are there — make sure you see that. Also, although both sums go from 0 to $N/2-1$, notice that for the first sum the first and last terms are $\underline{f}[0]$ and $\underline{f}[N-2]\,\omega[N,-(N-2)m]$, and for the second they are $\underline{f}[1]\omega[N,-m]$ and $\underline{f}[N-1]\,\omega[N,-(N-1)m]$.

Next, according to our observations on powers of ω we can also write $\underline{F}[m]$ as

$$\begin{aligned}\underline{F}[m] &= \sum_{n=0}^{N/2-1} \underline{f}[2n]\,\omega[N/2,-nm] \;+\; \sum_{n=0}^{N/2-1} \underline{f}[2n+1]\,\omega[N/2,-nm]\,\omega[N,-m]\\ &= \sum_{n=0}^{N/2-1} \underline{f}[2n]\,\omega[N/2,-nm] \;+\; \omega[N,-m] \sum_{n=0}^{N/2-1} \underline{f}[2n+1]\,\omega[N/2,-nm]\,.\end{aligned}$$

Let's study these sums more closely. There are $N/2$ even indices and $N/2$ odd indices, and we appear, in each sum, *almost* to be taking a DFT of order $N/2$ of the $N/2$-tuples $\underline{f}$[even indices] and $\underline{f}$[odd indices]. Why "almost?" The DFT of order $N/2$ accepts as input an $N/2$-length signal *and returns* an $N/2$-length signal. But the sums above give all N entries of the N-length $\underline{F}$ as m goes from 0 to $N-1$. We're going to do two things to bring in $\underline{\mathcal{F}}_{N/2}$.

- First, if we take m to go from 0 to $N/2-1$, then we get the first $N/2$ outputs $\underline{F}[m]$. We write, informally,

$$\underline{F}[m] = (\underline{\mathcal{F}}_{N/2}\,\underline{f}_{\text{even}})[m] \;+\; \omega[N,-m]\,(\underline{\mathcal{F}}_{N/2}\,\underline{f}_{\text{odd}})[m]\,, \qquad m = 0,1,\ldots,N/2-1\,.$$

That makes sense; $N/2$-tuples go in and $N/2$-tuples come out.

- Second, what is the story for an index m in the second half of the range, from $N/2$ to $N-1$? Instead of letting m go from $N/2$ to $N-1$ we can write these indices in the form $m+N/2$, where m goes from 0 to $N/2-1$. What happens with $\underline{F}[m+N/2]$?

We again have

$$\underline{F}[m+N/2] = \underline{\mathcal{F}}_{N/2}\underline{f}_{\text{even}}[m+N/2] + \omega[N,-(m+N/2)]\underline{\mathcal{F}}_{N/2}\underline{f}_{\text{odd}}[m+N/2].$$

But look: for $\underline{\mathcal{F}}_{N/2}$ *the outputs and inputs are periodic of period* $N/2$. That is,

$$\begin{aligned}\underline{\mathcal{F}}_{N/2}\underline{f}_{\text{even}}[m+N/2] &= \underline{\mathcal{F}}_{N/2}\underline{f}_{\text{even}}[m],\\ \underline{\mathcal{F}}_{N/2}\underline{f}_{\text{odd}}[m+N/2] &= \underline{\mathcal{F}}_{N/2}\underline{f}_{\text{odd}}[m].\end{aligned}$$

Now by a calculation,

$$\omega[N,-(m+N/2)] = -\omega[N,-m].$$

Thus

$$\underline{F}[m+N/2] = \underline{\mathcal{F}}_{N/2}\underline{f}_{\text{even}}[m] - \omega[N,-m]\underline{\mathcal{F}}_{N/2}\underline{f}_{\text{odd}}[m].$$

We are there.

The description of the FFT algorithm. It's really very significant what we've done here. Let's summarize:

- We start with an input $\underline{f}$ of length N and want to compute its output $\underline{F} = \underline{\mathcal{F}}_N\,\underline{f}$, also of length N.
- The steps we take serve to compute the component outputs $\underline{F}[m]$ for $m = 0, 1, \ldots, N-1$ by computing a DFT on two sequences, each of half the length, and arranging properly. That is:
  (1) Separate $\underline{f}[n]$ into two sequences, the even and odd indices (0 is even), each of length $N/2$.
  (2) Compute $\underline{\mathcal{F}}_{N/2}\,\underline{f}_{\text{even}}$ and $\underline{\mathcal{F}}_{N/2}\,\underline{f}_{\text{odd}}$.
  (3) The outputs $\underline{F}[m]$ are obtained by arranging the results of this computation according to

$$\begin{aligned}\underline{F}[m] &= (\underline{\mathcal{F}}_{N/2}\,\underline{f}_{\text{even}})[m] + \omega[N,-m]\,(\underline{\mathcal{F}}_{N/2}\,\underline{f}_{\text{odd}})[m],\\ \underline{F}[m+N/2] &= (\underline{\mathcal{F}}_{N/2}\,\underline{f}_{\text{even}})[m] - \omega[N,-m]\,(\underline{\mathcal{F}}_{N/2}\,\underline{f}_{\text{odd}})[m],\end{aligned}$$

  for $m = 0, 1, \ldots, N/2$.

Another look at $\underline{\mathcal{F}}_4$. Let's do the case $N = 4$ as an example, comparing it to our earlier calculation. The first step is to rearrange the inputs to group the even and odd indices. This is done by a *permutation matrix*

$$M = \begin{pmatrix} 1 & 0 & 0 & 0\\ 0 & 0 & 1 & 0\\ 0 & 1 & 0 & 0\\ 0 & 0 & 0 & 1\end{pmatrix}$$

whose effect is

$$\begin{pmatrix} 1 & 0 & 0 & 0\\ 0 & 0 & 1 & 0\\ 0 & 1 & 0 & 0\\ 0 & 0 & 0 & 1\end{pmatrix}\begin{pmatrix}\underline{f}[0]\\ \underline{f}[1]\\ \underline{f}[2]\\ \underline{f}[3]\end{pmatrix} = \begin{pmatrix}\underline{f}[0]\\ \underline{f}[2]\\ \underline{f}[1]\\ \underline{f}[3]\end{pmatrix}.$$

The matrix M is defined by what it does to the natural basis (discrete $\underline{\delta}$'s!) $\underline{\delta}_0$, $\underline{\delta}_1$, $\underline{\delta}_2$, $\underline{\delta}_3$ of $\mathbb{R}^4$; namely, $M\underline{\delta}_0 = \underline{\delta}_0$, $M\underline{\delta}_1 = \underline{\delta}_2$, $M\underline{\delta}_2 = \underline{\delta}_1$, and $M\underline{\delta}_3 = \underline{\delta}_3$.

Next, the even and odd indices are fed respectively to two DFTs of order $4/2 = 2$. This is the crucial reduction in the FFT algorithm.

$$\begin{pmatrix} 1 & 1 & 0 & 0\\ 1 & \omega_2^{-1} & 0 & 0\\ 0 & 0 & 1 & 1\\ 0 & 0 & 1 & \omega_2^{-1}\end{pmatrix}\begin{pmatrix}\underline{f}[0]\\ \underline{f}[2]\\ \underline{f}[1]\\ \underline{f}[3]\end{pmatrix} = \begin{pmatrix}\underline{f}[0] + \underline{f}[2]\\ \underline{f}[0] + \underline{f}[2]\omega_2^{-1}\\ \underline{f}[1] + \underline{f}[3]\\ \underline{f}[1] + \underline{f}[3]\omega_2^{-1}\end{pmatrix}.$$

On the left we have a block diagonal matrix. It's a 4×4 matrix with the 2×2 $\underline{\mathcal{F}}_2$ matrices down the diagonal and zeros everywhere else. We saw this step, but we didn't see the intermediate result written just above on the right because our formulas passed right away to the reassembly of the $\underline{F}[m]$'s. That reassembly is the final step.

So far we have

$$\underline{\mathcal{F}}_2\,\underline{f}_{\text{even}} = \begin{pmatrix} f_0 + \underline{f}[2] \\ f_0 + \underline{f}[2]\omega_2^{-1} \end{pmatrix}, \quad \underline{\mathcal{F}}_2\,\underline{f}_{\text{odd}} = \begin{pmatrix} \underline{f}[1] + \underline{f}[3] \\ \underline{f}[1] + \underline{f}[3]\omega_2^{-1} \end{pmatrix}$$

and in each case the indexing is $m = 0$ for the first entry and $m = 1$ for the second entry. The last stage, to get the $\underline{F}[m]$'s, is to recombine these half-size DFTs in accordance with the even and odd sums we wrote down earlier. In putting the pieces together we want to leave the even indices alone, put a $+\omega_4^{-m}$ in front of the mth component of the first half of the $\underline{\mathcal{F}}_2$ of the odds, and put a $-\omega_4^{-m}$ in front of the mth component of the $\underline{\mathcal{F}}_2$ of the second half of the odds. This is done by the matrix

$$B_4 = \begin{pmatrix} 1 & 0 & 1 & 0 \\ 0 & 1 & 0 & \omega_4^{-1} \\ 1 & 0 & -1 & 0 \\ 0 & 1 & 0 & -\omega_4^{-1} \end{pmatrix}.$$

It works like this:

$$\begin{aligned}
\begin{pmatrix} 1 & 0 & 1 & 0 \\ 0 & 1 & 0 & \omega_4^{-1} \\ 1 & 0 & -1 & 0 \\ 0 & 1 & 0 & -\omega_4^{-1} \end{pmatrix}
\begin{pmatrix} \underline{f}[0] + \underline{f}[2] \\ \underline{f}[0] + \underline{f}[2]\omega_2^{-1} \\ \underline{f}[1] + \underline{f}[3] \\ \underline{f}[1] + \underline{f}[3]\omega_2^{-1} \end{pmatrix}
&= \begin{pmatrix} \underline{f}[0] + \underline{f}[2] + \underline{f}[1] + \underline{f}[3] \\ \underline{f}[0] + \underline{f}[2]\omega_2^{-1} + \underline{f}[1]\omega_4^{-1} + \underline{f}[3]\omega_4^{-1}\omega_2^{-1} \\ \underline{f}[0] + \underline{f}[2] - \underline{f}[1] - \underline{f}[3] \\ \underline{f}[0] + \underline{f}[2]\omega_2^{-1} - \underline{f}[1]\omega_4^{-1} - \underline{f}[3]\omega_4^{-1}\omega_2^{-1} \end{pmatrix} \\
&= \begin{pmatrix} \underline{f}[0] + \underline{f}[1] + \underline{f}[2] + \underline{f}[3] \\ \underline{f}[0] + \underline{f}[1]\omega_4^{-1} - \underline{f}[2] - \underline{f}[3]\omega_4^{-1} \\ \underline{f}[0] - \underline{f}[1] + \underline{f}[2] - \underline{f}[3] \\ \underline{f}[0] - \underline{f}[1]\omega_4^{-1} - \underline{f}[2] + \underline{f}[3]\omega_4^{-1} \end{pmatrix},
\end{aligned}$$

where we (finally) used $\omega_2^{-1} = e^{-2\pi i/2} = -1$. This checks with what we got before.

One way to view this procedure is as a factorization of $\underline{\mathcal{F}}_4$ into simpler matrices. It works like this:

$$\begin{aligned}
\begin{pmatrix} 1 & 1 & 1 & 1 \\ 1 & \omega_4^{-1} & \omega_4^{-2} & \omega_4^{-3} \\ 1 & \omega_4^{-2} & \omega_4^{-4} & \omega_4^{-6} \\ 1 & \omega_4^{-3} & \omega_4^{-6} & \omega_4^{-9} \end{pmatrix}
&= \begin{pmatrix} 1 & 1 & 1 & 1 \\ 1 & \omega_4^{-1} & -1 & -\omega_4^{-1} \\ 1 & -1 & 1 & -1 \\ 1 & -\omega_4^{-1} & -1 & \omega_4^{-1} \end{pmatrix} \\
&= \begin{pmatrix} 1 & 0 & 1 & 0 \\ 0 & 1 & 0 & \omega_4^{-1} \\ 1 & 0 & -1 & 0 \\ 0 & 1 & 0 & -\omega_4^{-1} \end{pmatrix}
\begin{pmatrix} 1 & 1 & 0 & 0 \\ 1 & \omega_2^{-1} & 0 & 0 \\ 0 & 0 & 1 & 1 \\ 0 & 0 & 1 & \omega_2^{-1} \end{pmatrix}
\begin{pmatrix} 1 & 0 & 0 & 0 \\ 0 & 0 & 1 & 0 \\ 0 & 1 & 0 & 0 \\ 0 & 0 & 0 & 1 \end{pmatrix} \\
&= \begin{pmatrix} 1 & 0 & 1 & 0 \\ 0 & 1 & 0 & \omega_4^{-1} \\ 1 & 0 & -1 & 0 \\ 0 & 1 & 0 & -\omega_4^{-1} \end{pmatrix}
\begin{pmatrix} 1 & 1 & 0 & 0 \\ 1 & -1 & 0 & 0 \\ 0 & 0 & 1 & 1 \\ 0 & 0 & 1 & -1 \end{pmatrix}
\begin{pmatrix} 1 & 0 & 0 & 0 \\ 0 & 0 & 1 & 0 \\ 0 & 1 & 0 & 0 \\ 0 & 0 & 0 & 1 \end{pmatrix}.
\end{aligned}$$

Look at all the zeros! There are 48 entries total in the three matrices that multiply together to give $\underline{\mathcal{F}}_4$, but only 20 entries are nonzero.

In the same way, the general shape of the factorization to get $\underline{\mathcal{F}}_N$ via $\underline{\mathcal{F}}_{N/2}$ is

$$\underline{\mathcal{F}}_N = \begin{pmatrix} I_{N/2} & \Omega_{N/2} \\ I_{N/2} & -\Omega_{N/2} \end{pmatrix} \begin{pmatrix} \underline{\mathcal{F}}_{N/2} & 0 \\ 0 & \underline{\mathcal{F}}_{N/2} \end{pmatrix} \begin{pmatrix} \text{sort the even} \\ \text{and odd indices} \end{pmatrix}.$$

$I_{N/2}$ is the $N/2 \times N/2$ identity matrix. 0 is the zero matrix (of size $N/2 \times N/2$ in this case). $\Omega_{N/2}$ is the diagonal matrix with entries $1, \omega_N^{-1}, \omega_N^{-2}, \dots, \omega_N^{-(N/2-1)}$ down the diagonal.[13] $\underline{\mathcal{F}}_{N/2}$ is the DFT of half the order, and the permutation matrix puts the $N/2$ even indices first and the $N/2$ odd indices second.

Thus, the way this factorization works is:

- The inputs are $\underline{f}[0], \underline{f}[1], \dots, \underline{f}[N-1]$.
- The matrix on the right is a permutation matrix that puts the even indices in the first $N/2$ slots and the odd indices in the second $N/2$ slots.
  - Alternatively, think of the operation as first starting with $\underline{f}[0]$ and taking every other $\underline{f}[n]$ — this collects $\underline{f}[0]$, $\underline{f}[2]$, $\underline{f}[4]$, and so on — and then starting back with $\underline{f}[1]$ and taking every other $\underline{f}[n]$ — this collects $\underline{f}[1]$, $\underline{f}[3]$, $\underline{f}[5]$, and so on. As we iterate the process, this will be a more natural way of thinking about the way the first matrix chooses how to send the inputs on to the second matrix.
- The outputs of the first matrix operation are a pair of $N/2$-vectors. The matrix in the middle accepts these as inputs. It computes half-size DFTs on these half-size inputs and outputs two $N/2$-vectors, which are then passed along as inputs to the third matrix.
- The third matrix, on the left, reassembles the outputs from the half-size DFTs and outputs the $\underline{F}[0], \underline{F}[1], \dots, \underline{F}[N-1]$.
  - This is similar in spirit to a step in the "merge-sort" algorithm for sorting numbers. Operations (comparisons in that case, DFTs in this case) are performed on smaller lists that are then merged to longer lists.
- The important feature, as far as counting the multiplications go, is that suddenly there are a lot of zeros in the matrices.

As to this last point, we can already assess some savings in the number of operations when the even/odd splitting is used versus the straight evaluation of the DFT from its original definition. If we compute $\underline{F} = \underline{\mathcal{F}}_N \underline{f}$ just as a matrix product, there are N^2 multiplications and N^2 additions for a total of $2N^2$ operations. On the other hand, with the splitting, the computations in the inner block matrix of two DFTs of order $N/2$ require $2(N/2)^2 = N^2/2$ multiplications and $2(N/2)^2 = N^2/2$ additions. The sorting and recombining by the third matrix require another $N/2$ multiplications and N additions — and this is *linear* in N. Thus the splitting method needs on the order of N^2 operations while the straight DFT needs $2N^2$.

[13]The notation $\Omega_{N/2}$ isn't the greatest — it's written with $N/2$ because of the dimensions of the matrix, though the entries are powers of ω_N. Still, it will prove useful to us later on, and it appears in the literature.

We've cut the work in half, pretty much, though it's still of the same order. We'll get back to this analysis later.

Divide and conquer. At this point it's clear what we'd like to do: repeat the algorithm, each time halving the size of the DFT. The factorization from N to $N/2$ is the top level:

$$\underline{\mathcal{F}}_N = \begin{pmatrix} I_{N/2} & \Omega_{N/2} \\ I_{N/2} & -\Omega_{N/2} \end{pmatrix} \begin{pmatrix} \underline{\mathcal{F}}_{N/2} & 0 \\ 0 & \underline{\mathcal{F}}_{N/2} \end{pmatrix} \begin{pmatrix} \text{sort the even} \\ \text{and odd indices} \end{pmatrix}.$$

At the next level "down" we don't do anything to the matrices on the ends, but we factor each of the two $\underline{\mathcal{F}}_{N/2}$'s the same way, into a permutation matrix on the right, a block matrix of $\underline{\mathcal{F}}_{N/4}$'s in the middle, and a reassembly matrix on the left. (I'll come back to the sorting — it's the most interesting part.). That is,

$$\underline{\mathcal{F}}_{N/2} = \begin{pmatrix} I_{N/4} & \Omega_{N/4} \\ I_{N/4} & -\Omega_{N/4} \end{pmatrix} \begin{pmatrix} \underline{\mathcal{F}}_{N/4} & 0 \\ 0 & \underline{\mathcal{F}}_{N/4} \end{pmatrix} \begin{pmatrix} \text{sort } N/2\text{-lists to} \\ \text{two } N/4\text{-lists} \end{pmatrix},$$

and putting this into the top-level picture, the operations become nested (or recursive):

$$\underline{\mathcal{F}}_N = \begin{pmatrix} I_{N/2} & \Omega_{N/2} \\ I_{N/2} & -\Omega_{N/2} \end{pmatrix}$$
$$\cdot \begin{pmatrix} \begin{pmatrix} I_{N/4} & \Omega_{N/4} \\ I_{N/4} & -\Omega_{N/4} \end{pmatrix} \begin{pmatrix} \underline{\mathcal{F}}_{N/4} & 0 \\ 0 & \underline{\mathcal{F}}_{N/4} \end{pmatrix} \begin{pmatrix} N/2 \text{ to} \\ N/4 \text{ sort} \end{pmatrix} & 0 \\ 0 & \begin{pmatrix} I_{N/4} & \Omega_{N/4} \\ I_{N/4} & -\Omega_{N/4} \end{pmatrix} \begin{pmatrix} \underline{\mathcal{F}}_{N/4} & 0 \\ 0 & \underline{\mathcal{F}}_{N/4} \end{pmatrix} \begin{pmatrix} N/2 \text{ to} \\ N/4 \text{ sort} \end{pmatrix} \end{pmatrix}$$
$$\cdot \begin{pmatrix} \text{sort the } N/2\text{-even} \\ \text{and } N/2\text{-odd indices} \end{pmatrix}.$$

To be able to repeat this, to keep halving the size of the DFT, we now see that we need to take N to be a power of 2. The construction then continues, going down levels till we get from $\underline{\mathcal{F}}_N$ to $\underline{\mathcal{F}}_1$. Note that the DFT of order 1 takes a single input and returns it unchanged; i.e., it is the identity transform.

When the halving is all over, here's what happens. The work is in the initial sorting and in the reassembling, since the final DFT in the factorization is $\underline{\mathcal{F}}_1$, which leaves alone whatever it gets. Thus, reading from right to left, the initial inputs $\underline{f}[0], \dots, \underline{f}[N-1]$ are first sorted and then passed back up through a number of reassembly matrices, ultimately winding up as the outputs $\underline{F}[0], \dots, \underline{F}[N-1]$.

It's clear, with the abundance of zeros in the matrices, that there should be a savings in the total number of operations, though it's not clear how much. The *entire trip*, from $\underline{f}$'s to $\underline{F}$'s, is called the Fast Fourier Transform (FFT). It's fast because of the reduction in the number of operations. Remember, the FFT is not a new transform, *it is just computing the DFT of the initial inputs.*

7.11.2. Factoring the DFT matrix. Rather than trying now to describe the general process in more detail, let's look at an example more thoroughly and from the matrix point of view. One comment about the matrix factorization description versus other sorts of descriptions of the algorithm: Since the initial ideas of Cooley and Tukey there have been many other styles of FFT algorithms proposed and implemented, similar in some respects to Cooley and Tukey's formulation and different in others. It became a mess. In a 1992 book, *Computational Frameworks* *for the Fast Fourier Transform*, Charles van Loan showed how many of the ideas could be unified via a study of different matrix factorizations of the DFT. This is not the only way to organize the material, but it has been very influential.

Let's take the case $N = 16$, just to live it up. Once again, the initial input is a length 16 signal (or vector) $\underline{f}$ and the final output is another 16-tuple, $\underline{F} = \underline{\mathcal{F}}_{16}\underline{f}$. At the top level, we can write this as

$$\underline{F} = \underline{\mathcal{F}}_{16}\underline{f} = \begin{pmatrix} I_8 & \Omega_8 \\ I_8 & -\Omega_8 \end{pmatrix} \begin{pmatrix} \underline{\mathcal{F}}_8 & 0 \\ 0 & \underline{\mathcal{F}}_8 \end{pmatrix} \begin{pmatrix} \underline{f}_{\text{even}} \\ \underline{f}_{\text{odd}} \end{pmatrix} = \begin{pmatrix} I_8 & \Omega_8 \\ I_8 & -\Omega_8 \end{pmatrix} \begin{pmatrix} \underline{G} \\ \underline{H} \end{pmatrix},$$

where $\underline{G}$ and $\underline{H}$ are the results of computing $\underline{\mathcal{F}}_8$ on $\underline{f}_{\text{even}}$ and $\underline{f}_{\text{odd}}$, respectively. Write this as

$$\underline{F} = B_{16} \begin{pmatrix} \underline{G} \\ \underline{H} \end{pmatrix}, \quad B_{16} = \begin{pmatrix} I_8 & \Omega_8 \\ I_8 & -\Omega_8 \end{pmatrix},$$

where B is supposed to stand for "butterfly" — more on this, later.

But how, in the nesting of operations, did we get to $\underline{G}$ and $\underline{H}$? The next level down (or in) has

$$\underline{G} = \underline{\mathcal{F}}_8\underline{f}_{\text{even}} = \begin{pmatrix} I_4 & \Omega_4 \\ I_4 & -\Omega_4 \end{pmatrix} \begin{pmatrix} \underline{G}' \\ \underline{H}' \end{pmatrix} = B_8 \begin{pmatrix} \underline{G}' \\ \underline{H}' \end{pmatrix},$$

where $\underline{G}'$ and $\underline{H}'$ are the result of computing $\underline{\mathcal{F}}_4$'s on the even and odd subsets of $\underline{f}_{\text{even}}$. Got it? Likewise we write

$$\underline{H} = \underline{\mathcal{F}}_8\underline{f}_{\text{odd}} = \begin{pmatrix} I_4 & \Omega_4 \\ I_4 & -\Omega_4 \end{pmatrix} \begin{pmatrix} \underline{G}'' \\ \underline{H}'' \end{pmatrix} = B_8 \begin{pmatrix} \underline{G}'' \\ \underline{H}'' \end{pmatrix},$$

where $\underline{G}''$ and $\underline{H}''$ are the result of computing $\underline{\mathcal{F}}_4$'s on the even and odd subsets of $\underline{f}_{\text{odd}}$.

Combining this with what we did first, we have

$$\underline{F} = \underline{\mathcal{F}}_{16}\underline{f} = B_{16} \begin{pmatrix} B_8 & 0 \\ 0 & B_8 \end{pmatrix} \begin{pmatrix} \underline{G}' \\ \underline{H}' \\ \underline{G}'' \\ \underline{H}'' \end{pmatrix}.$$

Continue this for two more steps — it remains to find DFTs of order 4 and 2. The result then looks like

$$\begin{aligned} &\underline{F} = \underline{\mathcal{F}}_{16}\underline{f} \\ &= B_{16} \begin{pmatrix} B_8 & 0 \\ 0 & B_8 \end{pmatrix} \begin{pmatrix} B_4 & 0 & 0 & 0 \\ 0 & B_4 & 0 & 0 \\ 0 & 0 & B_4 & 0 \\ 0 & 0 & 0 & B_4 \end{pmatrix} \begin{pmatrix} B_2 & 0 & 0 & 0 & 0 & 0 & 0 & 0 \\ 0 & B_2 & 0 & 0 & 0 & 0 & 0 & 0 \\ 0 & 0 & B_2 & 0 & 0 & 0 & 0 & 0 \\ 0 & 0 & 0 & B_2 & 0 & 0 & 0 & 0 \\ 0 & 0 & 0 & 0 & B_2 & 0 & 0 & 0 \\ 0 & 0 & 0 & 0 & 0 & B_2 & 0 & 0 \\ 0 & 0 & 0 & 0 & 0 & 0 & B_2 & 0 \\ 0 & 0 & 0 & 0 & 0 & 0 & 0 & B_2 \end{pmatrix} \\ &\qquad \cdot \begin{pmatrix} 16 \times 16 \text{ permutation} \\ \text{matrix that sorts} \\ \text{the inputs} \end{pmatrix} \underline{f}. \end{aligned}$$

Note that

$$B_2 = \begin{pmatrix} 1 & 1 \\ 1 & -1 \end{pmatrix}.$$

Each B_2 receives a pair of inputs coming from a pair of $\underline{\mathcal{F}}_1$'s, and since the $\underline{\mathcal{F}}_1$'s don't do anything, each B_2 receives a pair of the original inputs $\underline{f}[m]$, but shuffled from the ordering $\underline{f}[0], \underline{f}[1], \ldots, \underline{f}[15]$. We'll get back to the question of sorting the indices, but first let's be sure that it's worth it.

And the point of this is, again? There are lots of zeros in the factorization of the DFT. After the initial sorting of the indices (also lots of zeros in that matrix) there are 4 reassembly stages. In general, for $N = 2^n$ there are $n = \log_2 N$ reassembly stages after the initial sorting. The count $\log_2 N$ for the number of reassembly stages follows in the same way as the count for the number of merge-sort steps in the sorting algorithm, but I want to be a little more precise this time.

We now consider the *computational complexity* of the FFT algorithm in general. Let $C(N)$ denote the number of elementary operations involved in finding the DFT via the FFT algorithm; these include additions and multiplications. We reassemble $\underline{\mathcal{F}}_N$ from two $\underline{\mathcal{F}}_{N/2}$'s by another set of elementary operations. From our earlier considerations, or from the factorization, the number of operations can easily be shown to be proportional to N. Thus the basic recursion relationship is

$$C(N) = 2C(N/2) + KN.$$

We can solve this recurrence equation as follows: let

$$n = \log_2 N$$

and let

$$T(n) = \frac{C(N)}{N}$$

so that

$$C(N) = NT(n).$$

Then $n - 1 = \log_2(N/2)$ and thus

$$T(n-1) = \frac{C(N/2)}{N/2} = 2\frac{C(N/2)}{N}, \quad \text{or} \quad NT(n-1) = 2C(N/2).$$

Substituting into the recurrence relationship for C then gives

$$NT(n) = NT(n-1) + KN$$

or simply

$$T(n) = T(n-1) + K.$$

This already implies that $T(n)$ is linear. But $C(1)$ is obviously 0, because there aren't any operations needed to compute the DFT of a signal of length 1. Hence $T(0) = C(1) = 0$, $T(1) = K$, and in general

$$T(n) = Kn.$$

In terms of C this says

$$C(N) = KN\log_2 N.$$

Various implementations of the FFT try to make the constant K as small as possible. The best one around now, I think, brings the number of multiplications down to $N \log_2 N$ and the number of additions down to $3 \log_2 N$. Remember that this is for complex inputs. Restricting to real inputs cuts the number of operations in half.

As I pointed out when talking about the problem of sorting, when N is large, the reduction in computation from N^2 to $N \log_2 N$ is an enormous savings. For example, take $N = 32{,}768 = 2^{15}$. Then $N^2 = 2^{30} = 1{,}073{,}741{,}824$, about a billion, while $2^{15} \log_2 2^{15} = 491{,}520$, a measly half a million. (Cut down to half that for real signals.) That's a substantial reduction.

7.11.3. Sorting the indices. If we think of recursively factoring the inner DFT matrix, then in implementing the whole FFT the first thing that's done is to sort and shuffle the inputs.[14] It's common to display a flow diagram for the FFT, and much of the pictorial splendor in many treatments of the FFT is in showing how the $\underline{f}[m]$'s are shuffled and passed on from stage to stage. The flow chart of the complete FFT algorithm is called a *butterfly diagram* — hence the naming of the matrices B. You can find butterfly diagrams in any of the standard works.

The principle of sorting the inputs is as stated earlier. Start with the first input, $\underline{f}[0]$, and take every other one. Then start with the second input, $\underline{f}[1]$ (which was skipped over in the first pass), and take every other one. This produces $\underline{f}_{\text{even}}$ and $\underline{f}_{\text{odd}}$. The next sorting repeats the process *for the subsequences* $\underline{f}_{\text{even}}$ and $\underline{f}_{\text{odd}}$, and so on.

For $N = 8$ (I don't have the stamina for $N = 16$ again) this looks like

$\underline{f}[0]$	$\underline{f}[0]$	$\underline{f}[0]$	$\underline{f}[0]$
$\underline{f}[1]$	$\underline{f}[2]$	$\underline{f}[4]$	$\underline{f}[4]$
$\underline{f}[2]$	$\underline{f}[4]$	$\underline{f}[2]$	$\underline{f}[2]$
$\underline{f}[3]$	$\underline{f}[6]$	$\underline{f}[6]$	$\underline{f}[6]$
$\underline{f}[4]$	$\underline{f}[1]$	$\underline{f}[1]$	$\underline{f}[1]$
$\underline{f}[5]$	$\underline{f}[3]$	$\underline{f}[5]$	$\underline{f}[5]$
$\underline{f}[6]$	$\underline{f}[5]$	$\underline{f}[3]$	$\underline{f}[3]$
$\underline{f}[7]$	$\underline{f}[7]$	$\underline{f}[7]$	$\underline{f}[7]$

Though there's no more shuffling from the third to the fourth column we've written the last column to indicate that the inputs go in, one at a time, in that order to the waiting B's.

---

[14] Shuffling is actually an apt description: see, D. Knuth, *The Art of Computer Programming*, Vol. 3, p. 237.

The FFT algorithm with $N = 8$ is thus

$$\underline{F} = B_8 \begin{pmatrix} B_4 & 0 \\ 0 & B_4 \end{pmatrix} \begin{pmatrix} B_2 & 0 & 0 & 0 \\ 0 & B_2 & 0 & 0 \\ 0 & 0 & B_2 & 0 \\ 0 & 0 & 0 & B_2 \end{pmatrix} \begin{pmatrix} \underline{f}[0] \\ \underline{f}[4] \\ \underline{f}[2] \\ \underline{f}[6] \\ \underline{f}[1] \\ \underline{f}[5] \\ \underline{f}[3] \\ \underline{f}[7] \end{pmatrix}.$$

The sorting can be described in a neat way via binary numbers. Each sort puts a collection of inputs into a "top" bin or a "bottom" bin. Let's write 0 for top and 1 for bottom (as in 0 for even and 1 for odd). Assigning digits from right to left, the least significant bit is the first sort, the next most significant bit is the second sort, and the most significant bit (for the three sorts needed when $N = 8$) is the final sort. We thus augment the table, above, to (read the top/bottom descriptions right to left):

$\underline{f}[0]$	$\underline{f}[0]$	0	top	$\underline{f}[0]$	00	top-top	$\underline{f}[0]$	000	top-top-top
$\underline{f}[1]$	$\underline{f}[2]$	0	top	$\underline{f}[4]$	00	top-top	$\underline{f}[4]$	100	bottom-top-top
$\underline{f}[2]$	$\underline{f}[4]$	0	top	$\underline{f}[2]$	10	bottom-top	$\underline{f}[2]$	010	top-bottom-top
$\underline{f}[3]$	$\underline{f}[6]$	0	top	$\underline{f}[6]$	10	bottom-top	$\underline{f}[6]$	110	bottom-bottom-top
$\underline{f}[4]$	$\underline{f}[1]$	1	bottom	$\underline{f}[1]$	01	top-bottom	$\underline{f}[1]$	001	top-top-bottom
$\underline{f}[5]$	$\underline{f}[3]$	1	bottom	$\underline{f}[5]$	01	top-bottom	$\underline{f}[5]$	101	bottom-top-bottom
$\underline{f}[6]$	$\underline{f}[5]$	1	bottom	$\underline{f}[3]$	11	bottom-bottom	$\underline{f}[3]$	011	top-bottom-bottom
$\underline{f}[7]$	$\underline{f}[7]$	1	bottom	$\underline{f}[7]$	11	bottom-bottom	$\underline{f}[7]$	111	bottom-bottom-bottom

The numbers in the final column are exactly the binary representations for 0, 4, 2, 6, 1, 5, 3, 7.

Now notice that we get from the initial natural ordering

$\underline{f}[0]$	000
$\underline{f}[1]$	001
$\underline{f}[2]$	010
$\underline{f}[3]$	011
$\underline{f}[4]$	100
$\underline{f}[5]$	101
$\underline{f}[6]$	110
$\underline{f}[7]$	111

to the ordering that we feed in,

$\underline{f}[0]$	000
$\underline{f}[4]$	100
$\underline{f}[2]$	010
$\underline{f}[6]$	110
$\underline{f}[1]$	001
$\underline{f}[5]$	101
$\underline{f}[3]$	011
$\underline{f}[7]$	111

by *reversing the binary representation* of the numbers in the first table.

What happens to $\underline{f}[6]$ in the sort, for example? 6 in binary is 110. That's bottom-bottom-top: the first sort puts $\underline{f}[6]$ in the top 4-list, and the second sort generates 4 "2-lists," two top "2-lists" and two bottom "2-lists," and puts $\underline{f}[6]$ (along with $\underline{f}[2]$) in the bottom of the two top 2-lists. The final sort puts $\underline{f}[6]$ just below $\underline{f}[2]$. The slot for $\underline{f}[6]$ in the final sort, corresponding to "bottom-bottom-top," is the fourth one down — that's 110, the reverse of binary 011 (the fourth slot in the original ordering).

This same procedure for sorting works for all N. In summary:

(1) Write the numbers 0 to $N-1$ in binary (with leading 0's so all numbers have the same length). That enumerates the slots from 0 to $N-1$.

(2) Reverse the binary digits of each slot number. For a binary number m call this reversed number $\overleftarrow{m}$.

(3) The input $\underline{f}[\overleftarrow{m}]$ goes in slot m.

This step in the FFT algorithm is called *bit reversal*, for obvious reasons. In fact, people have spent plenty of time coming up with efficient bit reversal algorithms.[15] In running an FFT routine, as in MATLAB, you don't do the sorting, of course. The program takes care of that. If, in the likely case, you don't happen to have 2^n samples in whatever data you've collected, then a common dodge is to add zeros to get up to the closest power of 2. This is referred to as *zero padding*, and FFT routines will automatically do it for you. But, like anything else, it can be dangerous if used improperly. We discuss zero padding below, and there are some problems on it.

Bit reversal via permutation matrices. To write down the permutation matrix that does this sorting, you perform the "every other one" algorithm to the rows (or columns) of the $N \times N$ identity matrix, reorder the rows according to that, and then repeat. Thus for $N = 8$ there are two steps:

$$\begin{pmatrix} 1&0&0&0&0&0&0&0\\ 0&1&0&0&0&0&0&0\\ 0&0&1&0&0&0&0&0\\ 0&0&0&1&0&0&0&0\\ 0&0&0&0&1&0&0&0\\ 0&0&0&0&0&1&0&0\\ 0&0&0&0&0&0&1&0\\ 0&0&0&0&0&0&0&1 \end{pmatrix} \text{ sort \& rearrange rows} \rightarrow \begin{pmatrix} 1&0&0&0&0&0&0&0\\ 0&0&1&0&0&0&0&0\\ 0&0&0&0&1&0&0&0\\ 0&0&0&0&0&0&1&0\\ 0&1&0&0&0&0&0&0\\ 0&0&0&1&0&0&0&0\\ 0&0&0&0&0&1&0&0\\ 0&0&0&0&0&0&0&1 \end{pmatrix}$$

$$\text{sort \& rearrange top \& bottom halves} \rightarrow \begin{pmatrix} 1&0&0&0&0&0&0&0\\ 0&0&0&0&1&0&0&0\\ 0&0&1&0&0&0&0&0\\ 0&0&0&0&0&0&1&0\\ 0&1&0&0&0&0&0&0\\ 0&0&0&0&0&1&0&0\\ 0&0&0&1&0&0&0&0\\ 0&0&0&0&0&0&0&1 \end{pmatrix}.$$

[15] See P. Rosel, Timing of some bit reversal algorithms, *Signal Processing* **18** (1989), 425–433, for a survey of 12 (!) different bit-reversal algorithms.

And sure enough

$$\begin{pmatrix} 1 & 0 & 0 & 0 & 0 & 0 & 0 & 0 \\ 0 & 0 & 0 & 0 & 1 & 0 & 0 & 0 \\ 0 & 0 & 1 & 0 & 0 & 0 & 0 & 0 \\ 0 & 0 & 0 & 0 & 0 & 0 & 1 & 0 \\ 0 & 1 & 0 & 0 & 0 & 0 & 0 & 0 \\ 0 & 0 & 0 & 0 & 0 & 1 & 0 & 0 \\ 0 & 0 & 0 & 1 & 0 & 0 & 0 & 0 \\ 0 & 0 & 0 & 0 & 0 & 0 & 0 & 1 \end{pmatrix} \begin{pmatrix} \underline{f}[0] \\ \underline{f}[1] \\ \underline{f}[2] \\ \underline{f}[3] \\ \underline{f}[4] \\ \underline{f}[5] \\ \underline{f}[6] \\ \underline{f}[7] \end{pmatrix} = \begin{pmatrix} \underline{f}[0] \\ \underline{f}[4] \\ \underline{f}[2] \\ \underline{f}[6] \\ \underline{f}[1] \\ \underline{f}[5] \\ \underline{f}[3] \\ \underline{f}[7] \end{pmatrix}.$$

7.11.4. Zero padding. As we have seen, the FFT algorithm for computing the DFT is set up to work with an input length that is a power of 2. While not all implementations of the FFT require an input to be of that length, many programs only accept inputs of certain lengths, and when this requirement is not met, it's common to add enough zeros to the end of the signal to bring the input up to the length required. That's zero padding. Let's spell it out. Let $\underline{f} = (\underline{f}[0], \underline{f}[1], \dots, \underline{f}[N-1])$ be the original input. For an integer $M > N$, define

$$\underline{g}[n] = \begin{cases} \underline{f}[n], & 0 \le n \le N-1\,, \\ 0, & N \le n \le M-1\,. \end{cases}$$

Then

$$\underline{G}[m] = \underline{\mathcal{F}}_M \underline{g}[m] = \sum_{n=0}^{M-1} \omega_M^{-mn} \underline{g}[n] = \sum_{n=0}^{N-1} \omega_M^{-mn} \underline{f}[n]\,.$$

Work a little bit with ω_M^{-mn}:

$$\omega_M^{-mn} = e^{-2\pi i m n/M} = e^{-2\pi i m n N/MN} = e^{-2\pi i n(mN/M)/N} = \omega_N^{n(mN/M)}\,.$$

Thus whenever mN/M is an integer, we have

$$\underline{G}[m] = \sum_{n=0}^{N-1} \omega_N^{-n(mN/M)} \underline{f}[n] = \underline{F}[mN/M]\,.$$

We could also write this equation for $\underline{F}$ in terms of $\underline{G}$ as

$$\underline{F}[m] = \underline{G}[mM/N]$$

whenever mM/N is an integer. This is what we're more interested in: the program computes the zero padded transform $\underline{G} = \underline{\mathcal{F}}_M \underline{g}$, and we'd like to know what the outputs $\underline{F}[m]$ of our original signal are in terms of the $\underline{G}$'s. The answer is that the mth component of $\underline{F}$ is the mM/Nth component of $\underline{G}$ whenever mM/N is an integer.

We pursue this, starting with getting rid of the ridiculous proviso that mM/N is an integer. We can *choose* M, so let's choose

$$M = kN$$

for some integer k; so M is twice as large as N, or 3 times as large as N, or whatever. Then $mM/N = km$, always an integer, and

$$\underline{F}[m] = \underline{G}[km]\,,$$

which is much easier to say in words:

- If $\underline{f}$ is zero padded to a signal $\underline{g}$ of length M, where $M = kN$, then the mth component of $\underline{F} = \underline{\mathcal{F}}\,\underline{f}$ is the kmth component of $\underline{G} = \underline{\mathcal{F}}\,\underline{g}$.

Time and frequency grids. Remember the two grids, in time and in frequency, that arise when we use a discrete approximation of a continuous signal? Zero padding the inputs has an important consequence for the spacing of the grid points in the frequency domain. Suppose that the discrete signal $\underline{f} = (\underline{f}[0], \underline{f}[1], \dots, \underline{f}[N-1])$ comes from sampling a continuous signal $f(t)$ at points t_n, so that $\underline{f}[n] = f(t_n)$. Suppose also that the N sample points in the time domain of $f(t)$ are spaced Δt apart. Then the length of the interval on which $f(t)$ is defined is $N\Delta t$, and the spectrum $\mathcal{F}f(s)$ is spread out over an interval of length $1/\Delta t$. Remember, knowing N and Δt determines everything. Going from N inputs to $M = kN$ inputs by padding with zeros lengthens the interval in the time domain to $M\Delta t$ but it doesn't change the spacing of the sample points. For the sample points associated with the discrete signals $\underline{f}$ and $\underline{F} = \underline{\mathcal{F}}_N\,\underline{f}$ we have

$$\Delta t \Delta\nu_{\text{unpadded}} = \frac{1}{N}$$

by the reciprocity relations (see Section 7.5), and for $\underline{g}$ and $\underline{G} = \underline{\mathcal{F}}_M\,\underline{g}$ we have

$$\Delta t \Delta\nu_{\text{padded}} = \frac{1}{M} = \frac{1}{kN}\,.$$

The Δt in both equations is the same, so

$$\frac{\Delta\nu_{\text{padded}}}{\Delta\nu_{\text{unpadded}}} = \frac{1}{k} \quad \text{or} \quad \Delta\nu_{\text{padded}} = \frac{1}{k}\Delta\nu_{\text{unpadded}}\,;$$

that is, the spacing of the sample points *in the frequency domain* for the padded sequence has decreased by the factor $1/k$. At the same time, the total extent of the grid in the frequency domain has not changed because it is $1/\Delta t$ and Δt has not changed. What this means is:

- Zero padding in the time domain *refines* the grid in the frequency domain.

There's a warning that goes along with this. Using zero padding to refine the grid in the frequency domain is a valid thing to do only if the original continuous signal f is already known to be zero outside of the original interval. If not, then you're killing off real data by filling $\underline{f}$ out with zeros. See the book by Briggs and Henson for this and for other important practical concerns.

Problems and Further Results

7.1. *Discrete triangle function*

We can obtain a discrete triangle function $\underline{\Lambda}[n]$ by sampling the usual triangle function $\Lambda(t)$. We need to specify the length, N, of $\underline{\Lambda}[n]$ and the sampling rate m. Define

$$\underline{\Lambda}[n] = \Lambda(n/m), \quad n = -N, \ldots, N\,.$$

For the plots below, what are the values of m and N?

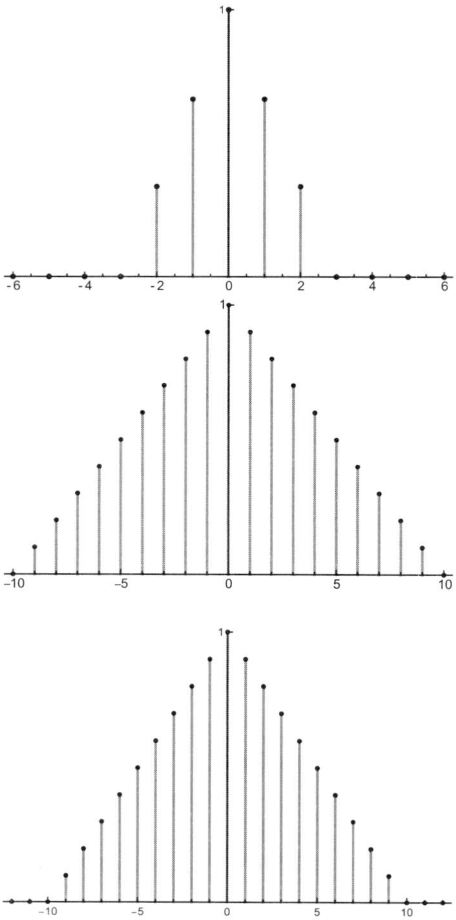

7.2. *Convolving discrete rectangle functions*

We're familiar with the result $\Pi * \Pi = \Lambda$ for the continuous-time rectangle and triangle functions. Use software to try this out with discrete signals. Take, e.g., $\underline{\Pi} = \sum_{n=-4}^{4} \underline{\delta}_n$ and plot $\underline{\Pi} * \underline{\Pi}$. Verify algebraically what you see. Now use $\underline{\Pi}' = (1/2)(\underline{\delta}_{-4} + \underline{\delta}_4) + \sum_{n=-3}^{3} \underline{\delta}_n$ (i.e., making the value 1/2 at the endpoints), and plot $\underline{\Pi}' * \underline{\Pi}'$. How does this compare to your first plot? Try some other experiments.

7.3. *Practice with the DFT*

Find the discrete Fourier transform of each of the sequences below. (You should *not* have to grind through all the calculations!) *Note*: You might check your results in MATLAB, or another package, but the point of the problem is to get some hands-on experience with calculating DFTs.

(a) $\underline{f} = (1, 1, 1, 1, 1, 1, 1, 1)$.
(b) $\underline{f} = (1, 1, 1, 1, 0, 0, 0, 0)$.
(c) $\underline{f} = (0, 0, 1, 1, 1, 1, 0, 0)$.
(d) $\underline{f} = (1, 1, -1, 1, -1, 1, -1, 1)$.
(e) $\underline{f} = (0, 0, 1, 0, 0, 0, 1, 0)$.

7.4. *DFT of real-valued signals*

Let $\underline{f}$ and $\underline{g}$ be real discrete signals of length N with DFTs $\underline{F}$ and $\underline{G}$, respectively. Consider the possibility of computing both of these DFTs simultaneously by using an N-point DFT on the complex signal $\underline{h}[n] = \underline{f}[n] + i\underline{g}[n]$. If $\underline{H} = \underline{\mathcal{F}}\,\underline{h}$, find separate expressions (if possible) for $\underline{F}$ and $\underline{G}$ in terms of $\underline{H}$. If it is not possible to separate $\underline{F}$ and $\underline{G}$ from $\underline{H}$, explain.

7.5. *A true story*

Professor Osgood and a graduate student were working on a discrete form of the sampling theorem. This included looking at the DFT of the discrete rect function

$$\underline{f}[n] = \begin{cases} 1, & |n| \le \frac{N}{6}, \\ 0, & -\frac{N}{2} + 1 \le n < -\frac{N}{6}, \quad \frac{N}{6} < n \le \frac{N}{2}. \end{cases}$$

The grad student, ever eager, said, "Let me work this out." A short time later the student came back saying, "I took a particular value of N and I plotted the DFT using MATLAB (their FFT routine). Here are plots of the real part and the imaginary part."

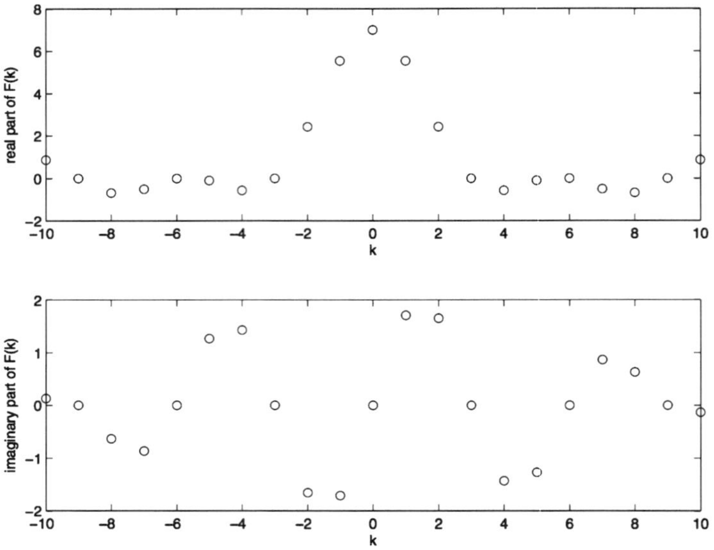

(a) Produce these figures.

Professor Osgood said, "That can't be correct."

(b) Is Professor Osgood right to object? If so, state what the basis of his objection is and produce the correct plot. If not, explain why the student is correct.[16]

7.6. *DFT variety*

Consider two sequences $\underline{f}[n]$ and $\underline{g}[n]$ of length $N = 6$. These two sequences are related by the DFT relationship $\underline{F}[m] = (-1)^m \underline{G}[m]$. The sequence $\underline{g}[n]$ is shown below.

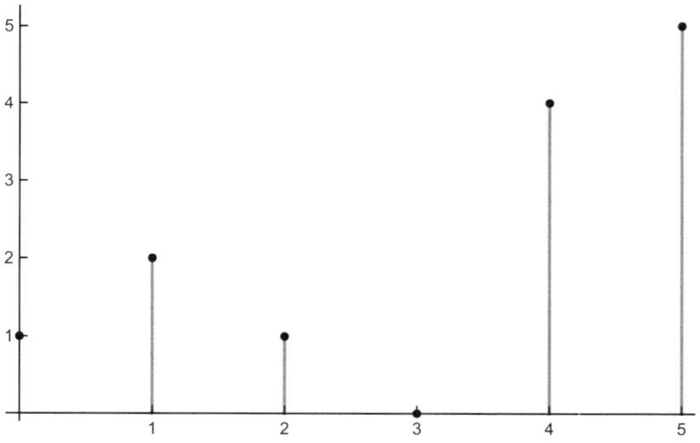

(a) Express the sequence $\underline{f}[n]$ in terms of $\underline{g}[n]$. Also write out the sequence $\underline{f}[n]$ as an N-tuple $(\underline{f}[0], \ldots, \underline{f}[5])$.
Hint: Write $(-1)^m$ as a complex exponential.

[16] Readers of this book will know that Professor Osgood has (on occasion) been wrong.

(b) Evaluate the summation $\sum_{m=0}^{5} |\underline{F}[m]|^2$.
(c) Let $\underline{h} = \underline{f} * \underline{g}$. Evaluate $\underline{h}[4]$, the fifth element of this convolution.

7.7. *Approximation of continuous Fourier transform*

We can view the DFT as a discrete approximation to the Fourier transform of a continuous-time signal, and we know how to take the DFT in MATLAB (fft, fftshift, ifftshift, ...) by sampling the continuous-time signal. In this problem, we want to approximate the Fourier transform of the timelimited signal $f(t)$ shown below.

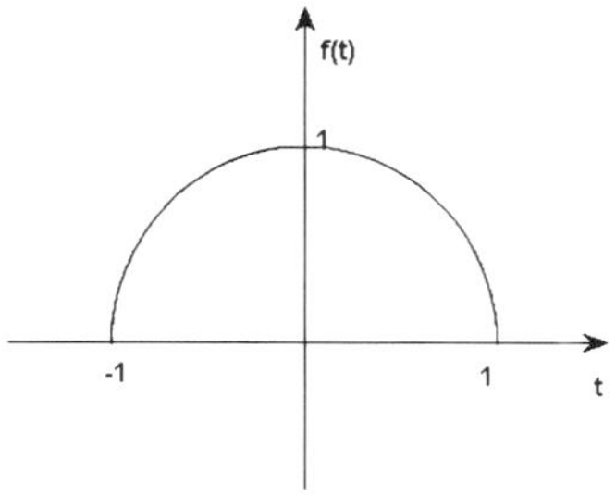

We sample $f(t)$ in three different versions, sample (a), sample (b), and sample (c):

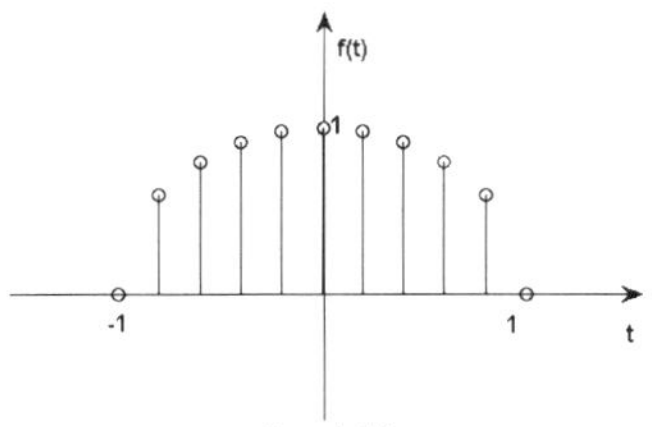

Sample (a)

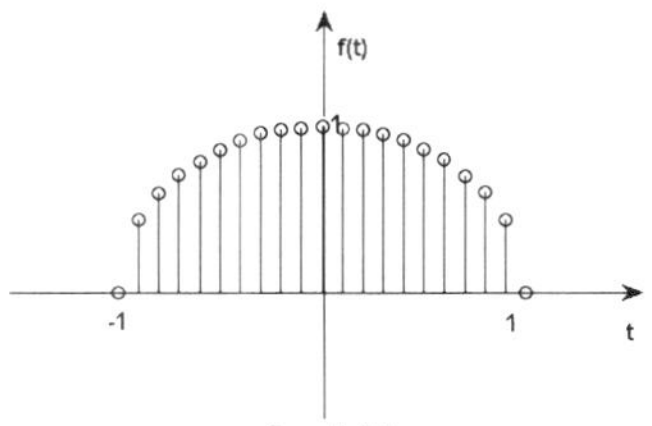

Sample (b)

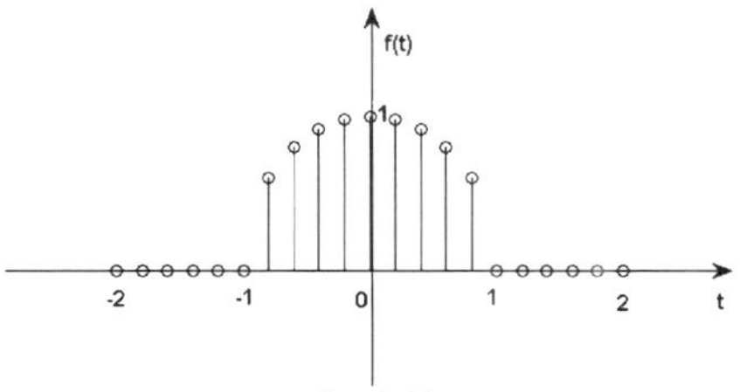

Sample (c)

Don't worry — you don't need to evaluate the Fourier transform of $f(t)$ or write a MATLAB code. Possible results, obtained using MATLAB, are shown below. What you need to do is match (a), (b), and (c) to the approximation of the Fourier transform in (A) through (F). *Hint*: Start with finding a plot that matches sample (a) first. Remember the relationship $2BL = N$ between the bandwidth and the limitation in time.

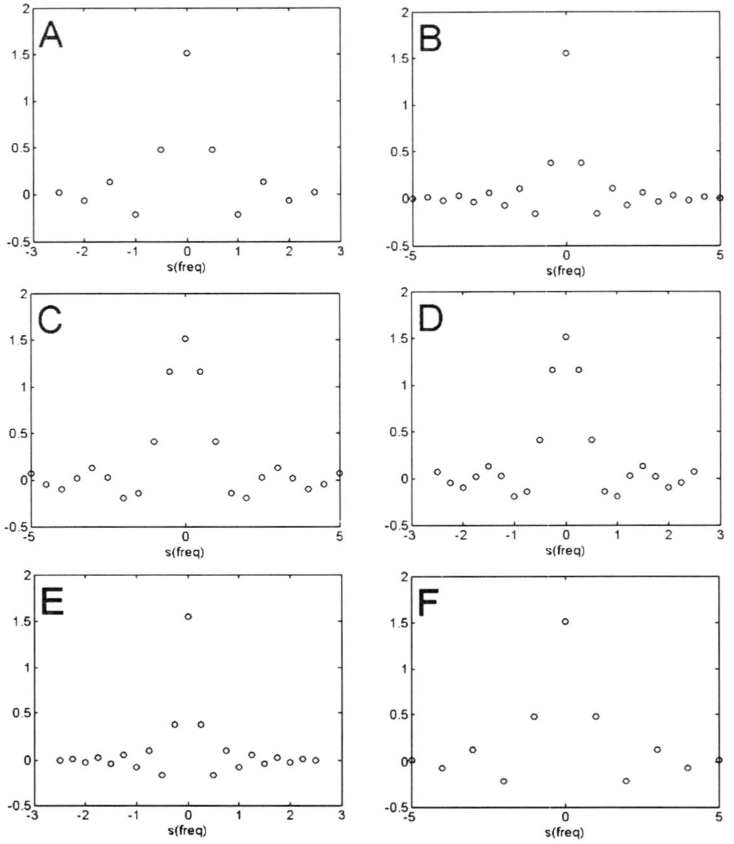

7.8. *Sampling a periodic signal: Another interpretation of the DFT*

Let $x(t)$ be a periodic signal of period 1. Let $N > 1$ and sample $x(t)$ by multiplying by a finite Ш-function of spacing $1/N$, forming

$$g(t) = x(t) \sum_{k=0}^{N-1} \delta(t - k/N).$$

Let c_n be the nth Fourier coefficient,

$$c_n = \int_{-1/2}^{1/2} e^{-2\pi i n t} g(t)\, dt = \int_0^1 e^{-2\pi i n t} g(t)\, dt.$$

(The integral from $-1/2$ to $1/2$ is more convenient to use for the calculations in this problem.)

Finally, let $\underline{f}$ be the discrete signal with values at the sample points,

$$\underline{f}[n] = x(n/N), \quad n = 0, 1, \ldots, N-1.$$

What is the relationship between c_n and the DFT of $\underline{f}[n]$?

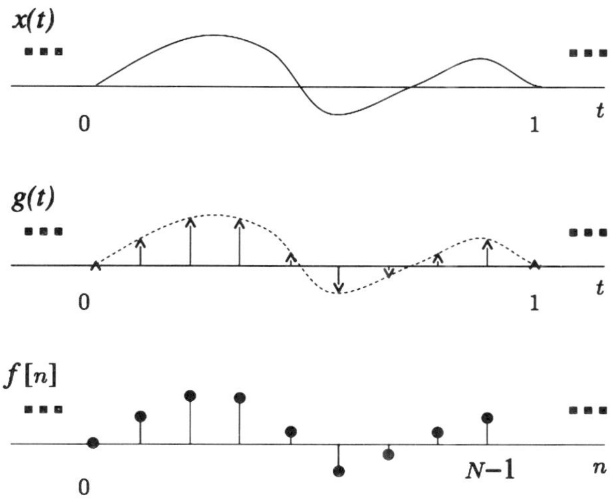

7.9. *Some particular cases*[17]

Let $\underline{x} = (p, q, r, s, t, u)$ be a vector of length 6 and assume that $\underline{x}$ has the 6-point DFT $\underline{\mathcal{F}}\,\underline{x} = (P, Q, R, S, T, U)$, where $p, \dots, u, P, \dots, U$ are complex scalars. For each signal (i)–(v), indicate its corresponding discrete Fourier transform from parts (A)–(J). Provide a *brief* explanation for each answer.

Note that the 3-point DFT is computed for the signal of length 3 in part (i) and the 2-point DFT is computed for the signal of length 2 in part (ii).

(i) $(p + s, q + t, r + u)$.
(ii) (p, s).
(iii) $(p, -q, r, -s, t, -u)$.
(iv) (s, t, u, p, q, r).
(v) $(p, q, r, s, t, u) * (1, \cos(2\pi/3), \cos(4\pi/3), \cos(6\pi/3), \cos(8\pi/3), \cos(10\pi/3))$.

(A) $\frac{1}{3}(P + R + T, Q + S + U)$.
(B) (P, S).
(C) $\frac{1}{2}(P + S, Q + T, R + U)$.
(D) (S, T, U, P, Q, R).
(E) (P, R, T).
(F) (T, U, P, Q, R, S).
(G) $(P, -Q, R, -S, T, -U)$.
(H) $3(0, 0, R, 0, T, 0)$.
(I) $(P + S, 0, Q + T, 0, R + U, 0)$.

7.10. *Duality relation in matrix form*

One of the duality relations for the DFT is $\underline{\mathcal{F}}\,\underline{\mathcal{F}}\,\underline{f} = N\underline{f}^{-}$. What is the matrix form of this; i.e., what is the matrix $\underline{\mathcal{F}}\,\underline{\mathcal{F}}$?

---

[17] From D. Kammler.

7.11. *DFT properties*

(a) Prove the shift theorem for the discrete Fourier transform:

$$\underline{\mathcal{F}}(\tau_p \underline{f}) = \underline{\omega}^{-p} \underline{\mathcal{F}}\, \underline{f},$$

where

$$\tau_p \underline{f}[n] = \underline{f}[n-p].$$

(b) *Replication.* Suppose that the signal $\underline{f} = (\underline{f}[0], \underline{f}[1], \dots, \underline{f}[N-1])$ has discrete Fourier transform $\underline{F}$. We create a new signal $\underline{g}[n]$, $n = 0, 1, \dots, 2N-1$, with *twice* the number of points defined by

$$\underline{g}[n] \quad = \quad \begin{cases} \underline{f}[n], & n = 0, 1, \dots, N-1\,, \\ \underline{f}[n-N], & n = N, N+1, \dots, 2N-1\,. \end{cases}$$

Find the DFT of $\underline{g}$ in terms of $\underline{F}$.

(c) *Simple upsampling.* Again suppose that the signal $\underline{f}$, of size N, has discrete Fourier transform $\underline{F}$. We create a new signal $\underline{h}$ of size $2N$, $n = 0, 1, \dots, 2N-1$, with twice the number of points, differently, by inserting 0's among the values $\underline{f}[n]$; i.e.,

$$\underline{h}[n] \quad = \quad \begin{cases} \underline{f}[n/2], & n \text{ even}\,, \\ 0, & n \text{ odd}\,. \end{cases}$$

Find the DFT of $\underline{h}$ in terms of $\underline{F}$.

(d) *Simple downsampling.* With $\underline{f}$ as above, we create another new signal $\underline{g}$, with *half* the number of points, $n = 0, 1, \dots, N/2 - 1$ (assume that N is even) by keeping only the values of $\underline{f}[n]$ at even indices; i.e.,

$$\underline{g}[n] = \underline{f}[2n]\,.$$

Find the DFT of $\underline{g}$ in terms of $\underline{F}$.

7.12. *The DFT, upsampling, and linear interpolation*

(a) Let $\underline{y}$ be the discrete signal, periodic of order M,

$$\underline{y} = \left(1\,, \frac{1}{2}\,, 0\,, \dots\,, 0\,, \frac{1}{2}\right).$$

Show that its DFT is

$$\underline{Y}[m] = 1 + \cos(2\pi m/M)\,.$$

(b) Let $\underline{f} = (\underline{f}[0]\,, \underline{f}[1]\,, \dots\,, \underline{f}[N-1])$ be a discrete signal and let $\underline{F} = (\underline{F}[0]\,, \underline{F}[1]\,, \dots\,, \underline{F}[N-1])$ be its DFT. Recall that the upsampled version of $\underline{f}$ is the signal $\underline{h}$ of order $2N$ obtained by inserting zeros between the values of $\underline{f}$; i.e.,

$$\underline{h} = (\underline{f}[0]\,, 0\,, \underline{f}[1]\,, 0\,, \underline{f}[2]\,, \dots\,, 0\,, \underline{f}[N-1]\,, 0)\,.$$

Show that $\underline{\widetilde{f}} = \underline{h} * \underline{y}$ is the linearly interpolated version of $\underline{f}$:

$$\left(\underline{f}[0]\,, \frac{\underline{f}[0] + \underline{f}[1]}{2}\,, \underline{f}[1]\,, \frac{\underline{f}[1] + \underline{f}[2]}{2}\,, \underline{f}[2]\,, \dots\,, \underline{f}[N-1]\,, \frac{\underline{f}[N-1] + \underline{f}[0]}{2}\right).$$

Hint: Here we take $M = 2N$ for the period of $\underline{y}$. Note that

$$\underline{y} = \underline{\delta}_0 + \frac{1}{2}\underline{\delta}_1 + \frac{1}{2}\underline{\delta}_{2N-1}$$

and remember the effect of convolving with a shifted discrete δ. Line up the $2N$-tuples.

(c) In Problem 7.11 you showed that the DFT of $\underline{h}$ is a replicated form of $\underline{F}$,

$$\underline{H}[m] = \underline{F}[m], \quad m = 0, 1, \ldots, 2N-1,$$

$$\underline{H} = (\overbrace{\underbrace{\underline{F}[0], \underline{F}[1], \ldots, \underline{F}[N-1]}_{\underline{F}}, \underbrace{\underline{F}[0], \underline{F}[1], \ldots, \underline{F}[N-1]}_{\underline{F}}}^{\underline{H}}).$$

Assuming this, find the DFT of $\widetilde{\underline{f}}$.

7.13. *Upsampling by a factor L*

This is the first of several problems generalizing and working with upsampling and downsampling from Problem 7.11. Upsampling first. Suppose $\underline{f}$ is a signal of length N and let L be a positive integer. From $\underline{f}$ we manufacture a signal $\underline{g}$ of length LN by inserting $L-1$ zeros after each $\underline{f}[m]$, $m = 0, \ldots, N-1$. For example, if $N = 4$ and $L = 5$, then

$$\underline{g} = (\underline{f}[0], 0, 0, 0, 0, \underline{f}[1], 0, 0, 0, 0, \underline{f}[2], 0, 0, 0, 0, \underline{f}[3], 0, 0, 0, 0),$$

a signal of length $20 = 4 \cdot 5$. There are 4 blocks of length 5 of the form $\underline{f}[k], 0, 0, 0, 0$ where k goes from 0 to 3.

More compactly, $\underline{g}$ is defined by

$$\underline{g}[m] = \begin{cases} \underline{f}[m/L], & \text{if } L \text{ divides } m, \\ 0, & \text{otherwise,} \end{cases}$$

for $m = 0, 1, \ldots, LN-1$. We say that $\underline{g}$ is obtained from $\underline{f}$ by upsampling by a factor L.

Show that $\underline{\mathcal{F}}\,\underline{g}$ is a concatenation of copies of $\underline{\mathcal{F}}\,\underline{f}$.

This more general upsampling is usually accompanied by some kind of interpolation (e.g., as in Problem 7.12) to replace the inserted zeros by values that more reasonably interpolate the given sample values. A simple scheme is to convolve the upsampled signal with a dinc whose DFT (a $\underline{\Pi}$) is wide enough to isolate one copy of $\underline{\mathcal{F}}\,\underline{f}$; the result is a digital low-pass filter. There's a lot that can be done, and thus one enters the world of digital filters, filter banks, etc.

7.14. *Downsampling by a factor P*

Here's a more general take on downsampling. As aways, let $\underline{f}$ be a signal of length N, and suppose that P divides N, say $N = MP$ for a positive integer M. To downsample $\underline{f}$ by the factor P is to form the signal $\underline{g}$ of length $N/P = M$ by taking every Pth sample of $\underline{f}$. Thus

$$\underline{g}[m] = \underline{f}[mP], \quad m = 0, 1, 2, \ldots, M-1.$$

For example, if $N = 20$ and $P = 4$, then $M = 5$ and

$$\underline{g} = (\underline{f}[0], \underline{f}[4], \underline{f}[8], \underline{f}[12], \underline{f}[16]) .$$

(a) Express $\underline{\mathcal{F}}\,\underline{g}$ as a convolution of a discrete Ш with $\underline{\mathcal{F}}\,\underline{f}$ (in general, not just for the example). Write the expression for $\underline{\mathcal{F}}\,\underline{g}[m]$ at an index m, and show that

$$\underline{\mathcal{F}}\,\underline{g}[0] = \frac{1}{P}\sum_{k=0}^{P-1} \underline{\mathcal{F}}\,\underline{f}\Big[\frac{kN}{P}\Big].$$

(b) What conditions must the spectrum $\underline{\mathcal{F}}\,\underline{f}$ satisfy to avoid overlaps in the spectrum $\underline{\mathcal{F}}\,\underline{g}$ due to the periodizing that you observed in (a)? How could you achieve this prior to downsampling?

Upsampling (together with interpolation) and downsampling are often combined to change the original sampling rate by a fractional amount. This area of digital signal processing is called, naturally, *sample-rate conversion.* Let me quote Wikipedia:

> Sample-rate conversion is the process of changing the sampling rate of a discrete signal to obtain a new discrete representation of the underlying continuous signal. Application areas include image scaling and audio/visual systems, where different sampling rates may be used for engineering, economic, or historical reasons.
>
> For example, Compact Disc Digital Audio and Digital Audio Tape systems use different sampling rates, and American television, European television, and movies all use different frame rates. Sample-rate conversion prevents changes in speed and pitch that would otherwise occur when transferring recorded material between such systems.

7.15. *Time domain multiplexing*

Let $\underline{f}$, $\underline{g}$, $\underline{h}$ be length N signals with discrete Fourier transforms $\underline{F}$, $\underline{G}$, $\underline{H}$, respectively.

(a) We combine $\underline{f}$ and $\underline{g}$ to form the $2N$-dimensional signal $\underline{w}$,

$$\underline{w}[n] = \begin{cases} \underline{f}[n/2], & \text{for } n \text{ even}, \\ \underline{g}[(n-1)/2], & \text{for } n \text{ odd}, \end{cases}$$

where $0 \le n \le 2N - 1$. Find the discrete Fourier transform $\underline{W} = \underline{\mathcal{F}}\,\underline{w}$ in terms of $\underline{F}$ and $\underline{G}$.

Hint: Consider constructing $\underline{w}$ from upsampled versions of $\underline{f}$ and $\underline{g}$.

(b) How does your answer to part (a) simplify if $\underline{g} = \underline{0}$? Explain why this makes sense given what you know about replication and upsampling.

(c) Suppose we defined $\underline{w}[n]$ for $0 < n < 3N - 1$ by

$$\underline{w}[n] = \begin{cases} \underline{f}[n/3], & \text{for } n = 0 \mod 3, \\ \underline{g}[(n-1)/3], & \text{for } n = 1 \mod 3, \\ \underline{h}[(n-2)/3], & \text{for } n = 2 \mod 3. \end{cases}$$

Describe how you expect $\underline{W}$ to depend on $\underline{F}$, $\underline{G}$, and $\underline{H}$. You do not need to do any calculations for this part; you may generalize from part (a). To keep everything simple and concrete, set $N = 4$ for this part.

7.16. *Summation rule of the DFT*

Suppose $\underline{f}$ is an mN-periodic function and $\underline{g}$ is defined as

$$\underline{g}[n] = \sum_{l=0}^{m-1} \underline{f}[n + lN], \quad n = 0, 1, \ldots, N-1.$$

(a) Show that $\underline{g}[n]$ is periodic of period N.
(b) Suppose $\underline{f}$ has the DFT $\underline{F}$. What is $\underline{G}[k]$ in terms of $\underline{F}$?
(c) Suppose $\underline{f} = (a, b, c, d, e, f)$ has the DFT $\underline{F} = (A, B, C, D, E, F)$. What is the DFT of $\underline{g} = (a + c + e, b + d + f)$?

7.17. *Discrete correlation*

Let $\underline{f}$ and $\underline{g}$ be discrete, real signals, each periodic of period N. Define their correlation by the formula

$$(\underline{f} \star \underline{g})[m] = \sum_{n=-\frac{N}{2}+1}^{\frac{N}{2}} \underline{f}[n]\underline{g}[n+m]\,.$$

(a) Show that

$$\underline{\mathcal{F}}(\underline{f} \star \underline{g}) = \overline{\underline{\mathcal{F}}\,\underline{f}}\,\underline{\mathcal{F}}\,\underline{g}\,.$$

Use the reversed signal to write the correlation in terms of convolution.
(b) Give an upper bound for $(\underline{f} \star \underline{g})[0]$.

7.18. *DFT's frequency response and spectral leakage*

Compute the DFT of the discrete signal

$$\underline{f}[k] = e^{2\pi i x k/N}\,;$$

i.e., find a closed form expression for $\underline{\mathcal{F}}\,\underline{f}$. This is the *frequency response* of the DFT.

Qualitatively, how is the spectrum different when x is an integer and when x is not an integer? Plot the magnitude of the frequency response for $x = 2.5$ and $N = 8$. Does the plot agree with what you expect?

7.19. *Leakage et al., again*[18]

Consider the periodic signal $f(t) = \cos\left(2\pi \frac{5}{32} t\right)$. The signal and its samples are shown below.

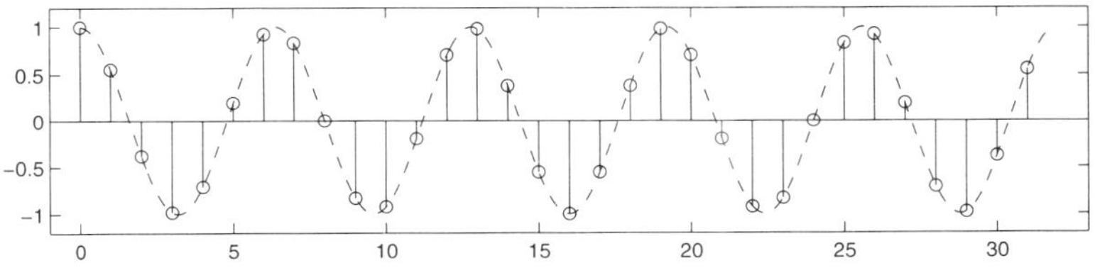

[18]From J. Gill.

(a) Sketch the continuous Fourier transform of $f(t)$.
(b) Sketch the DFT of the 32-point sample sequence $\big(f(0), f(1), \ldots, f(31)\big)$.
(c) Sketch the DFT of the 16-point sample sequence $\big(f(0), f(2), \ldots, f(30)\big)$.
(d) The DFTs sketched in parts (b) and (c) should be consistent with the continuous Fourier transform of $f(t)$. Explain why this is expected. In parts (a) and (b) the sampling rates (1 and 1/2, respectively) are larger than the Nyquist rate, 5/16. This means that there will be no aliasing in the sampled signal, so the DFT of the sampled signal will look like the frequency response of f.
(e) The DFT of the first 16 samples $\big(f(0), f(1), \ldots, f(15)\big)$ has the following plot (the DFT is purely imaginary).

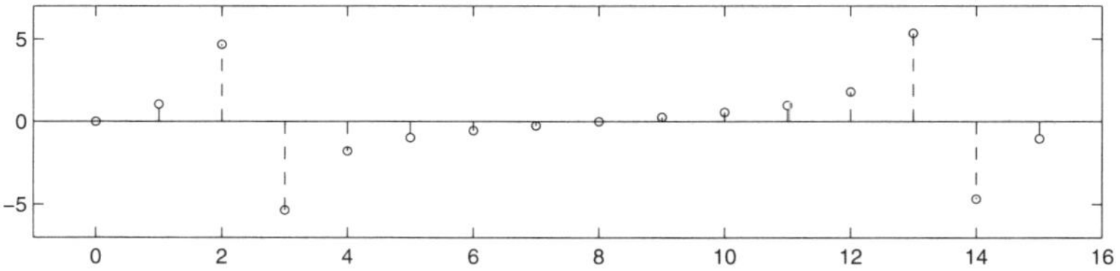

This DFT is qualitatively different from the DFTs of parts (b) and (c). Explain why the above DFT does *not* agree with the continuous Fourier transform of $f(t)$.

7.20. *A discrete rectangle function for an interval*

Here's a variation of the discrete rectangle function and its DFT: again we want a signal that's 1 over a block of consecutive indices (which we think of as a discrete version of an interval) and 0 off that block. Take a set $\mathcal{I}$ of at most N consecutive integers and let

$$\underline{\Pi}_{\mathcal{I}} = \sum_{k \in \mathcal{I}} \underline{\delta}_k \,.$$

Show that

$$\underline{\mathcal{F}\,\Pi}_{\mathcal{I}}[m] = \begin{cases} |\mathcal{I}|\,, & m \equiv 0 \mod N, \\ \omega^{-m\mathcal{I}_{\text{mid}}} \dfrac{\sin\left(\frac{\pi m |\mathcal{I}|}{N}\right)}{\sin\left(\frac{\pi m}{N}\right)}\,, & m \not\equiv 0 \mod N\,, \end{cases}$$

where $|\mathcal{I}|$ is the number of elements in $\mathcal{I}$ and

$$\mathcal{I}_{\text{mid}} = \frac{\alpha + \beta}{2} = \frac{\min \mathcal{I} + \max \mathcal{I}}{2}.$$

This is supposed to suggest the midpoint of $\mathcal{I}$, except that $\mathcal{I}_{\text{mid}}$ need not be an integer. Nevertheless, $\mathcal{I}_{\text{mid}}$ is a perfectly fine number and makes sense in the formula.

7.21. *A discrete uncertainty principle*[19]

In the chapter on convolution, we proved the uncertainty principle for continuous-time signals, the main idea being that time width and frequency width are inversely related. A similar relationship holds for discrete signals,

[19] From A. Siripuram.

as we will see in this problem. Let $\underline{f}$ be an N-length vector and let $F = \mathcal{F}\underline{f}$ be its DFT. Let $\phi(\underline{f})$ indicate the number of nonzero coordinates in $\underline{f}$, and similarly let $\phi(\mathcal{F}\underline{f})$ be the number of nonzero coordinates in $F = \mathcal{F}\underline{f}$. Then the discrete uncertainty principle states that

$$\phi(\underline{f})\phi(\mathcal{F}\underline{f}) \geq N.$$

In other words, the time width ($\phi(\underline{f})$) and frequency width ($\phi(\mathcal{F}\underline{f})$) are inversely related.

Below is an outline of the proof of this uncertainty principle. Complete the proof by briefly justifying the statements (a)–(f) in the proof.

Proof. Let the vector s denote the sign of $\underline{f}$:

$$s[n] := \begin{cases} 1 & \text{if } \underline{f}[n] > 0\,, \\ -1 & \text{if } \underline{f}[n] < 0\,, \\ 0 & \text{if } \underline{f}[n] = 0\,, \end{cases}$$

and let (f, s) denote the dot product of the vectors f and s. Assume that $f \neq 0$. Defining $s[n]$ gives us the advantage of expressing $|f[n]|$ as $|f[n]| = f[n]s[n]$.

Now

$$\begin{aligned} \max_m |F[m]| &\overset{(a)}{\leq} \sum_{n=0}^{N-1} |f[n]| \\ &\overset{(b)}{=} (f, s) \\ &\overset{(c)}{\leq} (f, f)^{1/2}(s, s)^{1/2} \\ &\overset{(d)}{=} \frac{1}{\sqrt{N}}(F, F)^{1/2}(s, s)^{1/2} \\ &\overset{(e)}{=} \frac{1}{\sqrt{N}}(F, F)^{1/2}\phi(f)^{1/2} \\ &\overset{(f)}{\leq} \frac{\max_m |F[m]|}{\sqrt{N}}\phi(F)^{1/2}\phi(f)^{1/2}. \end{aligned}$$

Canceling out $\max_m |F[m]|$ on both sides, we are left with the desired inequality

$$\phi(\underline{f})\phi(F) \geq N.$$

7.22. *Radix-3 FFT*

Let $f[n]$ be a 9-point sequence. Show that the order-9 DFT, $\underline{F}[m] = (\underline{\mathcal{F}}f)[m]$, can be expressed in terms of the order-3 DFTs of the following 3-point sequences:

$$\begin{aligned} \underline{f}_A &= (f[0],\, f[3],\, f[6])\,, \\ \underline{f}_B &= (f[1],\, f[4],\, f[7])\,, \\ \underline{f}_C &= (f[2],\, f[5],\, f[8])\,, \end{aligned}$$

as

$$\underline{F}[m] = \underline{F}_A[m] + \omega^{-m}\underline{F}_B[m] + \omega^{-2m}\underline{F}_C[m],$$

where $\omega = e^{2\pi i/9}$ and $m = 0, 1, , \dots, 8$. (To make sense of this expression you need to use the fact that $\underline{F}_A[m]$, $\underline{F}_B[m]$, and $\underline{F}_C[m]$ are periodic of period 3.)

This result is the first step in establishing a "radix-3" FFT algorithm.

7.23. *A version of the stretch theorem*

Does the stretch theorem have an analog for periodic signals $\underline{f}$ of length N? Since the signal is periodic, we can certainly define $\sigma_a f[n] = f[an]$ for any integer a (and we assume $a \neq 0$). But for the Fourier transform $\underline{\mathcal{F}}(\underline{f}[an])$ we might think the spectrum scales like m/a, and m/a is an integer only when a divides m. Any hope? Yes, but we need to know a few more facts about congruences mod N.

Instead of dividing, as in forming m/a, we think in terms of a multiplicative inverse, as in forming $a^{-1}m$ where a^{-1} is an element in $\mathbb{Z}_N$ for which $a^{-1}a \equiv 1 \mod N$. When does a number $a \in \mathbb{Z}_N$ have a multiplicative inverse? It does if a and N have no common factors, i.e., if their greatest common divisor is 1. (Check some examples.) The numbers in $\mathbb{Z}_N$ that have a multiplicative inverse are called the *units* of $\mathbb{Z}_N$, and the set of all units in $\mathbb{Z}_N$ is often denoted by $\mathbb{Z}_N^*$. For example, $\mathbb{Z}_4^* = \{1, 3\}$, $\mathbb{Z}_{10}^* = \{1, 3, 7, 9\}$, and if p is any prime, then $\mathbb{Z}_p^* = \{1, 2, \dots, p-1\}$. The second thing we need to know is that if a is any unit in $\mathbb{Z}_N$, then for $n = 0, 1, \dots, N-1$ the numbers an range over all of $\mathbb{Z}_N$; they may not go in order, but all the numbers in $\mathbb{Z}_N$ are hit. You can establish these results yourself or look them up.

Armed with these facts, derive the following discrete version of the stretch theorem: Let a be a unit in $\mathbb{Z}_N$. If $\underline{f}[n] \rightleftharpoons \underline{F}[n]$, then $\underline{f}[an] \rightleftharpoons \underline{F}[a^{-1}n]$. In terms of σ_a we can write this without the variable as $\underline{\mathcal{F}}(\sigma_a \underline{f}) = \sigma_{a^{-1}}(\underline{\mathcal{F}}\underline{f})$. Notice that there's no a^{-1} out front on the right as there is in the continuous stretch theorem.

7.24. What is $\underline{1} * \underline{f}$? What is $\underline{1} * \underline{1}$?, $\underline{1} * \underline{a}$, where $\underline{a} = (a, a, \dots, a)$?

7.25. In the chapter we saw that $\underline{\delta}_a * \underline{\delta}_b = \underline{\delta}_{a+b}$. This useful identity is actually a defining property of convolution if we throw in the algebraic properties that make convolution like multiplication. To make this precise, let $\mathcal{D}_N$ be the collection of discrete signals of period N. This is a (complex) vector space, because if $\underline{f}, \underline{g} \in \mathcal{D}_N$, so is $\underline{f} + \underline{g}$, and likewise if α is a complex number, then $\alpha\underline{f} \in \mathcal{D}_N$ whenever $\underline{f} \in \mathcal{D}_N$. Now show the following:

There is exactly one operation $C(\underline{f}, \underline{g})$, operating on pairs of signals in $\mathcal{D}_N$, and producing an element of $\mathcal{D}_N$, with the following properties:

(*i*) C is bilinear, meaning

$$C(\underline{f} + \underline{g}, \underline{h}) = C(\underline{f}, \underline{h}) + C(\underline{g}, \underline{h}), \quad C(\underline{f}, \underline{g} + \underline{h}) = C(\underline{f}, \underline{g}) + C(\underline{f}, \underline{h}),$$
$$C(\alpha\underline{f}, \underline{g}) = \alpha C(\underline{f}, \underline{g}), \quad C(\underline{f}, \alpha\underline{g}) = \alpha C(\underline{f}, \underline{g}).$$

(*ii*) $C(\underline{\delta}_a, \underline{\delta}_b) = \underline{\delta}_{a+b}$.

Hint: With statements of this type one always proves uniqueness first. That is, *suppose* we have an operator C satisfying the properties. Then what has to happen?

7.26. *Orthonormal discrete sincs*

Consider the discrete functions $\underline{f}_n$, $-\infty < n < \infty$, defined by

$$\underline{f}_n[m] = \operatorname{sinc}(m-n)\,, \quad -\infty < m < \infty.$$

Show that they form an orthonormal family, where we use the inner product of two infinite sequences a_n and b_n defined by

$$\sum_{n=-\infty}^{\infty} a_n \overline{b_n}\,.$$

For this, use the results in Problem 6.2.

7.27. *The discrete time Fourier transform*

In Fourier analysis, different transforms are used to go between the time and frequency domains depending on whether the signal is continuous or discrete in each domain. For example, the DFT is used to transform signals from discrete time to discrete frequency while the Fourier transform transforms a continuous-time signal into a continuous-frequency signal. The *discrete time Fourier transform*, or DTFT, takes a discrete-time signal to a continuous-frequency signal. The DTFT of a signal $\underline{f}[n]$ is defined as

$$F(s) = \mathcal{D}\underline{f}(s) = \sum_{n=-\infty}^{\infty} \underline{f}[n]e^{-2\pi i s n}\,, \quad s \in \mathbb{R}\,.$$

(a) What is the DTFT of $\overline{\underline{f}[n]}$, the complex conjugate of $\underline{f}[n]$, in terms of $F(s)$?

(b) Find a formula for the inverse DTFT transform $\mathcal{D}^{-1}$, a transform such that

$$\mathcal{D}^{-1}(\mathcal{D}\underline{f})[n] = \underline{f}[n].$$

Hint: $F(s) = \mathcal{D}\underline{f}(s)$ is periodic of period 1. Think Fourier coefficients, except with a plus sign in the exponential.

(c) What is $\mathcal{D}\underline{\delta}_0$, the DTFT of the discrete δ centered at 0? (*Note*: This is not the periodic discrete δ used in the DFT. This $\underline{\delta}_0$ is just 1 at 0 and 0 elsewhere.)

(d) What is $\mathcal{D}\underline{1}$, the DTFT of the discrete constant function $\underline{1}$? Your answer should not involve any complex exponentials.

(e) Let $\underline{f}[n]$ be a discrete-time signal of length N; i.e., $\underline{f}[n]$ is identically zero if n is outside $[0 : N-1]$. What is the relationship between the DTFT $\mathcal{D}\underline{f}(s)$ and the DFT $\underline{\mathcal{F}}\underline{f}[n]$ of $\underline{f}[n]$?

7.28. *Convolution theorem for the discrete time Fourier transform*

Let $\underline{f}[n]$ and $\underline{g}[n]$ be discrete signals defined for all integers $n = 0, \pm 1, \pm 2, \ldots$ but not assumed to be periodic. Their convolution is the discrete signal

$$(\underline{f} * \underline{g})[n] = \sum_{m=-\infty}^{\infty} \underline{f}[n-m]\underline{g}[m]\,,$$

also defined for all integers $n = 0, \pm 1, \pm 2, \ldots$.

Show that the DTFT from the previous problem satisfies

$$\mathcal{D}(\underline{f} * \underline{g})(s) = \mathcal{D}\underline{f}(s)\mathcal{D}\underline{g}(s)\,.$$

7.29. *Solving a quadratic, the hard way*[20]

The quadratic formula tells us how to find the roots x_0 and x_1 of a quadratic function $x^2 + bx + c$:

$$x^2 + bx + c = (x - x_0)(x - x_1) \qquad \Rightarrow \qquad \{x_0, x_1\} = \frac{1}{2}\left(-b \pm \sqrt{b^2 - 4c}\right).$$

In this problem, we'll rederive this formula using the DFT. Throughout parts (a)–(c), assume that the vector $\underline{x} = (x_0, x_1)$ has DFT $\underline{X} = (X_0, X_1)$.

(a) Express x_0 and x_1 in terms of X_0 and X_1.
(b) Express b and c in terms of X_0 and X_1.
(c) Combine parts (a) and (b) to express x_0 and x_1 in terms of b and c.

7.30. *Solving a cubic, the hard way*[21]

In this problem we will find the complex roots of the cubic

$$f(x) = x^3 - 3\alpha\beta x - (\alpha^3 + \beta^3) = 0.$$

It can be shown (not required for this problem) that any cubic can be written in this form.

(a) Find a 3×3 circulant matrix A such that $\det(xI - A) = f(x)$:

$$A = \begin{pmatrix} a_1 & a_3 & a_2 \\ a_2 & a_1 & a_3 \\ a_3 & a_2 & a_1 \end{pmatrix}, \quad \det(xI - A) = f(x).$$

Hint: Pick $a_1 = 0$ and compute $\det(xI - A)$. You should be able to read off a_2, a_3 by inspection.

(b) Use (a) to find all the roots of $f(x)$. Recall that the roots of $\det(xI - A)$ are the eigenvalues of A.

7.31. *The DFT and difference operators*

There isn't a derivative theorem for the DFT because there isn't a derivative of a discrete signal. Instead one forms differences, and there are some analogies to the continuous case.

[20] From Raj Bhatnagar.
[21] From Aditya Siripuram.

Let $\underline{f}$ be a (periodic) discrete signal of length N. Define the backward difference $\nabla \underline{f}$ to be the discrete signal

$$\nabla \underline{f}[n] = \underline{f}[n] - \underline{f}[n-1], \quad n = 0, 1, \ldots, N-1.$$

Similarly, the forward difference operator is

$$\Delta \underline{f}[n] = \underline{f}[n+1] - \underline{f}[n], \quad n = 0, 1, \ldots, N-1.$$

In both cases note how periodicity enters; e.g., $\nabla \underline{f}[0] = \underline{f}[0] - \underline{f}[-1] = \underline{f}[0] - \underline{f}[N-1]$.

(a) For practice, using the definition of the DFT show that

$$\underline{\mathcal{F}}(\nabla \underline{f})[n] = (1 - e^{-2\pi i n/N})\underline{\mathcal{F}}\,\underline{f}[n], \quad n = 0, 1, 2, \ldots, N-1.$$

(b) We can also get this result another way, and an analogous result for $\Delta \underline{f}$, via convolution. Express $\nabla \underline{f}$ and $\Delta \underline{f}$ in terms of convolution, and show that

$$\underline{\mathcal{F}}(\Delta \underline{f})[n] = (e^{2\pi i n/N} - 1)\underline{\mathcal{F}}\,\underline{f}[n].$$

Is it true that

$$\nabla \Delta \underline{f} = \Delta \nabla \underline{f}?$$

Finally, express $\nabla \underline{f}$ and $\Delta \underline{f}$ in terms of each other via convolution.

One can define higher-order differences (like higher-order derivatives) recursively. For example,

$$\nabla^{(0)} \underline{f}[n] = \underline{f}[n] \quad \text{(just the original signal — no differences)},$$
$$\nabla^{(1)} \underline{f}[n] = \underline{f}[n] - \underline{f}[n-1] \quad \text{(the first difference, as above)},$$
$$\nabla^{(2)} \underline{f}[n] = \nabla(\nabla \underline{f})[n] = \nabla \underline{f}[n] - \nabla \underline{f}[n-1] = \underline{f}[n] - 2\underline{f}[n-1] + \underline{f}[n-2],$$
$$\nabla^{(3)} \underline{f}[n] = \nabla(\nabla^{(2)})\underline{f}[n] = \nabla^{(2)} \underline{f}[n] - \nabla^{(2)} \underline{f}[n-1] = \cdots,$$

and so on. Similarly for $\Delta^{(m)} \underline{f}$. (Again, these are all defined because of periodicity.)

(c) Express $\nabla^{(m)} \underline{f}$ and $\Delta^{(m)} \underline{f}$ in terms of convolution and find their DFTs in terms of $\underline{\mathcal{F}}\,\underline{f}$.

(d) Clearly, if $\underline{f}$ is a constant signal, then all differences are identically 0; i.e., $\nabla^{(m)} \underline{f}[n] = 0$ and $\Delta^{(m)} \underline{f}[n] = 0$ for all n and for any m. Conversely, suppose there exists an m_0 such that

$$\nabla^{(m_0)} \underline{f}[n] = 0, \quad \text{or} \quad \Delta^{(m_0)} \underline{f}[n] = 0, \quad n = 0, 1, 2, \ldots, N-1.$$

Show that $\underline{f}$ is a constant signal, and say what the constant is. (*Hint*: The DFT of $\nabla^{(m_0)} \underline{f}$ or $\Delta^{(m_0)} \underline{f} = 0$ is zero. Use your result from part (b).)

7.32. *A discrete form of Wirtinger's inequality*

Glance back at Problem 1.26 on Wirtinger's inequality; it's an inequality between the integrals of a periodic function $f(t)$ and its derivative $f'(t)$. There's a discrete version of the inequality involving difference operators; see the previous problem.

Let $\underline{f}[n]$, $n = 0, \dots, N-1$, be a discrete, periodic function of length N with

$$\sum_{k=0}^{N-1} \underline{f}[k] = 0.$$

In terms of the DFT this condition means

$$\underline{\mathcal{F}}\underline{f}[0] = \sum_{k=0}^{N-1} \underline{f}[k] = 0.$$

(You'll have to see how this comes in.)

The problem is to establish the inequality

$$\sum_{k=0}^{N-1} |\underline{f}[k]|^2 \leq \frac{1}{4\sin^2(\pi/N)} \sum_{k=0}^{N-1} |\Delta \underline{f}[k]|^2.$$

This sure looks like some kind of application of Parseval's identity. It is, twice in fact — with some work.

Start with $\|\underline{\mathcal{F}}(\Delta \underline{f})\|^2 = N\|\Delta \underline{f}\|^2$, from Parseval, and show $|\underline{\mathcal{F}}(\Delta \underline{f})[k]|^2 = 4\sin^2(\pi k/N)$.

Chapter 8

Linear Time-Invariant Systems

8.1. We Are All Systemizers Now

An eminent former dean of my very own School of Engineering was heard to opine that we didn't really have separate Departments of Electrical Engineering, Mechanical Engineering, Chemical Engineering, and so on. Rather, we had very well mixed departments of large systems and small systems, electrical systems and mechanical systems, chemical systems and, most recently, biological systems. And so on.[1] If "system" is a catch-all phrase describing a process of going from inputs to outputs, then that's probably as good a description of our organization as any, maybe especially as a way of contrasting engineering with fields that seem to be stuck on input. This chapter is primarily a *very limited* treatment of some of the meeting points between Fourier analysis and *linear systems*, with just enough general observations thrown in to encourage you to find out more. In fact, it will take some time before the Fourier transform enters the scene, so please be patient.

For a taxonomy of linear systems, soon to be defined, several phrases come to mind: matrix theory, linear dynamical systems, integral operators. It's not clear that Fourier series or Fourier transforms, continuous or discrete, have anything to do with any of these. But in fact, Fourier analysis connects to linear systems in a deep way, and that's a connection you should take with you from our work here.

To begin with some general observations, for us a *system* is a correspondence — a mathematician would call it a mapping — from input signals to output signals. We'll typically write this as

$$w(t) = L(v(t)),$$

[1]The title of this section comes from the remark: "We are all Keynesians now." Look it up.

or without the variable as

$$w = L(v).$$

The system is "L" and it operates on the signal v in some way to produce another signal w. Upon this very general notion vast empires have been built, but not without being more specific.

We often think of the signals v and w as functions of time or of a spatial variable (or variables) or just as likely as functions of a discrete variable (or variables). As in the chapter on the DFT, the discrete case may arise via sampling continuous-time signals, or things might have started off discrete. As well, one might input a continuous-time signal and output a discrete-time signal, or vice versa. In any case, we have to specify the domain of L, i.e., the signals that we can feed into the system as inputs. For example, maybe it's appropriate to consider L as operating on finite energy signals (as in the square integrable functions from our study of Fourier series), or on bandlimited signals (as in sampling theory). Whatever, they have to be specified. Of course we also have to define just how it is that L operates to unambiguously produce an output, say integration, as in the Fourier transform, or finding Fourier coefficients of a periodic signal, the latter being an example of starting with a continuous-time input ($f(t)$) and ending with a discrete output ($\hat{f}(n)$). Defining how L operates might then involve continuity properties or convergence properties or boundedness properties. Finally, we often want to be able to say what the range of L is; what kind of signals come out.

Working with and thinking in terms of systems, all of these points included, is as much an adopted attitude as it is an application of a collection of results. I don't want to go so far as saying that any problem should be viewed through the lens of systems from start to finish, but it often provides, at the very least, a powerful organizing principle.

8.2. Linear Systems

With the extreme degree of generality described above one shouldn't expect to be able to say anything terribly interesting — a system is some kind of operation that relates an incoming signal to an outgoing signal. Someway. Somehow. So what. Imposing more structure can make "system" a more interesting notion, and the simplest nontrivial extra assumption is that the system is *linear*, where the sum of two inputs produces the corresponding sum of outputs and scaling an input results in the same scaling of the output.

Things become very interesting indeed under this natural and widely observed property, but unchecked enthusiasm leads quickly into deep waters. Broadly speaking, many of the mathematical problems come from working with infinite-dimensional vector spaces of signals and settling on appropriate definitions of limits, convergence, etc. In some ways the development parallels finite-dimensional linear algebra, but there's plenty more going on. The area of mathematics that comes into play is called *functional analysis*, a highly developed subject that originated in old applied problems (integral equations and partial differential equations) and in turn found great utility in new applied problems (quantum mechanics).[2] Rest easy, however. Front loading the mathematical issues hasn't been our style to this point and it won't be here. Just as we did for the infrastructure that supports the theory of distributions, we'll be setting much aside.

"Linearity" was first encountered in terms of the principle of superposition, which, when it happens, says something like:

> If u and v satisfy $\ldots X \ldots$, then $u + v$ and αu, α a number, also satisfy X.

This isn't a systems approach exactly, but that's how things started. Two examples that you know are:

(1) ODEs

As in, if u and v are each solutions to $y''(t) + p(t)y(t) = 0$, then the linear combination $\alpha u + \beta v$ is also a solution.

(2) Systems of linear equations

As in, if A is a matrix and if $\underline{u}$ and $\underline{v}$ are each solutions to $A\underline{x} = 0$, then the linear combination $\alpha\underline{u} + \beta\underline{v}$ is also a solution.

There is clearly a similarity in these two examples, at least in the language.

From "u and v satisfy $\ldots X \ldots$" the conditions and descriptions gradually shifted to an emphasis on the nature of the *operation*, as in "u and v are operated on by ..." and "superposition" shifted to "linearity" of the operation. The *system* $w = L(v)$ is *linear* if L as a mathematical operation is linear. This means exactly that for all signals v, v_1, v_2 in the domain of L and for all scalars α,

$$L(v_1(t) + v_2(t)) = L(v_1(t)) + L(v_2(t)) \qquad (L \text{ is } \textit{additive}),$$

$$L(\alpha v(t)) = \alpha L(v(t)) \qquad (L \text{ is } \textit{homogeneous}).$$

You have to say whether you're taking α to be real or complex. Instead of using the word "system," mathematicians tend to say that L is a *linear operator*.

Note that to define this notion we have to be able to add and scale inputs and outputs. Not all systems can be linear because, depending on the application, it just might not make sense to add inputs or to scale them. Ditto for the outputs. Think of a communication system where the inputs and outputs are code words in some alphabet. Does it make sense to add code words, or to scale a code word? A system where you can add and scale inputs and outputs but where one or both of the properties above do not hold is generically referred to as *nonlinear*.

One immediate comment. If the zero signal is the input to a linear system, then the output is also the zero signal, since

$$\begin{aligned} L(0) &= L(0 \cdot 0) \\ &\quad \text{(on the left-hand side that's the zero signal;} \\ &\qquad \text{on the right it's the scalar 0 times the signal 0)} \\ &= 0 \cdot L(0) \quad \text{(using homogeneity)} \\ &= 0\,. \end{aligned}$$

[2] Depending on your interests and stamina, I highly recommend skimming (if not devouring) the four volumes (!) of *Methods of Modern Mathematical Physics* by M. Reed and B. Simon.

If a system is nonlinear, it may not be that $L(0) = 0$; take, for example, $L(v(t)) = v(t) + 1$. This may be the quickest mental checkoff to see if a system is *not* linear. Does the zero signal go to the zero signal? If not, then the system *cannot* be linear.

Notational note. A common notational convention when dealing with linear systems is to drop the parentheses when L is acting on a single signal, so we write Lv instead of $L(v)$. This convention comes from the analogy of general linear systems to those given by multiplication by a matrix. More on that connection later.

In a phrase we've already used, an expression of the form $\alpha_1 v_1 + \alpha_2 v_2$ is called a *linear combination* of v_1 and v_2, or, occasionally, with a nod toward the older terminology, a superposition of v_1 and v_2. Here α_1 and α_2 are (typically) numbers, real, complex, binary, whatever makes sense in context. Then to say that L is linear is to say that

$$L(\alpha_1 v_1 + \alpha_2 v_2) = \alpha_1 L v_1 + \alpha_2 L v_2.$$

In words, a linear combination of inputs produces a corresponding linear combination of outputs. That's an alternate way of saying that L satisfies the principle of superposition. One can extend this directly to finite sums and, with proper assumptions of continuity of L and convergence of the sums, to infinite sums. That is,

$$L\left(\sum_{n=0}^{N} \alpha_n v_n(t)\right) = \sum_{n=0}^{N} \alpha_n L v_n(t) \quad \text{and} \quad L\left(\sum_{n=0}^{\infty} \alpha_n v_n(t)\right) = \sum_{n=0}^{\infty} \alpha_n L v_n(t).$$

Do not minimize the importance of these properties. If a signal can be written as a sum of its components in some basis (think Fourier series and complex exponentials) and if we know the action of L on the components (the complex exponentials), then we can find the action of L on the composite signal:

$$L\left(\sum_{k=-\infty}^{\infty} \hat{f}(k) e^{2\pi i k t}\right) = \sum_{n=-\infty}^{\infty} \hat{f}(k) L e^{2\pi i k t}\,.$$

We won't make an issue of convergence and continuity for the kinds of things we want to do, and we'll even allow a more general version of superposition using integrals, discussed later. This is part of suppressing the mathematical challenges.

How are linearity and (the older terminology of) superposition related to each other in, say, the ODE example of solutions to $y''(t) + p(t)y(t) = 0$? Define the operator L by

$$Ly = y'' + py.$$

Then, for sure, the domain of L better be included in the space of functions that are twice differentiable. We might also want to impose additional conditions on the functions, like initial conditions, boundary conditions, etc. You can check that L is linear, the key old properties from calculus being that the derivative of the sum is the sum of the derivatives and similarly for the derivative of a constant times a function. Properties that you probably mocked at the time. Solutions to the ODE are functions for which $Ly = 0$, known (from linear algebra) as the null space (or kernel) of L. If two functions u and v are in the null space, then, because L is linear, so is any linear combination of u and v. Voilà, superposition. You can make the same sort of translation for the example of systems of linear equations. Let's look at some others.

8.3. Examples

As we've done with other topics, we'll develop some general properties of linear systems and also consider some particular cases. As I said earlier, thinking in terms of linear systems is partly an attitude, and it's helpful to keep specific examples in mind that lead to more general considerations. In fact, the examples really aren't so neatly separated. There will be a fair amount of "this example has aspects of that example," an important aspect of the subject and something you should look for. You too can develop an attitude.

8.3.1. Direct proportion. The most basic example of a linear system is the relationship "A is directly proportional to B." Suitably interpreted (a slippery phrase), you see this in some form or function in all linear constructions. In fact, I'm willing to raise it to the level of an aphorism, or perhaps even a secret of the universe:

- Any linear system is a generalization of direct proportion.

In important cases the goal of an analysis of a linear system is to get it to look as much as possible like direct proportion and to understand when this can or cannot be done.

Usually one thinks of direct proportion in terms of a constant of proportionality, as in "the current is proportional to the voltage, $I = (1/R)V$," or "the acceleration is proportional to the force, $a = (1/m)F$." The conclusions are then of the type, "doubling the voltage results in doubling the current," or "tripling the force triples the acceleration."

The constant of proportionality can also be a function, and the key property of linearity is still present. Fix a function $a(t)$ and define

$$Lv(t) = a(t)v(t)\,.$$

Then, to be formal about it, L is linear because

$$\begin{aligned} L(\alpha v_1(t) + \beta v_2(t)) &= a(t)(\alpha v_1(t) + \beta v_2(t)) \\ &= \alpha a(t)v_1(t) + \beta a(t)v_2(t) = \alpha L v_1(t) + \beta L v_2(t)\,. \end{aligned}$$

This is such a common situation that you don't usually jump to considering it in the context of linearity per se. Nevertheless, here are two examples.

Switching on and off is a linear system. Suppose we have a system consisting of a single switch. When the switch is closed, a signal goes through unchanged, and when the switch is open, the signal doesn't go through at all (so by convention what comes out the other end is the zero signal). Suppose that the switch is closed for $-\frac{1}{2} \leq t \leq \frac{1}{2}$. Is this a linear system? Sure. It's described by

$$Lv(t) = \Pi(t)v(t),$$

i.e., multiplication by Π.

We could modify this any number of ways:

- Switching on and off at various time intervals.
  - This is modeled by multiplication by a sum of shifted and scaled Π's.
- Switching on and staying on, or switching off and staying off.
  - This is modeled by multiplication by the unit step $H(t)$ or by $1 - H(t)$.

All of these are linear systems — they are direct proportion — and you can come up with many other systems built on the same principle.

Sampling is a linear system. Sampling is multiplication by Ш. To sample a signal $v(t)$ with sample points spaced p apart is to form

$$Lv(t) = Ш_p(t)v(t) = \sum_{k=-\infty}^{\infty} v(kp)\delta(t - kp).$$

It doesn't hurt to say in words the consequences of linearity, namely: we can first add two signals and sample the sum, or we can sample each signal separately and add the sampled versions. We get the same result either way.

8.3.2. Matrix multiplication. Similar to simple direct proportion, but maybe a level up in sophistication, are linear systems defined by matrix multiplication:

$$\underline{w} = A\underline{v},$$

where $\underline{v} \in \mathbb{R}^n$, $\underline{w} \in \mathbb{R}^m$, and A is an $m \times n$ matrix. This *looks* like direct proportion. Written out in the usual way, if the entries of A are a_{ij} and if $\underline{v} = (v_1, \ldots, v_n)$, $\underline{w} = (w_1, \ldots, w_m)$, then

$$w_i = \sum_{j=1}^{n} a_{ij}v_j, \quad i = 1, \ldots, m.$$

This looks less like direct proportion, but with some sympathy it certainly looks like a reasonable generalization of direct proportion. You do know that matrix multiplication is linear:

$$A(\alpha\underline{u} + \beta\underline{v}) = \alpha A\underline{u} + \beta A\underline{v}.$$

The DFT, for example, can be thought of as such a linear system.

The linear system might have special properties according to whether A has special properties. As mentioned in the last chapter, for a square matrix these might be:

Symmetric: $A^{\mathsf{T}} = A$,

Hermitian: $A^* = A$,

Orthogonal: $AA^{\mathsf{T}} = I$,

Unitary: $AA^* = I$.

We'll come back to this.

Matrix multiplication as a discrete linear system. For reasons that will become clear in just a bit, I want to express matrix multiplication in a different way. Same operations, just a different mindset.

Think of the matrix A, with entries a_{ij}, as defining a discrete signal $\underline{A}$ that is a function of the two discrete variables i and j,

$$\underline{A}[i, j] = a_{ij}.$$

Likewise, and as we did with the DFT, think of the vectors $\underline{v}$ and $\underline{w}$ as defining discrete signals of length n and m, respectively. Then we can write the matrix product $\underline{w} = A\underline{v}$ as

$$\underline{w}[i] = \sum_{j=1}^{n} \underline{A}[i, j]\underline{v}[j], \quad i = 1, \ldots, m.$$

Feels different somehow.

In turn, rewriting those special properties of A (size $n \times n$) in terms of $\underline{A}$, we have:

Symmetric: $\underline{A}[i, j] = \underline{A}[j, i]$,

Hermitian: $\underline{A}[i, j] = \overline{\underline{A}[j, i]}$,

Orthogonal: $\sum_{k=1}^{n} \underline{A}[i, k]\underline{A}[j, k] = \underline{\delta}_0[i - j] = \begin{cases} 1, & i = j, \\ 0, & i \neq j, \end{cases}$

Unitary: $\sum_{k=1}^{n} \underline{A}[i, k]\overline{\underline{A}[j, k]} = \underline{\delta}_0[i - j] = \begin{cases} 1, & i = j, \\ 0, & i \neq j. \end{cases}$

You'll see why I wrote the latter two properties in terms of $\underline{\delta}$.

Linear dynamical systems. Speaking of matrix multiplication, for those who have taken a more advanced ODE class, the linear system

$$L\underline{v}(t) = e^{tA}\underline{v},$$

where A is an $n \times n$ matrix and $\underline{v} \in \mathbb{R}^n$, is the linear system associated with the initial value problem

$$\dot{\underline{x}}(t) = A\underline{x}, \quad \underline{x}(0) = \underline{v}.$$

Here the matrix e^{tA} varies in time, and the system describes how the initial value $\underline{v}$ evolves over time.

8.3.3. Integration against a kernel.

A linear system defined via matrix multiplication is a model for more complicated situations. The continuous, infinite-dimensional version of matrix multiplication is the linear system given by an operation of the form

$$w(x) = Lv(x) = \int_a^b k(x, y)v(y)\, dy.$$

Here, $k(x, y)$ is called the *kernel*[3] and one speaks of "integrating $v(y)$ against a kernel." I call this an "infinite-dimensional" system because, typically, the set of inputs (the domain of L) is an infinite-dimensional vector space.

[3] Not to be confused with kernel as in the null space of a linear operator. A simple clash of terminology.

This is certainly a linear system since

$$\begin{aligned} L(\alpha_1 v_1(x) + \alpha_2 v_2(x)) &= \int_a^b k(x,y)(\alpha_1 v_1(y) + \alpha_2 v_2(y))\,dy \\ &= \alpha_1 \int_a^b k(x,y)v_1(y)\,dy + \alpha_2 \int_a^b k(x,y)v_2(y)\,dy \\ &= \alpha_1 L v_1(x) + \alpha_2 L v_2(x). \end{aligned}$$

Naturally one has to state conditions on $k(x, y)$ and on the inputs to feel good about the integral existing, but we take feeling good for granted.

To take one quick example, the Fourier transform is of this form

$$\mathcal{F}f(s) = \int_{-\infty}^{\infty} e^{-2\pi i s t} f(t)\,dt = \int_{-\infty}^{\infty} k(s,t) f(t)\,dt\,.$$

In this case the kernel is $k(s, t) = e^{-2\pi i s t}$.

To imagine how this generalizes the (finite) matrix linear systems, think, if you dare, of the values of v as being listed in an infinite column, $k(x, y)$ as an infinite (square) matrix, $k(x, y)v(y)$ as the product of the (x, y)-entry of k with the yth entry of v, and the integral

$$w(x) = \int_a^b k(x,y)v(y)\,dy$$

as summing the products $k(x, y)$ across the xth row of k with the entries of the column v, resulting in the xth value of the output, $w(x)$. It was to bring out this analogy that I rewrote matrix multiplication as

$$\underline{w}[i] = \sum_{j=1}^{n} \underline{A}[i,j]\underline{v}[j]\,, \quad i = 1, \dots, m\,,$$

and matrix multiplication can thus be thought of as "summation against a kernel."

You won't be misleading yourself (much) thinking in terms of this analogy, though there are several lifetimes of work that went into pursuing the idea — it's an analogy, not a perfect correspondence. This is one entree to functional analysis.

Staying on this path, the system may have special properties according to whether the kernel $k(x, y)$ has special properties:

Symmetric: $k(x, y) = k(y, x)$.
One can define the transpose of the operator L with kernel $k(x, y)$ by

$$(L^{\mathsf{T}} v)(x) = \int_a^b k(y,x)v(y)\,dy.$$

This comes to us from functional analysis, again. Thus if $k(x, y) = k(y, x)$, then $L = L^{\mathsf{T}}$. Nice.

Hermitian: $k(x,y) = \overline{k(y,x)}$.

Same thing: when $k(x,y)$ et al. are complex, the adjoint of L is

$$(L^*v)(x) = \int_a^b \overline{k(y,x)}v(y)\,dy.$$

Thus if $k(x,y) = \overline{k(y,x)}$, then $L = L^*$. Nice, again.

Orthogonal and unitary operators are a little harder to define in terms of the kernel, and an alternate approach is taken in mainstream functional analysis in terms of preserving norms (which is what orthogonal and unitary matrices do). However, if we allow ourselves to borrow freely from distributions and δ-functions (don't tell anyone), we can write conditions that look like the matrix conditions.

Yielding to temptation, we have, for example,

$$\begin{aligned}
L(L^Tv)(x) &= \int_a^b k(x,y)(L^Tv)(y)\,dy \\
&= \int_a^b k(x,y)\left(\int_a^b k(z,y)v(z)\,dz\right)dy \\
&= \int_a^b \left(\int_a^b k(x,y)k(z,y)\,dy\right)v(z)\,dz
\end{aligned}$$

(combining and rearranging the integrals).

For L to be orthogonal we want $LL^Tv(x) = v(x)$, so the condition on k is

$$\int_a^b k(x,y)k(z,y)\,dy = \delta(x-z).$$

The condition for L to be unitary is the same except with a complex conjugate on one of the k's:

$$\int_{-\infty}^{\infty} k(x,y)\overline{k(x,z)}\,dx = \delta(y-z)\,.$$

What happens when the system is the Fourier transform and the kernel is $k(s,t) = e^{-2\pi ist}$? That's left as a problem.

8.3.4. Convolution: Continuous and discrete. Convolution is a special case of integrating (summing in the discrete case) against a kernel, but it's so important in applications that it deserves its own mention. In the continuous realm, fix a signal $h(x)$ and define

$$w(x) = Lv(x) = (h * v)(x)\,.$$

Then explicitly, to make this look like what we did above,

$$w(x) = (h*v)(x) = \int_{-\infty}^{\infty} h(x-y)v(y)\,dy = \int_{-\infty}^{\infty} k(x,y)v(y)\,dy,$$

where the kernel is

$$k(x,y) = h(x-y).$$

It's interesting here, and significant, that the kernel depends on the difference $x-y$ and not on x and y independently. This brings up the so-called time-invariant systems, a topic that gets its own section in a little bit.

For discrete signals. Convolution in the finite, discrete case naturally involves periodic discrete signals, and we know what to do. Fix a periodic discrete signal, say $\underline{h}$, of length N and define $\underline{w} = L\underline{v}$ by convolution with $\underline{h}$:

$$\underline{w}[m] = (\underline{h} * \underline{v})[m] = \sum_{n=0}^{N-1} \underline{h}[m-n]\underline{v}[n] \,.$$

We're summing against the kernel $\underline{h}[m-n]$.

Translation. Signals, whether functions of a continuous or discrete variable, can be delayed or advanced to produce new signals. As a system the operation of translating, or shifting, by an amount b is just

$$Lv(t) = v(t-b) \,.$$

This is very familiar to us, to say the least.

To add a little to what we've said before, think of t as "time now," where now starts at $t = 0$, and b as "time ago." Think of $t - b$ as a delay in time by an amount b; "delay" if $b > 0$, and "advance" if $b < 0$. To delay a signal 24 hours from current time ("now") is to consider the difference between current time and time 24 hours ago, i.e., $t - 24$. The signal v delayed 24 hours is $v(t-24)$ because it's not until $t = 24$ that the signal "starts."

We could show directly and easily enough that such translation in time (or in space, if that's the physical variable) is a linear system. But we can also observe that translation is nothing other than convolving with a translated δ, and convolution is a linear system. That is,

$$v(t-b) = (\delta_b * v)(t) \,,$$

and the same for a discrete signal:

$$\underline{v}[m-n] = (\underline{\delta}_n * \underline{v})[m] \,.$$

By the same token, we see that periodizing a signal is a linear system, since this amounts to convolution with a $Ш_p$. Thus

$$w(t) = Lv(t) = (Ш_p * v)(t)$$

is a linear system. In words, linearity means that the periodization of the sum of two signals is the sum of the periodizations, with a similar statement for scaling by a constant.

8.4. Cascading Linear Systems

An important operation is to *compose* or *cascade* two (or more) linear systems. If L and M are linear systems, then, as long as the combined operations make sense, ML is also a linear system. First applying L, followed by M,

$$\begin{aligned}(ML)(\alpha_1 v_1 + \alpha_2 v_2) &= M(L(\alpha_1 v_1 + \alpha_2 v_2)) \\ &= M(\alpha_1 Lv_1 + \alpha_2 Lv_2) = \alpha_1 MLv_1 + \alpha_2 MLv_2 \,.\end{aligned}$$

In general we do *not* have $ML = LM$.

The phrase "as long as the combined operations make sense" means paying attention to the domains and ranges of the individual systems. For example, if we start out with an integrable function $f(t)$, then $\mathcal{F}f$ makes sense but $\mathcal{F}f$ may not be integrable so, strictly speaking, $\mathcal{F}\mathcal{F}f$ may not be defined.[4] These concerns, too, fall under our doctrine of noting the issue but setting it aside.

Cascading linear systems defined by matrix multiplication amounts to multiplying the matrices — a short but important observation. There's a version of this in the case of cascading linear systems given by integration against a kernel.

First, if L is the linear system given by

$$Lv(x) = \int_a^b k(x,y)v(y)\,dy$$

and M is another linear system (not necessarily given by an integral), then the composition ML is the linear system

$$MLv(x) = \int_a^b M(k(x,y))v(y)\,dy\,.$$

This is important enough in itself.

What does this mean, and when is it true? For one thing, $k(x,y)$ has to be a signal upon which M can operate, and in writing $M(k(x,y))$ (and then integrating with respect to y) we intend that its operation on $k(x,y)$ is in its x-dependence. To see where the formula comes from, imagine approximating $Lv(x)$ by a Riemann sum:

$$Lv(x) \approx \sum_n k(x,y_n)v(y_n)\Delta y_n\,.$$

Linearity of M then implies

$$\begin{aligned} MLv(x) &\approx \sum_n M(k(x,y_n)v(y_n)\Delta y_n) \\ &= \sum_n M(k(x,y_n))v(y_n)\Delta y_n \end{aligned}$$

($v(y_n)\Delta y_n$ are constants when M is operating in the x-variable).

In the limit we get the result. This requires some continuity assumptions on M, etc. The restrictions are mild, and we can be safe in assuming that we can perform such operations for the applications we're interested in.

Now take the special case when M is also given by integration against a kernel. Say

$$Mv(x) = \int_a^b \ell(x,y)v(y)\,dy\,.$$

[4]Though this has never bothered us. If $f(t)$ is a Schwartz function, then we can keep applying $\mathcal{F}$. Duality results imply that $\mathcal{F}\mathcal{F}\mathcal{F}\mathcal{F}f = f$, so cascading $\mathcal{F}$ doesn't go on producing new signals forever.

Then

$$\begin{aligned} MLv(x) &= \int_a^b \ell(x,y)Lv(y)\,dy \\ &= \int_a^b \ell(x,y)\left(\int_a^b k(y,z)v(z)\,dz\right)dy \\ &\qquad \text{(we introduced a new variable of integration)} \\ &= \int_a^b \int_a^b \ell(x,y)k(y,z)v(z)\,dz\,dy\,. \end{aligned}$$

Under the ever-present banner of whatever-hypotheses-are-necessary, we can interchange the order of integration and write

$$\int_a^b \int_a^b \ell(x,y)k(y,z)v(z)\,dz\,dy = \int_a^b \left(\int_a^b \ell(x,y)k(y,z)\,dy\right)v(z)\,dz\,.$$

Thus the cascaded system MLv is also given by integration against a kernel:

$$MLv(x) = \int_a^b K(x,z)v(z)\,dz\,,$$

where

$$K(x,z) = \int_a^b \ell(x,y)k(y,z)\,dy\,.$$

The formula for the kernel $K(x,z)$ should call to mind a matrix product. We had a version of this, above, when considering orthogonal and unitary operators.

8.5. The Impulse Response, or the Deepest Fact in the Theory of Distributions Is Well Known to All Electrical Engineers

Actually, before dragging distributions into the picture, let's briefly revisit your linear algebra class.

8.5.1. Matrices and linear systems, redux. It's not just that a discrete (finite-dimensional) linear system given by matrix multiplication is a nice example. It's the *only* example. You learned this in linear algebra. You learned that if L is a linear transformation between two finite-dimensional vector spaces, then L can be represented as a matrix, and the operation of L becomes matrix multiplication. To do this you choose a basis of the domain and a basis of the range, you look at what L does to each basis vector in the domain in terms of the basis of the range, and then you put the matrix together column by column. Different bases give different matrices.

Typically, in the case when L is a linear transformation from $\mathbb{R}^n$ to $\mathbb{R}^m$ (or $\mathbb{C}^n$ to $\mathbb{C}^m$), you use the natural basis vectors in the domain and range. *But not always.* For square matrices, looking for good bases (so to speak) is tied up in the question of diagonalizing a matrix — it's the spectral theorem, which I'll remind you of, below.

Examples. Let's look at two quick examples to jog your memories. Define

$$L(x, y, z) = (2x + 3y - z, x - y + 4z).$$

We pass L a 3-vector and it returns a 2-vector. Yes, it's linear. For the natural basis vectors we have

$$L(1, 0, 0) = (2, 1), \quad L(0, 1, 0) = (3, -1), \quad L(0, 0, 1) = (-1, 4).$$

Assembling these as columns, the matrix for L is

$$L = \begin{pmatrix} 2 & 3 & -1 \\ 1 & -1 & 4 \end{pmatrix}.$$

Then, for example, using the original formula for L, we get $L(2, 1, -3) = (10, -11)$. Multiplying by the matrix gives

$$\begin{pmatrix} 2 & 3 & -1 \\ 1 & -1 & 4 \end{pmatrix} \begin{pmatrix} 2 \\ 1 \\ -3 \end{pmatrix} = \begin{pmatrix} 10 \\ -11 \end{pmatrix}.$$

Like there was any doubt.

Here's a second example, a little less standard. Take the vector space of polynomials of degree ≤ 2. This is a three-dimensional vector space with basis $\{1, t, t^2\}$; to say this is a basis is to say that any quadratic can be written uniquely as a linear combination $y(t) = a \cdot 1 + b \cdot t + c \cdot t^2$. Now take the linear transformation $Ly(t) = y''(t) + py(t)$ where p is a constant. Then

$$\begin{aligned} L1 &= p = p \cdot 1 + 0 \cdot t + 0 \cdot t^2, \\ Lt &= pt = 0 \cdot 1 + p \cdot t + 0 \cdot t^2, \\ Lt^2 &= 2 + pt^2 = 2 \cdot 1 + 0 \cdot t + p \cdot t^2. \end{aligned}$$

Identifying

$$L1 = \begin{pmatrix} p \\ 0 \\ 0 \end{pmatrix}, \quad Lt = \begin{pmatrix} 0 \\ p \\ 0 \end{pmatrix}, \quad Lt^2 = \begin{pmatrix} 2 \\ 0 \\ p \end{pmatrix},$$

we assemble the matrix representing L column by column:

$$L = \begin{pmatrix} p & 0 & 2 \\ 0 & p & 0 \\ 0 & 0 & p \end{pmatrix}.$$

Then, for example, what is $L(5 + 3t + 2t^2)$? We operate with the matrix

$$\begin{pmatrix} p & 0 & 2 \\ 0 & p & 0 \\ 0 & 0 & p \end{pmatrix} \begin{pmatrix} 5 \\ 3 \\ 2 \end{pmatrix} = \begin{pmatrix} 5p + 4 \\ 3p \\ 2p \end{pmatrix}$$

and (if we want) we re-identify the result with the quadratic $(5p + 4) + 3pt + 2pt^2$. This is probably the most complicated way of computing $y'' + py$ that you've seen, but you get the point.

8.5.2. The discrete impulse response. Let's translate this into engineering lingo. In different terminology, we're going to do in general what the preceding examples illustrated.

Suppose L is a linear system that accepts a signal $\underline{v} = (\underline{v}[0], \underline{v}[1], \dots, \underline{v}[N-1])$ and returns a signal $\underline{w} = (\underline{w}[0], \underline{w}[1], \dots, \underline{w}[M-1])$. In terms of the shifted, discrete $\underline{\delta}$'s we can write

$$\underline{v} = \sum_{n=0}^{N-1} \underline{v}[n]\underline{\delta}_n,$$

and then, since L is linear,

$$\underline{w} = L\underline{v} = \sum_{n=0}^{N-1} \underline{v}[n]L\underline{\delta}_n.$$

At an index m, $m = 0, \dots, M-1$,

$$\underline{w}[m] = \sum_{n=0}^{N-1} \underline{v}[n]L\underline{\delta}_n[m].$$

Let

$$\underline{k}[m,n] = L\underline{\delta}_n[m].$$

Then, on the one hand, we've written the linear system in the form

$$\underline{w}[m] = \sum_{n=0}^{N-1} \underline{k}[m,n]\underline{v}[n].$$

Summation against a kernel! On the other hand, we can introduce the $M \times N$ matrix K, whose mn-entry is

$$K_{mn} = \underline{k}[m,n]$$

and write the system as

$$\underline{w} = K\underline{v}.$$

In either case, the discrete kernel $\underline{k}[m,n]$ or the matrix K is called the (discrete) *impulse response* of the system L. The terminology comes, of course, from recording how the system "responds" to an impulse, just as in $\underline{k}[m,n] = L\underline{\delta}_n[m]$, above. And so:

- To know the impulse response of a system is to know the system.

Also in your linear algebra class. An important problem that you studied in your linear algebra course was that of diagonalizing a square matrix: under what conditions on an $N \times N$ matrix A is there a basis of $\mathbb{R}^N$, or of $\mathbb{C}^N$, for which A becomes diagonal? The answer is often called the *spectral theorem*. Let me state one version of it for complex matrices:

> **Spectral Theorem.** If A is Hermitian ($A^* = A$), then there is an orthonormal basis for $\mathbb{C}^N$ consisting of eigenvectors of A, i.e, an orthonormal basis $\{\underline{v}_1, \dots, \underline{v}_N\}$ and complex numbers $\lambda_1, \dots, \lambda_N$ (the eigenvalues, not necessarily distinct) for which
>
> $$A\underline{v}_n = \lambda_n \underline{v}_n.$$

In terms of the basis of eigenvectors, A is a diagonal matrix with diagonal entries $\lambda_1, \lambda_2, \dots, \lambda_N$.

I wanted you to recall this result to make the point that in the direction $\underline{v}_n$ the system A acts exactly as direct proportion, with constant of proportionality λ_n. *Direct proportion*, there it is again.

I'm not going to even bring up the situation for continuous-time systems, but suffice it to say that finding analogs to the finite-dimensional spectral theorem was a motivating problem, with serious applications, for the development of functional analysis. There are many sources for this, but I refer you again to Reed and Simon's *Methods of Mathematical Physics*.

We'll return to eigenvalues and eigenvectors later in the context of time-invariant linear systems.

8.5.3. And now distributions: The continuous-time case. It's not just that a continuous-time linear system given by integration against a kernel is a nice example. It's the *only* example. Under very minimal assumptions, all linear systems are of this form. That's too bold a statement, honestly, but here's what it means and why engineers of all stripes think of things this way.

It's time to bring back distributions. We'll use integration as a way of operating on continuous-time signals with shifted δ's. Thus for a signal $v(x)$ we have

$$v(x) = \int_{-\infty}^{\infty} \delta(x-y)v(y)\,dy.$$

Now we allow ourselves to operate on this with a linear system L according to

$$Lv(x) = \int_{-\infty}^{\infty} (L\delta(x-y))\,v(y)\,dy.$$

Define the function (distribution)

$$h(x,y) = L\delta(x-y).$$

This is the *impulse response* of the system L. With allowances, we have expressed the system as integration against the kernel $h(x,y)$. Invoking the classical language, we state this as:

Superposition Theorem. If L is a linear system with impulse response $h(x,y)$, then

$$w(x) = Lv(x) = \int_{-\infty}^{\infty} h(x,y)v(y)\,dy\,.$$

To know the impulse response is to know the system. The integral is usually called the *superposition integral*, naturally enough.

What this means in practice is that we see how the system responds to a very short, very peaked signal. The limit of such responses is the impulse response.

Credit for introducing the impulse response belongs to engineers, but you will note the usual mathematical modus operandi: we answer the question of how a system responds to an impulse via a definition.

You can't miss the analogy with matrices, but can this be made more precise? What does it mean for L to operate on δ? Is that a function or a distribution? And so on. The answer is yes, all of this can be made precise, and it has a natural home in the context of distributions. The superposition theorem of electrical engineering is known as the Schwartz kernel theorem in mathematics, and the impulse response (which *is* a distribution) is known as the Schwartz kernel. Moreover, there is a uniqueness statement. It says, in the present context, that for each linear system L there is a *unique* kernel $h(x, y)$ such that

$$Lv(x) = \int_{-\infty}^{\infty} h(x,y)v(y)\,dy\,.$$

In proper treatments, it's not integration but the *pairing* of the Schwartz kernel with the input.[5]

The uniqueness is good to know, for if you've somehow expressed a linear system as an integral with a kernel, then you have *found* the impulse response. Thus, for example, the impulse response of the Fourier transform is $h(s, t) = e^{-2\pi ist}$ since

$$\mathcal{F}f(s) = \int_{-\infty}^{\infty} e^{-2\pi ist} f(t)\,dt = \int_{-\infty}^{\infty} h(s,t)f(t)\,dt\,.$$

We conclude from this that

$$\mathcal{F}\delta(t-s) = e^{-2\pi ist}\,,$$

which checks with what we know from earlier work.[6]

The Schwartz kernel theorem is considered probably the deepest and hardest result in the theory of distributions. So, by popular demand, we won't take this any farther.

Example. Let's return to the simplest linear system, direct proportion. Say,

$$Lv(x) = a(x)v(x).$$

What is the impulse response? Directly, using properties of δ,

$$h(x,y) = L\delta(x-y) = a(x)\delta(x-y) = a(y)\delta(x-y).$$

[5]Want more? See, once again, *Introduction to the Theory of Distributions*, by G. Friedlander, or A. Zemanian's *Distribution Theory and Transform Analysis*. Not for the fainthearted.

[6]This isn't circular reasoning, but neither is it a good way to find the Fourier transform of a shifted δ-function. It's hard to prove the Schwartz kernel theorem, but it's easy (with distributions) to find the Fourier transform of δ.

Does the superposition integral give back the direct proportion formula for the system? For an input v,

$$\begin{aligned}\int_{-\infty}^{\infty} h(x,y)v(y)\,dy &= \int_{-\infty}^{\infty} (a(y)\delta(x-y))v(y)\,dy \\ &= \int_{-\infty}^{\infty} (a(y)v(y))\delta(x-y)\,dy \\ &= a(x)v(x).\end{aligned}$$

Yes, it works.

8.6. Linear Time-Invariant (LTI) Systems

Good news, the Fourier transform will soon make its entrance, though it's not so obvious at first that "time invariance," the topic of this section, should be the vehicle.

If I run a program tomorrow, I expect to get the same answers as I do when I run it today. Except I'll get them tomorrow. The circuits that carry the currents and compute the 0's and 1's will behave (ideally) today just as they did yesterday and into the past, and just as they should tomorrow and into the future. We know that's not true indefinitely, of course — components fail — but as an approximation this kind of time invariance is a natural assumption for many systems. When it holds, *for linear systems*, it has important consequences for the mathematical and engineering analysis of the system.

I want to emphasize this: the time-invariance property is that a *shift* in time of the inputs should result in an *identical shift* in time of the outputs. Invariance is in terms of "shifts in time," or "differences in time." "Absolute time" doesn't make sense for invariance, but differences between two times does. If this reminds you of relativity in physics, you're thinking correctly.

What is the mathematical expression of invariance? To say that a linear system $w(x) = Lv(x)$ is *time invariant*, or is an *LTI system*, is to say that a delay of the input signal by an amount y produces a delay of the output signal by the same amount, but no other changes. As a formula,

$$Lv(x-y) = w(x-y)\,.$$

Careful how to interpret this — it's the "variable problem" again. Writing $Lv(x-y)$ means operating on the delayed signal $v(x-y)$. Delaying the input signal by 24 hours, then operating by L, produces a delay in the output by 24 hours, but that's the only thing that happens. More below.

Sometimes LTI is translated to read "Linear Translation-Invariant" system, recognizing that the variable isn't always time but that the operation is always translation. You also see LSI in use, meaning "Linear Shift-Invariant." We'll speak most often in terms of "time invariance," but I'm using the more generic variables x and y instead of, say, t and τ, to maintain the generality.

What about the impulse response for an LTI system? For a general linear system the impulse response $h(x, y) = L\delta(x - y)$ depends independently on x and y. But this *isn't the case* for an LTI system. Let

$$h(x) = L\delta(x).$$

If L is LTI, then a shift of input produces an identical shift of output:

$$L\delta(x - y) = h(x - y).$$

Hence the impulse response does not depend on x and y independently but rather *only on their difference*, $x - y$. In this case we refer to $h(x)$ itself, rather than $h(x - y)$, as the impulse response.

Importantly, the character of the impulse response means that the superposition integral assumes the form of a convolution:

$$w(x) = Lv(x) = \int_{-\infty}^{\infty} (L\delta(x - y))v(y)\,dy = \int_{-\infty}^{\infty} h(x - y)v(y)\,dy = (h * v)(x).$$

Conversely, let's show that a linear system given by convolution is time invariant. Fix a function $g(x)$ and define

$$w(x) = Lv(x) = (g * v)(x) = \int_{-\infty}^{\infty} g(x - y)v(y)\,dy.$$

Then

$$\begin{aligned} L(v(x - x_0)) &= \int_{-\infty}^{\infty} g(x - y)v(y - x_0)\,dy \\ &\qquad \text{(make sure you understand how we substituted the shifted } v\text{)} \\ &= \int_{-\infty}^{\infty} g(x - x_0 - s)v(s)\,ds \quad \text{(substituting } s = y - x_0\text{)} \\ &= (g * v)(x - x_0) = w(x - x_0). \end{aligned}$$

Thus L is time invariant, as we wanted to show.

Furthermore, we see that when L is defined this way, by a convolution, $g(x - y)$ *must* be the impulse response, because for a time-invariant system the impulse response is determined by

$$L\delta(x) = (g * \delta)(x) = g(x);$$

that is,

$$L\delta(x - y) = g(x - y).$$

A grand summary. This is a very satisfactory state of affairs. Let's summarize:

If L is a linear system, then

$$Lv(x) = \int_{-\infty}^{\infty} k(x, y)v(y)\,dy,$$

where $k(x, y)$ is the impulse response,

$$k(x, y) = L\delta(x - y).$$

The system is time invariant *if and only if* it is a convolution. The impulse response is a function of the difference $x - y$, and

$$Lv(x) = \int_{-\infty}^{\infty} h(x-y)v(y)\,dy = (h * v)(x)\,, \quad h(x) = L\delta(x)\,.$$

This last result is another indication of how fundamental, and natural, the operation of convolution is.

Translating in time and plugging into L. From my own bitter experience I can report that knowing when and how to plug the time shifts into a formula is not always easy. So I'd like to return briefly to the definition of time invariance and offer another way of writing it and also a way to think of it in terms of cascading systems. Bring back the shift operator,

$$(\tau_b v)(x) = v(x-b)\,.$$

If $w(x) = Lv(x)$, then its shift by b is

$$w(x-b) = (\tau_b w)(x)\,,$$

and the time-invariance property then says that

$$L(\tau_b v) = \tau_b(Lv), \quad \text{without writing the variable } x,$$

or, writing the variable,

$$L(\tau_b v)(x) = \tau_b(Lv)(x) = (Lv)(x-b).$$

It's the placement of parentheses here that means everything — it says that translating by b and then applying L (the left-hand side) has the same effect as applying L and then translating by b (the right-hand side). One says that an LTI system L "commutes with translation." Most succinctly,

$$L\,\tau_b = \tau_b\,L\,.$$

In fact, "commuting" is just the way to look at time invariance from a second point of view. We already observed that "translation by b" is itself a linear system. Then the combination of τ_b and L, in that order, produces the output $L(v(x-b))$. To say that L is an LTI system is to say that the system τ_b followed by the system L produces the same result as the system L followed by the system τ_b.

Now go back to the plugging in we did earlier in the convolution:

$$Lv(x) = (g * v)(x) = \int_{-\infty}^{\infty} g(x-y)v(y)\,dy\,.$$

Then

$$L(v(x-x_0)) = \int_{-\infty}^{\infty} g(x-y)v(y-x_0)\,dy\,.$$

Expanding the explanation,

$$\begin{aligned}
L(\tau_{x_0}v)(x) &= \int_{-\infty}^{\infty} g(x-y)\tau_{x_0}v(y)\,dy \\
&\qquad \text{(it's the translated signal } \tau_{x_0}v \text{ that's getting convolved)} \\
&= \int_{-\infty}^{\infty} g(x-y)v(y-x_0)dy \\
&= \int_{-\infty}^{\infty} g(x-x_0-s)v(s)\,ds \quad \text{(substituting } s = y - x_0\text{)} \\
&= (g*v)(x-x_0) \;=\; \tau_{x_0}(g*v)(x) \;=\; \tau_{x_0}(Lv)(x)\,.
\end{aligned}$$

In fact, if you look back to problems in the chapter on convolution, you'll see that we just rederived the general property

$$(\tau_b v) * g = \tau_b(v * g).$$

In words, the convolution of the shift is the shift of the convolution.

Back to direct proportion. Does direct proportion, that most basic of all linear systems, fare well with regard to time invariance? Afraid not, except in the simplest case.

Suppose that

$$Lv(x) = h(x)v(x)$$

is time invariant. Then for any y, on the one hand,

$$L(v(x-y)) = h(x)v(x-y)\,,$$

and on the other hand, by time invariance,

$$L(v(x-y)) = (Lv)(x-y) = h(x-y)v(x-y)\,.$$

Thus,

$$h(x-y)v(x-y) = h(x)v(x-y)\,.$$

This is to hold for every input v and every value y, and that can only happen if $h(x)$ is *constant.* Hence the relationship of direct proportion will only define a time-invariant linear system when the proportionality factor is constant ("genuine" direct proportion).

So, suppose your boss comes to you and says: "I want to build a set of switches and I want you to model that for me by convolution, because although I don't know what convolution means I know it's an important idea." You will have to say: "That cannot be done because while switches can be modeled as a linear system, the simple (even for you, boss) relation of direct proportion that we would use does not define a time-invariant system, as convolution must." You win.

Later, however, your boss comes back and says: "OK, no convolution, but find the impulse response of my switch system and find it fast." This you can do. To be definite, take the case of a single-switch system modeled by

$$Lv(x) = \Pi(x)v(x)\,,$$

for which the impulse response is

$$h(x, y) = L\delta(x - y) = \Pi(x)\delta(x - y) = \Pi(y)\delta(x - y) .$$

Sure enough, the impulse response is *not* a function of $x - y$ only. You take the rest of the day off.[7]

8.6.1. Discrete LTI systems and circulant matrices. A discrete system given by a convolution

$$\underline{w} = L\underline{v} = \underline{h} * \underline{v}$$

is time invariant, as you can check. "Time" here is discrete, and to verify time invariance is to recall the identity

$$(\tau_k \underline{v}) * \underline{h} = \tau_k(\underline{v} * \underline{h})$$

for discrete convolution. We'll work with periodic signals, as usual.

Since the operation $\underline{w} = \underline{h} * \underline{v}$, going from the input $\underline{v}$ to the output $\underline{w}$, is a linear transformation of $\mathbb{C}^N$ to itself, it must be given by multiplication by some $N \times N$ matrix:

$$\underline{w} = \underline{h} * \underline{v} = L\underline{v} .$$

What is the matrix L?

Remember that L is assembled column by column by computing L on the shifted $\underline{\delta}$'s. We know what happens when we convolve with shifted $\underline{\delta}$'s: we shift the index. Thus the mth entry in the nth column of L is

$$L\underline{\delta}_n[m] = (\underline{h} * \underline{\delta}_n)[m] = \underline{h}[m - n] ,$$

written simply as

$$(L)_{mn} = \underline{h}[m - n] .$$

The matrix L is constant along the diagonals and is filled out, column by column, by the shifted versions of $\underline{h}$. Note that shifts make sense because of periodicity.

This goes the other way. As an exercise you can show:

- If L is a matrix for which L_{mn} depends only on the difference $m - n$, then $\underline{w} = L\underline{v}$ is an LTI system. In terms of convolution the system is $\underline{w} = \underline{h} * \underline{v}$, where $\underline{h}$ is the first column of L. (And $L_{mn} = \underline{h}[m - n]$.)

To take an example, if

$$\underline{h} = (\underline{h}[0], \underline{h}[1], \underline{h}[2], \underline{h}[3]),$$

then the matrix L for which

$$\underline{w} = \underline{h} * \underline{v} = L\underline{v}$$

has columns

$$\begin{pmatrix} \underline{h}[0] \\ \underline{h}[1] \\ \underline{h}[2] \\ \underline{h}[3] \end{pmatrix}, \quad \begin{pmatrix} \underline{h}[-1] \\ \underline{h}[0] \\ \underline{h}[1] \\ \underline{h}[2] \end{pmatrix} = \begin{pmatrix} \underline{h}[3] \\ \underline{h}[0] \\ \underline{h}[1] \\ \underline{h}[2] \end{pmatrix}, \quad \begin{pmatrix} \underline{h}[-2] \\ \underline{h}[-1] \\ \underline{h}[0] \\ \underline{h}[1] \end{pmatrix} = \begin{pmatrix} \underline{h}[2] \\ \underline{h}[3] \\ \underline{h}[0] \\ \underline{h}[1] \end{pmatrix}, \quad \begin{pmatrix} \underline{h}[-3] \\ \underline{h}[-2] \\ \underline{h}[-1] \\ \underline{h}[0] \end{pmatrix} = \begin{pmatrix} \underline{h}[1] \\ \underline{h}[2] \\ \underline{h}[3] \\ \underline{h}[0] \end{pmatrix} .$$

---

[7] Don't tell your boss that you already did this calculation a few pages back, more generally for $Lv(x) = a(x)v(x)$.

Easier to see the pattern with an explicit $\underline{h}$. Say $\underline{h} = (1, 2, 3, 4)$. Then

$$L = \begin{pmatrix} 1 & 4 & 3 & 2 \\ 2 & 1 & 4 & 3 \\ 3 & 2 & 1 & 4 \\ 4 & 3 & 2 & 1 \end{pmatrix}.$$

In general, square matrices that are constant along the diagonals (allowing for different constants along the different diagonals) are called *Toeplitz matrices*, after the mathematician Otto Toeplitz who singled them out for special study. They have all sorts of interesting properties.

There's even more structure when the columns of a Toeplitz matrix are periodic, as happens when the matrix comes from convolution. These special Toeplitz matrices are called *circulant matrices*. We'll pursue some of their features in the problems. See the books *Toeplitz and Circulant Matrices: A Review* by Robert Gray and *Circulant Matrices* by P. Davis. For computational aspects see also the book *Matrix Computation* by G. Golub and C. van Loan.

A grand summary, take two. To button up the analogy between the continuous and discrete:

If L is a discrete linear system, then

$$L\underline{v}[m] = \sum_{n=0}^{N-1} \underline{k}[m,n]\underline{v}[n],$$

where $\underline{k}[m,n]$ is the impulse response

$$\underline{k}[m,n] = L\underline{\delta}_n[m]\,.$$

The system is time invariant *if and only if* it is a convolution. The impulse response is a function of the difference $m - n$, and

$$L\underline{v}[m] = \sum_{n=0}^{N-1} \underline{h}[m-n]\underline{v}[n] = (\underline{h} * \underline{v})[m]\,, \quad \underline{h} = L\underline{\delta}_0\,.$$

See the problems for more.

8.7. The Fourier Transform and LTI Systems

At last. The fact that LTI systems are identical with linear systems defined by convolution should trip the Fourier switch in your head. Given the LTI system

$$w(t) = (h * v)(t),$$

we take Fourier transforms and write

$$W(s) = H(s)V(s)\,,$$

turning convolution in the time domain to multiplication in the frequency domain. This bears rephrasing:

- In the frequency domain, the relationship between input and output of an LTI system is *exactly* direct proportion.

Recall that $H(s)$ is called the *transfer* function of the system. We introduced the transfer function earlier, in Chapter 3, and I refer you there for a quick review. The additional terminology is that the system $w = h * v$ is called a *filter* (or an LTI filter), and sometimes the impulse response h is called the *filter function*.

One catch phrase used to describe LTI filters is that they "add no new frequencies." Sure. Multiplying $V(s)$ by $H(s)$ can't stretch out $V(s)$ any further. It might shrink it, as with a low-pass filter, or otherwise cut it down, as with a band-pass filter, but it can't make it any bigger. Here's an example of a system that *does* add new frequencies. Consider

$$Lv(t) = v(t)^2 .$$

This is nonlinear. And if, for example, we feed in $v(t) = \cos 2\pi t$, we get out

$$w(t) = Lv(t) = \cos^2 2\pi t = \tfrac{1}{2} + \tfrac{1}{2}\cos 4\pi t .$$

Although the input has a single frequency at 1 Hz, the output has a DC component of 1/2 and a frequency component at 2 Hz.

8.7.1. Complex exponentials are eigenfunctions. Now consider an LTI system's response to inputting the building blocks of signals, a complex exponential of frequency ν. This is called the *frequency response* of the system.

For an LTI system $w = Lv = h * v$, with transfer function H, we find $L(e^{2\pi i\nu t})$ via the frequency domain. Taking Fourier transforms,

$$\begin{aligned} W(s) &= H(s)\mathcal{F}(e^{2\pi i\nu t}) \\ &= H(s)\delta(s-\nu) \quad \text{(using the Fourier transform pairing } \delta(s-\nu) \rightleftharpoons e^{2\pi i\nu t}\text{)} \\ &= H(\nu)\delta(s-\nu) \quad \text{(using the sampling property of } \delta\text{).} \end{aligned}$$

Now take the inverse transform. Because $H(\nu)$ is a *constant* we find that

$$L(e^{2\pi i\nu t}) = H(\nu)e^{2\pi i\nu t} .$$

This is quite an important discovery. We already know that L is a linear operator; this equation says further that:

- The exponentials $e^{2\pi i\nu t}$ are *eigenfunctions* of L; the output is a scalar multiple of the input $e^{2\pi i\nu t}$.
- The corresponding eigenvalues are the values of the transfer function $H(\nu)$.

This is *the* reason that complex exponentials are fundamental for studying LTI systems. To return to a familiar theme, it says that in the direction of a complex exponential an LTI system acts as direct proportion. Earlier we commented that an LTI system can't be one of direct proportion for every input except in the simplest case; i.e., if $Lv(x) = a(x)v(x)$ for any $v(x)$, then $a(x)$ is constant. But an LTI system *does* act as direct proportion when inputting complex exponentials.

Contrast this to what happens to a sine or cosine signal under an LTI system. We assume that the impulse response is real valued, a natural assumption. Feed in a cosine signal $\cos(2\pi\nu t)$. What is the response? Writing

$$v(t) = \cos(2\pi\nu t) = \tfrac{1}{2}e^{2\pi i\nu t} + \tfrac{1}{2}e^{-2\pi i\nu t} ,$$

we compute

$$\begin{aligned}
Lv(t) &= \tfrac{1}{2}H(\nu)e^{2\pi i\nu t} + \tfrac{1}{2}H(-\nu)e^{-2\pi i\nu t} \\
&= \tfrac{1}{2}H(\nu)e^{2\pi i\nu t} + \tfrac{1}{2}\overline{H(\nu)}e^{-2\pi i\nu t} \quad (H(-\nu) = \overline{H(\nu)} \text{ because } h(t) \text{ is real valued}) \\
&= \tfrac{1}{2}(H(\nu)e^{2\pi i\nu t} + \overline{H(\nu)e^{2\pi i\nu t}}) \\
&= \mathrm{Re}\{H(\nu)e^{2\pi i\nu t}\} \\
&= |H(\nu)|\cos(2\pi\nu t + \phi_H(\nu)),
\end{aligned}$$

where

$$H(\nu) = |H(\nu)|e^{i\phi_H(\nu)}.$$

The response is a cosine of the same frequency, but with a changed amplitude *and phase*. We would find a similar result for the response to a sine signal.

Neither the cosine nor the sine are themselves eigenfunctions of L. It is only the *complex exponential* that is an eigenfunction. We are (sort of) back to where we started in the course, with complex exponentials as a basis for decomposing a signal, and now for decomposing an operator.

Let's take this one step further. Suppose again that L has a real-valued impulse response $h(t)$. Suppose also that we input a real periodic signal, which we represent as a Fourier series

$$v(t) = \sum_{n=-\infty}^{\infty} c_n e^{2\pi int}.$$

Recall that because $v(t)$ is real the Fourier coefficients $c_n = \hat{v}(n)$ satisfy

$$c_{-n} = \overline{c_n}.$$

If we apply L to $v(t)$, we find

$$Lv(t) = \sum_{n=-\infty}^{\infty} c_n L e^{2\pi int} = \sum_{n=-\infty}^{\infty} c_n H(n) e^{2\pi int}.$$

But the fact that $h(t)$ is real valued implies that

$$H(-n) = \overline{H(n)},$$

the same symmetry as the Fourier coefficients. That is, if

$$C_n = c_n H(n),$$

then $C_{-n} = \overline{C_n}$, and the output $w(t)$ is also a real, periodic function with Fourier series

$$w(t) = \sum_{n=-\infty}^{\infty} C_n e^{2\pi int}.$$

The discrete case. The situation for discrete LTI systems is entirely analogous. Suppose the system is given by

$$\underline{w} = L\underline{v} = \underline{h} * \underline{v}.$$

Take $\underline{v} = \underline{\omega}^k$, a discrete complex exponential, as input and take the discrete Fourier transform of both sides of $\underline{w} = \underline{h} * \underline{\omega}^k$:

$$\begin{aligned}\underline{\mathcal{F}}\,\underline{w}[m] &= \underline{\mathcal{F}}\,\underline{h}[m]\,\underline{\mathcal{F}}\,\underline{\omega}^k[m] \\ &= \underline{\mathcal{F}}\,\underline{h}[m]\,N\underline{\delta}[m-k] \quad \text{(remember the extra factor } N\text{)} \\ &= \underline{H}[m]\,N\underline{\delta}[m-k] \\ &= \underline{H}[k]\,N\underline{\delta}[m-k] \quad \text{(sampling property of } \underline{\delta}\text{)}\,.\end{aligned}$$

Now take the inverse DFT of both sides (remembering that $\underline{\mathcal{F}}\,\underline{\delta}_k = (1/N)\underline{\omega}^k$):

$$\underline{w} = \underline{H}[k]\,\underline{\omega}^k\,.$$

Hence

$$L\underline{\omega}^k = \underline{H}[k]\,\underline{\omega}^k$$

and we see that $\underline{\omega}^k$ is an eigenvector of L for $k = 0, 1, \ldots, N-1$ with eigenvalue $\underline{H}[k]$.

We already know that $\underline{1}$, $\underline{\omega}$, $\underline{\omega}^2$, $\ldots$, $\underline{\omega}^{N-1}$ are an orthogonal basis for $\mathbb{C}^N$. That's the orthogonality of the discrete complex exponentials, a foundation for much of the theory of the DFT. This new result says that if L is a discrete LTI system, then the complex exponentials are an orthogonal basis for $\mathbb{C}^N$ consisting of eigenvectors of L. They diagonalize L, meaning that if the matrix of L is expressed in this basis, it is a diagonal matrix with diagonal entries $\underline{H}[k]$. There are some problems on this theme.

8.7.2. Matched filters. LTI systems are used extensively in the study of communications systems. Makes sense; it shouldn't matter whether I send a message today or tomorrow. But sometimes there's a failure to communicate[8] and a fundamental concern is to distinguish the signal from noise, to design a filter that will do the job. The filter should respond strongly to one particular signal and only to that signal. This is *not* a question of recovering or extracting a particular signal from the noise; it's a question of first *detecting* whether the signal is present. (Think radar.) If the filtered signal rises above a certain threshold, an alarm goes off, so to speak, and we believe that the signal we want is there.

Here's a *highly* condensed discussion of this central problem, but even such a condensed version fits naturally with what we're doing. With $w(t) = (h * v)(t)$ and $W(s) = H(s)V(s)$, we'll try to design the transfer function $H(s)$ so that the system responds strongly to a particular signal $v_0(t)$. What it means to "respond strongly" emerges as a consequence of the analysis. This is a fine example from engineering of the enduring mathematical principle of turning the solution of a problem into a definition.

Let's begin with some general observations. Suppose an incoming signal is of the form $v(t) + p(t)$ where $p(t)$ is noise. Then the output is $h(t) * (v(t) + p(t)) = w(t) + q(t)$, $q = h * p$. The contribution of the noise to the output has total energy

$$\int_{-\infty}^{\infty} |q(t)|^2\,dt\,,$$

[8]Presumably every communications engineer has seen the movie *Cool Hand Luke*.

which, using Parseval's theorem and the transfer function, we can write as

$$\int_{-\infty}^{\infty} |q(t)|^2\, dt = \int_{-\infty}^{\infty} |Q(s)|^2\, ds = \int_{-\infty}^{\infty} |H(s)|^2 |P(s)|^2\, ds$$

(capital letters for the Fourier transforms).

Now we make an assumption about the nature of the noise. Take the special case of *white noise*. A reasonable definition of that term, one that translates into a workable condition, is that $p(t)$ should have *equal power in all frequencies.*[9] This means simply that $|P(s)|$ is *constant*, say $|P(s)| = C$, so from the equation above the output energy of the noise is

$$E_{\text{noise}} = C^2 \int_{-\infty}^{\infty} |H(s)|^2\, ds\,.$$

We want to compare the energy of the noise output to the strength of the output signal $w(t) = (h * v)(t)$. Using Fourier inversion,

$$w(t) = \int_{-\infty}^{\infty} W(s) e^{2\pi i s t}\, ds = \int_{-\infty}^{\infty} H(s)V(s) e^{2\pi i s t}\, ds\,.$$

Now, long dormant since our discussion of Fourier series, used briefly in a problem on autocorrelation, and just waiting for this moment, the Cauchy-Schwarz inequality makes its triumphant reappearance! It says, in the form we're going to use it,

$$\left| \int_{-\infty}^{\infty} f(t)\overline{g(t)}\, dt \right|^2 \le \left(\int_{-\infty}^{\infty} |f(t)|^2\, dt \right) \left(\int_{-\infty}^{\infty} |g(t)|^2\, dt \right),$$

with equality if and only if f is a constant multiple of $\overline{g}$.

According to Cauchy-Schwarz,

$$\begin{aligned} |w(t)|^2 &= \left| \int_{-\infty}^{\infty} H(s)V(s)e^{2\pi i s t}\, ds \right|^2 \\ &\le \int_{-\infty}^{\infty} |H(s)|^2\, ds \int_{-\infty}^{\infty} |V(s)|^2\, ds \quad \text{(we also used } |e^{2\pi i s t}| = 1\text{).} \end{aligned}$$

That is,

$$\frac{|w(t)|^2}{E_{\text{noise}}} \le \frac{1}{C^2} \int_{-\infty}^{\infty} |V(s)|^2\, ds\,.$$

By *definition*, the fraction $|w(t)|^2/E_{\text{noise}}$ is the *signal-to-noise* ratio, abbreviated SNR. The biggest the SNR can be is when there is equality in the Cauchy-Schwarz inequality. Thus the filter that gives the strongest response, meaning largest SNR, when a given signal $v_0(t)$, part of a combined noisy signal $v_0(t) + p(t)$, is the one with transfer function proportional to $\overline{V_0(s)e^{2\pi i s t}}$, where $V_0(s) = \mathcal{F}v_0(t)$.

This result is sometimes referred to as the *matched filter theorem*:

- To design a filter that has the strongest response to a particular signal $v_0(t)$, in the sense of having the largest signal-to-noise ratio, design it so the transfer function $H(s)$ has the same shape as $\overline{V_0(s)}$.

[9]There are other ways to get to this definition; e.g., the autocorrelation of $p(t)$ should be zero.

To recapitulate, when the filter is designed this way, then

$$\begin{aligned}|w_0(t)|^2 &= \left|\int_{-\infty}^{\infty} H(s)V_0(s)e^{2\pi ist}\,ds\right|^2 \\ &= \frac{1}{C^2}E_{\text{noise}}\int_{-\infty}^{\infty}|V_0(s)|^2\,ds = \frac{1}{C^2}E_{\text{noise}}\int_{-\infty}^{\infty}|v_0(t)|^2\,dt \quad \text{(by Parseval).}\end{aligned}$$

Thus the SNR is

$$\frac{|w_0(t)|^2}{E_{\text{noise}}} = \frac{1}{C^2}\int_{-\infty}^{\infty}|v_0(t)|^2\,dt = \frac{1}{C^2}(\text{energy of } v_0(t))\,.$$

There's no impediment to converting this reasoning to digital signals to derive a discrete form of the matched filter theorem. I'll let you work that out, or look it up.

8.8. Causality

In the words of my late, esteemed colleague Ron Bracewell: "It is well known that effects never precede their causes." This commonsense sentiment, put into action for inputs and outputs of systems, is referred to as *causality*. In words, we might describe causality as saying that the past determines the present but not vice versa. In math:

- If $v_1(t) = v_2(t)$ for $t < t_0$, then $Lv_1(t) = Lv_2(t)$ for $t < t_0$, and this holds for all t_0.

At first glance you may be puzzled by the statement, or wonder if there's any statement at all: if two signals are the same, then mustn't their outputs be the same? But the signals $v_1(t)$ and $v_2(t)$ are assumed to exist for all time, and the system L might produce outputs based not only on the values of the inputs up to a certain time t_0, but on values into the future as well. A classic example is the moving average operator (a subject of several problems) defined for an $a > 0$ by

$$\mathcal{A}_a v(t) = \frac{1}{2a}\int_{t-a}^{t+a} v(x)\,dx$$

(different a's, different moving averages). The value of the output $w(t_0) = \mathcal{A}_a v(t_0)$ at any t_0 depends on the values of the input $v(t)$ on the interval from $t_0 - a$ to $t_0 + a$. The system is evidently not causal.

The condition as we stated it above is the general definition for a system L to be causal, and it's a nontrivial requirement. We're not even assuming that the system is linear. One also sees the definition stated with "$<$" replaced by "$\leq$", i.e., to include "the present." I gave the definition with "$<$" because it's more standard, especially in connection with characterizing causality via the impulse response, which we'll get to.

You'll see other versions of the definition depending on additional assumptions on the system, and it's worthwhile sorting these out. For example, we can formulate the causality condition a little more compactly, and a little more conveniently, when L *is* linear.

- A linear system $w(t) = Lv(t)$ is causal if and only if $v(t) = 0$ for $t < t_0$ implies $w(t) = 0$ for $t < t_0$.

Here's a proof. First suppose that L is linear and causal. We have to show that if $v(t) = 0$ for $t < t_0$, then $w(t) = Lv(t) = 0$ for $t < t_0$. We apply the definition of causality when one of the signals is the zero signal. That is, letting $u(t) = 0$ be the zero signal, then $Lu(t)$ is also the zero signal because L is linear, and causality means that if $v(t) = u(t) = 0$ for $t < t_0$, then $w(t) = Lv(t) = Lu(t) = 0$ for $t < t_0$.

Conversely, suppose that L is linear and that we have the implication $v(t) = 0$ for $t < t_0$ implies $w(t) = 0$ for $t < t_0$. We claim that L is causal. For this, if $v_1(t) = v_2(t)$ for $t < t_0$, then $v(t) = v_1(t) - v_2(t) = 0$ for $t < t_0$, so, by the hypothesis on L, if $t < t_0$, then $0 = Lv(t) = L(v_1(t) - v_2(t)) = Lv_1(t) - Lv_2(t)$; i.e., $Lv_1(t) = Lv_2(t)$ for $t < t_0$. Done.

Finally, for an LTI system one t_0 is as good as another, one might say, and we can simplify the definition of causality even further:

- An LTI system is causal if and only if $v(t) = 0$ for $t < 0$ implies $w(t) = 0$ for $t < 0$.

One direction of the equivalence relies only on linearity and does not need time invariance. For if L is linear and causal, then, taking $t_0 = 0$ in the preceding statement, we know that $v(t) = 0$ for $t < 0$ implies $w(t) = 0$ for $t < 0$

In the other direction, suppose that $w(t) = Lv(t)$ is an LTI system for which $v(t) = 0$ for $t < 0$ implies $Lv(t) = 0$ for $t < 0$. We claim that L satisfies the causality condition for linear systems, above. For this, take a signal $\tilde{v}(t)$ with $\tilde{v}(t) = 0$ for $t < t_0$ and let $w(t) = L\tilde{v}(t)$. The signal $v(t) = \tilde{v}(t + t_0)$ is zero for $t < 0$, and hence $Lv(t) = 0$ for $t < 0$ by the hypothesis. But by time invariance $Lv(t) = L(\tilde{v}(t + t_0)) = w(t + t_0)$. Thus $w(t + t_0) = 0$ for $t < 0$; i.e., $w(t) = 0$ for $t < t_0$.

In many treatments of causal systems it is only this last condition, for LTI systems, that is presented; it is tacitly assumed that linearity and time invariance are in force. In fact, one often runs directly into the following pair of definitions, usually given without the preceding motivation:

- A function $v(t)$ is causal if it is zero for $t < 0$.
- An LTI system is causal if causal inputs go to causal outputs.

Fine definitions, but they seem a little stark.

Still another definition of causality for LTI systems can be given in terms of the impulse response. Since $\delta(t) = 0$ for $t < 0$ (allowing for evaluating δ at points), the impulse response $h(t) = L\delta(t)$ of a causal LTI system satisfies $h(t) = 0$ for $t < 0$. Conversely, if the impulse response for an LTI system satisfies this condition and if $v(t)$ is a signal with $v(t) = 0$ for $t < 0$, then the output $w(t) = (v * h)(t)$ is zero for $t < 0$, too. Thus L is causal. This is the definition one finds in Bracewell's book.

He says simply:

- An LTI system is causal if and only if its impulse response $h(t)$ is zero for $t < 0$.

Many systems are causal and many aren't, and as an exercise you should look over the examples we've already had and decide which is which. (We already had one example of a noncausal system, the averaging operator.) Causality is sometimes called the condition of *physical realizability* because of its "past determines the present" interpretation. Many systems governed by differential equations are causal (current in RC circuits, for example). Indeed, the idea that a differential equation plus initial conditions determines the solution uniquely is another way causality is often put to use.[10]

If causality seems naturally to apply to a system where time is running and "past" and "present" make sense, it's important to realize that causality may not apply where everything is there already, not meaning to be glib. For example, optical systems may not be causal. The variable is typically not time but rather may represent some kind of spatial relation. The input may be an object and the output an image. There is no past and present, all the information that there is, and ever will be, is present. Or to take an example when time is still the relevant variable, one may want to *design* a noncausal system to filter *prerecorded* music. In the case of prerecorded music the full signal is known, and one may *want* to take into account past, present, and future values of the input to decide on what the output should sound like.

8.9. The Hilbert Transform

To set the stage for the main work in this section, let's now consider a causal LTI system given to us as a convolution

$$w(t) = (h * v)(t)\,.$$

Taking the Fourier transform,

$$W(s) = H(s)V(s)\,.$$

Owing to the fact that $h(t)$ is causal, the transfer function $H(s)$ has some special properties that are important in many applications.

We'll analyze $H(s)$ using the unit step function

$$u(t) = \begin{cases} 0, & t < 0, \\ 1, & t \geq 0. \end{cases}$$

I've modified this slightly from how I first defined it. For one thing, I'm calling it u instead of H since we're using H for the transfer function, and for another, it's more natural to have $u(t) = 0$ for $t < 0$ instead of for $t \leq 0$ for use in connection with causality as we've defined it. No big deal.

Since $h(t) = 0$ for $t < 0$, we can write

$$h(t) = u(t)h(t)\,;$$

[10] I'm not even going to go anywhere near what happens to causality in quantum mechanics.

that's the key idea.[11] Recall that the Fourier transform of $u(t)$ is

$$U(s) = \tfrac{1}{2}\left(\delta(s) + \frac{1}{\pi i s}\right).$$

Therefore

$$\begin{aligned} H(s) &= U(s) * H(s) \\ &= \tfrac{1}{2}\left(\delta(s) + \frac{1}{\pi i s}\right) * H(s) \\ &= \tfrac{1}{2}(\delta * H)(s) + \frac{1}{2\pi i s} * H(s) = \tfrac{1}{2}H(s) + \frac{1}{2\pi i s} * H(s). \end{aligned}$$

Thus we have the weird looking result

$$\tfrac{1}{2}H(s) = \frac{1}{2\pi i s} * H(s) \quad \text{or} \quad H(s) = \frac{1}{\pi i s} * H(s).$$

Let's work with this a little more.

Split $H(s)$ into its real and imaginary parts, writing, for the time being,

$$H(s) = R(s) + iI(s),$$

where

$$R(s) = \operatorname{Re} H(s), \quad I(s) = \operatorname{Im} H(s).$$

Then

$$\begin{aligned} R(s) + iI(s) &= \frac{1}{\pi i s} * \big(R(s) + iI(s)\big) \\ &= \frac{1}{\pi i s} * R(s) + \frac{1}{\pi s} * I(s) = \frac{1}{\pi s} * I(s) - i\frac{1}{\pi s} * R(s) \quad (\text{using } 1/i = -i). \end{aligned}$$

Now equate real and imaginary parts on the two sides of the equation. We see that

$$\begin{aligned} \operatorname{Re} H(s) &= \frac{1}{\pi s} * \operatorname{Im} H(s), \\ \operatorname{Im} H(s) &= -\frac{1}{\pi s} * \operatorname{Re} H(s). \end{aligned}$$

The real and imaginary parts of the transfer function $H(s)$ are paired with each other in the sense that we get one from the other via a convolution with $1/\pi s$. The real and imaginary parts of an arbitrary complex-valued function need not be related, but in this case they are very strongly related.[12] If we know one, then we know the other. It's as if we get all of $H(s)$ by knowing only half of it. This merits some additional discussion.

---

[11] Remind you of the sort of trick in the sampling theorem (without the periodization)? There are only so many ideas.

[12] If you've taken a course in complex analysis, you've seen the Cauchy-Riemann equations relating the real and imaginary parts of an analytic function. This is an aspect of what's going on here, but it's a topic that we'll have to leave aside.

8.9.1. The Hilbert transform. Way back in Chapter 4, I threatened to look at a convolution with $1/x$. The time has come. The *Hilbert transform*[13] of a function $v(x)$ is defined by

$$\mathcal{H}v(x) = -\frac{1}{\pi x} * v(x) = -\frac{1}{\pi}\int_{-\infty}^{\infty} \frac{v(y)}{x-y}\,dy = \frac{1}{\pi}\int_{-\infty}^{\infty} \frac{v(y)}{y-x}\,dy\,.$$

A few comments. The integral is improper since the denominator in the integrand is zero when $y = x$.[14] As we discussed in Chapter 4, for functions $v(x)$ arising in applications the integral exists as a *Cauchy principal value*, defined via a *symmetric* limit:

$$\mathcal{H}v(x) = \frac{1}{\pi}\text{pr.v.}\int_{-\infty}^{\infty} \frac{v(y)}{y-x}\,dy = \frac{1}{\pi}\lim_{\epsilon\to 0}\left(\int_{-\infty}^{x-\epsilon} \frac{v(y)}{y-x}\,dy + \int_{x+\epsilon}^{\infty} \frac{v(y)}{y-x}\,dy\right).$$

Having duly noted this, I won't write "pr.v." in front of the integral anymore.

In Chapter 4 we found the Fourier transform of $1/x$ to be

$$\mathcal{F}\left(\frac{1}{x}\right) = -\pi i \operatorname{sgn} s\,,$$

where

$$\operatorname{sgn} x = \begin{cases} +1, & x > 0\,, \\ -1, & x < 0\,. \end{cases}$$

Thus

$$\mathcal{F}\left(-\frac{1}{\pi x}\right) = i \operatorname{sgn} s\,,$$

and

$$\mathcal{F}(\mathcal{H}v)(s) = \mathcal{F}\left(-\frac{1}{\pi x} * v(x)\right) = i \operatorname{sgn} s V(s)\,.$$

For the moment, we want to draw just one conclusion from the calculation of the Fourier transform. If we apply $\mathcal{H}$ *twice* to a function, then, on taking the Fourier transform,

$$\begin{aligned}\mathcal{F}(\mathcal{H}(\mathcal{H}v))(s) &= \mathcal{F}\left(\left(-\frac{1}{\pi x}\right) * \left(-\frac{1}{\pi x}\right) * v(x)\right) \\ &= (i \operatorname{sgn} s)(i \operatorname{sgn} s)V(s) = -\operatorname{sgn}^2 s\, V(s) = -V(s)\,.\end{aligned}$$

Taking the inverse Fourier transform, we see that

$$\mathcal{H}(\mathcal{H}v(x)) = -v(x)\,.$$

Hence the Hilbert transform is *invertible*, and its inverse is just its negative:

$$\mathcal{H}^{-1}v(x) = -\mathcal{H}v(x) = \frac{1}{\pi x} * v(x)\,,$$

a result that's far from obvious.

---

[13]Named after David Hilbert (1862–1943). Hilbert made fundamental contributions to a great many areas of mathematics, including number theory, geometry, logic, and differential and integral equations. The Hilbert transform comes up in several areas of Fourier analysis, but I think it may well be true that more engineers know more about its properties than many mathematicians.

In Chapter 2, in the problems, we used $\mathcal{H}$ for the Hartley transform, so there's a clash of notation. An unfortunate but not uncommon situation. Just pay attention to the context and don't let it bother you.

[14]It's also improper because it has limits $\pm\infty$, but that's the kind of improper integral we're used to working with.

Let's now use the Hilbert transform and its inverse to recast the results on the real and imaginary parts of the transfer function for a causal LTI system:

- If L is a causal LTI system with transfer function $H(s)$, then

$$\operatorname{Im} H(s) = \mathcal{H}(\operatorname{Re} H(s)),$$
$$\operatorname{Re} H(s) = \mathcal{H}^{-1}(\operatorname{Im} H(s)).$$

Sorry if I can't make this statement more thrilling at this point, but it turns out to be quite significant. The Hilbert transform comes up in many applications, especially in communications. We'll examine a few more of its properties.

8.9.2. The Hilbert transform as an LTI system. Since it's a convolution, the Hilbert transform is in particular an LTI system:

$$w(t) = \mathcal{H}v(t) = -\frac{1}{\pi t} * v(t).$$

However, though $\mathcal{H}$ is naturally associated with causal systems, it's *not a causal* LTI system since its impulse response is $-1/\pi t$.

As a filter, this convolution is a rather strange operation. In terms of Fourier transforms,

$$W(s) = i \operatorname{sgn} s \, V(s),$$

and bringing in the amplitude and phase,

$$|W(s)|e^{i\phi_W(s)} = e^{i\pi/2} \operatorname{sgn} s \, |V(s)| \, e^{i\phi_V(s)}.$$

The amplitudes in the spectrum are unchanged but the phases are rotated. The phases corresponding to positive frequencies are rotated by $+\pi/2$, since $\operatorname{sgn} s = 1$ for $s > 0$. The phases corresponding to negative frequencies are rotated by $-\pi/2$, since $\operatorname{sgn} s = -1$ for $s < 0$, and

$$-1 \cdot e^{i\pi/2} = e^{-i\pi} e^{i\pi/2} = e^{-i\pi/2}.$$

In summary

$$W(s) = \begin{cases} |V(s)|e^{i(\phi_V(s)+\pi/2)}, & s > 0, \\ |V(s)|e^{i(\phi_V(s)-\pi/2)}, & s < 0. \end{cases}$$

For this reason $\mathcal{H}v(t)$ is referred to as the *quadrature function* of $v(t)$ in some literature. Among the various definitions, the term "quadrature" means the process of making something square, and in astronomy it refers to "any configuration in which the angular separation of two celestial bodies, as measured from a third, is 90°."

As an example, suppose we pass a cosine through the Hilbert LTI filter, say $v(t) = \cos 2\pi a t$, $a > 0$. Then $V(s) = (1/2)(\delta(s-a) + \delta(s+a))$, and the Fourier transform of the output $w(t)$ is

$$\begin{aligned} W(s) &= \frac{i}{2}\operatorname{sgn} s(\delta(s-a)+\delta(s+a)) \\ &= \frac{i}{2}(\operatorname{sgn} a\,\delta(s-a)+\operatorname{sgn}(-a)\delta(s+a)) \\ &= \frac{i}{2}(\delta(s-a)-\delta(s+a)) = -\frac{1}{2i}(\delta(s-a)-\delta(s+a)). \end{aligned}$$

We recognize this last expression as the Fourier transform of $-\sin 2\pi at$. Thus

$$\mathcal{H}\cos 2\pi at = -\sin 2\pi at\,.$$

Using $\mathcal{H}^{-1} = -\mathcal{H}$ we then also see that

$$\mathcal{H}\sin 2\pi at = \cos 2\pi at\,.$$

Personally, I can't think of a more complicated way of turning sines into cosines than convolving with $1/\pi t$, but there you are.

For another example, direct integration gives the Hilbert transform of the rectangle function:

$$\mathcal{H}\Pi(x) = \frac{1}{\pi}\int_{-\infty}^{\infty}\frac{\Pi(y)}{y-x}\,dy = \frac{1}{\pi}\int_{-1/2}^{1/2}\frac{1}{y-x}\,dy = \frac{1}{\pi}\ln\left|\frac{x-\frac{1}{2}}{x+\frac{1}{2}}\right|.$$

Here's a plot. The transform becomes infinite at $\pm 1/2$. (The things that look like cusps are just the parts of the graph going asymptotically to $\pm\infty$.)

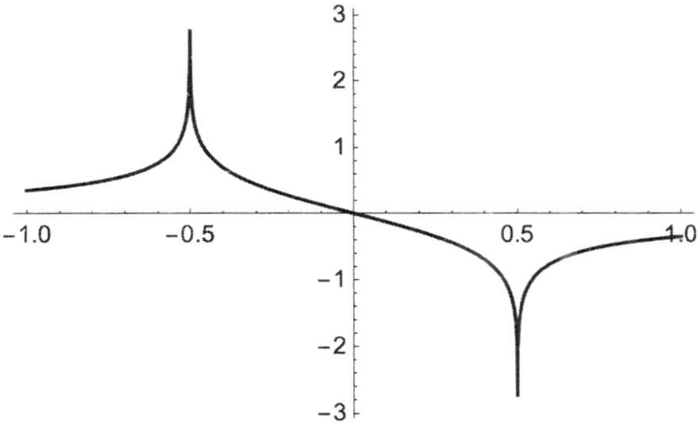

And in case you're wondering,

$$\mathcal{H}\operatorname{sinc} t = -\frac{\pi t}{2}\operatorname{sinc}^2\frac{t}{2}\,.$$

I'll put the calculations for this at the end of this section, together with a plot.

We won't produce any more examples. There's a table of Hilbert transforms in Bracewell's book, and you can also find them on the web.

8.9.3. The Hilbert transform and analytic signals. Those of you who have studied circuits are well aware of how much employing complex exponentials can streamline the analysis, and we have seen many benefits of representing real signals that way. Nevertheless, one drawback of introducing complex exponentials is that they seem inevitably to require that we consider negative and positive frequencies.

This manifests itself, for example, in the spectrum of a real signal having two sidebands, due to the relationship $\mathcal{F}f(-s) = \overline{\mathcal{F}f(s)}$. The use of the Hilbert transform allows one to modify a real signal to an *analytic signal*, which is a *complex-valued* form of the signal that involves only *positive* frequencies. This representation can simplify results and methods in some cases, especially in applications to communications systems. Here we'll just go through the basic construction, but take a course, any course, on communications to see this in daily use.

If we write the real signals $\cos 2\pi at$ and $\sin 2\pi at$ in terms of complex exponentials, as usual,

$$\cos 2\pi at = \tfrac{1}{2}(e^{2\pi iat} + e^{-2\pi iat}), \quad \sin 2\pi at = \frac{1}{2i}(e^{2\pi iat} - e^{-2\pi iat}),$$

then the right-hand sides display both the frequencies a and $-a$. It doesn't seem like that's saying much, but write instead

$$\begin{aligned} \cos 2\pi at &= \operatorname{Re}(\cos 2\pi at + i \sin 2\pi at) = \operatorname{Re} e^{2\pi iat}, \\ \sin 2\pi at &= \operatorname{Im}(\cos 2\pi at + i \sin 2\pi at) = \operatorname{Im} e^{2\pi iat}, \end{aligned}$$

and take note that the real and imaginary parts are a Hilbert transform pair, $\sin 2\pi at = -\mathcal{H} \cos 2\pi at$. The pairing of real and imaginary parts via the Hilbert transform is what's done in general. Here's how.

Take a real-valued signal $v(t)$. Since $v(t)$ is real valued, its Fourier transform, $V(s)$, has the symmetry property

$$V(-s) = \overline{V(s)},$$

so, in particular, $v(t)$ has positive and negative frequencies in its spectrum. Cut off the negative frequencies by multiplying $V(s)$ by the unit step function $u(s)$ (*not* the Fourier transform of u, just u). Actually, as we'll see, it's best to use $2u(s)V(s)$ for the cutoff.

Now form

$$\mathcal{Z}(t) = \mathcal{F}^{-1}(2u(s)V(s)).$$

By construction, $\mathcal{Z}(t)$ has no negative frequencies in its spectrum. Watch how it's related to the original signal $v(t)$. We can write $2u(s) = 1 + \operatorname{sgn} s$; hence

$$\begin{aligned} \mathcal{F}^{-1}(2u(s)V(s)) &= \mathcal{F}^{-1}((1 + \operatorname{sgn} s)V(s)) \\ &= \mathcal{F}^{-1}(V(s) + \operatorname{sgn} s V(s)) \\ &= \mathcal{F}^{-1}(V(s)) + \mathcal{F}^{-1}(\operatorname{sgn} s V(s)) \;=\; v(t) - i(\mathcal{H}v)(t). \end{aligned}$$

$\mathcal{Z}(t)$ is called the *analytic signal* associated with $v(t)$.

To summarize, let $v(t)$ be a real-valued signal with Fourier transform $V(s)$. Let $\mathcal{Z}(t) = \mathcal{F}^{-1}(2u(s)V(s))$. Then:

(1) $\mathcal{Z}(t) = v(t) - i(\mathcal{H}v)(t)$, so that $v(t) = \operatorname{Re} \mathcal{Z}(t)$.

(2) $\mathcal{Z}(t)$ has no negative frequencies in its spectrum.

(3) To get the spectrum of $\mathcal{Z}(t)$ cut off the negative frequencies of $v(t)$ and double the amplitudes of the positive frequencies.

Two instances of this, where we've already computed the Hilbert transform, are

$$v(t) = \cos 2\pi at \;\Rightarrow\; \mathcal{Z}(t) = \cos 2\pi at + i \sin 2\pi at\,,$$
$$v(t) = \operatorname{sinc} t \;\Rightarrow\; \mathcal{Z}(t) = \operatorname{sinc} t + \frac{i\pi t}{2}\operatorname{sinc}^2 \frac{t}{2}\,.$$

Example: Narrowband signals. Let's look at just one application of the analytic signal, to the representation of *narrowband signals*. Without giving a precise definition, a narrowband signal might look something like this:

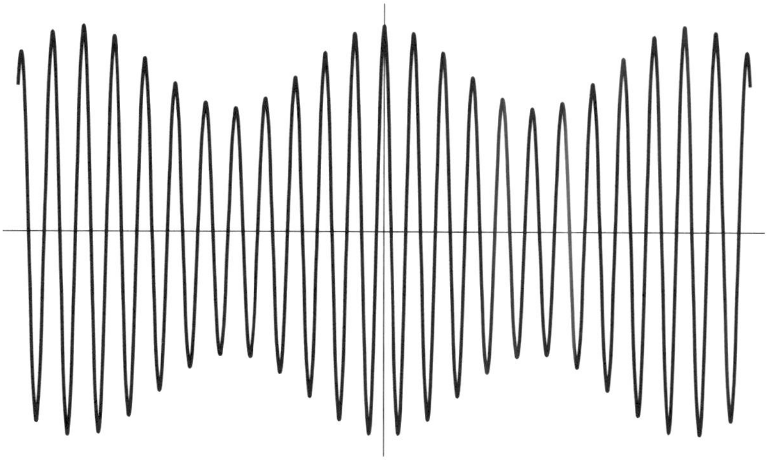

And its spectrum might look something like this:

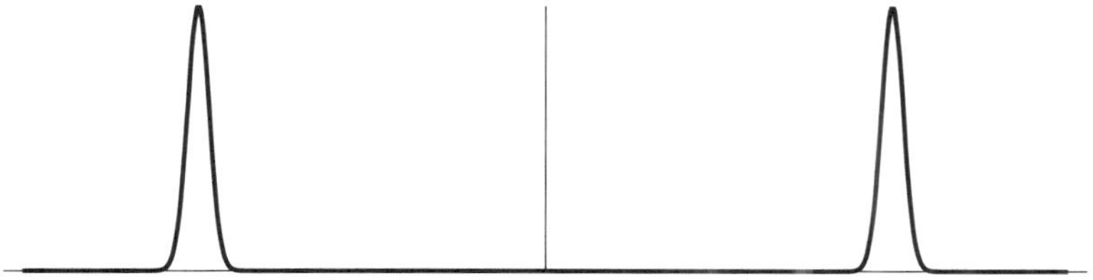

The idea is that the spectrum is concentrated in two islands, roughly centered about some frequencies $\pm s_0$, and that the spread about $\pm s_0$ is much less than s_0 itself. Often such a signal is of the form

$$v(t) = A(t)\cos(2\pi s_0 t + \phi(t))\,,$$

where the amplitude $A(t)$ and the phase $\phi(t)$ are *slowly varying*. Write $v(t)$ as

$$\begin{aligned} v(t) &= \tfrac{1}{2}A(t)\left(e^{2\pi i s_0 t + i\phi(t)} + e^{-(2\pi i s_0 t + i\phi(t))}\right) \\ &= \tfrac{1}{2}A(t)e^{i\phi(t)}e^{2\pi i s_0 t} + \tfrac{1}{2}A(t)e^{-i\phi(t)}e^{-2\pi i s_0 t}\,. \end{aligned}$$

Remember that we get the analytic signal corresponding to $v(t)$ by cutting off the negative frequencies and doubling the amplitudes of the positive frequencies. So to the extent that the spectral islands of $v(t)$ are really separated and the amplitude and phase are slowly varying functions, we are justified in saying that, approximately,

$$\mathcal{Z}(t) = A(t)e^{i\phi(t)}e^{2\pi i s_0 t}.$$

In terms of $\mathcal{Z}(t)$ the envelope of the signal $v(t)$ is

$$A(t) = |\mathcal{Z}(t)|$$

and the phase is

$$\psi(t) = \arg \mathcal{Z}(t) = 2\pi s_0 t + \phi(t).$$

In this context it is common to introduce the *instantaneous frequency*

$$s_{\text{inst}} = \frac{1}{2\pi}\frac{d\psi}{dt} = s_0 + \frac{1}{2\pi}\frac{d\phi}{dt}.$$

8.9.4. The Hilbert transform of sinc. To find the Hilbert transform of the sinc function we start in the frequency domain, with

$$\mathcal{F}(\mathcal{H}\,\text{sinc})(s) = i \operatorname{sgn} s\, \mathcal{F}\,\text{sinc}(s) = i \operatorname{sgn} s\, \Pi(s).$$

The graph of $\operatorname{sgn} s\, \Pi(s)$ looks like this:

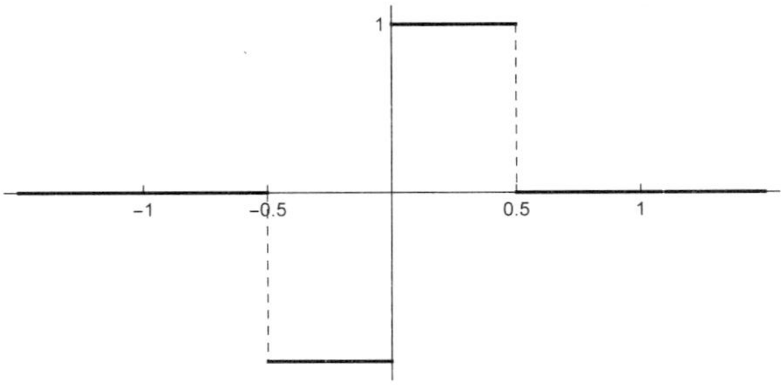

How could we find the inverse Fourier transform of something like that? Here's one way to do it. Remember the triangle function, defined by

$$\Lambda(x) = \begin{cases} 1 - |x|, & |x| \le 1, \\ 0, & \text{otherwise.} \end{cases}$$

Its Fourier transform is $\mathcal{F}\Lambda(s) = \text{sinc}^2 s$. If we scale to $\frac{1}{2}\Lambda(2x)$, which has the graph

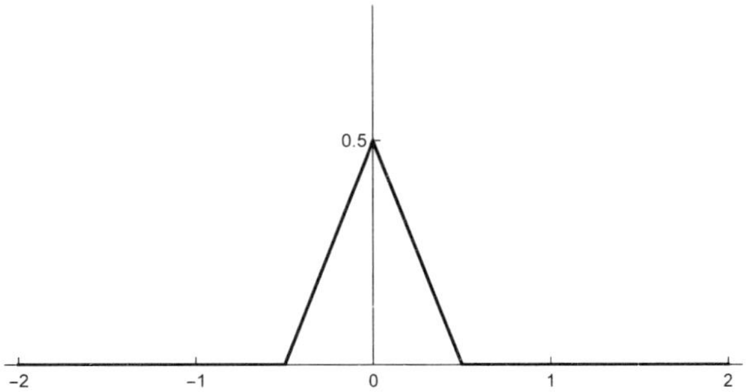

then

$$\operatorname{sgn} x\, \Pi(x) = -\frac{d}{dx}\frac{1}{2}\Lambda(2x).$$

Thus

$$\mathcal{F}^{-1}(i \operatorname{sgn} s\, \Pi(s)) = \mathcal{F}^{-1}\Big(-i\frac{d}{ds}\frac{1}{2}\Lambda(2s)\Big),$$

and we can find the right-hand side using the derivative theorem (and the stretch theorem). This gives

$$\mathcal{F}^{-1}\Big(-i\frac{d}{ds}\frac{1}{2}\Lambda(2s)\Big) = -i(-2\pi i t)\frac{1}{4}\operatorname{sinc}^2\frac{t}{2} = -\frac{\pi t}{2}\operatorname{sinc}^2\frac{t}{2}$$

and the final result

$$\mathcal{H}\operatorname{sinc} t = -\frac{\pi t}{2}\operatorname{sinc}^2\frac{t}{2}.$$

Here's a plot.

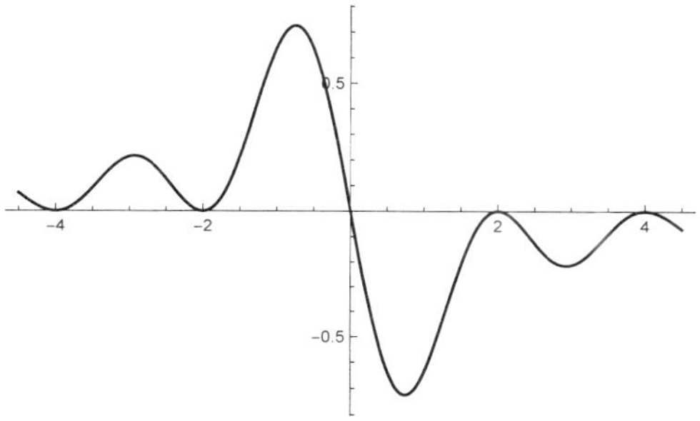

8.10. Filters Finis

We now want to live and work in the discrete world and discuss some typical digital filters and their uses. There are entire books and courses on digital filters, of course, not to mention the notable fraction of the world's economy that depends on them. We'll touch on only a few aspects, enough to give you a sense of how digital filters work and to encourage you to find out more. The presentation here draws on two excellent sources, the book *Digital Filters* by R. Hamming, certainly not new but still stimulating, and Briggs and Henson's book on the DFT, mentioned earlier.

The setup will be a discrete LTI system, given by

$$\underline{w} = \underline{h} * \underline{v},$$

where $\underline{v}$, $\underline{h}$, $\underline{w}$ are periodic signals of length N and $\underline{h}$ is the impulse response. Taking the DFT we write, as is customary,

$$\underline{W} = \underline{H}\,\underline{V}.$$

$\underline{H}$ is the transfer function.

8.10.1. Averaging in the time domain.

A common use of digital filters is to smooth data. Before convolution makes it's appearance we'll start the process as a least-squares fit of the data in the time domain. That turns into a running average, which is then realized as a discrete convolution and is finally refined through analysis in the frequency domain. We've spoken many times of convolution as a smoothing operation and here it is in the discrete setting, quite explicitly.

Suppose we have a list of data, typically arising as samples of some function $f(x)$, say $(x_0, f(x_0))$, $(x_1, f(x_1))$, $\ldots$, $(x_n, f(x_n))$. We suppose that the sample points $x_0, \dots, x_n$ are evenly spaced. We plot the data and join successive points to make a continuous curve — linear interpolation, in other words. That's a continuous fit but, say, it looks too jagged. We suspect too much noise in the measurements, for example. We set out to smooth the data.

Averaging and least-squares. A first attempt to define a smoothed version of $f(x)$, let's call it $g(x)$, is this:

- Instead of taking the points two at a time and joining successively, start by taking the first three data points $(x_0, f(x_0))$, $(x_1, f(x_1))$, $(x_2, f(x_2))$ and find the line that best fits these three points according to the method of least-squares.
- As the first new data point, take the midpoint of the least-squares fit. Since the x's are evenly spaced, the midpoint has x-coordinate x_1, and $g(x_1)$ is then defined to be the corresponding y-coordinate on the line; i.e., the midpoint of the least-squares fit is $(x_1, g(x_1))$.
- Shift one point over and repeat this procedure using the triple of original data points $(x_1, f(x_1))$, $(x_2, f(x_2))$, $(x_3, f(x_3))$. This gives the second new data point $(x_2, g(x_2))$.
- Then work with $(x_2, f(x_2))$, $(x_3, f(x_3))$, $(x_4, f(x_4))$ to get the third new data point, $(x_3, g(x_3))$, and so on through the list up to the triple $(x_{n-2}, f(x_{n-2}))$, $(x_{n-1}, f(x_{n-1}))$, $(x_n, f(x_n))$ producing the final new data point $(x_{n-1}, g(x_{n-1}))$.

Note that this procedure drops the original left endpoint and the original right endpoint; i.e., the new data points run from $(x_1, g(x_1))$ up to $(x_{n-1}, g(x_{n-1}))$.

There's a formula for $g(x_i)$, made much easier if we agree to write things in terms of discrete signals. Say we have N data points, which we represent as a discrete signal (or N-vector) $\underline{f}$. As we did when defining the DFT, we consider the samples to be tagged by the index m and the data points to be $(m, \underline{f}[m])$, rather than $(x_m, f(x_m))$. For reasons of symmetry, especially later when we switch to the frequency domain, it's a little more convenient to write

$$\underline{f} = \left(\underline{f}\left[-\frac{N}{2}+1\right], \dots, \underline{f}[0], \dots, \underline{f}\left[\frac{N}{2}\right]\right),$$

letting the index range from $-N/2+1$ to $N/2$. The formula for the new points $(m, \underline{g}[m])$ determined by the least-squares method as described is simply

$$\underline{g}[m] = \tfrac{1}{3}(\underline{f}[m-1] + \underline{f}[m] + \underline{f}[m+1]) .$$

In words, the new value $\underline{g}[m]$ is just the average of the old value $\underline{f}[m]$ with its adjacent values on either side, $\underline{f}[m-1]$ and $\underline{f}[m+1]$. I won't derive this here — sorry, but it's easily googled. The expression for $\underline{g}$ is naturally called a *running average* because the point $\underline{f}[m]$ is replaced by the average of itself and its neighbors, and the neighboring points overlap. Note, however, that it's always the original data that's getting averaged, not a combination of old and new data. This is the difference between a *nonrecursive filter*, such as we have here, and a *recursive filter*, for which the new values involve the old values together with previously computed new values.

Using shifted $\underline{\delta}$'s, it's very easy to express $\underline{g}$ as a convolution,

$$\underline{g} = \tfrac{1}{3}(\underline{\delta}_1 + \underline{\delta}_0 + \underline{\delta}_{-1}) * \underline{f},$$

since, to remind you,

$$\tfrac{1}{3}((\underline{\delta}_1 * \underline{f})[m] + (\underline{\delta}_0 * \underline{f})[m] + (\underline{\delta}_{-1} * \underline{f})[m]) = \tfrac{1}{3}(\underline{f}[m-1] + \underline{f}[m] + \underline{f}[m+1]).$$

The discrete signal

$$\underline{h} = \tfrac{1}{3}(\underline{\delta}_1 + \underline{\delta}_0 + \underline{\delta}_{-1})$$

is the impulse response of 3-point smoothing as an LTI system, $\underline{g} = \underline{h} * \underline{f}$.

Modified averaging. Convolution is an appealing and useful way to think of the operation and is very suitable to computation; the formula $\underline{g} = \underline{h} * \underline{f}$ is so easy to implement. But convolution has some consequences that we must sort out. If we let m run from $-N/2+2$ to $N/2-1$, dropping the left and right endpoints, then we have described algebraically what we originally described geometrically, no more no less. But if we let convolution be convolution, then it's natural (essential when the DFT comes into the picture) that the inputs be periodic, and with m running from $-N/2-1$ to $N/2$ the endpoints do come in and something funny happens there.

At the left endpoint $m = -N/2+1$ we have, using periodicity,

$$\begin{aligned}\underline{g}\Big[-\frac{N}{2}+1\Big] &= \tfrac{1}{3}\Big(\underline{f}\Big[-\frac{N}{2}\Big] + \underline{f}\Big[-\frac{N}{2}+1\Big] + \underline{f}\Big[-\frac{N}{2}+2\Big]\Big)\\ &= \tfrac{1}{3}\Big(\underline{f}\Big[\frac{N}{2}\Big] + \underline{f}\Big[-\frac{N}{2}+1\Big] + \underline{f}\Big[-\frac{N}{2}+2\Big]\Big)\\ &\qquad\Big(\text{the left index changed from } -\frac{N}{2} \text{ to } -\frac{N}{2}+N = \frac{N}{2}\Big).\end{aligned}$$

And at the right endpoint $m = N/2$, again using periodicity, we have

$$\begin{aligned}\underline{g}\Big[\frac{N}{2}\Big] &= \tfrac{1}{3}\Big(\underline{f}\Big[\frac{N}{2}-1\Big] + \underline{f}\Big[\frac{N}{2}\Big] + \underline{f}\Big[\frac{N}{2}+1\Big]\Big)\\ &= \tfrac{1}{3}\Big(\underline{f}\Big[\frac{N}{2}-1\Big] + \underline{f}\Big[\frac{N}{2}\Big] + \underline{f}\Big[-\frac{N}{2}+1\Big]\Big)\\ &\qquad\Big(\text{the right index changed from } \frac{N}{2}+1 \text{ to } \frac{N}{2}+1-N = -\frac{N}{2}+1\Big).\end{aligned}$$

We see that in computing the value of the new left endpoint the operation of convolution has averaged in the old value at the *right* endpoint, a value at the opposite end of the original data! The corresponding thing happens in computing the new right endpoint. This is certainly not what was called for in the original description of smoothing as the least-squares line fit of three adjacent points. Or put differently, it doesn't seem much like smoothing to average together data from opposite ends of the sampled values.

There are several things we could do:

- We could run the convolution as is and then just drop the computed values at the left and right endpoints. That would be in keeping with the original description as a moving least-squares fit of three adjacent data points.
- Or *before* convolving we could replace both the original left endpoint and original right endpoint values by the average of the two; that is, the first *and*

last values of the input $\underline{f}$ are modified to be

$$A = \tfrac{1}{2}\Big(\underline{f}\Big[-\frac{N}{2}+1\Big] + \underline{f}\Big[\frac{N}{2}\Big]\Big).$$

This is a theme in the Briggs and Henson book. It traces back to the phenomenon that the Fourier series converges to the average of the values at the endpoints at a jump discontinuity. (See Section 1.5.1.) The more you work with the DFT, the more you will run into the dictum: replace the values at the endpoints by their average.

What's most often done — and we'll do so here — is to modify the impulse response $\underline{h}$ rather than the data. In the original running average filter the coefficients of the filter function all have weight $1/3$. Instead, we take the neighbors, $\underline{f}[m-1]$ and $\underline{f}[m+1]$, to have half the weight of the midpoint $\underline{f}[m]$. This *modified running average filter* (or modified least-squares filter) is thus defined by

$$\underline{g} = \left(\tfrac{1}{4}\underline{\delta}_1 + \tfrac{1}{2}\underline{\delta}_0 + \tfrac{1}{4}\underline{\delta}_{-1}\right) * \underline{f}, \quad \underline{g}[m] = \tfrac{1}{4}\underline{f}[m-1] + \tfrac{1}{2}\underline{f}[m] + \tfrac{1}{4}\underline{f}[m+1],$$

so the impulse response is now

$$\underline{h} = \tfrac{1}{4}\underline{\delta}_1 + \tfrac{1}{2}\underline{\delta}_0 + \tfrac{1}{4}\underline{\delta}_{-1}.$$

We'll analyze the effects of this filter in the frequency domain, but first a confession. We did the 3-point running average because it was easy to describe and easy to write out. It's more common to take a 5-*point* running average. The least-squares interpretation is as before: take the least-squares fit to five consecutive data points and take the new data point to be the midpoint of this fit. With that description, the filter is

$$\underline{g} = \tfrac{1}{5}\left(\underline{\delta}_2 + \underline{\delta}_1 + \underline{\delta}_0 + \underline{\delta}_{-1} + \underline{\delta}_{-2}\right) * \underline{f},$$

so at an index m,

$$\underline{g}[m] = \tfrac{1}{5}(\underline{f}[m-2] + \underline{f}[m-1] + \underline{f}[m] + \underline{f}[m+1] + \underline{f}[m+2]).$$

Once again, to address the endpoint problem, this is usually modified so that the two end values are given half the weight of the interior values. This results in the modified 5-point running average filter:

$$\underline{g} = \left(\tfrac{1}{8}\underline{\delta}_2 + \tfrac{1}{4}\underline{\delta}_1 + \tfrac{1}{4}\underline{\delta}_0 + \tfrac{1}{4}\underline{\delta}_{-1} + \tfrac{1}{8}\underline{\delta}_{-2}\right) * \underline{f},$$

with impulse response

$$\underline{h} = \tfrac{1}{8}\underline{\delta}_2 + \tfrac{1}{4}\underline{\delta}_1 + \tfrac{1}{4}\underline{\delta}_0 + \tfrac{1}{4}\underline{\delta}_{-1} + \tfrac{1}{8}\underline{\delta}_{-2}.$$

Example. Let's have an example of the 5-point filter in action, a concocted example but one that hits the main points. Consider the signal

$$f(t) = 3\cos(2\pi t) + \cos(4\pi t) + 0.5\cos(12\pi t) + 0.3\cos(14\pi t),$$

for $-0.5 \le t \le 0.5$. You take samples of this signal, 30 of them, every 1/30th of a second. Here's a plot of the signal and the sampled values.

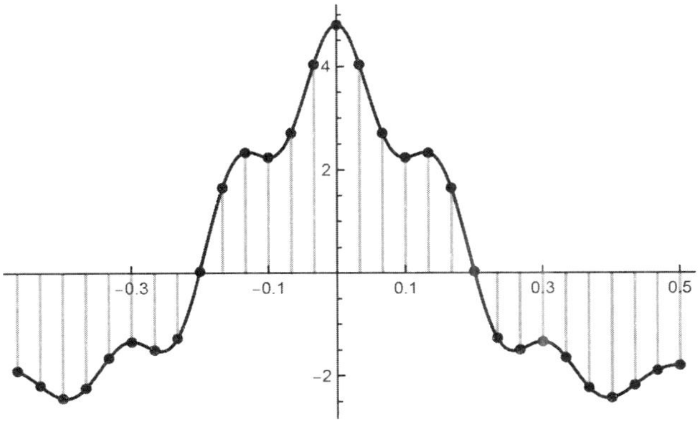

Of course, you plot only your measured values, not knowing the signal. Your discrete signal is $\underline{f}[m] = f(m/30)$, $m \in [-14 : 15]$. You see a low frequency pattern but you also see some regularly occurring bumps. You think (you hope) these bumps are some kind of noise that's not really indicative of the phenomenon that you're measuring, and you want to smooth them out. You form the discrete function $\underline{g}$ as above (which uses the periodicity of $\underline{f}$). Here's what you get for your plot.

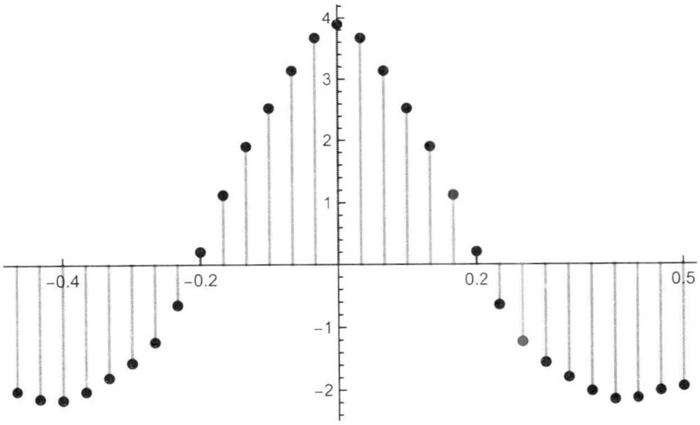

The bumps are gone; the data has been smoothed. Whether this smoothing has lost something essential in the data is up to you and your scientific conscience.

8.10.2. Analysis in the frequency domain. Sticking with the same example, how exactly have the higher frequencies that caused the bumps been suppressed? Take the DFT, of order 30, of $\underline{g} = \underline{h} * \underline{f}$ and, as usual, write it as

$$\underline{G} = \underline{H}\,\underline{F}\,.$$

Here's the plot of $\underline{F}$ showing spikes at the four frequencies (positive and negative) in the signal. Because $f(t)$ and its sampled version $\underline{f}[m]$ are real and even, the DFT is also real and even; you're looking at the plot of the actual DFT, not the magnitude of the DFT.

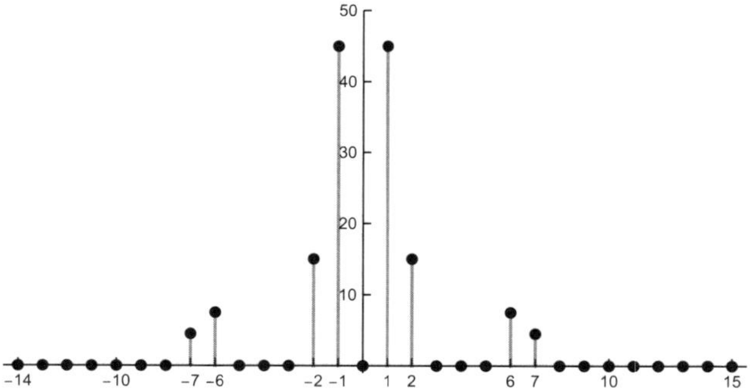

By the way, as a quick check (of the type you should get used to making) you might note that $\underline{F}[0] = 0$. That's as it should be, because

$$\underline{F}[0] = \sum_m \underline{f}[m],$$

and this is zero because the average value of $\underline{f}$ is zero.

The system $\underline{g} = \underline{h} * \underline{f}$ is *not* a simple low-pass filter, simply cutting the spectrum off beyond a specified frequency. It's a more gradual attenuation. The magic is in the transfer function $\underline{H}$. We found it, up to a scaling factor, back in Section 7.9; look back there. The impulse response $\underline{h}$ is a scaled discrete rectangle function. In the earlier notation ($p = 4$, $N = 30$),

$$\underline{h} = \frac{1}{4}\underline{\Pi}_{4,1/2} = \frac{1}{8}(\underline{\delta}_2 + \underline{\delta}_{-2}) + \frac{1}{4}\sum_{k=-1}^{1} \underline{\delta}_k$$

and

$$\underline{H}[m] = \underline{\mathcal{F}}\,\underline{h}[m] = \begin{cases} \frac{1}{4}\cdot 4, & m \equiv 0 \mod 30, \\ \frac{1}{4}\dfrac{\cos(\pi m/30)\sin(4\pi m/30)}{\sin(\pi m/30)}, & m \not\equiv 0 \mod 30\,. \end{cases}$$

Here's a plot.

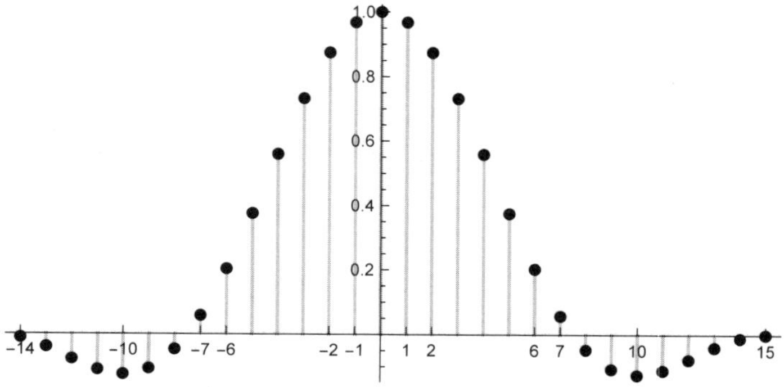

Taking the product $\underline{H}\,\underline{F}$ to produce $\underline{G}$ has the effect of diminishing the higher frequencies in the original signal $\underline{f}$ while leaving the magnitude of the lower frequency components largely unchanged. Here's a plot of $\underline{G}$, showing exactly this phenomenon.

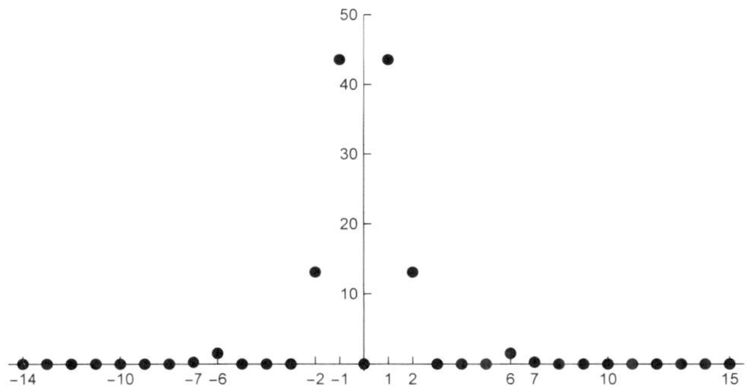

The pictures of the transfer functions in the frequency domain shed light on the difference between the straight running average, where all values are weighted equally, and the modified running average, where the end values are given half the weight of the interior values. Here are the transfer functions for running average filters of three points, five points, seven points, and nine points. Once again, we found these transfer functions in Section 7.9; they're of the form $\underline{\tilde{H}}_{p+1} = \frac{1}{p+1}\mathcal{F}\underline{\Pi}_{p,1}$. (I put the tilde on $\underline{H}$ to distinguish these transfer functions from the ones for the modified running average.) You can see the pattern:

$$\underline{\tilde{H}}_3[m] = \begin{cases} 1, & m \equiv 0, \\ \frac{1}{3}\frac{\sin(3\pi m/N)}{\sin \pi m/N}, & m \not\equiv 0, \end{cases} \qquad \underline{\tilde{H}}_5[m] = \begin{cases} 1, & m \equiv 0, \\ \frac{1}{5}\frac{\sin(5\pi m/N)}{\sin \pi m/N}, & m \not\equiv 0, \end{cases}$$

$$\underline{\tilde{H}}_7[m] = \begin{cases} 1, & m \equiv 0, \\ \frac{1}{7}\frac{\sin(7\pi m/N)}{\sin \pi m/N}, & m \not\equiv 0, \end{cases} \qquad \underline{\tilde{H}}_9[m] = \begin{cases} 1, & m \equiv 0, \\ \frac{1}{9}\frac{\sin(9\pi m/N)}{\sin \pi m/N}, & m \not\equiv 0. \end{cases}$$

Below are the plots. The 3-point filter is the thicker, dashed curve, followed by the others, thinner and squeezing in, in order (the 5-, 7- and 9-point filters). I replaced the discrete variable n/N with the continuous variable t to show more readily two important features. In each case the transfer function has a center lobe from its maximum down to its first minimum, and then there are sidelobes where the transfer function oscillates. On multiplying the transfer function times the DFT of an input to the filter, $\underline{G} = \underline{\tilde{H}}\underline{F}$, the frequencies close to the center are passed through pretty much unchanged, and the frequencies that are in the sidelobe regions are decreased. The sidelobes die out, so the farther out we go, the more the frequencies in those regions are eliminated, but it's a question of how fast the sidelobes are dying out.

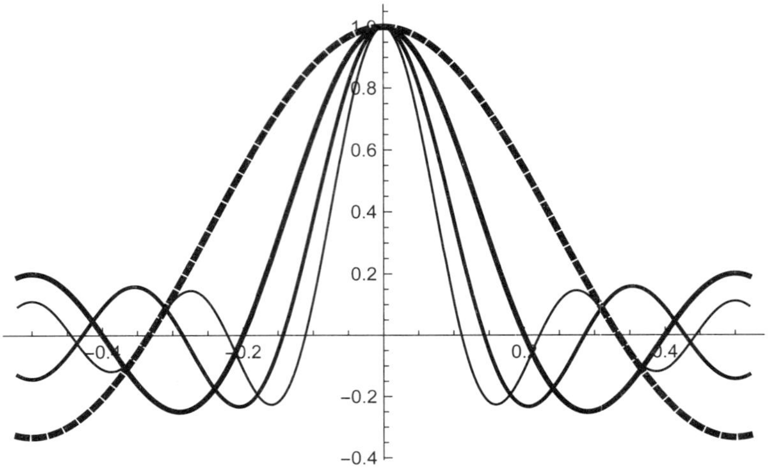

For the modified running average, $\underline{H}_{p+1} = \frac{1}{p}\underline{\mathcal{F}\Pi}_{p,1/2}$:

$$\underline{H}_3[m] = \begin{cases} 1, & m \equiv 0, \\ \frac{1}{2}\frac{\cos(\pi m/N)\sin(2\pi m/N)}{\sin \pi m/N}, & m \not\equiv 0, \end{cases}$$

$$\underline{H}_5[m] = \begin{cases} 1, & m \equiv 0, \\ \frac{1}{4}\frac{\cos(\pi m/N)\sin(4\pi m/N)}{\sin \pi m/N}, & m \not\equiv 0, \end{cases}$$

$$\underline{H}_7[m] = \begin{cases} 1, & m \equiv 0, \\ \frac{1}{6}\frac{\cos(\pi m/N)\sin(6\pi m/N)}{\sin \pi m/N}, & m \not\equiv 0, \end{cases}$$

$$\underline{H}_9[m] = \begin{cases} 1, & m \equiv 0, \\ \frac{1}{8}\frac{\cos(\pi m/N)\sin(8\pi m/N)}{\sin \pi m/N}, & m \not\equiv 0. \end{cases}$$

Here are the plots, again with a continuous variable and with the 3-point filter as the dashed, thicker curve on the outside, and the 9-point filter farthest inside.

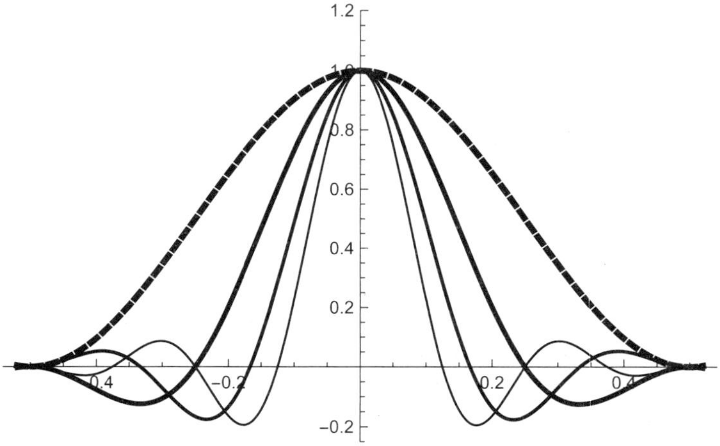

The center lobes are somewhat wider here than they are for the unmodified running average, so more frequencies in the center are passed through, but more important is that for the modified running average the sidelobes damp down much more quickly. Thus frequencies outside the center lobe are decreased much more dramatically. The result is smoother data (fewer high frequencies) back in the time domain.[15]

One word of caution. Although the modified running average as we've defined it is in widespread use, it's not the only way to smooth data. Having seen how these work you might imagine designing and analyzing your own filters.

8.10.3. Find that filter. Let's finish our very brief discussion of digital filters with the discrete form of low-pass and band-pass filters. You can get high-pass and notch filters directly from these.

Low-pass filters. The example of finding a running average had us convolving in the time domain and multiplying in the frequency domain, and the effect was to eliminate, or at least attenuate, the higher frequencies. If we start in the frequency domain instead and just cut off higher frequencies, we wind up with a low-pass filter. We discussed this in Chapter 3, and again in Chapter 5, strictly in the continuous case. One more time, here's the main idea.

An ideal low-pass filter sets all frequencies above a certain amount ν_c to zero and lets all frequencies below ν_c pass through unchanged. Writing, as usual,

$$w(t) = (h * v)(t), \quad W(s) = H(s)V(s)\,,$$

the transfer function is

$$H(s) = \Pi_{2\nu_c}(s) = \begin{cases} 1, & |s| \le \nu_c\,, \\ 0, & |s| > \nu_c\,. \end{cases}$$

The transfer function is just a scaled rect function. In the time domain the impulse response is

$$h(t) = 2\nu_c \operatorname{sinc}(2\nu_c t)\,.$$

In the discrete case with N-points, going, say, from $-N/2+1$ to $N/2$, the transfer function is defined to be

$$\underline{H}[m] = \begin{cases} 1, & |m| < m_c\,, \\ \frac{1}{2}, & |m| = m_c\,, \\ 0, & m_c < |m| \le \frac{N}{2}\,. \end{cases}$$

Here m_c is the index associated with the frequency where we want to cut off. We take the value to be $1/2$ at the endpoints. This choice comes from the "take the average at a jump discontinuity" principle.

It's a discrete rectangle function, of course, and while here we called it $\underline{H}$, if you look back (again) to Section 7.9 and to Section 7.10, there we called it $\underline{\Pi}$ and

---

[15]These plots are also in Hamming's book, and I refer you there for more information.

we found its DFT and inverse DFT explicitly (the dinc). With $\underline{h} = \underline{\mathcal{F}}^{-1}\underline{H}$ we have

$$\underline{h}[m] = \begin{cases} \dfrac{m_c}{N}, & m = 0, \\ \dfrac{1}{N}\dfrac{\cos(\pi m/N)\sin(2\pi m m_c/N)}{\sin(\pi m/N)}, & m \neq 0. \end{cases}$$

Here are plots of $\underline{H}[m]$ and $\underline{h}[m]$ with $N = 64$ and $m_c = 32$; this is an example from Briggs and Henson.

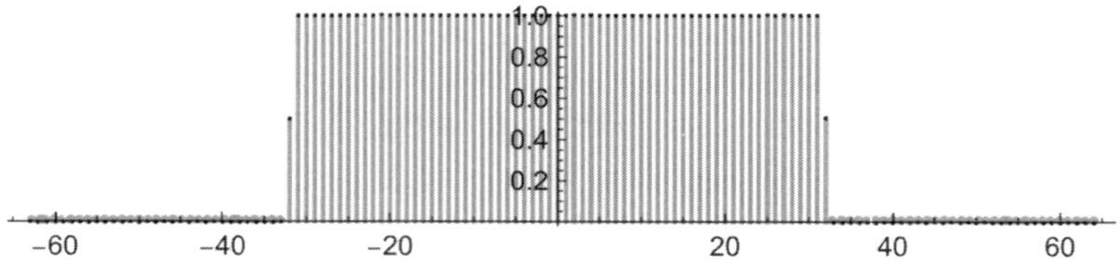

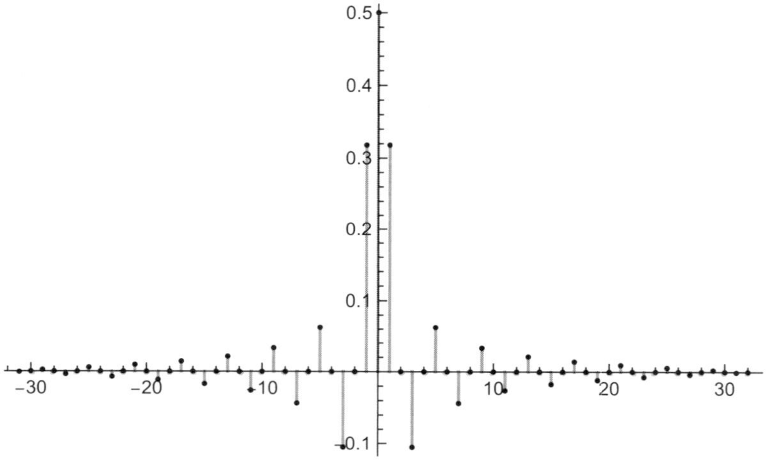

Remember, it's the function $\underline{h}$, in the second plot that gets convolved with an input. Notice again the sidelobes in the time domain, i.e., the many, small oscillations. These are less pronounced for this definition of $\underline{h}$, i.e., with $\underline{H}$ defined to be 1/2 at the endpoints, than they would be if $\underline{H}$ jumped straight from 0 to 1. But even so, using this filter with a signal that itself has "edges" can cause some unwanted effects. This is called *ringing*, a whole other topic. To counteract such effects, one sometimes brings the transfer function down to zero more gradually. One possible choice, suggested by Briggs and Henson, is

$$\underline{H}[m] = \begin{cases} 1, & |m| \leq m_c - m_0\,, \\ \sin\left(\frac{\pi(m_c - |m|)}{2m_0}\right), & m_c - m_0 < |m| \leq m_c\,, \\ 0, & m_c < |m|\,, \end{cases}$$

where, again, m_c is where you want the frequencies cut off and m_0 is where you start bringing the transfer function down.[16]

[16]This corrects a typo in Briggs and Henson on p. 268. They didn't have $|m|$ inside the sine function.

Here is the picture in frequency — the transfer function, $\underline{H}$.

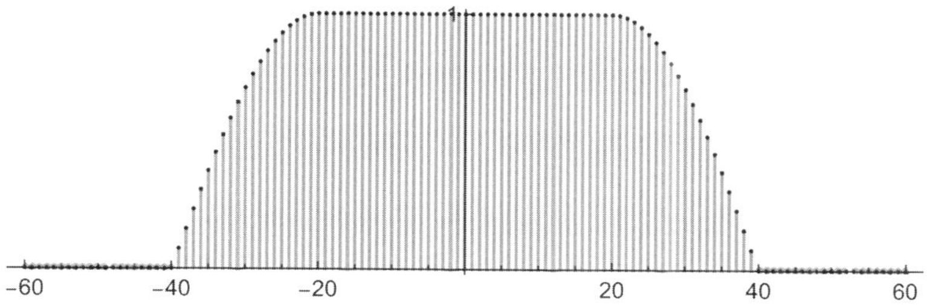

The cutoff frequency is $m_c = 32$ and $\underline{H}$ starts its descent at $m_0 = 20$.

And here is the picture in time — the impulse response, $\underline{h} = \underline{\mathcal{F}}^{-1}\underline{H}$.

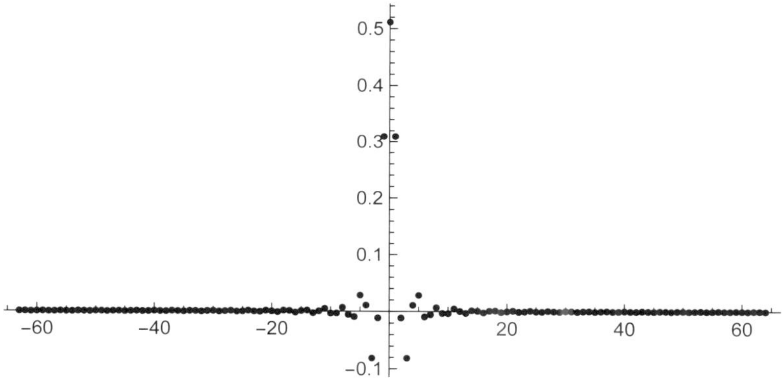

The sidelobes are definitely less pronounced.

Band-pass filters. We also looked earlier at *band-pass filters*, filters that pass a particular band of frequencies through unchanged and eliminate all others. The transfer function $B(s)$ for a band-pass filter can be constructed by shifting and combining the transfer function $H(s)$ for the low-pass filter.

We center our band-pass filter at $\pm\nu_0$ and cut off frequencies more than ν_c above and below $\pm\nu_0$. Thus we define

$$\begin{aligned} B(s) &= \begin{cases} 1, & \nu_0 - \nu_c \le |s| \le \nu_0 + \nu_c, \\ 0, & |s| < \nu_0 - \nu_c \text{ or } |s| > \nu_0 + \nu_c \end{cases} \\ &= \Pi_{2\nu_c}(s - \nu_0) + \Pi_{2\nu_c}(s + \nu_0)\,. \end{aligned}$$

From the representation of $B(s)$ in terms of Π's it's easy to find the impulse response, $b(t)$:

$$\begin{aligned} b(t) &= h(t)e^{2\pi i\nu_0 t} + h(t)e^{-2\pi i\nu_0 t} \quad \text{(using the shift theorem)} \\ &= 2h(t)\cos(2\pi\nu_0 t). \end{aligned}$$

Here's a picture of a discrete version of a band-pass filter in the frequency and time domains. The transfer function $\underline{H}$ is the sum of two discrete rectangle functions, with value 1/2 at the endpoints. The two halves are centered at $\pm m_0$ and cutoff frequencies more than m_c above and below $\pm m_0$.

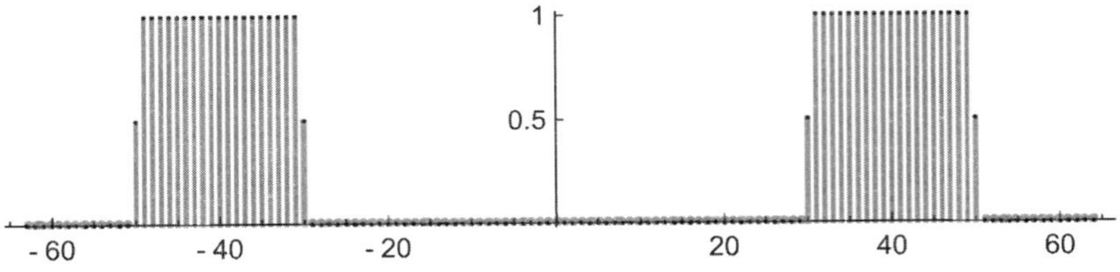

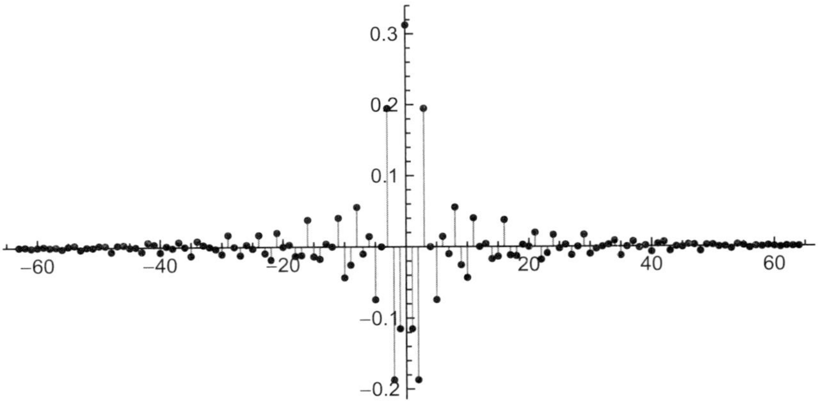

The sidelobes in the impulse response $\underline{h} = \underline{\mathcal{F}}^{-1}\underline{H}$ die off but are pretty wildly oscillating. As before, it's possible to mollify this by decreasing the transfer function to zero more gradually.

It's faster in frequency. It's worth pausing to remember the gains in computational efficiency provided by the FFT algorithm for calculating the DFT and realizing what this means for filtering. Calculating the convolution of two (periodic) sequences of length N requires on the order of N^2 operations. On the other hand, using the FFT on the inputs ($O(N \log_2 N)$ operations), multiplying *componentwise* the results (N operations), and then inverting (another $O(N \log_2 N)$ operations) requires a total of $O(N \log_2 N)$ operations. Filtering by passing to the frequency domain gives a considerable savings.

8.11. A Tribute: The Linear Millennium

I started writing these notes around the turn of the century, the most recent century, and at the time I felt an obligation to offer some millennial reminiscences. It's been quite a thousand years. I propose *linearity* as one of the most important themes of mathematics and its applications in times past and times present.

Why has linearity been so successful? I offer a few reasons. By all means, make your own list.

(1) On a small scale, smooth functions are approximately linear.

This is the basis of calculus, of course, but it's a very general idea and it has dominated the discussion for 300 or so years. Whenever one quantity changes smoothly (differentiably) with another, small changes in the one produce, approximately, directly proportional changes in the other.

Related to this in geometry is the notion of a linear approximation to a nonlinear object, e.g., the tangent space to a curved space (a manifold) and all the constructions that go with it, including tensors and differential forms.

(2) There's a highly developed, highly successful assortment of existence and uniqueness results for solving linear equations, and existence and uniqueness are related for linear problems. When does an equation have a solution? If it has one, does it have more than one? These are fundamental questions.

Think of solving systems of linear equations as the model here. Understanding the structure of the space of solutions to linear systems in the finite-dimensional, discrete case (i.e., matrices) is the starting point. Under this banner, the computational aspects of linearity are ubiquitous and profound. Take, for one example, *linear programming* — optimization under linear constraints. Hard to find a combination of two words that has had a greater impact.

Solving linear systems of equations has also served as the model of what to look for in the infinite-dimensional case. I've previously mentioned linear operators and the rise of functional analysis and its applications as an aspect of this.

In ancient times when people studied the classical differential equations of mathematical physics like the heat equation, the wave equation, and Laplace's equation, they all knew that they satisfied the principle of superposition and they all knew that this was important. The change in point of view was to emphasize the linear structure of the space of solutions, including in eigenvalue problems.

(3) Linearity can be associated wth groups, particularly through what has come to be known as representation theory. This mixes analysis, algebra, and geometry and contributes to all of those perspectives. Groups bring out symmetries in problems and symmetries in the solutions. This has turned out to be very important in sorting out and generalizing many phenomena in Fourier analysis. It has also been key in the description of the elementary particles in physics. An influential book by Persi Diaconis, *Group Representations in Probability and Statistics*, takes representation theory in another direction.

Finally, I'm willing to bet (but not large sums) that linearity won't necessarily hold sway too far into the new millennium and that nonlinear phenomena will become increasingly more central. Nonlinear problems have always been important, but it's the *computational* power now available that is making them more accessible to analysis.

Consider, for example, the field of convex optimization. Convexity is the next best thing to linearity. It's time has arrived. See, for example, the recent book *Convex Optimization* by Stephen Boyd and Lieven Vandenberghe. I got a quote from Stephen Boyd on this:

> Linear algebra studies linear equations. Convex optimization adds inequalities to linear equations.[17] Linearity is flatness, or zero curvature,

[17] OK, technically you get linear programming, but that's a stepping stone from linear equations to convex optimization.

geometrically. (And smoothness/calculus is approximate flatness.) Convexity is nonnegative curvature. Turns out lots of things you might want to analyze, or compute, end up involving nonnegative curvature; and with powerful computers, we're good to go.

Problems and Further Results

8.1. Are the following systems linear? Time invariant? If time invariant, what is the impulse response and the transfer function?
(a) $Lf = f^-$.
(b) $Lf = \tau_b f$.
(c) $Lf = \sigma_a f$.

8.2. *LTI input/output pairs*
Assuming that L is an LTI system, answer the following questions (each part is independent of the others; i.e., the system L in (a) is different from the system L in (b)).
(a) If $L\{\cos(2\pi t)\} = y(t)$, find $L\{\sin(2\pi t)\}$ in terms of $y(t)$.
(b) If $L\{\Pi(t)\} = y(t)$, find $L\{\Pi(t/2)\}$ in terms of $y(t)$.
(c) If $L\{\operatorname{sinc}(t)\} = \sin(\frac{2}{3}\pi t)$, find $L\{\operatorname{sinc}(t/2)\}$.
(d) If $L\{\Pi(t)\} = \cos(\pi t)$, find $L\{\Lambda(t)\}$.
Hint: For part (a), think first of how to relate $\sin(2\pi t)$ to $\cos(2\pi t)$, and then use that relationship to determine the unknown output. In fact, most of these problems can be solved by finding an appropriate connection between the two inputs.

8.3. A linear system L has the inputs (on the left) and outputs (on the right) shown below.

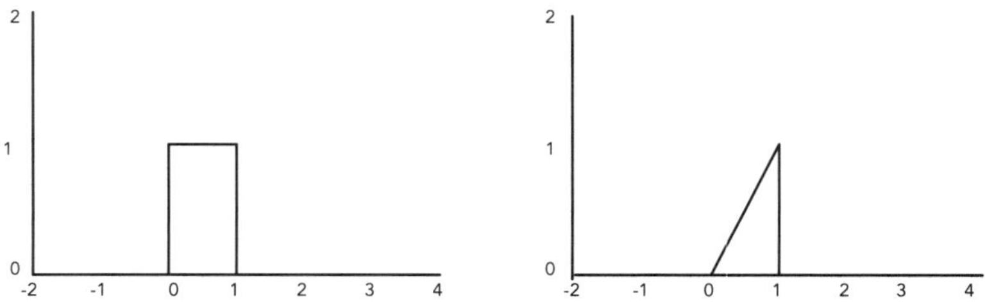

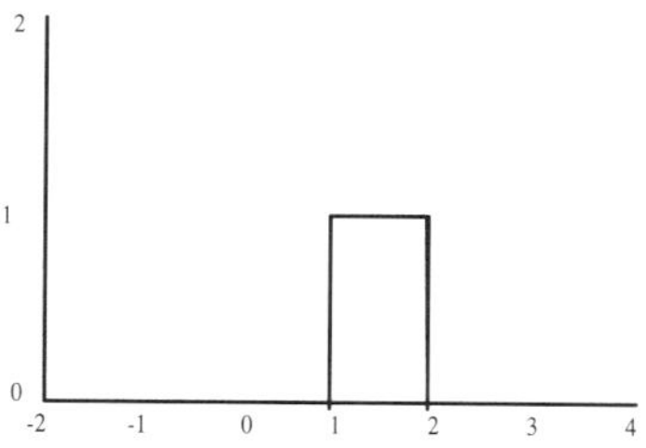

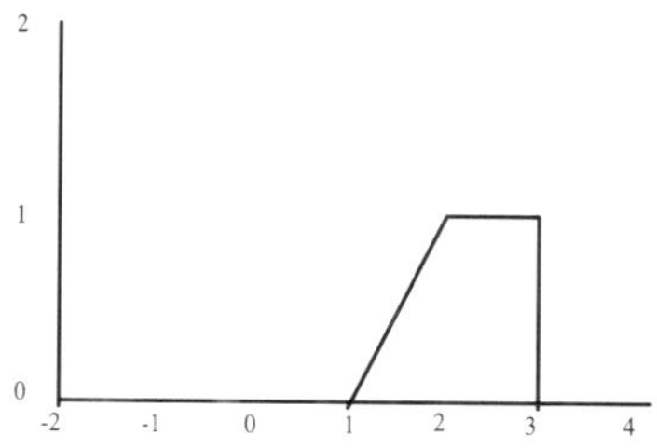

(a) Is L time invariant? Justify your answer.

(b) Sketch the output of L given the input below.

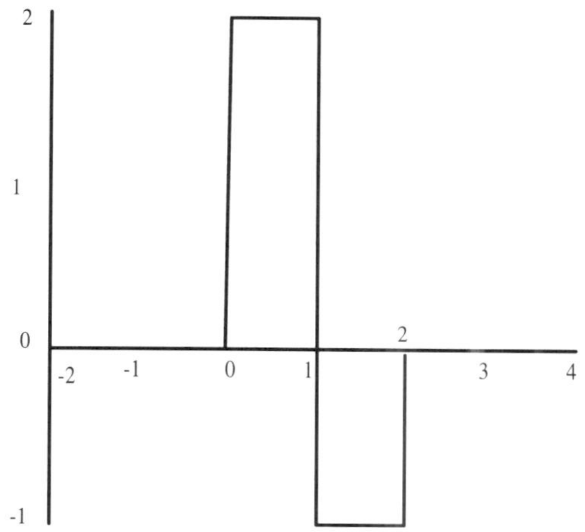

8.4. *LTI system input/outputs*

For each case below, determine whether the output function (with spectrum $Y(s)$) can be generated when the input function (with spectrum $X(s)$) is passed through a linear time-invariant system. Explain your reasons.

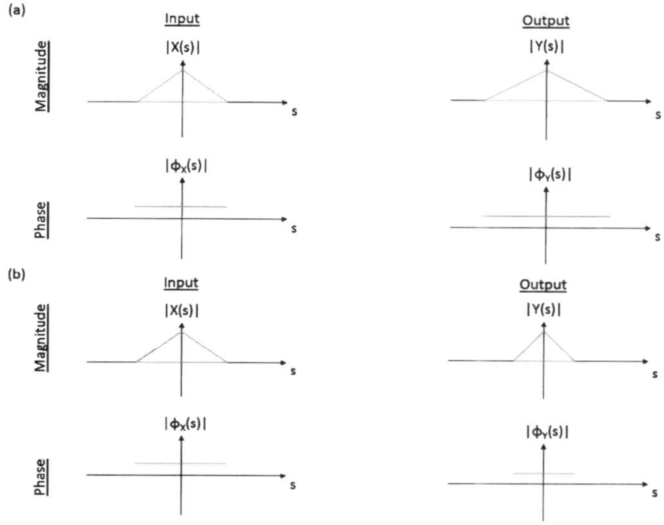

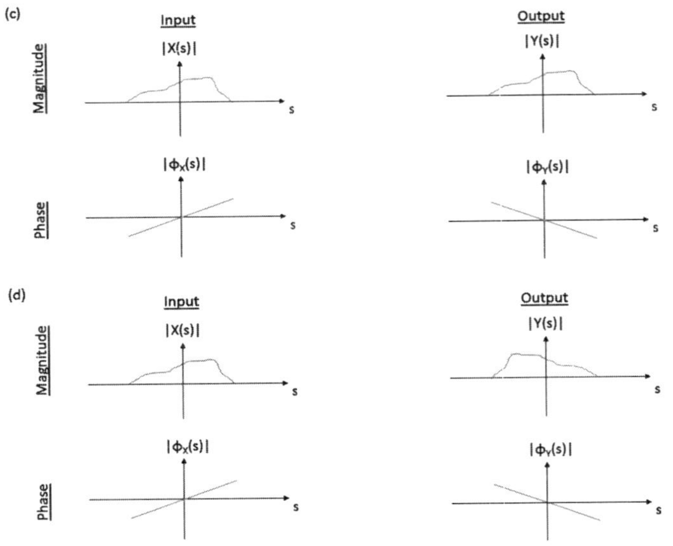

8.5. *Zero-order hold.*

The music on your CD (if you still use CDs) has been sampled at the rate of 44.1 kHz. This sampling rate comes from the sampling theorem together with experimental observations that your ear cannot respond to sounds with frequencies above about 20 kHz. (The precise value 44.1 kHz comes from the technical specs of the earlier audio tape machines that were used when CDs were first getting started.)

A problem with reconstructing the original music from samples is that interpolation based on the sinc function is not physically realizable — for one thing, the sinc function is not timelimited. Cheap CD players use zero-order hold. We've seen this before. Recall that the value of a given sample is held until the next sample is read, at which point that sample value is held, and so on.

Suppose the input is represented by a train of δ-functions, spaced $T = 1/44.1$ msec apart with strengths determined by the sampled values of the music, and suppose the output looks like a staircase function. The system for carrying out zero-order hold then looks like the diagram below. (The scales on the axes are the same for both the input and the output.)

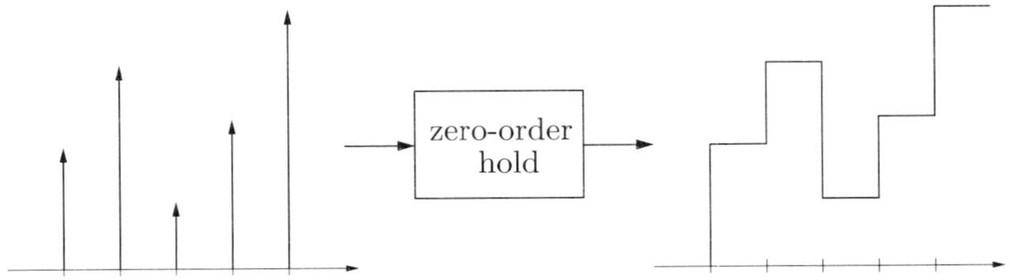

(a) Is this a linear system? Is it time invariant for shifts of integer multiples of the sampling period?
(b) Find the impulse response for this system.
(c) Find the transfer function.

8.6. Let $w(t) = Lv(t)$ be an LTI system with impulse response $h(t)$; thus $w(t) = (h * v)(t)$.
(a) If $h(t)$ is bandlimited, show that $w(t) = Lv(t)$ is bandlimited regardless of whether $v(t)$ is bandlimited.
(b) Suppose $v(t)$, $h(t)$, and $w(t) = Lv(t)$ are all bandlimited with spectrum contained in $-1/2 < s < 1/2$. According to the sampling theorem we can write

$$v(t) = \sum_{k=-\infty}^{\infty} v(k)\,\mathrm{sinc}(t-k)\,, \quad h(t) = \sum_{k=-\infty}^{\infty} h(k)\,\mathrm{sinc}(t-k)\,,$$
$$w(t) = \sum_{k=-\infty}^{\infty} w(k)\,\mathrm{sinc}(t-k)\,.$$

Express the sample values $w(k)$ in terms of the sample values of $v(t)$ and $h(t)$. (You will need to consider the convolution of shifted sincs.)

8.7. *LTI systems and sampling*[18]

A bandlimited signal $x(t)$ of bandwidth p passes through an LTI filter whose impulse response is $h(t)$. The input signal $x(t)$ and output signal $y(t) = h(t) * x(t)$ can be reconstructed from samples taken at any rate above p. The sampled values define discrete signals $\underline{x}[n] = x(n/p)$, $\underline{y}[n] = y(n/p)$. Note that $\underline{x}$ and $\underline{y}$ are defined for $n = 0, \pm 1, \pm 2, \ldots$ and are generally not periodic.

Find a discrete impulse response $\underline{k}$ such that $\underline{y} = \underline{k} * \underline{x}$. Convolution is defined by

$$(\underline{k} * \underline{x})[n] = \sum_{m=-\infty}^{\infty} \underline{k}[n-m]\underline{x}[m]\,.$$

8.8. *Unitary linear systems*

We said that a linear system L given by integration against a (complex-valued) kernel $k(x, y)$,

$$w(x) = Lv(x) = \int_{-\infty}^{\infty} k(x,y)v(y)\,dy\,,$$

is a *unitary* linear system if

$$\int_{-\infty}^{\infty} k(x,y)\overline{k(x,z)}\,dx = \delta(y-z)\,.$$

(This says that $L^*Lv = v$ where L is the adjoint of L. Needless to say, we assume integrals can be used to express δ-functions.)

[18] From J. Gill.

(a) Show that if L is unitary, then it preserves inner products:

$$(Lu, Lv) = (u, v)\,,$$

where the inner product of two signals u and v is

$$(u, v) = \int_{-\infty}^{\infty} u(x)\overline{v(x)}\,dx\,.$$

You'll want to use different variables of integration for $Lu(x)$ and $Lv(x)$, as in

$$Lu(x) = \int_{-\infty}^{\infty} k(x, y)u(y)\,dy, \quad Lv(x) = \int_{-\infty}^{\infty} k(x, z)v(z)\,dz\,.$$

(b) Parseval's identity for the Fourier transform says that

$$\int_{-\infty}^{\infty} \mathcal{F}u(x)\overline{\mathcal{F}v(x)}\,dx = (\mathcal{F}u, \mathcal{F}v) = (u, v) = \int_{-\infty}^{\infty} u(x)\overline{v(x)}\,dx\,.$$

So the Fourier transform preserves inner products. Show that $\mathcal{F}$ is a unitary linear system in the sense above.

8.9. *LTI systems, derivatives, step response, and impulse response*

(a) Let L be an LTI system with $w(t) = Lv(t)$. Show that $Lv'(t) = w'(t)$. *Hint*: Use that the derivative $v'(t)$ is approximated by its difference quotient $\frac{v(t+h)-v(t)}{h}$. Assume nothing goes wrong in taking limits as $h \to 0$.

(b) Let $u(t)$ be the unit step function. The output $Lu(t)$ is called the step response. Suppose that for the LTI system L when the input is a pulse $v(t) = u(t) - u(t-1)$, the output is $w(t) = e^{-t}u(t) - e^{-(t-1)}u(t-1)$. Find the impulse response of the system and the transfer function. If $H(s)$ is the transfer function, examine the behavior of $|H(s)|$ as $s \to 0$ and as $s \to \infty$. Is the system more like a low-pass or a high-pass filter?

8.10. *Accumulators as an LTI system*

Consider a continuous-time system L that takes input $v(t)$ and generates output $w(t)$ by

$$w(t) = Lv(t) = \int_{-\infty}^{t} v(\tau)\,d\tau\,.$$

This is sometimes called an accumulator.

(a) L is clearly a linear system because integration is a linear operation. Show that it is also a time-invariant system.

(b) What is the impulse response of L?

(c) A natural thought is that the input v can be recovered from the output w simply by taking the derivative of w, but some conditions have to be imposed. Show that if $\mathcal{F}v(0) = 0$, then $w' = v$. (Use part (b).)

8.11. *Derivative filter*[19]

Let $W > 0$ and let

$$H(s) = \begin{cases} 2\pi i s, & |s| < W, \\ 0, & |s| > W. \end{cases}$$

(a) Find the impulse response $h(t)$ of the filter with $H(s)$ as the transfer function. (No integral needed.)
(b) Find the output of the filter with input $\cos(2\pi f t)$. Consider the cases $0 < f < W$ and $f > W$.
(c) Find the output of the filter with input $\sin(2\pi f t)$.
(d) In words, what is the output of this filter for signals bandlimited to $< W$.

8.12. *Chirp transform*

The chirp signal is given by $f(t) = e^{i\pi t^2}$, and its Fourier transform is $F(s) = e^{i\pi/4} e^{-i\pi s^2}$. We consider the system described by the following block diagram, with input $v(t)$ and output $w(t)$, that involves multiplication and convolution with the chirp signal (so this is a linear system).

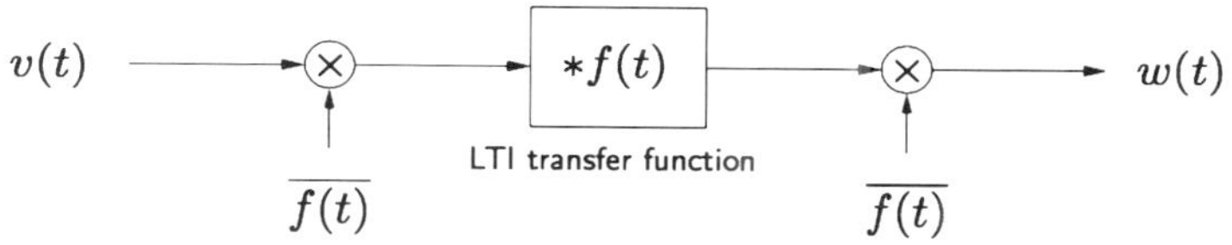

(a) Find the output $w(t)$ when the input is $v(t) = \delta(t)$.
(b) Find the output $w(t)$ when the input is $v(t) = \delta(t - \tau)$.
(c) Write out the superposition integral for this system. Do you recognize this system as a *familiar* operation?

8.13. Let $a > 0$ and, as earlier, define the averaging operator acting on a signal $f(t)$ to be

$$Lf(t) = \frac{1}{2a} \int_{t-a}^{t+a} f(x)\,dx\,.$$

(a) Show that L is an LTI system, and find its impulse response and transfer function.
(b) What happens to the system as $a \to 0$? Justify your answer based on properties of the impulse response.
(c) Find the impulse response and the transfer function for the cascaded system $M = L(Lf)$. Simplify your answer as much as possible; i.e., don't simply express the impulse response as a convolution.
(d) Three signals are plotted below. One is an input signal $f(t)$. One is $Lf(t)$ and one is $Mf(t)$. Which is which, and why?

[19]From J. Gill.

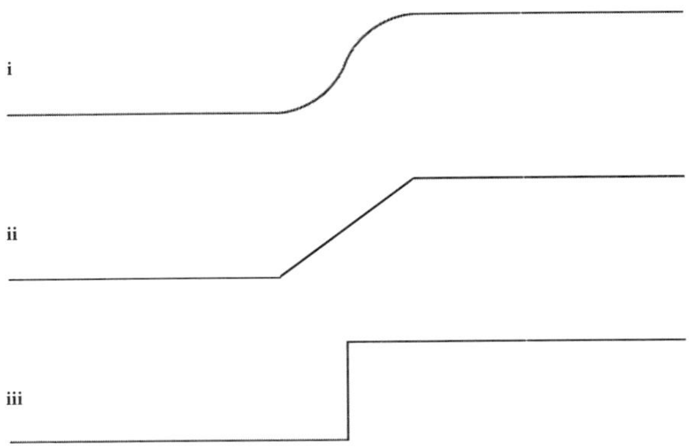

8.14. *System identification*

(a) Consider a filter with real and even transfer function $H(s)$. If the input to the filter is $\cos(2\pi\nu t)$, show that the output is $H(\nu)\cos(2\pi\nu t)$.

(b) Suppose you have an unknown filter. Download `identme.m` and `transferfcn.p`. The file `identme.m` is the filter that takes an input signal and computes the output by calling `transferfcn.p`. The file `transferfcn.p` is a "black box" because its code is obscured from the user (you). Using sinusoids of frequency $\nu_1 = 3$ and $\nu_2 = 8$, give separate examples of the filter satisfying superposition and homogeneity. Give an example of the filter satisfying time invariance. If you can only input signals and observe the output signals, is it possible to prove that the black box filter is LTI?

(c) Suppose you are told that the black box filter is indeed LTI, with an even transfer function $H(s)$. Let the input be a sinusoid with some frequency ν. Estimate $H(\nu)$ from the amplitude of the output. Sketch $H(s)$ by varying ν. What kind of filter is this? What is/are the cutoff frequency(ies)?
Hints: (1) Try frequencies in the range 0–20 Hz. (2) Ignore the fact that the output signal is not a perfect sinusoid. This is because the signal is not infinite duration.

8.15. *Radio receiver*[20]

The mixer section of a heterodyne receiver can be modeled as show below.

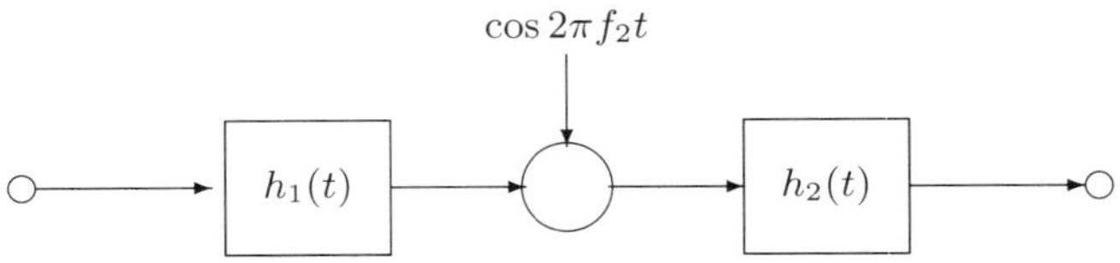

The first linear filter is a band-pass filter with center frequency f_1; it has the transfer function

$$H_1(f) = \begin{cases} \frac{1}{2} & \text{if } f_1 - \frac{1}{2}B \le |f| \le f_1 + \frac{1}{2}B, \\ 0 & \text{otherwise,} \end{cases}$$

[20] From J. Gill.

where $f_1 \gg B$. (The nonzero range of this filter can also be described by $\big||f| - |f_1|\big| < \frac{1}{2}B$.) This filter is followed by an ideal multiplier, which in turn is followed by an ideal low-pass filter with transfer function

$$H_2(f) = \begin{cases} 1 & \text{if } 0 \le |f| \le b, \\ 0 & \text{otherwise,} \end{cases}$$

where $|f_2 - f_1| + B/2 < b < f_1 + f_2 - B/2$.

(a) Is this system linear?

(b) What is its impulse response, that is, the response to the input $\delta(t)$?

(c) Is this system time invariant? If yes, what is its transfer function? If not, why not?

8.16. *A half-wave rectifier analysis*

A half-wave rectifier circuit converts AC into DC voltage. AC voltage is a sinusoid with frequency s_0 (and period $T_0 = 1/s_0$) and can be written as $v_{\text{ac}}(t) = A\sin(2\pi s_0 t)$. Ideal DC voltage is a constant waveform $v_{\text{dc}}(t) = V$. A circuit schematic for a half-wave rectifier implemented with an ideal diode and an RC low-pass filter is shown below.

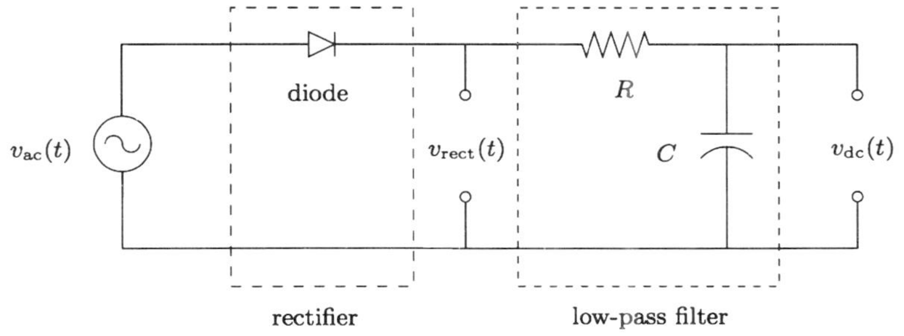

(a) The rectifier portion of the circuit will pass only the positive voltage waveform; i.e., $v_{\text{rect}}(t) = v_{\text{ac}}(t)$ when $v_{\text{ac}}(t) \ge 0$, and it is zero otherwise. Find the Fourier series coefficients for the rectified voltage waveform $v_{\text{rect}}(t)$.

(b) Ideally the low-pass filter would only pass the DC component of the $v_{\text{rect}}(t)$ and eliminate any other nonzero frequencies. However, we use a nonideal RC low-pass filter with the transfer function,

$$H(s) = \frac{1}{1 + i(2\pi sRC)}\,.$$

Compute the Fourier series coefficients for the output voltage $v_{\text{dc}}(t)$. (It is also a periodic waveform.)

8.17. *Stable LTI systems*

Let $w(t) = Lv(t)$ be an LTI system. One says that L is *stable* if a bounded input $v(t)$ results in a bounded output $w(t)$. (Recall that to say a function $f(t)$ is bounded means there is a constant M such that $|f(t)| \le M$ for all t.)

(a) Let $h(t)$ be the impulse response of the system. Show that if

$$\int_{-\infty}^{\infty} |h(t)|\, dt < \infty\,,$$

then the system is stable.

(b) You'll now show the converse to part (a). This has several steps. Given any function $h(t)$, define a function $v(t)$ by

$$v(t) = \begin{cases} \dfrac{h(-t)}{|h(-t)|}, & h(-t) \neq 0\,, \\ 0, & h(-t) = 0\,. \end{cases}$$

(i) Sketch $v(t)$ for the particular $h(t)$ shown below, and explain why $v(t)$ is bounded for a general $h(t)$.

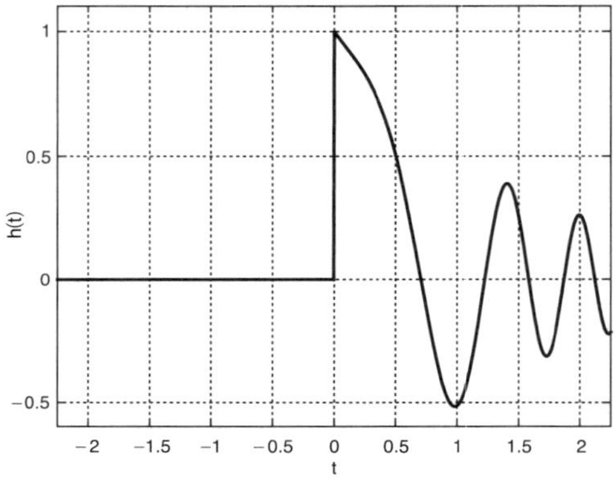

Now let $w(t) = Lv(t)$ be an LTI system with impulse response $h(t)$ and suppose that the integral of $|h(t)|$ is infinite:

$$\int_{-\infty}^{\infty} |h(t)|\, dt = \infty.$$

(ii) With $v(t)$ defined as above, show that the output $w(t) = Lv(t)$ is unbounded. (Consider the value of $w(t)$ at $t = 0$.)

(iii) Explain why this proves the converse to part (a).

(c) If $w(t) = Lv(t)$ is a stable LTI system, show that an input of finite energy results in an output of finite energy. (Use the frequency domain and show first that the condition of stability implies that the transfer function is bounded.)

8.18. *Derivative and Hilbert transform*

Recall the Hilbert transform,

$$\mathcal{H}f = -\frac{1}{\pi t} * f(t)\,.$$

(a) Derive the formula $\frac{d}{dt}\mathcal{H}f = \mathcal{H}\frac{df}{dt}$.

(b) Using this result, find $\mathcal{H}\delta''(t)$.

8.19. *Analytic signals*

(a) Show that for any two analytic signals $u(t)$ and $v(t)$

$$\int_{-\infty}^{\infty} u(t)v(t)\, dt = 0.$$

(b) Find the analytic signal representation of $\Pi(t)$.

8.20. *Linear modulation-invariant systems*[21]

Shifts and modulation are closely related operations via the Fourier transform. This problem gives an interpretation of this in terms of linear systems.

The *exponential modulation operator*, $\mathcal{E}_p$, operates on a signal f by $\mathcal{E}_p f(x) = e^{2\pi i p x} f(x)$. In terms of this operator, the usual shift theorems for the Fourier transform can be written $\mathcal{F}(\tau_p f) = \mathcal{E}_{-p}(\mathcal{F}f)$, and for the inverse Fourier transform we have $\mathcal{F}^{-1}(\mathcal{E}_p f) = \tau_{-p}(\mathcal{F}^{-1} f)$.

A linear operator L is said to be *modulation invariant* (LMI) provided that it commutes the exponential modulation operator $\mathcal{E}_p$ for every p; i.e.,

$$L(\mathcal{E}_p f) = \mathcal{E}_p L f.$$

Given a general linear system L, define a linear system $\tilde{L}$ by $\tilde{L} = \mathcal{F} L \mathcal{F}^{-1}$. Show that L is LTI if and only if $\tilde{L}$ is LMI.

8.21. *The Mellin transform*

A system L is said to be *linear scale invariant* (LSI) if L is linear and

$$L(\sigma_a v(t)) = \sigma_a (L v(t))$$

for all inputs $v(t)$ and for all $a > 0$. Thus just as LTI systems commute with the shift operator, LSI systems commute with the scaling operator.

Throughout this problem, we will assume that the input and output are *causal*, meaning that if $w = Lv$, then

$$v(t) = 0 \text{ for } t < 0 \qquad \text{and} \qquad w(t) = 0 \text{ for } t < 0.$$

(a) Show that if L is an LSI system, then it can be written as

$$w(t) = Lv(t) = \int_0^{\infty} v(y)\, m\Big(\frac{t}{y}\Big)\, \frac{dy}{y},$$

where $m(t) = L\{\delta(t-1)\}$. The function $m(t)$ is called the *Mellin impulse response*.

Hint: The argument for this is very much like the argument linking convolution and LTI systems, starting with

$$v(t) = \int_0^{\infty} v(y)\delta(t-y)\, dy, \quad v(t) = 0 \text{ for } t < 0.$$

You will need the scaling property of δ.

[21] From Akanksha Chowdhery.

Notation: The integral in part (a) is called the *Mellin convolution* and we denote it (just for this problem) by

$$(v \diamond m)(t) = \int_0^\infty v(y)\, m\Big(\frac{t}{y}\Big)\, \frac{dy}{y}.$$

(b) Show that if a system L can be written in the form $w(t) = Lv(t) = (v \diamond m)(t)$ for some function $m(t)$, then L is LSI.

(c) The *Mellin transform* is defined by

$$(\mathcal{M}f)(s) = \int_0^\infty f(t)t^{s-1}\, dt = \int_0^\infty f(t)t^s\, \frac{dt}{t},$$

assuming (as we will) that the integral converges. The Mellin transform is *not* an LSI system. Rather, show that

$$\mathcal{M}(\sigma_a f)(s) = a^{-s}\mathcal{M}f(s).$$

Notice that the "spectral variable" s is *not scaled* on taking the transform of the scaled function $\sigma_a f$. This property of the Mellin transform is used extensively by people working in computer vision for image identification and image registration. B. Mandelbrot used the Mellin transform in his study of fractals, objects that exhibit self-similarity at all scales.

(d) Show that if $w = v \diamond m$, then

$$\mathcal{M}w = (\mathcal{M}v)(\mathcal{M}m).$$

Assume that v, w, and m are causal.

Remark. The Mellin transform is related to the Fourier transform by $(\mathcal{M}f)(s) \equiv \mathcal{F}\{f(e^x)\}$. For those of you who have had a course in group theory, this connection to the Fourier transform explains how the properties of the two transforms are related. Namely, the exponential map e^x is an isomorphism of the additive group $(\mathbb{R}, +)$ with the multiplicative group $(\mathbb{R}^+, \times)$. Under this map, the Fourier transform, convolution, and LTI operators become the Mellin transform, Mellin convolution, and LSI operators, respectively.

8.22. *Discrete convolution, inner products, and reversed signals*

(a) Let $\underline{f}$, $\underline{g}$, $\underline{h}$ be discrete, real signals. Show that $(\underline{h} * \underline{f}, \underline{g}) = (\underline{f}, \underline{h}^- * \underline{g})$, where $(\cdot,\cdot)$ is the complex inner product.

(b) If L_h is the matrix corresponding to convolution with $\underline{h}$, as in $L_{\underline{h}}\underline{f} = \underline{h} * \underline{f}$, then to what does the transpose $L_{\underline{h}}^T$ correspond?

(c) What if the signals are complex valued?

8.23. *Savings account balance and LTI systems*

You never know where LTI systems might just come up. Suppose you just opened a new savings account with monthly interest rate α. Starting from the zeroth month, every month you deposit some money into the account. The fund deposit at the nth month is $x[n]$, and the balance at the nth month is $y[n]$. Since your new account has no money in it initially, $y[n] = 0$ for $n < 0$.

The relationship between $x[n]$ and $y[n]$ can be expressed as

$$y[n] = x[n] + (1 + \alpha)y[n-1]\,.$$

For this problem we consider $x[n]$ and $y[n]$ to be defined for all integers n, positive and negative, and not just from 0 to some $N-1$, but the principles we've used before are still valid.

(a) The system taking $x[n]$ as the input to $y[n]$ as the output is an LTI system. You need not show this. (It's easy to see, so think about it sometime.)
Show that the impulse response $h[n]$ is

$$h[n] = (1+\alpha)^n u[n],$$

where $u[n]$ is the discrete unit-step function:

$$u[n] = \begin{cases} 1, & n \geq 0, \\ 0, & n < 0. \end{cases}$$

(b) Suppose you deposit C dollars per month starting at month 0, so your input is $x[n] = Cu[n]$. Using the fact that an LTI system is given by convolution, show that after n months your balance is

(*) $$y[n] = Cu[n]\frac{(1+\alpha)^{n+1}-1}{\alpha}.$$

Note: Because we assume that the inputs and outputs are defined for all integers, the sum giving a convolution should go from $-\infty$ to ∞ instead of from 0 to $N-1$ as we write it for periodic signals. Nevertheless, it will reduce to a finite sum.

(c) Suppose you first deposit $1,000/month for 12 months, then you don't do anything for six months, then you withdraw $500/month for another 12 months, then you don't do anything.
If $x[n]$ is the input corresponding to this scenario, write

$$x[n] = \text{combination of scaled, shifted } u\text{'s}.$$

(d) Assuming the scenario in part (c), write an expression for your account balance after 36 months. In order to simplify the writing, write equation (*) in the form

$$y[n] = Cs[n] \quad \text{for an input } Cu[n].$$

Then write your answer in terms of $s[n]$. You should note how linearity and time invariance play into your solution.

8.24. *The DFT diagonalizes circulant matrices*

A discrete linear system with input $\underline{v} \in \mathbb{R}^n$ and output $\underline{w} \in \mathbb{R}^n$ is given by matrix multiplication,

$$\underline{w} = A\underline{v}$$

for an $n \times n$ matrix A. We have shown that the following three statements are equivalent (and are taken as known):

(a) The system is time invariant (or shift invariant).
(b) A is a circulant matrix.
(c) A acts by convolution, meaning there is an $\underline{h} \in \mathbb{R}^n$ such that

$$A\underline{v} = \underline{h} * \underline{v} \quad \text{for all } \underline{v} \in \mathbb{R}^n.$$

Show the following:

Theorem. A matrix A is circulant if and only if it is diagonalized by the discrete Fourier transform $\underline{\mathcal{F}}$; i.e.,

$$\underline{\mathcal{F}}\, A\, \underline{\mathcal{F}}^{-1} = \Lambda \quad \text{or equivalently} \quad \underline{\mathcal{F}}\, A = \Lambda \underline{\mathcal{F}}\,,$$

where Λ is a diagonal matrix.

8.25. A discrete LTI system is given by $\underline{w} = \underline{h} * \underline{v}$ where $\underline{h}[k] = k$ for $k = 0, 1, 2, 3$. This system is also given as $\underline{w} = L\underline{v}$ where L is the matrix

$$L = \begin{pmatrix} 0 & 3 & 2 & 1 \\ 1 & 0 & 3 & 2 \\ 2 & 1 & 0 & 3 \\ 3 & 2 & 1 & 0 \end{pmatrix}.$$

Find the eigenvalues of L by finding the transfer function of the system.

8.26. *Symmetric discrete signals*[22]

A real, discrete signal $\underline{f}$ of length N is *symmetric* if for some point k_0,

$$\underline{f}[k_0 + k] = \underline{f}[k_0 - k], \quad k = 0, \pm 1, \pm 2, \ldots.$$

(In particular, when $k_0 = 0$, then $\underline{f}$ is even.)

(a) The DFT $\mathcal{F}\underline{f}$ of N-periodic real signals $\underline{f}$ is given below. In which of these cases is $\underline{f}$ symmetric?
  (i) $\mathcal{F}\underline{f}[n] = (N-n)n$, $n = 0, 1, 2, \ldots, N-1$.
  (ii) $\mathcal{F}\underline{f}[n] = e^{2\pi i n}$.
  (iii) $\mathcal{F}\underline{f}[n] = n^2 e^{4\pi i n}$, $n = 0, 1, 2, \ldots, N-1$.
  (iv) $\mathcal{F}\underline{f}[n] = n(1-i) + Ni$, $n = 0, 1, 2, \ldots, N-1$.

(b) Let A be a real circulant $N \times N$ matrix with first column $\underline{h}$. Suppose that $\underline{h}$ is symmetric about k_0.
  (i) Show that A is orthogonal if and only if $\underline{h} * \underline{h} = \underline{\delta}_{k_0}$.
  (ii) Conclude that A is orthogonal if and only if all its eigenvalues are of the form $\pm \exp(\pi i k_0 n/N)$.

8.27. *Differences are LTI*

Recall the problems on the backward and forward difference operators, ∇ and Δ in the DFT chapter. We can think of them each as a system operating on length N discrete signals:

$$\underline{w}[n] = \nabla \underline{v}[n] = \underline{v}[n] - \underline{v}[n-1],$$
$$\underline{w}[n] = \Delta \underline{v}[n] = \underline{v}[n+1] - \underline{v}[n].$$

Show that ∇ and Δ are LTI systems. What are the respective impulse responses and transfer functions? What are the matrices representing ∇ and Δ, respectively, say, for $N = 5$?

[22]From A. Siripuram.

8.28. *Control engineering application*[23]

Consider a system described by the following equations (known as the state-space representation):

$$\frac{\partial u(t)}{\partial t} = a \cdot u(t) + b \cdot v(t),$$

$$w(t) = c \cdot u(t) + d \cdot v(t).$$

Here, $v(t)$ is the system input, $w(t)$ is the system output, and $u(t)$ is the hidden state variable. The constants a, b, c, and d are complex numbers.

(a) Prove the system is linear and time invariant. What is the impulse response, $h(t)$? *Hint*: Approach the problem in the frequency domain.

(b) In control theory, a system is called causal if the output depends on past and current inputs but not future inputs. For an LTI system, the condition for causality is $h(t) = 0$ for all $t < 0$. What are the conditions on a, b, c, and d constants for the above system to be causal?

(c) Another important system property is stability, more specifically bounded-input bounded-output (BIBO) stability. A system is BIBO stable if a bounded input results in a bounded output. For an LTI system, the condition for BIBO stability is $\int_{-\infty}^{\infty} |h(t)|\, dt < \infty$. What are the conditions on a, b, c, and d for the system above to be BIBO stable?

8.29. *Tape recorder*

A magnetic tape containing signal $f(x)$ is scanned by a point sensor (head) that moves uniformly and continuously in the x-direction. The gap between the head and the tape is a. Ideally, the output of the head, $g(x)$, would be proportional to $f(x)$. However, at any given position x_0, the output signal $g(x_0)$ is proportional to the total signal that the head "sees" from the tape. In fact, at any position, the head receives a signal from the entire tape. The incremental contribution by each tape segment to the output signal depends on $1/d^2$, where d is the distance from the tape segment to the head.

(a) Derive an expression for the output $g(x)$ in terms of the input $f(x)$. Assume a catch-all gain factor of K. (Write this as a convolution of f and another function h.)

(b) At what frequency will the response of the system be half that of the response to zero frequency?

[23]From Sercan Arik.

8.30. *Wireless communications in a tunnel*[24]

A transmitting antenna is placed at the entry of a tunnel. A receiving antenna is placed on a car inside the tunnel. The transmitting antenna transmits a signal in all directions. Among all the transmitted beams, some will reach the receiving antenna directly, and some will reach the receiving antenna after one or more reflections on the tunnel ceiling and/or floor.

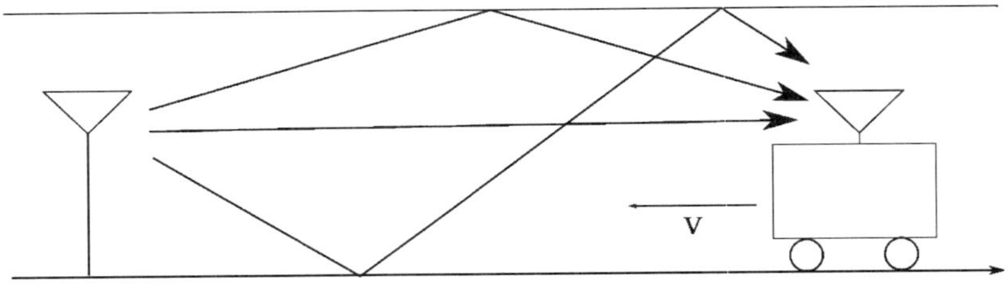

We make the simplifying assumption that there is only one beam going from the transmitting antenna to the receiving antenna for a given number of reflections; i.e., as illustrated, there is one direct beam, one beam with one reflection, one beam with two reflections, etc. (We label the beams according to the number of reflections, $0, 1, 2, \ldots$.) Each beam experiences a delay a_n, $n = 0, 1, 2, \ldots$, in going from the transmitter to the receiver.

(a) Suppose the car is not moving. The beams have a constant amplitude during the propagation, but any reflection against the floor or ceiling of the tunnel decreases their amplitude by a factor of α, one factor of α for each reflection. We assume that the propagation statistics are time invariant. The system is then

$$\text{received signal} = L(\text{transmitted signal}),$$

where the received signal is the sum of all the beams.
Is this system an LTI system? If yes, give its impulse response in terms of α and a_n. Sketch the impulse response. (The transmitted signal is a δ sent out in all directions. You can take arbitrary, increasing values for a_n.) Find the transfer function of the system.

(b) After making some measurements, you realize that the signal gets distorted because of nonlinear effects in air propagation. These effects are supposed to be independent of the propagation path. You manage to isolate the zero-order beam and find that the received signal to a sharp pulse (a δ) is

$$f(t) = \Lambda(t - a_0).$$

Is the system still time invariant? If yes, give its impulse response and sketch it. Find the transfer function of the system. (You can assume that the effects of distortion are smaller than the delay difference between two consecutive-order beams and that distortion does not affect the effects of attenuation.)

[24] From Damien Cerbelaud.

(c) We now send the signal

$$v(t) = \cos(2\pi\nu t)$$

into the tunnel. Give an expression for the signal received by the car.

(d) *Doppler effect*: If a receiver moving at speed v is sent a signal with frequency ν, then the receiver sees a modified sensed frequency, given by the formula

$$\tilde{\nu} = \nu\left(1 + \frac{v\cos(\theta)}{c}\right),$$

where θ is the angle between the speed vector and the signal propagation vector.

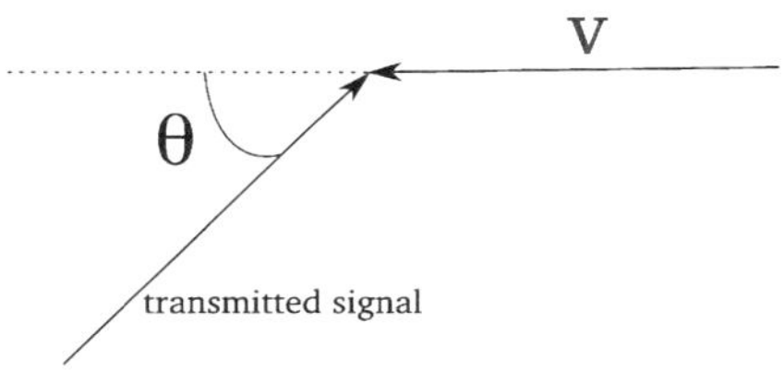

We now assume that the car starts moving slowly, with constant speed v. The same signal as in part (c) is transmitted through the tunnel. We assume that v is small compared to the transmission speed of the message and that the tunnel statistics/parameters are constant within the timeframe. Only the Doppler effect changes the received signal.
When the car moves, the delays experienced by each beam will be continuously changing, in which case the system is not an LTI system anymore. However, if we consider that the car is moving slowly compared to a transmission speed, then we can assume that these delays won't change during the message transmission, and the system can still be considered as an LTI system for this short period of time.
How is your answer to part (c) modified?
Give the spectrum bandwidth (upper bound) of the signal received at the car.

8.31. *Circulant matrices commute*
Let

$$c = \begin{pmatrix} c_0 \\ c_1 \\ \vdots \\ c_{N-1} \end{pmatrix}, \quad d = \begin{pmatrix} d_0 \\ d_1 \\ \vdots \\ d_{N-1} \end{pmatrix},$$

and let C, D be the corresponding $N \times N$ circulant matrices,

$$C = \begin{pmatrix} c_0 & c_{N-1} & c_{N-2} & \cdots & c_1 \\ c_1 & c_0 & c_{N-1} & \cdots & c_2 \\ c_2 & c_1 & c_0 & \cdots & c_3 \\ \vdots & \vdots & \vdots & \vdots & \vdots \\ c_{N-1} & c_{N-2} & c_{N-3} & \cdots & c_0 \end{pmatrix}, \quad D = \begin{pmatrix} d_0 & d_{N-1} & d_{N-2} & \cdots & d_1 \\ d_1 & d_0 & d_{N-1} & \cdots & d_2 \\ d_2 & d_1 & d_0 & \cdots & d_3 \\ \vdots & \vdots & \vdots & \vdots & \vdots \\ d_{N-1} & d_{N-2} & d_{N-3} & \cdots & d_0 \end{pmatrix}.$$

Recall that a discrete LTI system has two representations, as a circulant matrix and as convolution with the impulse response.

(a) What are the impulse responses for C, D, and CD?

(b) Show that any two circulant matrices commute; i.e., $CD = DC$. (The point is that this is unusual — matrices don't usually commute.)

8.32. *Invertible discrete LTI systems*

A discrete linear system L_1 is invertible if there is a second system L_2 such that

$$L_2 L_1 \underline{v} = \underline{v}$$

for all inputs $\underline{v}$. We call L_2 the inverse system, naturally. Regarding L_1 and L_2 as matrices, we have

$$L_2 L_1 = I,$$

the $N \times N$ identity matrix.

It is important, and fortunately easy to show, that if L_1 is an invertible LTI system, then the inverse is also LTI. In this case we can write the systems L_1 and L_2 as convolution with the respective impulse responses:

$$L_1 \underline{v} = \underline{h}_1 * \underline{v}, \quad L_2 \underline{v} = \underline{h}_2 * \underline{v}.$$

(a) What is $\underline{h}_1 * \underline{h}_2$? [This uses an algebraic property of convolution (analogous to multiplication). What is it?]

Recall that L_1 and L_2 are diagonalized by the DFT matrix; i.e.,

$$\underline{\mathcal{F}} L_1 \underline{\mathcal{F}}^{-1} = \Lambda_1, \quad \underline{\mathcal{F}} L_2 \underline{\mathcal{F}}^{-1} = \Lambda_2$$

where Λ_1 and Λ_2 are diagonal matrices.

(b) What is $\Lambda_2 \Lambda_1$?

Consider an LTI system that adds an echo to the input signal:

$$\underline{h_1}[n] = \underline{\delta}[n] + \alpha \underline{\delta}[n-m],$$

for some m. To be definite, let's take

$$\alpha = 0.5, \quad N = 4, \quad m = 2.$$

For the rest of the problem, you'll need the $N = 4$ DFT, so you'll also need

$$\underline{\omega} = (1, e^{2\pi i/4}, e^{2\pi i 2/4}, e^{2\pi i 3/4}) = (1, i, -1, -i),$$
$$\underline{\omega}^{-1} = (1, -i, -1, i).$$

(c) Find the eigenvalues of L_1, the matrix representing the system $L_1 \underline{v} = \underline{h}_1 * \underline{v}$.

(d) Find $\underline{h_2}$, the impulse response of the inverse system.

Chapter 9

n-Dimensional Fourier Transform

9.1. Space, the Final Frontier

To quote Ron Bracewell from his book *Two-Dimensional Imaging*, "In two dimensions phenomena are richer than in one dimension." True enough, working in two dimensions offers many new and rich possibilities. Contemporary applications of the Fourier transform are just as likely to come from problems in two, three, and even higher dimensions as they are from one. Imaging is one obvious and important example.

To capitalize on the work we've already done and to highlight the similarities and differences between the one-dimensional case and higher dimensions, we want to mimic the one-dimensional setting and arguments as much as possible. It is a measure of the naturalness of the fundamental concepts that the extension to higher dimensions of the basic ideas and the mathematical definitions that we've used so far proceeds almost automatically. However much we'll be able to do here, you should be able to read more on your own with some assurance that you won't be reading anything too different from what you've already read.

Vector notation. The higher-dimensional case looks most like the one-dimensional case when we use vector notation. For the sheer thrill of it, I'll give many of the definitions in n dimensions, but to raise the comfort level we'll usually look at the special case of two dimensions in more detail. Two and three dimensions are where most of our examples will come from.

We'll write a point in $\mathbb{R}^n$ as an n-tuple, say,

$$\underline{x} = (x_1, x_2, \ldots, x_n)\,.$$

As in earlier chapters, I'll be using the underline notation for vectors and note that we're going back to the usual indexing from 1 to n. We can try not to succumb to a particular interpretation, but it's still most common to think of the x_i's as coordinates in space and to call $\underline{x}$ a spatial variable. To define the Fourier transform we'll then also need an n-tuple of frequencies, and without saying yet what "frequency" means, we'll typically write

$$\underline{\xi} = (\xi_1, \xi_2, \dots, \xi_n)$$

for the *frequency variable dual to* $\underline{x}$ in the vernacular. The dimension of ξ_i is the reciprocal of the dimension of x_i.

Recall that the dot product of vectors in $\mathbb{R}^n$ is the real number defined by

$$\underline{x} \cdot \underline{\xi} = x_1\xi_1 + x_2\xi_2 + \cdots + x_n\xi_n .$$

The geometry of $\mathbb{R}^n$, orthogonality in particular, is governed by the dot product, and using it will help our understanding as well as streamline our notation.

Cartesian product. One comment on set notation. On a few occasions we'll want to indicate that we're assembling sets of higher dimension from sets of lower dimension. For example, $\mathbb{R}^n$ is assembled from n copies of $\mathbb{R}$, each element in $\mathbb{R}^n$ being an ordered n-tuple. The mathematical term for this is *Cartesian product*, a phrase you've probably heard, and there's a notation that goes along with it. Thus, for example,

$$\mathbb{R}^2 = \mathbb{R} \times \mathbb{R} = \{(x_1, x_2) : x_1 \in \mathbb{R}, , x_2 \in \mathbb{R}\} .$$

In general, if A and B are any two sets, their Cartesian product is the set of ordered pairs with elements coming from each:

$$A \times B = \{(a, b) : a \in A,\ b \in B\} .$$

If you haven't seen this before, you have now. From two sets you can iterate and consider Cartesian products of any finite number of sets.[1]

9.1.1. The Fourier transform.

We started in the first chapter with Fourier series and periodic phenomena and went on from there to define the Fourier transform. There's a place for Fourier series in higher dimensions, but, carrying all our hard won experience with us, we'll proceed directly to the higher-dimensional Fourier transform. I'll save higher-dimensional Fourier series for a later section, including a really interesting application to random walks where the dimension enters in a crucial way.

How shall we define the Fourier transform? We consider real- or complex-valued functions f defined on $\mathbb{R}^n$ and write $f(\underline{x})$ or $f(x_1, \dots, x_n)$, whichever is more convenient in context. The Fourier transform of $f(\underline{x})$ is the function $\mathcal{F}f(\underline{\xi})$ defined by

$$\mathcal{F}f(\underline{\xi}) = \int_{\mathbb{R}^n} e^{-2\pi i \underline{x}\cdot\underline{\xi}} f(\underline{x})\, d\underline{x} .$$

Alternate notations are $\hat{f}(\underline{\xi})$ and the uppercase version $F(\underline{\xi})$. The inverse Fourier transform of a function $g(\underline{\xi})$ is

$$\mathcal{F}^{-1}g(\underline{x}) = \int_{\mathbb{R}^n} e^{2\pi i \underline{x}\cdot\underline{\xi}} g(\underline{\xi})\, d\underline{\xi} .$$

[1] Defining the Cartesian product of an infinite number of sets is a vexing issue, foundationally speaking. It involves the Axiom of Choice. Google if you dare.

The Fourier transform, or the inverse transform, of a real-valued function is, in general, complex valued.

The complex exponential now features the dot product of the vectors $\underline{x}$ and $\underline{\xi}$. This is the key to extending the definitions from one dimension to higher dimensions and making them resemble the definitions in one dimension.[2] The integral is over all of $\mathbb{R}^n$, and as an n-fold multiple integral all the x_j's (or ξ_j's for $\mathcal{F}^{-1}$) go from $-\infty$ to ∞. The notation $d\underline{x}$ is shorthand for the n-dimensional volume element, $d\underline{x} = dx_1 dx_2 \cdots dx_n$. All parts of the integrand are scalar-valued functions; we're using vectors to write the integrand, particularly in the complex exponential, but we're not integrating a vector-valued function. It all looks pretty similar to the 1D case to me, but I also know how it works out.

Written out in coordinates, the definition of the Fourier transform reads

$$\mathcal{F}f(\xi_1, \xi_2, \ldots, \xi_n) = \int_{\mathbb{R}^n} e^{-2\pi i(x_1\xi_1 + \cdots + x_n\xi_n)} f(x_1, \ldots, x_n)\, dx_1 \ldots dx_n\,,$$

so for two dimensions,

$$\mathcal{F}f(\xi_1, \xi_2) = \int_{-\infty}^{\infty}\int_{-\infty}^{\infty} e^{-2\pi i(x_1\xi_1 + x_2\xi_2)} f(x_1, x_2)\, dx_1\, dx_2\,.$$

The coordinate expression is manageable in the two-dimensional case, but I hope to convince you that it's almost always *much* better to use the vector notation in writing formulas, deriving results, and so on.

The kinds of functions to consider and how they enter into the discussion — Schwartz functions, L^1, L^2, etc. — is also entirely analogous to the one-dimensional case, and so are the definitions of these types of functions. We don't have to redo distributions et al. (good news), and I'll only occasionally point out when this aspect of the general theory is (or must be) invoked and what extra has to be done.

Not surprisingly, there is not universal agreement on where to put the 2π, and in the case of n dimensions we're even treated to nth powers of 2π. Commonly, one sees

$$\mathcal{F}f(\underline{\xi}) = \int_{\mathbb{R}^n} e^{-i\underline{x}\cdot\underline{\xi}} f(\underline{x})\, d\underline{x}\,,$$

without the 2π in the complex exponential, leading to

$$\mathcal{F}^{-1}f(\underline{x}) = \frac{1}{(2\pi)^n}\int_{\mathbb{R}^n} e^{i\underline{x}\cdot\underline{\xi}} f(\underline{\xi})\, d\underline{\xi}$$

for the inverse Fourier transform. One also meets

$$\mathcal{F}f(\underline{\xi}) = \frac{1}{(2\pi)^{n/2}}\int_{\mathbb{R}^n} e^{-i\underline{x}\cdot\underline{\xi}} f(\underline{x})\, d\underline{x}, \quad \mathcal{F}^{-1}f(\underline{x}) = \frac{1}{(2\pi)^{n/2}}\int_{\mathbb{R}^n} e^{i\underline{x}\cdot\underline{\xi}} f(\underline{\xi})\, d\underline{\xi}\,.$$

There are others. You have been warned.

Arithmetic with vectors, including the dot product, is pretty much like arithmetic with numbers, and consequently the familiar algebraic properties of the

[2]In one dimension the definitions coincide. The dot product of two one-dimensional vectors is just the ordinary product of their (single) components.

Fourier transform are present in the higher-dimensional setting. We won't go through them all, but, for example,

$$\mathcal{F}f(-\underline{\xi}) = \int_{\mathbb{R}^n} e^{-2\pi i\underline{x}\cdot(-\underline{\xi})} f(\underline{x})\, d\underline{x} = \int_{\mathbb{R}^n} e^{2\pi i\underline{x}\cdot\underline{\xi}} f(\underline{x})\, d\underline{x} = \mathcal{F}^{-1}f(\underline{\xi}),$$

which is one way of stating the duality between the Fourier and inverse Fourier transforms. Here, if $\underline{\xi} = (\xi_1, \dots, \xi_n)$, then[3]

$$-\underline{\xi} = (-\xi_1, \dots, -\xi_n).$$

To be neater, we again use the notation

$$f^-(\underline{x}) = f(-\underline{x})$$

for the reversed signal, and with this the duality results read exactly as in the one-dimensional case:

$$\mathcal{F}f^- = (\mathcal{F}f)^-, \quad (\mathcal{F}f)^- = \mathcal{F}^{-1}f.$$

Fourier inversion gives us

$$\mathcal{F}\mathcal{F}f = f^-,$$

also just as in one dimension.

It's still the case that the complex conjugate of the integral is the integral of the complex conjugate, so when $f(\underline{x})$ is real valued,

$$\mathcal{F}f(-\xi) = \overline{\mathcal{F}f(\xi)}.$$

Finally, evenness and oddness are defined exactly as in the one-dimensional case:

- $f(\underline{x})$ is *even* if $f(-\underline{x}) = f(\underline{x})$, or, without writing the variables, if $f^- = f$.
- $f(\underline{x})$ is *odd* if $f(-\underline{\xi}) = -f(\underline{\xi})$, or if $f^- = -f$.

In three dimensions and up we no longer have the easy geometric interpretations of evenness and oddness in terms of a graph (because we can't draw a graph), but as algebraic properties of a function these conditions do have the familiar consequences for the higher-dimensional Fourier transform, in all dimensions. For example, if $f(\underline{x})$ is even, then $\mathcal{F}f(\underline{\xi})$ is even; if $f(\underline{x})$ is real and even, then $\mathcal{F}f(\underline{\xi})$ is real and even; etc. You could write them all out.

In connection with these formulas, and derivations to follow, I have to point out that changing the variable of integration can be more complicated for multiple integrals. Changing variables was one of our prized techniques in one dimension, and we'll handle the complications on a need-to-know basis.

Soon enough we'll calculate the Fourier transform of some model functions, but first let's look a little bit more at the complex exponentials in the definition and get a better sense of what the spectrum means in higher dimensions.

[3]It's bad practice and a bad habit to use the adjective "negative" for $-\underline{\xi}$ since we're talking about vectors, not numbers. Real numbers are positive or negative; vectors are not. (Neither are complex numbers.) Just say "minus $\underline{\xi}$." "Opposite frequency" is also suggestive. In any case the world won't end on this negative note.

9.1.2. Harmonics, periodicity, and spatial frequencies. The complex exponentials are again the building blocks — the harmonics — for the Fourier transform and its inverse. Now that they involve a dot product, what more do we need to know?

As mentioned, we tend to view $\underline{x} = (x_1, \dots, x_n)$ as a spatial variable and we call $\underline{\xi} = (\xi_1, \dots, \xi_n)$ a frequency variable, frequency vector, or *spatial frequency*; any term will do. It's not hard to imagine problems where one would want to specify n spatial directions each having the unit of length, but it's not so clear what an n-tuple of frequencies should mean. One thing we can say is that if the spatial variables $(x_1, \dots, x_n)$ do have the dimension of length, then the corresponding frequency variables $(\xi_1, \dots, \xi_n)$ have the dimension 1/length. For then

$$\underline{x} \cdot \underline{\xi} = x_1\xi_1 + \cdots + x_n\xi_n$$

is dimensionless and $\exp(\pm 2\pi i \underline{x} \cdot \underline{\xi})$ makes sense. There's your first reciprocal relationship in higher dimensions, and it corresponds to dimensions of time and 1/time in the one-dimensional time domain and frequency domain picture.

For some insight into the meaning of spatial frequencies let's look at the two-dimensional case in some detail; we'll often check things in 2D. The complex exponentials in $\mathcal{F}$ and $\mathcal{F}^{-1}$ are

$$\exp(\pm 2\pi i \underline{x} \cdot \underline{\xi}) = \exp(\pm 2\pi i (x_1\xi_1 + x_2\xi_2)),$$

and it doesn't matter for the following discussion whether we take + or − in the exponent. The exponential equals 1 whenever $\underline{x} \cdot \underline{\xi}$ is an integer, recalling that $e^{\pm 2\pi i \cdot \text{integer}} = 1$. With $\underline{\xi} = (\xi_1, \xi_2)$ fixed, asking for $\underline{x} \cdot \underline{\xi} = \text{integer}$ is a condition on $\underline{x} = (x_1, x_2)$, and there's a natural geometric interpretation that's very helpful in understanding the most important properties of the complex exponential. For a fixed $\underline{\xi}$ the equations

$$\xi_1 x_1 + \xi_2 x_2 = n, \quad n = 0, \pm 1, \pm 2, \dots,$$

describe a family of *evenly spaced, parallel lines* in the (x_1, x_2)-plane (the *spatial domain*, if you prefer that phrase). It's common to say that the complex exponential has *zero phase* along these lines, the terminology coming from optics.

Take $n = 0$. Then the condition on x_1 and x_2 is

$$\xi_1 x_1 + \xi_2 x_2 = 0$$

and we recognize this as the equation of a line through the origin with $\underline{\xi} = (\xi_1, \xi_2)$ as a normal vector to the line.[4] (Remember your vectors!) Furthermore $\underline{\xi}$ is a normal to *each* of the parallel lines in the family. One could also describe the geometry of the situation by saying that the lines each make an angle θ with the x_1-axis satisfying

$$\tan\theta = -\frac{\xi_1}{\xi_2},$$

but I think it's much better to think in terms of normal vectors to specify the direction. The vector point of view generalizes readily to higher dimensions, as we'll discuss.

[4]Note that $\underline{\xi} = (\xi_1, \xi_2)$ isn't assumed to be a unit vector, so it's not necessarily a unit normal.

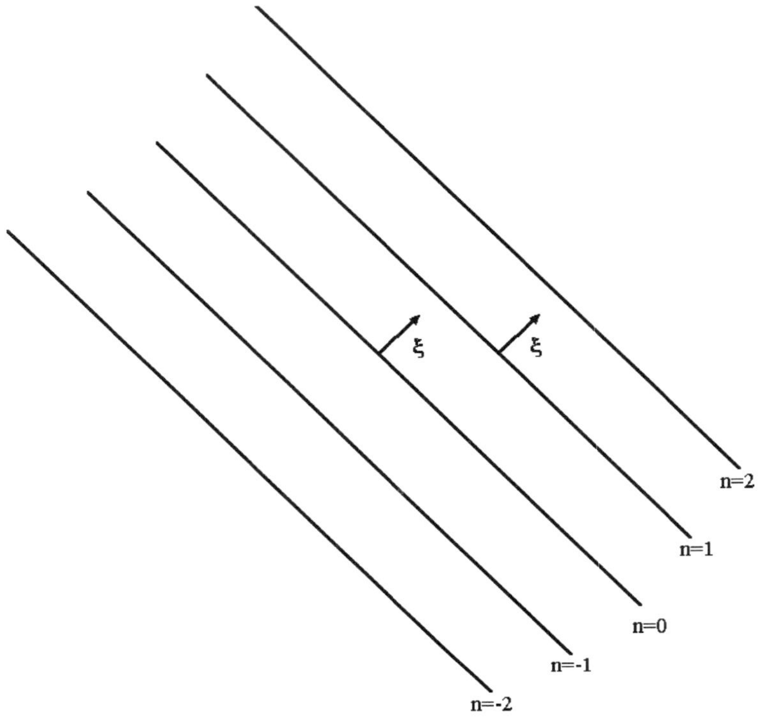

What is the spacing between adjacent lines $\xi_1 x_1 + \xi_2 x_2 = n$ and $\xi_1 x_1 + \xi_2 x_2 = n + 1$? The answer is

$$\text{distance} = \frac{1}{\|\underline{\xi}\|} = \frac{1}{\sqrt{\xi_1^2 + \xi_2^2}}\,.$$

I'll let you derive this. It's a nice exercise in vector geometry that you probably did in a course on vector calculus; or you can look up the formula for the distance from a point to a line and apply that.

This is a second reciprocal relationship, in two dimensions, between the spatial and frequency variables:

- The spacing of adjacent lines of zero phase is the reciprocal of the length of the frequency vector.

Drawing the family of parallel lines with a fixed normal $\underline{\xi}$ gives us some sense of the oscillatory nature of the harmonics $\exp(\pm 2\pi i\, \underline{x} \cdot \underline{\xi})$. The frequency vector $\underline{\xi} = (\xi_1, \xi_2)$, as a normal to the lines of zero phase, determines an orientation (for lack of a better word) for the harmonic $e^{2\pi i \underline{x}\cdot\underline{\xi}}$, and the magnitude $\|\underline{\xi}\|$, or rather its reciprocal $1/\|\underline{\xi}\|$, determines the period of the harmonic in the direction $\underline{\xi}$. To be precise, start at any point (a, b) and move in the direction of the *unit* normal, $\underline{\xi}/\|\underline{\xi}\|$. That is, move from (a, b) at unit speed along the line

$$\underline{x}(t) = (a, b) + t\frac{\underline{\xi}}{\|\underline{\xi}\|}\,, \quad \text{or} \quad x_1(t) = a + t\frac{\xi_1}{\|\underline{\xi}\|},\ x_2(t) = b + t\frac{\xi_2}{\|\underline{\xi}\|}\,.$$

The dot product of $\underline{x}(t)$ and $\underline{\xi}$ is

$$\underline{x}(t)\cdot\underline{\xi} = (x_1(t), x_2(t))\cdot(\xi_1,\xi_2) = a\xi_1 + b\xi_2 + t\frac{\xi_1^2+\xi_2^2}{\|\underline{\xi}\|} = a\xi_1 + b\xi_2 + t\|\underline{\xi}\|\,,$$

and the complex exponential is a function of t along the line,

$$\exp(\pm 2\pi i\,\underline{x}\cdot\underline{\xi}) = \exp(\pm 2\pi i(a\xi_1 + b\xi_2))\,\exp(\pm 2\pi i t\|\underline{\xi}\|)\,.$$

The factor $\exp(\pm 2\pi i(a\xi_1 + b\xi_2))$ doesn't depend on t and the factor $\exp(\pm 2\pi i t\|\underline{\xi}\|)$ is periodic in t with period $1/\|\underline{\xi}\|$. The period is the spacing between the lines of zero phase. Briefly, the harmonic $e^{\pm 2\pi i\underline{x}\cdot\underline{\xi}}$ is oscillating between the lines of zero phase with a period $1/\|\underline{\xi}\|$.

Now, if ξ_1 *or* ξ_2 is large, then the spacing of the lines of zero phase is close, because $1/\|\underline{\xi}\|$ is small, and if ξ_1 *and* ξ_2 are small, then the lines of zero phase are far apart because $1/\|\underline{\xi}\|$ is large. Thus although "frequency" is now a vector quantity, we still tend to speak in terms of a "high frequency" harmonic when the lines of zero phase are spaced close together, and a "low frequency" harmonic when the lines of zero phase are spaced far apart ("high" and "low" are relatively speaking, of course). Half-way between the lines of zero phase, when $t = 1/(2\|\underline{\xi}\|)$, we're on lines where the exponential is -1, hence 180° out of phase with the lines of zero phase.

One often sees pictures like the following.[5]

Here's what you're looking at. The function $e^{2\pi i\underline{x}\cdot\underline{\xi}}$ is complex valued, but consider its real part

$$\begin{aligned}\operatorname{Re} e^{2\pi i\underline{x}\cdot\underline{\xi}} &= \tfrac{1}{2}\left(e^{2\pi i\underline{x}\cdot\underline{\xi}} + e^{-2\pi i\underline{x}\cdot\underline{\xi}}\right)\\ &= \cos 2\pi i\,\underline{x}\cdot\underline{\xi} = \cos 2\pi(\xi_1 x_1 + \xi_2 x_2)\,,\end{aligned}$$

which has the same periodicity and same lines of zero phase as the complex exponential. Put down white stripes where $\cos 2\pi(\xi_1 x_1 + \xi_2 x_2) \geq 0$ and black stripes where $\cos 2\pi(\xi_1 x_1 + \xi_2 x_2) < 0$, or, if you want to get fancy, use a gray scale to go from pure white on the lines of zero phase, where the cosine is 1, down to pure black on the lines 180° out of phase, where the cosine is -1, and back up again. This gives a sense of a periodically varying intensity, and the slowness or rapidity of the changes in intensity indicates low or high spatial frequencies.

[5]One saw these pictures in Chapter 1 as examples of periodicity in space.

The spectrum. The set of all spatial frequencies where the Fourier transform exists is called the *spectrum*, just as in one dimension. Also, just as in one dimension, we often refer to the actual values of $\mathcal{F}f(\xi_1, \xi_2)$ as the spectrum. Ambiguous yes, but troublesome no. The Fourier transform of a function $f(x_1, x_2)$ finds the amount that the spatial frequencies (ξ_1, ξ_2) contribute to the function. Then Fourier inversion via the inverse transform recovers the function, adding together the corresponding spatial harmonics, each contributing an amount $\mathcal{F}f(\xi_1, \xi_2)$. Fourier inversion recovers the signal from its spectrum — that fundamental fact is intact in all dimensions. All of this raises the issue of finite sets of points, or more generally two-dimensional sets of measure zero, where $\mathcal{F}f(\xi_1, \xi_2) = 0$ counting as parts of the spectrum, since they don't affect Fourier inversion. Reread the discussion of this in Chapter 2 for one dimension and feel secure that the ideas transfer to higher dimensions. I in turn feel secure in saying that this issue will not come up in your working life.

As mentioned above, when $f(x_1, x_2)$ is real, we have

$$\mathcal{F}f(-\xi_1, -\xi_2) = \overline{\mathcal{F}f(\xi_1, \xi_2)},$$

so that if a particular $\mathcal{F}f(\underline{\xi}) = \mathcal{F}f(\xi_1, \xi_2)$ is not zero, then there is also a contribution from the opposite frequency $-\underline{\xi} = (-\xi_1, -\xi_2)$. Thus for a real signal, the spectrum, as a set of points in the (ξ_1, ξ_2)-plane, is symmetric about the origin.[6]

If we think of the exponentials at corresponding frequency vectors, $\underline{\xi}$ and $-\underline{\xi}$, combining to give the signal, then we're adding up (integrating) a bunch of cosines and the signal really does seem to be made of a bunch of stripes with different spacings, different orientations, and different intensities (the magnitudes $|\mathcal{F}f(\xi_1, \xi_2)|$). It may be hard to imagine that an image, for example, is such a sum of stripes, but, then again, why is music the sum of a bunch of sinusoidal pressure waves?

In the one-dimensional case we are used to drawing a picture of the magnitude of the Fourier transform to get some sense of how the energy is distributed among the different frequencies. We can do a similar thing in the two-dimensional case, putting a bright (or colored) dot at each point (ξ_1, ξ_2) that is in the spectrum, with a brightness proportional to, say, the square magnitude $|\mathcal{F}f(\xi_1, \xi_2)|^2$. If $\mathcal{F}f(\xi_1, \xi_2) = 0$, then (ξ_2, ξ_2) is colored black. This is the *energy spectrum* or the *power spectrum* of the function. For a real signal it's symmetric about the origin because

$$|\mathcal{F}f(\xi_1, \xi_2)| = |\overline{\mathcal{F}f(\xi_1, \xi_2)}| = |\mathcal{F}f(-\xi_1, -\xi_2)|.$$

Below are pictures of the respective spectra of the spatial harmonics, above. The origin is at the center of the images.

Which is which? The stripes have an orientation and a spacing determined by $\underline{\xi} = (\xi_1, \xi_2)$, which is normal to the stripes. The horizontal stripes have a normal of the form $(0, \xi_2)$, and they are of higher frequency so ξ_2 is (relatively) large. The vertical stripes have a normal of the form $(\xi_1, 0)$ and are of a lower frequency so ξ_1 is small. As for the oblique stripes, with a normal (ξ_1, ξ_2), for the upper right $\xi_1 > \xi_2 > 0$ and for the lower left ξ_1 is negative and $\xi_2 > |\xi_1|$.

[6] *N.b.* It's not the *complex values* $\mathcal{F}f(\xi_1, \xi_2)$ that are symmetric, just the set of points (ξ_1, ξ_2) of contributing frequencies.

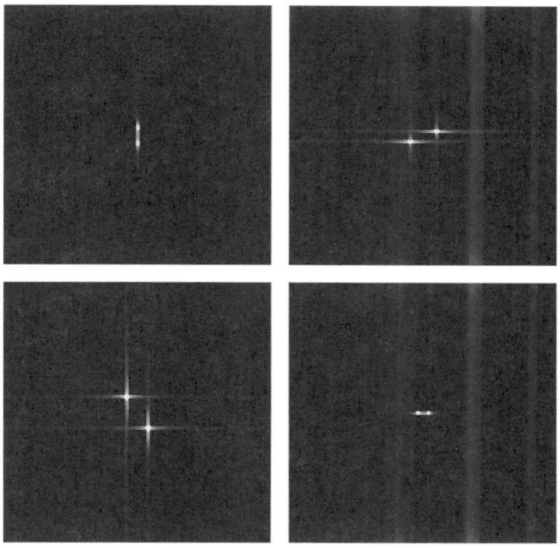

Here's a more interesting example.

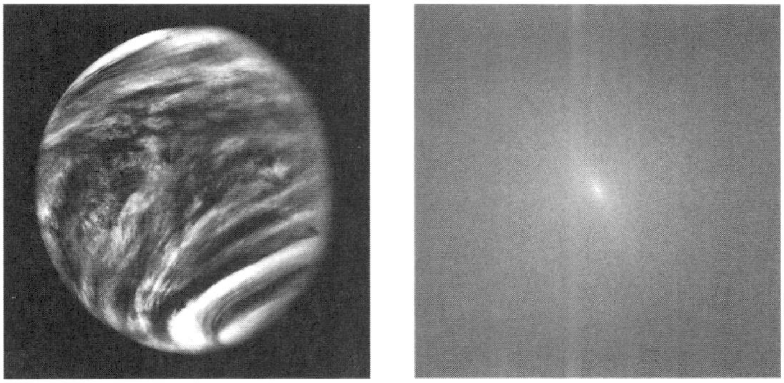

That's the first close-up image of Venus, together with its spectrum. The image was taken by NASA's Mariner 10 on February 5, 1974.[7]

For the picture of Venus, what function are we taking the Fourier transform *of*? The function, $f(x_1, x_2)$, is the intensity of light at each point (x_1, x_2) — that's what a grayscale image *is* for the purposes of Fourier analysis.

Incidentally, because the dynamic range (the range of intensities) in images can be so large it's common to light up the pixels in the spectral picture according to the *logarithm* of the intensity, sometimes with additional scaling that has to be specified. Check your local software package to see what it does.

Here's a natural application to an image of filtering the spectrum. The first picture shows an image with periodic noise; step back a little to see the background image a little more clearly. The noise, in the form of an oblique pattern of stripes, appears quite distinctly in the spectrum as the two bright dots in the second and fourth quadrants. If we eliminate those frequencies from the spectrum and take

[7]NASA makes many of their images available. Go to `https://www.nasa.gov/multimedia/imagegallery/index.html`.

the inverse transform, we see Stanford's Hoover Tower, and its cleaner spectrum. Denoising is a huge topic, of much current interest.

Pictures of Hoover Tower

There are reasons to *add* things to the spectrum as well as take them away. An important application of the Fourier transform in imaging is *digital watermarking*. Watermarking is an old technique to authenticate printed documents. Within the paper an image is imprinted (somehow — I don't know how this is done) that becomes visible when held up to a light. The idea is that someone trying to counterfeit the document will not know of or cannot replicate the watermark, but that someone who knows where to look can easily verify its presence and hence the authenticity of the document. The newer US currency now uses watermarks, as well as other anticounterfeiting techniques.

For electronic documents a *digital watermark* is typically incoporated by adding to the spectrum. Insert a few extra harmonics here and there and keep track of what you added. This is done in a way to make the changes in the image undetectable (you hope) and so that no one else could possibly tell what belongs in the spectrum and what you put there (you hope). If the recipients of the document know where to look in the spectrum, they can find your mark and verify that the document is legitimate.

Higher dimensions. In higher dimensions the words to describe the harmonics and the spectrum are pretty much the same, though we can't draw the pictures.[8] The harmonics are the complex exponentials $e^{\pm 2\pi i \underline{x}\cdot\underline{\xi}}$ and we have n spatial frequencies, $(\xi_1, \xi_2, \dots, \xi_n) = \underline{\xi}$, to go with the n spatial variables, $(x_1, \dots, x_n) = \underline{x}$. Again we single out where the complex exponentials are equal to 1 (zero phase),

[8]Any computer graphics experts out there care to add color and 3D-rendering to try to draw the spectrum?

which is when $\underline{\xi} \cdot \underline{x}$ is an integer. In three dimensions a given (ξ_1, ξ_2, ξ_3) determines a family

$$\underline{\xi} \cdot \underline{x} = n, \quad n = 0, \pm 1, \pm 2, \dots,$$

of *equally spaced, parallel planes* (of zero phase) in (x_1, x_2, x_3)-space. The normal to any of the planes is the vector $\underline{\xi} = (\xi_1, \xi_2, \xi_3)$, and adjacent planes are a distance $1/\|\underline{\xi}\|$ apart. The exponential is periodic in the direction $\underline{\xi}$ with period $1/\|\underline{\xi}\|$. In a similar fashion, in n dimensions we have families of parallel hyperplanes, i.e., $(n-1)$-dimensional subspaces, with normals $\underline{\xi} = (\xi_1, \dots, \xi_n)$, and distance $1/\|\underline{\xi}\|$ apart. See how nicely this works?

9.2. Getting to Know Your Higher-Dimensional Fourier Transform

You already know a lot about the higher-dimensional Fourier transform because you already know a lot about the one-dimensional Fourier transform — that's the whole point. Still, it's useful to collect a few of the basic facts. If some result or example corresponding to the one-dimensional case isn't mentioned here, that doesn't mean it doesn't hold or isn't worth mentioning; it only means that the following is a very quick and very partial survey. Sometimes we'll work in $\mathbb{R}^n$, for any n, and sometimes just in $\mathbb{R}^2$; nothing should be read into this for or against $n = 2$.

9.2.1. Finding a few Fourier transforms: Separable functions. There are times when a function $f(x_1, \dots, x_n)$ of n variables can be written as a product of n functions of one variable, as in

$$f(x_1, \dots, x_n) = f_1(x_1) f_2(x_2) \cdots f_n(x_n)\,.$$

Attempting to do this is a standard technique in finding special solutions of partial differential equations. There, it's called the method of *separation of variables*. Here, we call such a function *separable*.

Now, there's no theorem that says, "If $a, b, c, \dots$, then $f(x_1, x_2, \dots, x_n)$ is separable." But *when* a function is separable, which you have to see by inspection, its Fourier transform can be calculated as the product of the Fourier transform of the factors. Take $n = 2$ as a representative case:

$$\begin{aligned}
\mathcal{F}f(\xi_1, \xi_2) &= \int_{\mathbb{R}^n} e^{-2\pi i \underline{x} \cdot \underline{\xi}} f(\underline{x})\, d\underline{x} \\
&= \int_{-\infty}^{\infty} \int_{-\infty}^{\infty} e^{-2\pi i (x_1 \xi_1 + x_2 \xi_2)} f(x_1, x_2)\, dx_1\, dx_2 \\
&= \int_{-\infty}^{\infty} \int_{-\infty}^{\infty} e^{-2\pi i \xi_1 x_1} e^{-2\pi i \xi_2 x_2} f_1(x_1) f_2(x_2)\, dx_1\, dx_2 \\
&= \int_{-\infty}^{\infty} \left(\int_{-\infty}^{\infty} e^{-2\pi i \xi_1 x_1} f_1(x)\, dx_1 \right) e^{-2\pi i \xi_2 x_2} f_2(x_2)\, dx_2 \\
&= \mathcal{F}f_1(\xi_1) \int_{-\infty}^{\infty} e^{-2\pi i \xi_2 x_2} f_2(x_2)\, dx_2 \\
&= \mathcal{F}f_1(\xi_1)\, \mathcal{F}f_2(\xi_2).
\end{aligned}$$

In general, if $f(x_1, x_2, \dots, x_n) = f_1(x_1)f_2(x_2)\cdots f_n(x_n)$, then

$$\mathcal{F}f(\xi_1, \xi_2, \dots, \xi_n) = \mathcal{F}f_1(\xi_1)\mathcal{F}f_2(\xi_2)\cdots\mathcal{F}f_n(\xi_n).$$

Higher-dimensional rectangle functions. The simplest, useful example of a function that fits this description is a version of the rectangle function in higher dimensions. In two dimensions, for example, we want the function that has the value 1 on the square of side length 1 centered at the origin and has the value 0 outside this square. That is,

$$\Pi(x_1, x_2) = \begin{cases} 1, & |x_1| < \frac{1}{2} \quad \text{and} \quad |x_2| < \frac{1}{2}, \\ 0, & \text{otherwise.} \end{cases}$$

You can fight it out how you want to define things on the edges. A plot is below; I should probably call this the box function instead of the rectangle function, but we'll stick with rectangle.

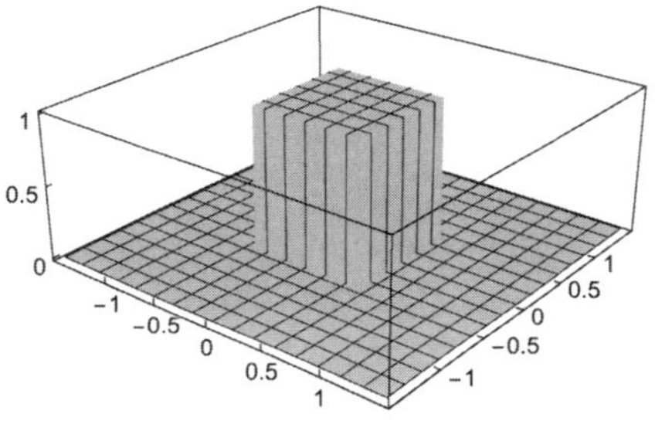

We can factor $\Pi(x_1, x_2)$ as the product of two one-dimensional rectangle functions:

$$\Pi(x_1, x_2) = \Pi(x_1)\Pi(x_2).$$

(I'm using the same notation for the rectangle function in one or more dimensions because, in this case, there's little chance of confusion.) We can write $\Pi(x_1, x_2)$ this way because it is identically 1 if *both* the coordinates are between $-1/2$ and $1/2$ and it is zero otherwise. So $\Pi(x_1, x_2)$ is zero if *either* of the coordinates is outside this range, and that's exactly what happens for the product $\Pi(x_1)\Pi(x_2)$.

For the Fourier transform of the two-dimensional Π we then have

$$\mathcal{F}\Pi(\xi_1, \xi_2) = \operatorname{sinc}\xi_1 \operatorname{sinc}\xi_2.$$

Here's what the graph looks like.

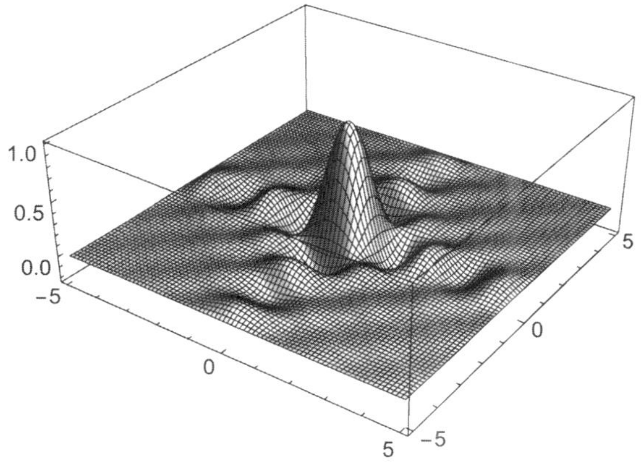

A helpful feature of factoring the rectangle function this way is the ability, easily, to change the widths in the different coordinate directions. For example, the function that is 1 in the rectangle $-a_1/2 < x_1 < a_1/2$, $-a_2/2 < x_2 < a_2/2$ and zero outside that rectangle is (in appropriate notation)

$$\Pi_{a_1 a_2}(x_1, x_2) = \Pi_{a_1}(x_1)\Pi_{a_2}(x_2)\,.$$

The Fourier transform of this is

$$\mathcal{F}\Pi_{a_1 a_2}(\xi_1, \xi_2) = (a_1 \operatorname{sinc} a_1\xi_1)(a_2 \operatorname{sinc} a_2\xi_2)\,.$$

Below we find plots of $\Pi_2(x_1)\Pi_4(x_2)$ and of its Fourier transform $(2 \operatorname{sinc} 2\xi_1)(4 \operatorname{sinc} 4\xi_2)$. You can see how the shapes have changed from what we had before.

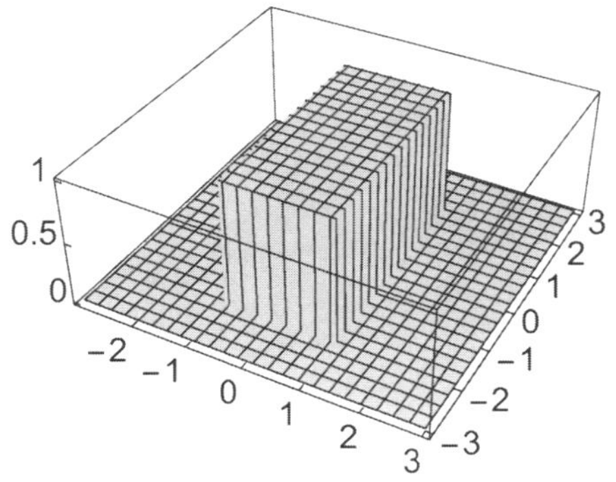

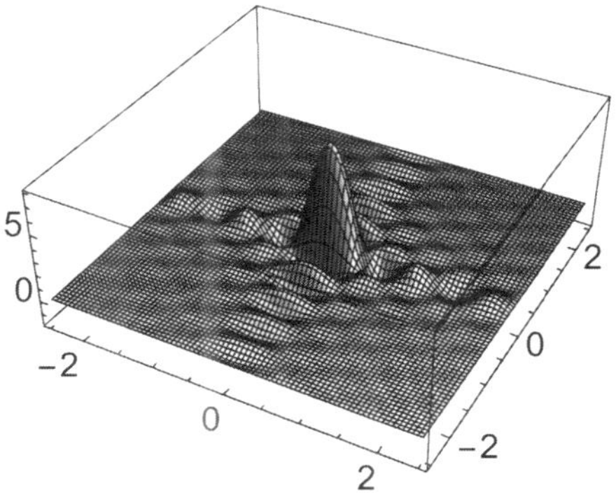

The direct generalization of the basic rectangle function to n dimensions is

$$\Pi(x_1, x_2, \ldots, x_n) = \begin{cases} 1, & |x_k| < \frac{1}{2}, \quad k = 1, \ldots, n\,, \\ 0, & \text{otherwise}, \end{cases}$$

which factors as

$$\Pi(x_1, x_2, \ldots, x_n) = \Pi(x_1)\Pi(x_2)\cdots\Pi(x_n)\,.$$

For the Fourier transform of the n-dimensional Π we then have

$$\mathcal{F}\Pi(\xi_1, \xi_2, \ldots, \xi_n) = \operatorname{sinc} \xi_1 \operatorname{sinc} \xi_2 \cdots \operatorname{sinc} \xi_n\,.$$

It's obvious how to modify higher-dimensional Π's to have different widths along different axes.

See the problems for a higher-dimensional generalization of the triangle function.

Gaussians. Another good example of a separable function, one that often comes up in practice, is a Gaussian. As for the one-dimensional case, the most natural Gaussian to use in connection with Fourier transforms is

$$g(\underline{x}) = e^{-\pi\|\underline{x}\|^2} = e^{-\pi(x_1^2+x_2^2+\cdots+x_n^2)}\,.$$

This factors as a product of n one-dimensional Gaussians:

$$g(x_1, \ldots, x_n) = e^{-\pi(x_1^2+x_2^2+\cdots+x_n^2)} = e^{-\pi x_1^2} e^{-\pi x_2^2} \cdots e^{-\pi x_n^2}\,.$$

Taking the Fourier transform and applying the one-dimensional result (and reversing the algebra that we did above) gets us

$$\mathcal{F}g(\underline{\xi}) = e^{-\pi\xi_1^2} e^{-\pi\xi_2^2} \cdots e^{-\pi\xi_n^2} = e^{-\pi(\xi_1^2+\xi_2^2+\cdots+\xi_n^2)} = e^{-\pi\|\underline{\xi}\|^2}\,.$$

As for one dimension, we see that g is its own Fourier transform.

Here's a plot of the two-dimensional Gaussian $g(x_1, x_2) = e^{-\pi(x_1^2+x_2^2)}$.

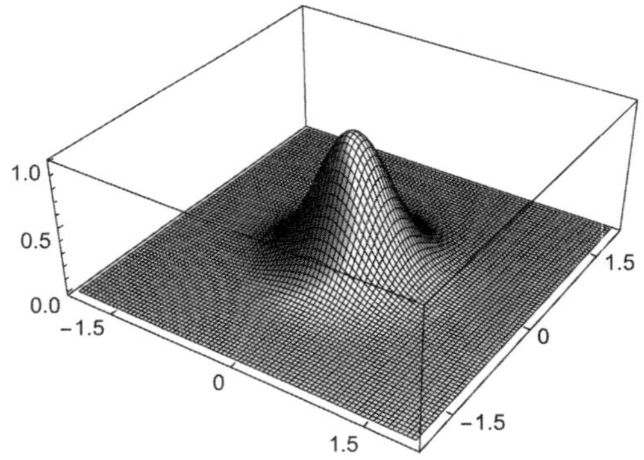

9.2.2. Tensor products. If you really want to impress your friends and confound your enemies, you can invoke *tensor products* in the context of separable functions. In mathematical parlance a separable signal f, say with

$$f(x_1, x_2, \ldots, x_n) = f_1(x_1) f_2(x_2) \cdots f_n(x_n),$$

is the tensor product of the functions f_i. One writes

$$f = f_1 \otimes f_2 \otimes \cdots \otimes f_n\,.$$

In terms of the variables, the rule to apply the tensor product is

$$(f_1 \otimes f_2 \otimes \cdots \otimes f_n)(x_1, x_2, \ldots, x_n) = f_1(x_1) f_2(x_2) \cdots f_n(x_n).$$

The formula for the Fourier transform (without pesky variables) is then, neatly,

$$\mathcal{F}(f_1 \otimes f_2 \otimes \cdots \otimes f_n) = \mathcal{F}f_1 \otimes \mathcal{F}f_2 \otimes \cdots \otimes \mathcal{F}f_n \,.$$

In words, the Fourier transform of the tensor product is the tensor product of the Fourier transforms. Not hard, but people often run in terror from the $\otimes$ symbol. Cool.

Note that in general the tensor product is *not* commutative. For example, let $f_1(x) = x^2$, $f_2(x) = e^x$. Then $(f_1 \otimes f_2)(x_1, x_2) = f_1(x_1)f_2(x_2) = x_1^2 e^{x_2}$, while $(f_2 \otimes f_1)(x_1, x_2) = f_2(x_1)f_1(x_2) = e^{x_1}x_2^2$.

Using the tensor product provides a formal way to talk about assembling a function of many variables as a product of functions of fewer variables. We can allow for greater generality, for example asking if a function of three variables can be separated into a product (a tensor product) of a function of one variable and a function of two variables. And so on. If a function factors in some such way, then the Fourier transform factors in the same way as the function. Do not run in terror from $\otimes$. Embrace $\otimes$.

9.2.3. Partial Fourier transforms. In the examples above the separability of the functions meant that we could find the Fourier transform one variable at a time, and the full Fourier transform is just the product of the individual Fourier transforms. I want to point out that there's a version of this one-variable-at-a-time thing even for functions that are not separable.

Write the Fourier transform as an iterated integral, which is what you'd do anyway if you were computing a specific example. In 2D,

$$\begin{aligned}
\mathcal{F}f(\xi_1, \xi_2) &= \int_{\mathbb{R}^2} e^{-2\pi i \underline{x}\cdot\underline{\xi}} f(\underline{x})\, d\underline{x} \\
&= \int_{-\infty}^{\infty}\int_{-\infty}^{\infty} e^{-2\pi i(x_1\xi_1 + x_2\xi_2)} f(x_1, x_2)\, dx_1 dx_2 \\
&= \int_{-\infty}^{\infty} e^{-2\pi i x_2\xi_2} \left(\int_{-\infty}^{\infty} e^{-2\pi i x_1\xi_1} f(x_1, x_2)\, dx_1\right) dx_2 \,.
\end{aligned}$$

The inner integral is the 1D Fourier transform of $f(x_1, x_2)$ with respect to x_1. To find the full transform $\mathcal{F}f(\xi_1, \xi_2)$ it then remains to take another 1D Fourier transform with respect to x_2. We could also do this in the other order. Now note: when the function is not separable, this doesn't lead to a product of 1D transforms.

We can make this a little more formal by introducing *partial Fourier transforms*, for lack of a better term. Say,

$$\mathcal{F}_1 f(\xi_1, x_2) = \int_{-\infty}^{\infty} e^{-2\pi i x_1\xi_1} f(x_1, x_2)\, dx_1 \,,$$

$$\mathcal{F}_2 f(x_1, \xi_2) = \int_{-\infty}^{\infty} e^{-2\pi i x_2\xi_2} f(x_1, x_2)\, dx_2 \,.$$

Then

$$\mathcal{F}f = \mathcal{F}_1(\mathcal{F}_2 f) = \mathcal{F}_2(\mathcal{F}_1 f).$$

This would not look as good if you tried to put in some variables.

The same idea extends to higher dimensions.

9.2.4. Linearity. The linearity property is obvious, but I'm obliged to state it:

$$\mathcal{F}(\alpha f + \beta g)(\underline{\xi}) = \alpha \mathcal{F}f(\underline{\xi}) + \beta \mathcal{F}g(\underline{\xi}) .$$

9.2.5. Shifts. In one dimension a shift in time corresponds to a phase change in frequency, and the statement of this is the shift theorem:

- If $f(x) \rightleftharpoons F(s)$, then $f(x-b) \rightleftharpoons e^{-2\pi isb}F(s)$.

It looks a little slicker (to me) if we use the delay operator $(\tau_b f)(x) = f(x-b)$, for then we can write

$$\mathcal{F}(\tau_b f)(s) = e^{-2\pi isb}\mathcal{F}f(s) .$$

Each to his or her own taste.

The shift theorem in higher dimensions can be made to look just like it does in the one-dimensional case. Suppose that a point $\underline{x} = (x_1, x_2, \dots, x_n)$ is shifted by a displacement $\underline{b} = (b_1, b_2, \dots, b_n)$ to $\underline{x} - \underline{b} = (x_1 - b_1, x_2 - b_2, \dots, x_n - b_n)$. Then the effect on the Fourier transform is:

- **Shift Theorem.** If $f(\underline{x}) \rightleftharpoons F(\underline{\xi})$, then $f(\underline{x} - \underline{b}) \rightleftharpoons e^{-2\pi i\underline{b}\cdot\underline{\xi}}F(\underline{\xi})$.

Vectors replace scalars and the dot product replaces multiplication, but the formulas look much the same.

Again we can introduce the delay operator, this time delaying by a vector,

$$\tau_{\underline{b}} f(\underline{x}) = f(\underline{x} - \underline{b}) ,$$

and the shift theorem then takes the form

$$\mathcal{F}(\tau_{\underline{b}} f)(\underline{\xi}) = e^{-2\pi i\underline{b}\cdot\underline{\xi}}\mathcal{F}f(\underline{\xi}) .$$

Each to his or her own taste, again.

If you're more comfortable writing things out in coordinates, the result, in two dimensions, would read

$$\mathcal{F}f(x_1 - b_1, x_2 - b_2) = e^{2\pi i(-\xi_1 b_1 - \xi_2 b_2)}\mathcal{F}f(\xi_1, \xi_2) .$$

The only advantage in this (and you certainly wouldn't do so for any dimension higher than two) is a more visible reminder that in shifting (x_1, x_2) to $(x_1 - b_1, x_2 - b_2)$ we shift the variables independently. This independence is also (more) visible in the Fourier transform if we break up the dot product and multiply the exponentials:

$$\mathcal{F}f(x_1 - b_1, x_2 - b_2) = e^{-2\pi i\xi_1 b_1}e^{-2\pi i\xi_2 b_2}\mathcal{F}f(\xi_1, \xi_2) .$$

The derivation of the shift theorem is pretty much as in the one-dimensional case (look that over), but let me show you how the change of variable works, in two ways. I'll do this for $n = 2$, and, yes, I'll write it out in coordinates. First off,

$$\mathcal{F}(f(x_1 - b_2, x_2 - b_2)) = \int_{-\infty}^{\infty}\int_{-\infty}^{\infty} e^{-2\pi i(x_1\xi_1 + x_2\xi_2)} f(x_1 - b_1, x_2 - b_2)\, dx_1\, dx_2 ,$$

with the usual problem of notation and variables on the left-hand side.

We want to make a change of variable, turning $f(x_1 - b_1, x_2 - b_2)$ into $f(u_1, u_2)$ by the substitutions $u_1 = x_1 - b_1$ and $u_2 = x_2 - b_2$. Because this substitution changes x_1 and x_2 separately we can work one variable at a time:

$$\begin{aligned}
\mathcal{F}(f(x_1 - b_1, x_2 - b_2)) &= \int_{-\infty}^{\infty}\int_{-\infty}^{\infty} e^{-2\pi i(x_1\xi_1 + x_2\xi_2)} f(x_1 - b_1, x_2 - b_2)\, dx_1\, dx_2 \\
&= \int_{-\infty}^{\infty} e^{2\pi i x_1\xi_1} \left(\int_{-\infty}^{\infty} e^{2\pi i x_2\xi_2} f(x_1 - b_1, x_2 - b_2)\, dx_2\right) dx_1 \\
&= \int_{-\infty}^{\infty} e^{2\pi i x_1\xi_1} \left(\int_{-\infty}^{\infty} e^{-2\pi i(u_2 + b_2)\xi_2} f(x_1 - b_1, u_2)\, du_2\right) dx_1 \\
&\quad \text{(substituting } u_2 = x_2 - b_2\text{)}
\end{aligned}$$

$$\begin{aligned}
&= e^{-2\pi i b_2\xi_2} \int_{-\infty}^{\infty} e^{-2\pi i x_1\xi_1} \left(\int_{-\infty}^{\infty} e^{-2\pi i u_2\xi_2} f(x_1 - b_1, u_2)\, du_2\right) dx_1 \\
&= e^{-2\pi i b_2\xi_2} \int_{-\infty}^{\infty} e^{-2\pi i u_2\xi_2} \left(\int_{-\infty}^{\infty} e^{-2\pi i x_1\xi_1} f(x_1 - b_1, u_2)\, dx_1\right) du_2 \\
&= e^{-2\pi i b_2\xi_2} \int_{-\infty}^{\infty} e^{-2\pi i u_2\xi_2} \left(\int_{-\infty}^{\infty} e^{-2\pi i(u_1 + b_1)\xi_1} f(u_1, u_2)\, du_1\right) du_2 \\
&\quad \text{(substituting } u_1 = x_1 - b_1\text{)}
\end{aligned}$$

$$\begin{aligned}
&= e^{-2\pi i b_2\xi_2} e^{-2\pi i b_1\xi_1} \int_{-\infty}^{\infty} e^{-2\pi i u_2\xi_2} \left(\int_{-\infty}^{\infty} e^{-2\pi i u_1\xi_1} f(u_1, u_2)\, du_1\right) du_2 \\
&= e^{-2\pi i b_2\xi_2} e^{-2\pi i b_1\xi_1} \int_{-\infty}^{\infty}\int_{-\infty}^{\infty} e^{-2\pi i(u_1\xi_1 + u_2\xi_2)} f(u_1, u_2)\, du_1 du_2 \\
&= e^{-2\pi i b_2\xi_2} e^{-2\pi i b_1\xi_1}\, \mathcal{F}f(\xi_1, \xi_2) \\
&= e^{-2\pi i(b_1\xi_1 + b_2\xi_2)}\, \mathcal{F}f(\xi_1, \xi_2)\,.
\end{aligned}$$

And there's our formula. Of course we allowed ourselves the usual slack in writing a double integral in terms of iterated integrals, swapping the order of integration, etc. Fine. The same sort of thing goes through in higher dimensions, just work one variable at a time.

Now, if you know the general change of variable formula for multiple integrals, you can make the argument look a little less cumbersome and, in fact, even more like the argument in one dimension. I want to put this on the record, as a warm up to using the change of variable formula when we do the generalization of the stretch theorem to higher dimensions.

We're still using $u_1 = x_1 - b_1$, $u_2 = x_2 - b_2$, which in vector form is $\underline{u} = \underline{x} - \underline{b}$. This is a rigid shift of coordinates on $\mathbb{R}^2$, and so the area element does not change,

$du_1 du_2 = dx_1 dx_2$. Write this in vector form as $d\underline{u} = d\underline{x}$. Then

$$\begin{aligned}\mathcal{F}(f(\underline{x}-\underline{b})) &= \int_{\mathbb{R}^2} e^{-2\pi i \underline{x}\cdot\underline{\xi}} f(\underline{x}-\underline{b})\,d\underline{x} \\ &= \int_{\mathbb{R}^2} e^{-2\pi i(\underline{u}+\underline{b})\cdot\underline{\xi}} f(\underline{u})\,d\underline{u}\end{aligned}$$

(*that's* what the general change of variables formula does for you)

$$\begin{aligned}&= e^{-2\pi i \underline{b}\cdot\underline{\xi}} \int_{\mathbb{R}^2} e^{-2\pi i \underline{u}\cdot\underline{\xi}} f(\underline{u})\,d\underline{u} \\ &= e^{-2\pi i \underline{b}\cdot\underline{\xi}} \mathcal{F}f(\underline{\xi})\,.\end{aligned}$$

There's our formula, again, and there's really nothing here that depended on working in two dimensions.

The good news is that we've certainly derived the shift theorem! The bad news is that you may be saying to yourself: "This is not what I had in mind when you said the higher-dimensional case is just like the one-dimensional case." I don't have a quick comeback to that, except to say that I'm trying to make honest statements about the similarities and the differences in the two cases. If you want, you can assimilate the formulas and just skip those derivations in the higher-dimensional case that bother your sense of simplicity. (I will too, mostly.)

9.2.6. Modulation. No surprise, nothing new. The modulation theorem works in higher dimensions just as it does in one dimension: a modulation in space corresponds to a shift in frequency.

- **Modulation Theorem.** $\mathcal{F}(e^{2\pi i \underline{\xi}_0\cdot\underline{x}} f(\underline{x})) = \mathcal{F}f(\underline{\xi} - \underline{\xi}_0)$.

You can verify this at your leisure.

9.2.7. Stretches. Surprise, something new. There's really only one stretch theorem in higher dimensions, but I'd like to give two versions of it. The first version, where the variables scale independently, can be derived in a manner similar to that used in the derivation of the one-dimensional shift theorem, i.e., making separate changes of variable. This case comes up often enough that it's worth giving it its own moment in the sun. The second version (which includes the first) needs the general change of variables formula for multiple integrals for the derivation.

- **Stretch Theorem, first version.**

$$\mathcal{F}(f(a_1x_1, a_2x_2)) = \frac{1}{|a_1|\,|a_2|}\mathcal{F}f\left(\frac{\xi_1}{a_1}, \frac{\xi_2}{a_2}\right), \quad a_1, a_2 \neq 0\,.$$

I'll skip the derivation — I've given enough clues. There is an analogous result in higher dimensions.

The reason there's a second version of the stretch theorem is that something more can be done by way of transformations in higher dimensions that doesn't come up in the one-dimensional setting. We can look at a *linear change of variables* in the spatial domain. In two dimensions we write this as

$$\begin{pmatrix} u_1 \\ u_2 \end{pmatrix} = \begin{pmatrix} a & b \\ c & d \end{pmatrix} \begin{pmatrix} x_1 \\ x_2 \end{pmatrix}, \quad \text{or} \quad \underline{u} = A\underline{x}, \quad A = \begin{pmatrix} a & b \\ c & d \end{pmatrix},$$

or, written out,

$$\begin{aligned} u_1 &= ax_1 + bx_2 , \\ u_2 &= cx_1 + dx_2 . \end{aligned}$$

The simple, independent stretch is the special case

$$\begin{pmatrix} u_1 \\ u_2 \end{pmatrix} = \begin{pmatrix} a_1 & 0 \\ 0 & a_2 \end{pmatrix} \begin{pmatrix} x_1 \\ x_2 \end{pmatrix}.$$

For a general linear transformation the coordinates can get mixed up together instead of simply changing independently.

A linear change of coordinates is not at all an odd thing to do. Think of linearly distorting an image, like shearing; some examples are in the problems. Think also of rotation, which we'll consider below.

The general stretch theorem answers the question of what happens to the spectrum when the spatial coordinates change linearly: what is

$$\mathcal{F}(f(u_1, u_2)) = \mathcal{F}(f(ax_1 + bx_2, cx_1 + dx_2))?$$

The nice answer is most compactly expressed in matrix notation, in fact just as easily for n dimensions as for two.

Two points: First, a linear transformation as a linear change of coordinates isn't much good if you can't change the coordinates back. Thus it's natural to work only with invertible transformations here, i.e., those matrices A for which $\det A \neq 0$. Second, we'll need the transpose of a matrix and its inverse. We introduce the notation

$$A^{-\mathsf{T}} = (A^{-1})^{\mathsf{T}}$$

for the transpose of the inverse of A. You can check that $A^{-\mathsf{T}} = (A^{\mathsf{T}})^{-1}$ as well; i.e., $A^{-\mathsf{T}}$ can be defined either as the transpose of the inverse or as the inverse of the transpose.

We can now state:

- **Stretch Theorem, general version.**

$$\mathcal{F}(f(A\underline{x})) = \frac{1}{|\det A|}\mathcal{F}f(A^{-\mathsf{T}}\underline{\xi}).$$

There's another way of writing this that you might prefer, depending (as always) on your tastes. Using $\det A^{\mathsf{T}} = \det A$ and $\det A^{-1} = 1/\det A$, we have

$$\frac{1}{|\det A|} = |\det A^{-\mathsf{T}}|$$

so the formula reads

$$\mathcal{F}(f(A\underline{x})) = |\det A^{-\mathsf{T}}|\,\mathcal{F}f(A^{-\mathsf{T}}\underline{\xi}).$$

Finally, I'm of a mind to introduce the general *scaling* operator defined by

$$(\sigma_A f)(\underline{x}) = f(A\underline{x}),$$

where A is an invertible $n \times n$ matrix. Then I'm of a mind to write

$$\mathcal{F}(\sigma_A f)(\underline{\xi}) = \frac{1}{|\det A|}\mathcal{F}f(A^{-\mathsf{T}}\underline{\xi}) = \frac{1}{|\det A|}\sigma_{A^{-\mathsf{T}}}(\mathcal{F}f)(\underline{\xi}),$$

or

$$\mathcal{F}(\sigma_A f) = \frac{1}{|\det A|}\sigma_{A^{-\mathsf{T}}}(\mathcal{F}f),$$

without the variable. Your choice. The derivation of the stretch theorem is coming right up, after a few comments.

Reciprocal means inverse transpose. In the one-dimensional setting we pointed out that the stretch theorem expressed a reciprocal relationship between the time and frequency domains, one of the early instances of this. Then it was simply that stretching in time corresponds to a reciprocal stretching in frequency. We've seen many examples of reciprocal relationships, and by now you should be treating "reciprocity" as an important organizing principle in the ever larger Fourier transform section of your brain.

The generalization of the stretch theorem to higher dimensions heralds something new, something not apparent in one dimension. If we want to keep using the word "reciprocal," and we do, then we should allow that *reciprocal means inverse transpose*. This reduces to the usual meaning of reciprocal in one dimension, since the matrix determining the stretch is 1×1: $Ax = ax$, $A^{-\mathsf{T}} = 1/a$. But it's a more encompassing notion in higher dimensions.

I'm not just making this up. We'll see an interesting and important example of this kind of reciprocity in action when we talk about X-ray diffraction and crystals. There $A^{-\mathsf{T}}$ comes up when we apply the Fourier transform to lattices and encounter the so-called *reciprocal lattices*.

Let's also look at the two-dimensional case in a little more detail. To recover the first version of the stretch theorem we apply the general version to the diagonal matrix

$$A = \begin{pmatrix} a_1 & 0 \\ 0 & a_2 \end{pmatrix}, \quad \text{with} \quad \det A = a_1 a_2 \neq 0.$$

Then

$$A^{-1} = \begin{pmatrix} 1/a_1 & 0 \\ 0 & 1/a_2 \end{pmatrix} \Rightarrow A^{-\mathsf{T}} = \begin{pmatrix} 1/a_1 & 0 \\ 0 & 1/a_2 \end{pmatrix}.$$

This gives

$$\mathcal{F}(f(a_1x_1, a_2x_2)) = \mathcal{F}(f(A\underline{x})) = \frac{1}{|\det A|}\mathcal{F}f(A^{-\mathsf{T}}\underline{\xi}) = \frac{1}{|a_1||a_2|}\mathcal{F}f\left(\frac{\xi_1}{a_1}, \frac{\xi_2}{a_2}\right).$$

Works like a charm.

An important special case of the stretch theorem is when A is a rotation matrix:

$$A = \begin{pmatrix} \cos\theta & -\sin\theta \\ \sin\theta & \cos\theta \end{pmatrix}.$$

For $\theta > 0$ this is a counterclockwise rotation by an angle θ:

$$\begin{pmatrix} \cos\theta & -\sin\theta \\ \sin\theta & \cos\theta \end{pmatrix}\begin{pmatrix} 1 \\ 0 \end{pmatrix} = \begin{pmatrix} \cos\theta \\ \sin\theta \end{pmatrix}, \quad \begin{pmatrix} \cos\theta & -\sin\theta \\ \sin\theta & \cos\theta \end{pmatrix}\begin{pmatrix} 0 \\ 1 \end{pmatrix} = \begin{pmatrix} -\sin\theta \\ \cos\theta \end{pmatrix}.$$

A rotation matrix is *orthogonal*, meaning that $AA^{\mathsf{T}} = I$:

$$\begin{aligned} AA^{\mathsf{T}} &= \begin{pmatrix} \cos\theta & -\sin\theta \\ \sin\theta & \cos\theta \end{pmatrix}\begin{pmatrix} \cos\theta & \sin\theta \\ -\sin\theta & \cos\theta \end{pmatrix} \\ &= \begin{pmatrix} \cos^2\theta + \sin^2\theta & 0 \\ 0 & \cos^2\theta + \sin^2\theta \end{pmatrix} = \begin{pmatrix} 1 & 0 \\ 0 & 1 \end{pmatrix}. \end{aligned}$$

Thus $A^{-1} = A^{\mathsf{T}}$ so that

$$A^{-\mathsf{T}} = (A^{-1})^{\mathsf{T}} = (A^{\mathsf{T}})^{\mathsf{T}} = A.$$

Also

$$\det A = \cos^2\theta + \sin^2\theta = 1.$$

The consequence of all of this for the Fourier transform is that if A is a rotation matrix, then

$$\mathcal{F}(f(A\underline{x})) = \mathcal{F}f(A\underline{\xi}).$$

In words:

- A rotation in the spatial domain corresponds to an identical rotation in the frequency domain.

This result is used all the time in image processing.

Finally, it's worth knowing that for a 2×2 matrix we can write down $A^{-\mathsf{T}}$ explicitly:

$$\begin{pmatrix} a & b \\ c & d \end{pmatrix}^{-1} = \frac{1}{\det A}\begin{pmatrix} d & -b \\ -c & a \end{pmatrix},$$

and the transpose is

$$\begin{pmatrix} a & b \\ c & d \end{pmatrix}^{-\mathsf{T}} = \frac{1}{\det A}\begin{pmatrix} d & -c \\ -b & a \end{pmatrix}.$$

This jibes with what we found for a rotation matrix.

Derivation of the general stretch theorem. Starting with

$$\mathcal{F}(f(A\underline{x})) = \int_{\mathbb{R}^n} e^{-2\pi i \underline{\xi}\cdot\underline{x}} f(A\underline{x})\, d\underline{x},$$

our object is to make a change of variable, $\underline{u} = A\underline{x}$, and this requires the change of variables formula for multiple integrals.[9] In the form we need it, we can state:

If A is an invertible $n \times n$ matrix and $\underline{u} = A\underline{x}$, then

$$\int_{\mathbb{R}^n} g(A\underline{x})\,|\det A|\,d\underline{x} = \int_{\mathbb{R}^n} g(\underline{u})\,d\underline{u}\,.$$

Want to feel good (or at least OK) about this in a familiar setting? Take the case $n = 1$. Then

$$\int_{-\infty}^{\infty} g(ax)\,|a|dx = \int_{-\infty}^{\infty} g(u)\,du\,,$$

making the substitution $u = ax$. The transformation $u = ax$ of $\mathbb{R}$ scales lengths, and the scaling factor is a ($du = a\,dx$). That's if a is positive. The absolute value of a is in there in case a is negative; thus $u = ax$ is *sense reversing*, as manifested in flipping $+\infty$ and $-\infty$.[10]

In n-dimensions the transformation $\underline{u} = A\underline{x}$ scales n-dimensional *volumes*, and the scaling factor is $\det A$ ($d\underline{u} = \det A\,d\underline{x}$). The absolute value $|\det A|$ is in there because a matrix A with $\det A > 0$ is sense preserving on $\mathbb{R}^n$, and it is sense reversing if $\det A < 0$.[11] Thus, in general,

$$d\underline{u} = |\det A|\,d\underline{x}\,,$$

so the substitution $\underline{u} = A\underline{x}$ leads right to the formula

$$\int_{\mathbb{R}^n} g(A\underline{x})\,|\det A|\,d\underline{x} = \int_{\mathbb{R}^n} g(\underline{u})\,d\underline{u}\,.$$

Since $\det A$ is a constant, it comes out of the integral. We can say, as we shall,

$$\int_{\mathbb{R}^n} g(A\underline{x})\,d\underline{x} = \frac{1}{|\det A|}\int_{\mathbb{R}^n} g(\underline{u})\,d\underline{u}\,.$$

To apply this to the Fourier transform of $f(A\underline{x})$ we have

$$\int_{\mathbb{R}^n} e^{-2\pi i\underline{\xi}\cdot\underline{x}} f(A\underline{x})\,d\underline{x} = \frac{1}{|\det A|}\int_{\mathbb{R}^n} e^{-2\pi i\underline{\xi}\cdot A^{-1}\underline{u}} f(\underline{u})\,d\underline{u}\,.$$

Next we use an identity for what happens to the dot product when there's a matrix operating on one of the vectors. Namely, for a matrix B and any vectors $\underline{\xi}$ and $\underline{u}$,

$$\underline{\xi}\cdot B\underline{u} = B^{\mathsf{T}}\underline{\xi}\cdot\underline{u}\,,$$

where B^{T} is the transpose of B. We take $B = A^{-1}$ and then

$$\underline{\xi}\cdot A^{-1}\underline{u} = A^{-\mathsf{T}}\underline{\xi}\cdot\underline{u}\,.$$

With this,

$$\frac{1}{|\det A|}\int_{\mathbb{R}^n} e^{-2\pi i\underline{\xi}\cdot A^{-1}\underline{u}} f(\underline{u})\,d\underline{u} = \frac{1}{|\det A|}\int_{\mathbb{R}^n} e^{-2\pi i A^{-\mathsf{T}}\underline{\xi}\cdot\underline{u}} f(\underline{u})\,d\underline{u}\,.$$

[9] We also used the change of variable formula in Chapter 3 in preparation for a derivation of the Central Limit Theorem. My feeling then, as now, is that if you're going to learn how to design chips having millions of transistors, you can certainly learn how to change variables in a multiple integral.

[10] If you trace out an interval from left to right and scale by $u = ax$ with $a < 0$, the scaled interval is traced out from right to left — that's sense reversal.

[11] This is a trickier notion in higher dimensions! Defining A to be sense reversing if $\det A < 0$ is another example of the mathematical modus operandi of turning the solution of a problem into a definition. It's hard to visualize "sense reversing" in higher dimensions, but operationally it involves $\det A < 0$. So make this the definition.

But this last integral is exactly $\mathcal{F}f(A^{-\mathsf{T}}\underline{\xi})$. We have shown that

$$\mathcal{F}(f(A\underline{x})) = \frac{1}{|\det A|}\mathcal{F}f(A^{-\mathsf{T}}\underline{\xi}),$$

as desired.

9.2.8. Shift and stretch. As an example of using the general stretch formula, let's combine a shift with a stretch and show that

$$\mathcal{F}(f(A\underline{x} - \underline{b})) = \exp(-2\pi i\underline{b}\cdot A^{-\mathsf{T}}\underline{\xi})\frac{1}{|\det A|}\mathcal{F}f(A^{-\mathsf{T}}\underline{\xi}).$$

(Here I think the exponential is a little crowded to write it as e to a power.) Combining shifts and stretches seems to cause a lot of problems for people (even in one dimension), so let me do this in several ways.

As a first approach and to keep the operations straight, write

$$g(\underline{x}) = f(\underline{x} - \underline{b})$$

and then

$$f(A\underline{x} - \underline{b}) = g(A\underline{x}).$$

Using the stretch theorem first,

$$\mathcal{F}(g(A\underline{x})) = \frac{1}{|\det A|}\mathcal{F}g(A^{-\mathsf{T}}\underline{\xi}).$$

Applying the shift theorem next gives

$$(\mathcal{F}g)(A^{-\mathsf{T}}\underline{\xi}) = \exp(-2\pi i\underline{b}\cdot A^{-\mathsf{T}}\underline{\xi})\mathcal{F}f(A^{-\mathsf{T}}\underline{\xi}).$$

Putting these together gives the final formula for $\mathcal{F}(f(A\underline{x} - \underline{b}))$.

Another way around is instead to write

$$g(\underline{x}) = f(A\underline{x})$$

and then

$$f(A\underline{x} - \underline{b}) = f(A(\underline{x} - A^{-1}\underline{b})) = g(\underline{x} - A^{-1}\underline{b}).$$

Now use the shift theorem first to get

$$\mathcal{F}(g(\underline{x} - A^{-1}\underline{b})) = \exp(-2\pi i A^{-1}\underline{b}\cdot\underline{\xi})\,(\mathcal{F}g)(\underline{\xi}) = \exp(-2\pi i\underline{b}\cdot A^{-\mathsf{T}}\underline{\xi})\,(\mathcal{F}g)(\underline{\xi}).$$

The stretch theorem comes next and it produces

$$\mathcal{F}g(\underline{\xi}) = \mathcal{F}(f(A\underline{x})) = \frac{1}{|\det A|}\mathcal{F}f(A^{-\mathsf{T}}\underline{\xi}).$$

This agrees with what we had before, as if there were any doubt.

By popular demand I'll do this one more time by expressing $f(A\underline{x} - \underline{b})$ using the delay and scaling operators. It's a question of which comes first, and parallel to the first derivation above we can write

$$f(A\underline{x} - \underline{b}) = \sigma_A(\tau_{\underline{b}}f)(\underline{x}) = (\sigma_A\tau_{\underline{b}}f)(\underline{x}),$$

which we verify by

$$(\sigma_A\tau_{\underline{b}}f)(\underline{x}) = (\tau_{\underline{b}}f)(A\underline{x}) = f(A\underline{x} - \underline{b}).$$

And now we have

$$\mathcal{F}(\sigma_A(\tau_{\underline{b}}f))(\underline{\xi}) = \frac{1}{|\det A|}\mathcal{F}(\tau_{\underline{b}}f)(A^{-\mathsf{T}}\underline{\xi}) = \frac{1}{|\det A|}\exp(-2\pi i A^{-\mathsf{T}}\underline{\xi}\cdot\underline{b})\mathcal{F}f(A^{-\mathsf{T}}\underline{\xi}).$$

I won't give a second version of the second derivation.

9.2.9. Parseval's identity. Yes, that works, too:

$$\int_{\mathbb{R}^n} f(\underline{x})\overline{g(\underline{x})}\,d\underline{x} = \int_{\mathbb{R}^n} \mathcal{F}f(\underline{\xi})\overline{\mathcal{F}g(\underline{\xi})}\,d\underline{\xi},$$

$$\int_{\mathbb{R}^n} |f(\underline{x})|^2\,d\underline{x} = \int_{\mathbb{R}^n} |\mathcal{F}f(\underline{\xi})|^2\,d\underline{\xi}.$$

Derivation suppressed. Same as one dimension. Splendid use of underlines and overlines here.

9.2.10. Derivatives. One variable, one ordinary derivative. Many variables, many partial derivatives. For the Fourier transform it's the same principle regardless: the Fourier transform swaps differentiation and multiplication. With $\underline{x} = (x_1, x_2, \ldots, x_n)$ and $\underline{\xi} = (\xi_1, \xi_2, \ldots, \xi_n)$, as usual, we have:

- **Derivative theorems.**

$$\frac{\partial}{\partial \xi_k}\mathcal{F}f(\underline{\xi}) = \mathcal{F}(-2\pi i x_k f(\underline{x})),$$

$$\mathcal{F}\left(\frac{\partial}{\partial x_k}f\right)(\underline{\xi}) = 2\pi i \xi_k \mathcal{F}f(\underline{\xi}).$$

From these you can find formulas for higher and mixed partial derivatives.

Once again, I am pleased to suppress the derivations. You just have to know how to take partial derivatives of $e^{-2\pi i \underline{x}\cdot\underline{\xi}} = e^{-2\pi i(x_1\xi_1 + x_2\xi_2 + \cdots + x_n\xi_n)}$. The derivation of the second formula uses a difference quotient, as in the derivation in one dimension.

9.2.11. Convolution. What about convolution? For two functions f and g defined on $\mathbb{R}^n$ the definition is

$$(f * g)(\underline{x}) = \int_{\mathbb{R}^n} f(\underline{x} - \underline{y})g(\underline{y})\,d\underline{y}.$$

Written out in coordinates, this looks more complicated. For $n = 2$, for example,

$$(f * g)(x_1, x_2) = \int_{-\infty}^{\infty}\int_{-\infty}^{\infty} f(x_1 - y_1, x_2 - y_2)g(y_1, y_2)\,dy_1\,dy_2.$$

The intelligent person would not write out the corresponding coordinatized formula for higher dimensions unless absolutely pressed. The intelligent person would also not try too hard to flip, drag, or otherwise manually visualize a convolution in higher dimensions. The intelligent person would be happy to learn that once again

$$\mathcal{F}(f * g)(\underline{\xi}) = \mathcal{F}f(\underline{\xi})\mathcal{F}g(\underline{\xi}) \quad \text{and} \quad \mathcal{F}(fg)(\underline{\xi}) = (\mathcal{F}f * \mathcal{F}g)(\underline{\xi}).$$

The algebraic properties of convolution continue to hold, as do the typical interpretations — smoothing, averaging, etc. — when applied by an intelligent person. Yes, convolution is ubiquitous and extremely important, but you know that.

Filters. Just as in one dimension, convolution is the heart and soul of filtering. And again as in one dimension, the basic examples (low-pass, high-pass, band-pass, notch) are defined through the higher-dimensional rectangle functions by what they do in the frequency domain, i.e., by defining the transfer function of the filter to pass some frequencies and eliminate others. If you take a course in image processing, you'll be spending time with each of these, as well as with other transformations.

The two-dimensional version of the aphorism "it takes high frequencies to make sharp corners" is "it takes high frequencies to make *edges*." A low-pass filter eliminates high frequencies and so softens edges. The effect is to blur the image. A high-pass filter passes the high frequencies (rapid changes in grayscale levels) and so emphasizes edges. In particular, high-pass filters are used for edge detection. Any standard software package that includes basic image processing will allow you to experiment with different frequency cutoffs.

Here I'll just show two filtered versions of the picture of Venus we had earlier. The one on the left is a result of a low-pass filter and the one on the right a high-pass filter — no formulas for the filters. I used the same cutoff frequency for the low-pass and high-pass filtering.

Here's what the spectra look like, respectively.

9.2.12. The Fourier transform of a radial function. For its use in many applications, we're going to consider one further aspect of the two-dimensional case. A function on $\mathbb{R}^2$ is *radial* (also called *radially symmetric* or *circularly symmetric*) if its values depend only on the distance from the origin. In polar coordinates the distance from the origin is denoted by r, so to say that a function is radial is to say that its values depend only on r (and that the values do not depend on θ, writing the usual polar coordinates as (r, θ)).

To take a simple example, the Gaussian $g(x_1, x_2) = e^{-\pi(x_1^2+x_2^2)}$ is radial; it can be written $g(r) = e^{-\pi r^2}$, $r^2 = x_1^2 + x_2^2$, and there's no θ dependence. On the other hand, if we scale the Gaussian in one direction, for example $h(x_1, x_2) = e^{-\pi((2x_1)^2+x_2^2)}$, then it is no longer radial.

The two-dimensional rectangle function $\Pi(x_1, x_2)$ is also not radial. In polar coordinates

$$\Pi(x_1, x_2) = \Pi(x_1)\Pi(x_2) = \Pi(r\cos\theta)\Pi(r\sin\theta)\,.$$

As you can verify this depends on both r and θ.

The definition of the Fourier transform is set up in Cartesian coordinates, but we'll be better off writing it in polar coordinates if we work with radial functions. This is actually *not* so straightforward, or, at least, it involves introducing some special functions to write the formulas in a compact way.

We have to convert

$$\int_{\mathbb{R}^2} e^{-2\pi i \underline{x}\cdot\underline{\xi}} f(\underline{x})\,d\underline{x} = \int_{-\infty}^{\infty}\int_{-\infty}^{\infty} e^{-2\pi i(x_1\xi_1+x_2\xi_2)} f(x_1, x_2)\,dx_1\,dx_2$$

to polar coordinates (r, θ) on the (x_1, x_2)-plane. There are several steps: To say that $f(\underline{x})$ is a radial function means that it becomes $f(r)$. To describe all of $\mathbb{R}^2$ in the limits of integration, we take r going from 0 to ∞ and θ going from 0 to 2π. The area element $dx_1\,dx_2$ becomes $r\,dr\,d\theta$.

The problem is the inner product $\underline{x}\cdot\underline{\xi} = x_1\xi_1 + x_2\xi_2$ in the exponential and how to write it in polar coordinates. For (x_1, x_2), varying over the (x_1, x_2)-plane, we have, as usual, $x_1 = r\cos\theta$, $x_2 = r\sin\theta$. Similarly, we write (ξ_1, ξ_2), which is fixed in the integral, in terms of polar coordinates on the (ξ_1, ξ_2)-plane; call them (ρ, ϕ), with $\xi_1 = \rho\cos\phi$, $\xi_2 = \rho\sin\phi$. Then

$$\underline{x}\cdot\underline{\xi} = r\rho(\cos\theta\cos\phi + \sin\theta\sin\phi) = r\rho\cos(\theta-\phi)\,,$$

using the addition formula for the cosine. The Fourier transform of f is thus

$$\int_{-\infty}^{\infty}\int_{-\infty}^{\infty} e^{-2\pi i \underline{x}\cdot\underline{\xi}} f(\underline{x})\,d\underline{x} = \int_0^{2\pi}\int_0^{\infty} f(r) e^{-2\pi i r\rho\cos(\theta-\phi)}\,r\,dr\,d\theta\,.$$

There's more to be done. First of all, because $e^{-2\pi i r\rho\cos(\theta-\phi)}$ is periodic in θ of period 2π, the integral

$$\int_0^{2\pi} e^{-2\pi i r\rho\cos(\theta-\phi)}\,d\theta$$

does not depend on ϕ.[12] Consequently,

$$\int_0^{2\pi} e^{-2\pi i r\rho\cos(\theta-\phi)}\,d\theta = \int_0^{2\pi} e^{-2\pi i r\rho\cos\theta}\,d\theta\,.$$

The next step is to define ourselves out of trouble. We introduce the function

$$J_0(a) = \frac{1}{2\pi}\int_0^{2\pi} e^{-ia\cos\theta}\,d\theta\,.$$

We give this integral a name, $J_0(a)$, because, try as you might, there is no simple closed-form expression for it. So we take the integral as defining a new function. It is called the zero-order Bessel function of the first kind. Sorry, but Bessel functions, of whatever order and kind, always seem to come up in problems involving circular symmetry — ask any physicist.

Incorporating J_0 into what we've done,

$$\int_0^{2\pi} e^{-2\pi i r\rho\cos\theta}\,d\theta = 2\pi J_0(2\pi r\rho)$$

and the Fourier transform of $f(r)$ is

$$2\pi\int_0^{\infty} f(r)J_0(2\pi r\rho)\,r\,dr\,.$$

Let's summarize:

- If $f(\underline{x})$ is a radial function (that is, $f(\underline{x})$ is really $f(r)$), then its Fourier transform is
$$F(\rho) = 2\pi\int_0^{\infty} f(r)J_0(2\pi r\rho)\,r dr\,.$$
- In words, the important conclusion to take away from this is that the Fourier transform of a radial function is also radial.

The formula for $F(\rho)$ in terms of $f(r)$ is sometimes called the zero-order *Hankel transform* of $f(r)$ but, again, we understand that it is nothing other than the Fourier transform of a radial function.

Circ and Jinc. A useful radial function to introduce, sort of a radially symmetric analog of the rectangle function, is

$$\operatorname{circ}(r) = \begin{cases} 1, & r < 1,\\ 0, & r \geq 1.\end{cases}$$

[12] We've applied this general fact implicitly or explicitly on earlier occasions when working with periodic functions; namely, if g is periodic with period 2π, then

$$\int_0^{2\pi} g(\theta-\phi)\,d\theta = \int_0^{2\pi} g(\theta)\,d\theta\,.$$

Convince yourself of this; for instance let $G(\phi) = \int_0^{2\pi} g(\theta-\phi)\,d\theta$ and show that $G'(\phi) \equiv 0$. Hence $G(\phi)$ is constant, so $G(\phi) = G(0)$.

(And one can argue about the value at the rim, $r = 1$.) Here's the graph.

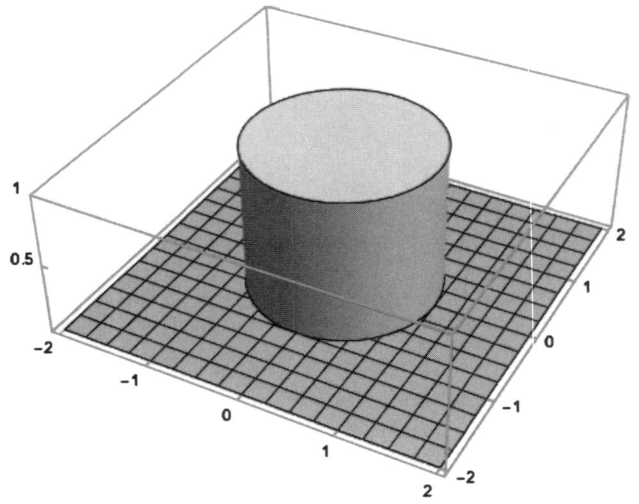

For its Fourier transform the limits of integration on r go only from 0 to 1, and so we have simply

$$\mathcal{F}\text{circ}(\rho) = 2\pi \int_0^1 J_0(2\pi r\rho)\, r\, dr\,.$$

We make a change of variable, $u = 2\pi r\rho$. Then $du = 2\pi\rho dr$ and the limits of integration go from $u = 0$ to $u = 2\pi\rho$. The integral becomes

$$\mathcal{F}\text{circ}(\rho) = \frac{1}{2\pi\rho^2}\int_0^{2\pi\rho} uJ_0(u)\, du\,.$$

We write the integral this way because, you will now be ecstatic to learn, there is an identity that brings in the first-order Bessel function of the first kind. That identity goes like this:

$$\int_0^x uJ_0(u)\, du = xJ_1(x)\,.$$

In terms of J_1 we can now write

$$\mathcal{F}\text{circ}(\rho) = \frac{J_1(2\pi\rho)}{\rho}\,.$$

It is customary to introduce the jinc function, defined by

$$\text{jinc}(\rho) = \frac{J_1(\pi\rho)}{2\rho}\,.$$

In terms of this,

$$\mathcal{F}\text{circ}(\rho) = 4\,\text{jinc}(2\rho)\,.$$

The graph of $\mathcal{F}$circ is

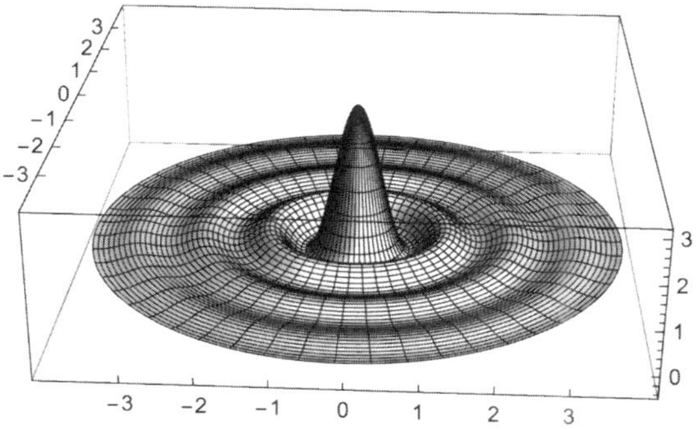

I could plot this (easily) because Bessel functions are so common (really) that they are built into many mathematical software packages, such as MATLAB or Mathematica. Looks vaguely like some kind of radially symmetric version of the $2D$ sinc function, but don't take that too seriously.

If you're wondering if these ideas extend to three dimensions and higher for functions that are spherically symmetric, the answer is yes. It's too much to go into here, but you should be able to track down the formulas you might need. More Bessel functions, oh joy.

Let me also mention, but only mention, the vast subject of *spherical harmonics*. These are harmonic functions (solutions of Laplace's equation) on the sphere. They allow for a development analogous to Fourier series for functions defined on a sphere, i.e., expansions in spherical harmonics.[13] They're big in physics, for instance in solutions of the Schrödinger equation in quantum mechanics, and they also find applications in computer graphics. Again I encourage you to poke around, but be prepared for a lot of new notation and new terminology.

9.3. A Little δ Now, More Later

We'll see that things get more interesting in higher dimensions for δ-functions, but the definition of the plain vanilla δ is the same as before, and its Fourier transform also works out just like before. To give the distributional definition, I'll pause, just for a moment, to define what it means for a function of several variables to be a Schwartz function. We don't actually need Schwartz functions to give the definition of δ as a distribution (we could use smooth functions of compact support), but they're needed for the generalized Fourier transform and this seems as good a time as any to say what they are.

[13]In one dimension we can consider periodic functions as being defined on a circle, and one way to generalize this is to consider functions defined on a sphere. But a sphere doesn't have a group structure (like a circle does, via rotations) and this complicates the attempts to extend Fourier analysis to spherical functions.

9.3.1. Schwartz functions. The theory and practice of tempered distributions works the same in higher dimensions as it does in one. The basis of the treatment is via the Schwartz functions as the class of test functions. The condition that a function of several variables be rapidly decreasing is that all partial derivatives (including mixed partial derivatives) decrease faster than any power of any of the coordinates. This can be stated in any number of equivalent forms. One way is to require that[14]

$$|\underline{x}|^p \, |\partial^q \varphi(\underline{x})| \to 0 \quad \text{as } |\underline{x}| \to \infty .$$

I'll explain the funny notation — it's an example of the occasional awkwardness that sets in when writing formulas in higher dimensions. Here p is a positive integer, so that just gives a power of $|\underline{x}|$, while q is a *multi-index.* This means that $q = (q_1, \dots, q_n)$, each q_i a positive integer, and writing ∂^q is supposed to stand in for

$$\frac{\partial^{q_1+\cdots+q_n}}{(\partial x_1)^{q_1}(\partial x_2)^{q_2}\cdots(\partial x_n)^{q_n}} .$$

There's no special font used to indicate multi-indices, though people more often use lowercase Greek letters. You just have to intuit it.

From here, the definitions of tempered distributions, the Fourier transform of a tempered distribution, and everything else goes through just as before. Shall we leave it alone? I think so.

9.3.2. δ in higher dimensions. The δ-function is the distribution defined by the pairing

$$\langle \delta, \varphi \rangle = \varphi(0, \dots, 0), \quad \text{or} \quad \langle \delta, \varphi \rangle = \varphi(\underline{0}) \quad \text{in vector notation,}$$

where $\varphi(x_1, , \dots, x_n)$ is a Schwartz function.[15] As is customary, we also write this in terms of integration as

$$\int_{\mathbb{R}^n} \varphi(\underline{x})\delta(\underline{x})\, d\underline{x} = \varphi(\underline{0}) .$$

You can show that δ is even as a distribution, once you've reminded yourself what it means for a distribution to be even.

As before, one has

$$f(\underline{x})\delta(\underline{x}) = f(\underline{0})\delta(\underline{x}) ,$$

when f is a smooth function, and for convolution

$$(f * \delta)(\underline{x}) = f(\underline{x}) .$$

I will not hesitate to write

$$(f * \delta)(x) = \int_{\mathbb{R}^n} f(\underline{y})\delta(\underline{x} - \underline{y})\, d\underline{y} .$$

The shifted δ-function $\delta(\underline{x} - \underline{b}) = \delta(x_1 - b_1, x_2 - b_2, , \dots, x_n - b_n)$, or $\delta_{\underline{b}} = \tau_{\underline{b}}\delta$, has the corresponding properties

$$f(\underline{x})\delta(\underline{x} - \underline{b}) = f(\underline{b})\delta(\underline{x} - \underline{b}) \quad \text{and} \quad f * \delta(\underline{x} - \underline{b}) = f(\underline{x} - \underline{b}), \quad \text{or} \quad f * \tau_{\underline{b}}\delta = \tau_{\underline{b}} f .$$

[14]Alternate definitions are possible, analogous to those in one dimension.

[15]δ is in a larger class of distributions than the tempered distributions. It is defined by the pairing $\langle \delta, \varphi \rangle = \varphi(\underline{0})$ when φ is any smooth function of compact support.

In some cases it is useful to know that we can factor the δ-function into one-dimensional deltas, as in

$$\delta(x_1, x_2, \dots, x_n) = \delta_1(x_1)\delta_2(x_2)\cdots\delta_n(x_n)\,.$$

I've put subscripts on the δ's on the right-hand side just to tag them with the individual coordinates. There are some advantages in doing this. Though it remains true, as a general rule, that multiplying distributions is not (and cannot be) defined, this is one case where it makes sense. The formula holds because of how each side acts on a Schwartz function.[16] Let's just check this in the two-dimensional case and play a little fast and loose by writing the pairing as an integral. Then, on the one hand,

$$\int_{\mathbb{R}^2} \varphi(\underline{x})\delta(\underline{x})\,d\underline{x} = \varphi(0,0)$$

by definition of the two-dimensional δ-function. On the other hand,

$$\begin{aligned}\int_{\mathbb{R}^2} \varphi(x_1,x_2)\delta_1(x_1)\delta_2(x_2)\,dx_1\,dx_2 &= \int_{-\infty}^{\infty}\left(\int_{-\infty}^{\infty}\varphi(x_1,x_2)\delta_1(x_1)\,dx_1\right)\delta_2(x_2)\,dx_2\\ &= \int_{-\infty}^{\infty}\varphi(0,x_2)\delta_2(x_2)\,dx_2 = \varphi(0,0).\end{aligned}$$

So $\delta(x_1, x_2)$ and $\delta_1(x_1)\delta_2(x_2)$ have the same effect when integrated against a test function.

9.3.3. The Fourier transform of δ.

And finally, the Fourier transform of the δ-function is, of course, 1 (that's the constant function 1):

$$\mathcal{F}\delta = 1\,.$$

The argument is the same as in the one-dimensional case, but it's been awhile. So, to reminisce, for any Schwartz function φ,

$$\begin{aligned}\langle \mathcal{F}\delta, \varphi\rangle &= \langle \delta, \mathcal{F}\varphi\rangle \quad \text{(the definition of the generalized Fourier transform)}\\ &= \mathcal{F}\varphi(\underline{0}) \quad \text{(the definition of how } \delta \text{ pairs with a test function, } \mathcal{F}\varphi \text{ in this case)}\\ &= \int_{-\infty}^{\infty} 1\cdot\varphi(x)\,dx\\ &\quad \text{(because in the classical Fourier transform of } \varphi \text{ the integrand has } e^{-2\pi i\underline{0}\cdot\underline{x}} = 1)\\ &= \langle 1, \varphi\rangle \quad \text{(how the smooth function 1 pairs with a test function } \varphi).\end{aligned}$$

(Now say it to yourself: where did we start; where did we finish?)

Appealing to duality, the Fourier transform of 1 is δ,

$$\mathcal{F}1 = \delta.$$

One can then shift to get

$$\delta(\underline{x}-\underline{b}) \rightleftharpoons e^{-2\pi i\underline{b}\cdot\underline{\xi}} \quad \text{or} \quad \mathcal{F}\delta_{\underline{b}}(\underline{\xi}) = e^{-2\pi i\underline{b}\cdot\underline{\xi}}\,,$$

and also

$$\mathcal{F}e^{-2\pi i\underline{b}\cdot\underline{x}} = \delta_{-\underline{b}}\,.$$

[16] The precise way to do this is through the use of tensor products of distributions, something we have not discussed, and will not.

You can now see (again) where those symmetrically paired dots come from in looking at the spectral picture for alternating black and white stripes. They come from the Fourier transforms of $\cos(2\pi\,\underline{x}\cdot\underline{\xi}_0) = \mathrm{Re}\{\exp(2\pi i\,\underline{x}\cdot\underline{\xi}_0)\}$, since

$$\mathcal{F}\cos(2\pi\,\underline{x}\cdot\underline{\xi}_0) = \tfrac{1}{2}(\delta(\underline{\xi}-\underline{\xi}_0) + \delta(\underline{\xi}+\underline{\xi}_0))\,.$$

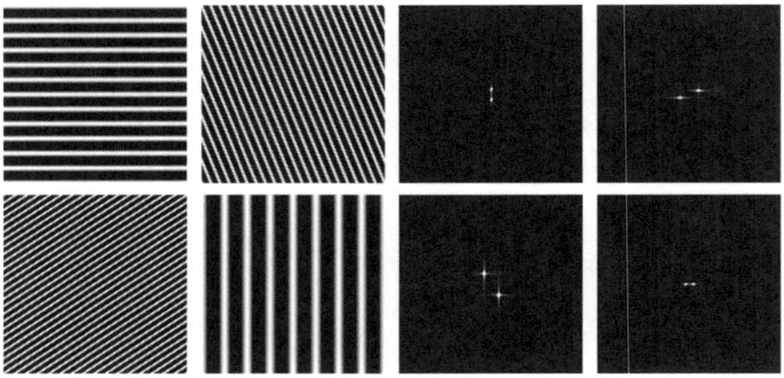

9.3.4. Scaling δ-functions. Recall how a one-dimensional δ-function scales:

$$\delta(ax) = \frac{1}{|a|}\delta(x)\,.$$

Writing a higher-dimensional δ-function as a product of one-dimensional δ-functions, we get a corresponding formula. In two dimensions,

$$\begin{aligned}\delta(a_1x_1, a_2x_2) &= \delta_1(a_1x_1)\delta_2(a_2x_2)\\ &= \frac{1}{|a_1|}\delta_1(x_1)\frac{1}{|a_2|}\delta_2(x_2)\\ &= \frac{1}{|a_1|\,|a_2|}\delta_1(x_1)\delta_2(x_2) = \frac{1}{|a_1a_2|}\delta(x_1,x_2),\end{aligned}$$

and in n-dimensions,

$$\delta(a_1x_1,\ldots,a_nx_n) = \frac{1}{|a_1\cdots a_n|}\delta(x_1,\ldots,x_n)\,.$$

It's also possible (and useful) to consider $\delta(A\underline{x})$ when A is an invertible matrix. The result is

$$\delta(A\underline{x}) = \frac{1}{|\det A|}\delta(\underline{x})\,.$$

This formula bears the same relationship to the preceding formula as the general stretch theorem bears to the first version of the stretch theorem.

Here's the derivation: from scratch, and for practice, let's consider pairing by integration and use the change of variables formula:

$$\begin{aligned}
\int_{\mathbb{R}^n} \delta(A\underline{x})\varphi(\underline{x})\,d\underline{x} &= \frac{1}{|\det A|}\int_{\mathbb{R}^n} \delta(\underline{u})\varphi(A^{-1}\underline{u})\,d\underline{u} \\
&\qquad \text{(making the change of variables } \underline{u} = A\underline{x}) \\
&= \frac{1}{|\det A|}\varphi(A^{-1}\underline{0}) \quad \text{(by how the } \delta\text{-function acts)} \\
&= \frac{1}{|\det A|}\varphi(\underline{0}) \quad (A^{-1}\underline{0} = \underline{0} \text{ because } A^{-1} \text{ is linear}).
\end{aligned}$$

Thus $\delta(A\underline{x})$ has the same effect as $(1/|\det A|)\delta$ when paired with a test function, so they must be equal. Suffice it to say, it's possible to give a derivation that doesn't resort (explicitly) to writing the pairing as integration.

9.4. Higher-Dimensional Fourier Series

The ideas and constructions for Fourier series that we studied in Chapter 1 carry over pretty much directly to periodic functions in two, three, and higher dimensions. Here we want to give just the basic setup so you can see that the situation is very similar to what we've already encountered, even as far as making the notations correspond. Then we'll look at a fascinating problem — random walks — where higher-dimensional Fourier series are central to the solution in a very surprising way.

Periodic functions. The definition of periodicity for real-valued functions of several variables is much the same as for functions of one variable, except that we allow for different periods in different slots. To take the two-dimensional case, we say that a function $f(x_1, x_2)$ is (p_1, p_2)-periodic if

$$f(x_1 + p_1, x_2) = f(x_1, x_2) \quad \text{and} \quad f(x_1, x_2 + p_2) = f(x_1, x_2)$$

for all x_1 and x_2. It follows that

$$f(x_1 + p_1, x_2 + p_2) = f(x_1, x_2)$$

and more generally that

$$f(x_1 + n_1p_1, x_2 + n_2p_2) = f(x_1, x_2)$$

for all integers n_1, n_2. One also says that $f(x_1, x_2)$ has two independent periods.

There's a small but important point having to do with periodicity of $f(x_1, x_2)$ one variable at a time or both variables together. The condition

$$f(x_1 + n_1p_1, x_2 + n_2p_2) = f(x_1, x_2)$$

for all integers n_1, n_2 *can* be taken as the definition of periodicity, but the condition

$$f(x_1 + p_1, x_2 + p_2) = f(x_1, x_2)$$

alone is *not* the appropriate definition. The former implies that $f(x_1 + p_1, x_2) = f(x_1, x_2)$ and $f(x_1, x_2 + p_2) = f(x_1, x_2)$ by taking (n_1, n_2) to be $(1, 0)$ and $(0, 1)$, respectively, and this independent periodicity is what we want. The latter condition does not imply independent periodicity.

As we did for functions of one variable, we need to settle on a standard case, and for our work now, it's enough to assume that the period in each variable is 1. So the condition is

$$f(x_1 + 1, x_2) = f(x_1, x_2) \quad \text{and} \quad f(x_1, x_2 + 1) = f(x_1, x_2),$$

or

$$f(x_1 + n_1, x_2 + n_2) = f(x_1, x_2) \quad \text{for all integers } n_1,\ n_2 .$$

If we use vector notation and write $\underline{x}$ for (x_1, x_2) and $\underline{n}$ for the pair (n_1, n_2) of integers, then we can write the condition as

$$f(\underline{x} + \underline{n}) = f(\underline{x}),$$

and, except for the typeface, it looks like the one-dimensional case.

Where is $f(x_1, x_2)$ defined? For a periodic function (of period 1) it is enough to know the function for $x_1 \in [0, 1]$ and $x_2 \in [0, 1]$, which we write as $(x_1, x_2) \in [0, 1]^2 = [0, 1] \times [0, 1]$ (Cartesian product). But we extend $f(x_1, x_2)$ to be defined on all of $\mathbb{R}^2$ via the periodicity condition.

We can consider periodicity in any dimension. To avoid conflicts with other notation, in this discussion I'll write the dimension as d rather than n. Let $\underline{x} = (x_1, x_2, \ldots, x_d)$ be a vector in $\mathbb{R}^d$ and let $\underline{n} = (n_1, n_2, \ldots, n_d)$ be a d-tuple of integers. Then $f(\underline{x}) = f(x_1, x_2, \ldots, x_d)$ is periodic of period 1 in each variable if

$$f(\underline{x} + \underline{n}) = f(\underline{x}) \quad \text{for all } \underline{n} .$$

In this case we consider the natural domain of $f(\underline{x})$ to be

$$[0, 1]^d = [0, 1] \times [0, 1] \times \cdots \times [0, 1] \quad (d\text{-fold Cartesian product}),$$

meaning the set of points $(x_1, x_2, \ldots, x_d)$ where $0 \le x_j \le 1$ for each $j = 1, 2, \ldots, d$, and we extend $f(\underline{x})$ to be defined on all of $\mathbb{R}^d$ by periodicity.

Complex exponentials, again. What are the building blocks for periodic functions in higher dimensions? We simply multiply complex exponentials of one variable. Taking again the two-dimensional case as a model, the function

$$e^{2\pi i x_1} e^{2\pi i x_2}$$

is periodic with period 1 in each variable. Note that once we get beyond one dimension it's not so helpful to think of periodicity "in time" and to force yourself to write the variable as t.

In d dimensions the corresponding exponential is

$$e^{2\pi i x_1} e^{2\pi i x_2} \cdots e^{2\pi i x_d} .$$

You may be tempted to use the usual rules and write this as

$$e^{2\pi i x_1} e^{2\pi i x_2} \cdots e^{2\pi i x_d} = e^{2\pi i (x_1 + x_2 + \cdots + x_d)} .$$

Don't do that quite yet.

Harmonics, Fourier series, et al. Can a periodic function $f(x_1, x_2, \ldots, x_d)$ be expressed as a Fourier series using multidimensional complex exponentials? The answer is yes and the formulas and theorems are virtually identical to those of the one-dimensional case. First of all, the natural setting is $L^2([0,1]^d)$. This is the space of square integrable functions:

$$\int_{[0,1]^d} |f(\underline{x})|^2 \, d\underline{x} < \infty\,;$$

e.g., in the case $d = 2$ the condition is

$$\int_0^1 \int_0^1 |f(x_1, x_2)|^2 \, dx_1 \, dx_2 < \infty\,.$$

The inner product of two (complex-valued) functions is

$$(f, g) = \int_{[0,1]^d} f(\underline{x})\overline{g(\underline{x})} \, d\underline{x}\,.$$

Orthogonality is defined as before: f and g are orthogonal if $(f, g) = 0$.

I'm not going to relive the greatest hits of Fourier series in the higher-dimensional setting. The only thing I want us to know now is *what the expansions look like*. It's nice — just watch.

Let's do the two-dimensional case as an illustration. The harmonics are of the form

$$e^{2\pi i n_1 x_1} e^{2\pi i n_2 x_2}\,,$$

where n_1 and n_2 are integers. Reflecting the periodicity in each variable, we would then imagine writing the Fourier series expansion as

$$\sum_{n_1, n_2} c_{n_1 n_2} e^{2\pi i n_1 x_1} e^{2\pi i n_2 x_2}\,,$$

where the sum is over all integers n_1, n_2. More on the coefficients in a minute, but first let's find a more attractive way of writing such sums.

Instead of working with the product of separate exponentials, it's *now* time to combine them and see what happens:

$$\begin{aligned} e^{2\pi i n_1 x_1} e^{2\pi i n_2 x_2} &= e^{2\pi i (n_1 x_1 + n_2 x_2)} \\ &= e^{2\pi i \, \underline{n} \cdot \underline{x}} \quad \text{(dot product in the exponent!)}, \end{aligned}$$

where we use vector notation and write $\underline{n} = (n_1, n_2)$. The Fourier series expansion then looks like

$$\sum_{\underline{n}} c_{\underline{n}} e^{2\pi i \underline{n} \cdot \underline{x}}\,.$$

The dot product in two dimensions has replaced ordinary multiplication in the exponent in one dimension, but the formula *looks the same*.

The sum has to be understood to be over all points (n_1, n_2) with integer coefficients. This set of points in $\mathbb{R}^2$ is called the two-dimensional *integer lattice*, written $\mathbb{Z}^2$, and we'll come back to it in a later section. What are the coefficients?

The argument we gave in one dimension extends easily to two dimensions (and more) and one finds that the coefficients are given by

$$\begin{aligned} c_{\underline{n}} &= \int_0^1 \int_0^1 e^{-2\pi i n_1 x_1} e^{-2\pi i n_2 x_2} f(x_1, x_2)\, dx_1\, dx_2 \\ &= \int_0^1 \int_0^1 e^{-2\pi i (n_1 x_1 + n_2 x_2)} f(x_1, x_2)\, dx_1\, dx_2 = \int_{[0,1]^2} e^{-2\pi i\, \underline{n}\cdot\underline{x}} f(\underline{x})\, d\underline{x}\,. \end{aligned}$$

Switching to d dimensions, we thus introduce the Fourier coefficients $\hat{f}(\underline{n})$ defined by the integral

$$\hat{f}(\underline{n}) = \int_{[0,1]^d} e^{-2\pi i\, \underline{n}\cdot\underline{x}} f(\underline{x})\, d\underline{x}\,.$$

Owing to periodicity, the Fourier coefficients can be computed by integrating over any d-dimensional cube of side length 1.

It should now come as no shock that the Fourier series for a periodic function $f(\underline{x})$ in $\mathbb{R}^d$ is

$$\sum_{\underline{n}} \hat{f}(\underline{n}) e^{2\pi i\, \underline{n}\cdot\underline{x}},$$

where the sum is over all points $\underline{n} = (n_1, n_2, \ldots, n_d)$ with integer entries. This set of points is the integer lattice in $\mathbb{R}^d$, written $\mathbb{Z}^d$.

The role of orthogonality is as before. The complex exponentials $e^{2\pi i\, \underline{n}\cdot\underline{x}}$ are orthonormal and the Fourier coefficient $\hat{f}(\underline{n})$ is the projection of f in the direction $e^{2\pi i\, \underline{n}\cdot\underline{x}}$. Visualize that? I don't think so.

I am also pleased to report that Rayleigh's identity continues to hold:

$$\int_{[0,1]^d} |f(\underline{x})|^2\, d\underline{x} = \sum_{\underline{n}} |\hat{f}(\underline{n})|^2\,.$$

That's it. We won't talk about convergence, Dirichet kernels, and so on. Instead, coming up next is an *extremely* interesting and important example of higher-dimensional Fourier series *in action*. Later we'll come back to higher-dimensional Fourier series in studying the higher-dimensional Ш, on the way to applications to crystallography and to sampling.

9.4.1. The eternal recurrence of the same, with apologies to F. Nietzsche. You've likely heard the term *random walk*. Here's the setup in one dimension, a random walk along a line:

- You're at home at the origin at time $n = 0$ and you take a step, left or right chosen with equal probability. Flip a coin — heads you move right, tails you move left. So at time $n = 1$ you're at one of the points $+1$ or -1. Again you take a step, left or right, chosen with equal probability. You're either back home at the origin or at ± 2. And so on. Keep flipping; keep stepping. Here's the question at the heart of the subject:
  - As you take more and more steps, will you get home (to the origin)?
  - If so, how often?

Perfectly natural questions, and we can formulate the same questions in two, three, or any number of dimensions, allowing for additional degrees of freedom in which directions to move. We can also tinker with the probabilities and assume that steps in some directions are more probable than in others, but we'll stick with the equally probable case.

The terms associated with answering the basic questions are *recurrent* (yes, you'll return to the origin) and *transient* (no, you won't). More later. The lesson of this section is that, surprisingly, recurrence or transience depends on the dimension.

Random walks are related to Brownian motion and can also be described as a Markov process. There are *many* areas of applications, queuing problems, to cite one example, and recently they have been applied in mathematical finance. A really interesting treatment is the book *Random Walks and Electrical Networks* by P. Doyle and J. L. Snell. We won't go into any of this, but if — or rather, *when* — you encounter these ideas in other courses, you have been so advised.

We need to make some use of notions from probability, but nothing beyond what we used in discussing the Central Limit Theorem in Chapter 3, and here we'll be working in the discrete setting. For this excursion and for your safe return, if you do return, you will need the following:

- The notion of a random variable, its expected value (which we also called the *mean*), and some of the properties of the expected value.[17]
- To remember what probability means.
- To know that for independent events the probabilities multiply, i.e.,

$$\operatorname{Prob}(A, B) = \operatorname{Prob}(A) \operatorname{Prob}(B)\,,$$

  meaning that the probability of A *and* B occurring (together) is the product of the separate probabilities of A and B occurring.

Think first of the space case $d = 3$ as an example. Start at the origin and start stepping. Each step is by a unit amount in one of six possible directions, and the directions are chosen with equal probability; e.g., throw a single die and have each number correspond to one of six directions. Wherever you go, you get there by adding to wherever you are one of the six unit steps $(\pm 1, 0, 0)$, $(0, \pm 1, 0)$, $(0, 0, \pm 1)$.

More generally, in d dimensions your steps are

$$(\pm 1, 0, \dots, 0), (0, \pm 1, 0, \dots, 0), \dots, (0, 0, \dots, \pm 1)\,.$$

Denote any of these "elementary" steps by $\underline{s}$ and the random process of choosing any of these steps by $\underline{S}$. So to repeat, to take a step is to choose one of the elementary steps and each choice is made with probability $1/d$. $\underline{S}$ is a random variable — it's a step you haven't taken yet — with values $\underline{s}$.

[17]Recall Sam Savage's definition of a random variable as a number you don't know yet.

Since we're interested in walks more than we are in individual steps, let's add an index to $\underline{S}$ and write $\underline{S}_1$ for the choice in taking the first step, $\underline{S}_2$ for the choice in taking the second step, and so on. We assume that each step is a new adventure — the choice at the nth step is made independently of the previous $n-1$ steps.

Then, again, the process $\underline{S}_n$ is a discrete random variable.

- The domain of $\underline{S}_n$ is the set of all possible walks and the value of $\underline{S}_n$ on a particular walk is the nth step in that walk.

(Many people would call $\underline{S}_n$ a *random vector* since its values are d-tuples.)

We're assuming that the distribution of values of $\underline{S}_n$ is uniform (each particular step is taken with probability $1/2d$, in dimension d) and that the steps are independent. In the language we've used in connection with the Central Limit Theorem, $\underline{S}_1, \underline{S}_2, \ldots, \underline{S}_n$ are independent, identically distributed random variables.[18]

- The possible random walks of n steps are described exactly as

$$\underline{W}_n = \underline{S}_1 + \underline{S}_2 + \cdots + \underline{S}_n.$$

A particular walk is

$$\underline{w}_n = \underline{s}_1 + \underline{s}_2 + \cdots + \underline{s}_n\,.$$

$\underline{W}_n$ is a random variable (random vector) — it's a walk you haven't taken yet.

Here's a picture of a random walk in $\mathbb{R}^3$.

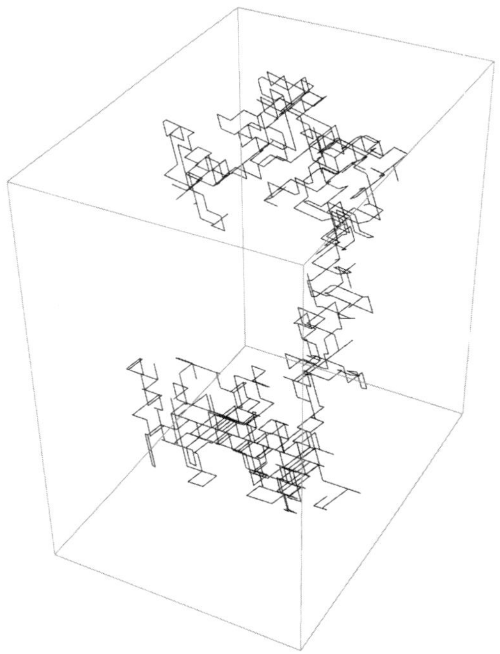

[18]We won't be appealing to the CLT; it's just that here you see again the same assumptions.

After a walk of n steps, $n \geq 1$, you are at a lattice point in $\mathbb{R}^d$, i.e., a point with integer coordinates.

In a famous paper in 1921 that pretty much founded the subject, the mathematician G. Pólya[19] proved the following:

Theorem. In dimensions 1 and 2, with probability 1, the walker visits the origin infinitely often; in symbols

$$\text{Prob}(\underline{W}_n = \underline{0} \text{ infinitely often}) = 1\,.$$

In dimensions ≥ 3, with probability 1, the walker escapes to infinity:

$$\text{Prob}\left(\lim_{n\to\infty} \|\underline{W}_n\| = \infty\right) = 1\,.$$

One says that a random walk along a line or in the plane is *recurrent* and that a random walk in higher dimensions is *transient*. The statements on probabilities are the precise definitions of those terms, once again turning the solution of a problem into a definition.

To get started, let V_n be the random variable that is 1 if the walker returns to the origin in n steps and zero otherwise. The expected value of V_n is then $\text{Prob}(\underline{W}_n = \underline{0}) \cdot 1$, the value of the function, 1, times the probability of that value occurring. Now set $V = \sum_{n=0}^{\infty} V_n$. The expected value of V is what we want and it is the sum of the expected values of the V_n; i.e.,

$$\text{expected visits to the origin} = \sum_{n=0}^{\infty} \text{Prob}(\underline{W}_n = \underline{0})\,.$$

How can we get a handle on this? It was Pólya's idea to realize the numbers $\text{Prob}(\underline{W}_n = \underline{0})$ as *Fourier coefficients* in a Fourier series. Of all things.

Here's the key object we'll work with. For each $\underline{x} \in \mathbb{R}^d$, fixed for the time being, consider

$$\Phi_n = e^{2\pi i\, \underline{W}_n \cdot \underline{x}}$$

as a function of n. For each n the possible values of $\underline{W}_n$, as a sum of n steps corresponding to different walks, lie among the lattice points, and if a particular walk $\underline{w}_n$ lands on a particular lattice point $\underline{l}$, then the value of Φ_n for that walk is $e^{2\pi i\, \underline{l}\cdot\underline{x}}$. What is the expected value of Φ_n over all walks of n steps? It is the mean, i.e., the weighted average of the values of Φ_n over the possible random walks of n steps, each value weighted by the probability of its occurrence. That is,

$$\text{expected value of } \Phi_n = \sum_{\underline{l}} \text{Prob}(\underline{W}_n = \underline{l}) e^{2\pi i\, \underline{l}\cdot\underline{x}}\,.$$

This is actually a finite sum because in n steps we can have reached only a finite number of lattice points. Put another way, $\text{Prob}(\underline{W}_n = \underline{l})$ is zero for all but finitely many lattice points $\underline{l}$.

---

[19] Pólya was born in Hungary and emigrated to the US after WWII. He worked in a great many areas of mathematics and was well known for both his theorems and his philosophy of theorem proving and problem solving.

From this expression you can see (finite) Fourier series coming into the picture, but put that off for the moment.[20] We can compute this expected value based on our assumption that steps are equally probable and independent of each other. First of all, we can write

$$\Phi_n = e^{2\pi i \underline{W}_n \cdot \underline{x}} = e^{2\pi i(\underline{S}_1+\underline{S}_2+\cdots+\underline{S}_n)\cdot \underline{x}} = e^{2\pi i \underline{S}_1 \cdot \underline{x}} e^{2\pi i \underline{S}_2 \cdot \underline{x}} \cdots e^{2\pi i \underline{S}_n \cdot \underline{x}}.$$

So we want to find the expected value of the product of exponentials. At this point we could appeal to a standard result in probability, stating that the expected value of the product of *independent* random variables is the product of their expected values. We can also think about this directly, however. The expected value of $e^{2\pi i \underline{S}_1 \cdot \underline{x}} e^{2\pi i \underline{S}_2 \cdot \underline{x}} \cdots e^{2\pi i \underline{S}_n \cdot \underline{x}}$ is, as above, the weighted average of the values that the function assumes, weighted by the probabilities of those values occurring. In this case we'd be summing over all steps $\underline{s}_1, \underline{s}_2, \ldots, \underline{s}_n$ of the values $e^{2\pi i \underline{s}_1 \cdot \underline{x}} e^{2\pi i \underline{s}_2 \cdot \underline{x}} \cdots e^{2\pi i \underline{s}_n \cdot \underline{x}}$ weighted by the appropriate probabilities. But now the fact that the steps are independent means

$$\begin{aligned}\mathrm{Prob}(&\underline{S}_1 = \underline{s}_1, \underline{S}_2 = \underline{s}_2, \ldots, \underline{S}_n = \underline{s}_n) \\ &= \mathrm{Prob}(\underline{S}_1 = \underline{s}_1)\,\mathrm{Prob}(\underline{S}_2 = \underline{s}_2)\cdots \mathrm{Prob}(\underline{S}_n = \underline{s}_n) = \frac{1}{(2d)^n} \\ &\text{(probabilities } \textit{multiply} \text{ for independent events)},\end{aligned}$$

and then,

$$\begin{aligned}&\text{expected value of } \Phi_n = \text{expected value of } e^{2\pi i \underline{S}_1 \cdot \underline{x}} e^{2\pi i \underline{S}_2 \cdot \underline{x}} \cdots e^{2\pi i \underline{S}_n \cdot \underline{x}} \\ &= \sum_{\underline{s}_1}\sum_{\underline{s}_2}\cdots\sum_{\underline{s}_n} \mathrm{Prob}(\underline{S}_1 = \underline{s}_1, \underline{S}_2 = \underline{s}_2, \ldots, \underline{S}_n = \underline{s}_n) e^{2\pi i \underline{s}_1 \cdot \underline{x}} e^{2\pi i \underline{s}_2 \cdot \underline{x}} \cdots e^{2\pi i \underline{s}_n \cdot \underline{x}} \\ &= \sum_{\underline{s}_1}\sum_{\underline{s}_2}\cdots\sum_{\underline{s}_n} \frac{1}{(2d)^n} e^{2\pi i \underline{s}_1 \cdot \underline{x}} e^{2\pi i s_2 \cdot \underline{x}} \cdots e^{2\pi i \underline{s}_n \cdot \underline{x}}.\end{aligned}$$

The sums go over all possible choices of $\underline{s}_1$, $\underline{s}_2, \ldots,$ $\underline{s}_n$. Now, these sums are uncoupled, and so the nested sum is the product

$$\sum_{\underline{s}_1} \frac{1}{2d} e^{2\pi i \underline{s}_1 \cdot \underline{x}} \sum_{\underline{s}_2} \frac{1}{2d} e^{2\pi i \underline{s}_2 \cdot \underline{x}} \cdots \sum_{\underline{s}_n} \frac{1}{2d} e^{2\pi i \underline{s}_n \cdot \underline{x}}.$$

But the sums are, respectively, the expected values of $e^{2\pi i \underline{s}_j \cdot \underline{x}}$, $j = 1, \ldots, n$, and *these expected values are all the same.* (The steps are independent and identically distributed.) So all the sums are equal, say, to the first sum, and we may write

$$\text{expected value of } \Phi_n = \left(\frac{1}{2d}\sum_{\underline{s}_1} e^{2\pi i \underline{s}_1 \cdot \underline{x}}\right)^n.$$

A further simplification is possible. The first step $\underline{s}_1$, as a d-tuple, has exactly one slot with a ± 1 and the rest 0's. Summing over these $2d$ possibilities allows us

[20] Also, though it's not in the standard form, i.e., a power series, I think of Pólya's idea here as writing down a *generating function* for the sequence of probabilities $\mathrm{Prob}(\underline{W}_n = \underline{l})$. For an appreciation of this kind of approach to a great variety of problems, pure and applied, see the book *generatingfunctionology* by H. Wilf. The first sentence of Chapter One reads: "A generating function is a clothesline on which we hang up a sequence of numbers for display." Seems pretty apt for the problem at hand.

to combine positive and negative terms, as follows. Check the case $d = 2$, for which the choices of s_1 are

$$(1,0)\,,\quad (-1,0)\,,\quad (0,1)\,,\quad (0,-1)\,.$$

This leads to a sum with four terms:

$$\begin{aligned}\sum_{\underline{s}_1} \frac{1}{2\cdot 2} e^{2\pi i\,\underline{s}_1\cdot\underline{x}} &= \sum_{\underline{s}_1} \frac{1}{2\cdot 2} e^{2\pi i\,\underline{s}_1\cdot(x_1,x_2)} \\ &= \tfrac{1}{2}\big(\tfrac{1}{2}e^{2\pi i x_1} + \tfrac{1}{2}e^{-2\pi i x_1} + \tfrac{1}{2}e^{2\pi i x_2} + \tfrac{1}{2}e^{-2\pi i x_2}\big) \\ &= \tfrac{1}{2}(\cos 2\pi x_1 + \cos 2\pi x_2).\end{aligned}$$

The same thing happens in dimension d, and our final formula is

$$\sum_{\underline{l}} \operatorname{Prob}(\underline{W}_n = \underline{l}) e^{2\pi i\,\underline{l}\cdot\underline{x}} = \left(\frac{1}{d}(\cos 2\pi x_1 + \cos 2\pi x_2 + \cdots + \cos 2\pi x_d)\right)^n .$$

Let us write

$$\phi_d(\underline{x}) = \frac{1}{d}(\cos 2\pi x_1 + \cos 2\pi x_2 + \cdots + \cos 2\pi x_d)\,.$$

Observe that $|\phi_d(\underline{x})| \le 1$, since $\phi_d(\underline{x})$ is the sum of d cosines divided by d, and $|\cos 2\pi x_j| \le 1$ for $j = 1, 2, \ldots, d$.

Back to Fourier series. Consider the sum of probabilities times exponentials, above, *as a function of* $\underline{x}$; i.e., let

$$f(\underline{x}) = \sum_{\underline{l}} \operatorname{Prob}(\underline{W}_n = \underline{l})\, e^{2\pi i\,\underline{l}\cdot\underline{x}}\,.$$

This is a (finite) Fourier series for $f(\underline{x})$ and therefore its coefficients *must be the Fourier coefficients*,

$$\operatorname{Prob}(\underline{W}_n = \underline{l}) = \hat{f}(\underline{l})\,.$$

But according to our calculation, $f(\underline{x}) = \phi_d(\underline{x})^n$, and so this must also be the Fourier coefficient of $\phi_d(\underline{x})^n$; that is,

$$\operatorname{Prob}(\underline{W}_n = \underline{l}) = \hat{f}(\underline{l}) = \widehat{(\phi_d)^n}(\underline{l}) = \int_{[0,1]^d} e^{-2\pi i\,\underline{l}\cdot\underline{x}} \phi_d(\underline{x})^n \, d\underline{x}\,.$$

In particular, the probability that the walker visits the origin, $\underline{l} = \underline{0}$, in n steps is

$$\operatorname{Prob}(\underline{W}_n = 0) = \int_{[0,1]^d} \phi_d(\underline{x})^n \, d\underline{x}\,.$$

We can put this together with $\sum_{n=0}^{\infty} \operatorname{Prob}(\underline{W}_n = \underline{0})$. We'd like to say that

$$\begin{aligned}\sum_{n=0}^{\infty} \operatorname{Prob}(\underline{W}_n = \underline{0}) &= \sum_{n=0}^{\infty} \int_{[0,1]^d} \phi_d(\underline{x})^n \, d\underline{x} \\ &= \int_{[0,1]^d} \left(\sum_{n=0}^{\infty} \phi_d(x)^n\right) d\underline{x} = \int_{[0,1]^d} \frac{1}{1-\phi_d(\underline{x})}\, d\underline{x}\end{aligned}$$

using the formula for the sum of a geometric series. The final answer is correct, but the derivation isn't quite legitimate. The formula for the sum of a geometric series is

$$\sum_{n=0}^{\infty} r^n = \frac{1}{1-r}$$

provided that $|r|$ is strictly less than 1. In our application we know only that $|\phi_d(\underline{x})| \leq 1$. To get around this difficulty let $\alpha < 1$ and write

$$\begin{aligned}\sum_{n=0}^{\infty} \operatorname{Prob}(\underline{W}_n = \underline{0}) &= \lim_{\alpha \to 1} \sum_{n=0}^{\infty} \alpha^n \operatorname{Prob}(\underline{W}_n = \underline{0}) = \lim_{\alpha \to 1} \int_{[0,1]^d} \Big(\sum_{n=0}^{\infty} \alpha^n \phi_d(x)^n \Big) \, d\underline{x} \\ &= \lim_{\alpha \to 1} \int_{[0,1]^d} \frac{1}{1 - \alpha\phi_d(\underline{x})} \, d\underline{x} = \int_{[0,1]^d} \frac{1}{1-\phi_d(\underline{x})} \, d\underline{x} \,.\end{aligned}$$

Pulling the limit inside the integral is justified by the Lebesgue dominated convergence theorem (Section 4.3). Not to worry.

- The crucial question now concerns the integral

$$\int_{[0,1]^d} \frac{1}{1-\phi_d(\underline{x})} \, d\underline{x} \,.$$

Is it finite or infinite? *This depends on the dimension of the space where the walk takes place.* This is exactly where the dimension d enters the picture.

Using some calculus (think Taylor series) it is not difficult to show (I won't) that if $|\underline{x}|$ is small, then

$$1 - \phi_d(\underline{x}) \sim c|\underline{x}|^2 \,,$$

for a constant c. Thus

$$\frac{1}{1-\phi_d(\underline{x})} \sim \frac{1}{c|\underline{x}|^2} \,,$$

and the convergence of the integral we're interested in depends on that of the power integral

$$\int_{\underline{x} \text{ small}} \frac{1}{|\underline{x}|^2} \, d\underline{x} \quad \text{in dimension } d \,.$$

It is an important mathematical fact of nature (something you should file away for future use) that:

- The power integral diverges for $d = 1, 2$.
- The power integral converges for $d \geq 3$.

Let me illustrate why this is so for $d = 1, 2, 3$. For $d = 1$ we have an ordinary improper integral (I mean one with finite limits; the improperness is coming from the denominator being zero),

$$\int_0^a \frac{dx}{x^2}, \quad \text{for some small } a > 0 \,.$$

This diverges, by direct integration. For $d = 2$ we have a double integral, and to check its properties we introduce polar coordinates (r, θ) and write

$$\int_{|\underline{x}| \text{ small}} \frac{dx_1 \, dx_2}{x_1^2 + x_2^2} = \int_0^{2\pi} \int_0^a \frac{r \, dr \, d\theta}{r^2} = \int_0^{2\pi} \left(\int_0^a \frac{dr}{r} \right) d\theta \,.$$

The inner integral diverges. In three dimensions we introduce spherical coordinates (ρ, θ, φ), and something different happens. The integral becomes

$$\int_{|\underline{x}| \text{ small}} \frac{dx_1\, dx_2\, dx_3}{x_1^2 + x_2^2 + x_3^2} = \int_0^{\pi}\int_0^{2\pi}\int_0^{a} \frac{\rho^2 \sin\phi\, d\rho\, d\theta\, d\varphi}{\rho^2}.$$

This time the ρ^2 in the denominator cancels with the ρ^2 in the numerator and the ρ-integral is *finite*. This phenomenon — convergence of the integral — persists in higher dimensions, for the same reason (introducing higher-dimensional spherical coordinates).

Let's take stock. We have shown that

$$\text{expected number of visits to the origin} = \sum_{n=0}^{\infty} \text{Prob}(\underline{W}_n = \underline{0}) = \int_{[0,1]^d} \frac{1}{1 - \phi_d(\underline{x})}\, d\underline{x}$$

and that this number is infinite in dimensions 1 and 2 and finite in dimension 3. From here we can go on to prove Pólya's theorem as he stated it:

$\text{Prob}(\underline{W}_n = \underline{0} \text{ infinitely often}) = 1$ in dimensions 1 and 2.

$\text{Prob}(\lim_{n\to\infty} \|\underline{W}_n\| = \infty) = 1$ in dimensions ≥ 3.

For the case $d \geq 3$, we know that the expected number of times that the walker visits the origin is finite. This can only be true if the actual number of visits to the origin is finite with probability 1. But the origin is no different from any other lattice point, so this must hold for any lattice point. In turn, this means that for any $K > 0$ the walker eventually stops visiting the ball $\|\underline{x}\| \leq K$ of radius K with probability 1, and this is exactly saying that $\text{Prob}(\lim_{n\to\infty} \|\underline{W}_n\| = \infty) = 1$.

To settle the case $d \leq 2$ we formulate a lemma. We haven't had many lemmas in this book, but I think I can get away with one or two.

> **Lemma.** Let p_n be the probability that a walker visits the origin *at least* n times and let q_n be the probability that a walker visits the origin *exactly* n times. Then $p_n = p_1^n$ and $q_n = p_1^n(1 - p_1)$.

We argue as follows: note first that $p_0 = 1$ since the walker starts at the origin. Then

$$\begin{aligned}
p_{n+1} &= \text{Prob(visit origin at least } n+1 \text{ times)} \\
&= \text{Prob(visit origin at least } n+1 \text{ times given visit at least } n \text{ times)} \\
&\qquad \times \text{Prob(visit at least } n \text{ times)} \\
&= \text{Prob(visit origin at least 1 time given visit at least 0 times)} \cdot p_n \\
&\qquad \text{(using independence and the definition of } p_n) \\
&= \text{Prob(visit at least 1 time)} \cdot p_n \\
&= p_1 \cdot p_n\,.
\end{aligned}$$

From $p_0 = 1$ and $p_{n+1} = p_1 \cdot p_n$ it follows (by induction) that $p_n = p_1^n$.

For the second part,

$$\begin{aligned} q_n &= \text{Prob(exactly } n \text{ visits to origin)} \\ &= \text{Prob(visits at least } n \text{ times)} - \text{Prob(visits at least } n+1 \text{ times)} \\ &= p_n - p_{n+1} = p_1^n(1-p_1). \end{aligned}$$

Now, if p_1 were less than 1, then the expected number of visits to the origin would be

$$\begin{aligned} \sum_{n=0}^{\infty} nq_n &= \sum_{n=0}^{\infty} np_1^n(1-p_1) = (1-p_1)\sum_{n=0}^{\infty} np_1^n \\ &= (1-p_1)\frac{p_1}{(1-p_1)^2} \\ &\qquad \left(\text{check the identity for the sums by differentiating } \frac{1}{1-x} = \sum_{n=0}^{\infty} x^n\right) \\ &= \frac{p_1}{1-p_1} < \infty\,. \end{aligned}$$

But this contradicts the fact we established earlier, namely

$$\text{expected visits to the origin} = \int_{[0,1]^2} \frac{1}{1-\phi_2(\underline{x})}\, d\underline{x} = \infty\,.$$

Thus we must have $p_1 = 1$, that is, the probability of returning to the origin is 1, and hence $\underline{W}_n$ must equal 0 infinitely often with probability 1.

We are done walking. I hope you enjoyed the trip.

There are other proofs of Pólya's theorem and there are *many* generalizations. I encourage you to do some poking around. There are also explanations/motivations for his idea of using Fourier series, but on this point I still think his approach is another example of *genius is as genius does.*

An anecdote. I arrived at Stanford in 1985 just after Pólya passed away at the age of 97. His friend, colleague, and collaborator M. Schiffer offered a very moving remembrance. He said that late in life Pólya liked to reread some of his papers, especially some of his early papers. Not out of pride, but to experience again the joy of discovery. What discovery. What joy.

9.5. III, Lattices, Crystals, and Sampling

That's a lot of topics for one section, but it all hangs together. Let's start with what we're aiming for, a generalization of the sampling theorem.

Our derivation of the sampling formula in Chapter 6 was a direct application of the important properties of

$$\text{Ш}_p(t) = \sum_{k=-\infty}^{\infty} \delta(t - kp)\,.$$

Without redoing the whole argument here, short as it is, let me remind you of what those properties are that made things work.

- Ш_p is *the* tool to use for periodizing and for sampling:

$$(f * \text{Ш}_p)(t) = \sum_{k=-\infty}^{\infty} f(t - kp)\,,$$

$$f(t)\text{Ш}_p(t) = \sum_{k=-\infty}^{\infty} f(kp)\delta(t - kp)\,.$$

- For the Fourier transform,

$$\mathcal{F}\text{Ш}_p = \frac{1}{p}\,\text{Ш}_{1/p}\,.$$

It is through the result on the Fourier transform that periodizing in one domain corresponds to sampling in the other domain. Pay particular attention here to the reciprocity in spacing between Ш_p and its Fourier transform.

The sampling formula itself says that if $\mathcal{F}f(s)$ is identically 0 for $|s| \geq p/2$, then

$$f(t) = \sum_{k=-\infty}^{\infty} f\left(\frac{k}{p}\right) \operatorname{sinc} p\left(t - \frac{k}{p}\right).$$

Everything we do now will rest on a generalization of Ш and its Fourier transform. There isn't much difference in substance between the two-dimensional case and higher dimensions, so we'll stick pretty much to the plane.

9.5.1. The two-dimensional Ш. The formula $\mathcal{F}\text{Ш}_p = (1/p)\text{Ш}_{1/p}$ depends crucially on the fact that Ш_p is a sum of impulses at evenly spaced points; this is an aspect of periodicity. We've already defined a two-dimensional δ, so to introduce a Ш that goes with it we need to define what "evenly spaced" means for points in $\mathbb{R}^2$. One way of spacing points evenly in $\mathbb{R}^2$ is to take all pairs (k_1, k_2) where k_1 and k_2 integers. The corresponding Ш-function is then defined to be

$$\text{Ш}(x_1, x_2) = \sum_{k_1, k_2=-\infty}^{\infty} \delta(x_1 - k_1, x_2 - k_2)\,.$$

Bracewell, and others, sometimes refer to this, evocatively, whimsically, as the "bed of nails." Pictured.

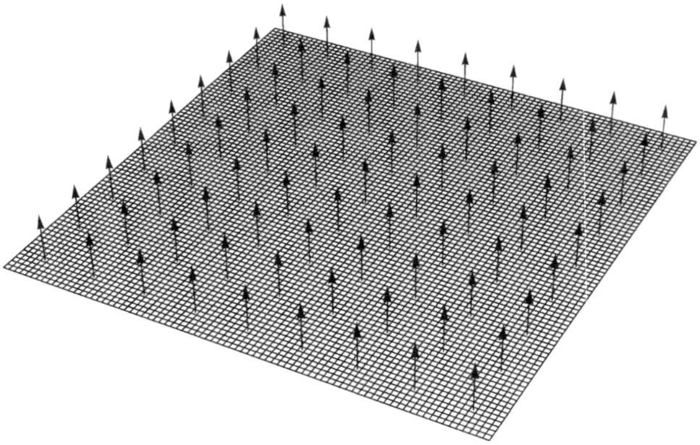

The points $\underline{k} = (k_1, k_2)$ with integer coordinates are said to form a *lattice* in the plane, an idea we met in the earlier section on Fourier series. As before, we denote this particular lattice, the *integer lattice*, by $\mathbb{Z}^2$. We'll have more general lattices in a short while. As a model of a physical system, you can think of such an array as a two-dimensional (infinite) crystal, where there's an atom at every lattice point.

Since we prefer to write things in terms of vectors, another way to describe $\mathbb{Z}^2$ is to use the standard basis of $\mathbb{R}^2$, the vectors $\underline{e}_1 = (1, 0)$, $\underline{e}_2 = (0, 1)$, and write the points in the lattice as

$$\underline{k} = k_1\underline{e}_1 + k_2\underline{e}_2 \,.$$

We can thus think of the elements of a lattice either as points or as vectors. Observe that the sum of two lattice points is another lattice point and that an integer multiple of a lattice point is another lattice point. The Ш-function can be written

$$Ш_{\mathbb{Z}^2}(\underline{x}) = \sum_{k_1,k_2=-\infty}^{\infty} \delta(\underline{x} - k_1\underline{e}_1 - k_2\underline{e}_2) = \sum_{\underline{k}\in\mathbb{Z}^2} \delta(\underline{x} - \underline{k}) \,,$$

the latter formula looking especially like the one-dimensional case. It is easy to show that $Ш_{\mathbb{Z}^2}$ is even.

Periodicity and sampling on $\mathbb{Z}^2$, and $\mathcal{F}Ш_{\mathbb{Z}^2}$, too. As it was in one dimension, $Ш_{\mathbb{Z}^2}$ is *the* tool to use to work with periodicity and sampling. If we form

$$\Phi(\underline{x}) = (\varphi * Ш_{\mathbb{Z}^2})(\underline{x}) = \sum_{\underline{k}\in\mathbb{Z}^2} \varphi(\underline{x} - \underline{k}) \,,$$

assuming that the sum converges, then Φ is periodic *on the lattice* $\mathbb{Z}^2$, or briefly, is $\mathbb{Z}^2$-periodic. This means that

$$\Phi(\underline{x} + \underline{n}) = \Phi(\underline{x})$$

for all $\underline{x}$ and for any lattice point $\underline{n} \in \mathbb{Z}^2$. Sure, that's true:

$$\Phi(\underline{x} + \underline{n}) = \sum_{\underline{k}\in\mathbb{Z}^2} \varphi(\underline{x} + \underline{n} - \underline{k}) = \sum_{\underline{k}'\in\mathbb{Z}^2} \varphi(\underline{x} - \underline{k}') = \Phi(\underline{x}) \,.$$

The sum (or difference) of two lattice points, $\underline{k} - \underline{n} = \underline{k}'$, is a lattice point, so we're still summing over $\mathbb{Z}^2$ and we get back Φ.

Sampling on $\mathbb{Z}^2$ also works as in one dimension. We sample a function $f(\underline{x})$ via multiplication with $Ш_{\mathbb{Z}^2}$:

$$f(\underline{x})Ш_{\mathbb{Z}^2}(\underline{x}) = \sum_{\underline{k}\in\mathbb{Z}^2} f(\underline{k})\delta(\underline{x}-\underline{k}).$$

Finally, periodicity leads to the important and remarkable formula

$$\mathcal{F}Ш_{\mathbb{Z}^2} = Ш_{\mathbb{Z}^2}\,,$$

corresponding precisely to the one-dimensional case. I'll put the details of the derivation of this in Section 9.5.4; it's Poisson summation. It's also true that

$$\mathcal{F}^{-1}Ш_{\mathbb{Z}^2} = Ш_{\mathbb{Z}^2}$$

because $Ш_{\mathbb{Z}^2}$ is even (use duality).

Already the most basic version of the two-dimensional sampling formula is within easy reach. However, it's much more interesting, as well as ultimately much more useful, to allow for some greater generality.

9.5.2. Lattices in general. The integer lattice $\mathbb{Z}^2$ isn't the only example of a set of evenly spaced points in the plane, though as we'll see there are reasons for considering it the most evenly spaced points. It's easy to imagine oblique lattices, too, and we'll want to be able to use such lattices to model crystals. As observed in nature, not all crystals are square, or even rectangular.

We'll consider such oblique arrangements, but be cautioned that the subject of lattices, as simple as it may seem, can go on and on. The effort here is just to get a few facts that we need.

We adopt the vector point of view for defining a general lattice. Take any basis $\underline{u}_1$, $\underline{u}_2$ of $\mathbb{R}^2$ and consider all the points (or vectors) that are *integer* linear combinations of $\underline{u}_1$ and $\underline{u}_2$. These constitute

$$\text{lattice points} = \underline{p} = p_1\underline{u}_1 + p_2\underline{u}_2, \quad p_1, p_2 = 0, \pm1, \pm2, \ldots.$$

We'll denote such a lattice by $\mathcal{L}$. The sum and difference of two lattice points is again a lattice point, as is any integer times a lattice point.[21]

The vectors $\underline{u}_1$ and $\underline{u}_2$ are said to be a *basis* for the lattice. Other vectors can also serve as a basis, and two bases for the same lattice are related by a 2×2 matrix with integer entries having determinant 1. (This is one of those interesting facts about lattices that we won't cover.) The parallelogram determined by the basis vectors (any basis vectors) is called a *fundamental parallelogram* for the lattice, or, in crystallographers' terms, a *unit cell*. A fundamental parallelogram for $\mathbb{Z}^2$ is the square $0 \le x_1 < 1$, $0 \le x_2 < 1$.[22]

[21] In mathematical terminology a lattice is a *module* over $\mathbb{Z}$; a module is like a vector space except that you can't divide by the scalars (the integers in this case) only add them and multiply them. For a module, as opposed to a vector space, the scalars form a ring, not a field.

[22] It's a common convention to define a fundamental parallelogram to be "half open", including two sides ($x_1 = 0$ and $x_2 = 0$ in this case) and excluding two ($x_1 = 1$ and $x_2 = 1$). This won't be an issue for our work.

By convention, one speaks of the *area of a lattice* in terms of the area of a fundamental parallelogram for the lattice, and, by convention, writing

$$\text{Area}(\mathcal{L}) = \text{area of a fundamental parallelogram.}$$

Two fundamental parallelograms *for the same lattice* have the same area since the bases are related by a 2×2 integer matrix with determinant 1, and the area scales by the determinant.

If we take the natural basis vectors $\underline{e}_1 = (1, 0)$ and $\underline{e}_2 = (0, 1)$ for $\mathbb{R}^2$, we get the integer lattice $\mathbb{Z}^2$. Then *any* lattice $\mathcal{L}$ can be *obtained* from $\mathbb{Z}^2$ via an invertible linear transformation A, the one that takes $\underline{e}_1$ and $\underline{e}_2$ to a basis $\underline{u}_1 = A\underline{e}_1$ and $\underline{u}_2 = A\underline{e}_2$ that defines $\mathcal{L}$. This is so precisely because A is linear: if

$$\underline{p} = p_1\underline{u}_1 + p_2\underline{u}_2, \quad p_1, p_2 \text{ integers},$$

is any point in $\mathcal{L}$, then

$$\underline{p} = p_1(A\underline{e}_1) + p_2(A\underline{e}_2) = A(p_1\underline{e}_1 + p_2\underline{e}_2),$$

showing that $\underline{p}$ is the image of the point $p_1\underline{e}_1 + p_2\underline{e}_2$ in $\mathbb{Z}^2$. We write

$$\mathcal{L} = A(\mathbb{Z}^2).$$

A fundamental parallelogram for $\mathcal{L}$ is determined by $\underline{u}_1$ and $\underline{u}_2$, and so

$$\text{Area}(\mathcal{L}) = \text{area of the parallelogram determined by } \underline{u}_1 \text{ and } \underline{u}_2 = |\det A|.$$

Below, for example, is the lattice $\mathcal{L}$ obtained from $\mathbb{Z}^2$ by applying

$$A = \begin{pmatrix} 3 & -1 \\ 1 & 2 \end{pmatrix}.$$

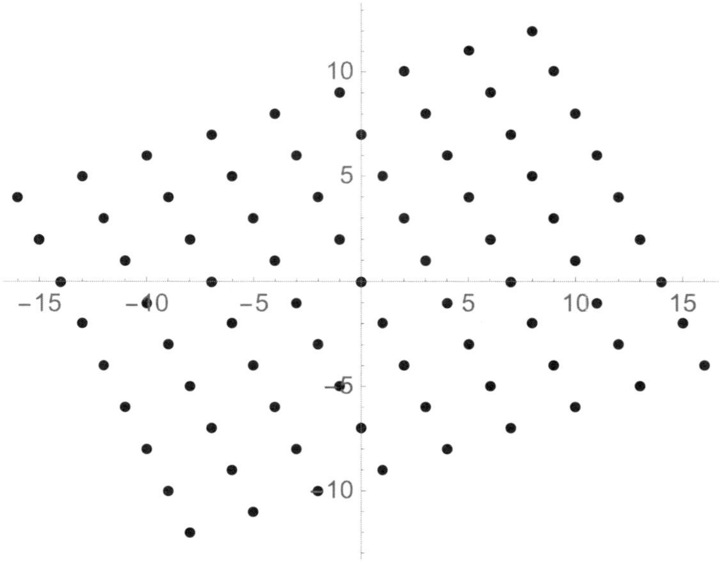

A basis for $\mathcal{L}$ is $\underline{u}_1 = (3, 1) = A\underline{e}_1$, $\underline{u}_2 = (-1, 2) = A\underline{e}_2$ (Draw the basis on the lattice!) The area of $\mathcal{L}$ is 7.

9.5.3. Ш for a lattice. It doesn't take a great leap in imagination to introduce Ш for a general lattice. If $\mathcal{L}$ is a lattice in $\mathbb{R}^2$, then the Ш-function associated with $\mathcal{L}$ is

$$Ш_{\mathcal{L}}(\underline{x}) = \sum_{\underline{p}\in\mathcal{L}} \delta(\underline{x} - \underline{p}) .$$

There's your general sum of δ-functions at evenly spaced points. We could also write the definition as

$$Ш_{\mathcal{L}}(\underline{x}) = \sum_{k_1,k_2=-\infty}^{\infty} \delta(\underline{x} - k_1\underline{u}_1 - k_2\underline{u}_2) .$$

You can check that $Ш_{\mathcal{L}}$ is even.

Periodizing and sampling. Periodizing with $Ш_{\mathcal{L}}$ via convolution results in a function that is periodic with respect to $\mathcal{L}$. If

$$\Phi(\underline{x}) = (\varphi * Ш_{\mathcal{L}})(\underline{x}) = \sum_{\underline{p}\in\mathcal{L}} \varphi(\underline{x} - \underline{p}),$$

then

$$\Phi(\underline{x} + \underline{p}) = \Phi(\underline{x})$$

for all $\underline{x} \in \mathbb{R}^2$ and all $\underline{p} \in \mathcal{L}$. Another way of saying this is that Φ has two independent periods, one each in the directions of any pair of basis vectors for the lattice. Thus if $\underline{u}_1$, $\underline{u}_2$ are a basis for $\mathcal{L}$, then

$$\Phi(\underline{x} + k_1\underline{u}_1) = \Phi(\underline{x}) \quad \text{and} \quad \Phi(\underline{x} + k_2\underline{u}_2) = \Phi(\underline{x}), \quad k_1, k_2 \text{ any integers.}$$

$Ш_{\mathcal{L}}$ is also the tool to use for sampling on a lattice, for

$$\varphi(\underline{x})Ш_{\mathcal{L}}(\underline{x}) = \sum_{\underline{p}\in\mathcal{L}} \varphi(\underline{p})\delta(\underline{x} - \underline{p}) .$$

Relating $Ш_{\mathcal{L}}$ and $Ш_{\mathbb{Z}^2}$. As $\mathcal{L}$ can be obtained from $\mathbb{Z}^2$ via a linear transformation, so too can $Ш_{\mathcal{L}}$ be expressed in terms of $Ш_{\mathbb{Z}^2}$. If $\mathcal{L} = A(\mathbb{Z}^2)$, then

$$Ш_{\mathcal{L}}(\underline{x}) = \sum_{\underline{p}\in\mathcal{L}} \delta(\underline{x} - \underline{p}) = \sum_{\underline{k}\in\mathbb{Z}^2} \delta(\underline{x} - A\underline{k}) .$$

Next, using the formula for $\delta(A\underline{x})$ from Section 9.3.4,

$$\delta(\underline{x} - A\underline{k}) = \delta(A(A^{-1}\underline{x} - \underline{k})) = \frac{1}{|\det A|}\delta(A^{-1}\underline{x} - \underline{k}) .$$

Therefore

$$Ш_{\mathcal{L}}(\underline{x}) = \frac{1}{|\det A|}Ш_{\mathbb{Z}^2}(A^{-1}\underline{x}) .$$

Incidentally, the evenness of $Ш_{\mathcal{L}}$ follows from the evenness of $Ш_{\mathbb{Z}^2}$ via this expression. Nice to have these little consistency checks.

We'll use this formula in just a moment. Compare it to our earlier formulas in Section 5.4.2 on how the one-dimensional Ш-function scales. Using the one-dimensional linear transformation $A(x) = px$, we have, just as above,

$$\text{Ш}_p(x) = \sum_{k=-\infty}^{\infty} \delta(x-kp) = \sum_{k=-\infty}^{\infty} \delta(x-A(k)) = \frac{1}{|\det A|} \sum_{n=-\infty}^{\infty} \delta(A^{-1}x-k) = \frac{1}{p}\text{Ш}(x/p)\,.$$

Dual lattices and $\mathcal{F}\text{Ш}_{\mathcal{L}}$. Of the many (additional) interesting things to say about lattices, the one that's most important for our concerns is how the Fourier transform of $\text{Ш}_{\mathcal{L}}$ depends on $\mathcal{L}$. This question leads to a fascinating phenomenon, one that is realized physically in X-ray diffraction images of crystals.

We mentioned earlier (derivation still forthcoming!) that for the integer lattice we have

$$\mathcal{F}\text{Ш}_{\mathbb{Z}^2} = \text{Ш}_{\mathbb{Z}^2}\,.$$

What about the Fourier transform of $\text{Ш}_{\mathcal{L}}$? We appeal to the general similarity theorem to obtain, for $\mathcal{L} = A(\mathbb{Z}^2)$,

$$\begin{aligned}
\mathcal{F}\text{Ш}_{\mathcal{L}}(\underline{\xi}) &= \frac{1}{|\det A|}\mathcal{F}(\text{Ш}_{\mathbb{Z}^2}(A^{-1}\underline{x})) \\
&= \frac{1}{|\det A|}\frac{1}{|\det A^{-1}|}\mathcal{F}\text{Ш}_{\mathbb{Z}^2}(A^{\mathsf{T}}\underline{\xi}) \\
&\qquad \text{(we just get } A^{\mathsf{T}} \text{ on the inside because we're already working with } A^{-1}\text{)} \\
&= \mathcal{F}\text{Ш}_{\mathbb{Z}^2}(A^{\mathsf{T}}\underline{\xi}) \\
&= \text{Ш}_{\mathbb{Z}^2}(A^{\mathsf{T}}\underline{\xi})\,.
\end{aligned}$$

There's a much neater version of this last result, of genuine physical importance, but we need a new idea.

In crystallography it is common (necessary!) to introduce the *reciprocal lattice* associated with a given lattice. Given a lattice $\mathcal{L}$, the reciprocal lattice is the lattice $\mathcal{L}^*$ consisting of all points (or vectors) $\underline{q}$ such that

$$\underline{q}\cdot\underline{p} = \text{an integer for every } \underline{p} \text{ in the lattice } \mathcal{L}.$$

In some other areas of applications, and in mathematics, the reciprocal lattice is known as the *dual lattice*. I find myself showing my mathematical heritage by generally using the latter term. Sorry for the unmotivated definition — where did that come from? — but read along.

> **Warning.** People in crystallography, those in materials science for example, use the reciprocal lattice all the time and define it this way. However, in some fields and for some applications the reciprocal lattice is normalized differently to require that $\underline{q}\cdot\underline{p}$ be an integer multiple of 2π. This alternate normalization is exactly tied up with the alternate ways of defining the Fourier transform; i.e., while we use $e^{-2\pi i\,\underline{\xi}\cdot\underline{x}}$, putting the 2π in the exponential, others do not put the 2π there and have to put a factor in front of the integral, and so on. I can do no more than to issue this warning and wish us all luck in sorting out the inconsistencies.

To develop the notion of the dual lattice a little and to explain the terminology "reciprocal," suppose we get the lattice $\mathcal{L}$ from $\mathbb{Z}^2$ by applying an invertible matrix A to $\mathbb{Z}^2$. We'll show that the reciprocal lattice $\mathcal{L}^*$ of $\mathcal{L}$ is given by

$$\mathcal{L}^* = A^{-\mathsf{T}}(\mathbb{Z}^2)\,.$$

In fact, one could take this as an alternate definition of the dual lattice.

There's a maxim lurking here, one that I want to reiterate from the discussion of the general stretch formula. Use of the Fourier transform always brings up reciprocal relations of some sort. In one dimension "reciprocal" means "one over" while in higher dimensions (to repeat):

- "Reciprocal" means inverse transpose.

Notice, by the way, that $(\mathbb{Z}^2)^* = \mathbb{Z}^2$, since A in this case is the identity; i.e., $\mathbb{Z}^2$ is *self-dual* as a lattice. This, coupled with the discussion to follow, is a reason for saying that $\mathbb{Z}^2$ wins the award for *most* evenly spaced points in $\mathbb{R}^2$.

Here's why $\mathcal{L}^* = A^{-\mathsf{T}}(\mathbb{Z}^2)$. Let $\underline{m} = (m_1, m_2)$ in $\mathbb{Z}^2$ and let $\underline{q} = A^{-\mathsf{T}}\underline{m}$. Let $\underline{p}$ be a point in $\mathcal{L}$. Because $\mathcal{L} = A(\mathbb{Z}^2)$, we have $\underline{p} = A\underline{m}'$ for some $\underline{m}' = (m_1', m_2')$ in $\mathbb{Z}^2$. For the dot product $\underline{q}\cdot\underline{p}$ we compute

$$\begin{aligned}\underline{q}\cdot\underline{p} &= A^{-\mathsf{T}}\underline{m}\cdot A\underline{m}'\\ &= \underline{m}\cdot A^{-1}(A\underline{m}') \quad \text{(because of how matrices operate with the dot product)}\\ &= \underline{m}\cdot\underline{m}' = m_1m_1' + m_2m_2' \quad \text{(an integer)}.\end{aligned}$$

According to the original definition, this says exactly that $\underline{q}$ is an element of the dual lattice $\mathcal{L}^*$.

We want to draw two conclusions from the result that $\mathcal{L}^* = A^{-\mathsf{T}}(\mathbb{Z}^2)$. First, we see that

$$\text{Area}(\mathcal{L}^*) = |\det \mathrm{A}^{-\mathsf{T}}| = \frac{1}{|\det \mathrm{A}|} = \frac{1}{\text{Area}(\mathcal{L})}\,.$$

Thus the areas of $\mathcal{L}$ and $\mathcal{L}^*$ are reciprocals. This is probably the crystallographer's main reason for using the term reciprocal: the areas of the unit cells are reciprocals.

The second conclusion, and the second reason to use the term reciprocal, has to do with bases of $\mathcal{L}$ and of $\mathcal{L}^*$. With $\mathcal{L} = A(\mathbb{Z}^2)$ let

$$\underline{u}_1 = A\underline{e}_1, \quad \underline{u}_2 = A\underline{e}_2$$

be a basis for $\mathcal{L}$. Since $\mathcal{L}^* = A^{-\mathsf{T}}(\mathbb{Z}^2)$, the vectors

$$\underline{u}_1^* = A^{-\mathsf{T}}\underline{e}_1, \quad \underline{u}_2^* = A^{-\mathsf{T}}\underline{e}_2$$

are a basis for $\mathcal{L}^*$. They have a special property with respect to $\underline{u}_1$ and $\underline{u}_2$, namely,

$$\underline{u}_i\cdot\underline{u}_j^* = \delta_{ij} \quad \text{(Kronecker delta)}\,.$$

This is simple to show, after all we've been through:

$$\underline{u}_i\cdot\underline{u}_j^* = A\underline{e}_i\cdot A^{-\mathsf{T}}\underline{e}_j = \underline{e}_i\cdot A^{\mathsf{T}}A^{-\mathsf{T}}\underline{e}_j = \underline{e}_i\cdot\underline{e}_j = \delta_{ij}\,.$$

Now, in linear algebra — independent of any connection with lattices — bases $\{\underline{u}_1, \underline{u}_2\}$ and $\{\underline{u}_1^*, \underline{u}_2^*\}$ of $\mathbb{R}^2$ are called *dual* (or, sometimes, *reciprocal*) if they satisfy

$$\underline{u}_i\cdot\underline{u}_j^* = \delta_{ij} \quad \text{(Kronecker delta)}\,.$$

We can therefore summarize what we've found by saying:

- If $\{\underline{u}_1, \underline{u}_2\}$ is a basis for a lattice $\mathcal{L}$ and if $\{\underline{u}_1^*, \underline{u}_2^*\}$ is the dual basis to $\{\underline{u}_1, \underline{u}_2\}$, then $\{\underline{u}_1^*, \underline{u}_2^*\}$ is a basis for the dual lattice $\mathcal{L}^*$.

(This could be taken as another alternate way of defining the dual lattice.)

Lots of words here, true, but it's worth your while understanding what we've just done. You're soon to see all of it in action.

Below is a picture of the dual lattice to the lattice pictured earlier, $\mathcal{L} = A(\mathbb{Z}^2)$ with

$$A = \begin{pmatrix} 3 & -1 \\ 1 & 2 \end{pmatrix}.$$

$\mathcal{L}^*$ is obtained from $\mathbb{Z}^2$ by applying

$$A^{-\mathsf{T}} = \begin{pmatrix} 2/7 & -1/7 \\ 1/7 & 3/7 \end{pmatrix}.$$

As the scales on the axes show, the dual lattice in this case is much more compressed, if that's the right word, than the original lattice. Its area is $1/7$.

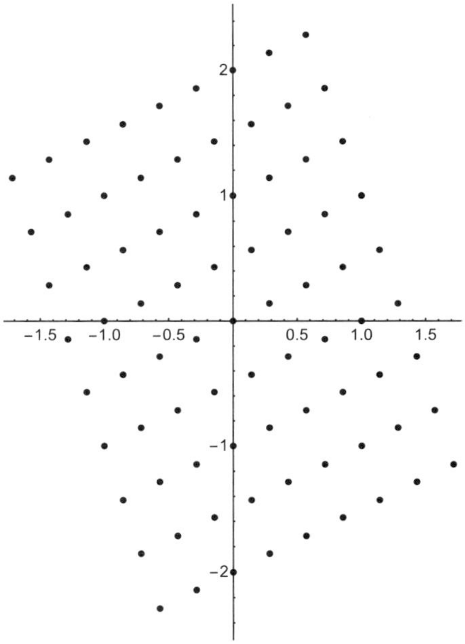

Back to the Fourier transform. We showed that if $\mathcal{L} = A(\mathbb{Z}^2)$, then

$$\mathcal{F}Ш_{\mathcal{L}}(\underline{\xi}) = Ш_{\mathbb{Z}^2}(A^{\mathsf{T}}\underline{\xi}).$$

We want to call forth the dual lattice and write this differently. To that end,

$$\begin{aligned}
Ш_{\mathbb{Z}^2}(A^{\mathsf{T}}\underline{\xi}) &= \sum_{\underline{n}\in\mathbb{Z}^2} \delta(A^{\mathsf{T}}\underline{\xi} - \underline{n}) \\
&= \sum_{\underline{n}\in\mathbb{Z}^2} \delta(A^{\mathsf{T}}(\underline{\xi} - A^{-\mathsf{T}}\underline{n})) \\
&= \frac{1}{|\det A^{\mathsf{T}}|}\sum_{\underline{n}\in\mathbb{Z}^2} \delta(\underline{\xi} - A^{-\mathsf{T}}\underline{n}) = \frac{1}{|\det A|}\sum_{\underline{n}\in\mathbb{Z}^2} \delta(\underline{\xi} - A^{-\mathsf{T}}\underline{n})\,.
\end{aligned}$$

But this last expression is exactly a sum over points $A^{-\mathsf{T}}\underline{n}$ in the dual lattice $\mathcal{L}^*$. We thus have the great formula

$$\mathcal{F}Ш_{\mathcal{L}}(\underline{\xi}) = \frac{1}{|\det A|}Ш_{\mathcal{L}^*}(\underline{\xi})\,.$$

In words, the Fourier transform of the Ш-function of a lattice is the Ш-function of the *dual* lattice, with some additional scaling.

Bringing in the areas of fundamental parallelograms for $\mathcal{L}$ and $\mathcal{L}^*$, we can write the result either in the form

$$\mathcal{F}Ш_{\mathcal{L}}(\underline{\xi}) = \text{Area}(\mathcal{L}^*)Ш_{\mathcal{L}^*}(\underline{\xi}) \quad \text{or} \quad \text{Area}(\mathcal{L})\mathcal{F}Ш_{\mathcal{L}}(\underline{\xi}) = Ш_{\mathcal{L}^*}(\underline{\xi})\,.$$

Interchanging the roles of $\mathcal{L}$ and $\mathcal{L}^*$, we likewise have

$$\mathcal{F}Ш_{\mathcal{L}^*}(\underline{\xi}) = \text{Area}(\mathcal{L})Ш_{\mathcal{L}}(\underline{\xi}) \quad \text{or} \quad \text{Area}(\mathcal{L}^*)\mathcal{F}Ш_{\mathcal{L}^*}(\underline{\xi}) = Ш_{\mathcal{L}}(\underline{\xi})\,.$$

Formulas for the inverse Fourier transforms look just like these because the Ш's are even.

Compare these results to the formula in one dimension,

$$\mathcal{F}Ш_p = \frac{1}{p}Ш_{1/p}\,,$$

and now you'll see why I said at the beginning of this section, "Pay particular attention here to the reciprocity in spacing between $Ш_p$ and its Fourier transform."

Higher dimensions. Everything in the preceding discussion goes through in higher dimensions with *no significant changes*; e.g., "area" becomes "volume." The only reason for stating definitions and results in two dimensions was to picture the lattices a little more easily. But, certainly, lattices in three dimensions are common in applications, maybe more common since, for example, they provide the natural framework for understanding crystals.

9.5.4. The Poisson summation formula, again, and $\mathcal{F}Ш_{\mathbb{Z}^2}$. As promised, we'll now show that $\mathcal{F}Ш_{\mathbb{Z}^2} = Ш_{\mathbb{Z}^2}$.

Back in Chapter 5 we derived the Poisson summation formula: if φ is a Schwartz function, then

$$\sum_{k=-\infty}^{\infty} \mathcal{F}\varphi(k) = \sum_{k=-\infty}^{\infty} \varphi(k)\,.$$

It's a remarkable identity and it's the basis for showing that

$$\mathcal{F}Ш = Ш$$

for the one-dimensional Ш. In fact, the Poisson summation formula is *equivalent* to the Fourier transform identity.[23] The situation in higher dimensions is completely analogous. All we need is a little bit on higher-dimensional Fourier series, which we'll bring in here without fanfare. Suppose φ is a Schwartz function on $\mathbb{R}^2$. We periodize φ to be periodic on the integer lattice $\mathbb{Z}^2$ via

$$\Phi(\underline{x}) = (\varphi * Ш_{\mathbb{Z}^2})(\underline{x}) = \sum_{\underline{n}\in\mathbb{Z}^2} \varphi(\underline{x} - \underline{n})\,.$$

Then Φ has a two-dimensional Fourier series

$$\Phi(\underline{x}) = \sum_{\underline{k}\in\mathbb{Z}^2} \widehat{\Phi}(\underline{k}) e^{2\pi i \underline{k}\cdot\underline{x}}\,.$$

Because $\varphi(\underline{x})$ is rapidly decreasing, there is no issue with convergence of the sum defining $\Phi(\underline{x})$ and no issue with the convergence of the Fourier series for $\Phi(\underline{x})$.

Let's see what happens with the Fourier coefficients:

$$\begin{aligned}
\widehat{\Phi}(k_1,k_2) &= \int_0^1\int_0^1 e^{-2\pi i(k_1x_1+k_2x_2)}\Phi(x_1,x_2)\,dx_1\,dx_2\\
&= \int_0^1\int_0^1 e^{-2\pi i(k_1x_1+k_2x_2)} \sum_{n_1,n_2=-\infty}^{\infty} \varphi(x_1-n_1,x_2-n_2)\,dx_1\,dx_2\\
&= \sum_{n_1,n_2=-\infty}^{\infty} \int_0^1\int_0^1 e^{-2\pi i(k_1x_1+k_2x_2)}\varphi(x_1-n_1,x_2-n_2)\,dx_1\,dx_2
\end{aligned}$$

(no issue with interchanging integrals and sums, either).

Now we make the change of variables $u = x_1 - n_1$, $v = x_2 - n_2$. We can either do this separately (because the variables are changing separately) or together using the general change of variables formula.[24] Either way, the result is

$$\begin{aligned}
&\sum_{n_1,n_2=-\infty}^{\infty} \int_0^1\int_0^1 e^{-2\pi i(k_1x_1+k_2x_2)}\varphi(x_1-n_1,x_2-n_2)\,dx_1\,dx_2\\
&\quad= \sum_{n_1,n_2=-\infty}^{\infty} \int_{-n_1}^{1-n_1}\int_{-n_2}^{1-n_2} e^{-2\pi i(k_1(u+n_1)+k_2(v+n_2))}\varphi(u,v)\,du\,dv\\
&\quad= \sum_{n_1,n_2=-\infty}^{\infty} \int_{-n_1}^{1-n_1}\int_{-n_2}^{1-n_2} e^{-2\pi i(k_1n_1+k_2n_2)}e^{-2\pi i(k_1u+k_2v)}\varphi(u,v)\,du\,dv\\
&\quad= \sum_{n_1,n_2=-\infty}^{\infty} \int_{-n_1}^{1-n_1}\int_{-n_2}^{1-n_2} e^{-2\pi i(k_1u+k_2v)}\varphi(u,v)\,du\,dv\\
&\qquad(\text{because } e^{-2\pi i(k_1n_1+k_2n_2)} = 1)\\
&\quad= \int_{-\infty}^{\infty}\int_{-\infty}^{\infty} e^{-2\pi i(k_1u+k_2v)}\varphi(u,v)\,du\,dv\\
&\quad= \mathcal{F}\varphi(k_1,k_2)\,.
\end{aligned}$$

[23] You showed in a problem that it's even equivalent to duality, and hence to Fourier inversion.

[24] Right here is where the property of $\mathbb{Z}^2$ as the simplest lattice comes in. If we were working with an oblique lattice, we could not make such a simple change of variables. We would have to make a more general linear change of variables. This would lead to a more complicated result.

The Fourier coefficients of $\Phi(\underline{x})$ are values of the Fourier *transform* of $\varphi(\underline{x})$. We have found, just as we did in one dimension, that the Fourier series for the $\mathbb{Z}^2$-periodization of φ is

$$\Phi(\underline{x}) = \sum_{\underline{k}\in\mathbb{Z}^2} \mathcal{F}\varphi(\underline{k})e^{2\pi i\,\underline{k}\cdot\underline{x}}\,.$$

We now evaluate $\Phi(\underline{0})$ in two ways, plugging $\underline{x} = \underline{0}$ into its definition as the periodization of φ and also into its Fourier series. The result is

$$\sum_{\underline{k}\in\mathbb{Z}^2} \mathcal{F}\varphi(\underline{k}) = \sum_{\underline{k}\in\mathbb{Z}^2} \varphi(\underline{k})\,.$$

This is the Poisson summation formula.

To wrap it all up, here's the derivation of

$$\mathcal{F}Ш_{\mathbb{Z}^2} = Ш_{\mathbb{Z}^2}\,,$$

as quick as can be. For any Schwartz function ψ,

$$\langle \mathcal{F}Ш_{\mathbb{Z}^2}, \psi\rangle = \langle Ш_{\mathbb{Z}^2}, \mathcal{F}\psi\rangle = \sum_{\underline{k}\in\mathbb{Z}^2} \mathcal{F}\psi(\underline{k}) = \sum_{\underline{k}\in\mathbb{Z}^2} \psi(\underline{k}) = \langle Ш_{\mathbb{Z}^2}, \psi\rangle\,.$$

Beautiful. And yes, it all works in higher dimensions, too.

9.5.5. Crystals, take two. In a few paragraphs, here's one reason why the meet and greet we just finished on dual lattices is so interesting and worthwhile. The physical model of a crystal is a three-dimensional lattice, possibly oblique, with atoms at the lattice points. An X-ray diffraction experiment scatters X-rays off the atoms in the crystal and results in spots of varying intensity on the X-ray detector, also located at lattice points. From this and other information the crystallographer attempts to deduce the structure of the crystal. The first thing the crystallographer has to know is that the lattice of spots on the detector arising from diffraction comes from the *dual of the crystal lattice*. In fact, it's more complicated, for the spots are the projections of the three-dimensional dual lattice onto the two-dimensional plane of the X-ray detector.

We can explain this phenomenon — atoms on a lattice, spots on the dual lattice — via the Fourier transform. What the crystallographer ultimately wants to find is the electron density distribution for the crystal. The mathematical model for crystals puts a delta at each lattice point, one for each atom. If we describe the electron density distribution of a single atom by a function $\rho(\underline{x})$, then the electron density distribution of the crystal with atoms at points of a (three-dimensional) lattice $\mathcal{L}$ is

$$\rho_{\mathcal{L}}(\underline{x}) = \sum_{\underline{p}\in\mathcal{L}} \rho(\underline{x} - \underline{p}) = (\rho * Ш_{\mathcal{L}})(\underline{x})\,.$$

This is a periodic function with three independent periods, one in the direction of each of the three basis vectors that determine $\mathcal{L}$. We worked with a one-dimensional version of this in Chapter 5.

The basic fact in X-ray crystallography is that the scattered amplitude of the X-rays diffracting off the crystal is proportional to the square magnitude of the Fourier transform of the electron density distribution.[25] This data, the results of X-ray diffraction, thus comes to us *in the frequency domain.* Using the convolution theorem and the formula for $\mathcal{F}\text{III}_{\mathcal{L}}$, we have

$$\mathcal{F}\rho_{\mathcal{L}}(\underline{\xi}) = \mathcal{F}\rho(\underline{\xi})\mathcal{F}\text{III}_{\mathcal{L}}(\underline{\xi}) = \mathcal{F}\rho(\underline{\xi})\left(\text{Volume}(\mathcal{L}^*)\,\text{III}_{\mathcal{L}^*}(\underline{\xi})\right),$$

where $\mathcal{L}^*$ is the dual lattice. Now from the sampling property of δ,

$$\mathcal{F}\rho_{\mathcal{L}}(\underline{\xi}) = \text{Volume}(\mathcal{L}^*) \sum_{\underline{q}\in\mathcal{L}^*} \mathcal{F}\rho(\underline{q})\delta(\underline{\xi} - \underline{q}).$$

The important conclusion is that the diffraction pattern has peaks at the lattice points of the *reciprocal* lattice. The picture is not complete, however. The intensities of the spots are related to the square magnitude of the Fourier transform of the electron density distribution, and for a description of the crystal it is also necessary to determine the phases. This is called phase retrieval and is a hard problem.

Here's a picture of an actual diffraction experiment kindly provided by my colleague James Harris. In fact, let me quote directly from Jim, for those of you who may be familiar with what he's describing.

> The three images on the left are Reflection High Energy Electron Diffraction (RHEED) patterns of GaAs surfaces taken with the electron beam at grazing incidence — $\approx 2°$ — and show that when the surface is slightly rough — this is on a nanoscale — the beam goes through the small surface features and gives a three-dimensional spot diffraction pattern, but as the surface smooths to atomic scale smoothness, the pattern becomes two dimensional — streaked in the bottom image with the middle one in an intermediate stage. The other thing to note is that as the surface smooths, the surface crystal lattice becomes twice as large as the bulk lattice, so you see fainter lines that are halfway between the brighter streaks and, also, right in the middle of the 3D rows of spots so that the reciprocal lattice is half that of the top pattern. The photos on the right side are optical microscope images of the surface at each stage.

[25] In short, a diffraction pattern comes from the Fourier transform of the apertures. The apertures are the spaces between atoms. The spaces between the atoms are determined by the electron density distribution.

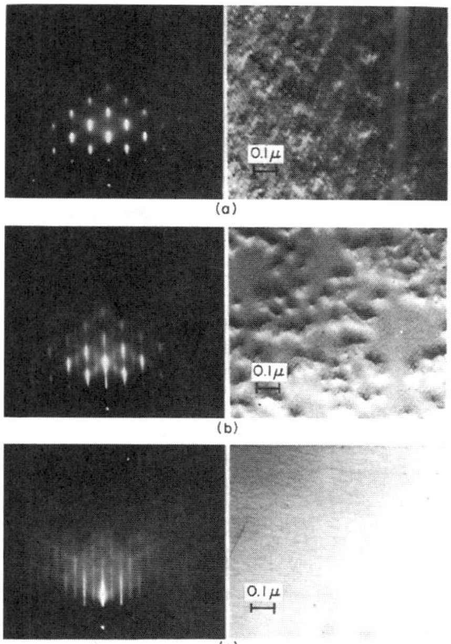

(a)
(b)
(c)

Some references. A major area of applications of X-ray crystallography is unraveling the structure of important biological molecules. There are many good sources to read about this. For further technical reading I recommend *Crystals, X-rays and Proteins* by D. Sherwood and J. Cooper, mentioned back in Chapter 5. I also highly recommend *The Eighth Day of Creation: Makers of the Revolution in Biology* by H. Judson. It's a book for educated laymen (those coveted readers) about the growth of molecular biology, the discovery of the double helix structure of DNA, etc. In addition are two recent biographies of crystallographic heroes, *Dorothy Hodgkins: A Life* and *Max Perutz and the Secret of Life*, both by G. Ferry. Dorothy Hodgkins, at Oxford, worked on the structures of vitamin B12 and insulin. Max Perutz, at Cambridge, worked on the structure of hemoglobin. They were close friends and each received a Nobel Prize for their discoveries.

9.5.6. Bandlimited functions on $\mathbb{R}^2$ and sampling on a lattice. Talk about excitement. Using many of the ideas above let's now develop some sampling theorems in two dimensions. To state the terminology and assumptions: A function $f(\underline{x})$ on $\mathbb{R}^2$ is bandlimited if $\mathcal{F}f(\underline{\xi})$ is identically zero outside of some bounded region. ($\mathcal{F}f(\underline{\xi})$ has compact support.) We will always assume that $f(\underline{x})$ is real valued, and hence $\mathcal{F}f(-\underline{\xi}) = \overline{\mathcal{F}f(\underline{\xi})}$. Thus, as we've pointed out before, if $\mathcal{F}f(\underline{\xi}) \neq 0$, then $\mathcal{F}f(-\underline{\xi}) \neq 0$ and so, as a point set in $\mathbb{R}^2$, the spectrum is symmetric about the origin.

We want to derive an interpolation formula for bandlimited signals that is associated with a lattice $\mathcal{L}$. We'll follow the recipe of first periodizing $\mathcal{F}f$ via $Ш_{\mathcal{L}}$, then cutting off, and then taking the inverse Fourier transform. The result will be a sinc reconstruction of $f(\underline{x})$ from its sampled values, but just *where those sampled values are* is what's especially interesting and relevant to what we've just done.

To get the argument started we assume that the support of $\mathcal{F}f(\underline{\xi})$ lies in a parallelogram. This parallelogram determines a fundamental parallelogram for a lattice $\mathcal{L}$, and the spectrum gets shifted parallel to itself and off itself through convolution with $\mathrm{III}_{\mathcal{L}}$.

This periodization is the first step, and it's analogous to the one-dimensional case, where the spectrum lies in an interval, say from $-p/2$ to $p/2$, and the spectrum gets shifted off itself through convolution with III_p. To remind you, in 1D the crucial point is that the spectrum lies between $\pm p/2$, while III_p has δ's spaced p apart. The spacing of the δ's is big enough to shift the spectrum off itself. Cutting off by Π_p then gives back the original Fourier transform. It's all about the equation

$$\mathcal{F}f = \Pi_p(\mathcal{F}f * \mathrm{III}_p)\,.$$

Correspondingly, in two dimensions, the parallelogram containing the spectrum determines a lattice with, we assume, big enough spacing of the lattice points so that convolving with a III based on the lattice shifts the spectrum off itself.

Using the general stretch theorem, we'll be able to get a general result by first deriving a special case when the spectrum lies in a square. Suppose, then, that $\mathcal{F}f(\underline{\xi})$ is identically zero for $|\xi_1| \geq 1/2$, $|\xi_2| \geq 1/2$ (or put another way, that $\mathcal{F}f(\underline{\xi}) = 0$ outside the open square $|\xi_1| < 1/2$, $|\xi_2| < 1/2$). We work with the integer lattice $\mathbb{Z}^2$ with basis $\underline{e}_1$ and $\underline{e}_2$. The (open) fundamental parallelogram for $\mathbb{Z}^2$ is $0 < \xi_1 < 1$, $0 < \xi_2 < 1$, and the spectrum is inside the center fourth of four copies of it, as pictured.

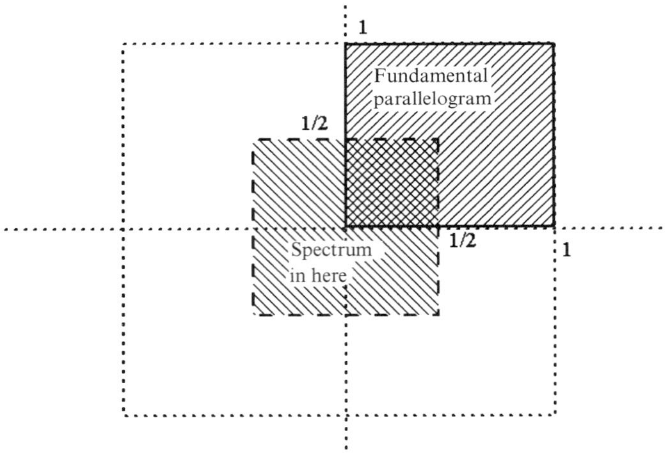

Periodizing $\mathcal{F}f$ by $\mathrm{III}_{\mathbb{Z}^2}$ shifts the spectrum off itself, and no smaller rectangular lattice will do for this. Draw a picture; convince yourself. We then cut off by the two-dimensional rect function $\Pi(\xi_1, \xi_2) = \Pi(\xi_1)\Pi(\xi_2)$ and this gives back $\mathcal{F}f$. The magic equation is

$$\mathcal{F}f(\underline{\xi}) = \Pi(\underline{\xi})(\mathcal{F}f * \mathrm{III}_{\mathbb{Z}^2})(\underline{\xi})\,,$$

just as in the one-dimensional case. And now it's time to take the inverse Fourier transform. Using $\mathcal{F}\mathrm{III}_{\mathbb{Z}^2} = \mathrm{III}_{\mathbb{Z}^2}$, or rather $\mathcal{F}^{-1}\mathrm{III}_{\mathbb{Z}^2} = \mathrm{III}_{\mathbb{Z}^2}$, and invoking the convolution theorem we obtain

$$\begin{aligned} f(\underline{x}) = f(x_1, x_2) &= (\operatorname{sinc} x_1 \ \operatorname{sinc} x_2) * (f(\underline{x}) \cdot Ш_{\mathbb{Z}^2}(\underline{x})) \\ &= (\operatorname{sinc} x_1 \ \operatorname{sinc} x_2) * \Big(f(\underline{x}) \cdot \sum_{k_1,k_2=-\infty}^{\infty} \delta(\underline{x} - k_1\underline{e}_1 - k_2\underline{e}_2)\Big) \\ &= (\operatorname{sinc} x_1 \ \operatorname{sinc} x_2) * \sum_{k_1,k_2=-\infty}^{\infty} f(k_1,k_2)\delta(x_1 - k_1, x_2 - k_2) \\ &= \sum_{k_1,k_2=-\infty}^{\infty} f(k_1,k_2) \operatorname{sinc}(x_1 - k_1) \operatorname{sinc}(x_2 - k_2)\,. \end{aligned}$$

Convolving the product of sincs with the shifted δ shifts each slot separately; convince yourself of that, too.

In solidarity with the general case soon to follow, let's write this square sampling formula in vector notation as

$$f(\underline{x}) = \sum_{k_1,k_2=-\infty}^{\infty} f(k_1\underline{e}_1 + k_2\underline{e}_2) \operatorname{sinc}(\underline{x}\cdot\underline{e}_1 - k_1) \operatorname{sinc}(\underline{x}\cdot\underline{e}_2 - k_2)\,.$$

Now suppose more generally that the spectrum of $\mathcal{F}f$ lies in an open parallelogram, as pictured, with $\underline{u}_1$ and $\underline{u}_2$ parallel to the sides and as long as the sides.

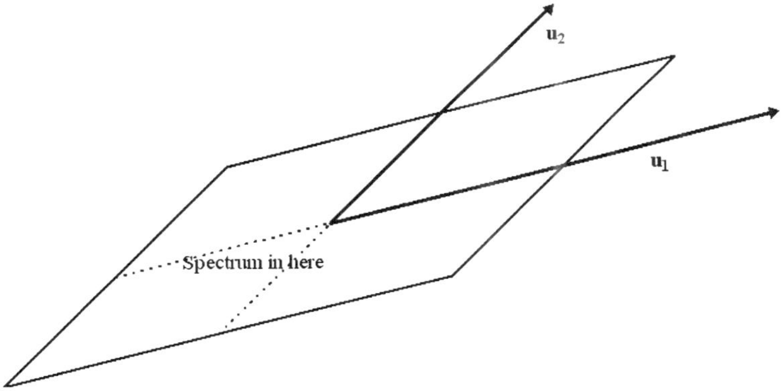

Let A be the 2×2 matrix that takes $\underline{e}_1$ to $\underline{u}_1$ and $\underline{e}_2$ to $\underline{u}_2$, so that A maps the lattice $\mathbb{Z}^2$ to the lattice $\mathcal{L}$ with basis $\underline{u}_1$ and $\underline{u}_2$. Let $B = A^{-\mathsf{T}}$ (hence $B^{-\mathsf{T}} = A$) and remember that B takes $\mathbb{Z}^2$ to the dual lattice $\mathcal{L}^*$ of $\mathcal{L}$. A basis for $\mathcal{L}^*$ (the dual basis to $\underline{u}_1$ and $\underline{u}_2$) is

$$\underline{u}_1^* = B\underline{e}_1, \quad \underline{u}_2^* = B\underline{e}_2\,.$$

Next let

$$g(\underline{x}) = f(B\underline{x})\,.$$

According to the general stretch theorem,

$$\mathcal{F}g(\underline{\xi}) = \frac{1}{|\det B|}\mathcal{F}f(B^{-\mathsf{T}}\underline{\xi}) = |\det A|\,\mathcal{F}f(A\underline{\xi})\,.$$

The determinant factor out front doesn't matter; what's important is that the spectrum of g is in the square $-1/2 < \xi_1 < 1/2$, $-1/2 < \xi_2 < 1/2$, since the corresponding points $A\underline{\xi}$ lie in the parallelogram containing the spectrum of f. That is, $\mathcal{F}g$ is identically zero outside the open square.

We now apply the square sampling formula to g to write

$$g(\underline{x}) = \sum_{k_1,k_2=-\infty}^{\infty} g(k_1\underline{e}_1 + k_2\underline{e}_2)\,\mathrm{sinc}(\underline{x}\cdot\underline{e}_1 - k_1)\,\mathrm{sinc}(\underline{x}\cdot\underline{e}_2 - k_2)\,.$$

With $\underline{y} = B\underline{x}$ we can then say

$$\begin{aligned} f(\underline{y}) &= \sum_{k_1,k_2=-\infty}^{\infty} f(B(k_1\underline{e}_1 + k_2\underline{e}_2))\,\mathrm{sinc}(B^{-1}\underline{y}\cdot\underline{e}_1 - k_1)\,\mathrm{sinc}(B^{-1}\underline{y}\cdot\underline{e}_2 - k_2) \\ &= \sum_{k_1,k_2=-\infty}^{\infty} f(k_1B\underline{e}_1 + k_2B\underline{e}_2)\,\mathrm{sinc}(A^{\mathsf{T}}\underline{y}\cdot\underline{e}_1 - k_1)\,\mathrm{sinc}(A^{\mathsf{T}}\underline{y}\cdot\underline{e}_2 - k_2) \\ &= \sum_{k_1,k_2=-\infty}^{\infty} f(k_1\underline{u}_1^* + k_2\underline{u}_2^*)\,\mathrm{sinc}(\underline{y}\cdot A\underline{e}_1 - k_1)\,\mathrm{sinc}(\underline{y}\cdot A\underline{e}_2 - k_2) \\ &= \sum_{k_1,k_2=-\infty}^{\infty} f(k_1\underline{u}_1^* + k_2\underline{u}_2^*)\,\mathrm{sinc}(\underline{y}\cdot\underline{u}_1 - k_1)\,\mathrm{sinc}(\underline{y}\cdot\underline{u}_2 - k_2)\,. \end{aligned}$$

We're done. Change $\underline{y}$ to $\underline{x}$ for psychological comfort, and the lattice sampling formula says that

$$f(\underline{x}) = \sum_{k_1,k_2=-\infty}^{\infty} f(k_1\underline{u}_1^* + k_2\underline{u}_2^*)\,\mathrm{sinc}(\underline{x}\cdot\underline{u}_1 - k_1)\,\mathrm{sinc}(\underline{x}\cdot\underline{u}_2 - k_2).$$

This is a sinc reconstruction formula giving the function in terms of sample values on a lattice. But it's the dual lattice! Here's how to remember the highlights:

- The spectrum of f lies in a parallelogram, which determines a lattice with basis $\underline{u}_1$ and $\underline{u}_2$.
- That lattice determines a dual lattice (in the spatial domain) with dual basis $\underline{u}_1^*$ and $\underline{u}_2^*$.
- The sincs use data from the lattice, while the sample points are exactly the lattice points in the *dual* lattice.

Look back at the one-dimensional sampling formula and tell yourself what you see there of this picture.

Now think about the following: what should we mean by "sampling rate" vis à vis the two-dimensional lattice sampling formula?

The next stops on this path would be to deliver the higher-dimensional versions of the sampling formulas (straightforward), to investigate aliasing, and to consider the case of a finite spectrum and finite sampling. With regret, we won't do that.

9.6. The Higher-Dimensional DFT

A brief discussion of the DFT in higher dimensions is in order, the goal being just to derive the formula. This goes like the one-dimensional case, so look that over. Or you can skip down to the final result if you'd like. I'll leave some properties of the DFT for the problems.

As has been our custom, we'll work in two dimensions with the assurance that all goes through smoothly above and beyond. And as in the 1D case we start by making an assumption that is false. We suppose that a function $f(x_1, x_2)$ is identically zero outside the rectangle

$$[0, L_1] \times [0, L_2] = \{(x_1, x_2) : 0 \le x_1 \le L_1\,, 0 \le x_2 \le L_2\}$$

(spatially limited) *and* that the Fourier transform $\mathcal{F}f(\xi_1, \xi_2)$ is identically zero outside the rectangle

$$[0, B_1] \times [0, B_2] = \{(\xi_1, \xi_2) : 0 \le \xi_1 \le B_1\,, 0 \le \xi_2 \le B_2\}$$

(bandlimited). That's right; this is false because, just as in one dimension, a function cannot be both limited in space and in frequency, unless it's identically zero. But play along.

The first step is to replace $f(x_1, x_2)$ by a sampled version. In the x_i slot of the spatial variables, $i = 1, 2$, we sample at a rate of $2B_i$, the bandwidth of the ith frequency. The sample points are spaced $1/2B_i$ apart, starting at 0 and ending at $\frac{N_i - 1}{2B_i}$, if, say, there are N_i of them. So the sample points in the x_i variable are

$$\mathcal{K}_i = \left\{0, \frac{1}{2B_i}, \frac{2}{2B_i}, \dots, \frac{N_i - 1}{2B_i}\right\},$$

and observe that

$$\frac{N_i}{2B_i} = L_i, \quad \text{or} \quad N_i = 2B_i L_i\,.$$

Now let

$$\mathcal{K} = \mathcal{K}_1 \times \mathcal{K}_2 = \left\{\left(\frac{k_1}{2B_1}, \frac{k_2}{2B_2}\right) : 0 \le k_1 \le N_1 - 1,\ 0 \le k_2 \le N_2 - 1\right\}$$

be the set of pairs of the sample points in the two variables. We can think of $\mathcal{K}$ as a finite rectangular lattice, and the sampled version of $f(x_1, x_2)$ on $\mathcal{K}$ is

$$\begin{aligned} f_{\text{sampled}}(\underline{x}) &= f(\underline{x}) \sum_{\underline{k} \in \mathcal{K}} \delta(\underline{x} - \underline{k}) \\ &= \sum_{\underline{k} \in \mathcal{K}} f(\underline{k}) \delta(\underline{x} - \underline{k}) \\ &= \sum_{k_1, k_2} f\left(\frac{k_1}{2B_1}, \frac{k_2}{2B_2}\right) \delta\left(x_1 - \frac{k_1}{2B_1}\right) \delta\left(x_2 - \frac{k_2}{2B_2}\right). \end{aligned}$$

Then for the Fourier transform

$$\mathcal{F}f_{\text{sampled}}(\underline{\xi}) = \sum_{\underline{k} \in \mathcal{K}} f(\underline{k}) \mathcal{F}\delta(\underline{x} - \underline{k}) = \sum_{\underline{k} \in \mathcal{K}} f(\underline{k}) e^{-2\pi i \underline{\xi} \cdot \underline{k}}\,.$$

In the ξ_i-variable we in turn now sample $\mathcal{F}f_{\text{sampled}}(\xi_1, \xi_2)$ at a rate of L_i, dictated by the support of $f(x_1, x_2)$ in the x_i-variable. The sample points are spaced $1/L_i$ apart, starting at 0 and ending at $(M_i - 1)/L_i$, if, say, there are M_i of them. Here

$$\frac{M_i}{L} = 2B_i \quad \text{or} \quad M_i = 2B_i L_i = N_i,$$

and, just as in the 1D case, we see that the number of sample points in the ith frequency variable is the same as in the ith spatial variable.

Let

$$\mathcal{L}_i = \left\{0, \frac{1}{L_i}, \frac{2}{L_i}, \dots, \frac{N_i - 1}{L_i}\right\}$$

be the sample points for $\mathcal{F}f_{\text{sampled}}(\xi_1, \xi_2)$ in the ξ_i-variable, and let

$$\mathcal{L} = \mathcal{L}_1 \times \mathcal{L}_2$$

be the finite, rectangular lattice of sampling points in frequency. The discrete Fourier transform of f comes from $\mathcal{F}f_{\text{sampled}}$ evaluated at the frequency lattice points,

$$\mathcal{F}f_{\text{sampled}}(\underline{l}) = \sum_{\underline{k} \in \mathcal{K}} f(\underline{k}) e^{-2\pi i \underline{l} \cdot \underline{k}}, \quad \underline{l} \in \mathcal{L},$$

or

$$\begin{aligned} \mathcal{F}f_{\text{sampled}}\left(\frac{l_1}{L_1}, \frac{l_2}{L_2}\right) &= \sum_{k_1, k_2} f\left(\frac{k_1}{2B_1}, \frac{k_2}{2B_2}\right) e^{-2\pi i \left(\frac{k_1}{2B_1}\frac{l_1}{L_1} + \frac{k_2}{2B_2}\frac{l_2}{L_2}\right)} \\ &= \sum_{k_1, k_2} f\left(\frac{k_1}{2B_1}, \frac{k_2}{2B_2}\right) e^{-2\pi i \left(\frac{k_1 l_1}{N_1} + \frac{k_2 l_2}{N_2}\right)}, \end{aligned}$$

using $2B_j L_j = N_j$ in the complex exponential.

The final step is to identify the sample points with their indices

$$\frac{k_j}{2B_n} \leftrightarrow k_j, \quad \frac{\ell_j}{L_j} \leftrightarrow \ell_j,$$

forget that everything came from sampling $f(x_1, x_2)$, and give the definition solely in terms of discrete functions. Anticipating, as in 1D, that various calculations depend only on integers mod N_1 and mod N_2, let $\underline{f}$ be a discrete signal defined on $\mathbb{Z}_{N_1} \times \mathbb{Z}_{N_2}$. The discrete Fourier transform of $\underline{f}$ is the discrete signal $\underline{\mathcal{F}}\,\underline{f}$, also defined on $\mathbb{Z}_{N_1} \times \mathbb{Z}_{N_2}$, defined by

$$\underline{\mathcal{F}}\,\underline{f}[l_1, l_2] = \sum_{k_1 \in \mathbb{Z}_{N_1}, k_2 \in \mathbb{Z}_{N_2}} \underline{f}[k_1, k_2] e^{-2\pi i \left(\frac{k_1 l_1}{N_1} + \frac{k_2 l_2}{N_2}\right)}, \quad l_1 \in \mathbb{Z}_{N_1}, l_2 \in \mathbb{Z}_{N_2}.$$

The next order of business would be periodicity, reindexing, orthogonality of the complex exponentials, etc. I'll pass, except for a few comments.

The inputs and outputs are periodic in each variable, having period N_i in variable i. Thus, if we wish, we can consider the input and its DFT to be defined on $\mathbb{Z} \times \mathbb{Z}$ and periodic in each slot, or, as we said above, to be defined on $\mathbb{Z}_{N_1} \times \mathbb{Z}_{N_2}$. The periodicity means that reindexing in each slot is possible. In imaging applications it's common to index the respective slots from $-N_1/2+1$ to $N_1/2$ and from $-N_2/2+1$ to $N_2/2$ (assuming N_1 and N_2 are even, which we are assuming), so the center of the image and of the spectrum is at the origin.

The separate complex exponentials associated with each slot are orthogonal but not orthonormal (that's just the old, 1D calculation). This means that the higher-dimensional case again suffers from the scourge of extra multiplicative factors. For example, the inverse DFT is

$$\underline{\mathcal{F}}^{-1}\underline{g}[k_1, k_2] = \frac{1}{N_1 N_2} \sum_{l_1 \in \mathbb{Z}_{N_1}, l_2 \in \mathbb{Z}_{N_2}} \underline{g}[l_1, l_2] e^{2\pi i \left(\frac{k_1 l_1}{N_1} + \frac{k_2 l_2}{N_2}\right)}, \quad k_1 \in \mathbb{Z}_{N_1}, k_2 \in \mathbb{Z}_{N_2}.$$

The $N_1 N_2$ factor appears in other formulas. Try to remain calm.

Also note that it's not really a simple dot product that occurs in the complex exponentials because of the divisions by N_1 and N_2, respectively. If $N_1 = N_2$, which is a common situation (so $\underline{f}$ and $\underline{\mathcal{F}}\underline{f}$ are defined on a square grid of discrete points), we can write a little more simply

$$\underline{\mathcal{F}}\underline{f}[\underline{l}] = \sum_{\underline{k} \in \mathbb{Z}_N^2} \underline{f}[\underline{k}] e^{-\frac{2\pi i}{N} \underline{k} \cdot \underline{l}}, \quad \underline{\mathcal{F}}^{-1}\underline{g}[\underline{k}] = \frac{1}{N^2} \sum_{\underline{k} \in \mathbb{Z}_N^2} \underline{g}[\underline{k}] e^{\frac{2\pi i}{N} \underline{k} \cdot \underline{l}}.$$

As you'll see in the problems, the 2D DFT can be computed by iterated partial DFTs, in the same sense that the 2D Fourier transform can be computed by iterated partial Fourier transforms. For computation, the 1D FFT algorithm is applied to each of the partial transforms.

Finally, it bears repeating that all goes through in the same manner for higher dimensions.

9.7. Naked to the Bone

We now turn to a brief development of the use of the two-dimensional Fourier transform in medical imaging. Specifically, we'll see how the Fourier transform works together with another transform, the Radon transform, to solve the *problem of tomography*, stated, one way, as:

- Can we reconstruct a two-dimensional image from one-dimensional sections of that image?

"Tomography" derives from the Greek word "*tomos*," which means cut or slice. The image, in the case we're interested in, is a two-dimensional slice through your body. The one-dimensional sections of that slice are X-rays that pass through it. If we have information on what happens to the X-rays en route, can we reconstruct what's inside?

Our problem is one of computed tomography (CT) and *not* magnetic resonance imaging (MRI). Interestingly, while these are two very different modalities of medical imaging, the Fourier transform comes to the rescue in similar ways for each.

For an account of the history of medical imaging I recommend the book *Naked to the Bone: Medical Imaging in the Twentieth Century* by Bettyann Kevles, from which I stole the title of this section. Other sources are *Medical Imaging Systems* by my colleague A. Macovski,[26] and *Introduction to the Mathematics of Medical Imaging* by C. Epstein. The book by Briggs and Hensen on the DFT also has a very good discussion.

9.7.1. Dimmer and dimmer. What happens when light passes through murky water? It gets dimmer and dimmer the farther it goes, of course — this is not a trick question. If the water is the same murkiness throughout, meaning, for example, uniform density of stuff floating around in it, then it's natural to assume that the intensity of light decreases by the same *percent amount* per length of path traveled. Through absorption, scattering, etc., whatever intensity goes in, a certain percentage of that intensity goes out. So over a given distance, the murky water removes a percentage of light, and this percentage depends only on the distance traveled and not on where the starting and stopping points are. We're assuming here that light is traveling in a straight line through the water.

By the way, optical fibers provide an interesting and important example of the progress in making something — glass in this case — less murky. There's an interesting graph in the book *Signals* by J. Pierce and A. Noll showing how the clarity of glass has improved over the ages, with some poetic license in estimating the clarity of the windows of ancient Egypt.[27] The big jump in clarity going to optical fibers was achieved largely by eliminating water in the glass.

Constant percent change characterizes exponential growth, or decay, so the attenuation of the intensity of light passing through a homogeneous medium is modeled by

$$I = I_0 e^{-\mu x},$$

where I_0 is the initial intensity, x is the distance traveled, and μ is a (positive) "murkiness constant." The variable x has dimension of length, and μ has dimension of 1/length and units "murkiness/length." The number μ is constant because we assume that the medium is homogeneous. We know the value of I_0, and one measurement of x and I will determine μ. In fact, what we do is to put a detector at a known distance x and measure the intensity when the light arrives at the detector.

Now suppose the water is not uniformly murky, but rather the light passes through a number of layers, each layer of uniform murkiness. If the ith layer has murkiness constant μ_i and is of length Δx_i and if there are n layers, then the

[26] I'm hoping that my colleague D. Nishimura will publish his lecture notes based on the course he teaches at Stanford. Look for them.

[27] Unfortunately, I can't reproduce the graph here. Costs too much. But it's worth the effort to track down the book, for all sorts of interesting topics. John Pierce was a very eminent engineer at Bell Labs. He pretty much invented the telecommunications satellite. He retired and moved to Stanford where, together with Max Matthews, another Bell Labs alum, he worked at the Center for Computer Research in Music and Acoustics.

intensity of light that reaches the detector can be modeled by

$$I = I_0 \exp\left(-\sum_{i=1}^{n} \mu_i \Delta x_i\right).$$

In the limit (invoking that wonderful catchall phrase), if the murkiness is described by a density function $\mu(x)$, over an infinitesimal layer dx, then the intensity arriving at the detector is modeled by

$$I = I_0 \exp\left(-\int_L \mu(x)\,dx\right),$$

where L is the line the light travels along. It's common to call the $\mu(x)$ the *linear attenuation coefficient*, and measurements of the incoming and outgoing intensities give us its line integral

$$p = \int_L \mu(x)\,dx = -\ln\left(\frac{I}{I_0}\right).$$

Can we recover the density function $\mu(x)$ from knowledge of the intensities? Not so easily. Certainly not from a single reading; many arrangements of murkiness along the path could result in the same final intensity at the detector.

If we could vary the detector along the path and record the results, then we would be able to determine $\mu(x)$. That is, if we could form

$$p(\xi) = \int_{\xi_0}^{\xi} \mu(x)\,dx$$

as a *function* of a variable position ξ along the line (ξ_0 is some fixed starting point, the source), then we could find μ from p by finding the derivative p'. The trouble is moving the detector through the murky water along the path.

Tomography. X-rays are light, too, and when they pass through murky stuff (your body) along a straight line, they are attenuated and reach a detector on the other end at a reduced intensity. We can continue to assume that the attenuation, the decrease in intensity, is exponentially decreasing with the path length. The exponential of what? What do the X-rays pass through?

From the start we set this up as a two-dimensional problem. Take a planar slice through your body. The gunk in this two-dimensional slice — bones, organs, other tissue — is of *variable density*. Let's say it's described by a function $\mu(x_1, x_2)$. To know $\mu(x_1, x_2)$ is to know what's inside. That's the goal.

Take a line L passing through this slice, in the plane of the slice, the path that an X-ray would follow. We need to define the integral of $\mu(x_1, x_2)$ along L. For this, parameterize the line by $x_1(s)$, $x_2(s)$, where s is the arclength parameter going, say, from s_0 to s_1.[28] Then the density along the line is $\mu(x_1(s), x_2(s))$, and

[28] The "arclength parameter" means that we trace out the line at unit speed. It seems to be a custom to call the arclength parameter s. You'll see that in math books. The only reason I can think of for this is that in mechanics s is often used for distance traveled. I have no idea where that comes from.

the integral is

$$\int_{s_0}^{s_1} \mu(x_1(s), x_2(s))\, ds\,.$$

The attenuation of the X-ray intensity along the line is

$$I = I_0 \exp\left(- \int_{s_0}^{s_1} \mu(x_1(s), x_2(s))\, ds \right).$$

We know I_0, we know the line, and we measure I.

Instead of writing out the parameters and limits, we often write the integral simply as

$$\int_L \mu(x_1, x_2)\, ds\,.$$

We'll refer to this as a *line integral of* μ *along* L. The line integrals are, in effect, what we're measuring, and we do so for many lines. Let's restate:

- The fundamental problem of tomography is to determine the function $\mu(x_1, x_2)$ from the line integrals.

For example, what's $\mu(x_1, x_2)$ for your head?

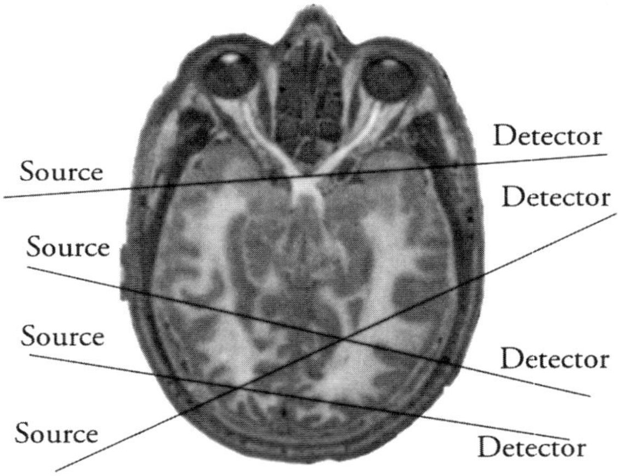

In trying to solve this problem, what's not allowed is to move the detector through the body — that's not covered by health insurance plans. What is done, contrary to the random lines in the picture, is to send a group of parallel X-rays through the region and then rotate the source (and the detector) to get families of parallel X-rays circling around the two-dimensional cross-section of the body.[29]

9.7.2. The Radon transform. Let's get organized. For each line L, passing through the slice, the integral

$$\int_L \mu(x_1, x_2)\, ds$$

[29]Other trajectories are also used, but we'll stick with parallel lines.

is a number. The operation "line L determines a number" thus defines a real-valued function of L. The whole subject of tomography is about this function, and the first, key idea — not obvious — is to view it as a *transform* of μ. Let

$$\mathcal{R}\mu(L) = \int_L \mu(x_1, x_2)\, ds\,.$$

This is called the *Radon transform* of μ, introduced by Johann Radon way back in 1917. The fundamental question of tomography can then be further restated as:

- From knowledge of the values $\mathcal{R}\mu(L)$ for many lines L, can we recover μ? In other words, can we invert the Radon transform?

I feel obliged to add that X-rays and the application to medical imaging were *not* Radon's motivation. He was just solving a math problem.

Parametrizing the collection of lines. To make $\mathcal{R}\mu(L)$ amenable to calculations, we need to describe the collection of all lines. Not just the (Cartesian) equation of a *particular* line, but some choice of parameters that will, as they vary, describe the *collection* of lines. Then we'll write $\mathcal{R}\mu$ as a function of these parameters.

There are many ways to describe the collection of all lines in the plane. One that may seem most natural to you is to use the slope-intercept form for the equation of a line from your early childhood. That's the equation $y = mx + b$ where m is the slope and b is the y-intercept. A line can thus be associated with a unique pair of parameters (m, b), and as the parameters m and b vary, separately, we get different lines. But there's a catch. Vertical lines (x = constant, infinite slope) are left out of this description.

Another approach, one that allows us to describe *all* lines and that (it turns out) is also well suited for the application to tomography goes as follows: first, a line *through the origin* is determined by its unit normal vector $\underline{n}$. Now, $\underline{n}$ and $-\underline{n}$ determine the same line, so we represent all the (distinct) normal vectors as $\underline{n} = (\cos\phi, \sin\phi)$ for an angle ϕ satisfying $0 \leq \phi < \pi$, measured counterclockwise from the x_1-axis. There is a one-to-one correspondence between the ϕ's with $0 \leq \phi < \pi$ and the collection of lines through the origin; note the strict inequality $\phi < \pi$.

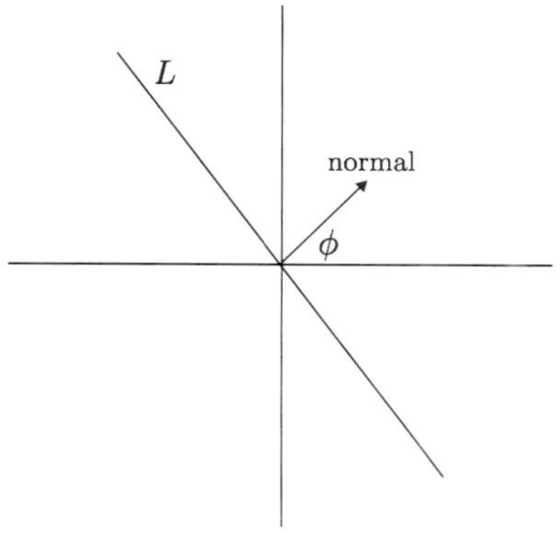

A line not through the origin can then be described by its unit normal vector together with the *directed distance* of the line from the origin, a positive number if measured in the direction of $\underline{n}$ and a negative number if measured in the direction $-\underline{n}$. Call this directed distance ρ. Thus $-\infty < \rho < \infty$.

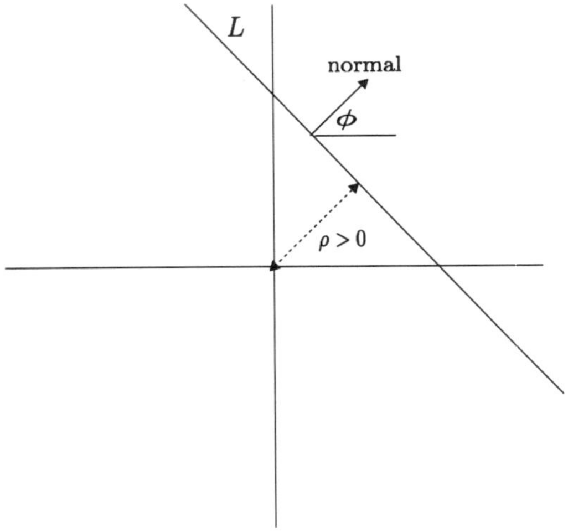

The set of pairs (ρ, ϕ) provides a parameterization for the set of all lines in the plane. Once again:

- A pair (ρ, ϕ) means, in this context, the unique line with normal vector $\underline{n} = (\cos\phi, \sin\phi)$ which is at a directed distance ρ from the origin, measured in the direction $\underline{n}$ if $\rho > 0$ and in the direction $-\underline{n}$ if $\rho < 0$. (These aren't polar coordinates, which is one reason we're calling them (ρ, ϕ), not to be confused with (r, θ) as you often write polar coordinates.)

OK, this parametrization of the collection of lines in the plane is probably not one that would spring to mind. But it has many advantages that we'll come to appreciate.

Anytime you're confronted with a new coordinate system you should ask yourself what the situation is when one of the coordinates is fixed and the other is free to vary. If ϕ is fixed and ρ varies, we get a family of parallel lines, all with unit normal $\underline{n} = (\cos\phi, \sin\phi)$. This is the main reason why the (ρ, ϕ)-coordinates are useful. For a fixed ϕ setting on your CT machine the family of parallel lines is a family of parallel X-rays that are sent through the region. Then change ϕ and send through the next family of parallel X-rays.

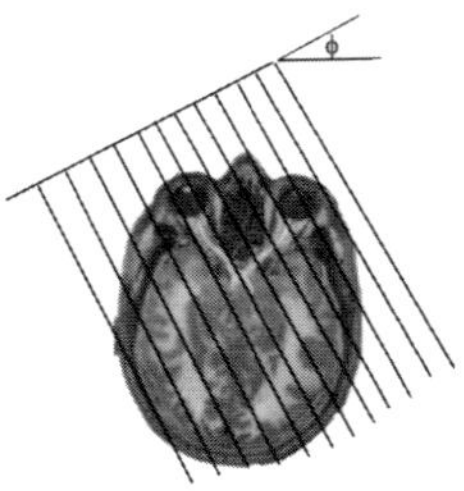

On the other hand, if ρ is fixed and ϕ varies, we have to distinguish some cases. When $\rho = 0$, we just get the family of all lines through the origin. When ρ is fixed and positive and ϕ varies from 0 to π (including 0, excluding π), we get the family of lines tangent to the upper semicircle of radius ρ, including the tangent at $(\rho, 0)$ and excluding the tangent at $(-\rho, 0)$. When ρ is fixed and negative, we get lines tangent to the lower semicircle, including the tangent at $(-|\rho|, 0)$ and excluding the tangent at $(|\rho|, 0)$. The two situations are pictured below. No particular CT application here.

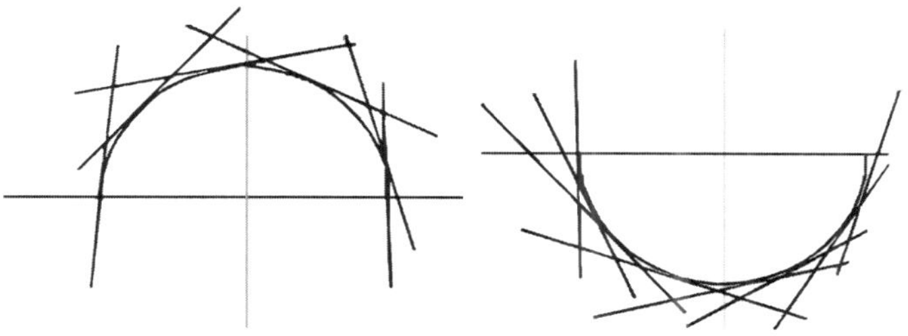

The Radon transform in coordinates. We need two more facts to write the Radon transform in terms of the coordinates (ρ, ϕ), one easy and one not so easy. The easy fact is the Cartesian equation of the line with given coordinates (ρ, ϕ). Using the dot product, the line determined by (ρ, ϕ) has equation

$$\rho = \underline{x} \cdot \underline{n} = (x_1, x_2) \cdot (\cos\phi, \sin\phi) = x_1 \cos\phi + x_2 \sin\phi\,.$$

You can verify this. We'll write the equation as

$$\rho - x_1 \cos\phi - x_2 \sin\phi = 0\,, \quad -\infty < x_1 < \infty,\ -\infty < x_2 < \infty\,.$$

Next, the not-so-easy fact is to consider the 1D δ-function *along the line*,

$$\delta(\rho - x_1 \cos\phi - x_2 \sin\phi)\,,$$

as a function of x_1, x_2. This is called a *line impulse*, and it's an example of the greater variety one has in defining different sorts of δ's in two dimensions. Many more details are provided in Section 9.8.

With some interpretation and argument one can show that integrating a function $f(x_1, x_2)$ against the line impulse associated with a line L results precisely in the line integral of f along L,

$$\iint_{\mathbb{R}^2} f(x_1, x_2)\delta(\rho - x_1 \cos\phi - x_2 \sin\phi)\, dx_1 dx_2 = \int_L f(x_1, x_2)\, ds\,.$$

This is all we'll need here (it's derived in Section 9.8) and with it the Radon transform of $\mu(x_1, x_2)$ can be expressed as

$$\mathcal{R}\mu(\rho, \phi) = \int_{-\infty}^{\infty}\int_{-\infty}^{\infty} \mu(x_1, x_2)\delta(\rho - x_1 \cos\phi - x_2 \sin\phi)\, dx_1\, dx_2\,.$$

We'll most often work with this expression for $\mathcal{R}\mu$. One also sees the Radon transform written in vector form

$$\mathcal{R}\mu(\rho, \underline{n}) = \int_{\mathbb{R}^2} \mu(\underline{x})\delta(\rho - \underline{x} \cdot \underline{n})\, d\underline{x}\,.$$

This expression suggests generalizations to higher dimensions, for example with lines replaced by planes in 3D and line impulses replaced by plane impulses.[30] Interesting, but we won't pursue it.

Projections. We'll study $\mathcal{R}\mu(\rho, \phi)$ by first fixing ϕ and letting ρ vary. Then, as we saw in the earlier figure, we're looking at parallel lines passing through the domain of μ, all perpendicular to a particular line making an angle ϕ with the x_1-axis. For such a family of lines, the collection of values $\mathcal{R}\mu(\rho, \phi)$ is often referred to as a *projection* of μ, the idea being that the line integrals over parallel lines, at the fixed angle ϕ, are giving some kind of profile, or projection, of μ in that direction. Then varying ϕ gives a family of projections parametrized by ϕ, and one speaks of the inversion problem as "determining $\mu(x_1, x_2)$ from its projections."[31]

9.7.3. Getting to know your Radon transform.

Before pushing ahead with inverting the Radon transform, I feel an obligation to list a few of its properties. Just enough to get some sense of how to work with it.

First, some comments on what kinds of functions $\mu(x_1, x_2)$ one wants to use. Inspired by honest medical applications, we would *not* want to require that the cross-sectional density $\mu(x_1, x_2)$ be smooth, or even continuous. Jump discontinuities in $\mu(x_1, x_2)$ correspond naturally to changes from bone to muscle, etc. Moreover, although, mathematically speaking, the lines extend infinitely, in practice the paths are finite. The easiest thing is just to assume that $\mu(x_1, x_2)$ is zero outside of some bounded region ($\mu(x_1, x_2)$ has compact support); it's describing the density of a slice of a finite-extent body, after all.

Circ function. There aren't too many cases where one can compute the Radon transform explicitly. One example where it's possible is the circ function introduced earlier. Expressed in polar coordinates it's

$$\operatorname{circ}(r) = \begin{cases} 1, & r < 1, \\ 0, & r \geq 1. \end{cases}$$

We have to integrate the circ function along any line. Think in terms of projections, as defined above. From the circular symmetry, it's clear that the projections are independent of ϕ; i.e., in the figures below the Radon transform has the same values on the two families of lines, and the values depend only on ρ.

[30] The coordinates describing the collection of all planes in 3D are analogous the 2D coordinates for lines: directed distance of the plane from the origin together with a unit normal to the plane, the latter specified by two angles (longitude and latitude). The (ρ, ϕ)-coordinates thus generalize nicely, whereas, for example, the slope-intercept coordinates don't generalize as directly. Plane impulses are more involved because integrals over planes are more involved than integrals over lines. Higher dimensions? Sure.

[31] But don't read too much into the term "projection." You are *not* geometrically projecting the shape of the two-dimensional cross section that the lines are cutting through. You are looking at the attenuated, parallel X-rays that emerge.

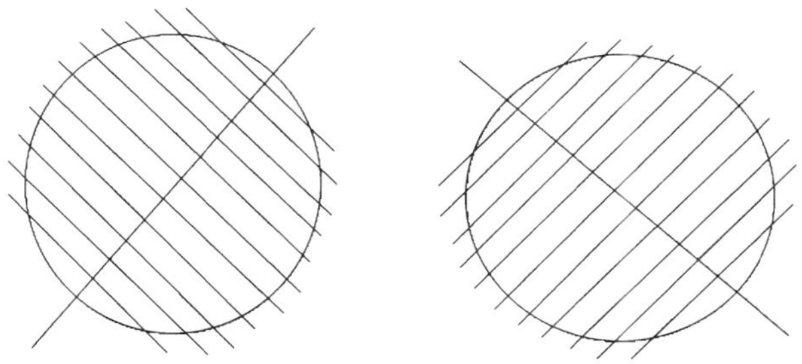

Because of this we can take any convenient value of ϕ, say $\phi = 0$, and find the integrals over the parallel lines in this family (vertical lines). The circ function is 0 outside the unit circle, so we need only to find the integral of the constant function 1 over any chord of the unit circle parallel to the x_2-axis. This is easy. If the chord is at a distance ρ from the origin, $|\rho| < 1$, then

$$\mathcal{R}(1)(\rho, 0) = \int_{-\sqrt{1-\rho^2}}^{\sqrt{1-\rho^2}} 1\, dx_2 = 2\sqrt{1-\rho^2}\,.$$

Thus for any (ρ, ϕ),

$$\mathcal{R}\operatorname{circ}(\rho, \phi) = \begin{cases} 2\sqrt{1-\rho^2}, & |\rho| < 1, \\ 0, & |\rho| \geq 1. \end{cases}$$

Gaussians again. Another example where we can compute the Radon transform exactly is for a Gaussian,

$$g(x_1, x_2) = e^{-\pi(x_1^2+x_2^2)}\,.$$

Any guesses as to what $\mathcal{R}g$ is? Let's do it.

Using the representation in terms of the line impulse, we can write

$$\mathcal{R}g(\rho, \phi) = \int_{-\infty}^{\infty}\int_{-\infty}^{\infty} e^{-\pi(x_1^2+x_2^2)}\delta(\rho - x_1\cos\phi - x_2\sin\phi)\, dx_1\, dx_2\,.$$

We now make a change of variables in this integral, putting

$$\begin{aligned} u_1 &= x_1\cos\phi + x_2\sin\phi, \\ u_2 &= -x_1\sin\phi + x_2\cos\phi. \end{aligned}$$

This is a rotation of coordinates through an angle ϕ, making the u_1-axis correspond to the x_1-axis. The area element is the same, and we also have that

$$u_1^2 + u_2^2 = x_1^2 + x_2^2\,.$$

In the new coordinates the integral becomes

$$\begin{aligned}
\mathcal{R}g(\rho,\phi) &= \int_{-\infty}^{\infty}\int_{-\infty}^{\infty} e^{-\pi(u_1^2+u_2^2)}\delta(\rho-u_1)\,du_1du_2 \\
&= \int_{-\infty}^{\infty}\left(\int_{-\infty}^{\infty} e^{-\pi u_1^2}\delta(\rho-u_1)\,du_1\right)e^{-\pi u_2^2}\,du_2 \\
&= \int_{-\infty}^{\infty} e^{-\pi\rho^2}e^{-\pi u_2^2}\,du_2 \quad \text{(the usual property of the one-dimensional } \delta\text{)} \\
&= e^{-\pi\rho^2}\int_{-\infty}^{\infty} e^{-\pi u_2^2}\,du_2 \\
&= e^{-\pi\rho^2} \text{ (the integral is 1 because the Gaussian is normalized to have area 1).}
\end{aligned}$$

Writing this in polar coordinates, $r = x_1^2 + x_2^2$, we have shown that

$$\mathcal{R}(e^{-\pi r^2}) = e^{-\pi\rho^2}\,.$$

How about that!

Linearity. $\mathcal{R}(\alpha f+\beta g) = \alpha\mathcal{R}f+\beta\mathcal{R}g$. Of course. Integration is a linear function of the integrand.

Shifts. This is a little easier to write (and to derive) in vector form. Let $\underline{n} = (\cos\phi, \sin\phi)$. The result is

$$\mathcal{R}(\mu(\underline{x}-\underline{b})) = \mathcal{R}\mu(\rho-\underline{b}\cdot\underline{n},\phi)\,.$$

In words, shifting $\underline{x}$ by $\underline{b}$ has the effect of shifting each projection a distance $\underline{b}\cdot\underline{n}$ in the ρ-variable.

To derive this we write the definition as

$$\mathcal{R}(\mu(\underline{x}-\underline{b})) = \int_{\mathbb{R}^2}\mu(\underline{x}-\underline{b})\delta(\rho-\underline{x}\cdot\underline{n})\,d\underline{x}\,.$$

If $\underline{b} = (b_1, b_2)$, then the change of variable $u_1 = x_1 - b_1$ and $u_2 = x_2 - b_2$, or simply $\underline{u} = \underline{x} - \underline{b}$ with $\underline{u} = (u_1, u_2)$, converts this integral into

$$\begin{aligned}
\mathcal{R}(\mu(\underline{x}-\underline{b})) &= \int_{\mathbb{R}^2}\mu(\underline{u})\delta(\rho-(\underline{u}+\underline{b})\cdot\underline{n})\,d\underline{u} \\
&= \int_{\mathbb{R}^2}\mu(\underline{u})\delta(\rho-\underline{u}\cdot\underline{n}-\underline{b}\cdot\underline{n})\,d\underline{u} \\
&= \mathcal{R}\mu(\rho-\underline{b}\cdot\underline{n},\phi)\,.
\end{aligned}$$

Evenness. Finally, the Radon transform *always* has a certain symmetry. It is always an even function of ρ and ϕ. This means that

$$\mathcal{R}\mu(-\rho,\phi+\pi) = \mathcal{R}\mu(\rho,\phi)\,.$$

Convince yourself that this makes sense in terms of the projections. The derivation goes as follows:

$$\begin{aligned}
\mathcal{R}\mu(-\rho,\phi+\pi) &= \int_{-\infty}^{\infty}\int_{-\infty}^{\infty} \mu(x_1,x_2)\delta(-\rho - x_1\cos(\phi+\pi) - x_2\sin(\phi+\pi))\,dx_1\,dx_2\\
&= \int_{-\infty}^{\infty}\int_{-\infty}^{\infty} \mu(x_1,x_2)\delta(-\rho - x_1(-\cos\phi) - x_2(-\sin\phi))\,dx_1\,dx_2\\
&= \int_{-\infty}^{\infty}\int_{-\infty}^{\infty} \mu(x_1,x_2)\delta(-\rho + x_1\cos\phi + x_2\sin\phi)\,dx_1\,dx_2\\
&= \int_{-\infty}^{\infty}\int_{-\infty}^{\infty} \mu(x_1,x_2)\delta(\rho - x_1\cos\phi - x_2\sin\phi)\,dx_1\,dx_2 \quad \text{(because } \delta \text{ is even)}\\
&= \mathcal{R}\mu(\rho,\phi)
\end{aligned}$$

9.7.4. Inverting the Radon transform. The inversion problem is solved by a result that equates the *two-dimensional* Fourier transform of μ to a *one-dimensional Fourier transform* of $\mathcal{R}\mu(\rho,\phi)$, *taken with respect to* ρ. Once $\mathcal{F}\mu$ is known, μ can be found by Fourier inversion.

The formulation of this equation between the Fourier transforms of μ and its projections is called the *projection-slice theorem*[32] and is the cornerstone of tomography. We'll go through the derivation, but it must be said at once that for practical applications all of this has to be implemented *numerically*, i.e., with the DFT (and the FFT). Much of the early work in computed tomography was in finding efficient algorithms for doing just this.

Starting with

$$\mathcal{R}\mu(\rho,\phi) = \int_{-\infty}^{\infty}\int_{-\infty}^{\infty} \mu(x_1,x_2)\delta(\rho - x_1\cos\phi - x_2\sin\phi)\,dx_1\,dx_2\,,$$

we want to find its Fourier transform with respect to ρ, regarding ϕ as fixed. Seems rather ambitious, if not preposterous, but watch. For lack of a better notation, we write this as $\mathcal{F}_\rho\mathcal{R}\mu$. Calling the frequency variable r, dual to ρ, we then have

$$\begin{aligned}
\mathcal{F}_\rho\mathcal{R}\mu(r,\phi) &= \int_{-\infty}^{\infty} e^{-2\pi i r\rho}\mathcal{R}\mu(\rho,\phi)\,d\rho\\
&= \int_{-\infty}^{\infty} e^{-2\pi i r\rho}\int_{-\infty}^{\infty}\int_{-\infty}^{\infty} \mu(x_1,x_2)\delta(\rho - x_1\cos\phi - x_2\sin\phi)\,dx_1\,dx_2\,d\rho\\
&= \int_{-\infty}^{\infty}\int_{-\infty}^{\infty} \mu(x_1,x_2)\left(\int_{-\infty}^{\infty} \delta(\rho - x_1\cos\phi - x_2\sin\phi)e^{-2\pi i r\rho}\,d\rho\right)dx_1\,dx_2\\
&= \int_{-\infty}^{\infty}\int_{-\infty}^{\infty} \mu(x_1,x_2)e^{-2\pi i r(x_1\cos\phi + x_2\sin\phi)}\,dx_1\,dx_2\\
&= \int_{-\infty}^{\infty}\int_{-\infty}^{\infty} \mu(x_1,x_2)e^{-2\pi i(x_1 r\cos\phi + x_2 r\sin\phi)}\,dx_1\,dx_2\,.
\end{aligned}$$

Check out what happened here: by interchanging the order of integration we wind up integrating the one-dimensional δ against the complex exponential $e^{-2\pi i r\rho}$. For

[32] Also called the central slice theorem, or the center slice theorem.

that integration we can regard $\delta(\rho - x_1 \cos\phi - x_2 \sin\phi)$ as a shifted δ-function, and the integration with respect to ρ produces $e^{-2\pi i(x_1 r\cos\phi + x_2 r \sin\phi)}$. Now if we let

$$\xi_1 = r\cos\phi\,,$$
$$\xi_2 = r\sin\phi\,,$$

the remaining double integral is

$$\int_{-\infty}^{\infty}\int_{-\infty}^{\infty} e^{-2\pi i(x_1\xi_1 + x_2\xi_2)}\mu(x_1, x_2)\,dx_1\,dx_2 = \int_{\mathbb{R}^2} e^{-2\pi i\underline{x}\cdot\underline{\xi}}\,\mu(\underline{x})\,d\underline{x}\,.$$

This is the two-dimensional Fourier transform of μ at (ξ_1, ξ_2)!

Observe that

$$r^2 = \xi_1^2 + \xi_2^2 \quad \text{and} \quad \tan\phi = \frac{\xi_2}{\xi_1}\,.$$

This means that (r, ϕ) is the set of polar coordinates for the (ξ_1, ξ_2)-frequency plane. With ϕ fixed and ρ varying, what we've found is $\mathcal{F}\mu(\xi_1, \xi_2)$ for (ξ_1, ξ_2) along the line in the (ξ_1, ξ_2)-plane described in polar coordinates by $\phi =$ constant and $-\infty < r < \infty$. So now change ϕ and do it all again. Then as ϕ varies between 0 and π (including 0, excluding π) and as r varies between $-\infty$ and ∞, we get all the points in the (ξ_1, ξ_2)-plane and all values of $\mathcal{F}\mu(\xi_1, \xi_2)$.

We have shown:

- Projection-Slice Theorem.

$$(\mathcal{F}_\rho \mathcal{R}\mu)(r, \phi) = \mathcal{F}\mu(\xi_1, \xi_2), \quad \xi_1 = r\cos\phi,\ \xi_2 = r\sin\phi\,.$$

From this we can find $\mu(x_1, x_2)$!

A summary. That derivation happened pretty fast. Let's slowly unpack the steps in using the projection-slice theorem to reconstruct a density μ from its projections, adding a little notation that we'll use in the next section. It's so impressive.

(1) We have a source and a detector that rotate together about some center. The angle of rotation is ϕ, where $0 \le \phi < \pi$.

(2) A family of parallel X-rays pass from the source through a (planar) region of unknown, variable density, $\mu(x_1, x_2)$, and are registered by the detector when they emerge.

For each ϕ the readings at the detector thus give a function $g_\phi(\rho)$ of ρ, where ρ is the (directed) distance that a particular X-ray is from the center of the beam of parallel X-rays.

Each such function $g_\phi(\rho)$, for different ϕ's, is called a projection.

(3) For each ϕ we compute $\mathcal{F}g_\phi(r)$, i.e., the Fourier transform of $g_\phi(\rho)$ with respect to ρ.

(4) Since $g_\phi(\rho)$ also depends on ϕ, so does its Fourier transform. Thus we have a function of two variables, $G(r, \phi)$. The projection-slice theorem tells us that this is the Fourier transform of μ:

$$\mathcal{F}\mu(\xi_1, \xi_2) = G(r, \phi), \quad \text{where } \xi_1 = r\cos\phi,\ \xi_2 = r\sin\phi\,.$$

Letting ϕ vary, $\mathcal{F}\mu(\xi_1, \xi_2)$ *is known* for all (ξ_1, ξ_2).

(5) Now take the inverse two-dimensional Fourier transform to recover μ:

$$\mu(\underline{x}) = \int_{\mathbb{R}^2} e^{2\pi i \underline{x}\cdot\underline{\xi}}\, \mathcal{F}\mu(\underline{\xi})\, d\underline{\xi}\,.$$

Running the numbers. Very briefly, let's go through how one might set up a numerical implementation of the procedure we've just been through. This comes from Briggs and Henson's all powerful DFT book, and you can track down many other references.

The function that we know is $g_\phi(\rho)$. That's what the sensor gives us, at least in discrete form. To normalize things we suppose that $g_\phi(\rho)$ is zero for $|\rho| \geq 1$. This means, effectively, that the region we're passing X-rays through is contained within the circle of radius one; the region is bounded so we can assume that it lies within some disk and we scale to assume the region lies within the unit disk.

Suppose we have M equal angles, $\phi_j = j\pi/M$, for $j = 0, \dots, M-1$. Suppose next that for each angle we send through N X-rays. We're assuming that $-1 \leq \rho \leq 1$, so the rays are spaced $\Delta\rho = 2/N$ apart and we index them to be

$$\rho_n = \frac{2n}{N}\,, \quad n = -\frac{N}{2}+1, \dots, \frac{N}{2}\,.$$

Then our projection data are the MN values

$$g_{nj} = g_{\phi_j}(\rho_n)\,, \quad j = 0, \dots, M-1\,,\; n = -\frac{N}{2}+1, \dots, \frac{N}{2}\,.$$

The first step in applying the projection-slice theorem is to find the one-dimensional Fourier transform of $g_{\phi_j}(\rho)$ with respect to ρ, which, since the function is zero for $|\rho| \geq 1$, is the integral

$$\mathcal{F}g_{\phi_j}(r) = \int_{-1}^{1} e^{-2\pi i r\rho} g_{\phi_j}(\rho)\, d\rho\,.$$

We have to approximate and discretize the integral. One approach to this is very much like the one we took in obtaining the DFT (Chapter 7). First, we're integrating with respect to ρ, and we already have sample points at the $\rho_n = 2n/N$; evaluating g at those points gives exactly $g_{nj} = g_{\phi_j}(\rho_n)$. We'll use these for a trapezoidal rule approximation.

We also have to discretize in r, the frequency variable dual to ρ. According to the sampling theorem, if we want to reconstruct $\mathcal{F}g_{\phi_j}(r)$ from its samples in r, the sampling rate is determined by the extent of $g_{\phi_j}(\rho)$ in the spatial domain, where the variable ρ is limited to $-1 \leq \rho \leq 1$. So the sampling rate in r is 2 and the sample points are spaced $1/2$ apart:

$$r_m = \frac{m}{2}\,, \quad m = -\frac{N}{2}+1, \dots, \frac{N}{2}\,.$$

The result of the trapezoidal approximation using $\rho_n = 2n/N$ and of discretizing in r using $r_m = m/2$ is

$$\mathcal{F}g_{\phi_j}(r_m) \approx \frac{2}{N} \sum_{n=-N/2+1}^{N/2} e^{-2\pi i \rho_n r_m} g_{nj}$$
$$= \frac{2}{N} \sum_{n=-N/2+1}^{N/2} e^{-2\pi i n m/N} g_{nj}\,.$$

(The 2 in $2/N$ comes in from the form of the trapezoidal rule.) Up to the constant out front, this is a DFT of the sequence (g_{nj}), $n = -N/2+1, \ldots, N/2$. (Here n is varying, while j indexes the projection.) That is,

$$\mathcal{F}g_{\phi_j}(r_m) \approx \frac{2}{N}\underline{\mathcal{F}}(g_{nj})[m]\,.$$

Computing this DFT for each of the M projections ϕ_j $(j = 0, \ldots, M-1)$ gives the data $\mathcal{F}g_{\phi_j}(r_m)$. Call this

$$G_{mj} = \underline{\mathcal{F}}(g_{nj})[m]\,.$$

The next step is to take the two-dimensional *inverse* Fourier transform of the data G_{mj}. Now, there's an interesting problem that comes up in implementing this efficiently. The G_{mj} are presented as data points based on a *polar coordinate* grid in the frequency domain, something like this:

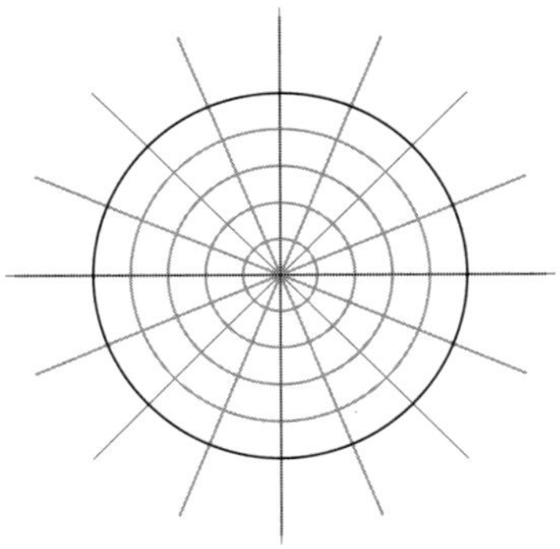

The vertices in this picture are the points (r_m, ϕ_j) and that's where the data points G_{mj} live. However, efficient FFT algorithms depend on the data being presented on a *Cartesian* grid. One way to reconcile this is to manufacture data at Cartesian grid points by taking a weighted average of the G_{mj} at the polar grid points which are nearest neighbors:

$$G_{\text{Cartesian}} = w_a G_a + w_b G_b + w_c G_c + w_d G_d\,.$$

This is called, appropriately enough, *gridding*.

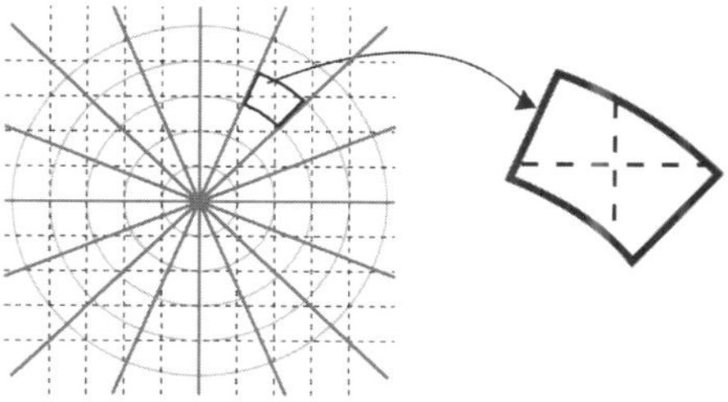

Choosing the weighting factors w_a, w_b, w_c, and w_d is part of the art, but *the most significant introductions of error in the whole process come from this step.* In the medical imaging business the errors introduced by approximating the transforms are termed *artifacts*, a better sounding word than "errors" when dealing with medical matters.

The final picture is then created by

$$\mu(\text{grid points in spatial domain}) = \underline{\mathcal{F}}^{-1}(G_{\text{Cartesian}})\,.$$

This is your brain. This is your brain on Fourier transforms. For a Fourier reconstruction of a model brain see the very accessible early survey paper: L. A. Shepp and J. B. Kruskal, Computerized tomography: The new medical X-ray technology, American Mathematical Monthly **85** (1978), 420–439. (Briggs and Henson also discuss this.) Their brain is modeled by a high density elliptical shell (the skull) with lower density elliptical regions inside. Here's the image.

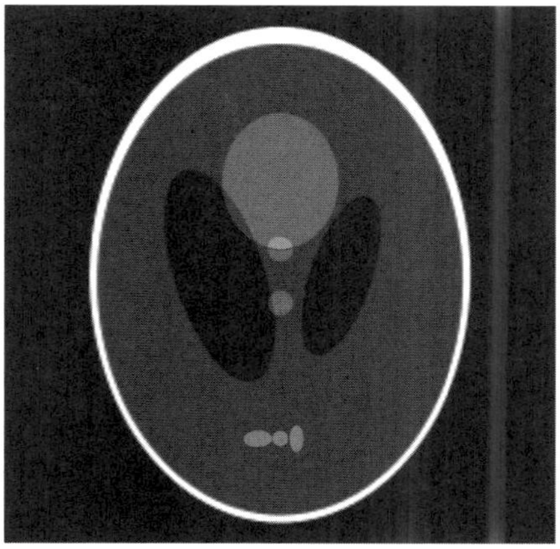

You can find this by searching for "Shepp-Logan phantom" (it's in the public domain). It's possible to compute explicitly the Radon transform for lines going through an elliptical region and the sampling can be carried out based on these formulas. Both MATLAB and Mathematica have implementations of the Radon transform and inversion as we've described it, so you can experiment numerically with reconstructions.

9.8. Appendix: Line Impulses

In the previous section we used a δ-function along a line, known as a line impulse, in connection with the Radon transform. I want to provide some further details. I don't often see line impulses covered in textbooks, and I've always found them to be interesting and attractive topic. We'll define line impulses in several different ways, though the definitions will all be in terms of pairing with a test function. We do need a little background, and I also apologize in advance for not motivating just why things turn out as they do.

It's possible to define impulses along circles, ellipses, and more generally along level curves $g(x_1, x_2) =$ constant of a function $g(x_1, x_2)$. The definition uses a construction for distributions — pullbacks — that we mentioned briefly in Chapter 5 and that I'll treat in more detail in Section 9.9.

9.8.1. Lines. First things first. We have to standardize how we describe lines in $\mathbb{R}^2$. We'll use the same idea as earlier, in Section 9.7.2, when we introduced coordinates on the set of all lines in $\mathbb{R}^2$. To recall, geometrically a line is determined by a unit vector $\underline{n}$ normal to the line and a signed distance c (previously called ρ) from the origin to the line. If $c > 0$, then the line is at a distance c from $\underline{0}$ in the direction of $\underline{n}$, and if $c < 0$, it's at a distance $|c|$ from $\underline{0}$ in the direction $-\underline{n}$. Thus in all cases $c\underline{n}$ goes from the origin to the line, and the point on the line closest to the origin is $c\underline{n}$. If $c = 0$, the line passes through $\underline{0}$, and then $\underline{n}$ and $-\underline{n}$ determine the same line. The vector equation of the line is

$$\underline{n} \cdot \underline{x} = c,$$

or in Cartesian coordinates (x_1, x_2),

$$ax_1 + bx_2 = c,$$

with

$$\underline{n} = (a, b), \quad a^2 + b^2 = 1.$$

We let

$$\underline{n}^\perp = (b, -a).$$

(Read this as "$\underline{n}$ perp.") We get $\underline{n}^\perp$ from $\underline{n}$ by a *clockwise* rotation of 90°. Be careful: despite the $\perp$ sign, $\underline{n}^\perp$ is *parallel* to the line $\underline{n} \cdot \underline{x} = c$. Also note that

$$(\underline{n}^\perp)^\perp = -\underline{n}.$$

I admit that it takes some remembering to keep straight which vector points which way relative to the line, but just try labeling things the other way and see the kinds of trouble you get into.

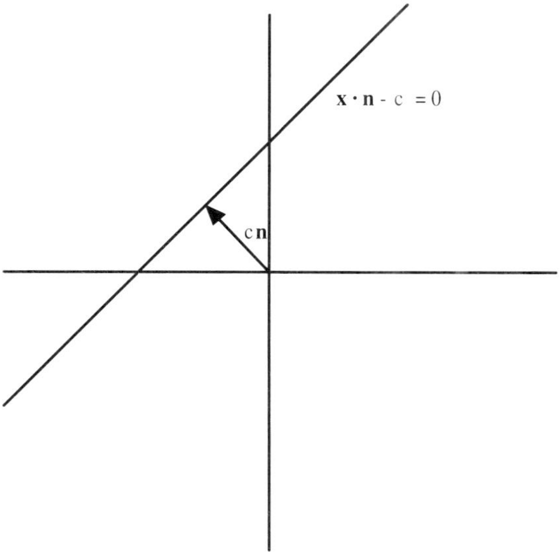

One other fact. The line $\underline{n} \cdot \underline{x} = c_1$ meets the line $\underline{n}^{\perp} \cdot \underline{x} = c_2$ at the point $c_1\underline{n} + c_2\underline{n}^{\perp}$, a distance $\sqrt{c_1^2 + c_2^2}$ from the origin.

Line coordinates. Some formulas and many calculations, especially integrals and integrations, will be simplified by using $\underline{n}^{\perp}$ and $\underline{n}$ as a basis for $\mathbb{R}^2$, writing a point $\underline{x} \in \mathbb{R}^2$ as

$$\underline{x} = s\underline{n}^{\perp} + t\underline{n}.$$

Observe, then, that it's the ordered basis $\{\underline{n}^{\perp}, \underline{n}\}$ that is positively oriented, in the sense that you go from $\underline{n}^{\perp}$ to $\underline{n}$ by a *counterclockwise* rotation of 90°.[33]

We can also think of $\underline{n}^{\perp}$ and $\underline{n}$ as providing a change of coordinates through

$$\begin{pmatrix} x_1 \\ x_2 \end{pmatrix} = \begin{pmatrix} b & a \\ -a & b \end{pmatrix} \begin{pmatrix} s \\ t \end{pmatrix},$$

or

$$\begin{aligned} x_1 &= bs + at\,, \\ x_2 &= -as + bt\,. \end{aligned}$$

This is a clockwise rotation of the (s,t)-coordinates to the (x_1, x_2)-coordinates through an angle θ, where $\tan\theta = -a/b$. The inverse is

$$\begin{pmatrix} s \\ t \end{pmatrix} = \begin{pmatrix} b & -a \\ a & b \end{pmatrix} \begin{pmatrix} x_1 \\ x_2 \end{pmatrix},$$

or

$$\begin{aligned} s &= bx_1 - ax_2\,, \\ t &= ax_1 + bx_2\,. \end{aligned}$$

The coordinate change is an orthogonal transformation of $\mathbb{R}^2$, so the area elements are equal, $dsdt = dx_1dx_2$.

[33] In the same way that you go from $\underline{e}_1 = (1,0)$ to $\underline{e}_2 = (0,1)$ by a counterclockwise rotation of 90°.

A little more generally, we can allow for an affine change

$$\underline{x} = s\underline{n}^{\perp} + t\underline{n} + c\underline{n},$$

or

$$\begin{pmatrix} x_1 \\ x_2 \end{pmatrix} = \begin{pmatrix} b & a \\ -a & b \end{pmatrix} \begin{pmatrix} s \\ t \end{pmatrix} + c\begin{pmatrix} a \\ b \end{pmatrix},$$

or

$$\begin{aligned} x_1 &= bs + at + ac, \\ x_2 &= -as + bt + bc\,. \end{aligned}$$

It will be helpful to have the s-axis, that's $t = 0$, correspond to the line $\underline{n} \cdot \underline{x} = c$, so the parametric equations of $\underline{n} \cdot \underline{x} = c$ are

$$\underline{x} = s\underline{n}^{\perp} + c\underline{n}\,, \quad \text{or} \quad x_1 = bs + ac, \quad x_2 = -as + bc\,, \quad -\infty < s < \infty\,.$$

Since $\|\underline{n}^{\perp}\| = 1$, this is the arclength parametrization, meaning that the line is traced out at unit speed.

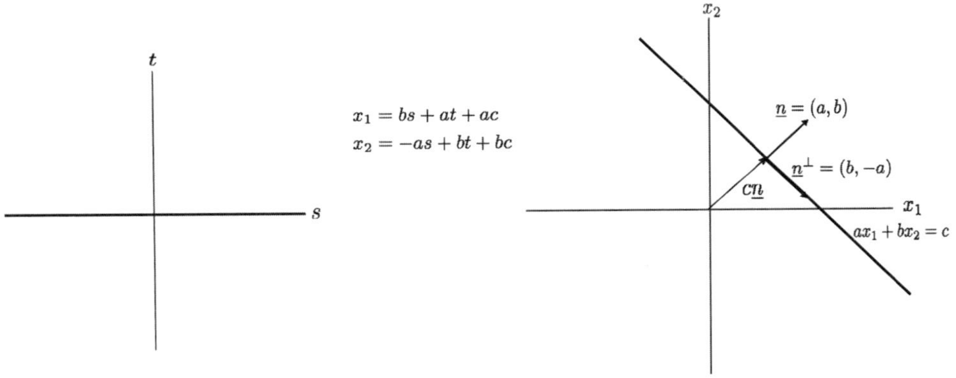

In many cases, using these coordinates will turn a two-dimensional calculation into a one-dimensional calculation imbedded in two dimensions. Be open minded.

9.8.2. Π-functions for strips. We'll need a generalization for strips of the rectangle function Π_p for intervals. Just as an interval is determined by its center and its width, a strip is determined by its center line and its width. A strip centered on a line through the origin of width p is thus bounded by two parallel lines, $\underline{n} \cdot \underline{x} = p/2$ and $\underline{n} \cdot \underline{x} = -p/2$. The strip is the region in between (excluding the boundary lines[34]) and is described by $|\underline{n} \cdot \underline{x}| < p/2$. We define

$$\Pi_{p\underline{n}}(\underline{x}) = \begin{cases} 1, & |\underline{n} \cdot \underline{x}| < p/2, \\ 0, & |\underline{n} \cdot \underline{x}| \geq p/2. \end{cases}$$

[34] I invite vigorous debate on assigning a value on the boundary lines.

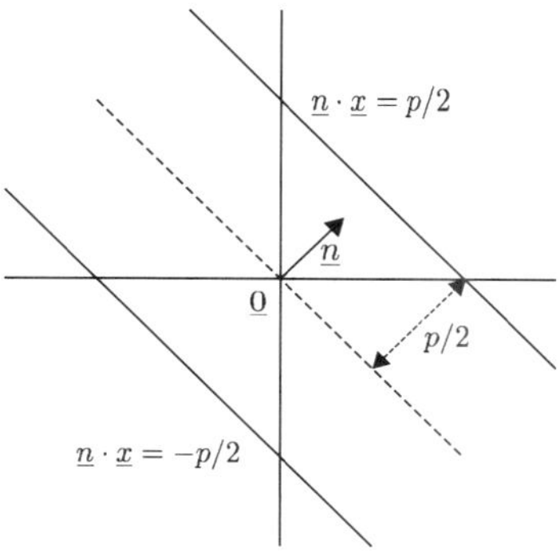

Observe that $\Pi_{p\underline{n}}$ is even,

$$\Pi_{p\underline{n}}(-\underline{x}) = \Pi_{p\underline{n}}(\underline{x}),$$

and that

$$\Pi_{p\underline{n}}(\underline{x}) = \Pi_{\underline{n}}\Big(\frac{1}{p}\underline{x}\Big).$$

The Π-function for a strip centered on a line $\underline{n} \cdot \underline{x} = c$ is $\Pi_{p\underline{n}}(\underline{x} - c\underline{n})$.

9.8.3. Line impulses: Definitions. We'll use the notation $\delta_{\underline{n}}(\underline{x} - c\underline{n})$ for the soon to be defined line impulse for the line $\underline{n}\cdot\underline{x} = c$. There are several ways one may choose to define $\delta_{\underline{n}}(\underline{x} - c\underline{n})$, and each has its advantages. The most direct approach is to use a sequence of approximating functions, much as in the one-dimensional case.

$\delta_{\underline{n}}$ *via approximating functions.* In one dimension we took on δ via a sequence of scaled rectangle functions $(1/\epsilon)\Pi_\epsilon(x)$, obtaining the pairing of δ with a test function $\varphi(x)$ as a limit,

$$\langle \delta, \varphi \rangle = \lim_{\epsilon \to 0} \int_{-\infty}^{\infty} \frac{1}{\epsilon}\Pi_\epsilon(x)\varphi(x)\, dx = \lim_{\epsilon \to 0} \frac{1}{\epsilon} \int_{-\epsilon/2}^{\epsilon/2} \varphi(x)\, dx = \varphi(0)\,.$$

To define a line impulse for the line $\underline{n} \cdot \underline{x} = c$ we use the sequence

$$\frac{1}{\epsilon}\Pi_{\epsilon\underline{n}}(\underline{x} - c\underline{n})\,.$$

Then for a test function $\varphi(\underline{x})$,

$$\begin{aligned} \langle \delta_{\underline{n}}(\underline{x} - c\underline{n}), \varphi(\underline{x}) \rangle &= \lim_{\epsilon \to 0} \int_{\mathbb{R}^2} \frac{1}{\epsilon}\Pi_{\epsilon\underline{n}}(\underline{x} - c\underline{n})\varphi(\underline{x})\, d\underline{x} \\ &= \lim_{\epsilon \to 0} \frac{1}{\epsilon} \int_{|\underline{n}\cdot\underline{x}| < \epsilon/2} \varphi(x_1, x_2)\, dx_1 dx_2. \end{aligned}$$

This has every appearance of turning into a line integral, and it will. Watch. Employ the coordinates

$$x_1 = bs + at + ac, \quad x_2 = -as + bt + bc\,.$$

Then

$$\frac{1}{\epsilon} \int_{|\underline{n}\cdot\underline{x}| < \epsilon/2} \varphi(x_1, x_2)\, dx_1 dx_2 = \frac{1}{\epsilon} \int_{-\infty}^{\infty} \left(\int_{-\frac{\epsilon}{2}}^{\frac{\epsilon}{2}} \varphi(bs + at + ac, -as + bt + bc)\, dt \right) ds\,.$$

As $\epsilon \to 0$ this becomes

$$\int_{-\infty}^{\infty} \varphi(bs + ac, -as + bc)\, ds\,.$$

That's exactly the line integral of $\varphi(x_1, x_2)$ along the line $ax_1 + bx_2 = c$: you parametrize the line by arclength, here it's s, and integrate $\varphi(x_1(s), x_2(s))$.

This is the basic definition,

$$\langle \delta_{\underline{n}}(\underline{x} - c\underline{n}), \varphi(\underline{x})\rangle = \int_{-\infty}^{\infty} \varphi(bs + ac, -as + bc)\, ds\,,$$

against which other expressions will be compared.

Pullback. Speaking of other expressions, there's another way of expressing this line integral that will be useful in some derivations. Introduce the function $\psi(s,t)$ by

$$\psi(s,t) = \varphi(bs + at + ac, -as + bt + bc)\,.$$

Mathematicians refer to this as a *pullback* of the function $\varphi(x_1, x_2)$ from the (x_1, x_2)-plane to the (s,t)-plane. We met this notion in Section 5.6.1, and we'll revisit it in Section 9.9. When $t = 0$, we have $\psi(s, 0) = \varphi(bs + ac, -as + bc)$. The function φ is evaluated along the line $ax_1 + bx_2 = c$, whose equation in the (s,t)-plane is $t = 0$, and

$$\langle \delta_{\underline{n}}(\underline{x} - c\underline{n}), \varphi(\underline{x})\rangle = \int_{-\infty}^{\infty} \psi(s, 0)\, ds\,.$$

Finally, from a course in vector calculus you may be familiar with still another way of writing line integrals, say, as an expression of the form

$$\int_C P(x_1, x_2)dx_1 + Q(x_1, x_2)dx_2\,,$$

where C is the path of integration.[35] One also says that the integrand $P(x_1, x_2)dx_1 + Q(x_1, x_2)dx_2$ is a *1-form.* Using this notation, you can check that

$$\langle \delta_{\underline{n}}(\underline{x} - c\underline{n}), \varphi(\underline{x})\rangle = \int_{ax_1 + bx_2 = c} \varphi(x_1, x_2)(b dx_1 - a dx_2)\,.$$

See Section 9.8.6 for additional discussion.

Line impulses and the 1D δ. For a little formal reassurance on all this let's look at the line impulses in the x_1- and x_2-coordinate directions, respectively, and the 1D deltas $\delta(x_1)$ and $\delta(x_2)$ in the coordinate directions. By this I mean that

$$\int_{-\infty}^{\infty} \delta(x_1)\varphi(x_1, x_2)\, dx_1 = \varphi(0, x_2)\,, \quad \int_{-\infty}^{\infty} \delta(x_2)\varphi(x_1, x_2)\, dx_2 = \varphi(x_1, 0)\,.$$

We want to see that $\delta(x_1)$ is the line impulse for the x_2-axis ($x_1 = 0$) and that $\delta(x_2)$ is the line impulse for the x_1-axis ($x_2 = 0$). Treating the pairing of $\delta(x_1)$ and

---

[35]People still call this a line integral even if the path is a curve.

$\varphi(x_1, x_2)$ as integration and operating as we would with one-dimensional δ's, we have

$$\begin{aligned}\langle\delta(x_1), \varphi(x_1, x_2)\rangle &= \int_{-\infty}^{\infty}\int_{-\infty}^{\infty} \delta(x_1)\varphi(x_1, x_2)\, dx_1 dx_2 \\ &= \int_{-\infty}^{\infty}\left(\int_{-\infty}^{\infty} \delta(x_1)\varphi(x_1, x_2)\, dx_1\right) dx_2 \\ &= \int_{-\infty}^{\infty} \varphi(0, x_2)\, dx_2\,.\end{aligned}$$

That's the line integral of $\varphi(x_1, x_2)$ along the x_2-axis. In the same manner

$$\langle\delta(x_2), \varphi(x_1, x_2)\rangle = \int_{-\infty}^{\infty} \varphi(x_1, 0)\, dx_1.$$

Similarly, and more generally, the line $\underline{n}\cdot\underline{x} = c$, $\underline{n} = (a, b)$, is described by $t = c$ in the line coordinates $\underline{x} = s\underline{n}^{\perp} + t\underline{n} = (bs + at, -as + bt)$. The 1D δ that we want to pair with is $\delta(t - c)$, and loading the expression up with variables,

$$\begin{aligned}\langle\delta(t - c), \varphi(x_1(s, t), x_2(s, t))\rangle &= \int_{-\infty}^{\infty}\int_{-\infty}^{\infty} \delta(t - c)\varphi(bs + at, -as + bt)\, dsdt \\ &= \int_{-\infty}^{\infty}\left(\int_{-\infty}^{\infty} \delta(t - c)\varphi(bs + at, -as + bt)\, dt\right) ds \\ &= \int_{-\infty}^{\infty} \varphi(bs + ac, -as + bc)\, ds\,.\end{aligned}$$

We're back to our earlier result.

If you look back at the derivation of the projection-slice theorem in Section 9.7.4 you'll see this kind of use of a line impulse.

$\delta_{\underline{n}}$ as the derivative of the unit cliff. In one dimension we have the familiar formula

$$\frac{d}{dx}H(x) = \delta(x),$$

and more generally

$$\frac{d}{dx}H(x - c) = \delta(x - c)\,,$$

where

$$H(x) = \begin{cases} 0, & x \leq 0\,, \\ 1, & x > 0\,, \end{cases}$$

is the Heaviside unit step.

We can generalize this result to line impulses using a *unit cliff* instead of a unit step. Define

$$H_{\underline{n}}(\underline{x}) = H(\underline{n}\cdot\underline{x}) = \begin{cases} 0, & \underline{n}\cdot\underline{x} \leq 0\,, \\ 1, & \underline{n}\cdot\underline{x} > 0\,. \end{cases}$$

The cliff is along the line $\underline{n}\cdot\underline{x} = 0$, and the unit normal $\underline{n}$ points into the half-plane where $H_{\underline{n}}(\underline{x}) = 1$, where the graph of $H_{\underline{n}}$ has a plateau of height 1. To put a cliff along $\underline{n}\cdot\underline{x} = c$ we shift to $H_{\underline{n}}(\underline{x} - c\underline{n})$, and again $\underline{n}$ points into the upper plateau.

Let $\nabla_{\underline{n}}$ denote the directional derivative in the direction $\underline{n} = (a, b)$. Thus for a function $f(x_1, x_2)$,

$$\nabla_{\underline{n}} f = \operatorname{grad} f \cdot \underline{n} = a\frac{\partial f}{\partial x_1} + b\frac{\partial f}{\partial x_2}.$$

Let's show that

$$\nabla_{\underline{n}} H_{\underline{n}}(\underline{x} - c\underline{n}) = \delta_{\underline{n}}(\underline{x} - c\underline{n}).$$

This formula makes it clear that $\delta_{\underline{n}}(\underline{x} - c\underline{n})$ is zero in the two open half-planes bounded by $\underline{n} \cdot \underline{x} - c = 0$. The action is at the edge.

To verify the formula, take a test function $\varphi(\underline{x})$ and appeal to the definition of differentiating a distribution (integration by parts) to write

$$\begin{aligned}
\langle \nabla_{\underline{n}} H_{\underline{n}}(\underline{x} - c\underline{n}), \varphi(\underline{x})\rangle &= \int_{\mathbb{R}^2} \nabla_{\underline{n}} H_{\underline{n}}(\underline{x} - c\underline{n})\varphi(\underline{x})\, dx_1 dx_2 \\
&= -\int_{\mathbb{R}^2} H_{\underline{n}}(\underline{x} - c\underline{n})\nabla_{\underline{n}}\varphi(\underline{x})\, dx_1 dx_2.
\end{aligned}$$

Now introduce the coordinates

$$x_1 = bs + at + ac, \quad x_2 = -as + bt + bc$$

and the pullback

$$\psi(s, t) = \varphi(bs + at + ac, -as + bt + bc).$$

Then

$$\frac{\partial}{\partial t}\psi(s, t) = \nabla_{\underline{n}}\varphi(x_1, x_2)$$

for as we can see by the chain rule,

$$\begin{aligned}
\frac{\partial}{\partial t}\psi(s, t) &= \frac{\partial}{\partial t}\varphi(x_1(s, t), x_2(s, t)) \\
&= \frac{\partial \varphi}{\partial x_1}\frac{\partial x_1}{\partial t} + \frac{\partial \varphi}{\partial x_2}\frac{\partial x_2}{\partial t} \\
&= a\frac{\partial \varphi}{\partial x_1} + b\frac{\partial \varphi}{\partial x_2} \\
&= \nabla_{\underline{n}}\varphi(x_1, x_2).
\end{aligned}$$

Changing coordinates and using the definition of $H_{\underline{n}}$ then yields

$$\begin{aligned}
-\int_{\mathbb{R}^2} H(\underline{n} \cdot \underline{x} - c)\nabla_{\underline{n}}\varphi(\underline{x})\, dx_1 dx_2 &= -\int_{-\infty}^{\infty}\int_0^{\infty} \frac{\partial}{\partial t}\psi(s, t)\, dt ds \\
&= \int_{-\infty}^{\infty} \psi(s, 0)\, ds \\
&= \int_{-\infty}^{\infty} \varphi(bs + ac, -as + bc)\, ds \\
&= \langle \delta_{\underline{n}}(\underline{x} - c\underline{n}), \varphi(\underline{x})\rangle,
\end{aligned}$$

which is what we had to show. Pretty good.

9.8.4. Properties of line impulses. We'll only do two properties of line impulses, symmetry and scaling. There are others, sampling and convolution to name two, but this is supposed to be a short section.

Symmetries of $\delta_{\underline{n}}$. The one-dimensional δ is even as a distribution, and one usually writes simply

$$\delta(-x) = \delta(x), \quad \text{and also} \quad \delta(-(x-c)) = \delta(x+c).$$

Similar results hold for a line impulse. To derive them, recall the notation

$$\varphi^-(x) = \varphi(-x)$$

for a function, and for a distribution S the distribution S^- defined by

$$\langle S^-, \varphi\rangle = \langle S, \varphi^-\rangle.$$

Now compute

$$\begin{aligned}
\langle \delta_{\underline{n}}^-(\underline{x} - c\underline{n}), \varphi(\underline{x})\rangle &= \langle \delta_{\underline{n}}(\underline{x} - c\underline{n}), \varphi^-(\underline{x})\rangle \\
&= \int_{-\infty}^{\infty} \varphi^-(bs + ac, -as + bc)\, ds \\
&= \int_{-\infty}^{\infty} \varphi(-(bs + ac), -(-as + bc))\, ds \\
&= \int_{-\infty}^{\infty} \varphi(-bs - ac, , as - bc)\, ds \\
&= \int_{-\infty}^{\infty} \varphi(bs + a(-c), -as + b(-c))\, ds \\
&= \langle \delta_{\underline{n}}(\underline{x} + c\underline{n}), \varphi(\underline{x})\rangle.
\end{aligned}$$

Thus

$$\delta_{\underline{n}}^-(\underline{x} - c\underline{n}) = \delta_{\underline{n}}(\underline{x} + c\underline{n}),$$

or as we might then write

$$\delta_{\underline{n}}(-\underline{x}) = \delta_{\underline{n}}(\underline{x}) \quad \text{and} \quad \delta_{\underline{n}}(-(\underline{x} - c\underline{n})) = \delta_{\underline{n}}(\underline{x} + c\underline{n})\,.$$

Scaling line impulses. Allowing for some shameless integrations we can find a formula for a scaled line impulse $\delta_n(p\underline{x})$, $p > 0$. It goes like this:

$$\begin{aligned}
\langle \delta_{\underline{n}}(p\underline{x}), \varphi(\underline{x})\rangle &= \int_{\mathbb{R}^2} \delta_{\underline{n}}(p\underline{x})\varphi(\underline{x})\, d\underline{x} \\
&= \int_{\mathbb{R}^2} \delta_{\underline{n}}(\underline{u})\frac{1}{p^2}\varphi\left(\frac{1}{p}\underline{u}\right)\, d\underline{u} \quad \text{(substituting } \underline{u} = p\underline{x}\text{)} \\
&= \frac{1}{p^2}\int_{-\infty}^{\infty} \varphi\left(\frac{bs}{p}, -\frac{as}{p}\right)\, ds \quad \text{(using the parametrization of } \underline{n}\cdot\underline{x} = 0\text{)} \\
&= \frac{1}{p}\int_{-\infty}^{\infty} \varphi(bs', -as')\, ds' \quad \text{(substituting } s' = s/p\text{)} \\
&= \frac{1}{p}\langle \delta_{\underline{n}}(\underline{x}), \varphi(\underline{x})\rangle.
\end{aligned}$$

So, just as in the 1D case, we have the formula

$$\delta_{\underline{n}}(p\underline{x}) = \frac{1}{p}\delta_{\underline{n}}(\underline{x}).$$

Since a general invertible linear transformation A maps lines to lines, we could also express $\delta_{\underline{n}}(A\underline{x})$ as a line impulse. The formula above is for

$$A = \begin{pmatrix} p & 0 \\ 0 & p \end{pmatrix}.$$

I invite you to find the general formula!

9.8.5. Fourier transform of a line impulse. The main order of business in this section is to obtain the Fourier transform of a line impulse. We'll start by assuming that the line passes through the origin and then get the more general result via a shift. We need to invoke the definition of the Fourier transform for tempered distributions: for a test function $\varphi(\underline{x})$,

$$\langle \mathcal{F}\delta_{\underline{n}}, \varphi \rangle = \langle \delta_{\underline{n}}, \mathcal{F}\varphi \rangle .$$

The right-hand side is something we can work with, and doing so depends on partial Fourier transforms, from Section 9.2.3, and their inverses.

We'll again use the pullback, $\psi(s,t) = \varphi(bs + at, -as + bt)$ (here $c = 0$), but we need dual variables to s and t since in $\langle \delta_{\underline{n}}, \mathcal{F}\varphi \rangle$ we're working with the Fourier transform of φ. Call the dual variables σ and τ, so that $\tau = 0$ corresponds to the line $a\xi_1 + b\xi_2 = c$. Then

$$\langle \delta_{\underline{n}}, \mathcal{F}\varphi \rangle \quad = \quad \int_{-\infty}^{\infty} \mathcal{F}\psi(\sigma, 0)\, d\sigma .$$

Next,

$$\mathcal{F}\psi(\sigma, 0) \quad = \quad \int_{-\infty}^{\infty}\int_{-\infty}^{\infty} \psi(s,t) e^{2\pi i s\sigma + t\cdot 0}\, dsdt = \int_{-\infty}^{\infty}\int_{-\infty}^{\infty} \psi(s,t) e^{2\pi i s\sigma}\, dsdt,$$

and hence

$$\begin{aligned} \int_{-\infty}^{\infty} \mathcal{F}\psi(\sigma, 0)\, d\sigma &= \int_{-\infty}^{\infty}\int_{-\infty}^{\infty} \left(\int_{-\infty}^{\infty} \psi(s,t) e^{2\pi i s\sigma}\, ds \right) dtd\sigma \\ &= \int_{-\infty}^{\infty} \left(\int_{-\infty}^{\infty} \mathcal{F}_1\psi(\sigma,t)\, d\sigma \right) dt. \end{aligned}$$

That's the partial Fourier transform in the first slot coming in. But now,

$$\begin{aligned} \int_{-\infty}^{\infty} \mathcal{F}_1(\sigma,t)\, d\sigma &= \int_{-\infty}^{\infty} e^{2\pi i \sigma\cdot 0} \mathcal{F}_1\psi(\sigma,t)\, d\sigma \\ &= \mathcal{F}_1^{-1}\mathcal{F}_1\psi(0;t) \\ &= \psi(0,t) \\ &= \varphi(at, bt) . \end{aligned}$$

Putting all this together and recalling that (at, bt) is the arclength parametrization of the line $\underline{n}^{\perp} \cdot \underline{x} = 0$ gives

$$\begin{aligned} \langle \mathcal{F}\delta_n, \varphi \rangle &= \int_{-\infty}^{\infty} \psi(0,t)\, dt \\ &= \int_{-\infty}^{\infty} \varphi(at,bt)\, dt \\ &= \langle \delta_{\underline{n}^{\perp}}, \varphi \rangle . \end{aligned}$$

Lo and behold, we have obtained the very attractive result

$$\mathcal{F}\delta_{\underline{n}} = \delta_{\underline{n}^\perp}\,.$$

From here we can get the Fourier transform of $\delta_n(\underline{x} - c\underline{n})$ straight from the shift formula (which we'll just gratefully accept):

$$\mathcal{F}\delta_n(\underline{x} - c\underline{n}) = e^{-2\pi i(\underline{\xi}\cdot c\underline{n})}\delta_{\underline{n}^\perp}(\underline{\xi}).$$

9.8.6. For those who are comfortable with differential forms, et al. For those who know about differential forms, wedge products, etc., here is a more direct definition of a line impulse without an explicit limiting process. Again I am leaving this unmotivated, with apologies and the comment that it follows from a more general construction of the pullback of a distribution. (There it is again.) See Friedlander, *Introduction to the Theory of Distributions*, to which I have referred previously, for further information.

Introduce the 1-form

$$\omega_g = b\,dx_1 - a\,dx_2\,,$$

where $g(x_1, x_2) = c - ax_1 - bx_2$ is the line we're impulsing along. The form is chosen so that

$$dg \wedge \omega_g = dx_1 \wedge dx_2.$$

That's the wedge product, $\wedge$, but if you're reading this section, you know that. The pairing of $\delta_{\underline{n}}(\underline{x} - c\underline{n})$ with φ is then *defined* to be the line integral

$$\langle \delta(\underline{x} - c\underline{n}), \varphi(\underline{x})\rangle = \int_{g=0} \varphi\omega_g = \int_{\underline{n}\cdot\underline{x}=c} \varphi(x_1, x_2)\,(b\,dx_1 - a\,dx_2).$$

This definition is invariant in the sense that while the pairing depends on $g(x_1, x_2)$ it is independent of any particular parametrization for the line. We do have to use a parametrization (any parametrization) to compute the integral, however. The arclength parametrization for the line is

$$x_1(s) = bs + ac, \quad x_2(s) = -as + bc,$$

and in terms of this parametrization

$$\langle \delta_{\underline{n}}(\underline{x} - c\underline{n}), \varphi(\underline{x})\rangle = \int_{-\infty}^{\infty} \varphi(bs + ac, -as + bc)\,ds,$$

the same result we obtained previously via a limit of approximating functions.

Other line impulses. We have standardized the way we describe lines by using the unit normal and the distance from the origin. When we use the notation $\delta_{\underline{n}}$ or $\delta_{\underline{n}}(\underline{x} - c\underline{n})$, it is understood that $\|\underline{n}\| = 1$; this is the only line impulse that has been defined thus far. At least we can say, at this point, that $\delta_{-\underline{n}} = \delta_{\underline{n}}$. For if $\underline{n} = (a, b)$, then

$$\begin{aligned}
\langle \delta_{-\underline{n}}, \varphi\rangle &= \int_{-\infty}^{\infty} \varphi(-bs, as)\,ds \\
&= \int_{-\infty}^{\infty} \varphi(bs', -as')\,ds' \quad \text{(making a change of variable } s' = -s) \\
&= \langle \delta_{\underline{n}}, \varphi\rangle\,.
\end{aligned}$$

You can view this as another evenness property of line impulses.

Suppose we describe a line by

$$Ax_1 + Bx_2 = C,$$

or as the level set $0 = G(x_1, x_2) = C - Ax_1 - Bx_2$, where this time $A^2 + B^2 \neq 1$. What is the line impulse $\delta(G = 0)$ (for lack of a better notation), meaning, what is the pairing of $\delta(G = 0)$ with a test function $\varphi(\underline{x})$?

Most directly, we introduce

$$\omega_G = \frac{B}{A^2 + B^2} dx_1 - \frac{A}{A^2 + B^2} dx_2 = \frac{1}{A^2 + B^2} (B dx_1 - A dx_2) .$$

As before, ω_G is chosen so that

$$dG \wedge \omega_G = dx_1 \wedge dx_2 .$$

Then $\delta(G = 0)$ is defined by the pairing

$$\langle \delta(G = 0), \varphi(\underline{x}) \rangle = \int_{G=0} \varphi \omega_G = \frac{1}{A^2 + B^2} \int_{Ax_1 + Bx_2 = 0} \varphi(x_1, x_2) \, (B dx_1 - A dx_2).$$

The line integral (the definition of $\delta(G = 0)$) depends on $G(x_1, x_2)$ but is independent of the way the line $G = 0$ is parametrized. Computing the integral requires a parametrization, however. For example, if $B \neq 0$, we can use the parametrization

$$x_1(t) = t, \quad x_2(t) = \frac{C}{B} - \frac{A}{B} t .$$

Then

$$\begin{aligned} \langle \delta(G = 0), \varphi(\underline{x}) \rangle &= \frac{1}{A^2 + B^2} \int_{G=0} \varphi(x_1, x_2) \, (B dx_1 - A dx_2) \\ &= \frac{1}{A^2 + B^2} \int_{-\infty}^{\infty} \varphi\Big(t, \frac{C}{B} - \frac{A}{B} t\Big) \Big(B dt + \frac{A^2}{B} dt\Big) \\ &= \frac{1}{B} \int_{-\infty}^{\infty} \varphi\Big(t, \frac{C}{A} - \frac{B}{A} t\Big) \, dt . \end{aligned}$$

We could also introduce arclength parametrization after the fact, writing

$$x_1(s) = \frac{Bs}{\sqrt{A^2 + B^2}} + \frac{CA}{A^2 + B^2}, \quad x_2(s) = \frac{-As}{\sqrt{A^2 + B^2}} + \frac{CB}{A^2 + B^2} .$$

We then find that

$$\langle \delta(G = 0), \varphi(\underline{x}) \rangle = \frac{1}{\sqrt{A^2 + B^2}} \int_{-\infty}^{\infty} \varphi(x_1(s), x_2(s)) \, ds .$$

Another take on scaling a line impulse. With newfound flexibility in defining line impulses there's a little more we can say on scaling. Let $p > 0$ and let $\underline{n} = (a, b)$ be a unit vector. Apply the preceding calculation to the line $pax_1 + pax_2 = 0$, with $A = pa$ and $B = pb$, to define the line impulse $\delta_{p\underline{n}}$:

$$\begin{aligned} \langle \delta_{p\underline{n}}(\underline{x}), \varphi(\underline{x}) \rangle &= \frac{1}{p} \int_{-\infty}^{\infty} \varphi(bs, -as) \, ds \\ &= \frac{1}{p} \langle \delta_{\underline{n}}(\underline{x}), \varphi(\underline{x}) \rangle. \end{aligned}$$

This shows that

$$\delta_{p\underline{n}} = \frac{1}{p}\delta_{\underline{n}}\,,$$

and it means we can add to our earlier scaling formula:

$$\delta_{\underline{n}}(p\underline{x}) = \frac{1}{p}\delta_{\underline{n}}(\underline{x}) = \delta_{p\underline{n}}(\underline{x})\,.$$

Nice.

Similarly, for the line $pax_1 + pbx_2 = pc$, a distance c from the origin, we have

$$\delta_{p\underline{n}}(\underline{x} - pc\underline{n}) = \frac{1}{p}\delta_{\underline{n}}(\underline{x} - c\underline{n}).$$

9.9. Appendix: Pullback of a Distribution

I kept threatening to talk about this, so here's a little bit on the general notion of the *pullback* of a distribution. I think it's safe to say that the chief application of this construction is to define an impulse along a curve, a surface, etc., something like $\delta(g(\underline{x})) = \delta(g(x_1, \dots, x_n))$, where δ is the usual one-dimensional δ and $g(x_1, \dots, x_n)$ is a *real-valued function.*[36] A line impulse and its use in inverting the Radon transform is one example, but δ's along curves or surfaces also come up in fundamental solutions of partial differential equations. There's even a "Feynman fundamental solution" that uses such δ's in the case of a wave equation, so you know it's not just mathematicians who play with these things.

For a function $g(x_1, x_2)$ of two variables, $\delta(g(x_1, x_2))$ is an impulse along the set $g(x_1, x_2) = 0$, which is typically one or more curves in the (x_1, x_2)-plane. A line impulse corresponds to the case when $g(x_1, x_2)$ is an affine function; e.g., $g(x_1, x_2) = Ax_1 + Bx_2 - C$. A circle impulse uses, e.g., $g(x_1, x_2) = (x_1 - a)^2 + (x_2 - b)^2 - c^2$. For a function $g(x_1, x_2, x_3)$ we'd be looking at an impulse $\delta(g(x_1, x_2, x_3))$ along the surface (or surfaces) $g(x_1, x_2, x_3) = 0$ in $\mathbb{R}^3$. In general, with an assumption on g, below, the equation $g(x_1, x_2, \dots, x_n) = 0$ defines a smooth $(n-1)$-dimensional manifold in $\mathbb{R}^n$ (often called a *hypersurface*), and $\delta(g(x_1, x_2, \dots, x_n))$ is an impulse along it.

Just as we saw for the particular case of line impulses, the ultimate effect of the general definition is to make the pairing of $\delta(g(\underline{x}))$ with a test function $\varphi(\underline{x})$ (using vector notation) work out to be the integral of φ along the set defined by $g(\underline{x}) = 0$:

$$\langle \delta(g(\underline{x})), \varphi(\underline{x}) \rangle = \int_{g(\underline{x})=0} \varphi(\underline{x}) \frac{1}{\|\nabla g(\underline{x})\|}\, dA(\underline{x})\,.$$

Here

$$\|\nabla g(\underline{x})\| = \left(\left(\frac{\partial g}{\partial x_1}\right)^2 + \cdots + \left(\frac{\partial g}{\partial x_n}\right)^2 \right)^{1/2}.$$

The formula for $\langle \delta(g(\underline{x})), \varphi(\underline{x}) \rangle$ only makes sense if $\nabla g(\underline{x})$ is never $\underline{0}$. In fact, the condition $\nabla g(\underline{x}) \neq \underline{0}$ is what's needed to guarantee that the sets $g(\underline{x}) = 0$ are smooth curves, surfaces, or higher-dimensional surfaces, so we'll make that assumption in

[36] Look back to Section 5.6 where we found a formula for $\delta(g(x))$ when $g(x)$ is a function of one variable. What we do here can be adapted to that case.

what follows.[37] Note that in the case of a line impulse, if $g(x_1, x_2) = ax_1 + bx_2 - c$ with $a^2 + b^2 = 1$ (our standard assumption), then $\|\nabla g(x_1, x_2)\| = \sqrt{a^2 + b^2} = 1$.

I'm being deliberately vague on how I've written the integral over the set $g(\underline{x}) = 0$. For a function $g(x_1, x_2)$ it's a line integral along a curve, for $g(x_1, x_2, x_3)$ it's a surface integral along a surface, and for higher dimensions it's a generalization of a surface integral. Writing $dA(\underline{x})$ in the formula is supposed to connote integrating "with respect to surface area" in the appropriate context (arclength along a curve in the case of a function $g(x_1, x_2)$). You probably did surface integrals in $\mathbb{R}^3$ in a course on multivariable calculus, or in a class on electromagnetics. Remember Gauss's law for electrostatics? You learned about the divergence theorem, Stokes's theorem, etc. You probably didn't like it. *So you can quit now,* if you'd rather, and just put the formula above into your head for intuition, or maybe as something you'd really have to compute for a particular application. If you want to learn more about surface integrals in higher dimensions et al., that's usually taught in a math course with the word "manifolds" in the title.

For those few who are proceeding, since we don't really evaluate δ at points (though, heaven knows, we have lapsed), how can we define $\delta(g(x_1, x_2))$, a two-dimensional distribution, via a pairing? (We'll work just for a function $g(x_1, x_2)$ of two variables, but the argument carries over to higher dimensions.) We'll ask this more generally:

- For a distribution T how do we define $T(g(x_1, x_2))$?

It's somehow the composition of a distribution with a function that we want to define, though "composition" doesn't formally make sense because we don't evaluate distributions at points, or at least we try not to. In this context, in particular, it's more common to use the term *pullback* and the notation that goes along with it.

Pullbacks for functions, redux. Here's the definition of pullback, first for functions in the setting we need, complete with special notation. We introduced this in Section 5.6 and we also used it in the previous section. It's really just composition in this case, but let this play out. If f is a function of one variable and $g(x_1, x_2)$ is a function of two, then the composition $(f \circ g)(x_1, x_2) = f(g(x_1, x_2))$ makes sense. One says that $(f \circ g)(x_1, x_2)$ is the *pullback* of f by g, and this is written

$$(g^* f)(x_1, x_2) = f(g(x_1, x_2)).$$

The function f of one variable has been "pulled back" by g in forming $g^* f$ to operate on two variables.[38] It's easy to think of examples. Say $I = g(x_1, x_2)$ is the intensity (e.g., grayscale level) of an image at the point (x_1, x_2) and $f(I)$ multiplies intensity by a fixed constant, $f(I) = kI$. Then $(g^* f)(x_1, x_2)$ is the "intensity multiplier function" *on the image*.

Pullback of a distribution. We're aiming to define the pullback of a distribution, $g^* T$. Here T operates on a test function of one variable and, sticking with the 2D case, the function g is a function of (x_1, x_2) for which ∇g is never $\underline{0}$. Thus $g^* T$

[37] This mathematical fact follows from what is usually called the implicit function theorem.

[38] To repeat the warning in Section 5.6, the $*$ is a superscript and the notation $g^* f$ has nothing to do with convolution.

pairs with a test function of two variables. How? To see what to do, as always we take our lead from what happens if T comes from a function f and the pairing $\langle g^* f, \varphi\rangle$ with a test function φ is via integration. Things look a little better if we incorporate vector notation, writing $g(\underline{x})$ for $g(x_1, x_2)$. Then $(g^* f)(\underline{x}) = (f \circ g)(\underline{x})$ and the pairing is

$$\langle g^* f, \varphi\rangle = \int_{-\infty}^{\infty}\int_{-\infty}^{\infty} f(g(x_1, x_2))\varphi(x_1, x_2)\, dx_1 dx_2 = \int_{\mathbb{R}^2} (g^* f)(\underline{x})\varphi(\underline{x})\, d\underline{x}\,.$$

We'll also suppose that f has compact support, as we often did in previous arguments, because for this argument we're going to integrate by parts and we want boundary terms at $\pm\infty$ to go away. Before moving on, note once again that f is a function of one variable, but $g^* f$ pairs with a test function of two variables.

We want to shift the action from $f(g(x_1, x_2))$ to $\varphi(x_1, x_2)$ in the integral, but this can't be accomplished by a simple change of variable, as in trying a substitution $u = g(x_1, x_2)$. Why? Because the dimensions have to match up to make a change of variable; we can't solve for x_1 and x_2 in terms of u to change variables in $\varphi(x_1, x_2)$.

Tricks are needed, and there are two of them. First, because f has compact support we have

$$(g^* f)(\underline{x}) = f(g(\underline{x})) = -\int_{g(\underline{x})}^{\infty} f'(y)\, dy \quad \text{(remember, } g(\underline{x}) \text{ is a number).}$$

Next,

$$\begin{aligned}\langle g^* f, \varphi\rangle &= \int_{\mathbb{R}^2} (g^* f)(\underline{x})\varphi(\underline{x})\, d\underline{x} \\ &= -\int_{\mathbb{R}^2} \left(\int_{g(\underline{x})}^{\infty} f'(y)\, dy\right) \varphi(\underline{x})\, d\underline{x}\,.\end{aligned}$$

This last integral can be written

$$-\int_{-\infty}^{\infty} \left(\int_{C(y)} \varphi(\underline{x})\, d\underline{x}\right) f'(y)\, dy\,,$$

where we put

$$C(y) = \{\underline{x} : g(\underline{x}) < y\}\,.$$

Where does this second sleight-of-hand come from? It's a matter of describing the region of integration in two ways.[39] I'll draw an explanatory picture in a simpler case, where $g(x)$ and $\varphi(x)$ are functions of only one variable, and from that I'll let you consider the situation as it is above.

Here's a plot of a 2D region R bounded below by a curve $y = g(x)$ and a second plot illustrating how you'd compute the $dydx$ integral over the region,

$$\iint_R f'(y)\varphi(x)\, dydx = \int_{-\infty}^{\infty} \left(\int_{g(x)}^{\infty} f'(y)\, dy\right) \varphi(x)\, dx\,.$$

[39] You would have discovered this when you invented the Lebesgue integral back in Chapter 4.

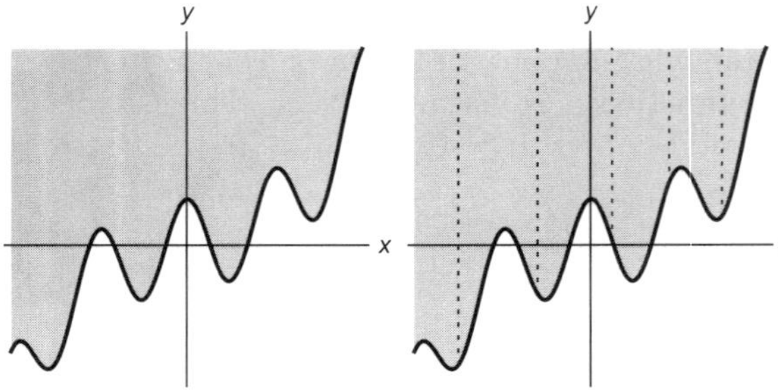

Now how do you interchange the order of integration to $dxdy$? For each y you integrate first over horizontal slices, which are exactly the segment, or segments, $C(y) = \{x : g(x) < y\}$, and then you integrate dy for $-\infty < y < \infty$. The picture is

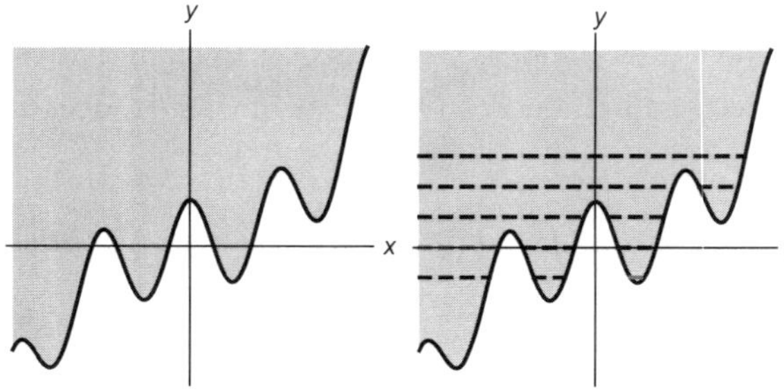

and the integral is

$$\iint_R f'(y)\varphi(x)\, dxdy = \int_{-\infty}^{\infty} \left(\int_{C(y)} \varphi(x)\, dx \right) f'(y)\, dy\,.$$

I'll leave it to you to see how to generalize this to 2D. It's the same idea. In that case you're integrating over a solid bounded below by the surface $y = g(x_1, x_2)$ (y is the vertical coordinate) and you change the order of integration from $dydx_1dx_2$ to dx_1dx_2dy. The horizontal slices $C(y) = \{\underline{x} : g(\underline{x}) < y\}$ are 2D regions.

Back to the 2D case, where we now have

$$\langle g^* f, \varphi \rangle = -\int_{-\infty}^{\infty} \left(\int_{C(y)} \varphi(\underline{x})\, d\underline{x} \right) f'(y)\, dy\,, \quad C(y) = \{\underline{x} : g(\underline{x}) < y\}\,.$$

Integrate by parts!

$$-\int_{-\infty}^{\infty} \left(\int_{C(y)} \varphi(\underline{x})\, d\underline{x} \right) f'(y)\, dy = \int_{-\infty}^{\infty} f(y) \frac{d}{dy} \left(\int_{C(y)} \varphi(\underline{x})\, d\underline{x} \right) dy\,.$$

Putting all this together,

$$\begin{aligned}\langle g^* f, \varphi\rangle &= -\int_{-\infty}^{\infty}\left(\int_{C(y)} \varphi(\underline{x})\, d\underline{x}\right) f'(y)\, dy \\ &= \int_{-\infty}^{\infty} f(y)\frac{d}{dy}\left(\int_{C(y)} \varphi(\underline{x})\, d\underline{x}\right) dy \\ &= \left\langle f, \frac{d}{dy}\left(\int_{C(y)} \varphi(\underline{x})\, d\underline{x}\right)\right\rangle .\end{aligned}$$

This tells us how to define the pullback g^*T for a distribution T. For a test function $\varphi(\underline{x})$ we set

$$\Phi_g(t) = \frac{d}{dy}\left(\int_{C(y)} \varphi(\underline{x})\, d\underline{x}\right), \quad C(y) = \{\underline{x} : g(\underline{x}) < y\} .$$

Then

$$\langle g^*T, \varphi\rangle = \langle T, \Phi_g\rangle .$$

This is a general definition. The argument works the same way in higher dimensions, pulling back T by a real-valued function $g(\underline{x}) = g(x_1, x_2, \ldots, x_n)$, for which $\nabla g \neq \underline{0}$. The resulting definition looks the same.

9.9.1. Pulling back δ. Directly from the definition, the pullback of δ by $g(x_1, x_2)$ is

$$\langle g^*\delta, \varphi\rangle = \langle \delta, \Phi_g\rangle = \Phi_g(0) = \frac{d}{dy}\left(\int_{C(y)} \varphi(\underline{x})\, d\underline{x}\right)\bigg|_{y=0} .$$

This is how the impulse along $g(x_1, x_2) = 0$ pairs with a test function. Further labor, which, alas, is too much to include, would recover the integral formula

$$\langle \delta(g(\underline{x})), \varphi(\underline{x})\rangle = \int_{g(\underline{x})=0} \varphi(\underline{x})\frac{1}{\|\nabla g(\underline{x})\|}\, dA(\underline{x})$$

from the beginning of the section. (Uses the divergence theorem among other things.) Instead we'll work with the definition as we derived it, the issue being to find an expression for $\Phi_g(0)$ for a given $g(x_1, x_2)$. Let's see how this proceeds for a line impulse, where we have earlier calculations to go by.

Take a simple case, say $g(x_1, x_2) = x_2$, so $g(x_1, x_2) = 0$ describes the x_1-axis. Then $C(t) = \{(x_1, x_2) : x_2 < y\}$, which is the half-plane under the horizontal line $x_2 = y$. We have

$$\int_{C(y)} \varphi(\underline{x})\, d\underline{x} = \int_{-\infty}^{y}\left(\int_{-\infty}^{\infty} \varphi(x_1, x_2)\, dx_1\right) dx_2 .$$

Then

$$\Phi_g(y) = \frac{d}{dy}\int_{-\infty}^{y}\left(\int_{-\infty}^{\infty} \varphi(x_1, x_2)\, dx_1\right) dx_2 = \int_{-\infty}^{\infty} \varphi(x_1, y)\, dx_1$$

and

$$\Phi_g(0) = \int_{-\infty}^{\infty} \varphi(x_1, 0)\, dx_1 .$$

That's the integral of φ along the x_1-axis; just what we obtained previously. See how taking the y-derivative stripped the double integral down to a single integral? That's the sort of thing that happens in general.

More generally, let's find (one last time) the impulse along the line $0 = g(x_1, x_2) = ax_1 + bx_2 - c$, with $a^2 + b^2 = 1$. We have to find

$$\Phi_g(y) = \iint_{ax_1+bx_2<c+y} \varphi(x_1, x_2)\, dx_1 dx_2$$

and differentiate with respect to y. With a change of coordinates it's the same calculation we just did. The region we're integrating over is the half-plane under the (oblique) line $ax_1 + bx_2 = c + y$. Use line coordinates (s, t) from Section 9.8.1:

$$x_1 = bs + at + ac, \qquad x_2 = -as + bt + bc\,,$$

where, recall, $t = 0$ and $-\infty < s < \infty$ (the s-axis) corresponds to the line $ax_1 + bx_2 = c$ and then $t =$ const. to a parallel line. In terms of these coordinates

$$\iint_{ax_1+bx_2<c+y} \varphi(x_1, x_2)\, dx_1 dx_2 = \int_{-\infty}^{y} \left(\int_{-\infty}^{\infty} \varphi(bs + at + ac, -as + bt + bc)\, ds \right) dt\,.$$

Differentiating with respect to y,

$$\Phi_g(y) = \int_{-\infty}^{\infty} \varphi(bs + ay + ac, -as + by + bc)\, ds\,,$$

$$\Phi_g(0) = \int_{-\infty}^{\infty} \varphi(bs + ac, -as + bc)\, dt\,.$$

That's the line integral of φ along $ax_1 + bx_2 = c$ with its arclength parametrization (the $a^2 + b^2 = 1$ assumption). Just as we got before. I'll let you redo the earlier calculation when the line is described by $G(x_1, x_2) = Ax_1 + Bx_2 - C = 0$ not assuming that $A^2 + B^2 = 1$. You'll get a factor of $1/\sqrt{A^2 + B^2}$, which is $1/\|\nabla G\|$.

Now screw your courage to the sticking place and find some impulses along other curves and maybe some surfaces. Circles. Ellipses. Spheres. Ellipsoids. Fourier transforms, too. Go wild. Good words upon which to end this book.

Problems and Further Results

9.1. *Some 2D Fourier transforms*

Find the 2D Fourier transform of the 2D signals below. Simplify your answers as much as possible. You may use the fact that $\int_0^\infty J_0(x)dx = 1$ for the Bessel function $J_0(x)$.

(a) Four rectangles, each of height 1, as positioned below:

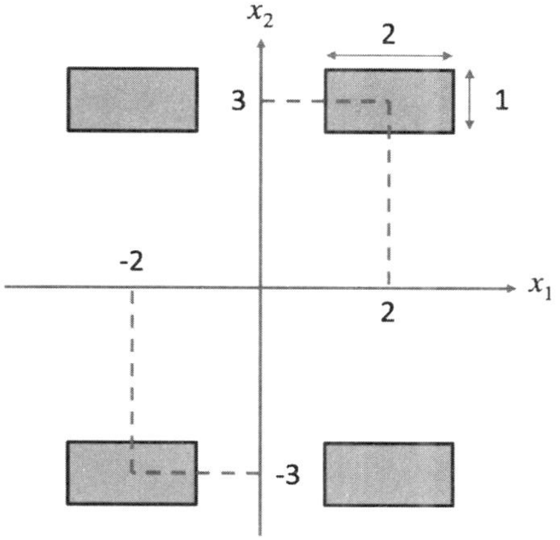

(b) $f(x_1, x_2) = 1/\sqrt{x_1^2 + x_2^2}$.

9.2. *More 2D Fourier transforms*

Find the 2D Fourier transforms of

(a) $\sin 2\pi a x_1 \sin 2\pi b x_2$,
(b) $e^{-2\pi i(ax+by)} \cos(2\pi cx)$,
(c) $\cos(2\pi(ax + by))$. *Hint*: Use the addition formula for the cosine.

9.3. Given a function $f(x_1, x_2)$ define

$$g(x_1, x_2) = f(-x_1, x_2)\,, \qquad h(x_1, x_2) = f(x_1, -x_2)\,,$$
$$k(x_1, x_2) = f(x_2, x_1)\,, \qquad m(x_1, x_2) = f(-x_2, -x_1)\,.$$

Find $\mathcal{F}g$, $\mathcal{F}h$, $\mathcal{F}k$, and $\mathcal{F}m$ in terms of $\mathcal{F}f$. In each case interpret your result geometrically in terms of an image and its spectrum.

9.4. *Bowtie*[40]

(a) Find the Fourier transform of the function

$$f(x_1, x_2) = \begin{cases} 1, & \text{inside the shaded square,} \\ 0, & \text{outside the shaded square.} \end{cases}$$

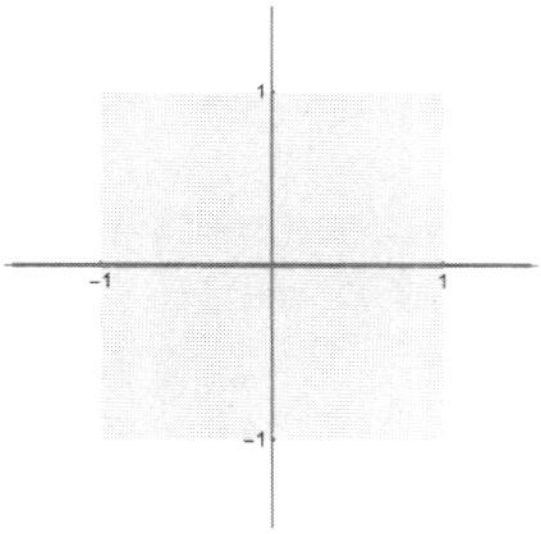

[40]From J. Gill.

(b) Find the Fourier transform of the function

$$g(x_1, x_2) = \begin{cases} 1, & \text{inside the shaded square,} \\ 0, & \text{outside the shaded square.} \end{cases}$$

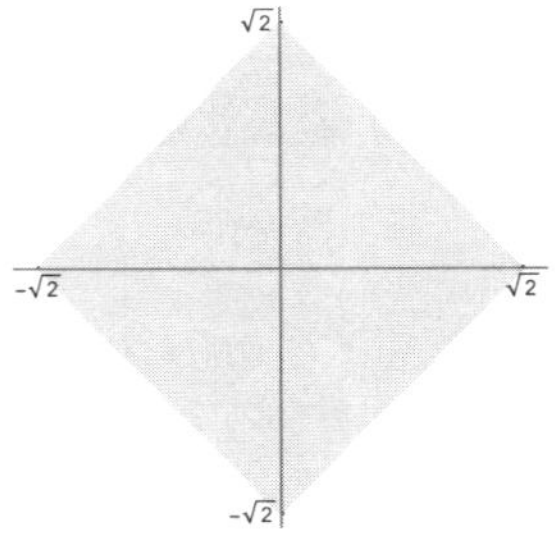

The shaded square is a rotation of the square in part (a). Recall that the matrix

$$A = \begin{pmatrix} \cos\theta & -\sin\theta \\ \sin\theta & \cos\theta \end{pmatrix}$$

gives a rotation by θ, counterclockwise if $\theta > 0$ and clockwise if $\theta < 0$.
(c) Find the Fourier transform of the function in the bowtie:

$$h(x_1, x_2) = \begin{cases} 1, & \text{inside the shaded squares,} \\ 0, & \text{outside the shaded squares.} \end{cases}$$

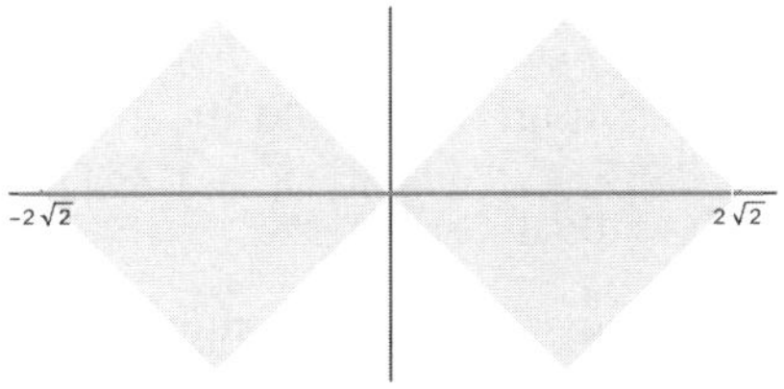

9.5. *Linear transformations and the 2D Fourier transform*

Consider the 2D rectangular function $\Pi(x, y)$. This is depicted by a 2D image, where white corresponds to 1, and black to 0. The power spectrum is also shown.

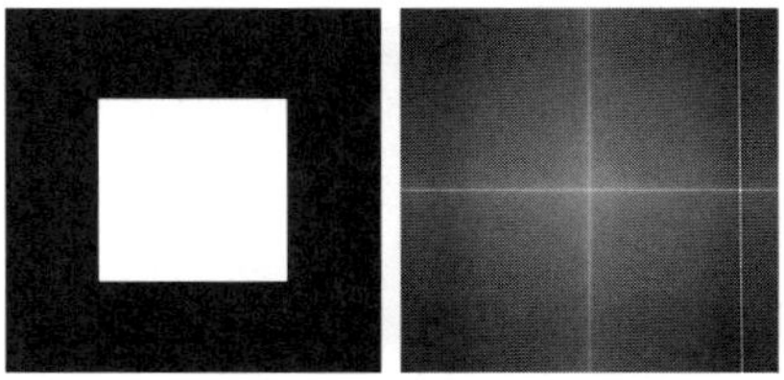

The square is subjected to three different linear transformations and the images A, B, and C are obtained.

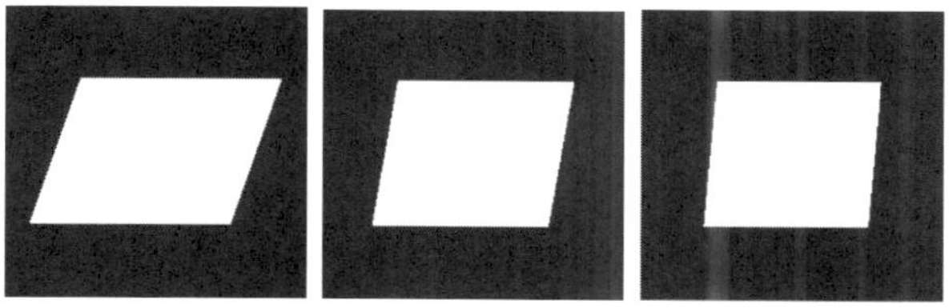

A B C

(a) Each of the figures is a result of a horizontal shear. If $|k_1| > |k_2| > |k_3|$, match the following linear transformations with figures A, B, and C:

$$\begin{pmatrix} 1 & k_1 \\ 0 & 1 \end{pmatrix}, \quad \begin{pmatrix} 1 & k_2 \\ 0 & 1 \end{pmatrix}, \quad \begin{pmatrix} 1 & k_3 \\ 0 & 1 \end{pmatrix}.$$

Here are plots of the power spectra for the different shears. Match the shear to the power spectrum. (Plots courtesy of Mathematica.)

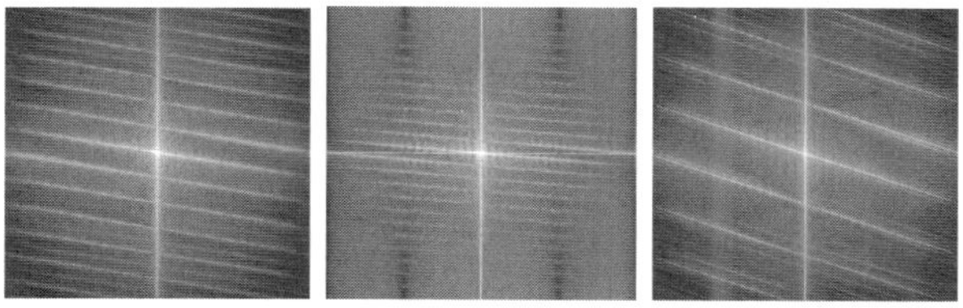

9.6. *Matching 2D Fourier transforms, again*

The pictures below show a two-dimensional function $f(\underline{x})$ (on the left) and the magnitude of its spectrum, $|F(\underline{s})|$ (on the right). The spectrum was obtained by MATLAB's two-dimensional DFT routine. The images are normalized so that white pixels correspond to 1 and black pixels to 0.

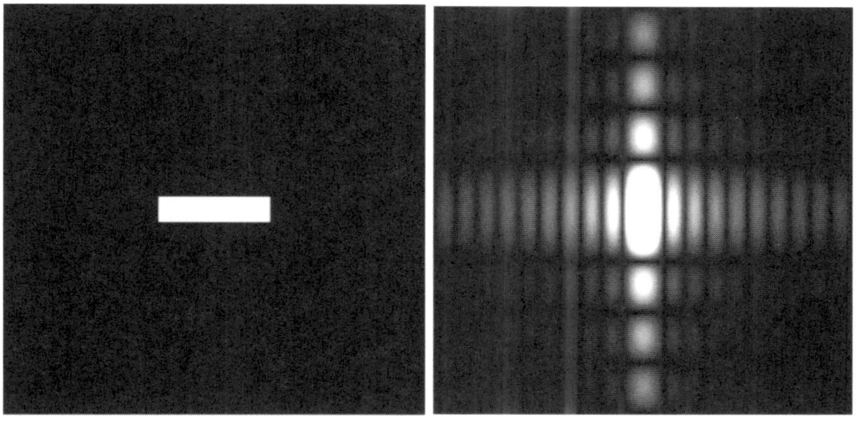

Here are magnitude plots of four modified versions of $f(\underline{x})$ followed by five possible Fourier transform magnitude plots. Match them.

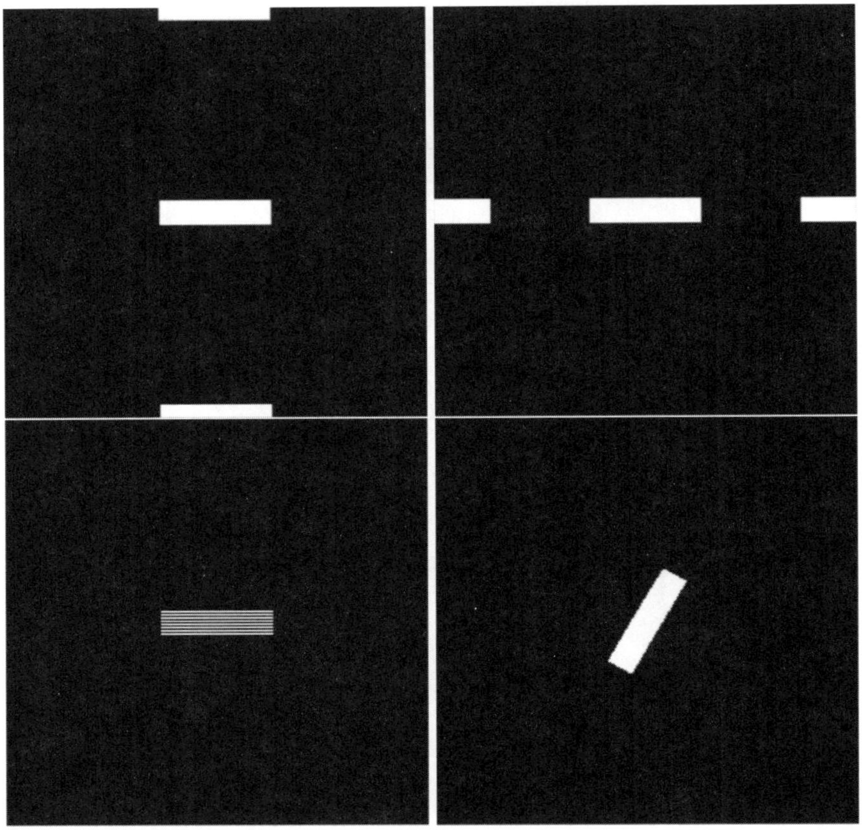

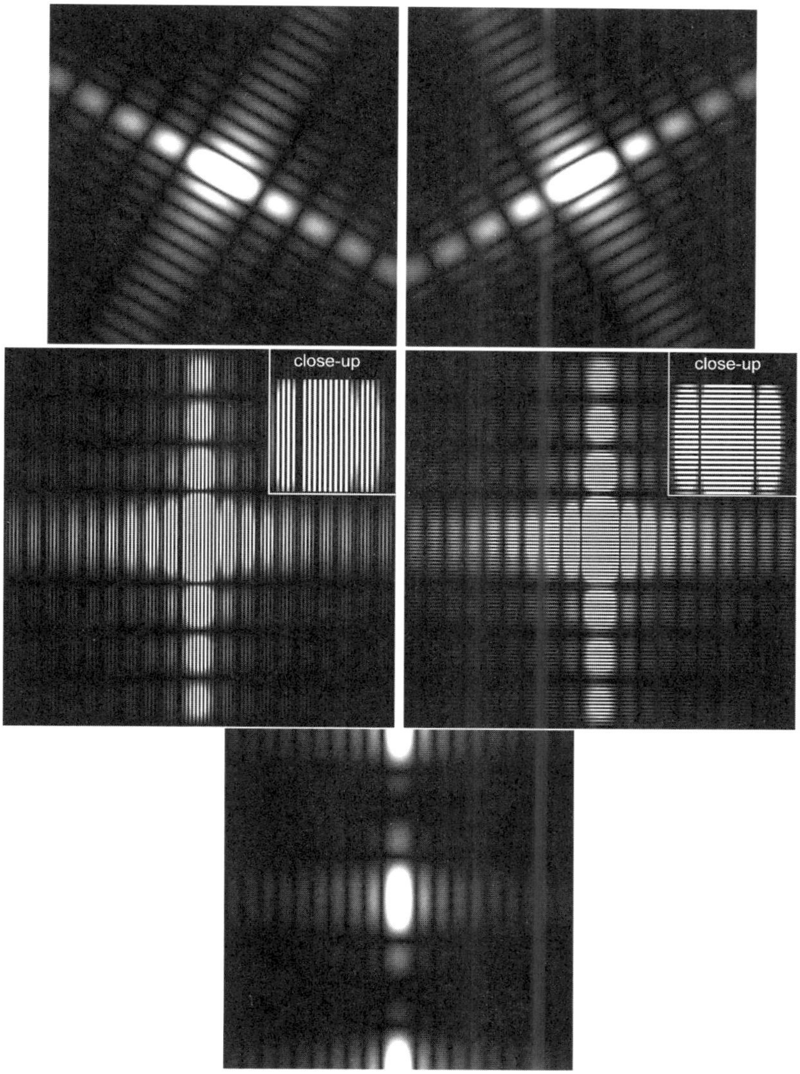
close-up
close-up

9.7. *Reflection across a line*

A common operation in image processing is to reflect an image across a line. Say the line passes through the origin at an angle θ. A point p and its reflection $R(p)$ are as illustrated.

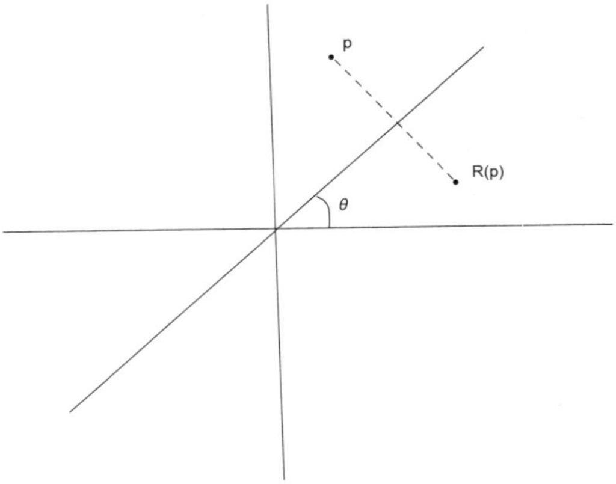

Recall that for $\theta > 0$

$$A_\theta = \begin{pmatrix} \cos\theta & -\sin\theta \\ \sin\theta & \cos\theta \end{pmatrix}$$

gives a counterclockwise rotation of the plane through an angle θ.

(a) Explain why

$$A_\theta^{-1} = A_{-\theta} = \begin{pmatrix} \cos\theta & \sin\theta \\ -\sin\theta & \cos\theta \end{pmatrix}.$$

(b) Explain why

$$B = \begin{pmatrix} 1 & 0 \\ 0 & -1 \end{pmatrix}$$

gives a reflection across the x-axis (the horizontal axis).

(c) Explain how to multiply A_θ, $A_{-\theta}$, and B *in the appropriate order* to obtain the reflection R. You should get

$$R = \begin{pmatrix} \cos 2\theta & \sin 2\theta \\ \sin 2\theta & -\cos 2\theta \end{pmatrix}.$$

Explain, geometrically, why $RR = I$.

(d) If $f(\underline{x})$ is an image, then $f(R\underline{x})$ is the reflected image. How do the spectra compare; i.e., how do $\mathcal{F}(f(\underline{x}))$ and $\mathcal{F}(f(R\underline{x}))$ compare? Explain this in words as well as in formulas.

9.8. *Indicator function of a triangle*

Let T be the triangular region in the x_1x_2-plane shown below.

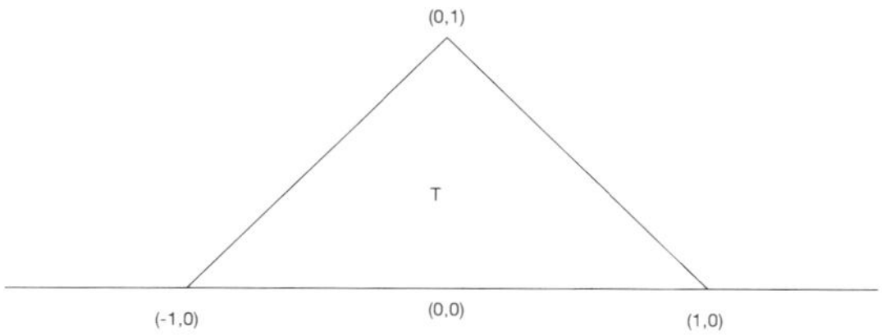

The indicator function is

$$I_T(x_1, x_2) = \begin{cases} 1, & (x_1, x_2) \text{ inside } T, \\ 0, & (x_1, x_2) \text{ outside } T. \end{cases}$$

(Fights might break out over the boundary values.)

An explicit calculation gives the 2D Fourier transform of $I_T(x_1, x_2)$ as

$$\mathcal{F}I_T(\xi_1, \xi_2) = \frac{1}{2\pi i \xi_1}(e^{\pi i(\xi_1 - \xi_2)} \operatorname{sinc}(\xi_1 + \xi_2) - e^{-\pi i(\xi_1+\xi_2)} \operatorname{sinc}(\xi_1 - \xi_2)).$$

(This isn't the 2D triangle function and its Fourier transform — see a later problem — it's the indicator function *of* the 2D triangular region.)

(a) Why should we have $\mathcal{F}I_T(-\xi_1, \xi_2) = \mathcal{F}I_T(\xi_1, \xi_2)$? Show that this is the case from the formula.

(b) From the formula, we should in principle be able to find the Fourier transform of the indicator function for *any* triangular region. Why?

(c) Obtain the formula

$$\mathcal{F}I_{T'}(\xi_2, \xi_2) = \frac{1}{\pi i \xi_1}(e^{-2\pi i \xi_2} \operatorname{sinc}(2(\xi_1 + \xi_2)) - e^{-2\pi i(\xi_1+\xi_2)} \operatorname{sinc}(2\xi_2))$$

for the triangle T' shown below.

Hint: What does the matrix $A = \begin{pmatrix} 1 & -\frac{1}{2} \\ 0 & \frac{1}{2} \end{pmatrix}$ do to T'?

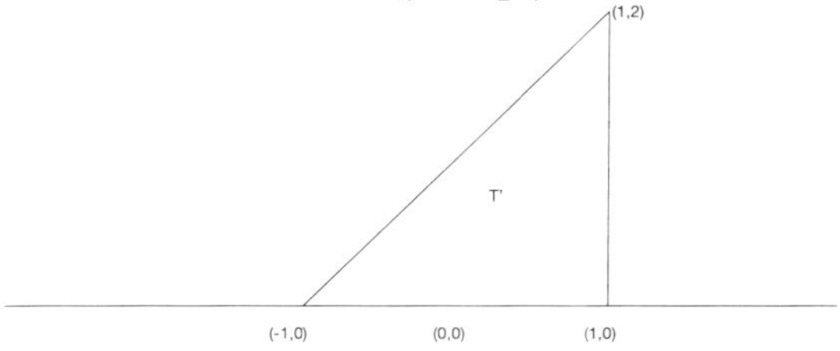

9.9. *The indicator function for a parallelogram*

One set of data that describes a parallelogram is the distances between sides, p and q, and the vectors that give the directions of the sides. Let $\underline{u}$ be a unit vector in the direction of the sides that are p apart and let $\underline{v}$ be a unit vector in the direction of the sides that are q apart.

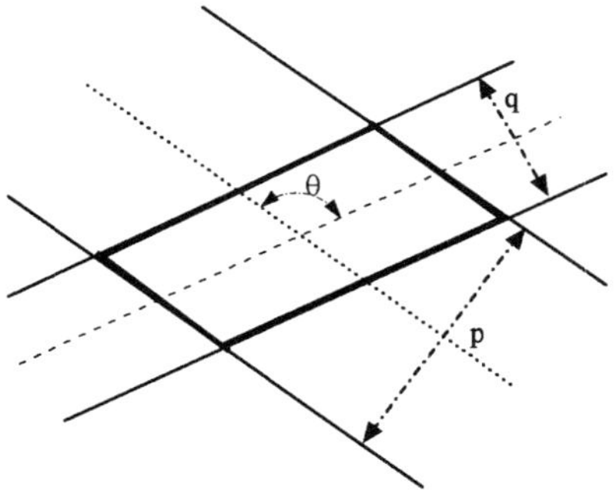

Consider a parallelogram P centered at $(0,0)$ and show that

$$\mathcal{F}I_P(\underline{\xi}) = \frac{pq}{|\sin\theta|}\operatorname{sinc}\left(\frac{p(\underline{u}\cdot\underline{\xi})}{\sin\theta}\right)\operatorname{sinc}\left(\frac{q(\underline{v}\cdot\underline{\xi})}{\sin\theta}\right),$$

where I_P is the indicator function for P.

9.10. *Convolution and stretch*

We had a problem on scaling and convolution in one dimension, namely,

$$(\sigma_a f) * g = \frac{1}{|a|}\sigma_a(f * \sigma_{1/a} g), \quad (\sigma_a f) * (\sigma_a g) = \frac{1}{|a|}\sigma_a(f * g).$$

The second formula (which follows from the first) is more memorable in that it says "the convolution of the stretched signals is the stretched convolution of the signals."

The natural generalization of these formulas to higher dimensions replaces the scaling operator $(\sigma_a f)(x) = f(ax)$ by $(\sigma_A f)(\underline{x}) = f(A\underline{x})$. Show that

$$((\sigma_A f) * g)(\underline{x}) = \frac{1}{|\det A|}(f * (\sigma_{A^{-1}} g)(A\underline{x}))$$

and deduce that also

$$(\sigma_A f * \sigma_A g)(\underline{x}) = \frac{1}{|\det A|}(f * g)(A\underline{x}) = \frac{1}{|\det A|}\sigma_A(f * g)(\underline{x}).$$

9.11. *Projections and 3D Fourier transforms*

Let $f(x_1, x_2, x_3)$ be a three-dimensional function whose Fourier transform is $\mathcal{F}f(\xi_1, \xi_2, \xi_3)$.

(a) Let

$$g(x_1, x_2) = \int_{-\infty}^{\infty} f(x_1, x_2, x_3)\, dx_3 .$$

One says that g is the projection of f *along* the x_3-direction. Find $\mathcal{F}g(\xi_1, \xi_2)$ in terms of $\mathcal{F}f$.

(b) Let

$$h(x_3) = \int_{-\infty}^{\infty}\int_{-\infty}^{\infty} f(x_1, x_2, x_3)\, dx_1 dx_2 .$$

One says that h is the projection of f *onto* the x_3-direction. Find $\mathcal{F}h(\xi_3)$ in terms of $\mathcal{F}f$.

9.12. *Convolution and separable functions*

Suppose that $f(x_1, x_2)$ and $g(x_1, x_2)$ are separable, say, $f(x_1, x_2) = f_1(x_1)f_2(x_2)$ and $g(x_1, x_2) = g_1(x_1)g_2(x_2)$. Show that $f * g$ is also separable with

$$(f * g)(x_1, x_2) = (f_1 * g_1)(x_1)(f_2 * g_2)(x_2)\,.$$

Hint: Use the Fourier transform.

This looks kind of slick in terms of tensor products:

$$(f_1 \otimes f_2) * (g_1 \otimes g_2) = (f_1 * g_1) \otimes (f_2 * g_2)\,.$$

Naturally, this generalizes to functions of more than two variables.

9.13. *2D triangle function*

Let's define a 2D triangle as a separable function by

$$\begin{aligned}\Lambda(x_1, x_2) &= \Lambda(x_1)\Lambda(x_2) \quad \text{(1D triangle functions on the right)}\\ &= \begin{cases} (1 - |x_1|)(1 - |x_2|), & |x_1| < 1 \quad \text{and} \quad |x_2| < 1,\\ 0, & \text{otherwise.}\end{cases}\end{aligned}$$

(a) Plot the graph of $\Lambda(x_1, x_2)$. You should see a sort of pyramid (so it may be better to call Λ a pyramid function). Are the lateral faces flat? Why or why not?
(b) You may also find it interesting to plot a contour diagram and relate it to what you saw in your 3D plot.
(c) Show that, just as in the 1D case, $\Lambda(x_1, x_2) = (\Pi * \Pi)(x_1, x_2)$ (a 2D convolution of 2D Π's on the right). Use Problem 9.12.
(d) To wrap up, find $\mathcal{F}\Lambda(\xi_1, \xi_2)$ and plot the graph.

We won't do it, but it's clear how to extend the definition of Λ to higher dimensions and to scale the pyramid differently along different axes.

9.14. *Two-dimensional heat equation*

In this problem, we will consider the problem of heat conduction within a material that has different temperature conductivity properties along the two dimensions x and y. The general two-dimensional version of the heat equation can be written as

$$\frac{\partial f}{\partial t} = \lambda_x^2 \frac{\partial^2 f}{\partial x^2} + \lambda_y^2 \frac{\partial^2 f}{\partial y^2}\,,$$

where the positive constants λ_x and λ_y are the heat conductivity in the x- and y-direction, respectively.

(a) Find the solution $f(x, y, t)$ of this equation with the initial condition that $f(x, y, 0) = \delta(x, y)$.
(b) Describe the isothermal curves (i.e., the curves of constant temperature, $f(x, y) = K$, in the (x, y)-plane) and sketch them for the cases $\lambda_x > \lambda_y$, $\lambda_x = \lambda_y$, and $\lambda_x < \lambda_y$.

9.15. *Electrostatics and the Hankel transform*[41]

Electrostatics is the study of how a charge density produces an electric potential. The two quantities are related by Poisson's equation,

$$\Delta V(\underline{x}) = -\frac{q(\underline{x})}{\epsilon_0} \quad \left(\Delta = \frac{\partial^2}{\partial x^2} + \frac{\partial^2}{\partial y^2} + \frac{\partial^2}{\partial z^2}\right),$$

where V is the electric potential, q is the charge density, and ϵ_0 is the vacuum dielectric constant. (A similar equation describes how a mass density produces a gravitational potential.) We will assume that $V(\underline{x})$ and $q(\underline{x})$ are *radial* functions; that is, they depend only on the distance to the origin $r = |\underline{x}|$. To emphasize this, we write $V(\underline{x}) = V(r)$ and $q(\underline{x}) = q(r)$. As the Fourier transform of a radial function is radial, we will have a similar simplification in the frequency domain, $\rho = |\underline{s}|$.

In three dimensions with radial symmetry, Poisson's equation is

$$\frac{1}{r^2}\frac{d}{dr}\left[r^2\frac{dV}{dr}\right] = -\frac{q(r)}{\epsilon_0}.$$

The three-dimensional Fourier transform of a radial function is

$$\mathcal{H}f(\rho) = \frac{2}{\rho}\int_0^\infty \sin(2\pi r\rho)f(r)\, r\, dr\,.$$

This is called the *3D Hankel transform* of $f(r)$. The 3D Hankel transform has the following properties which you may freely use without proof:

Property (i). The convolution of two radial functions f and g is defined as

$$(f * g)(r) = \iiint_{\mathbb{R}^3} f(\underline{y})g(\underline{x}-\underline{y})\, d\underline{y} \qquad \text{where } r = |\underline{x}|,$$

and the Hankel transform of a product is a convolution:

$$\mathcal{H}\{fg\} = \mathcal{H}f(\rho) * \mathcal{H}g(\rho).$$

Property (ii). The Hankel transform satisfies the derivative property

$$\mathcal{H}\left\{\frac{1}{r^2}\frac{d}{dr}\left[r^2\frac{df}{dr}\right]\right\} = -4\pi^2\rho^2\mathcal{H}f(\rho).$$

Property (iii). The Hankel transform is its own inverse:

$$\mathcal{H}^{-1}f = \mathcal{H}f.$$

(a) Show that

$$\mathcal{H}\left\{\frac{1}{r^2}\right\} = \frac{\pi}{\rho}.$$

(b) Show that the solution to the 3D radial Poisson equation is *Coloumb's Law*,

$$V(\underline{x}) = \frac{1}{4\pi\epsilon_0}\iiint \frac{q(\underline{y})}{|\underline{x}-\underline{y}|}d\underline{y}.$$

You may use the result of part (a) even if you were unable to show the formula.

[41] From Raj Bhatnager.

9.16. *2D Gaussian vectors*[42]

Let $X = (X_1, X_2)$ be a random 2D vector whose distribution is of the form

$$(*) \qquad p(x_1, x_2) = \frac{1}{2\pi\sigma^2\sqrt{1-\rho^2}} \exp\left(-\frac{x_1^2 + x_2^2 + 2\rho x_1 x_2}{2\sigma^2(1-\rho^2)}\right),$$

where $-1 < \rho < 1$ is the correlation between X_1 and X_2 and σ is the standard deviation (assumed equal) for both X_1 and X_2. This can also be written in the (more convenient) vector form. Let $\underline{x} = (x_1, x_2)$ and $\rho = \cos\theta$. Then

$$p(\underline{x}) = \frac{1}{2\pi\sigma^2 \sin\theta} \exp\left(-\underline{x}^T A^T A x\right),$$

where

$$A = \frac{1}{\sqrt{2}\sigma \sin\theta} \begin{pmatrix} \cos(\theta/2) & \cos(\theta/2) \\ \sin(\theta/2) & -\sin(\theta/2) \end{pmatrix}.$$

Note that $\det A = -1/2\sigma^2 \sin\theta$.

(a) Assume $\sigma = 1$, and find $\mathcal{F}p$, the 2D Fourier transform of p, by using the generalized stretch theorem.

(b) Suppose $\underline{X} = (X_1, X_2)$ and $\underline{Y} = (Y_1, Y_2)$ are two independent 2D vectors whose distributions are of the form (*) with correlation coeffecients ρ_1 and ρ_2, respectively. Let $\underline{Z} = (Z_1, Z_2)$ be the sum of $\underline{X}$ and $\underline{Y}$,

$$\underline{Z} = \underline{X} + \underline{Y}.$$

As before, the distribution of $\underline{Z}$ is given by a 2D convolution of the distributions of $\underline{X}$ and $\underline{Y}$.

Find the distribution of $\underline{Z}$ using the Fourier transform computed in part (a). What is the correlation between Z_1 and Z_2?

9.17. *Scaling radial functions*[43]

Let $\underline{x}$ be an n-dimensional vector and define

$$f(\underline{x}) = \frac{1}{|\underline{x}|^a},$$

where $0 < a < n$. Further, let $F(\underline{s})$ be the n-dimensional Fourier transform of $f(\underline{x})$ and assume that $F(\underline{s})$ exists.

(a) Show that $F(t\underline{s}) = t^{a-n} F(\underline{s})$, where t is a real positive constant.

(b) Explain why this implies that

$$F(\underline{s}) = \frac{C}{|\underline{s}|^{n-a}},$$

where C is a constant that you don't need to find.

[42] From A. Siripuram.

[43] From Raj Bhatnagar.

9.18. *2D discrete Fourier transform*

Let f be an $M \times N$ matrix (you can think of f as an $M \times N$ image). The 2D DFT of f is given by the following formula:

$$\mathcal{F}f[k,l] = \sum_{m=0}^{M-1} \sum_{n=0}^{N-1} f[m,n]\omega_N^{-ln}\omega_M^{-km},$$

where

$$\omega_N = e^{2\pi i/N}, \quad \omega_M = e^{2\pi i/M}.$$

Independent of the problems to follow, we comment that the 2D DFT is "separable" in the following sense. Let f_m be the mth row of the matrix f: it's a vector of length N. Let $\mathcal{F}f_m$ be the 1D DFT of f_m. Then

$$\mathcal{F}f_m[l] = \sum_{n=0}^{N-1} f_m[n]\omega_N^{-ln} = \sum_{n=0}^{N-1} f[m,n]\omega_N^{-ln}.$$

Then, from the given formula for the 2D DFT, we can easily see that

$$\mathcal{F}f[k,l] = \sum_{m=0}^{M-1} \mathcal{F}f_m[l]\omega_M^{-km}.$$

In other words, this is a 1D DFT of the vector of length M, which consists of the lth entry in each vector $\mathcal{F}f_m$.

We can take advantage of the separability by first computing the 1D DFT of the rows of f and then doing another 1D DFT to find the $[k,l]$ entry in $\mathcal{F}f$.

(a) *Modulation.* Let g be an $M \times N$ image obtained from f in the following way: $g[m,n] = f[m,n]\omega_M^{mk_0}\omega_N^{nl_0}$, where k_0 and l_0 are integers. What is $\mathcal{F}g$ in terms of $\mathcal{F}f$?

(b) *2D convolution.* Given two matrices f and g of size $M \times N$, their 2D convolution is

$$(f * g)[m,n] = \sum_{u=0}^{M-1} \sum_{v=0}^{N-1} f[u,v]g[m-u, n-v].$$

Show that $\mathcal{F}(f * g) = \mathcal{F}f \times \mathcal{F}g$, where on the right we mean the $M \times N$ matrix whose elements are the products of the corresponding elements of $\mathcal{F}f$ and $\mathcal{F}g$.

Hint: Write out the expression for $\mathcal{F}(f * g)$. You'll get a nasty looking expression with four nested sums. But swap the order of summation, change the variable of summation (analogous to changing the variable of integration), and use the fact that the 1D DFT can be computed by summing over any N (or M) consecutive indices.

9.19. *Accelerated MRI*[44]

In MRI the data is collected in the 2D frequency domain. Acquiring data takes time and it's important to find ways to speed things up. This problem explores the principle behind one method for doing this, called *sensitivity encoding* (referred to as SENSE), a breakthrough in MRI. See the paper *SENSE:*

[44] From Moosa Zaidi.

Sensitivity Encoding for Fast MRI by Klaas P. Pruessmann, Markus Weiger, Markus B. Scheidegger, and Peter Boesiger, from 1999.

Consider an image N_1 pixels long in the vertical direction and N_2 pixels long in the horizontal direction, and assume that N_1 and N_2 are both even. Represent the image by the discrete function $\underline{f}[k_1, k_2]$ wth $k_1 = 0, \ldots N_1 - 1$ and $k_2 = 0, \ldots, N_2 - 1$.

If we acquire all the values $\underline{\mathcal{F}}\underline{f}[\ell_1, \ell_2]$, $\ell_1 = 0, \ldots N_1 - 1$ and $\ell_2 = 0, \ldots, N_2 - 1$, of the DFT of $\underline{f}$, we can obtain $\underline{f}$ by Fourier inversion. Instead, we acquire the values of $\underline{\mathcal{F}}\underline{f}$ for *all* values of ℓ_1 but only for *even* values of ℓ_2. That is, we obtain all values of the function

$$\underline{G}[u_1, u_2] = \underline{\mathcal{F}}\underline{f}[u_1, 2u_2],$$

for $u_1 = 0, \ldots, N_1 - 1$ and $u_2 = 0, \ldots, \frac{N_2}{2} - 1$.

(a) The inverse 2D DFT of $\underline{G}$ is

$$\underline{g}[v_1, v_2] = \frac{1}{N_1(N_2/2)} \sum_{u_1=0}^{N_1-1} \sum_{u_2=0}^{N_2/2-1} \underline{G}[u_1, u_2] e^{2\pi i \left(\frac{v_1 u_1}{N_1} + \frac{v_2 u_2}{N_2/2} \right)},$$

for $v_1 = 0, \ldots, N_1 - 1$ and $v_2 = 0, \ldots, \frac{N_2}{2} - 1$. Show that

$$\underline{g}[v_1, v_2] = \underline{f}[v_1, v_2] + \underline{f}\left[v_1, v_2 + \frac{N_2}{2}\right].$$

(b) If $\underline{f}$ represents the image below, sketch the image represented by $\underline{g}$.

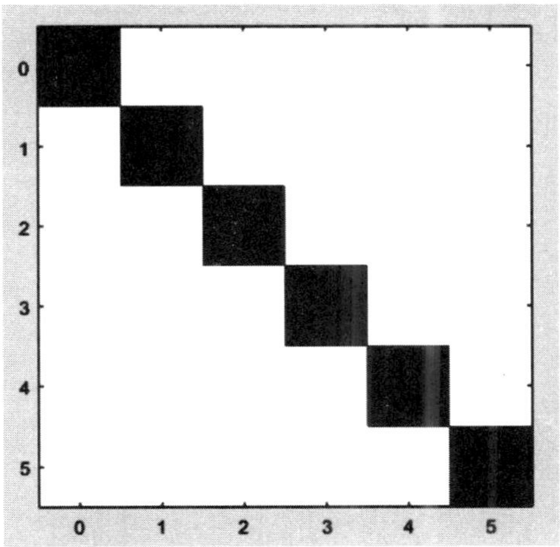

With no other information, we cannot reconstruct the original information from just $\underline{G}$, and the acceleration we have attempted by acquiring only the frequency data seems not to work. The idea is that we may be able to reconstruct the signal if we collect data from two different coils that have different sensitivities. (This is hardware! The "coils" in an MRI machine detect the signals. You'll find more if you search for Radiofrequency coils.) Say coil 1 has sensitivity $s_1[k_1, k_2]$ and coil 2 has sensitivity $s_1[k_1, k_2]$, so that the object as seen by coil 1 is $s_1[k_1, k_2]\underline{f}[k_1, k_2]$ and the object as seen by coil 1 is $s_2[k_1, k_2]\underline{f}[k_1, k_2]$. For each coil we acquire frequency data in the same accelerated way as above, so that we now have the values of *two* functions:

$$\underline{G}_1[u_1, u_2] = \underline{\mathcal{F}}(\underline{f}s_1)[u_1, 2u_2]\,,$$
$$\underline{G}_2[u_1, u_2] = \underline{\mathcal{F}}(\underline{f}s_2)[u_1, 2u_2]\,,$$

for $u_1 = 0, \dots, N_1 - 1$ and $u_2 = 0, \dots, \frac{N_2}{2} - 1$. If we compute the inverse 2D DFTs as before, we obtain

$$g_1[v_1, v_2] = s_1[v_1, v_2]f[v_1, v_2] + s_1[v_1, v_2 + N_2/2]f[v_1, v_2 + N_2/2],$$
$$g_2[v_1, v_2] = s_2[v_1, v_2]f[v_1, v_2] + s_2[v_1, v_2 + N_2/2]f[v_1, v_2 + N_2/2]\,.$$

For each v_1, v_2 regard this as a system of two linear equations in the two unknowns $f[v_1, v_2]$ and $f[v_1, v_2 + N_2/2]$. We can solve these equations and recover the original image if for every v_1, v_2 the vectors

$$(s_1[v_1, v_2], s_1[v_1, v_2 + N_2/2]), \quad (s_2[v_1, v_2], s_2[v_1, v_2 + N_2/2])$$

are linearly independent for all v_1 and v_2. This is the basic idea behind SENSE.

9.20. *Watermarking an image*

A common application of the 2D Fourier transform is the compression of images. The first JPEG compression standard heavily quantizes the high-frequency components within an image because the human eye is poor at discriminating between signals with high spatial frequencies.

Another application of this property is hiding one signal inside another without experiencing a decrease in the overall quality of the initial signal (for example, the French television standard SECAM used to encode the sound information in the high frequencies of the image spectrum to save transmission bandwidth). This procedure is called *digital watermarking*, which is what this problem is about.

The transform commonly used in image processing is a modified version of the DFT, called the *discrete cosine transform* (DCT); we'll use it in this problem. The basic principle of the DCT is to replicate the initial signal in all dimensions and obtain an even signal that has twice the size of the original signal in all dimensions and take the DFT of this. The output of the DFT is then truncated to keep only the same amount of samples as initially present. Given that we take the DFT on real and even signals, the output is then also real and even. Truncating the signal does not delete any information, and this technique thus has the advantage of outputting only real numbers. (You can find *many* descriptions of the DCT with a little googling.)

(a) Download the image `watermarked`. This image has been watermarked (i.e., another smaller image has been hidden in it) using the following scheme:
   - The image was decomposed in 8×8 blocks.
   - We computed the DCT on each of these blocks.
   - In each block, the bottom-right 2×2 block was erased and replaced by the amplitudes of the corresponding 2×2 block of the hidden image, divided by 10. Thus, the dimensions of the hidden image are equal to a quarter of those of the original image.

Write a MATLAB program that reads the original image, and reconstructs the hidden image inside it. Use the functions `dct, idct, imread,` and `imwrite.` Go through MATLAB help to get some information about those functions.
If we define the matrix

$$D = dct(I_8),$$

then the 2D discrete cosine transform of an 8×8 block A is given by the following equation:

$$\hat{A} = DAD^T.$$

What/who/where is the hidden signal?

(b) What happens if you reiterate the process on the image you just discovered?

9.21. *Echo planar MRI*[45]

CT and MRI scans are ubiquitous in society today. What does the doctor order when you have a head injury? CT scan. Knee injury? MRI. Neither of these imaging modalities would be possible without Fourier transforms. In this problem, we will look at a simple MR image reconstruction example.

It turns out that the data acquired from an MRI scan are samples of the Fourier transform of the image. The Fourier transform of the image is referred to as k-space, and different readout techniques sample k-space along different trajectories. To recover the image, the trajectory must cover all of k-space, but one of the beauties of MRI is the flexibility in how to traverse k-space. A simple trajectory is echo planar imaging (EPI), which alternately samples k-space from left to right and then from right to left as it moves down each row.

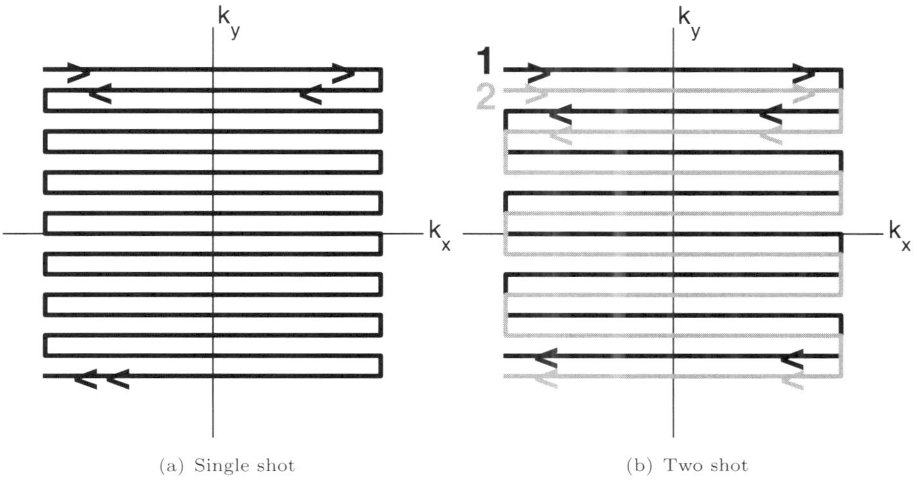

Figure 9.1. The EPI trajectory zig-zags from top to bottom, filling up k-space.

[45]From Raj Bhatnagar.

(a) Download `epiMRI.mat` and load it into MATLAB. The single shot EPI readout data `k` is a vector of length $256 \times 256 = 65{,}536$ samples. Populate a matrix K of size 256×256 with the readout data so that the matrix K is the Fourier transform of the image, which is 256×256 pixels. Reconstruct the MR image (can you tell what it is a scan of?). Remember to use `fftshift` so that MATLAB takes the inverse 2DFT (`ifft2`) properly.

(b) Suppose your trajectory is a two-shot *interleaved* EPI sequence. That is, you first acquire trajectory (1) and then you acquire trajectory (2). Reconstruct the second image from the readout data `ki`, where the first row is the data for trajectory (1) and the second row is the data for trajectory (2). What happens if you only use the first shot (and use zeros for the second shot)? What about if you only use the second shot (and use zeros for the first shot)? Why is the first shot image nonnegative whereas the second shot image is positive and negative? What happens when you add these two images together, and what basic property of the FT gives us this result?

Hint: To explain the artifact in the image from the first shot, look back at the DFT problems on upsampling and downsampling. You may also find the command `colorbar` useful.

Note: There are many other trajectories, including spiral, interleaved spiral, and radial projections. Because the sampling is not on a Cartesian grid, one way to reconstruct the image is to map the acquired data to the Cartesian grid through an operation called *gridding*. We encountered this in the discussion of numerically implementing CT. Then the inverse 2DFT can be applied. Projection sampling is very important for CT imaging because a CT scan gives you the data to fill in k-space along the radial spokes. In addition to gridding, there is an alternative reconstruction technique that involves filtering each projection (spoke) and backprojecting (i.e., smearing) each filtered projection across a blank image. Believe it or not, the sum of these filtered, backprojected images reconstructs the image! If you are interested in medical imaging:

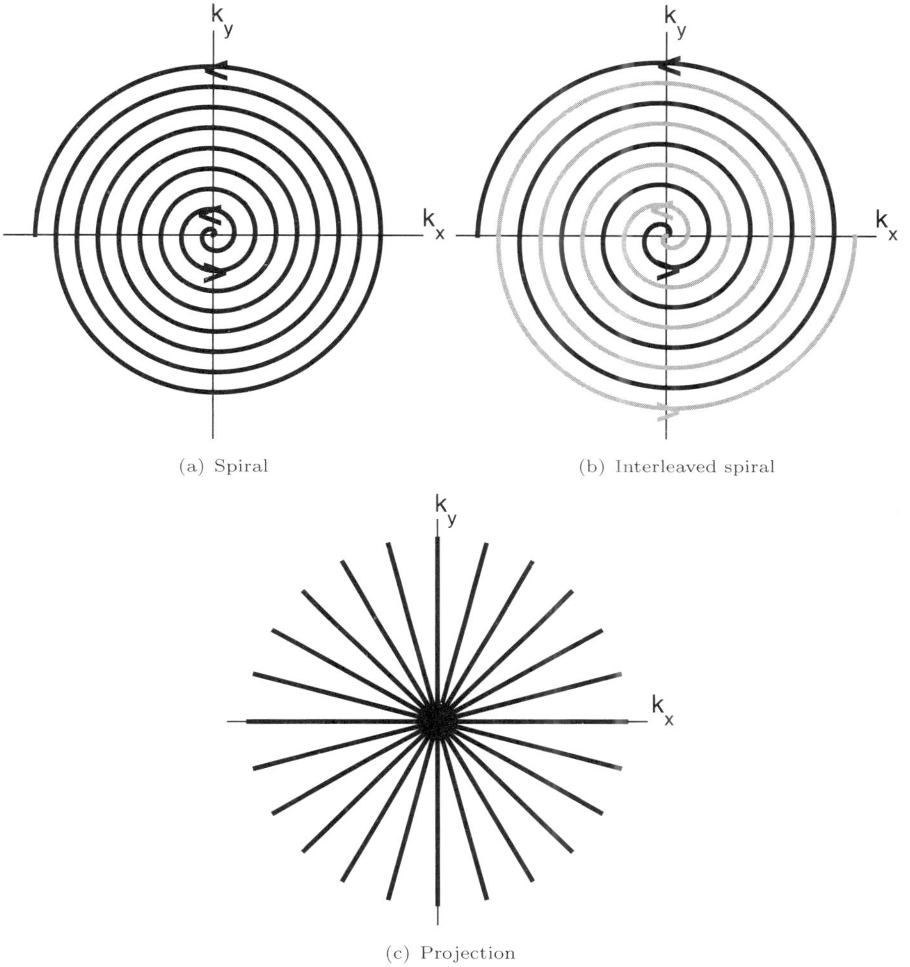

Figure 9.2. Various other trajectories that cover k-space.

A List of Mathematical Topics that Are Fair Game

There's a danger in setting down at the outset a list of specific mathematical terms and techniques that will be in play. It may seem to set the starting line beyond where you imagined the finish line to be. In Paul Halmos's book *Measure Theory* (*definitely* a math book, and by a master writer) there is the following memorable quote, referring to §0, a section on prerequisites:

> ... he should not be discouraged if, on a first reading of §0, he finds that he does not have the prerequisites for reading the prerequisites.

So don't worry too much about the following list. We can and will deal with particular mathematical definitions, theorems, formulas, and odds and ends when they come up and in context. With that understanding, here is a collection of things you can expect.

General know-how on notation: It will help if you can calmly reflect on the general idea of a function and how variables can be used, and misused. Notation is often an issue in this subject. You wouldn't think that plugging into a formula (e.g., convolution) could be a challenge, but you'll be surprised.

A further comment on terminology and notation for functions. First, we use the terms "signal" and "function" interchangeably. Second, mathematicians tend to be careful in writing $f\colon A \to B$ to indicate that f is a function (or mapping — that's the other standard term) with domain the set A and range contained in the set B (B, which can be bigger than the range, is sometimes called the codomain). They, the mathematicians, are also careful to distinguish f as the name for the function from $f(a)$, the value of the function f at a. You've probably run across this and it was certainly a part of my education. Nevertheless, we will be less fastidious and we'll refer to "the function $f(x)$" as well as "the function f." I feel like I should issue an apology to my mathematician colleagues, and you know who you are, but no.

Complex numbers: I'm assuming that the basic definitions and properties of complex numbers are known to you, as well as how the arithmetic of complex numbers works. But some people (especially non-EEs) tend to be a little rusty on these things. For that reason, there's an appendix collecting some basic facts on complex numbers. This should serve as a review for those who need one or want one. It covers material through the complex exponentials (Euler's formula), known to EEs as phasors, with a few applications to signals.

Calculus: Elementary differentiations and integrations you should be able to do with your spine, not your brain, and both integration by parts and integration by substitution are absolute musts. That applies particularly to definite integrals. You should be able to handle improper integrals and understand some issues of convergence, and you might even think about how you first learned about the definite integral in calculus as a limit of sums. To a certain extent we'll also be interested in functions whose properties are *defined* in terms of integration and differentiation, e.g., rates of growth or decay, degrees of smoothness, the property of being bandlimited. You should not blanch at the sight of Taylor series.

Multivariable calculus: Here it's mostly the basic notions on functions of several variables and how they're represented that you'll find helpful. We'll present some applications to differential equations, especially partial differential equations. We'll use multiple integrals — we'll even change variables in multiple integrals — when we talk about the Fourier transform in higher dimensions. Do not run from this. We won't use Green's theorem, Stokes's theorem, or the divergence theorem, which is probably not a disappointment unless you study electrodynamics.

Fourier series: This is our first topic, so it's not quite fair to call it a prerequisite. We won't go too far or too deep. It would be helpful if some of the material looked familiar. Certainly the idea of a periodic function and properties of sines and cosines will be central, and you should be able to tinker with trig functions, as in adjusting frequency, amplitude, and phase.

Linear algebra: There are several ways that ideas from linear algebra come up. I'll appeal to geometric ideas like vectors, dot products, and orthogonality, and we'll work with complex numbers as well as with real numbers. In particular, we will use orthogonality in studying Fourier series. It's very important to believe that you can correctly transfer your intuition from situations where you can draw pictures (two dimensions, geometric vectors) to situations where you can't (higher dimensions, functions). This is one of the great strengths of descriptions that use ideas and vocabulary from linear algebra.

When we talk about the discrete Fourier transform (DFT) and the Fast Fourier Transform algorithm (FFT), we'll do some of it in terms of vectors, matrices, and linear transformations. It will help if you are familiar with symmetric, Hermitian, orthogonal, and unitary matrices — not much, just the definitions.

Also important is the general idea of linearity as in linear operators, which we'll see in the guise of linear systems. Engineers and scientists are used to dealing with superposition and its consequences in this context. Happy with eigenvalues and eigenvectors? I'd like to think so.

Finite sums: We'll be tossing around finite sums of various descriptions, mostly of complex exponentials, especially when we work with the discrete Fourier transform. Confidence in changing the index of summation (like changing the variable of integration) will help you. You should be able to recognize a geometric series and find a closed form expression for its sum. See Appendix C.

Probability and statistics: We'll do some applications of the Fourier transform to some fundamental results in probability, specifically the Central Limit Theorem. The basic notions are random variables and probability distributions, along with mean and variance. You'll see these ideas in many other classes. We'll absolutely talk about Gaussians, so I hope you believe they're worth talking about, even if you have only some vague notion that "it's the bell shaped curve that comes up in prob and stat."

δ-functions: Actually, we'll spend some time developing δ-functions from scratch, and all that goes with them, but if you've seen them before all the better.

Appendix B

Complex Numbers and Complex Exponentials

B.1. Complex Numbers

This appendix is intended as a reference and a quick review of complex numbers. I'm assuming that the definition, notation, and arithmetic of complex numbers are known to you, but we'll put the basic facts on the record. In the course of our work we'll also use calculus operations involving complex numbers, usually complex-valued functions of a real variable. For what we'll do, this will *not* involve the area of mathematics referred to as "complex analysis." For our purposes, the extensions of the formulas of calculus to complex numbers and complex-valued functions are straightforward and can be reliably employed.

First a matter of fundamental importance:

> **Declaration of principles.** Without apology I will write
>
> $$i = \sqrt{-1}\,.$$
>
> In many areas of science and engineering it's common to use j for $\sqrt{-1}$. If you want to use j in your own work, I won't try to talk you out of it. But I'll use i.

Before we plunge into notation and formulas, there are two points to keep in mind:

- Using complex numbers *greatly* simplifies the algebra we'll be doing. This isn't the only reason they're used, but it's a good one.
- We'll use complex numbers to represent real quantities — real signals, for example. At this point in your life this should not cause a metaphysical crisis, but if it does, my only advice is to *get over it.*

Let's go to work.

Complex numbers, real and imaginary parts, complex conjugates. A *complex number* is determined by two real numbers, its *real* and *imaginary* parts. We write

$$z = x + iy$$

where x and y are real. x is the real part and y is the imaginary part, and we write $x = \operatorname{Re} z$, $y = \operatorname{Im} z$. *Note*: It's y that is the imaginary part of $z = x + iy$, not iy. One says that iy is an *imaginary number* or is *purely imaginary*. One says that z has positive real part (resp. positive imaginary part) if x (resp. y) is positive. The set of all complex numbers is denoted by $\mathbb{C}$. (The set of all real numbers is denoted by $\mathbb{R}$.)

Elementary operations on complex numbers are defined according to what happens to the real and imaginary parts. For example, if $z = a + ib$ and $w = c + di$, then their sum and product are

$$\begin{aligned} z + w &= (a + c) + (b + d)i, \\ zw &= (ac - bd) + i(ad + bc). \end{aligned}$$

Get the formula for the product by multiplying out and using $i^2 = -1$, and grouping the real and imaginary parts. I'll come back to the formula for the general quotient z/w, but here's a particular little identity that's used often: since $i \cdot i = i^2 = -1$, we have

$$\frac{1}{i} = -i \quad \text{and} \quad i(-i) = 1\,.$$

By the way, read "$-i$" as "minus i" not "negative i." To say $-i$ is "negative i" would imply that i is positive, and it isn't.

The *complex conjugate* of $z = x + iy$ is

$$\bar{z} = x - iy\,.$$

Other notations in use for the complex conjugate are z^* and sometimes even $z^\dagger$. It's useful to observe that

$$z = \bar{z} \quad \text{if and only if } z \text{ is real, i.e., } y = 0.$$

Note also that

$$\overline{z + w} = \overline{z} + \overline{w}\,, \quad \overline{zw} = \overline{z}\,\overline{w}\,, \quad \overline{\overline{z}} = z\,.$$

We can find expressions for the real and imaginary parts of a complex number using the complex conjugate. If $z = x + iy$, then $\overline{z} = x - iy$ so that in the sum $z + \overline{z}$ the imaginary parts cancel. That is, $z + \overline{z} = 2x$, or

$$x = \operatorname{Re} z = \frac{z + \overline{z}}{2}\,.$$

Similarly, in the difference, $z - \bar{z}$, the real parts cancel and $z - \bar{z} = 2iy$, or

$$y = \operatorname{Im} z = \frac{z - \overline{z}}{2i}\,.$$

Don't forget the i in the denominator here.

The formulas $\overline{z+w} = \bar{z} + \bar{w}$ and $\overline{zw} = \bar{z}\bar{w}$ extend to sums and products of more than two complex numbers, and to integrals (being limits of sums), leading to formulas like

$$\overline{\int_a^b f(t)g(t)\,dt} = \int_a^b \overline{f(t)}\,\overline{g(t)}\,dt \quad \text{(here } dt \text{ is a real quantity).}$$

This overextended use of the overline notation for complex conjugates shows why it's useful to have alternate notations, such as

$$\left(\int_a^b f(t)g(t)\,dt\right)^* = \int_a^b f(t)^* g(t)^*\,dt\,.$$

It's best not to mix stars and bars in a single formula, so please be mindful of this.

The *magnitude* of $z = x + iy$ is

$$|z| = \sqrt{x^2 + y^2}\,.$$

Multiplying out the real and imaginary parts gives

$$z\bar{z} = (x + iy)(x - iy) = x^2 - i^2 y^2 = x^2 + y^2 = |z|^2\,.$$

The formula $|z|^2 = z\bar{z}$ comes up all the time. Note that

$$|z| = |\bar{z}|.$$

Also,

$$|z + w|^2 = |z|^2 + 2\,\mathrm{Re}\{z\bar{w}\} + |w|^2.$$

This likewise comes up a lot. To verify it,

$$\begin{aligned} |z + w|^2 &= (z + w)(\bar{z} + \bar{w}) \\ &= z\bar{z} + z\bar{w} + w\bar{z} + w\bar{w} \\ &= |z|^2 + (z\bar{w} + \overline{z\bar{w}}) + |w|^2 \\ &= |z|^2 + 2\,\mathrm{Re}\{z\bar{w}\} + |w|^2. \end{aligned}$$

On the right-hand side it looks like z and w don't enter symmetrically, but since $\mathrm{Re}\{z\bar{w}\} = \mathrm{Re}\{\overline{z\bar{w}}\} = \mathrm{Re}\{\bar{z}w\}$, we also have

$$|z + w|^2 = |z|^2 + 2\,\mathrm{Re}\{\bar{z}w\} + |w|^2.$$

The quotient z/w. For people who really need to find the real and imaginary parts of a quotient z/w, here's how it's done. Write $z = a + bi$ and $w = c + di$. Then

$$\begin{aligned} \frac{z}{w} &= \frac{a + bi}{c + di} \\ &= \frac{a + bi}{c + di}\,\frac{c - di}{c - di} \\ &= \frac{(a + bi)(c - di)}{c^2 + d^2} = \frac{(ac + bd) + (bc - ad)i}{c^2 + d^2}\,. \end{aligned}$$

Thus

$$\operatorname{Re} \frac{a+bi}{c+di} = \frac{ac+bd}{c^2+d^2}, \quad \operatorname{Im} \frac{a+bi}{c+di} = \frac{bc-ad}{c^2+d^2}.$$

Do *not* memorize this. Remember the "multiply the top and bottom by the conjugate" sort of thing.

Polar form. Since a complex number is determined by two real numbers, it's natural to associate $z = x + iy$ with the pair $(x, y) \in \mathbb{R}^2$ and hence to identify z with the point in the plane with Cartesian coordinates (x, y). One also then speaks of the "real axis" and the "imaginary axis."

We can also introduce polar coordinates r and θ and relate them to the complex number $z = x + iy$ through the equations

$$r = \sqrt{x^2 + y^2} = |z| \quad \text{and} \quad \theta = \tan^{-1} \frac{y}{x}.$$

The angle θ is called the *argument* or the *phase* of the complex number. One sees the notation

$$\theta = \arg z \quad \text{and also} \quad \theta = \angle z.$$

Going from polar to Cartesian through $x = r\cos\theta$ and $y = r\sin\theta$, we have the *polar form* of a complex number:

$$x + iy = r\cos\theta + ir\sin\theta = r(\cos\theta + i\sin\theta).$$

B.2. The Complex Exponential and Euler's Formula

The real workhorse for us is the complex exponential function. The exponential function e^z for a complex number z is defined, just as in the real case, by the Taylor series:

$$e^z = 1 + z + \frac{z^2}{2!} + \frac{z^3}{3!} + \cdots = \sum_{n=0}^{\infty} \frac{z^n}{n!}.$$

This converges for all $z \in \mathbb{C}$, but we won't check that.

Notice also that

$$\begin{aligned} \overline{e^z} &= \overline{\left(\sum_{n=0}^{\infty} \frac{z^n}{n!}\right)} \quad \text{(a heroic use of the bar notation)} \\ &= \sum_{n=0}^{\infty} \frac{\bar{z}^n}{n!} \\ &= e^{\bar{z}}. \end{aligned}$$

Also, e^z satisfies the differential equation $f' = f$ with initial condition $f(0) = 1$ (this is often taken as a definition, even in the complex case). By virtue of this, one can verify the key algebraic properties:

$$e^{z+w} = e^z e^w.$$

Here's how this goes. Thinking of w as fixed,

$$\frac{d}{dz} e^{z+w} = e^{z+w};$$

hence e^{z+w} must be a constant multiple of e^z:

$$e^{z+w} = ce^z \,.$$

What is the constant? At $z = 0$ we get

$$e^w = ce^0 = c \,.$$

Done. Using similar arguments one can show the other basic property of exponentiation,

$$(e^z)^r = e^{zr}$$

if r is real. It's actually a tricky business to define $(e^z)^w$ when w is complex (and hence to establish when $(e^z)^w = e^{zw}$). This requires introducing the *complex* logarithm, and special considerations are necessary. We will not go into this.

The most remarkable thing happens when the exponent is purely imaginary. The result is called Euler's formula and reads

$$e^{i\theta} = \cos\theta + i\sin\theta \,.$$

I want to emphasize that the left-hand side has only been defined via a series. The exponential function in the real case has nothing to do with the trig functions sine and cosine, and why it should have anything to do with them in the complex case is a true wonder. Euler's formula is usually proved by substituting into and manipulating the Taylor series for $\cos\theta$ and $\sin\theta$. Here's another, more elegant, way of seeing it. It relies on results for differential equations, but the proofs of those are no more difficult than the proofs of the properties of Taylor series that one needs in the usual approach.

Let $f(\theta) = e^{i\theta}$. Then $f(0) = 1$ and $f'(\theta) = ie^{i\theta}$, so that $f'(0) = i$. Moreover

$$f''(\theta) = i^2 e^{i\theta} = -e^{i\theta} = -f(\theta)\,;$$

i.e., f satisfies

$$f'' + f = 0, \quad f(0) = 1, \ f'(0) = i \,.$$

On the other hand, if $g(\theta) = \cos\theta + i\sin\theta$, then

$$g''(\theta) = -\cos\theta - i\sin\theta = -g(\theta), \quad \text{or} \quad g'' + g = 0$$

and also

$$g(0) = 1, \quad g'(0) = i \,.$$

Thus f and g satisfy the same differential equation with the same initial conditions; so f and g must be equal. Slick.

I prefer using the second-order ordinary differential equation here since that's the one naturally associated with the sine and cosine. We could also do the argument with the first-order equation $f' = f$. Indeed, if $f(\theta) = e^{i\theta}$, then $f'(\theta) = ie^{i\theta} = if(\theta)$ and $f(0) = 1$. Likewise, if $g(\theta) = \cos\theta + i\sin\theta$, then $g'(\theta) = -\sin\theta + i\cos\theta = i(\cos\theta + i\sin\theta) = ig(\theta)$ and $g(0) = 1$. This implies that $f(\theta) = g(\theta)$ for all θ.

Plugging $\theta = \pi$ into Euler's formula gives $e^{i\pi} = \cos\pi + i\sin\pi = -1$, better written as

$$e^{i\pi} + 1 = 0 \,.$$

This is sometimes referred to as the most famous equation in mathematics. It expresses a simple relationship — and why should there be any at all? — between the fundamental numbers e, π, 1, and 0, not to mention i. We'll probably never see this most famous equation again, but now we've seen it once.

Consequences of Euler's formula. The polar form $z = r(\cos\theta + i\sin\theta)$ can now be written as

$$z = re^{i\theta},$$

where $r = |z|$ is the magnitude and θ is the phase of the complex number z. Using the arithmetical properties of the exponential function, we also have that if $z_1 = r_1 e^{i\theta_1}$ and $z_2 = r_2 e^{i\theta_2}$, then

$$z_1 z_2 = r_1 r_2 e^{i(\theta_1+\theta_2)}.$$

The magnitudes multiply and the arguments (phases) add.

Euler's formula also gives a dead easy way of deriving the addition formulas for the sine and cosine. On the one hand,

$$\begin{aligned} e^{i(\alpha+\beta)} &= e^{i\alpha}e^{i\beta} \\ &= (\cos\alpha + i\sin\alpha)(\cos\beta + i\sin\beta) \\ &= (\cos\alpha\cos\beta - \sin\alpha\sin\beta) + i(\sin\alpha\cos\beta + \cos\alpha\sin\beta). \end{aligned}$$

On the other hand,

$$e^{i(\alpha+\beta)} = \cos(\alpha+\beta) + i\sin(\alpha+\beta).$$

Equating the real and imaginary parts gives

$$\begin{aligned} \cos(\alpha+\beta) &= \cos\alpha\cos\beta - \sin\alpha\sin\beta, \\ \sin(\alpha+\beta) &= \sin\alpha\cos\beta + \cos\alpha\sin\beta. \end{aligned}$$

I went through this derivation because it expresses in a simple way an extremely important principle in mathematics and its applications.

- If you can compute the same thing two different ways, chances are you've done something significant.

Take this seriously. This maxim appears throughout the course.

Symmetries of the sine and cosine: Even and odd functions. Using the identity

$$\overline{e^{i\theta}} = e^{\overline{i\theta}} = e^{-i\theta}$$

we can express the cosine and the sine as the real and imaginary parts, respectively, of $e^{i\theta}$:

$$\cos\theta = \frac{e^{i\theta} + e^{-i\theta}}{2} \quad \text{and} \quad \sin\theta = \frac{e^{i\theta} - e^{-i\theta}}{2i}.$$

Once again this is a simple observation. Once again there is something more to say.

You are very familiar with the symmetries of the sine and cosine function: $\cos\theta$ is an *even function*, meaning

$$\cos(-\theta) = \cos\theta,$$

and $\sin\theta$ is an *odd function*, meaning

$$\sin(-\theta) = -\sin\theta\,.$$

Why is this true? There are many ways of seeing it (Taylor series, differential equations), but here's one you may not have thought of before, and it fits into a general framework of evenness and oddness that we'll find useful when discussing symmetries of the Fourier transform.

If $f(x)$ is *any* function, then the function defined by

$$f_e(x) = \frac{f(x) + f(-x)}{2}$$

is *even*. Check it out:

$$f_e(-x) = \frac{f(-x) + f(-(-x))}{2} = \frac{f(-x) + f(x)}{2} = f_e(x)\,.$$

Similarly, the function defined by

$$f_o(x) = \frac{f(x) - f(-x)}{2}$$

is *odd*. Moreover

$$f_e(x) + f_0(x) = \frac{f(x) + f(-x)}{2} + \frac{f(x) - f(-x)}{2} = f(x)\,.$$

The conclusion is that any function can be written as the sum of an even function and an odd function. Or, to put it another way, $f_e(x)$ and $f_o(x)$ are, respectively, the *even and odd parts* of $f(x)$, and $f(x)$ is the sum of its own even and odd parts. We can find some symmetries in a function even if it's not symmetric.

And what are the even and odd parts of the function $e^{i\theta}$? For the even part we have

$$\frac{e^{i\theta} + e^{-i\theta}}{2} = \cos\theta$$

and for the odd part we have

$$\frac{e^{i\theta} - e^{-i\theta}}{2} = i\sin\theta\,.$$

Nice.

Algebra and geometry. To wrap up this review I want to say a little more about the complex exponential and its use in representing sinusoids. To set the stage for this we'll consider the mix of algebra and geometry — one of the reasons why complex numbers are often so handy.

We not only think of a complex number $z = x + iy$ as a point in the plane, we also think of it as a vector with tail at the origin and tip at (x, y). In polar form, either written as $re^{i\theta}$ or as $r(\cos\theta + i\sin\theta)$, we recognize $|z| = r$ as the length of the vector and θ as the angle that the vector makes with the x-axis (the real axis). Note that

$$|e^{i\theta}| = |\cos\theta + i\sin\theta| = \sqrt{\cos^2\theta + \sin^2\theta} = 1\,.$$

Many are the times you will use

$$|e^{i(\text{something real})}| = 1 .$$

Once we make the identification of a complex number with a vector, we have an easy back-and-forth between the algebra of complex numbers and the geometry of vectors. Each point of view can help the other.

Take addition. The sum of $z = a + bi$ and $w = c + di$ is the complex number $z + w = (a + c) + (c + d)i$. Geometrically, this is the vector sum; if z and w are regarded as vectors from the origin, then $z + w$ is the vector from the origin that is the diagonal of the parallelogram determined by z and w.

Similarly, as a vector, $z - w = (a - c) + (b - d)i$ is the vector that goes from the tip of w to the tip of z, i.e., along the other diagonal of the parallelogram determined by z and w. Notice here that we allow for the customary ambiguity in placing vectors. On the one hand we identify the complex number $z - w$ with the vector with tail at the origin and tip at $(a - c, b - d)$. On the other hand we allow ourselves to place the (geometric) vector anywhere in the plane as long as we maintain the same magnitude and direction of the vector.

It's possible to give a geometric interpretation of zw (where, you will recall, the magnitudes multiply and the arguments add) in terms of similar triangles, but we won't need this.

Complex conjugation also has a simple geometric interpretation. If $z = x + iy$, then the complex conjugate $\bar{z} = x - iy$ is the mirror image of z in the x-axis. Think either in terms of reflecting the point (x, y) to the point $(x, -y)$ or reflecting the vector. This gives a natural geometric reason why $z + \bar{z}$ is real — since z and $\bar{z}$ are symmetric about the real axis, the diagonal of the parallelogram determined by z and $\bar{z}$ obviously goes along the real axis. In a similar vein, $-\bar{z} = -(x - iy) = -x + iy$ is the reflection of $z = x + iy$ in the y-axis, and now you can see why $z - \bar{z}$ is purely imaginary.

There are plenty of examples of the interplay between the algebra and geometry of complex numbers, and the identification of complex numbers with points in the plane (Cartesian or polar coordinates) often leads to some simple approaches to problems in analytic geometry. Equations in x and y (or in r and θ) can often be recast as equations in complex numbers, and having access to the arithmetic of complex numbers frequently simplifies calculations.

B.3. Further Applications of Euler's Formula

We've already done some work with Euler's formula $e^{i\theta} = \cos\theta + i\sin\theta$, and we agree it's a fine thing to know. For additional applications we'll replace θ by t and think of

$$e^{it} = \cos t + i \sin t$$

as describing a point in the plane that is moving in time. How does it move? Since $|e^{it}| = 1$ for every t, the point moves along the unit circle. In fact, from looking at the real and imaginary parts separately,

$$x = \cos t, \quad y = \sin t,$$

we see that e^{it} is a (complex-valued) parametrization of the circle: the circle is traced out exactly once in the *counterclockwise* direction as t goes from 0 to 2π. We can also think of the vector from the origin to z as rotating counterclockwise about the origin, like a backwards moving clock hand.

For our efforts I prefer to work with

$$e^{2\pi it} = \cos 2\pi t + i \sin 2\pi t$$

as the "basic" complex exponential. Via its real and imaginary parts the complex exponential $e^{2\pi it}$ contains the sinusoids $\cos 2\pi t$ and $\sin 2\pi t$, each of frequency 1 Hz. If you like, including the 2π or not is the difference between working with frequency in units of Hz, or cycles per second, and "angular frequency" in units of radians per second. With the 2π, as t goes from 0 to 1 the point $e^{2\pi it}$ traces out the unit circle exactly once (one cycle) in a counterclockwise direction. The units in the exponential $e^{2\pi it}$ are (as they are in $\cos 2\pi t$ and $\sin 2\pi t$)

$$e^{(2\pi \text{ radians/cycle}) \cdot i \cdot 1 \text{ (cycles/sec)} \cdot (t \text{ sec})}.$$

Without the 2π the units in e^{it} are

$$e^{i \cdot (1 \text{ radians/sec}) \cdot (t \text{ sec})}.$$

We can always pass easily between the complex form of a sinusoid as expressed by a complex exponential and the real signals as expressed through sines and cosines. But for many, many applications, calculations, prevarications, etc., it is *far* easier to stick with the complex representation. To repeat what I said earlier in this appendix, if you have philosophical trouble using complex entities to represent real entities, the best advice I can give you is to *get over it.*

We can now feel free to change the amplitude and frequency and to include a phase shift. The general (real) sinusoid is of the form, say, $A \sin(2\pi\nu t + \phi)$; the amplitude is A, the frequency is ν (in Hz), and the phase is ϕ. (We'll take A to be positive for this discussion.) The general complex exponential that includes this information is then

$$Ae^{i(2\pi\nu t + \phi)}.$$

Note that i is multiplying the entire quantity $2\pi\nu t + \phi$. The term *phasor* is often used to refer to the complex exponential $e^{2\pi i\nu t}$.

And what is $Ae^{i(2\pi\nu t+\phi)}$ describing as t varies? The magnitude is $|Ae^{i(2\pi i\nu t+\phi)}| = |A| = A$ so the point is moving along the circle of radius A. Assume for the moment that ν is positive — we'll come back to negative frequencies later. Then the point traces out the circle in the counterclockwise direction at a rate of ν cycles per second; 1 second is ν times around, including the possibility of a fractional, or even an irrational, number of times around. The phase ϕ determines the starting point on the circle, for at $t = 0$ the point is $Ae^{i\phi}$. In fact, we can write

$$Ae^{i(2\pi\nu t+\phi)} = e^{2\pi i\nu t} Ae^{i\phi}$$

and think of this as the (initial) vector $Ae^{i\phi}$ set rotating at a frequency ν Hz through multiplication by the time-varying phasor $e^{2\pi i\nu t}$.

What happens when ν is negative? That simply reverses the direction of motion around the circle from counterclockwise to clockwise. The catch phrase is just so: positive frequency means counterclockwise rotation and negative frequency means clockwise rotation. Now, we can write a cosine, say, as

$$\cos 2\pi\nu t = \frac{e^{2\pi i\nu t} + e^{-2\pi i\nu t}}{2},$$

and one sees this formula interpreted through statements like "a cosine is the sum of phasors of positive and negative frequency," or similar phrases. The fact that a cosine is made up of a positive and negative frequency, in this sense, is important for some analytical considerations, particularly having to do with the Fourier transform (and we'll see this phenomenon more generally), but I don't think there's a geometric interpretation of negative frequencies without appealing to the complex exponentials that go with real sines and cosines. "Negative frequency" is clockwise rotation of a phasor, period.

Sums of sinusoids. As a brief, final application of these ideas we'll consider the sum of two sinusoids of the same frequency.[1] In real terms, the question is what one can say about the superposition of two signals

$$A_1 \sin(2\pi\nu t + \phi_1) + A_2 \sin(2\pi\nu t + \phi_2).$$

Here the frequency is the same for both signals but the amplitudes and phases may be different.

If you answer too quickly, you might say that a phase shift between the two terms is what leads to beats. Wrong. Perhaps physical considerations (up to you) can lead you to conclude that the frequency of the sum is again ν. That's right, but it's not so obvious looking at the graphs of the individual sinusoids and trying to imagine what the sum looks like, e.g., for these sinusoids:

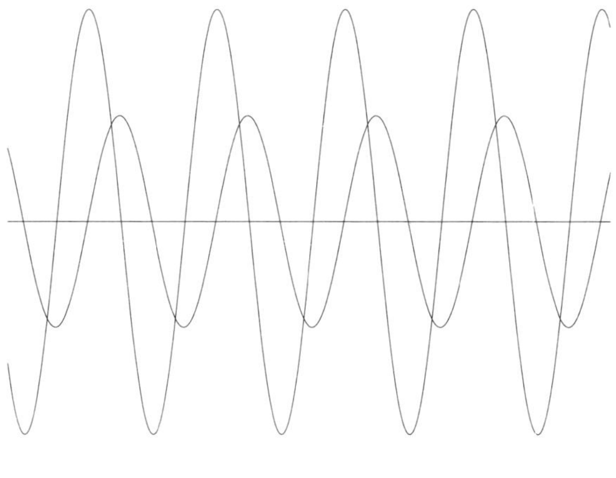

[1]The idea for this example comes from *A Digital Signal Processing Primer* by K. Stieglitz.

An algebraic analysis based on the addition formulas for the sine and cosine does not look too promising either. But it's easy to see what happens if we use complex exponentials.

Consider

$$A_1 e^{i(2\pi\nu t+\phi_1)} + A_2 e^{i(2\pi\nu t+\phi_2)},$$

whose imaginary part is the sum of sinusoids, above. Before messing with the algebra, think geometrically in terms of rotating vectors. At $t = 0$ we have the two vectors from the origin to the starting points, $z_0 = A_1 e^{i\phi_1}$ and $w_0 = A_2 e^{i\phi_2}$. Their sum $z_0 + w_0$ is the starting point (or starting vector) for the sum of the two motions. But how do those two starting vectors move? They rotate together at the same rate, the motion of each described by $e^{2\pi i\nu t} z_0$ and $e^{2\pi i\nu t} w_0$, respectively. Thus their sum also rotates at that rate — think of the whole parallelogram (vector sum) rotating rigidly about the vertex at the origin. Now mess with the algebra and arrive at the same result:

$$A_1 e^{i(2\pi\nu t+\phi_1)} + A_2 e^{i(2\pi\nu t+\phi_2)} = e^{2\pi i\nu t}(A_1 e^{i\phi_1} + A_2 e^{i\phi_2}).$$

What is the situation if the two exponentials are "completely out of phase"?

Of course, the simple algebraic manipulation of factoring out the common exponential does not work if the frequencies of the two terms *are* different. If the frequencies of the two terms are different ... now *that* gets interesting.

Problems and Further Results

B.1. If $0 \leq \arg w - \arg z < \pi$, show that the area of the triangle whose vertices are 0, z, and w is given by $\frac{1}{2}\mathrm{Im}\,\{\bar{z}w\}$.

B.2. Show that the equation of a line can be written as $az + \bar{a}\bar{z} + b = 0$, where a is a complex number and b is real (and $z = x + iy$ is the variable). What is the slope of the line?

B.3. Recall the identity $|z+w|^2 = |z|^2 + |w|^2 + 2\,\mathrm{Re}\{z\bar{w}\}$. As a generalization of this, show that

$$|z+w|^2 + |z+\bar{w}|^2 = 2(|z|^2 + |w|^2) + 4\,\mathrm{Re}\{z\}\,\mathrm{Re}\{w\},$$

$$|z+w|^2 + |z-\bar{w}|^2 = 2(|z|^2 + |w|^2) + 4\,\mathrm{Im}\{z\}\,\mathrm{Im}\{w\}.$$

Geometric Sums

Most of this appendix is on geometric sums, but before running that down I want to make a few comments on a matter that has brought many students to grief. Say we have two sums,

$$A = \sum_n a_n \quad \text{and} \quad B = \sum_n b_n \,.$$

How do we write their product AB in summation notation? On purpose, I haven't specified the ranges of the indices of summation nor whether the sums are finite or infinite. That's not the point. The point is that many people write

$$AB = \left(\sum_n a_n\right)\left(\sum_n b_n\right) = \sum_n a_n b_n \,,$$

and that's wrong. You can see that it's wrong by writing out a few terms, a practice I encourage, and comparing. On the one hand,

$$\begin{aligned}\left(\sum_n a_n\right)\left(\sum_n b_n\right) &= (a_1 + a_2 + a_3 + \cdots)(b_1 + b_2 + b_3 + \cdots) \\ &= a_1b_1 + a_1b_2 + a_1b_3 + a_2b_1 + a_2b_2 + a_2b_3 + \cdots \,.\end{aligned}$$

On the other hand,

$$\sum_n a_n b_n = a_1b_1 + a_2b_2 + a_3b_3 + \cdots \,.$$

See, we're missing the cross terms, the terms with different indices. How do you write this correctly? You have to use *different indices of summation* for the two sums and then write

$$AB = \left(\sum_n a_n\right)\left(\sum_m b_m\right) = \sum_{n,m} a_n b_m \,.$$

Convince yourself. Write out a few terms. They're all there. Some people prefer to write the right-hand side as nested sums,

$$\sum_n \sum_m a_n b_m\,, \quad \text{which is equal to} \quad \sum_m \sum_n a_n b_m\,.$$

This is better if A and B don't have the same number of terms, in which case you specify the runs of n and m separately.

I am suppressing all questions about convergence in the case of infinite sums. Meaning that I'm skating around the details on swapping $\sum_n \sum_m$ and $\sum_m \sum_n$, much as I skate around the details on swapping $dxdy$ and $dydx$ in double integrals when it comes up.

On to the main event.

We'll be running into *geometric sums* in various settings, mostly in connection with the discrete Fourier transform but in some other places as well. You've certainly seen such sums in your past, maybe your ancient past, but you might be rusty and I thought it would be helpful to collect some formulas and techniques. This will be mostly in the form of problems for you to work out.

A geometric sum is (most often) of the form

$$\sum_{n=0}^{N} x^n = 1 + x + \cdots + x^N.$$

Typically, x is a real or complex number, but keep your mind open to even more general possibilities.[1] The sum gets the adjective "geometric" because the ratio of two consecutive terms is constant, $x^{n+1}/x^n = x$, and ratios are fundamental for much of Euclidian geometry. In fact, anytime that happens, and you'll see examples, you call the sum "geometric."

We can find a closed form expression for the sum by the following trick: first, give the sum a name, say

$$S = 1 + x + x^2 + \cdots + x^N.$$

Multiply by x:

$$xS = x + x^2 + x^3 + \cdots + x^{N+1}.$$

Subtract xS from S. On the one hand, this is $(1-x)S$. On the other hand, all that remains of the sums are the first and last terms:

$$(1-x)S = S - xS = (1 + x + x^2 + \cdots + x^N) - (x + x^2 + x^3 + \cdots + x^{N+1}) = 1 - x^{N+1}.$$

Solve for S:

$$S = \frac{1 - x^{N+1}}{1 - x}.$$

Slick. Just one thing: if $x = 1$, we get $0/0$, which is bad form. But, of course, if $x = 1$, then S is just the sum of $N + 1$ 1's so $S = N + 1$. You'll see this case distinction come up anytime we have a geometric sum.

---

[1] When $x = 0$, the first term in the sum is 0^0, an undefined quantity, so we ignore that little problem and just declare the sum to be 0.

This formula for S is the basic result. There are many variations and consequences. For one, if $|x| < 1$, then $x^{N+1} \to 0$ as $N \to \infty$, which allows us to say:

$$\text{if } |x| < 1, \text{ then } \sum_{n=0}^{\infty} x^n = \frac{1}{1-x}.$$

For another, sometimes you want a sum that starts with x instead of with 1, as in $x + x^2 + \cdots + x^N$. You can rederive a formula for the sum using the "multiply by x and subtract" trick, which amounts to

$$x + x^2 + \cdots + x^N = x(1 + x + \cdots + x^{N-1}) = \frac{x(1-x^N)}{1-x},$$

or you can say:

$$x + x^2 + \cdots + x^N = (1 + x + \cdots + x^N) - 1 = \frac{1-x^{N+1}}{1-x} - 1 = \frac{x(1-x^N)}{1-x}.$$

Take your pick.

In most of our applications we'll be looking at a geometric sum of complex exponentials, consecutive powers of $e^{\pm 2\pi i t}$ or of $e^{\pm 2\pi i/N}$, say. Below are some practice problems.

Problems and Further Results

C.1. If w is a real or complex number, $w \neq 1$, and p and q are any integers, show that

$$\sum_{n=p}^{q} w^n = \frac{w^p - w^{q+1}}{1-w}.$$

(As above, if $w = 1$, then the sum is $q + 1 - p$.)

Discuss the cases when $p = -\infty$ or $q = \infty$. What about $p = -\infty$ *and* $q = +\infty$?

C.2. Find the sum

$$\sum_{n=0}^{N-1} e^{2\pi i n/N}$$

and explain your answer geometrically.

C.3. Derive the formula

$$\sum_{k=-N}^{N} e^{2\pi i k t} = \frac{\sin(2\pi t(N+1/2))}{\sin(\pi t)}.$$

Do this problem. This is an important formula.

Index

Published Titles in This Series

33 **Brad G. Osgood,** Lectures on the Fourier Transform and Its Applications, 2019
32 **John M. Erdman,** A Problems Based Course in Advanced Calculus, 2018
31 **Benjamin Hutz,** An Experimental Introduction to Number Theory, 2018
30 **Steven J. Miller,** Mathematics of Optimization: How to do Things Faster, 2017
29 **Tom L. Lindstrøm,** Spaces, 2017
28 **Randall Pruim,** Foundations and Applications of Statistics: An Introduction Using R, Second Edition, 2018
27 **Shahriar Shahriari,** Algebra in Action, 2017
26 **Tamara J. Lakins,** The Tools of Mathematical Reasoning, 2016
25 **Hossein Hosseini Giv,** Mathematical Analysis and Its Inherent Nature, 2016
24 **Helene Shapiro,** Linear Algebra and Matrices, 2015
23 **Sergei Ovchinnikov,** Number Systems, 2015
22 **Hugh L. Montgomery,** Early Fourier Analysis, 2014
21 **John M. Lee,** Axiomatic Geometry, 2013
20 **Paul J. Sally, Jr.,** Fundamentals of Mathematical Analysis, 2013
19 **R. Clark Robinson,** An Introduction to Dynamical Systems: Continuous and Discrete, Second Edition, 2012
18 **Joseph L. Taylor,** Foundations of Analysis, 2012
17 **Peter Duren,** Invitation to Classical Analysis, 2012
16 **Joseph L. Taylor,** Complex Variables, 2011
15 **Mark A. Pinsky,** Partial Differential Equations and Boundary-Value Problems with Applications, Third Edition, 1998
14 **Michael E. Taylor,** Introduction to Differential Equations, 2011
13 **Randall Pruim,** Foundations and Applications of Statistics, 2011
12 **John P. D'Angelo,** An Introduction to Complex Analysis and Geometry, 2010
11 **Mark R. Sepanski,** Algebra, 2010
10 **Sue E. Goodman,** Beginning Topology, 2005
9 **Ronald Solomon,** Abstract Algebra, 2003
8 **I. Martin Isaacs,** Geometry for College Students, 2001
7 **Victor Goodman and Joseph Stampfli,** The Mathematics of Finance, 2001
6 **Michael A. Bean,** Probability: The Science of Uncertainty, 2001
5 **Patrick M. Fitzpatrick,** Advanced Calculus, Second Edition, 2006
4 **Gerald B. Folland,** Fourier Analysis and Its Applications, 1992
3 **Bettina Richmond and Thomas Richmond,** A Discrete Transition to Advanced Mathematics, 2004
2 **David Kincaid and Ward Cheney,** Numerical Analysis: Mathematics of Scientific Computing, Third Edition, 2002
1 **Edward D. Gaughan,** Introduction to Analysis, Fifth Edition, 1998